World Sustainability Series

Series Editor

Walter Leal Filho, European School of Sustainability Science and Research, Research and Transfer Centre "Sustainable Development and Climate Change Management", Hamburg University of Applied Sciences, Hamburg, Germany

Due to its scope and nature, sustainable development is a matter which is very interdisciplinary, and draws from knowledge and inputs from the social sciences and environmental sciences on the one hand, but also from physical sciences and arts on the other. As such, there is a perceived need to foster integrative approaches, whereby the combination of inputs from various fields may contribute to a better understanding of what sustainability is, and means to people. But despite the need for and the relevance of integrative approaches towards sustainable development, there is a paucity of literature which address matters related to sustainability in an integrated way.

Notes on the quality assurance and peer review of this publication

Prior to publication, the works published in this book are initially assessed and reviewed by an in-house editor. If suitable for publication, manuscripts are sent for further review, which includes a combined effort by the editorial board and appointed subject experts, who provide independent peer-review. The feedback obtained in this way was communicated to authors, and with manuscripts checked upon return before finally accepted. The peer-reviewed nature of the books in the "World Sustainability Series" means that contributions to them have, over many years, been officially accepted for tenure and promotion purposes.

Salem Al-Naemi · Rachid Benlamri ·
Walter Leal Filho · Jeffrey McDonnell ·
Rehan Sadiq · Haruna Musa Moda
Editors

Water and Food Security in the Face of Climate Change: Challenges and Opportunities for Resilience

Proceeding of the International Conference Water and Food Security in the Face of Climate Change: Challenges and Opportunities for Resilience (WFCC2025)

Volume 1

Springer

Editors
Salem Al-Naemi
UNEVOC Centre
University of Doha for Science
and Technology
Doha, Qatar

Rachid Benlamri
Academic Food Security Research Cluster
University of Doha for Science
and Technology
Doha, Qatar

Walter Leal Filho
Research and Transfer Centre
"Sustainability and Climate Change
Management"
Hamburg University of Applied Sciences
Hamburg, Germany

Jeffrey McDonnell
National Hydrology Research Centre
University of Saskatchewan
Saskatoon, SK, Canada

Rehan Sadiq
Academic University of British Columbia
Vancouver, BC, Canada

Haruna Musa Moda
Environmental Health and Safety
University of Doha for Science
and Technology
Doha, Qatar

ISSN 2199-7373 ISSN 2199-7381 (electronic)
World Sustainability Series
ISBN 978-3-032-00097-2 ISBN 978-3-032-00098-9 (eBook)
https://doi.org/10.1007/978-3-032-00098-9

This Springer imprint is published by the registered company Springer Nature Switzerland AG
The registered company address is: Gewerbestrasse 11, 6330 Cham, Switzerland

Foreword

What we eat, drink, and breathe should be safe, clean, and good for our health. My interest in the topics explored in this excellent collection of scholarly articles is reflected in the concept of elemental metabolomics that my colleagues and I described a decade ago. We defined elemental metabolomics as the measurement and analysis of the overall levels of chemical elements in biological samples, along with the observation of how these levels vary under different biological or environmental conditions. This method uses advanced mass spectrometry and other techniques for high-throughput elemental analysis, making it an invaluable tool in medical diagnostics, nutrition, food science, agriculture, and environmental monitoring of pollutants.

However, we should not focus only on food, nutrition, and human health. Our concerns must expand to the entire planet—we have only one Earth. We must take care of it and leave it to future generations in better shape than it was when we received it. That is why I care deeply about environmental issues, and I am pleased to see that the environment plays a big part in this book.

The main topics—food, water, health, and the environment—are all connected. The One Health approach recognizes that the health of people, animals, and environment are closely interconnected. We cannot protect human health without considering the health of animals and the ecosystems we all share. Diseases, pollutants, and food systems cross these boundaries, so the solutions must be systemic. One Health encourages collaboration across fields of medicine, veterinary science, agriculture, and environmental science to create healthier communities, ecosystems, and a more sustainable planet.

Science has a big role to play. It gives us facts, not guesses or wishful thinking. Scientifically verified facts help leaders make informed decisions—including laws, policies, and strategies for managing our shared resources.

This book also reflects the values of the University of Doha for Science and Technology. We proudly support the United Nations Sustainable Development Goals (SDGs) and Qatar's Vision 2030. The research and discussions in this book are well-aligned with all 17 SDGs neatly grouped across the four main tracks of the book. Through research, education, and responsible action, we are working together to build a future that is fair, healthy, and sustainable for everyone.

May 2025

Dr. Vladimir Brusic
University of Doha for Science and Technology
Doha, Qatar

Preface

In an era defined by unprecedented climatic shifts, the interplay between water resources and food security has become increasingly critical. As we face the many interrelated challenges posed by climate change, we are compelled to explore how alterations in weather patterns, water availability, and agricultural productivity threaten the delicate balance that sustains our global food systems. This book, *Water and Food Security in the Face of Climate Change: Challenges and Opportunities*, part of scientific papers presented at the international conference on water and food security, University of Doha for Science and Technology, Qatar, seeks to illuminate these pressing issues while offering insights into viable pathways for resilience and adaptation.

Water is the lifeblood of agriculture; it is essential for crop growth, livestock health, and food production. However, as climate change accelerates, we witness rising temperatures, altered precipitation patterns, and increased frequency of extreme weather events, all of which strain our water resources. In this context, understanding the nexus between water and food security is paramount, as communities worldwide strive to ensure access to nutritious food amidst growing uncertainties.

This volume assembles contributions from a diverse array of experts across disciplines ranging from environmental science to agricultural economics and public policy. Together, they tackle a spectrum of themes, including the implications of dwindling freshwater supplies on agricultural practices, the role of innovative technologies in improving water use efficiency, and strategies for fostering sustainable practices that bolster both water and food security.

Moreover, this book highlights the opportunities that arise from collaborative efforts and innovative approaches to adapt to climate challenges. By fostering sustainable water management, promoting resilient agricultural practices, and implementing policies that prioritize food security, we can create a more secure future for all.

As you delve into the chapters that follow, we invite you to reflect on the profound interconnectedness of water and food systems. Our collective ability to respond to the challenges of climate change will depend on our understanding of these systems and our commitment to building adaptive capacity across communities and regions.

In the face of uncertainty, we have the power to transform challenges into opportunities. It is our hope that this book serves as a valuable resource for policymakers, researchers, practitioners, and students who envision a world where water and food security are secured for future generations, even amidst the changing climate.

Doha, Qatar Salem Al-Naemi
Doha, Qatar Rachid Benlamri
Hamburg, Germany Walter Leal Filho
Saskatoon, Canada Jeffrey McDonnell
Vancouver, Canada Rehan Sadiq
Doha, Qatar Haruna Musa Moda

Reviewers

Ravi Rangarajan, University of Doha for Science and Technology, Qatar, e-mail: ravi.rangarajan@udst.edu.qa
Amin Esmaeili Khalil Saraei, University of Doha for Science and Technology, Qatar, e-mail: amin.esmaeili@udst.edu.qa
Farhat Abbas, University of Doha for Science and Technology, Qatar, e-mail: farhat.abbas@udst.edu.qa
Bahman Boostani, University of Doha for Science and Technology, Qatar, e-mail: bahman.boostani@udst.edu.qa
Aparna Krishna, University of Doha for Science and Technology, Doha, Qatar, e-mail: aparna.krishna@udst.edu.qa
Raj Kumar Dahal, University of Doha for Science and Technology, Qatar, e-mail: raj.dahal@udst.edu.qa
Norr Hassan, University of Doha for Science and Technology, Qatar, e-mail: norr.hassan@udst.edu.qa
Ashraf Chamseddine, University of Doha for Science and Technology, Doha, Qatar, e-mail: ashraf.chamseddine@udst.edu.qa
Haruna Musa Moda, University of Doha for Science and Technology, Qatar, e-mail: haruna.moda@udst.edu.qa
Azad Ashraf, University of Doha for Science and Technology, Qatar, e-mail: azad.ashraf@udst.edu.qa
Hassan Hijazi, University of Doha for Science and Technology, Qatar, e-mail: hassan.hijazi@udst.edu.qa
Zahida Bano, University of Doha for Science and Technology, Qatar, e-mail: Zahida.qau@gmail.com
Christopher Fellows, The University of New England, Australia, e-mail: cfellows@une.edu.au
Nourhan Ghoneim, International Maritime College Oman, IMCO, Oman, e-mail: norhan@imco.edu.om
Venkatesh Meda, University of Saskatchewan, Saskatoon, Canada, e-mail: vem281@campus.usask.ca

Majid Ibrahim, University of Doha for Science and Technology, Qatar, e-mail: majid.ibraheem@udst.edu.qa
Ghoson Al-Obaidly, University of Doha for Science and Technology, Qatar, e-mail: ghoson.alobaidly@udst.edu.qa
Mahmoud Bataineh, University of Doha for Science and Technology, Qatar, e-mail: mahmoud.bataineh@udst.edu.qa
Monther Tarawneh, Tafila Technical University, Jordan, e-mail: mtarawneh@ttu.edu.jo
Tahira Makda, University of Doha for Science and Technology, Qatar, e-mail: tahira.makda@udst.edu.qa
Ahmed Badawi, University of Doha for Science and Technology, Qatar, e-mail: dr.ahmedbadawi100@gmail.com
Ahmad Alqudah, Department of Biological and Environmental Sciences, College of Arts and Sciences Qatar University, Qatar, e-mail: aalqudah@qu.edu.qa
Hassan Ali, University of Doha for Science and Technology, Qatar, e-mail: hassan.ali@udst.edu.qa
Vajiheh Akbarzadeh, University of Doha for Science and Technology, Qatar, e-mail: vajiheh.akbarzadeh@udst.edu.qa
Muhammad Zubair, University of Doha for Science and Technology, Qatar, e-mail: muhammad.zubair@udst.edu.qa
Yasen Harrye, University of Doha for Science and Technology, Qatar, e-mail: yasen.harrye@udst.edu.qa
Kashif Rasool, Qatar Environment and Energy Research Institute (QEERI), Hamad Bin Khalifa University, Qatar Foundation, Doha, Qatar, e-mail: krasool@hbku.edu.qa
Muhammad Zafar, Quaid-i-azam University Islamabad, Pakistan, e-mail: zafar@qau.edu.pk
Nour El Islam Bachari, Laboratory of Biological Oceanography and Marine Environment (LOBEM), Faculty of Biology, University of Sciences and Technology Houari Boumediene, Algeria, e-mail: bacharinouri@gmail.com
Noor Yashooa, Center for Environmental studies, University of Kurdistan Hewlêr, Erbil City, Kurdistan Region, Iraq, e-mail: noor.khalid@ukh.edu.krd
Dr. Adenipekun Shitu, Federal University, Oye-Ekiti, Ekiti State, Nigeria, e-mail: gabriel.shitu@fuoye.edu.ng
Dr. Yamen Idelbi, Yarmouk Private University, Syrian Arab Republic, e-mail: yamen6@hotmail.com
Lumu Emmanuel Roy Muyingo, Uganda Technology and Management University, Uganda, e-mail: lumuemmanuel@utamu.ac.ug
Aiyad Gannan, University of Doha for Science and Technology, Qatar, e-mail: aiyad.gannan@udst.edu.qa
sihem arris, University Constantine3-Salah Boubnider, Algeria, e-mail: sihem.arris@univ-constantine3.dz
Dr. Emmanuel Amase, Federal University Gashua, Nigeria, e-mail: emmaamase60@gmail.com

Mr. Valentin Tembe Mwela, Red Cross Higher Institute of Health Sciences of Bandundu, Congo, the Democratic Republic of the, e-mail: tembevalentin@gmail.com
Walid Kh Alqaisi, University of Doha for Science and Technology, Qatar, e-mail: walid.alqaisi@udst.edu.qa
Adefunke Odetoye, Federal University, Oye-Ekiti, Ekiti State, Nigeria, e-mail: adefunke.odetoye@fuoye.edu.ng
Djamal Abdessemed, University of the Sciences and Technology Houari Boumediene, Algeria, e-mail: adjamal@yahoo.com
Matthew Ayeni, Landmark University, OmuAran, Kwara State, Nigeria, e-mail: ayeni.matthew@lmu.edu.ng
Hakeem A. Ajonbadi, University of Doha for Science and Technology, Doha, Qatar, e-mail: hakeem.ajonbadi@udst.edu.qa
Sarah Higgins, University of Doha for Science and Technology, Qatar, e-mail: sarah.higgins@udst.edu.qa
Yusuf Bicer, Hamad Bin Khalifa University, Qatar, e-mail: ybicer@hbku.edu.qa
Mario Elzein, University of Doha for Science and Technology, Qatar, e-mail: mario.elzein@udst.edu.qa
Ali El-Keblawy, University of Sharjah, United Arab Emirates, e-mail: akeblawy@sharjah.ac.ae
Muhammad Farrukh Syed, National University of Computer and Emerging Science, Pakistan, e-mail: mfarrukh.shahid@nu.edu.pk
Prof. Harender Raj, Dr. Y. S. Parmar University of Horticulture and Forestry, India, e-mail: hrg_mpp@yahoo.com
Fahim Ullah Khan, University of Doha for Science and Technology, Doha, Qatar, e-mail: fahim.khan@udst.edu.qa
Hassan Mahasneh, University of Doha for Science and Technology, Qatar, e-mail: hassan.mahasneh@udst.edu.qa
Adnan Akhunzada, University of Doha for Science and Technology, Doha, Qatar, e-mail: adnan.adnan@udst.edu.qa
Hafiz Mohkum Hammad, Department of Agronomy Muhammad Nawaz Shareef University of Agriculture Multan, Pakistan, e-mail: hafiz.mohkum@mnsuam.edu.pk
Dr. Muhammad Riaz, Government College University Faisalabad, Pakistan, e-mail: mr548@ymail.com
Jean-Luc Bertrand-Krajewski, INSA Lyon, France, e-mail: jean-luc.bertrand-krajewski@insa-lyon.fr
Dr. Eugene Mohareb, University of Reading, UK, e-mail: e.mohareb@reading.ac.uk
Abdelhadi Ammari, Nationale Higher School of Hydraulics (ENSH), Algeria, e-mail: a.ammari@ensh.dz
Ayoub Zeroul, Ecole Nationale Supérieure d'Hydraulique, Algeria, e-mail: a.zeroual@ensh.dz
Felicia Olooto, Kwara State University Malete, Nigeria, e-mail: feliciamolooto@gmail.com

Susan Ajiere, University of Port Harcourt, Nigeria, e-mail: ajieresusan@gmail.com
Mohit Sharma, Central University of Jammu, India, e-mail: mohitcuj@gmail.com
Shipan Das Gupta, Noakhali Sceince and Technology University, Bangladesh, e-mail: shipan.bge@nstu.edu.bd
Mahwish Ejaz, University of Cambridge, United Kingdom, e-mail: me514@cam.ac.uk
Shams Ali Baig, Welcome to Abdul Wali Khan University Mardan, Pakistan, e-mail: drbaig@awkum.edu.pk
Riadh Bechir, IRA Medenine, Tunisia, e-mail: riadh113med@yahoo.fr
Serap Tepe, University of Health Sciences, Türkiye, e-mail: serap.tepe@sbu.edu.tr
Omar Ouda, Ministry of Environment, Water and Agriculture, Saudi Arabia, e-mail: oouda@mewa.gov.sa
Rukhsora Kholikova, Tashkent State University of Economics, Uzbekistan, e-mail: r.xoliqova@tsue.uz
Uthra R., SRM Institute of Science and Technology, India, e-mail: uthrar@srmist.edu.in
Maha Dewidar, University of Doha for Science and Technology, Doha, Qatar, e-mail: maha.dewidar@udst.edu.qa
Mohammed Noori Farhan, Diyala University, Iraq, e-mail: mohamedeco@uodiyala.edu.iq
Dr. Saeed Rauf, University of Sargodha, Pakistan, e-mail: saeedbreeder@hotmail.com
Safwan Mohammed, University of Debrecen, Hungary, e-mail: safwan@agr.unideb.hu
Kachalla Malabe, University of Maiduguri, Nigeria, e-mail: kmalabemohammed@gmail.com
Matthew Olufemi Adio, Federal University, Oye-Ekiti, Ekiti State, Nigeria, e-mail: matthew.adio@fuoye.edu.ng
Anthony Kola-Olusanya, Osun State University, Nigeria, e-mail: anthony.olusanya@gmail.com
Murad Yaghi, Al Hussein Technical University, Jordan, e-mail: murad.yaghi@htu.edu.jo
Dr. Mulugeta Genanu Kebede, Addis Ababa University, Ethiopia, e-mail: mule.genanu@gmail.com
Norbert Francois Tchouaffe Tchiadje, University of Ebolowa, Cameroon, e-mail: ntchoua@gmail.com
Nagi Abdussamie, University of Doha for Science and Technology, Qatar, e-mail: nagi.abdussamie@udst.edu.qa
Sidra Pervez, Shaheed Benazir Bhutto Women University Peshawar, Pakistan, e-mail: drsidrapervez@sbbwu.edu.pk
Gianluca Viscusi, University of Salerno, Italy, e-mail: gviscusi@unisa.it
Dr. Funmilola Aduke Akinola, Federal University Gashua, Nigeria, e-mail: faakinola@yahoo.com
Khaled Hassan, Egyptian Society for Migration Studies (EGYMIG), Egypt, e-mail: kehassan@gmail.com
Abdulaziz G. Alghamdi, Soil Sciences Department, College of Food and Agricultural Sciences, King Saud University, Saudi Arabia, e-mail: agghamdi@ksu.edu.sa

Imeldalyn Pacada, Phillipine Rice Research Institute, Philippines, e-mail: imeldalyn.pacada@gmail.com
Monidip Mondal, Jawaharlal Nehru University, India, e-mail: monidip33@gmail.com
Ahmed A. Abdalla, University of Doha for Science and Technology, Qatar, e-mail: ahmed.abdalla@udst.edu.qa
Monirul Mirza, University of Toronto, Canada, e-mail: monirul.mirza@utoronto.ca
Mamdouh Elmallah, Arab Academy for Science, Technology and Maritime Transport, Egypt, e-mail: mamdouhelmallah664@gmail.com
Eunice Moriyike Ogunbusola, Federal University, Oye-Ekiti, Ekiti State, Nigeria, e-mail: eunice.ogunbusola@fuoye.edu.ng
Asma Ayat, University of Salah Boubnider, Constantine, Algeria, e-mail: ayat_asma@hotmail.fr
Mohammad Javad Maghsoodi Tilaki, Universiti Sains Malaysia, e-mail: maghsoodi.tilaki@gmail.com
Prasanna K., Birbal Sahni Institute of Paleosciences, India, e-mail: prasanna@bsip.res.in
Solomon Mwampikita, Centre for Trade Policy and Development (CTPD), Zambia, e-mail: researcher.cce@ctpd.org.zm
Peter Fejer, Debrecen University, Hungary, e-mail: fejerp@agr.unideb.hu
Dr. Kehinde Oyeniran, Department of Biological Sciences, Bamidele Olumilua University of Education Science and Technology, Ikere-Ekiti, Ekiti State Nigeria, Nigeria, e-mail: oyeniran.kehinde@bouesti.edu.ng
Kurobuza Eddy Tukamushaba, University of Doha for Science and Technology, Qatar, e-mail: kurobuza.tukamushaba@udst.edu.qa
Fatima AlHamedi, Department of Integrative Agriculture, College of Agriculture and Veterinary Medicine, United Arab Emirates University, United Arab Emirates, e-mail: 200270204@uaeu.ac.ae
Faith Mugisha Ahabyoona, Cavendish University, Uganda, e-mail: ahabyoonaf@gmail.com
Anteneh Yeneneh, International Maritime College Oman, IMCO, Oman, e-mail: anteneh@imco.edu.om
Mohamed Himrane, University of Jijel, Algeria, e-mail: m_himrane@univ-jijel.dz
Kevi Govender, University of Doha for Science and Technology, Qatar, e-mail: pregalathan.govender@udst.edu.qa
Jafar Momani, University of Doha for Science and Technology, Qatar, e-mail: jafar.momani@udst.edu.qa
Abooali Golzary, University of Tehran, Iran, Islamic Republic of, e-mail: abooaligolzary1@gmail.com
Seyedmasnour Bidoki, Yazd University, Iran, Islamic Republic of, e-mail: smbidoki@yahoo.com
Dr. Mehdi Ghaffari, Oil Crops Research Department, Seed and Plant Improvement Institute (SPII), Iraq, e-mail: mghaffari69@gmail.com
Abdulkarim Assalem, Al-Baath University, Syrian Arab Republic, e-mail: assalem1@gmail.com

Contents

Polices, Legislations and Society

Matthew Olufemi Adio Department of Agricultural Economics and Farm Management, Faculty of Agriculture, Federal University, Oye-Ekiti, Ekiti State, Nigeria

Faith Mugisha Ahabyoona Department of Accounting, Finance and Microfinance, Kyambogo University, Kampala, Uganda

Ghufran Ahmed National University of Computer and Emerging Sciences, Karachi, Pakistan

Mourad Taleb Ahmed Laboratory Valorization and Recycling of Matters and Sustainable Development, University of Science and Technology Houari Boumediene, Bab Ezzouar, Algeria

Halima Aissaoui Laboratory of Environmental Engineering Process, University Constantine 3 Salah Boubnider, Ali Mendjeli, Constantine, Algeria

Oyepeju Abosede Ajetunmobi Department of Agricultural Economics, Faculty of Agricultural Sciences, Ladoke Akintola University of Technology, Ogbomoso, Nigeria

Hakeem A. Ajonbadi University of Science and Technology, Doha, Qatar

Adnan Akhunzada University of Doha for Science and Technology, Doha, Qatar

Akindele James Akintola Department of Agricultural Economics and Farm Management, Faculty of Agriculture, Federal University, Oye-Ekiti, Ekiti State, Nigeria

Ali Akyol Business Informatics University of Oldenburg, Oldenburg, Germany

Marwa Al-Ani Department of Chemical Engineering, College of Engineering, Qatar University, Doha, Qatar

Mustafa Khaled Al-Ani Ministry of Water Resources—National Water Resources Center, New Delhi, India

Ala Al-Ardah Department of Materials Science and Physics, College of Arts and Sciences, Qatar University, Doha, Qatar

Meera Alazem University of Doha for Science and Technology, Doha, Qatar

Sara Al-Baker Department of Chemical Engineering, College of Engineering, Qatar University, Doha, Qatar

Ali Hussein Ibraheem Al-Bayati Department of Biology, College of Education for Woman, University of Anbar, Ramadi, Iraq

Alzulfa Albulushi International Maritime College Oman, National University of Science and Technology, Suhar, Oman

Reem Al-Haidose Biotechnology Center, Agricultural Research Department, Ministry of Municipality, Doha, Qatar

Mohamed Saleh Al-Haj Faculty of Humanities, Sana'a University, Sanaa, Republic of Yemen

Naser Alhakmani International Maritime College Oman, National University of Science and Technology, Suhar, Oman

Ali Hamad Alhamadi Qatar Science and Technology Secondary School for Boys, Doha, Qatar

Naval Al-Hamdan Department of Chemistry and Earth Sciences, College of Arts and Sciences, Qatar University, Doha, Qatar

Ali A. Al-Hamzah Water Technologies Innovation Institute and Research Advancement, Saudi Water Authority, Jubail, Kingdom of Saudi Arabia;
School of Science and Technology, The University of New England, Armidale, Australia

Maathir Alhasani International Maritime College Oman, National University of Science and Technology, Suhar, Oman

Salim Alhashmi International Maritime College Oman, National University of Science and Technology, Suhar, Oman

Yousef Alhorr Gulf Organisation for Research and Development, Ar-Rayyan, Qatar

Hassan Ali College of Engineering and Technology, University of Doha for Science and Technology, Doha, Qatar;
Centre of Excellence for Sustainability and Food Security, University of Doha for Science and Technology, Doha, Qatar;
Department of Electrical Engineering, University of Doha for Science and Technology, Doha, Qatar

Lamia Ali Laboratory of Biological Oceanography and the Marine Environment (LOBEM), University of Science and Technology of Houari Boumedien (USTHB), Algiers, Algeria

Suzan Ali College of Computing and Information Technology, University of Doha for Science and Technology, Doha, Qatar

Saad Alkentar Homs University, Homs, Syria

Shahed AlNawaiseh School of Computing and Informatics, Al Hussein Technical University, Amman, Jordan

Ghoson Abdulaziz AL-Obaidly University of Doha, Doha, Qatar

Huthaifa Al-Omari School of Computing and Informatics, Al Hussein Technical University, Amman, Jordan

Noora Al-Qahtani Center for Advanced Materials (CAM), Office of the Vice President for Research and Graduate Studies (VPRGS), Qatar University, Doha, Qatar

Mansour Al Qubeissi College of Engineering and Technology, University of Doha for Science and Technology, Doha, Qatar

Walid Kh Alqaisi College of Engineering and Technology, University of Doha for Science and Technology, Doha, Qatar

Ahmad M. Alqudah Department of Biological and Environmental Sciences, College of Art and Science, Qatar University, Doha, Qatar

Bassem Alsahwa HIAST, Damascus, Syria

Rashid Alsawai International Maritime College Oman, National University of Science and Technology, Suhar, Oman

Ali Alseari Qatar Science and Technology Secondary School for Boys, Doha, Qatar

Abdulaziz Al Shandoudi International Maritime College Oman, National University of Science and Technology, Suhar, Oman

Mohammad Alsheyab National Planning Council, Doha, Qatar

Fahad Jassim Altamimi Qatar Science and Technology Secondary School for Boys, Doha, Qatar

Alshayma N. Al-Thani Department of Biological and Environmental Sciences, College of Art and Science, Qatar University, Doha, Qatar

Faisal Alzyoud Department of Computer Science, Faculty of IT, Isra University, Amman, Jordan

Abdulla Ansari Environmental Health and Safety, University of Doha for Science and Technology, Doha, Qatar

Sihem Arris Laboratory of Environmental Process Engineering, Faculty of Process Engineering, University of Constantine 3 Salah Boubnider, Constantine, Algeria

Summayya Asghar Department of Biochemistry, Shaheed Benazir Bhutto Women University (SBBWU), Peshawar, Pakistan

Hasna Ashraf University of Doha for Science and Technology, Doha, Qatar

Selin Aslantaş University of Health Sciences, Istanbul, Turkey

Abdulkareem Assalem Homs University, Homs, Syria

Asma Ayat Environmental Process Engineering Laboratory, University of Constantine3 Salah Boubnider, Constantine, Algeria;
University Setif 1; Ferhat ABBAS, Faculty of Technology, Setif, Algeria

Matthew Durojaiye Ayeni Department of Agricultural Economics and Extension, College of Agricultural Sciences, Landmark University, Omuaran, Kwara State, Nigeria

Nour El Islam Bachari Laboratory of Biological Oceanography and the Marine Environment (LOBEM), University of Science and Technology of Houari Boumedien (USTHB), Algiers, Algeria

Ahmed Badawi College of Engineering and Technology, University of Doha for Science and Technology, Doha, Qatar;
Department of Electrical Engineering, University of Doha for Science and Technology, Doha, Qatar

Evelyn H. Bandonill Rice Chemistry and Food Science Division, Science City of Munoz, Philippines;
Phillipine Rice Research Institute, Science City of Munoz, Nueva Ecija, Philippines

Zahida Bano University of Doha for Science and Technology, Doha, Qatar

Mohd Basheer University of Jordan, Amman, Jordan

Mahmoud Bataineh College of General Education, University of Doha for Science and Technology, Doha, Qatar

Scott Daniel Beard AtkinsRéalis, Epsom, Surrey, UK

Riadh Bechir Laboratoire LESOR Institut des Régions Arides de Médenine, Université de Gabès, Medenine, Tunisia

Jesusa C. Beltran Socio Economics Division, Science City of Munoz, Philippines;
Phillipine Rice Research Institute, Science City of Munoz, Nueva Ecija, Philippines

Amel Benhadji Laboratory of Reaction Engineering, University of Science and Technology Houari Boumediene, Algiers, Algeria

Farouk Benhamouda University Constantine 3, Ali Mendjeli, Constantine, Algeria

Rachid Benlamri Centre of Excellence for Sustainability and Food Security, University of Doha for Science and Technology (UDST), Doha, Qatar

Aqeela Bibi Department of Biochemistry, Shaheed Benazir Bhutto Women University (SBBWU), Peshawar, Pakistan

Asifa Bibi Bahria University, Islamabad, Pakistan

Seyedmasnour Bidoki Department of Textile Engineering, Yazd University, Yazd, Iran

Abdullah bin Bilal Al-Subhi Ministry of Environment Water and Agriculture, Riyadh, Kingdom of Saudi Arabia

Bahman Boostani College of Engineering Technology, University of Doha for Science and Technology, Doha, Qatar

Nabil Bougherira Laboratoire Ressource en Eau and Développement Durable, Université Badji Mokhtar Annaba, Annaba, Algeria

Hamza Bouguerra Laboratoire Ressource en Eau and Développement Durable, Université Badji Mokhtar Annaba, Annaba, Algeria

Yasser Boumezough Laboratory of Environmental Process Engineering, Faculty of Process Engineering, University of Constantine 3 Salah Boubnider, Constantine, Algeria

Ahmet Çabuk Karabük University, Karabük, Turkey

Julianne C. Caday University of Doha for Science and Technology, Doha, Qatar

Mohamed Chabi College of Engineering, Qatar University, Doha, Qatar

Hicham Chaffai Laboratoire Ressource en Eau and Développement Durable, Université Badji Mokhtar Annaba, Annaba, Algeria

Rami Chahin Business Informatics University of Oldenburg, Oldenburg, Germany

Shuhong Chai Centre for Maritime Engineering and Hydrodynamics, Australian Maritime College, University of Tasmania, Launceston, TAS, Australia

Supriyo Chakraborty Ministry of Earth Sciences, Indian Institute of Tropical Meteorology, Pune, India

Roneta Chaliulina Gulf Organisation for Research and Development, Ar-Rayyan, Qatar

Madhurima Chatterjee Department of Biological and Environmental Sciences, College of Art and Science, Qatar University, Doha, Qatar

Anis Chekirbane National Institute of Agronomy (INAT), Tunis, Tunisia

Chris Chin Australian Maritime College, Launceston, TAS, Australia

Hafsa Chohan University of Doha for Science and Technology, Doha, Qatar

Raj Kumar Dahal University of Doha for Science and Technology, Doha, Qatar

Larbi Dajbri Laboratoire Ressource en Eau and Développement Durable, Université Badji Mokhtar Annaba, Annaba, Algeria

Shaimaa R. Dalal Mansoura University, Mansoura, Egypt

Ali Demir Department of Textile Technologies and Design Engineering, Istanbul Technology University, Istanbul, Turkey

Nenita V. Desamero Plant Breeding and Biotechnology Division, Science City of Munoz, Philippines;
Phillipine Rice Research Institute, Science City of Munoz, Nueva Ecija, Philippines

Maha A. Dewidar University of Doha for Science and Technology, Doha, Qatar

Israr Ud Din Institute of Biotechnology and Genetic Engineering, The University of Agriculture Peshawar, Peshawar, Pakistan

Mouad Djelloudi Laboratory of Environmental Engineering Process, University Constantine 3, Ali Mendjeli, Constantine, Algeria

Animesh Dutta University of Guelph, Guelph, ON, Canada

Taghreed El-Kasaby Mansoura University, Mansoura, Egypt

Ali El-Keblawy Department of Applied Biology, College of Sciences, University of Sharjah, Sharjah, United Arab Emirates;
Faculty of Pharmacy, Al Salam University, Tanta, Egypt

Sohayla El Mokadem College of Engineering and Technology, University of Doha for Science and Technology, Doha, Qatar

Attiat Elnaggar Research Institute for Science and Engineering, University of Sharjah, Sharjah, United Arab Emirates;
Department of Botany and Microbiology, Faculty of Science, Alexandria University, Alexandria, Egypt

I. M. Elzein College of Engineering and Technology, University of Doha for Science and Technology, Doha, Qatar

Amin Esmaeili College of Engineering Technology, University of Doha for Science and Technology (UDST), Doha, Qatar

Anika Fajutagana College of Health Sciences, University of Doha for Science and Technology, Doha, Qatar

Olufemi Oyedokun Fakunle Department of Agricultural Economics and Farm Management, Faculty of Agriculture, Federal University, Oye-Ekiti, Ekiti State, Nigeria

Wajid Farhad University College Dera Murad Jamali, Dera Murad Jamali, Pakistan;
Lasbela University of Agriculture, Water and Marine Sciences, Uthal, Pakistan

Ahmed Faris Qatar Science and Technology Secondary School for Boys, Doha, Qatar

Aitazaz A. Farooque Canadian Center for Climate Change and Adaptation University of Prince Edward Island, St Peter's Bay, PEI, Canada;
Faculty of Sustainable Design Engineering, University of Prince Edward Island, Charlottetown, PEI, Canada

Christopher M. Fellows Water Technologies Innovation Institute and Research Advancement, Saudi Water Authority, Jubail, Kingdom of Saudi Arabia;
School of Science and Technology, The University of New England, Armidale, Australia

Valeria Freundt University of Doha for Science and Technology (UDST), Doha, Qatar

Milad Sabouri Gasgarei Carleton University, Ottawa, ON, Canada

Nourhan I. Ghoneim International Maritime College Oman, National University of Science and Technology, Suhar, Oman

Jorge Marx Gómez Business Informatics University of Oldenburg, Oldenburg, Germany

Abooali Golzary School of Environment, College of Engineering, Faculty of Environment, University of Tehran, Tehran, Iran

Giuliana Gorrasi Department of Industrial Engineering, University of Salerno, Fisciano, Italy

Ala Gouissem University of Doha for Science and Technology (UDST), Doha, Qatar

Eric Gubesch Centre for Maritime Engineering and Hydrodynamics, Australian Maritime College, University of Tasmania, Launceston, TAS, Australia

Yasmin A. Hageer Department of International Affairs, College of Arts and Sciences, Qatar University, Doha, Qatar

Fahima Halimi Laboratoire de Recherhe en Géologie, Université Badji Mokhtar Annaba, Annaba, Algeria

Hafiz Mohkum Hammad Department of Agronomy, MNS University of Agriculture Multan, Multan, Pakistan;
Department of Agricultural and Biological Engineering, University of Florida, Gainesville, FL, USA

Azzedine Hani Laboratoire Ressource en Eau and Développement Durable, Université Badji Mokhtar Annaba, Annaba, Algeria

Andrew Harris BMT Australia, Adeliade, SA, Australia

Yasen Harrye College of Engineering and Technology, University of Doha for Science and Technology, Doha, Qatar

Khaled El-Sayed Hassan Former Demographic Consultant at UNDP-Kuwait, Kuwait City, Kuwait;
Vice-president of the Egyptian Society for Migration Studies (EGYMIG), Cairo, Egypt

Norr Hassan Department of Environmental Health and Safety, University of Doha for Science and Technology, Doha, Qatar

Hocine Hassani University of Chlef, Chlef, Algeria

Aneela Hayder University of Guelph, Guelph, ON, Canada

Massoomeh Hedayati Marzbali Department of Urban Planning, School of Housing, Building and Planning, Universiti Sains Malaysia, Penang, Malaysia

Hassan Y. Hijazi University of Doha for Science and Technology, Doha, Qatar

Mohammed Himrane University of Jijel, Jijelv, Algeria

Woon Hong Centre for Maritime Engineering and Hydrodynamics, Australian Maritime College, University of Tasmania, Launceston, TAS, Australia

Fouzia Houma Marine and Coastal Ecosystems Laboratory (ECOSYSMarL), National Higher School of Marine Sciences and Coastal Management (ENSSMAL), Algiers, Algeria

Majid Ibrahim College of General Education, Doha University for Science and Technology, Doha, Qatar

Masoud Irannezhad Department of Civil Engineering, College of Engineering, Australian University, West Mishref, Safat, Kuwait;
Water, Energy and Environmental Engineering Research Unit, Faculty of Technology, University of Oulu, Oulu, Finland

Bachari Nour El Islam Laboratory of Biological Oceanography and the Marine Environment (LOBEM), University of Science and Technology of Houari Boumedien (USTHB), El Alia, Bab Ezzouar, Algiers, Algeria

Najma Nur Islam Department of Applied Biology, College of Sciences, University of Sharjah, Sharjah, United Arab Emirates

Seyed Behbood Issa-Zadeh School of Maritime Science, University of Gibraltar, Campus Europa Point, Gibralta GX11 1AA, UK

Asma Ghafoor Jan Department of Biological and Environmental Sciences, College of Art and Science, Qatar University, Doha, Qatar

Mohamed Jaoued Laboratoire LESOR Institut des Régions Arides de Médenine, Université de Gabès, Medenine, Tunisia

Abitha John University of Doha for Science and Technology (UDST), Doha, Qatar

Matilda Kaliba University of Zambia, Lusaka, Zambia

Muhammad Nawaz Kandhro Department of Agronomy, Sindh Agriculture University Tandojam, Tandojam, Pakistan

Akano Kayode Lagos State University of Education (LASUED), Oto-Ijanikin (Epe Campus), Lagos State, Nigeria;
Department of Agricultural Education, College of Vocational and Entrepreneurship Education (COVED), Lagos State, Nigeria

Carol Khadra QAFCO (Qatar Fertiliser Company), Mesaieed, Qatar

Fahim Ullah Khan University of Doha for Science and Technology, Doha, Qatar;
Department of Agriculture, Hazara University Dodhial Mansehra, Khyber Pakhtunkhwa, Pakistan

Khalaf Khatatneh Prince Abdullah bin Ghazi for Communication and IT College, AlBalqa Applied University, Amman, Jordan

Chaima Khayri Environmental Process Engineering Laboratory, University of Constantine3 Salah Boubnider, Constantine, Algeria

Djihane Khechine Environmental Process Engineering Laboratory, University of Constantine3 Salah Boubnider, Constantine, Algeria

Khaoula Khellaf Environmental Process Engineering Laboratory, University of Constantine3 Salah Boubnider, Constantine, Algeria

Anthony Kola-Olusanya Osun State University, Osogbo, Nigeria

Aparna Krishna Applied Research, Innovation and Economic Development Directorate, University of Doha for Science and Technology, Doha, Qatar

Anurag Kumar Birbal Sahni Institute of Palaeosciences, Lucknow, India

Jamal Kurdi Chemical Engineering, University of Doha for Science and Technology, Doha, Qatar

Ahmet Kurklu Department of Agricultural Mechanization and Technologies Engineering, Akdeniz University, Antalya, Turkey

Ayaz Ahmed Lashari Department of Agronomy, Sindh Agriculture University Tandojam, Tandojam, Pakistan

Latif-ur-Rahman Institute of Chemical Sciences, University of Peshawar, Peshawar, Pakistan

Mohamed Said Lebad Laboratory of the Industrial Science Process Engineering, University of the Sciences and Technology Houari Boumediene (USTHB), Algiers, Algeria

Sherzod Madrahimov Department of Sciences, Texas A&M University, Doha, Qatar

Mohammad Javad Maghsoodi Tilaki Department of Geography, School of Humanities, Universiti Sains Malaysia, Penang, Malaysia

Hassan Mahasneh College of Engineering and Technology, University of Doha for Science and Technology, Doha, Qatar

Tahira Jehan Makda University of Doha for Science and Technology (UDST), Doha, Qatar

K. M. Malabe Department of Agricultural Extension Services, University of Maiduguri, Maiduguri, Nigeria

Hind Mamlouk Department of Sciences, Texas A&M University, Doha, Qatar

Dana Khider Mawlood Centre for Environmental Studies, University of Kurdistan Hewlêr, Erbil City, Kurdistan Region, Iraq

Mohamed Meddi Ecole Nationale Supérieure d'Hydraulique (GEE), Guerrouaou, Algeria

Danish Mehmood SZABIST University, Islamabad, Pakistan

Fatima Mekhlouf University of Relizane, Relizane, Algeria

A.-H. Meniai Laboratory of Environmental Engineering Process, University Constantine 3, Ali Mendjeli, Constantine, Algeria

Saeed Mohajernasab Delmar Systems, Perth, WA, Australia

Asma Mohamed Department of Biological and Environmental Sciences, College of Arts and Sciences, Qatar University, Doha, Qatar

Raneem Mohamed Department of Biological and Environmental Sciences, College of Arts and Sciences, Qatar University, Doha, Qatar

Bryant Montenegro Gulf Organisation for Research and Development, Ar-Rayyan, Qatar

Kareem A. Mosa Department of Applied Biology, College of Sciences, University of Sharjah, Sharjah, United Arab Emirates;
Department of Biotechnology, Faculty of Agriculture, Al-Azhar University, Cairo, Egypt

Haruna Musa Moda Department of Environmental Health and Safety, University of Doha for Science and Technology, Doha, Qatar

S. B. Mustapha Department of Agricultural Extension Services, University of Maiduguri, Maiduguri, Nigeria

Lumu Emmanuel Roy Muyingo Business, Law, Communication and Accounting, INTI International University, Kuala Lumpur, Malaysia

Solomon Mwampikita Centre for Trade Policy and Development (CTPD), Olympia Park, Lusaka, Zambia

Lamri Nacef Laboratory of Biological Oceanography and the Marine Environment (LOBEM), University of Science and Technology of Houari Boumedien (USTHB), El Alia, Bab Ezzouar, Algiers, Algeria

Muhammad Asif Nawaz Department of Biotechnology, Shaheed Benazir Bhutto University (SBBU), Sheringal, Upper Dir, Pakistan

Jonathan M. Niones Genetic Resource Division, Science City of Munoz, Philippines;
Phillipine Rice Research Institute, Science City of Munoz, Nueva Ecija, Philippines

Oluwaseun Oyebisi Odetola Osun State University, Osogbo, Nigeria

Ayodele Olarewaju Department of Agricultural Economics and Farm Management, Faculty of Agriculture, Federal University, Oye-Ekiti, Ekiti State, Nigeria

Mohammed Ouda Electrical Engineering, University of Doha for Science and Technology, Doha, Qatar

Omar K. M. Ouda Ministry of Environment Water and Agriculture, Riyadh, Kingdom of Saudi Arabia

Imeldalyn G. Pacada Plant Breeding and Biotechnology Division, Science City of Munoz, Philippines;
Phillipine Rice Research Institute, Science City of Munoz, Nueva Ecija, Philippines

Loida M. Perez Plant Breeding and Biotechnology Division, Science City of Munoz, Philippines;
Phillipine Rice Research Institute, Science City of Munoz, Nueva Ecija, Philippines

Sidra Pervez Department of Biochemistry, Shaheed Benazir Bhutto Women University (SBBWU), Peshawar, Pakistan

Michael Phillips University of Doha for Science and Technology, Doha, Qatar

K. Prasanna Birbal Sahni Institute of Palaeosciences, Lucknow, India

Muhammad Qasim SZABIST University, Islamabad, Pakistan

Hadia Qureshi Department of Biochemistry, Shaheed Benazir Bhutto Women University (SBBWU), Peshawar, Pakistan

Ravi Rangarajan Department of Environmental Health and Safety, College of Health Sciences, University of Doha for Science and Technology (UDST), Doha, Qatar

Attiq ur Rehman Department of Agricultural Sciences, Faculty of Agriculture and Forestry, The University of Helsinki, Helsinki, Finland

Juliet P. Rillon Crop Protection Division, Science City of Munoz, Philippines;
Phillipine Rice Research Institute, Science City of Munoz, Nueva Ecija, Philippines

Harid Roumaissa Marine and Coastal Ecosystems Laboratory (ECOSYSMarL), National Higher School of Marine Sciences and Coastal Management (ENSSMAL), Algiers, Algeria

Nassila Sabba Laboratory Valorization and Recycling of Matters and Sustainable Development, University of Science and Technology Houari Boumediene, Bab Ezzouar, Algeria;
Laboratory of Recovery and Recycling of Matter for Sustainable Development, USTHB, University of Science and Technology Houari Boumediene, Algiers, Algeria

Milad Sabouri Department of Civil and Environmental Engineering, Carleton University, Ottawa, Canada

Lamine Salim Laboratory of Biological Oceanography and the Marine Environment (LOBEM), University of Science and Technology of Houari Boumedien (USTHB), El Alia, Bab Ezzouar, Algiers, Algeria;
Higher School of Saharan Agriculture Adrar, Adrar, Algeria

Haitem Sekhri Laboratory of Environmental Engineering Process, University Constantine 3, Ali Mendjeli, Constantine, Algeria

Farah Seray Laboratory Valorization and Recycling of Matters and Sustainable Development, University of Science and Technology Houari Boumediene, Bab Ezzouar, Algeria

Farzin Shabani Department of Biological and Environmental Sciences, College of Art and Science, Qatar University, Doha, Qatar

Anupam Sharma Birbal Sahni Institute of Palaeosciences, Lucknow, India

Chris Shearer BMT Australia, Melbourne, VIC, Australia

Yusuf Shogbesan Osun State University, Osogbo, Nigeria

Mohamed R. Shouman Faculty of Maritime Transport and Technology, Arab Academy for Science, Technology and Maritime Transport, Alexandria, Egypt

Mostafa H. Sliem Center for Advanced Materials (CAM), Office of the Vice President for Research and Graduate Studies (VPRGS), Qatar University, Doha, Qatar

Mostafa Soliman College of Engineering and Technology, University of Doha for Science and Technology, Doha, Qatar

Sara Soussy College of Engineering Technology, University of Doha for Science and Technology, Doha, Qatar

Roel R. Suralta Crop Biotech Center, Santo Domingo, Philippines;
Phillipine Rice Research Institute, Science City of Munoz, Nueva Ecija, Philippines

Salah Eddine Tachi Laboratoire Ressource en Eau and Développement Durable, Université Badji Mokhtar Annaba, Annaba, Algeria

Mourad Taleb Ahmed Laboratory of Recovery and Recycling of Matter for Sustainable Development, USTHB, University of Science and Technology Houari Boumediene, Algiers, Algeria

Monther Tarawneh Department of IT, Faculty of IT and Communications, Tafila Technical University, Tafila, Jordan

Omar Tarawneh Software Engineering Department, Amman Arab University, Amman, Jordan

Michel Tchotsoua Department of Geography and Geomatic, University of Ngaoundéré, Ngaoundéré, Cameroon

Norbert F. Tchouaffe-Tchiadje Forestry, Water and Environment, Higher Institute for Agriculture, University of Ebolowa, Ebolowa, Cameroon;
Kesmonds International University, Ngaoundéré, Cameroon

Serap Tepe University of Health Sciences, Istanbul, Turkey

Nick Tighe BMT Australia, Melbourne, VIC, Australia

Eddy Kurobuza Tukamushaba Department of Management, College of Business, University of Doha for Science and Technology, Doha, Qatar

R. Uthra SRM Institute of Science and Technology, Kattankulathur, Tamil Nadu, India

Gianluca Viscusi Department of Industrial Engineering, University of Salerno, Fisciano, Italy

Murad A. Yaghi School of Computing and Informatics, Al Hussein Technical University, Amman, Jordan

Raziq Yaqub Department of Electrical Engineering and Computer Science, A&M University, Huntsville, AL, USA

Noor Kh. Yashooa Centre for Environmental Studies, University of Kurdistan Hewlêr, Erbil City, Kurdistan Region, Iraq

Anteneh Mesfin Yeneneh International Maritime College Oman, National University of Science and Technology, Suhar, Oman

Rokia Youcef Laboratory Valorization and Recycling of Matters and Sustainable Development, University of Science and Technology Houari Boumediene, Bab Ezzouar, Algeria;
Laboratory of Recovery and Recycling of Matter for Sustainable Development, USTHB, University of Science and Technology Houari Boumediene, Algiers, Algeria

Florence Adeoti Yusuf Osun State University, Osogbo, Nigeria

Muhammad Zubair University of Doha for Science and Technology, Doha, Qatar

Aiste Zukaityte Gulf Organisation for Research and Development, Ar-Rayyan, Qatar

Alhareth Zyoud Department of Electrical and Computer Engineering, Faculty of Engineering, Birzeit University, Birzeit, Ramallah, Palestine

Climate Change Impact

Minimizing the Impact of Acid Rain on Crops and Plants by Reducing the Sulfur Oxide Emissions from the Combustion of Organic Sulfur Compounds in Petroleum Fuels

Hassan Y. Hijazi, Julianne C. Caday, Noor Abdelaziz, and Meera Alazem

Abstract A molecularly imprinted (MIP) film was made by photo radical polymerization on a plastic substrate and used to remove organic sulfur compounds from gasoline fuel. The removal of these compounds reduces the emission of sulfur oxide gases, which are responsible for the formation of acid rain. Optimization of the polymeric films included the selection of the functional monomers and solvents. The adsorption studies of the film, including adsorption kinetics, were investigated, and analysis was performed using a gas chromatography pulse flame photometric detector (GC-PFPD). Benzothiophene (BT) was selected as an example model of the organic sulfur compound in gasoline. BT-1-Vim-MIP film exhibited the highest binding capacity and imprinting factor compared to NIP among the selected monomers. 1-octanol produces a stable film and is chosen as the most suitable solvent for film preparation compared with other solvents. The adsorption equilibrium for 1-Vim-MIP was achieved within 60 min, and the adsorption capacity (Q) reached 34.2 mg g^{-1} for the BT compound.

Keywords Polymer · Sulfur · Plants · Acid rain · Food security · Petroleum products

1 Introduction

Acid rain is a phenomenon resulting mainly from the emission of gas pollutants such as sulfur oxides (SO_x) into the atmosphere, produced mainly from petroleum products such as gasoline. SO_x gases emitted into the atmosphere chemically transform to form sulfuric acid (H_2SO_4), causing acid rain [1]. Acid rain has several detrimental

H. Y. Hijazi (✉) · J. C. Caday · N. Abdelaziz · M. Alazem
University of Doha for Science and Technology, Doha, Qatar
e-mail: hassan.hijazi@udst.edu.qa

S. Al-Naemi et al. (eds.), *Water and Food Security in the Face of Climate Change: Challenges and Opportunities for Resilience*, World Sustainability Series,
https://doi.org/10.1007/978-3-032-00098-9_1

effects on crop growth, including soil pH alteration, nutrient deficiency, leaf and root damage, microbial activity, water uptake, and crop yield and quality reduction. These impacts negatively affect the amount of crops grown, posing a threat to food availability. pH of acid rain can vary between 3.0 and 5.5, changing soil acidity and reducing levels of calcium, magnesium, and potassium that are vital for plant growth. Soil acidity can result in nutrient deficiency in plants. Nutrients may become less available or even toxic at low pH levels [2, 3]. Aluminum and manganese, for example, tend to be more soluble and, hence, harmful to plants when the soils are acidic [4]. Acid environments also affect plant roots, limiting their ability to absorb water as well as nutrients; this can hinder fodder production and reduce crop yields. Similarly, leaves and stems on plants can get hurt, too, leading to leaf yellowing, tissue death, and reduced photosynthesis, harming plant productivity or health [5]. Such bacteria on earth break down organic matter into usable form, thus recycling nutrients. Such important organisms may be adversely affected by changes in pH level, especially towards acidic conditions; hence, nutrients available for crops decrease [6]. Rainwater acidification alters water absorption by plants, changing the soil structure and making it lose its water retention capacity [7]. In sandy soils, this leads to poor water uptake by crops, which worsens drought effects on them. Nutrient deficiency, root damage, and direct leaf damage together can significantly reduce crops' amount and quality [8, 9]. This can impact farmers economically and affect the food stock [10]. Not to mention that acid rain is also responsible for water pollution, infrastructure and cultural heritage damage, and human health risks [11–13].

Removal of sulfur compounds from petroleum products produces fuels containing low amounts of sulfur, which helps reduce the emission of SO_x gases that result from the combustion of sulfur-containing fuels. This aligns with worldwide governmental regulation of the sulfur content in petroleum products in order to protect the environment. The maximum allowable concentration for gasoline fuel in the United States is 15 ppm, while in Europe, it is 10 ppm [14, 15]. The most common method to remove sulfur from crude oil is hydrodesulfurization (HDS). Unfortunately, it remains less effective for removing certain refractory organic sulfur compounds, such as thiophenes and their derivatives [16, 17]. Thiophenes and their derivatives account for more than 70% of the total amount of sulfur in gasoline and diesel [18]. Therefore, the development of new absorptive materials for the removal of refectory sulfur compounds is necessary. Molecular imprinted polymer is a promising material due to its selectivity, high adsorption capacity, mechanical and thermal stability, relative ease of preparation and minimal cost compared to other adsorptive materials.

Molecularly imprinted polymers (MIPs) are artificial materials prepared selectively to recognize and bind to targeted molecules, mimicking the binding sites of natural receptors (Fig. 1). The process involves preparing a polymer in the existence of an imprint (template), monomers, cross-linkers, photo-initiator, and porogen (solvent). First, the imprint and monomers form selective bonds, followed by polymerization with cross-linkers. After polymerization, the polymer releases the template molecule by washing using an appropriate solvent, leaving a cavity.

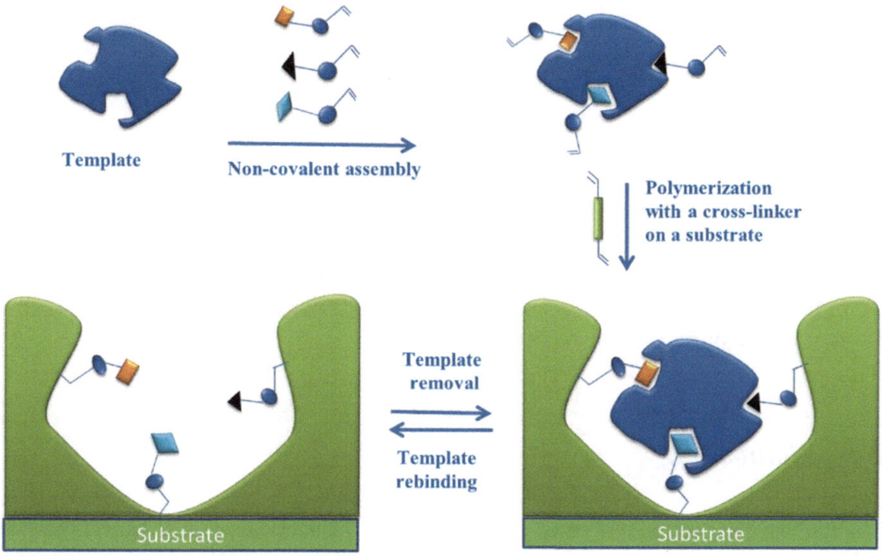

Fig. 1 Process for the synthesis of molecularly imprinted polymer

Durable cavities left in the polymer have size, shape, and functionality complementary to the original template and can function as target re-binding sites. This means that MIPs can be used for specific high-affinity molecular recognition, an approach that can be applied to various applications [19–21]. MIPs can be synthesized using different methods, including bulk polymerization, which produces selective polymers with high affinity. However, this method can result in low adsorption capacity due to small or discontinuous pores. Additionally, grinding these polymers produces particles with random shapes and sizes, affecting the adsorption of targets [21].

Uniform particles can be prepared more effectively through precipitation polymerization or by grafting onto silica or solid core particles. However, precipitation polymerization requires a substantial amount of template due to the high dilution factor, while the grafting method is both intricate and time-intensive [22, 23]. Alternatively, MIPs prepared as thin films exhibit high selectivity and enhance surface area [24–26], which offers high surface adsorption compared to other formats, such as bulk materials or powders. This is due to their large surface area-to-volume ratio and the accessibility of adsorption sites. Molecularly imprinted polymer films were prepared in our previous work [24] and successfully demonstrated the ability to extract organic sulfur compounds from water samples selectively.

2 Experimental

2.1 Materials

Methacrylic acid (99%), 4-vinylpyridine (95%), 1-vinylimidazol (≥ 99%), ethylene glycol dimethacrylate (98%), 2.2-dimethoxy-2-phenylacetophenone (99%), benzothiophene (≥ 95%), 3-methyl benzothiophene (96%), were all purchased from Sigma Aldrich. Methanol, acetonitrile, chloroform, hexane, toluene, and 1-octanol were ACS reagent grades purchased from ACP chemicals. Micro cover slides (18 × 18 mm^2) were purchased from Fisher Scientific.

2.2 Methodology

Molecularly imprinted polymer films were synthesized as a thin film. The pre-polymerization mixture was accurately prepared for each MIP film by combining the appropriate components, including monomer, template, crosslinker, photoinitiator, and porogen (solvent), in a glass vial (Table 1). The mixtures undergo degassing in a sonicator for five minutes to eliminate dissolved air, which may impede photo-polymerization. A precise volume of the pre-polymerization mixture was then dispensed onto the surface of a plastic slide. Once the mixture was evenly distributed on the slide, the polymer mixture was exposed to UV light (254 nm) for 45 min. Subsequently, the slide's cover was removed, and the MIP-coated plastic slides were immersed in a washing solution (methanol) to extract and eliminate any unreacted components. This washing step was repeated multiple times, as necessary, to remove the template entirely. The template removal was monitored using a UV-Vis spectrophotometer to ensure complete elimination from the polymer film. Similarly, non-imprinted polymer (NIP) was synthesized following the same procedure but without the template for comparative analysis. The mass of the polymer was determined by comparing weights using an analytical balance. An illustration depicting the steps of MIP film preparation steps is provided in Fig. 2.

Table 1 MIP film compositions with different monomers

Polymer	Template	Monomer	Crosslinker	Initiator	Porogen (1-octanol)
MMA-MIP	BT 0.149 mmol	MAA 0.596 mmol	EDGMA 2.98 mmol	DMPA 0.046 mmol	900 μL
4-VP-MIP		4-VP 0.596 mmol			
1-Vim-MIP		1-Vim 0.596 mmol			

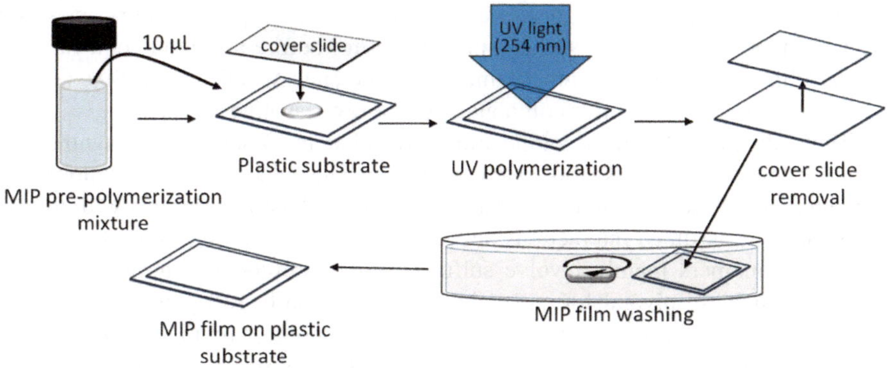

Fig. 2 Preparation steps of MIP film on a plastic substrate

The gas chromatography pulse flame photometric detector (GC-PFPD) was the primary analysis tool in optimizing the MIP film. Additionally, this instrument facilitated the detailed analysis of organic sulfur compounds present in gasoline samples due to its exceptional chromatographic separation capabilities and selective PFPD detection for sulfur compounds. The selection of monomer and porogen, accompanied by the kinetic and adsorption capacity study of MIP films, was carried out by introducing two MIP films (~ 3 mg each) into glass vials. Each vial contained 10 mL of raw gasoline or hexane solution spiked with an initial concentration of 0.50 mmol L^{-1} BT. The samples were stirred using magnetic stirring at room temperature for various durations. Subsequently, the residual BT content in the samples was quantified using GC-PFPD. The adsorption capacity (Q) of the BT compound was then calculated according to the following equation:

$$Q = \frac{(C_0 - C)V}{m_{film}}$$

In the equation, V (L) represents the volume of the gasoline sample, C_0 represents the concentration of sulfur compound(s) before removal, C represents the concentration of the sulfur compound(s) at equilibrium after removal, and m_{film} (g) represents the mass of MIP film.

3 Results and Discussion

Gasoline samples were collected and placed in gas canisters from various fuel stations using proper storage protocols. Benzothiophene-MIP was selected in this study to illustrate the ability of MIP to remove sulfur compounds from gasoline samples. The analysis involved injecting 1 µL of each sample into a gas chromatography pulse flame photometric detector (GC-PFPD). A calibration curve was constructed for BT by analyzing varied concentrations ranging from 5 to 100 ppm. The curve was used to determine the amount of the compound present before and after removal from the gasoline. A Chromatogram obtained from the analysis of raw gasoline (Fig. 3) reveals the presence of sulfur compounds in the raw gasoline samples.

Three distinct MIP film compositions (detailed in Table 1) were synthesized using different monomers (illustrated in Fig. 4). The selection of these monomers was predicted based on their potential to establish robust interactions with the BT template. The assumed interactions between the aromatic thiophene ring of the BT and the monomers mainly involve sulfur-nitrogen, sulfur–oxygen and π–π interactions, which are pivotal for enhancing the formation of recognition cavities via the polymerization process. An excess of these monomers relative to the BT was employed to favour the formation of BT-monomer complexes, thereby facilitating the creation of highly selective recognition sites within the MIPs. Various porogens were tested to find the best solvent that forms the most stable film on the plastic substrate. Among the six porogens (methanol, acetonitrile, chloroform, hexane, toluene, and 1-octanol) examined in this study, 1-octanol was found to be the most suitable solvent for MIP film preparation; other solvents produced films that appeared to be unstable and flaky.

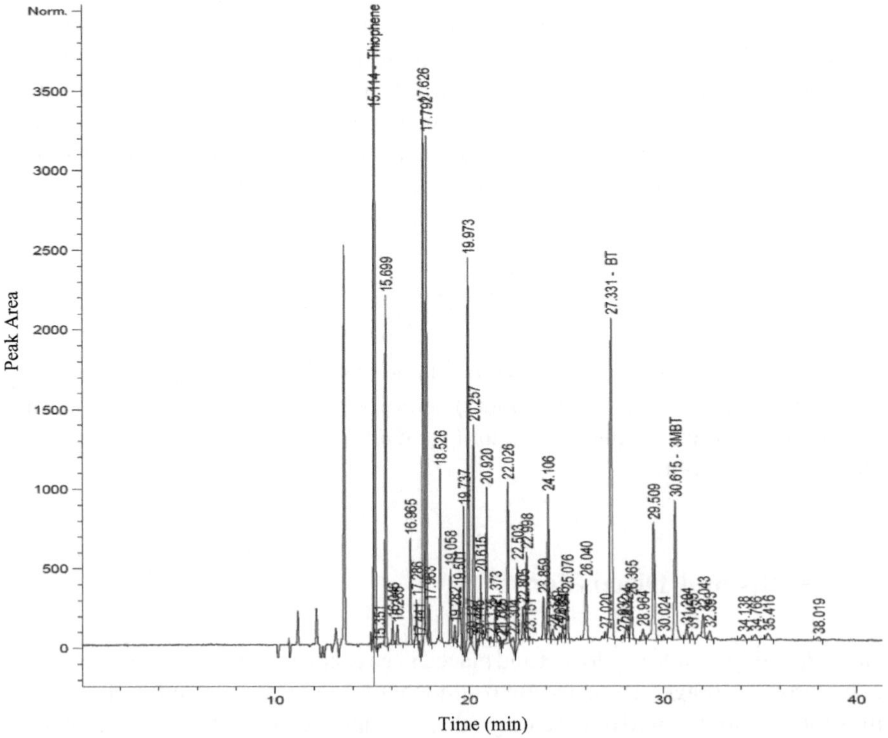

Fig. 3 A GC chromatogram was obtained for a gasoline sample of sulfur compounds

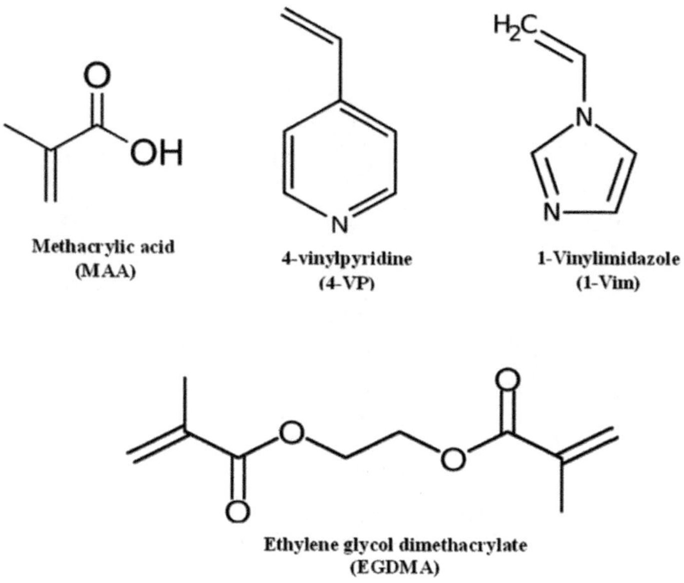

Fig. 4 Selected monomers and cross-linkers

Based on the data presented in Fig. 5, BT-1-Vim-MIP film exhibited the highest binding capacity and imprinting factor ($IF = Q_{MIP}/Q_{NIP}$) among the selected monomers, demonstrating an adsorption capacity of 34.2 mg g^{-1} during a 60-min removal period for the BT compound in gasoline sample. Therefore, 1-vinylimidazole emerges as the optimal monomer for removing BT from gasoline samples.

Figure 6 shows the adsorption kinetics of BT on the MIP film. The adsorption capacity of BT increased rapidly within 30 min and then slowed down until it reached equilibrium at 60 min. The initial rapid adsorption occurred due to the higher number of binding sites available on the polymeric film, which was gradually reduced later. The adsorption equilibrium for the BT compound was achieved after 60 min. Reusing the films is possible after the desorption of BT with a proper solvent before repeating the adsorption process. Unlike other adsorption materials like powders or beads that need filtration or further separation, MIP films can be easily removed from gasoline after adsorption.

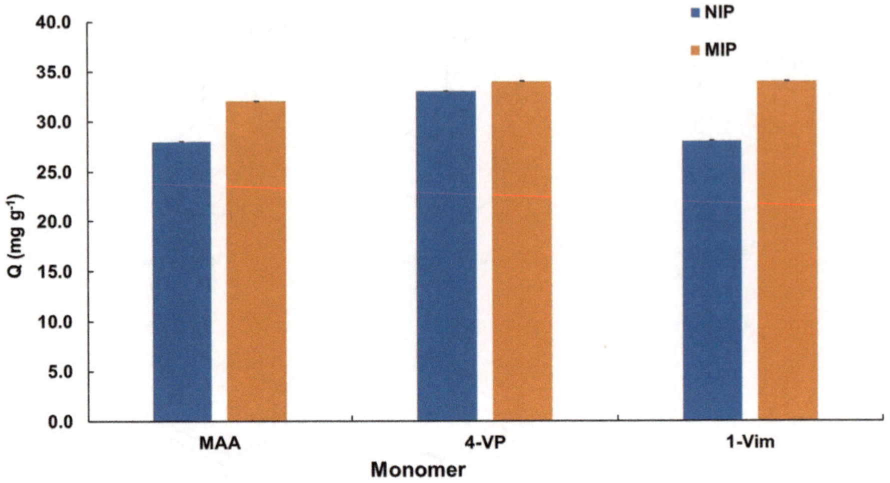

Fig. 5 Comparison of three BT-MIP films prepared as adsorbents for removing benzothiophene compounds from gasoline samples at 60 min (equilibration time)

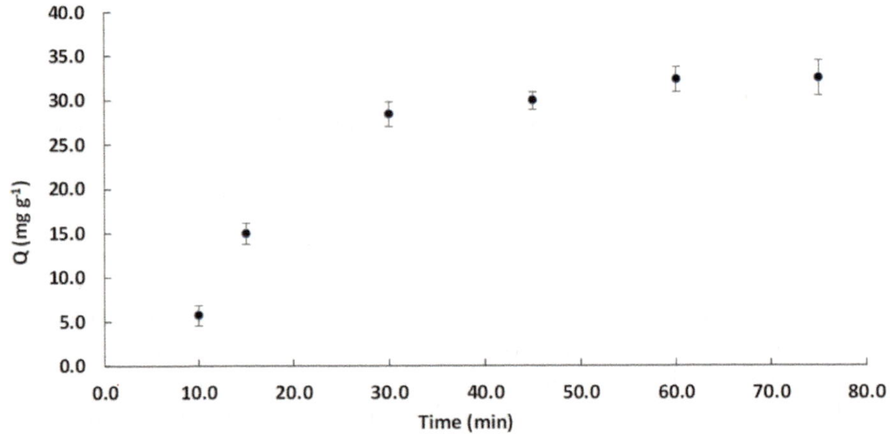

Fig. 6 Kinetic curve for adsorption of BT in 10 mL of hexane solution with an initial concentration of 0.50 mmol

4 Conclusion

The research outcomes indicate that the MIP film exhibited notable adsorption capacity (34.2 mg g^{-1}) and capability for removing sulfur compounds from gasoline samples. Removing these compounds is key to reducing the emissions of SO_x gases and mitigating the impact of acid rain on crops, which safeguards agricultural productivity and economic stability. MIP films for other sulfur compounds can be prepared and used simultaneously to remove sulfur compounds from gasoline.

Regeneration of the used MIP films using a proper solvent is possible, lowering the polymer preparation cost. This research supports sustainable agriculture, aids food security targets, and contributes to a cleaner and healthier environment.

Acknowledgements This work was supported by the Qatar National Research Fund (a member of the Qatar Foundation) and the University of Doha for Science and Technology (UDST).

Conflict of Interest No conflict of interest.

References

1. Grennfelt P et al (2020) Acid rain and air pollution: 50 years of progress in environmental science and policy 49(4). https://doi.org/10.1007/S13280-019-01244-4
2. Climate change: unpacking the burden on food safety. FAO (2020). https://doi.org/10.4060/ca8185en
3. EPA Homepage (2024). https://www.epa.gov/acidrain/effects-acid-rain. Last accessed 08 Aug 2024
4. Van Antwerpen R et al (2022) A review of field management practices impacting root health in sugarcane. In: Advances in agronomy, vol 173
5. Rodríguez-Sánchez VM, Rosas U, Calva-Vásquez G, Sandoval-Zapotitla E (2020) Does acid rain alter the leaf anatomy and photosynthetic pigments in urban trees? Plants 9:1–16
6. Zhang Y et al (2023) An overview of the direct and indirect effects of acid rain on plants: relationships among acid rain, soil, microorganisms, and plants. Sci Total Environ 873. https://doi.org/10.1016/j.scitotenv.2023.162388
7. Wei H et al (2021) Leaching of simulated acid rain deteriorates soil physiochemical and mechanical properties in three agricultural soils. Catena 206
8. Li HR et al (2020) Acid rain increases the impact of rice blast on crop health via inhibiting resistance enzymes. Plants 9:1–16
9. Pham HTT, Nguyen AT, Do ATN, Hens L (2021) Impacts of simulated acid rain on the growth and the yield of soybean (Glycine max (l.) merr.) in the mountains of northern Vietnam. Sustainability (Switzerland) 13
10. Zhang Q et al (2023) High environmental costs behind rapid economic development: evidence from economic loss caused by atmospheric acid deposition. J Environ Manag 334
11. Zhang L et al (2024) Chemical characteristics of long-term acid rain and its impact on lake water chemistry: a case study in Southwest China. J Environ Sci (China) 138:121–131
12. Esteban-Cantillo OJ, Menendez B, Quesada B (2024) Climate change and air pollution impacts on cultural heritage building materials in Europe and Mexico. Sci Total Environ 921
13. Barreca AI, Neidell M, Sanders NJ (2021) Long-run pollution exposure and mortality: evidence from the acid rain program. J Public Econ 200
14. Many Northeastern states require lower-sulfur heating oil while some embrace biodiesel blends—National Oil heat Research Alliance. https://noraweb.org/2014/11/many-northeastern-sates-require-lower-sulfur-heating-oil-while-some-embrace-biodiesel-blends/. Last accessed 08 Aug 2024
15. EU: fuels: diesel and gasoline|transport policy. https://www.transportpolicy.net/standard/eu-fuels-diesel-and-gasoline/. Last accessed 08 Aug 2024
16. Hossain MN, Park HC, Choi HS (2019) A comprehensive review on catalytic oxidative desulfurization of liquid fuel oil. Catalysts 9. https://doi.org/10.3390/catal9030229
17. Zhu W et al (2013) Catalytic kinetics of oxidative desulfurization with surfactant-type polyoxometalate-based ionic liquids. Fuel Process Technol 106:70–76

18. Yang YZ, Liu XG, Xu BS (2014) Recent advances in molecular imprinting technology for the deep desulfurization of fuel oils. New Carbon Mater 29:1–14. https://doi.org/10.1016/S1872-5805(14)60121-9
19. Belbruno JJ (2019) Molecularly imprinted polymers. Chem Rev 119:94–119
20. Ndunda EN (2020) Molecularly imprinted polymers—a closer look at the control polymer used in determining the imprinting effect: a mini review. J Mol Recognit 33:e2855
21. Awokoya KN, Okoya AA, Elujulo O (2021) Preparation, characterization and evaluation of a styrene-based molecularly imprinted polymer for capturing pyridine and pyrrole from crude oil. Sci African 13:e00947
22. Yan H, Ho Row K (2006) Characteristic and synthetic approach of molecularly imprinted polymer. Int J Mol Sci 7:155–178
23. Vasapollo G et al (2011) Molecularly imprinted polymers: present and future prospective. Int J Mol Sci 12:5908–5945. https://doi.org/10.3390/ijms12095908
24. Hijazi HY, Bottaro CS (2020) Molecularly imprinted polymer thin-film as a micro-extraction adsorbent for selective determination of trace concentrations of polycyclic aromatic sulfur heterocycles in seawater. J Chromatogr A 1617:460824
25. Boysen RI, Schwarz LJ, Nicolau Dv, Hearn MTW (2017) Molecularly imprinted polymer membranes and thin films for the separation and sensing of biomacromolecules. J Sep Sci 40:314–335
26. Kryscio DR, Peppas NA (2012) Surface imprinted thin polymer film systems with selective recognition for bovine serum albumin. Anal Chim Acta 718:109–115

The Social Impacts of Climate Change and Its Reflections on Working Life in Terms of Occupational Safety

Serap Tepe, Ahmet Çabuk, and Selin Aslantaş

Abstract The risks posed by climate change are critical elements that threaten the social and economic structures globally. To effectively address these threats, it is essential to comprehensively analyze the short, medium and long-term impacts of climate change. The economic effects of climate change can be examined across a broad spectrum, from changes in production processes to increases in energy costs and the economic costs of natural disasters. Climate change also significantly affects the agricultural sector; changes in temperature and precipitation patterns negatively impact agricultural productivity, threatening food security. In the energy sector, climate change necessitates a reduction in dependence on fossil fuels and an increased shift towards renewable energy sources. The effects on social events and migration include accelerated local and international migration movements, increased social conflicts, and deepening social inequalities. In the health sector, climate change leads to a decline in air quality, reduced water resources, and the spread of vector-borne diseases, adversely affecting public health and increasing disease and mortality rates, especially among vulnerable groups. Workplace accidents may also rise due to extreme weather conditions and changing work environments. This study provides a detailed analysis of the impacts of climate change across various sectors and discusses potential solutions to mitigate these effects. Combating climate changes requires significant cooperation and coordination at both national and international levels. Policymakers, scientists, and non-governmental organizations must work together to develop effective and sustainable solution strategies, playing a critical role in this endeavor.

Keywords Climate changes · Migration · Occupational health · Occupational safety · Social impact

S. Tepe (✉) · S. Aslantaş
University of Health Sciences, Istanbul, Turkey
e-mail: serap.tepe@sbu.edu.tr

A. Çabuk
Karabük University, Karabük, Turkey

S. Al-Naemi et al. (eds.), *Water and Food Security in the Face of Climate Change: Challenges and Opportunities for Resilience*, World Sustainability Series,
https://doi.org/10.1007/978-3-032-00098-9_2

1 Introduction

Climate can be defined as the combination of the frequency, temporal distribution, and variability of weather conditions observed over many years in any region of the Earth. The main factor determining weather conditions is the amount of energy received from the Sun. Although the climate is defined as the entirety of weather conditions, seas, lands, and the living beings inhabiting these areas also play a significant role within the climate system. The interactions among these factors constitute the climate system [1]. Products are obtained using land, labor, and energy as inputs for production. The use of energy in production, in particular, results in the creation of waste materials and CO_2. Atmospheric CO_2, which is considered a global pollution stock and also defined as a greenhouse gas, leads to an increase in temperature [2]. The rise in temperature due to the greenhouse gas effect is the primary cause of climate change, a significant threat to humanity [3]. Climate change, which poses a danger to the existence and continuity of all living entities on Earth, leads to the emergence and increased frequency of various hazards and risks [4]. As a result of climate change, the temperature of the Earth is rising, and the perceived temperature is increasing accordingly. This situation leads to variable and extreme weather events (extremely hot and cold weather, heavy precipitation) characterized by their high frequency, intensity, and duration [5, 6]. Due to climate change, ice and glaciers are melting [7], sea levels are rising [8], and the acidity levels of seas and oceans are increasing. Additionally, droughts, fires, heavy rainfall, floods, landslides, storms, heatwaves, and hurricanes are occurring more frequently. These occurrences due to climate change result in natural disasters and emergencies, with an increase in their number, severity, and duration. Climate change, seen as a global threat and an obstacle to development, is a determinant factor affecting many aspects of human development and is full of uncertainties. The scale and direction of climate change impacts vary according to the geographical and socio-economic conditions of the affected communities. Climate change disrupts social and political structures through effects such as migration, conflict, war, and changes in population and demographics [9]. The ecosystem and biodiversity are disrupted due to damage or extinction of plant and animal species [10]. The economy and financial system are weakened by changes in interest rates, prices, and exchange rates, supply and demand fluctuations, production and delivery delays, and negative impacts on labor productivity and business practices [11, 12]. Human and community health is adversely affected by changing human profiles, loss of housing and livelihoods, food and water shortages, heat strokes, air pollution, malnutrition, displacement, and respiratory and cardiovascular diseases, infectious diseases, physical and mental disorders, mental illnesses, and deaths resulting from suicide due to water and vectors (mosquitoes, ticks, flies, etc.) [13–15]. Terrestrial and aquatic agriculture are limited due to the deterioration of natural resources and the loss of agricultural lands. Overall, climate change is a hazard negatively impacting the existence and continuity of living and non-living entities, societies, and states. The new dangers and risks posed by this hazard cause irreversible and irreparable damage to the economy, health, social and political structures, agriculture, ecosystems, biodiversity, and occupational health and safety.

2 The Social Impacts of Climate Change and Its Reflections on Working Life

2.1 The Impact of Climate Change on Economic

Climate change, which has long been evident, possesses the potential to create hazards and risks that adversely affect our world. A substantial portion of the global community, particularly scientists, acknowledges the significant and escalating impacts of climate change on societies and economies worldwide. Considering factors such as annual income, labor productivity and supply, migration, food security, and energy consumption, climate change exerts a profound influence on the economic and financial systems [16]. Disruptions and deteriorations in the economy and finance result in rising food prices, increased income inequality, alterations in supply chains, decreased fertility rates [17], and migration from unproductive farmlands to urban slums, particularly in developing countries, leading to escalations in housing prices. The imbalance within a nation's economy disproportionately affects the poor and wealthy segments, with lower-income groups suffering more severe consequences compared to higher-income groups. Similarly, developing countries experience more significant economic and financial damage than developed countries.

2.2 The Impact of Climate Change on the Agricultural Sector

Changes in the patterns, frequency, and intensity of precipitation due to climate change result in phenomena such as floods, erosion, landslides, reduced snow cover, droughts, coastal flooding, coastal erosion, decreased rainfall, storms, diseases, and invasive species, all of which contribute to the decline or cessation of terrestrial and aquatic agriculture. The reduction in animal and plant production leads to decreased yields of agricultural products like fruits, grains, and aquatic products. This decline in productivity raises agricultural product prices, restricts access to food for low-income communities, prompts migration, causes spoilage of products, and creates storage issues, weakens rural economies, reduces forest products, and diminishes worker productivity and participation, thereby harming worker health. Furthermore, the scarcity of agricultural products and the rising prices of animal and plant products affect education and fertility, causing regional variations in resource allocation for these areas. In some regions, the resources dedicated to education and fertility are significantly altered [18, 19].

2.3 The Impact of Climate Change on the Energy Sector

The energy required for the production of climate control services to adapt to extreme temperatures is a primary source of greenhouse gases that contribute to climate change. To meet both the production needs and the energy required for adapting to temperature variations, increased energy production is necessary. The growing energy demand, particularly directed towards low- and middle-income countries, exacerbates the issue of energy poverty, highlighting the adverse effects of climate change on the energy sector [20].

2.4 The Impact of Climate Change on Social Events

Long-term crises, conflicts, natural disasters, persistent poverty, pandemics, famine, and economic challenges, coupled with irregular weather patterns, high temperatures and evaporation, glacial melting, and rising sea levels due to climate change, diminish hopes for peace, security, stability, and sustainable development, rendering the world an increasingly "unsafe space." These issues, due to their diversity and complexity, are compounding exponentially, transcending national borders and affecting individuals, communities, and states in multifaceted ways [21].

2.5 The Impact of Climate Change on Migration

Climate change is a global issue requiring both local and international cooperation. It is crucial for every individual, community, and government to fulfill their responsibilities in addressing this challenge. The relationship between climate change and migration is complex and multifaceted. Various effects of climate change can compel people to migrate from their homes. Climate change can lead to more frequent and severe natural disasters (floods, droughts, hurricanes, fires), causing people to lose their homes, agricultural lands, and livelihoods. For instance, individuals living in coastal areas may lose habitable land due to rising sea levels, forcing them to migrate [22].

Climate change can also lead to reduced water resources and water scarcity. People living in arid regions might have to migrate to access water sources. Agricultural productivity can decline due to climate change, making it difficult for farmers and agricultural workers to sustain their livelihoods. Consequently, people in agrarian economies may migrate to more fertile areas. Climate change can jeopardize livelihoods in sectors like agriculture, fishing, and tourism, prompting people to migrate in search of economic opportunities. Extreme weather events can damage infrastructure (roads, bridges, buildings), disrupting economic activities and compelling people to move to safer and more stable regions.

Furthermore, climate change can facilitate the spread of vector-borne diseases such as malaria and dengue fever, causing people to migrate in search of healthier living conditions. Extreme temperatures, cold waves, and air pollution can reduce the quality of life and lead to health problems, prompting migration for better living conditions. Climate change-induced resource scarcity and deteriorating living conditions can also lead to social tensions and conflicts, causing people to migrate to secure and peaceful areas.

The increase in migration flows due to climate change can alter social and political dynamics in both sending and receiving countries, necessitating revisions in migration policies and humanitarian aid strategies. Communities and countries with low adaptive capacities are more likely to experience higher levels of migration. Developed countries and communities can be more resilient to climate change through adaptive measures (e.g., infrastructure strengthening, improved agricultural techniques). However, in areas lacking such capacities, people may be forced to migrate in search of better living conditions.

Climate change is increasingly becoming a significant driver of migration, necessitating integrated and comprehensive strategies to address both climate change and the needs of migrants and host communities. Current research underscores that migration often represents a strategic response rather than a consequence of poverty or unbearable vulnerability [23]. Understanding how secondary effects of climate change, such as weakened agriculture and increased competition for water and food resources, contribute to instability and migration decisions is crucial. Despite the complexity of migration decisions, climate change is becoming an increasingly important factor.

Climate-induced migrations pose threats to common human interests and collective security worldwide. The cumulative effects of these trends can have serious implications for the stability of nations lacking sufficient resources, good governance, and response flexibility [24]. Policymakers and analysts are increasingly viewing climate-induced migrations as a security concern [25]. Adverse conditions resulting from climate change directly impact human lives, forcing people to migrate when they can no longer sustain themselves in their current regions.

These migration movements can be categorized into rapid and sudden migrations due to quick-onset events (e.g., storms, hurricanes, floods) and gradual migrations due to slow-onset events (e.g., drought, ocean acidification, soil degradation, sea-level rise). Rapid-onset events typically lead to immediate evacuation, while slow-onset events result in long-term environmental degradation. Although slow-onset events cause displacement gradually, they are expected to affect more people over time compared to rapid-onset events [26].

2.6 The Impact of Climate Change on Health

According to the World Health Organization (WHO), climate-related diseases are expected to cause approximately 250,000 additional deaths annually between 2030 and 2050. Climate change and associated weather events pose a significant threat to human health and public Health. Rising temperatures have adverse effects on human health, which are intensifying over time. These impacts disproportionately affect countries and populations living in poverty and inequality, including children, pregnant women, the elderly, individuals with disabilities, those with chronic illnesses, and outdoor workers, leaving them particularly vulnerable [27, 28]. Climate change affects human health and well-being worldwide. It manifests through changes in extreme weather conditions, resulting in storms, floods, droughts, and heatwaves; air pollution, land use changes; the distribution of disease burden and vectors; decreased food availability and distribution, migration; conflict, economic disruptions; infrastructure damage, and inadequate water resources [29]. These direct and indirect effects lead to physical, psychological, and social disabilities, deteriorating individual and public health.

The worsening health conditions due to climate change can be categorized as follows:

- Heat-Related Illnesses: Increased temperatures lead to more frequent and severe heatwaves, causing heatstroke, dehydration, and other heat-related illnesses.
- Vector-Borne Diseases: Changes in climate affect the distribution and prevalence of vector-borne diseases like malaria and dengue fever, as warmer temperatures and altered precipitation patterns create more favorable conditions for vectors such as mosquitoes.
- Respiratory Issues: Poor air quality, exacerbated by climate change, leads to respiratory problems, including asthma and other chronic respiratory diseases.
- Waterborne Diseases: Floods and increased rainfall can contaminate water supplies, leading to outbreaks of waterborne diseases such as cholera and dysentery.
- Food Security and Nutrition: Climate change affects agricultural productivity, reducing food availability and quality, which can lead to malnutrition and food insecurity.
- Mental Health: The stress and trauma associated with extreme weather events and displacement due to climate change can have significant mental health impacts, including anxiety, depression, and post-traumatic stress disorder (PTSD).

Addressing the health impacts of climate change requires comprehensive strategies, including strengthening healthcare systems, improving public health infrastructure, promoting sustainable agricultural practices, enhancing early warning systems for extreme weather events, and implementing policies to reduce greenhouse gas emissions. By taking these measures, we can mitigate the adverse health effects of climate change and protect vulnerable populations.

Infectious Diseases

Climate change has expanded the range of certain vectors (mosquitoes, sandflies, ticks, etc.), causing the spread of infectious diseases that adversely affect allergy sufferers, infants, children, people with poor health, and those in need of care [5]. Changing weather conditions, water contamination, variations in temperature, precipitation, humidity, and wind speed, droughts, and floods create environments conducive to the growth, reproduction, and spread of pathogens. This situation intensifies the development and transmissibility of pathogens. Vector-borne infectious diseases occur worldwide, especially in areas with inadequate healthcare services and conditions, spreading rapidly and disproportionately. Diseases such as dengue fever, malaria, Lyme disease, Zika, chikungunya, and West Nile Virus (WNV) can cause physical and mental health problems and may lead to death [5, 13, 15, 28, 30].

Chronic Diseases and Air Pollution

Injuries from storms, floods, and wildfires account for 88%, the increase in the severity of chronic diseases due to air pollution accounts for 88%, and the rise in allergic symptoms from long-term exposure to plants, molds, and fungi is around 80% [27]. The increase in pollution levels is due to the formation, transportation, dispersion, and accumulation of pollutants caused by climate change. In 2012, fine particulate matter caused an additional 7 million deaths globally. Increases in ambient temperature, rainfall frequency, and atmospheric stagnation due to climate change significantly affect the formation, transport, dispersion, and accumulation of air pollutants [27, 29].

Extreme Weather Events and Health

The intensity, frequency, and duration of weather events are changing due to climate change. These changes result in pollution, extreme heat and cold, and fires, leading to respiratory diseases and their severity from pollen dissemination, fungi, and infectious agents. Additionally, winter hypothermia cases occur due to drops in ambient temperature and inadequate shelter conditions [27].

Water and Food Security

Governments and health systems have responsibilities in managing the health risks arising from decreased food availability and distribution and potable water due to climate change. Temperature affects plant and animal production in terrestrial and aquatic environments. The health, productivity, and survival abilities of plants and animals are directly related to temperature. The sensitivity of plants and animals to weather conditions leads to increased food prices, thereby restricting food access for the poor and causing malnutrition. Furthermore, fluctuations in the global food system will have health consequences for laborers, farmers, and consumers. Another adverse effect of climate change is water security. Reduced rainfall and increased drought due to weather conditions affect groundwater sources, water supply, and potable water quality [5, 7, 15].

Mental Health

Increased temperatures due to climate change lead to a rise in crime and aggressive behavior, suicides, organic diseases, dementia, anxiety, stress, psychological developmental disorders, mental and behavioral disorders, physical and psychological fatigue, a threefold increase in the risk of death from any cause, and trauma. The increased frequency and severity of floods, storms, and hurricanes lead to epidemics, negatively impacting human and community well-being and mental health. These effects manifest in both the short and long term [5, 15].

2.7 The Impact of Climate Change on Occupational Accidents

Climate change, driven by global warming, leads to extreme weather events such as rising temperatures, droughts, floods, and storms. These extreme weather events increase the frequency, duration, and severity of emergencies and disasters, thereby enlarging the affected areas and the number of impacted living and non-living entities. Electrical facilities, often located outdoors, are more susceptible to hazards caused by these events. Weather-related damage to electrical equipment, such as transformer explosions and damages caused by salt, dust, freezing, fires, and short circuits, increase the incidence of electrical accidents. These accidents typically manifest as electrical fires, electric shocks, and incidents involving electrical equipment [31, 32]. High metabolic heat production, high ambient and radiant heat, low air flow, and high humidity induce heat stress in humans, which can reduce work capacity and motor-cognitive performance [33]. Similarly, changing weather conditions increase fog and smoke formation, temperature, rainfall, and sandstorms [34, 35]. Factors affecting the frequency and severity of accidents include environmental conditions (wet and muddy ground, frozen ground, deformed ground, deformation in bridge connections, etc.), weather conditions (rain, snow, fog, etc.), vehicles (overheating, tire deformations, etc.), and humans [36–38]. The combined effects of extreme weather events on environmental conditions, weather, vehicles, and humans due to climate change lead to an increase in traffic accidents resulting in injuries and fatalities. Exposure to heat in work environments increases body temperature. If fluid intake is inadequate to compensate for increased sweating, dehydration occurs. Increased dehydration reduces a person's capacity to perform physical and mental tasks, thereby increasing accident risk. Injuries and fatalities from accidents reduce work capacity, income, and lead to psychological losses [5]. Therefore, it is believed that climate change significantly impacts occupational accidents. High temperatures can cause heat stress among workers, leading to heat stroke, fatigue, and loss of concentration. Severe storms and floods pose significant dangers for workers in construction and outdoor jobs, potentially increasing occupational accidents. Lightning strikes pose a serious risk, particularly for workers in open fields. Climate change can exacerbate air pollution, leading to increased incidence of asthma and other respiratory diseases, which adversely affect outdoor workers. Rising sea levels endanger workers in coastal areas, leading to an increase in occupational accidents due to floods and inundation. Climate change may expand the range of disease-carrying vectors such as mosquitoes and ticks, increasing the risk of illness for agricultural and forestry workers. The uncertainty and stress caused by climate change can negatively affect workers' mental health, leading to attention and concentration issues that may cause accidents. Hazards caused by climate change (increased or decreased ambient temperature, air pollution, ultraviolet radiation, extreme weather conditions, vector-borne diseases, and expanded living areas) lead to changes in working environments and conditions, and activities [39, 40]. These changes physically and psychologically impact workers, increasing the likelihood of stress-related occupational accidents [41–43]. Occupational accidents causing fatalities and injuries due to climate change generally result from falls (in both summer and winter), electrical shocks, and hyperthermia [44, 45]. Sectors most affected include agriculture, construction, hospitality, emergency services, and healthcare. The risk of injury is higher on hot days than on cold days. For instance, construction workers are most often injured on hot days [46], whereas fishing, transportation, and electricity–gas–water distribution workers are most frequently injured in cold weather [47]. Extreme hot and cold weather causes fatigue, reduced reaction and decision-making ability, sweaty palms, dehydration, heat stroke, kidney diseases, heat cramps, musculoskeletal disorders, frostbite, asthma, depression, and anxiety among workers [48, 49]. Employers and occupational safety specialists must monitor these conditions and other potential hazards due to climate change, taking appropriate measures to minimize risks by considering the potential impacts of climate change.

3 Responses to Climate Change

Many countries have signed the Paris Agreement to reduce greenhouse gas emissions and keep global warming below 2 °C. This agreement requires countries to set emission reduction targets and develop strategies to achieve these goals [50]. Governments are creating policies to promote renewable energy sources like solar, wind, and hydroelectric power. These measures aim to reduce carbon emissions by decreasing reliance on fossil fuels. Cities are implementing sustainability initiatives to increase energy efficiency, encourage public transportation, and preserve green spaces. Local agriculture initiatives promote community gardens and local production to enhance food security and reduce the carbon footprint. Additionally, individuals can take measures to combat climate change. Individuals can take steps such as using energy-efficient appliances, reducing energy consumption, and opting for renewable energy sources. Changing consumption habits is also a crucial factor. Individual habits such as consuming less meat, recycling, and choosing sustainable products can also contribute to the fight against climate change. Controlling climate change-related risks requires international cooperation. Implementing new international policies will make the fight against climate change-related risks more meaningful.

3.1 International Agreements and Policies

The Paris Agreement, adopted in 2015, marks a significant milestone in global climate policy. It is a legally binding international treaty involving 196 countries, aiming to limit global warming to well below 2 °C, preferably to 1.5 °C, compared to pre-industrial levels. Each country must submit Nationally Determined Contributions (NDCs), outlining their plans for reducing greenhouse gas emissions and adapting to climate change impacts. These NDCs are reviewed every five years, with the expectation that countries will increase their ambition over time. The agreement also emphasizes the need for financial support for developing countries to help them reduce emissions and adapt to climate change effects [50].

3.2 Renewable Energy Initiatives

Governments worldwide are investing in renewable energy sources to reduce dependence on fossil fuels. Countries like Germany and Denmark have made significant strides in wind energy, while countries like China and the United States are leading in solar energy investments. These initiatives are supported by policies that provide subsidies, tax incentives, and grants for renewable energy projects. The transition to renewable energy not only reduces carbon emissions but also creates job opportunities and fosters economic growth [51].

3.3 Urban Sustainability Projects

Cities play a crucial role in addressing climate change due to their high energy consumption and carbon emissions. Urban sustainability projects focus on improving energy efficiency in buildings, expanding public transportation networks, and increasing green spaces. For example, Copenhagen aims to become the world's first carbon-neutral capital by 2025 through measures such as promoting cycling, improving public transport, and investing in green infrastructure. Similarly, cities like Singapore and Vancouver are implementing comprehensive sustainability plans to reduce their carbon footprint and enhance the quality of life for their residents [52].

3.4 Local Agriculture and Food Security

Promoting local agriculture is essential for reducing the carbon footprint associated with food production and transportation. Community gardens and local farms can enhance food security, especially in urban areas. Initiatives such as urban farming in Detroit and community-supported agriculture (CSA) programs in various cities encourage local food production and consumption. These initiatives not only reduce greenhouse gas emissions but also provide fresh, healthy food options and strengthen community bonds [53].

3.5 *Individual Actions*

Individuals can significantly impact climate change by adopting more sustainable lifestyles. Simple actions like using energy-efficient appliances, reducing water and energy consumption, and choosing renewable energy options can make a difference. Additionally, changing dietary habits, such as reducing meat consumption and opting for plant-based diets, can lower greenhouse gas emissions. Recycling, reducing waste, and supporting sustainable products and companies also contribute to environmental conservation [54].

3.6 *International Cooperation*

Addressing climate change effectively requires robust international cooperation. Collaborative efforts through organizations like the United Nations, the World Bank, and the International Monetary Fund (IMF) provide financial and technical support to developing countries. These organizations help implement climate resilience projects, promote sustainable development, and facilitate knowledge sharing. Strengthening international policies and agreements will enhance global efforts to mitigate climate change impacts and build a sustainable future for all [55].

In summary, combating climate change necessitates a multi-faceted approach involving international agreements, government policies, urban sustainability projects, local agriculture initiatives, individual actions, and international cooperation. By taking comprehensive and coordinated actions, we can mitigate the adverse effects of climate change and build a resilient and sustainable future.

References

1. Kurnaz L (2019) Son Buzul Erimeden, Doğan Kitap, İstanbul
2. Desmet K, Rossi-Hansberg E (2015) On the spatial economic impact of global warming. J Urban Econ 88:16–37. https://doi.org/10.1016/j.jue.2015.04.004
3. Dablander F, Sachisthal MS, Cologna V, Strahm N, Bosshard A, Grüning N-M, Green AJK, Brick C, Aron AR, B JM (2024) Climate change engagement of scientists. Nat Clim Change. https://doi.org/10.1038/s41558-024-02091-2

4. Carroll N, Frijters P, Shields MA (2009) Quantifying the costs of drought: new evidence from life satisfaction data. J Popul Econ 22(2):445–461. https://doi.org/10.1007/s00148-007-0174-3
5. Berry HL, Bowen K, Kjellstrom T (2010) Climate change and mental health: a causal pathways framework. Int J Public Health 55(2):123–132. https://doi.org/10.1007/s00038-009-0112-0
6. Hathaway J, Maibach EW (2018) Health implications of climate change: a review of the literature about the perception of the public and health professionals. Curr Environ Health Rep 5(1):197–204. https://doi.org/10.1007/s40572-018-0190-3
7. Habtie TT, Teferi E, Guta F (2024) Climate change in Tigray, Ethiopia: assessing impacts on smallholders' risk perception and food security. Environ Challenges 16:100972. https://doi.org/10.1016/j.envc.2024.100972
8. Trinh VQ, Trinh HH, Li T, Vo XV (2024) Climate change exposure, financial development, and the cost of debt: evidence from EU countries. J Financ Stab 101315. https://doi.org/10.1016/j.jfs.2024.101315
9. Dasgupta S, Emmerling J, Shayegh S (2023) Inequality and growth impacts of climate change—insights from South Africa. Environ Res Lett 18(12):124005. https://doi.org/10.1088/1748-9326/ad0448
10. Diffenbaugh NS, Burke M (2019) Global warming has increased global economic inequality. Proc Natl Acad Sci 116(20):9808–9813. https://doi.org/10.1073/pnas.1816020116
11. Bejarano J, Rodríguez D (2024) The effects of climate change on a small and open economy: economic and monetary perspectives. Lat Am J Central Banking 100144. https://doi.org/10.1016/j.latcb.2024.100144
12. Cao Z, Chen SX, Dong T, Lee E (2024) Climate change uncertainty and supply chain financing. Br Account Rev 101423. https://doi.org/10.1016/j.bar.2024.101423
13. Gizaw Z, Salubi E, Pietroniro A, Schuster-Wallace CJ (2024) Impacts of climate change on water-related mosquito-borne diseases in temperate regions: a systematic review of literature and meta-analysis. Acta Trop 258:107324. https://doi.org/10.1016/j.actatropica.2024.107324
14. Loud ES, Woolf GR, Ralph A, Gadhia T, Navaratnam AMD, Watson L, McGushin A, Clissold E, Hilton B (2024) Climate change and public health: an evaluation framework for local government. Public Health Pract 8:100524. https://doi.org/10.1016/j.puhip.2024.100524
15. Watts N, Adger WN, Agnolucci P (2015) Health and climate change: Policy responses to protect public health. Environnement, Risques et Sante, 14(6):466–468. John Libbey Eurotext. https://doi.org/10.1016/S0140-6736(15)60854-6
16. Gilli M, Calcaterra M, Emmerling J, Granella F (2024) Climate change impacts on the within-country income distributions. J Environ Econ Manag 127:103012. https://doi.org/10.1016/j.jeem.2024.103012
17. Casey G, Shayegh S, Moreno-Cruz J, Bunzl M, Galor O, Caldeira K (2019) The impact of climate change on fertility. Environ Res Lett 14(5):054007. https://doi.org/10.1088/1748-9326/ab0843
18. Kundzewicz ZW, Matczak P (2012) Climate change regional review: Poland. WIREs Clim Change 3(4):297–311. https://doi.org/10.1002/wcc.175
19. Neokye EO, Wang X, Thakur KK, Quijón PA, Nawaz RA (2024) Climate change impacts on oyster aquaculture—Part II: impact assessment and adaptation measures. Environ Res 259:119535. https://doi.org/10.1016/j.envres.2024.119535
20. De Cian E, Sue Wing I (2019) Global energy consumption in a warming climate. Environ Resour Econ 72(2):365–410. https://doi.org/10.1007/s10640-017-0198-4
21. Özkan KM (2024) Bir Güvenlik Sorunu Olarak İklim Değişikliğinin Çatişma ile Olan İlişkisi. Güvenlik ve İstihbarat Çalişmalari Dergisi 2(1):51–77
22. Bilben MS (2019) Dünyadan Örnekler Işiğinda İklim Değişikliği Kaynakli Göçleri Anlamak. Mediterr J Human 9(2). https://doi.org/10.13114/MJH.2019.494
23. Morrissey J (2014) Environmental change and human migration in Sub-Saharan Africa. In: Piguet E, Laczko F (eds) People on the move in a changing climate (global migration issues), vol 2. Springer, Dordrecht. https://doi.org/10.1007/978-94-007-6985-4_4
24. Werz M, Hoffman M (2016) Europe's twenty-first century challenge: climate change, migration and security. Eur View 15. https://doi.org/10.1007/s12290-016-0385-7

25. White G (2011) Climate change and migration: security and borders in a warming world, Online edn, Oxford Academic, 19 Jan 2012. https://doi.org/10.1093/acprof:oso/9780199794829.001.0001. Accessed 19 July 2024
26. Gemenne F, Brucker P, Lonesco D (2014) The state of enviromental migration 2014-a review of 2013. International Organization for Migration
27. Ponte N, Alves F, Vidal DG (2024) Exploring Portuguese physicians' perceptions of climate change impacts on health: a qualitative study. J Clim Change Health 100333. https://doi.org/10.1016/j.joclim.2024.100333
28. Winkler MS, Röösli M, Ragettli MS, Cissé G, Müller P, Utzinger J, Perez L (2015) Mitigating and adapting to climate change: a call to public health professionals. Int J Public Health 60(6):631–632. https://doi.org/10.1007/s00038-015-0722-7
29. Sarfaty M, Mitchell M, Bloodhart B, Maibach E (2014) A Survey of African American physicians on the health effects of climate change. Int J Environ Res Public Health 11(12):12473–12485. https://doi.org/10.3390/ijerph111212473
30. Alves MFP, Leal CMS, Vidal DG (2023) Doenças Transmitidas Por Vetores Num Contexto De Alterações Climáticas: Antecipando Riscos Para Uma Melhor Preparação Dos Territórios. Estudo De Caso Da Região De Coimbra, Portugal. Recima21—Revista Científica Multidisciplinar 4(10):e4104181. https://doi.org/10.47820/recima21.v4i10.4181
31. Jeong M-C, Kim J (2019) Prediction and analysis of electrical accidents and risk due to climate change. Int J Environ Res Public Health 16(16):2984. https://doi.org/10.3390/ijerph16162984
32. Yates D, Luna BQ, Rasmussen R, Bratcher D, Garre L, Chen F, Tewari M, Friis-Hansen P (2014) Stormy weather: assessing climate change hazards to electric power infrastructure: a sandy case study. IEEE Power Energ Mag 12(5):66–75. https://doi.org/10.1109/MPE.2014.2331901
33. Ebi KL, Capon A, Berry P, Broderick C, de Dear R, Havenith G, Honda Y, Kovats RS, Ma W, Malik A, Morris NB, Nybo L, Seneviratne SI, Vanos J, Jay O (2021) Hot weather and heat extremes: health risks. The Lancet 398(10301):698–708. https://doi.org/10.1016/S0140-6736(21)01208-3
34. Ambanattu ET, Umar AC, Akolaa EA, Sreejith A, Sreedharan J, Muttappallymyalil J (2023) Effect of climate change on road traffic accidents in the UAE: a narrative review. Int J Commun Med Public Health 10(7):2626–2628. https://doi.org/10.18203/2394-6040.ijcmph20232063
35. Islam MdM, Alharthi M, Alam MdM (2019) The impacts of climate change on road traffic accidents in Saudi Arabia. Climate 7(9):103. https://doi.org/10.3390/cli7090103
36. Alirezaei M, Onat N, Tatari O, Abdel-Aty M (2017) The climate change-road safety-economy nexus: a system dynamics approach to understanding complex interdependencies. Systems 5(1):6. https://doi.org/10.3390/systems5010006
37. Shahid S, Minhans A (2016) Climate change and road safety: a review to assess impacts in Malaysia. Jurnal Teknologi 78(4). https://doi.org/10.11113/jt.v78.7991
38. Zou Y, Zhang Y, Cheng K (2021) Exploring the impact of climate and extreme weather on fatal traffic accidents. Sustainability 13(1):390. https://doi.org/10.3390/su13010390
39. Lan T, Goh YM, Jensen O, Asmone AS (2022) The impact of climate change on workplace safety and health hazard in facilities management: an in-depth review. Saf Sci 151:105745. https://doi.org/10.1016/j.ssci.2022.105745
40. Schulte PA, Chun H (2009) Climate change and occupational safety and health: establishing a preliminary framework. J Occup Environ Hyg 6(9):542–554. https://doi.org/10.1080/15459620903066008
41. Adam-Poupart A, Labreche F, Smargiassi A, Duguay P, Busque M-A, Gagne C, Rintamäki H, Kjellstrom T, Zayed J (2013) Climate change and occupational health and safety in a temperate climate: potential impacts and research priorities in Quebec, Canada. Ind Health 51(1):68–78. https://doi.org/10.2486/indhealth.2012-0100
42. Clayton S (2021) Climate change and mental health. Curr Environ Health Rep 8(1):1–6. https://doi.org/10.1007/s40572-020-00303-3
43. Santurtún A, Shaman J (2023) Work accidents, climate change and COVID-19. Sci Total Environ 871:162129. https://doi.org/10.1016/j.scitotenv.2023.162129

44. Jackson LL, Rosenberg HR (2010) Preventing heat-related illness among agricultural workers. J Agromed 15(3):200–215. https://doi.org/10.1080/1059924X.2010.487021
45. Liao C-W (2012) Pattern analysis of seasonal variation in occupational accidents in the construction industry. Proc Eng 29:3240–3244. https://doi.org/10.1016/j.proeng.2012.01.473
46. Santurtún A, Moraes SL, Fdez-Arroyabe P, Obregón M, Almendra R (2023) Descriptive analysis of occupational accidents in Spain and their relationship with heatwaves. Prev Med 175:107697. https://doi.org/10.1016/j.ypmed.2023.107697
47. Marinaccio A, Scortichini M, Gariazzo C, Leva A, Bonafede M, de' Donato FK, Stafoggia M, Viegi G, Michelozzi P, Carla A, Paola A, Stefania A, Sandra B, Lucia B, Sergio B, Laura B, Serena B, Giuseppe B, Simone B, Francesco U (2019) Nationwide epidemiological study for estimating the effect of extreme outdoor temperature on occupational injuries in Italy. Environ Int 133:105176. https://doi.org/10.1016/j.envint.2019.105176
48. Ferrari GN, Ossani PC, de Souza RCT, Leal GCL, Galdamez EVC (2023) Impact of rising temperatures on occupational accidents in Brazil in the period 2006 to 2019: a multiple correspondence analysis. Saf Sci 161:106078. https://doi.org/10.1016/j.ssci.2023.106078
49. Karthick S, Kermanshachi S, Pamidimukkala A, Namian M (2023) A review of construction workforce health challenges and strategies in extreme weather conditions. Int J Occup Saf Ergon 29(2):773–784. https://doi.org/10.1080/10803548.2022.2082138
50. Paris Agreement, UNFCCC (2015). https://www.wipo.int/wipolex/en/text/581795
51. Renewable Energy Statistics, IRENA (2021). https://www.irena.org/publications/2021/Aug/Renewable-energy-statistics-2021
52. C40 Cities (2020). https://www.c40.org/?gad_source=1&gclid=Cj0KCQjwtsy1BhD7ARIsAHOi4xZKXidGwDd7svMykaPt8rTDAe4YdFNVgZfOdGgPWcDvRftHmXRgTcIaAgSYEALw_wcB
53. The State of Food and Agriculture (2019). https://openknowledge.fao.org/server/api/core/bitstreams/11f9288f-dc78-4171-8d02-92235b8d7dc7/content
54. IPCC (2018). UN Environment Programme
55. World Bank (2020). https://www.worldbank.org/en/home

Impact of Sowing Methods on Growth and Yield of Wheat Cultivars in Diverse Environments

Wajid Farhad, Ayaz Ahmed Lashari, Hafiz Mohkum Hammad, Muhammad Nawaz Kandhro, and Aitazaz A. Farooque

Abstract Weather conditions and sowing methods are key aspects affecting wheat productivity. This study was conducted over two consecutive growing seasons to investigate the impact of sowing methods on the morpho-yield performance of wheat cultivars under diverse ecological conditions, using Arid (Tandojam, Sindh) and Semi-arid (Quetta, Balochistan) as experimental sites. Six wheat cultivars were sown using broadcast, furrow, and drilling methods. The crop response to environmental conditions showed that the grain yield was 3642 kg ha^{-1} at Tandojam and 3809 kg ha^{-1} at Quetta sites. Among the various sowing methods, the furrow method proved to be more effective and efficient than the broadcast and drilling methods, with optimal agronomic observations. The maximum grain yield of 3905 kg ha^{-1} was observed under the furrow sowing method. The varietal performance, based on grain yield per hectare, was superior in the cultivar Mehran-89, with a grain yield of 4110 kg ha^{-1}. Hence, the furrow sowing method, along with site-specific wheat cultivars, is the best crop management practice for achieving maximum wheat growth and yield.

W. Farhad
University College Dera Murad Jamali, Dera Murad Jamali, Pakistan

Lasbela University of Agriculture, Water and Marine Sciences, Uthal, Pakistan

A. A. Lashari · M. N. Kandhro
Department of Agronomy, Sindh Agriculture University Tandojam, Tandojam, Pakistan

H. M. Hammad (✉)
Department of Agronomy, MNS University of Agriculture Multan, Multan, Pakistan
e-mail: hafizmohkum@gmail.com

Department of Agricultural and Biological Engineering, University of Florida, Gainesville, FL, USA

A. A. Farooque
Canadian Center for Climate Change and Adaptation University of Prince Edward Island, St Peter's Bay, PEI, Canada

Faculty of Sustainable Design Engineering, University of Prince Edward Island, Charlottetown, PEI, Canada

S. Al-Naemi et al. (eds.), *Water and Food Security in the Face of Climate Change: Challenges and Opportunities for Resilience*, World Sustainability Series,
https://doi.org/10.1007/978-3-032-00098-9_3

Keywords Morpho-yield · Sowing methods · Multi-environments

1 Introduction

In Pakistan, wheat is grown on an area of 9.60 million ha with total production 31.4 million tons and an average yield 3.27 t ha^{-1} [1] which is too low as compared to other countries such as China, USA, India etc. Pakistan is among top ten wheat producing countries of the world [2] but the average yield of the wheat is much lower than their potential yield [3].

Since, the improvement in wheat yield has been associated with accumlization, selection and hybridization dates back and management practices [4]. The genetic variation may be evaluated under different environment and induced mutations or a combination of all these methods for the improvement of wheat crop [5]. Its productivity can be enhanced with the development of high yielding, well adapted cultivars through hybridization genetically diversified genotypes [6]. While, the yield potential per unit area has been increased by cross breeding of the most productive strains [7, 8]. The wheat breeders are concentrating their efforts to improve the yield potential of wheat by developing new cultivars with desirable genetic make-up; efforts are being made to develop modern cultivars of wheat with high yield potential [9].

Multi-environment trials are important in the plant breeding and agronomy for studying yield stability and predicting yield performance of various genotypes and agronomic treatments across environments [10]. Similarly, the performance of wheat genotypes varies in different environment regarding to their characteristics yield and its component [11], agronomic trait [12], correlation and path analysis [13], additive main effects and multiplicative interaction [14], and multiple cluster [15]. The evaluation of elite wheat cultivars and advanced lines is essential for the further improvement of wheat [16].

Sowing method is another major factor that determines crop vigor and yield [17, 18]. In Pakistan, wheat is mostly cultivated as broadcast. [19]. Drilling method is considered suitable because of its uniform seed distribution and sowing at desired depth and resulted higher germination and uniform plant [20]. Sindh Province is a major contributor of wheat after Punjab in Pakistan where most of the area under wheat comes after cotton crop. Usually low yield is obtained by conventional methods of wheat planting. Wheat cultivation by drilling on raised beds has been investigated for its suitability in cotton-wheat cropping system [21] and obtained better grain yields over flat beds and broadcast seeding. Keeping in view the above factors, the study was carried out on morpho-yield evaluation of wheat cultivars against various sowing methods under two different ecological conditions.

2 Methodology

2.1 Study Region

Two field experiments were conducted at multi-location (i) Agronomic Research Farm at Sindh Agriculture University, Tandojam and (ii) Agriculture Research Institute, Quetta to evaluate the performance of six wheat cultivars under different environment conditions. The cultivars were tested against various sowing methods at both locations. The experiment site Tandojam is located at 25°25′ 60N 68°31′ 60E and Quetta located at 30°12′ 0° N, 67°0′ 0E. According to the chemical analysis of experimental soil relatively more organic matter and total nitrogen was recorded in Quetta region while the soil pH was greater in Tandojam region (Table 1). All other agronomic practices were kept standard.

3 Observational Procedure of the Experiments

Plant height (cm) at maturity	Plant height (cm) was measured at physiological maturity. Five representative plants were selected in each experimental unit and the height measured from ground level to the tip of the spike
Number of effective tillers plant^{-1}	Two representative rows were selected in each experimental unit. Numbers of effective tillers were counted in the selected rows Effective tillers m^{-2} calculated from the number of effective tillers counted and the area sampled
Grains spike-^{1}	Number of grains spike^{-1} was recorded from each spike from 10 randomly selected plants in each treatment
Grain weight per spike (g)	All seeds were collected from each spike and weighed on top loading digital balance

(continued)

Table 1 Chemical properties of the soil before sowing and after harvesting of crop

Parameters	Tandojam		Quetta	
	Before sowing	After sowing	Before sowing	After sowing
EC (dSm^{-1})	4.42	4.30	3.93	3.89
pH	7.50	7.40	7.13	7.08
Organic matter (O.M%)	0.596	0.542	0.71	0.69
Total nitrogen (N%)	0.026	0.022	0.034	0.031
Available phosphorous (P mg kg^{-1})	4.37	4.31	4.87	4.84
Exchangeable potassium (K mg kg^{-1})	128.31	127.20	144.36	143.19

(continued)

1000 grain weight (g)	A sample of thousand grains were taken from each plot and weighed on an electric balance after drying at 70 °C for 24 h in an oven
Grain yield (t ha^{-1})	The 1 m^{-2} rows area harvested for biological yield was threshed. The grains were cleaned, dried and weighed to record the grain yield and then grain yield kg ha^{-1} was calculated

3.1 *Statistical Analysis*

The effects sowing method on the various parameters were evaluated by ANOVA using the SAS statistical software [22]. The year effect was non-significant, so data of both years were pooled together. When F-values were significant, the least significant difference (LSD) test was used for comparing treatment means.

4 Results

4.1 *Plant Height*

Plant height of wheat cultivars sown at research centers under Tandojam (Sindh) and Quetta (Balochistan) varied significantly ($P < 0.05$) under various sowing methods, environmental conditions as well as by their interactions. Table 2 showed at Tandojam, the plants height was greater produced by cultivars Mehran-89 (88.29 cm), Kiran-2004 (84.57 cm) and Zardana (83.88 cm); while the of relatively shorter in height was found for cultivars Sarsabz (79.81 cm), Rasco-2005 (78.81 cm) and minimum was observed in Sariab-2005 (71.49 cm) as indicated in Table 2. Under Quetta ecology plant height was maximum (82.35 cm) in Zardana, followed by Rasco-2005 (81.49 cm) and Sariab-95 (78.57 cm); while lower plant height was observed in cultivars Kiran-2004 (70.78 cm) and Sarsabz (70.77 cm). On average, the plant height was maximum (83.11 cm) in Zardana, followed by Rasco-2005 (80.15 cm) and Mehran-89 (80.00 cm); while lower plant height was observed in cultivar Sariab-95 (75.03 cm).

The effect of sowing methods indicates that at Tandojam wheat plants grew taller under furrow sowing (87.88 cm) than drilling (82.61 cm) and broadcasting (72.94 cm); while under Quetta conditions also, furrow sowing produced plants of maximum height (77.88 cm) followed by drilling method (76.33 cm) and broadcast (73.73 cm).

The crop response to environmental conditions indicated that the plant height was significantly ($P < 0.05$) higher (81.14 cm) under Tandojam environmental as compared to Quetta (75.94 cm); while the overall average plant height was 78.54 cm.

Table 2 Impact of environmental dynamics on plant height (cm) of wheat cultivars sown under various sowing methods

Locations	Cultivars	Sowing methods			
		Broadcast	Furrow	Drilling	Mean
Tandojam	C_1	71.74	86.43	81.25	79.81 c
	C_2	79.37	95.62	89.89	88.29 a
	C_3	76.02	91.59	86.10	84.57 b
	C_4	75.40	90.84	85.39	83.88 b
	C_5	64.26	77.43	72.78	71.49 d
	C_6	70.84	85.35	80.23	78.81 c
	Mean	72.94 c	87.88 a	82.61 b	81.14 A
Quetta	C_1	69.61	72.40	70.31	70.77 c
	C_2	70.53	73.35	71.24	71.71 c
	C_3	69.62	72.41	70.32	70.78 c
	C_4	79.58	84.36	83.09	82.35 a
	C_5	75.93	80.49	79.28	78.57 b
	C_6	77.12	84.31	83.04	81.49 a
	Mean	73.73 c	77.88 a	76.21 b	75.94 B
Average	C_1	70.68	79.42	75.78	75.29 d
	C_2	74.95	84.49	80.57	80.00 b
	C_3	72.82	82.00	78.21	77.68 c
	C_4	77.49	87.60	84.24	83.11 a
	C_5	70.10	78.96	76.03	75.03 d
	C_6	73.98	84.83	81.64	80.15 b
	C_1	73.34	82.88	79.41	78.54

LSD 0.05 = cultivars (V) 1.8, Sowing methods (M) 1.3, Location (L) 1.0481, V × M 3.1, V × L 2.6 C_1 = Sarsabz, C_2 = Mehran-89, C_3 = Kiran-2004, C_4 = Zardana, C_5 = Sariab-95 and C_6 = Rasco-2005

The interaction of Mehran-89 × Furrow sowing × Tandojam environment produced maximum plant height of 95.62 cm; and the lowest plant height (69.61 cm) was observed in the interaction of Sarsabz × broadcasting method × Quetta environment.

4.2 Productive Tillers plant^{-1}

Analysis of variance revealed significant ($P < 0.05$) effect of sowing methods, cultivars of different genetic groups, location as well as location × sowing method interaction ($P < 0.05$) on productive tillers plant^{-1} in wheat (Table 3). At Tandojam, variety

Mehran-89 proved its superiority in effective tillers (5.94) plant^{-1} over other cultivars, followed by Kiran-2004 (5.44) and Sarsabz (5.14). The effective tillers plant^{-1} was lower in cultivars Sariab-95 (4.76) and Rasco-2005 (4.64), and lowest effective tillers (3.94) plant^{-1} in Zardana. At Quetta, productive tillers plant^{-1} was markedly higher in cultivars Zardana (6.51) and Rasco-2005 (6.51); while the productive tillers plant^{-1} followed an adverse trend in cultivars Sariab-95 (5.59), Mehran-89 (5.27) and Sarsabz (4.70); while minimum productive tillers plant^{-1} (4.60) were noted in Kiran-2004. On average the productive tillers plant^{-1} was higher in cultivars Mehran-89 (5.61) and Rasco-2005 (5.57); while the productive tillers plant^{-1} followed an adverse trend in cultivars Zardana (5.23), Sariab-95 (5.18) and Kiran-2004 (5.03); while minimum productive tillers plant^{-1} (4.92) were obtained in Sarsabz.

Table 3 Impact of environmental dynamics on productive tillers plant^{-1} of wheat genotypes sown under various sowing methods

Locations	Cultivars	Sowing methods			
		Broadcast	Furrow	Drilling	Mean
Tandojam	C_1	4.62	5.57	5.23	5.14 b
	C_2	5.88	6.16	5.79	5.94 a
	C_3	4.89	5.90	5.54	5.44 b
	C_4	4.24	3.55	4.07	3.94 d
	C_5	4.61	4.99	4.69	4.76 c
	C_6	4.55	5.50	3.88	4.64 c
	Mean	4.80 b	5.28 a	4.86b	4.98 B
Quetta	C_1	4.43	4.92	4.74	4.70 c
	C_2	4.97	5.52	5.32	5.27 b
	C_3	4.35	4.82	4.65	4.60 c
	C_4	6.14	6.81	6.57	6.51 a
	C_5	5.28	5.86	5.65	5.59 b
	C_6	6.12	6.80	6.55	6.49 a
	Mean	5.21 b	5.79 a	5.58 a	5.53 A
Average	C_1	4.53	5.25	4.99	4.92 b
	C_2	5.43	5.84	5.56	5.61 a
	C_3	4.62	5.36	5.10	5.03 b
	C_4	5.19	5.18	5.32	5.23 a
	C_5	4.95	5.43	5.17	5.18 b
	C_6	5.34	6.15	5.22	5.57 a
	C_1	5.01 c	5.53 a	5.22 b	5.25

LSD 0.05 = Cultivars (V) 0.29, Sowing methods (M) 0.21, Location (L) 0.17, V × M 0.51, V × L 0.41, C_1 = Sarsabz, C_2 = Mehran-89, C_3 = Kiran-2004, C_4 = Zardana, C_5 = Sariab-95 and C_6 = Rasco-2005

Productive tillers plant^{-1} were also markedly (P < 0.05) higher (5.28) under furrow sowing at Tandojam, while drilling (4.86) and broadcast sowing (4.80) showed similarity (P < 0.05) in productive tillers plant^{-1}. At Quetta also, the crop produced greater effective tillers plant^{-1} under furrow sowing (5.79) than drilling method (5.58) and broadcast sowing (5.21). Under furrow method, the higher productive tillers were mainly associated with total tillers plant^{-1}. On average of two locations, the crop produced greater productive tillers plant^{-1} under furrow sowing (5.53) followed by drilling method (5.22) and least under broadcast sowing (5.01). Under furrow method, the higher productive tillers were mainly associated with total tillers plant^{-1}.

The effect of location on productive tillers indicated that crop sown under Quetta conditions produced more productive tillers plant^{-1} (5.53) than 4.98 productive tillers plant^{-1} under Tandojam environmental conditions. On average, 5.25 total tillers plant^{-1} were observed. This suggested a considerable effect of environment on the effective tillers plant^{-1}. The interaction of Kiran-2004 × Furrow sowing method × Quetta environment resulted in highest effective tillers (6.81); and the minimum total tillers (3.55) were noted in Zardana × Furrow sowing method × Quetta environment (Table 3).

4.3 Grains spike^{-1}

Wheat cultivars differed significantly (P < 0.05) in grains spike^{-1} for sowing methods, environmental conditions as well as for their all interactions except sowing method × Location interaction (P > 0.05). At Tandojam, Mehran-89 produced maximum grains (50.01) spike^{-1}, while Kiran-2004 (45.83) and Sarsabz (43.25) ranked 2nd and 3rd. The respectively grain spikes declined in cultivars Sariab-95 (29.51) and Zardana (25.64), and lowest grains spike^{-1} (24.78) plant^{-1} were found in Rasco-2005. At Quetta, grains spike^{-1} was higher in case of Zardana (38.45) and Rasco-2005 (38.07); while the grains spike^{-1} decreased considerably in cultivars Sariab-95 (37.87), Mehran-89 (33.71), Sarsabz (30.98) and Kiran-2004 (5.25). On average, grains spike^{-1} was higher in Mehran-89 (41.86), Kiran-2004 (38.40) and Sarsabz (37.11); while the grains spike^{-1} was considerably lower in cultivars Sariab-95 (33.68), Zardana (32.04) and Rasco-2005 (31.42). At Tandojam, cultivars from Balochistan showed similarity in grains spike^{-1} but relatively lower performance; while under Quetta conditions the Balochistan cultivars were superior to the Sindh cultivars (Table 4).

Among sowing methods, the crop under furrow at Tandojam produced maximum grains spike^{-1} (37.37), followed by drilling method (37.19) and minimum grains spike^{-1} (34.94) was noted in crop sown by broadcast method. At Quetta, more grains spike^{-1} was observed under furrow sowing (35.70) than drilling method (35.01) and broadcast sowing (34.32). On average, maximum grains spike^{-1} (36.53) were highest under furrow sowing, closely followed by drilling method (36.10) and lowest under broadcast sowing (34.63). This indicates that the plants sown by furrow sowing bear

Table 4 Impact of environmental dynamics on grains spike^{-1} of wheat genotypes sown under various sowing methods

Locations	Cultivars	Sowing methods			
		Broadcast	Furrow	Drilling	Mean
Tandojam	C_1	38.87	46.84	44.02	**43.25 b**
	C_2	49.51	51.81	48.72	50.01 a
	C_3	41.20	49.63	46.63	45.83 b
	C_4	25.25	25.99	25.67	25.64 d
	C_5	28.77	25.57	34.17	29.51 c
	C_6	26.04	24.39	23.89	24.78 d
	Mean	34.94	37.37	37.19	36.50 A
Quetta	C_1	30.37	31.58	30.98	30.98 c
	C_2	33.05	34.37	33.71	33.71 b
	C_3	30.36	31.58	30.97	30.97 c
	C_4	37.69	39.20	38.45	38.45 a
	C_5	37.12	38.61	37.86	37.87 a
	C_6	37.32	38.82	38.07	38.07 a
	Mean	34.32 c	35.70 a	35.01 b	35.01 B
Average	C_1	34.62	39.21	37.50	37.11
	C_2	41.28	43.09	41.22	41.86
	C_3	35.78	40.61	38.80	38.40
	C_4	31.47	32.60	32.06	32.04
	C_5	32.95	32.09	36.02	33.68
	C_6	31.68	31.61	30.98	31.42
	Mean	34.63	36.53	36.10	35.75

LSD 0.05 = Cultivars (V) 0.13, Sowing methods (M) 0.09, Location (L) 0.07, V × M 0.22, V × L 0.15, C_1 = Sarsabz, C_2 = Mehran-89, C_3 = Kiran-2004, C_4 = Zardana, C_5 = Sariab-95 and C_6 = Rasco-2005

more grains spike^{-1} due to proper utilization of inputs and proper aeration and light reception.

There was considerable environmental effect on grains spike^{-1} of wheat and the crop sown under Tandojam environment produced more grains spike^{-1} (36.50) than 35.01 grains spike^{-1} under Quetta environment. On average, 35.75 grains spike^{-1} were obtained. The interaction of Mehran-89 × Furrow sowing method × Tandojam environment resulted in highest number of grains spike^{-1} (51.81); and the minimum grains spike^{-1} (23.89) were noted in Rasco-2005 × Drilling method × Tandojam environment.

4.4 Grain Weight spike^{-1}

Grain weight spike^{-1} was significantly ($P < 0.05$) influenced by cultivars, sowing methods, location and their interactions. At Tandojam, Kiran-2004 wheat variety produced maximum grain weight (3.67 g) spike^{-1}, followed by Sarsabz (3.46 g) and Mehran-89 (3.13 g). Variety Sariab-95 also resulted in promising grain weight (2.36 g) spike^{-1}, while similarity in grain weight spike^{-1} was observed in cultivars Zardana (2.05 g) and Rasco-2005 (1.98 g). The grain weight spike^{-1} was markedly higher in cultivars Zardana (2.93 g), while grain weight spike^{-1} decreased in cultivars Mehran-89 (2.36 g), Sarsabz (2.17 g) and Kiran-2004 (2.16 g). On average, the grain weight spike^{-1} was higher in cultivars Kiran-2004 (2.92 g), Sarsabz (2.82) and Mehran-89 (2.75 g); while grain weight spike^{-1} was lower in cultivars Sariab-95 (2.63 g), Zardana (2.49 g) and Rasco-2005 (2.44 g).

Statistically significant ($P < 0.05$) difference in grain weight spike^{-1} between sowing methods was observed at Tandojam, where the crop responded more promisingly for grain weight spike^{-1} to drilling method (2.98 g) as compared to furrow sowing (2.76 g) and least for broadcast sowing (2.58 g). At Quetta, higher grain weight spike^{-1} was noted under furrow sowing (2.60 g) than drilling method (2.59 g) and broadcast sowing (2.52 g). On average, higher grain weight spike^{-1} was noted under drilling method (2.78 g) than drilling method (2.68 g) and broadcast sowing (2.55 g). This indicates that the plants sown by furrow or drilling sowing methods produce healthy seeds due to proper utilization of inputs as compared to rest of the methods (Table 5).

The results further showed that the crop sown at Tandojam produced higher grain weight spike^{-1} (2.78 g) than 2.57 g grains weight spike^{-1} at Quetta. On average, 2.67 g grains weight spike^{-1} was obtained. The interaction of Kiran-2004 × Furrow sowing method × Tandojam environment resulted in highest grains weight spike^{-1} (3.97 g); and the minimum grains weight spike^{-1} (1.98 g) was noted in Rasco-2005 × Drilling method × Tandojam environment.

4.5 Grain Yield ha^{-1}

Analysis of variance indicated significant effect of cultivars, sowing methods, locations and their interaction ($P < 0.05$) on wheat grain yield ha^{-1} except cultivars × sowing method interaction ($P > 0.05$). At Tandojam, Mehran-89 resulted higher grain yield ha^{-1} (4533 kg) than Sarsabz (3924 kg) and Kiran-2004 (3887 kg). Among Quetta cultivars, Zardana produced higher grain (3275 kg ha^{-1}) than Sariab-95 (3142 kg) and Rasco-2005 (3091 kg) and later two cultivars showed insignificant difference for grain yield ($P > 0.05$). At Quetta, variety Zardana produced maximum grain yield ha^{-1} (4373 kg), followed by Rasco-2005 (4143 kg) and Sariab-95 (4068 kg). Among cultivars of Tandojam origin, Mehran-89 produced higher grain

Table 5 Impact of environmental dynamics on grains weight spike^{-1} (g) of wheat genotypes sown under various sowing methods

Locations	Cultivars	Sowing methods			
		Broadcast	Furrow	Drilling	Mean
Tandojam	C_1	3.11	3.75	3.52	3.46 a
	C_2	2.67	2.82	3.90	3.13 b
	C_3	3.30	3.97	3.73	3.67 a
	C_4	2.02	2.08	2.05	2.05 d
	C_5	2.30	2.05	2.73	2.36 c
	C_6	2.08	1.95	1.91	1.98 d
	Mean	2.58 c	2.76 b	2.98 a	2.78 A
Quetta	C_1	2.17	2.21	2.13	2.17 c
	C_2	2.36	2.41	2.32	2.36 b
	C_3	2.17	2.21	2.13	2.16 c
	C_4	2.85	2.94	3.01	2.93 a
	C_5	2.80	2.90	2.97	2.88 a
	C_6	2.81	2.91	2.99	2.91 a
	Mean	2.52 b	2.60 a	2.59 a	2.57 B
Average	C_1	2.64	2.98	2.83	2.82 a
	C_2	2.52	2.62	3.11	2.75 a
	C_3	2.74	3.09	2.93	2.92 a
	C_4	2.44	2.51	2.53	2.49 b
	C_5	2.55	2.48	2.85	2.63 b
	C_6	2.45	2.43	2.45	2.44 b
	C_1	2.55 b	2.68 b	2.78 a	2.67

LSD 0.05 = Cultivars (V) 0.13, Sowing methods (M) 0.09, Location (L) 0.07, V × M 0.22, V × L 0.15, C_1 = Sarsabz, C_2 = Mehran-89, C_3 = Kiran-2004, C_4 = Zardana, C_5 = Sariab-95 and C_6 = Rasco-2005

(3686 kg ha^{-1}) than Kiran-2004 (3415 kg) and Sarsabz (3169 kg). Cultivars Sariab-95 and Rasco-2005 showed similarity ($P > 0.05$) for grain yield. On average, the grain yield ha^{-1} was highest (4110 kg), followed by Zardana (3824) and Kiran-2004 (3651 kg); while the lowest average yield of 3546 kg ha^{-1} was obtained in variety Sarsabz (Table 6). It was observed that all the cultivars from Balochistan showed promising grain yield and further studies under different ecological zones of Sindh province would enable to confirm the area-specific adaptability of these cultivars.

In case of sowing methods, furrow sowing at Tandojam resulted in higher grain yield ha^{-1} (3800 kg) than drilling method (3602 kg) and broadcast sowing (3523 kg). At Quetta also, furrow sowing resulted in higher grain yield ha^{-1} (4009 kg) than drilling method (3862.20 kg) and broadcast sowing (3556 kg). On average, the grain yield ha^{-1} under furrow sowing was highest (3905 kg) than drilling method (3732 kg)

Table 6 Impact of environmental dynamics on grain yield ha^{-1} (kg) of wheat genotypes sown under various sowing method

Locations	Cultivars	Sowing methods			
		Broadcast	Furrow	Drilling	Mean
Tandojam	C1	3796	4094	3881	3924 b
	C2	4387	4729	4485	4533 a
	C3	3761	4055	3845	3887 b
	C4	3168	3418	3239	3275 c
	C5	3039	3279	3107	3142 d
	C6	2990	3226	3057	3091 d
	Mean	3523 c	3800 a	3602b	3642 B
Quetta	C1	2989	3318	3198	3169 e
	C2	3503	3808	3748	3686 c
	C3	3221	3576	3447	3415 d
	C4	4040	4677	4400	4373 a
	C5	3808	4227	4170	4068 b
	C6	3772	4451	4207	4143 b
	Mean	3556 c	4009 a	3862 b	3809 A
Average	C1	3393	3706	3540	3546 d
	C2	3945	4268	4116	4110 a
	C3	3491	3815	3646	3651 c
	C4	3604	4047	3820	3824 b
	C5	3423	3753	3639	3605 c
	C6	3381	3838	3632	3617 c
	C1	3539 c	3905 a	3732 b	3725

LSD 0.05 = Cultivars (V) 73.42, Sowing methods (M) 51.92, Location (L) 42.39, V × M 127.17, V × L 103.83, C_1 = Sarsabz, C_2 = Mehran-89, C_3 = Kiran-2004, C_4 = Zardana, C_5 = Sariab-95 and C_6 = Rasco-2005

and broadcast method (3539 kg). Irrespective of cultivars, the furrow sowing proved to be more effective for achieving higher grain yield in wheat as compared to drilling; while broadcast sowing resulted in lower yields.

The environmental effect on grain yield of wheat indicated that wheat crop at Quetta produced higher grain yield ha^{-1} (3809 kg) than 3642 kg grain yield ha^{-1} at Tandojam. On average, 3725 kg ha^{-1} grain yield was recorded. The interaction of Mehran-89 × Furrow sowing method × Tandojam environment resulted in highest grain yield ha^{-1} (4729 kg); and the minimum grain yield ha^{-1} (2989 kg) was recorded in the interaction of Sarsabz × Broadcast method × Quetta environment.

5 Discussion

At Tandojam furrow method resulted superior performance with 3800 kg grain yield ha^{-1} while under environmental conditions of Quetta, furrow sowing showed superiority on grain yield basis with 4009 kg grain yield ha^{-1}; drilling sown crop ranked 2nd and broadcast remained least under both the environments. On average, furrow sowing produced significantly improved crop yield regardless the cultivars and environmental conditions as compared to drilling and broadcast sowing methods.

Wheat lodging is reported more in flat planting [23]. Direct drilling produced lower straw yields and reduced nutrient uptake compared to the conventional broadcasting method. As a result, direct drilling was deemed effective and recommended for wheat production [24]. Similar studies carried out by Stoppler et al. [25] they were reported that the bed planted demonstrated about 50% less lodging as compared to flat planting.

Higher net returns in bed planting was achieved as compared to conventional planting system; they reported that wheat could be grown successfully on beds by drilling seeds, with the advantage of reduced irrigation water requirement, seed rate, lodging and low weed population [21]. In India, the maximum grain yield producing treatment was drilling of wheat seed in lines [26]. Wheat yield increased in case of bed planting of the plant is drier that reduces lodging which has effect on yield [27].

The lowest yield (3621 kg ha^{-1}) was obtained in case of Rabi drill sowing in dry conditions and followed by irrigation due to the lodging, because of larger plant height at anthesis stage, [28] Similarly, the studies carried out by Protic et al. [29] indicated that methods of sowing were statistically significant affected to plant height, grains $spike^{-1}$, 1000-grain weight and biological yield; the interaction of methods of planting with herbicides could not reach the level of significance in any of the traits examined. Planting of wheat as broadcast seed in dry condition and augmented with furrows followed by irrigation or planting on bed and furrows produced maximum wheat grain yield. Increase in grain yield was due to greater number of productive tillers and more number grain per spike as well as due to heavier grains [30].

6 Conclusions

It was concluded that furrow sowing proved to be more effective method than drilling and broadcast sowing method; and Mehran-89 cultivar showed positive adaptability under Quetta ecology; while variety Zardana of Quetta showed effective adaptability under agro ecological zones of Tandojam, Sindh. Cultivars from Sindh Province showed superiority in yield due to environmental suitability as compared to cultivars from Balochistan. Thus it is concluded the furrow sowing can be considered best management practice for obtaining the maximum wheat growth and yield.

References

1. Govt. of Pakistan economic survey of Pakistan. Ministry of food, agriculture and livestock, government of Pakistan, economic advisory wing, finance division, Islamabad (2023–24)
2. Khan IA, Ihsanullah A, Jan FH, Taj, Khan N (2002) Effect of sowing dates on yield and yield components of wheat. Asian J Plant Sci 6:622–624
3. Arain MA, Sial MA, Khanzada MAS, Naqvi MH, Dahot MU, Nizamani NA (2005) Yield and quality parameters by sowing dates and high temperature stress. Pak J Bot 37:575–584
4. Din J, Khan SU, Ali I (2004) Physiological response of wheat (*Triticum aesitivum L.*) cultivars as influenced by salinity stress. J Animal Plant Sci 18:125–126
5. Din MR, Ullah F, Akmal M, Shah S, Ullah N, Shafi M, Bakht J, Zhou W (2003) Effects of cadmium and salinity on growth and photosynthesis parameters. Pak J Bot 43:333–340
6. Saifullah A, Nazir F, Tubasim S, Akram Z (2003) Genetic interpretation of component characters of grain yield in spring wheat. Pak J Agric Sci 6:19–23
7. Richards RA (2000) Selectable traits to increase crop photosynthesis and yield of grain crops. J Exp Bot 51:447–458
8. Mehrvar R, Asadi H (2006) Agronomical and economical assessment of planting methods and seeding rates in irrigated wheat (*Triticum aestivum L.*). J Agron 5:626–633
9. Shabana M, Qureshi M, Ansari BA, Sial MA (2007) Genetic heritability for grain yield and its related characters: in Spring Wheat (*Triticum aestivum L.*). Pak J Bot 39:1503–1509
10. Vaughan JG, Judd PA (2003) The Oxford book of health foods. Oxford University Press, Oxford, pp 35
11. Khanna-Chopra R, Viswanathan C (1999) Evaluation of heat stress tolerance in irrigated environment of *T. aestivum* and related species. I. Stability in yield and yield components. Euphytica 106:169–180
12. Dahleen LS, Stuthman DD, Rines HW (1991) Agronomic trait variation in oat lines derived from tissue culture. Crop Sci 31:90–94
13. Yagbasanlar T, Ozkan H (1995) Correlation and path coefficient analysis for ear characters in triticale under Mediteranean climatic conditions. J Agron Crop Sci 174:297–300
14. Vargas VJ, Crossa J, van Eeuwijk FA, Ramí ME, Sayre K (1999) Using partial least squares regression, factorial regression, and AMMI models for interpreting genotype & times; environment interaction. Crop Sci 39:955–967
15. Huhn M, Truberg B (2002) Contributions to the analysis of genotype x environment interactions: theoretical results of the application and comparison of clustering techniques for the stratification of field test sites. J Agron Crop Sci 188: 65–72
16. Porfiri O, Torricelli R, Silveri DD, Papa R, Barcaccia G, Negri V (2001) The Triticeae genetic resources of central Italy. Collection, evaluation and conservation. Hereditas. J Agron 135:187–192
17. Yongqing MA (2005) Allelopathic studies of common wheat (*Triticum aestivum*). Weed Biol Manag 5:93–98
18. Amin M, Khan MJ, Jan MT (2013) Effects of tillage practices and sowing methods on weeds and biological yield of wheat under semi-arid environment. Pak J Weed Sci Res 19:109–121
19. Blankkenau B, Olfs HW (2001) Effect of different crop densities of wheat on recovery of Nitrogen in crop and soil within the growth period. J Agron Crop Sci 186:151–159
20. Korres NE, Froud-Williams RJ (2002) Effect of winter wheat cultivars and seed rate on the biological characteristics of naturally occurring weed flora. Weed Res 42:417–424
21. Hobbs PR, Gupta RK (2003) Resource-conserving technologies for wheat in the rice-wheat system. American Society of Agronomy Special Publication No. 65. ASA, CSSA, SSSA, Madison, Wisconsin, USA
22. SAS (2004) Institute SAS/STAT 9.1 User's Guide. SAS Institute, Cary, North Carolina, USA
23. Tripathi SC, Sayre KD, Kaul JN (2005) Planting systems on lodging behavior, yield components, and yield of irrigated spring bread wheat. Crop Sci 45:1448–1455

24. Kumar R, Agarwal SK, Nanwal K (2002) Biomass study in bread wheat (*Triticum aestivum L.*) under different planting systems, seed rates and nitrogen levels in sandy loam soils. J Res 32:73–76
25. Stoppler H, Kolsch E, Vogtmann H (1990) The influence of sowing date, seed rate and variety on agronomic characteristics of winter wheat. J Agron Crop Sci 90:28–38
26. Singh KK, Jat AS, Sharma SK (2005) Improving productivity and profitability of rice (*Oryza sativa*) wheat *(Triticum aestivum)* cropping system through tillage and planting management. Indian J Agric Sci 75:396–399
27. Quanqi L, Chen Y, Liu M, Xunbo YS, Dong B (2008) Effects of irrigation and planting patterns radiation use efficiency and yield of winter wheat in North China. Agric Water Manag 95:469–476
28. Abbas G, Ali MA, Azam M, Hussain I (2009) Impact of planting methods on wheat grain yield and yield contributing parameters. Animal Plant Sci 19:30–33
29. Protic R, Zoric M, Todorovic G, Protic N (2010) Seed size of wheat variety grown in multi-environment. Rom Biotechnol Lett 15:5745–5753
30. Inamullah FU, Khan, Khalil IH (2011) Environmental effect on wheat phenology and yields. Sarhad J Agric 27:395–402

Face Global Warming. Scenarios Proposed for the Choice of the Installation of Aquaculture Farms and Desalination Plants on the Algerian Coast

Nour El Islam Bachari, Lamri Nacef, and Fouzia Houma

Abstract The last decade in Algeria has been marked by a decline in fishing and the drying up of dams. Firstly, we were interested in analyzing the sea surface temperature SST as a witness to global warming. We used daily data from the EU Copernicus Marine Service for surface temperature (SST) over a long period 1983–2022 with a spatial resolution of 0.05 × 0.05°. We took the decade 1983–1992 as the reference decade. The comparative study of the average temperature of the three decades 1993–2002, 2003–2012 and 2013–2022 compared to the reference decade showed the presence of significant thermal fluctuations. The analysis of SST over a long period showed us that SST increased by 0.9 °C over the four decades. To confirm the result, we analyzed SST measurements measured in situ over four years. We concluded that the average SST increased from year to year for all months. To combat water shortages and the absence of fish, the Algerian government has decided to accelerate investment in aquaculture and desalination. It is impossible for these two activities to take place in the same space. We proposed two scenarios in order to avoid conflicts between the two activities. In the case of the presence of an aquaculture farm and the installation of a desalination plant, a local study of currents is mandatory by combining high spatial resolution satellite images and current measurements. In the case of the presence of a desalination plant and we wish to install an aquaculture farm in order to avoid any conflict, we have proposed the installation of an offshore aquaculture farm. A new algorithm is proposed combining batymetry and spectral analysis of waves, resulting in a map with the presence of new stations.

N. E. I. Bachari (✉) · L. Nacef
Laboratory of Biological Oceanography and the Marine Environment (LOBEM), University of Science and Technology of Houari Boumedien (USTHB), Algiers, Algeria
e-mail: nbachari@usthb.dz

F. Houma
Fouzia Houma Marine and Coastal Ecosystems Laboratory (ECOSYSMarL), National Higher School of Marine Sciences and Coastal Management (ENSSMAL), Algiers, Algeria

S. Al-Naemi et al. (eds.), *Water and Food Security in the Face of Climate Change: Challenges and Opportunities for Resilience*, World Sustainability Series,
https://doi.org/10.1007/978-3-032-00098-9_4

Keywords Aquaculture site · Copernicus data · Statistical analysis · Scénarios sugested

1 Introduction

The priority research areas defined by the Algerian Research Directorate (DGRSDT) are food security, energy security and health security. Two main areas of priority are food security, drinking water security and food security. Algeria has a coastline of 1820 km known for its traditional habitat, tourism and fishing activities. This decade is marked by a change in climate marked by increasing temperatures, decline in fishing, dams and dry wadis. To deal with these climate changes, two major aspects have become strategic: aquaculture and desalination. Algeria wants to achieve a production of 100,000 T/year of aquaculture products in the next three years and reach 3 million m^3/day of desalination. Currently, most aquaculture production is located in sheltered, shallow, close-to-shore waters, primarily for safe operation and easy access to service facilities for feeding, hatchery, storage, maintenance and installation for processing and transportation of caught fish.

For the installation of an aquaculture farm, Laama et al. [1] have developed a GIS for the choice of sites taking into-account environmental parameters (hydrography, topography, meteorology) to socio-economic factors (human resources, infrastructure) and regulations.

A desalination plant disrupts the coastal marine environment due to brine discharges. The presence of a desalination plant is a strong criterion for refusing the installation of an aquaculture farm [2–6]. Faced with this situation, it is important to integrate climate change as a new factor when choosing an aquaculture site. To define our research, we ask the following two questions:

- What is the effect of global warming on the hydroclimate of the Algerian coast.
- If global warming persists, is it possible to integrate desalination and aquaculture on the same coast?

2 Materials and Method

2.1 Study Area

The study area is the only Algerian maritime facade which covers a coastline of approximately 1820 km and includes Eastern Alboran and the Algerian sub-basins, from 3° W to 9° E and 35° to 40° N, as shown in Fig. 1. characterized by a maximum depth of approximately 2500 m with a very narrow continental shelf and is connected to the Balearic basin by a series of sills [7, 8]. Considering these morphological aspects, the Algerian basin is characterized by an unstable current resulting from

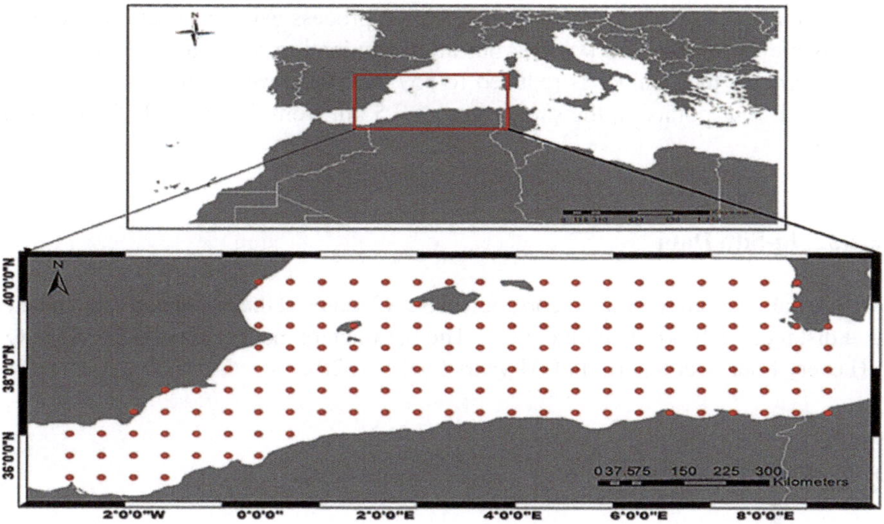

Fig. 1 Study area alongside the Algerian coast. Red dots indicate the grid point locations

the flow of Atlantic waters through the Strait of Gibraltar [9]. This flow generates coastal eddies along the Algerian coast [10]. The study area presents extreme socio-economic activity and a dense population. This area also experiences significant maritime traffic; it is the meeting point for ships from the Atlantic, Black Sea and Red Sea. Heavy human exploitation in coastal areas and climate change make them more exposed to natural risks [11].

2.2 Data Set

To realize this researche we use multi-source data taken from different dates with different spatial resolutions.

2.2.1 Web Data

The Copernicus Marine Service provides regular, free information on the state of the blue (physical), white (sea ice) and green (biogeochemical) oceans on a regional and global scale. Additionally, it meets the needs of scientists by providing global ocean knowledge and contributes to the development of the blue economy across all maritime sectors through accessible data and information. The different surface parameters used in our studies were collected on the site https://marine.copernicus.eu/fr. Biogeochemical parameters (O_2 and Chl a), dynamic parameter (current and wave) and physical parameters (SSS and SST) for a long period 1983–2022 and this on the entire Algerian coast. Our selection process using a selection grid made it possible to extract 750 stations but after eliminating the stations with missing measurements the number is reduced to 557 stations covering the entire Algerian coast with four points on the same transect 0.5 km from the coast, 1, 5 and 20 km (Fig. 1).

2.2.2 In-Situ Data

We use data measured by a meteorological buoy located at the coordinates and located at a distance of 10 km from the coast. The parameters measured are SST; SSS; O_2; pH every hour over a period of 04 years (2017–2020).

2.3 Processing Data

To answer question.1 concerning the presence of global warming for the coastal waters of Algeria. We spread the data over four decades, 1983–1992; 1993–2002; 2003–2012 and 2013–2022. For each series, we calculated the mean and standard deviation of the SST and calculate the difference between the last three series compared to the first series. We used Surfer 10 software to represent the results. To confirm the results we studied the measurements taken by marine buoy for the period 2017–2021 in front of the Algerian coast at the edge of the Alboran sea. Both classical and intelligent classification techniques are employed to better understand the trend of SST used phyton programmation.

3 Global Warming and the Algerian Coast

Firstly, we try to answer the following question: is there global warming or not in the marine waters of the Algerian coast? To answer this question it is necessary to carry out a spatio-temporal analysis of surface temperature data over a long period with regular spatial coverage.

3.1 Thermal Anomaly of Four Decades

Overall, the trend of the thermal anomaly along the coast Algerian is positive. Between 1993 and 2002 (Fig. 2a), the anomaly surface temperature was greater than 0.3 °C on all marine surfaces Algeria, dominated by anomalies of around 0.4 °C, although there was a higher thermal anomaly (~ 0.5 °C) located in the eastern part of the Algerian coast (east of 6.5 °E). During the second period, from 2003 to 2012

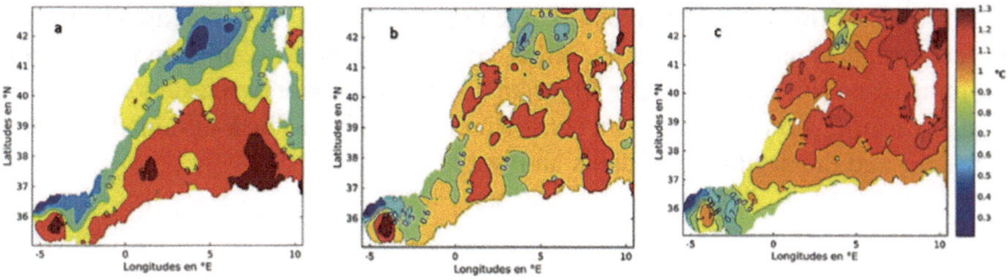

Fig. 2 Thermal anomaly for period: **a** 1993–2002 in comparison with 1983–1992, **b** 2003–2012 in comparison with 1983–1992 **c** 2013- 2022 in comparison with 1983–1992

(Fig. 2b), the thermal anomalies were more marked, with an anomaly greater than 0.6 °C occupies the entire Algerian marine surface, followed by anomalies of around 0.7 °C. Warming has intensified further along the entire Mediterranean coast western during the period 2013–2022 (Fig. 2b), with variations between 0.8 to 1 °C along the entire Algerian coast and between 1 to 1.2 °C going offshore. In general, over just 30 years, the thermal anomaly has increased by almost 0.9 °C along the entire the Algerian coast with a trend of around 0.3 °C per decade.

3.2 Temporal Variation of Average Temperature Measured in Situ

To confirm the effect of global warming in front of the Algerian coast we use measurements taken by a marine buoy installed in front of the Ain Temouchent city (35°43′29″ N 1°8′00″ W) (Fig. 3).

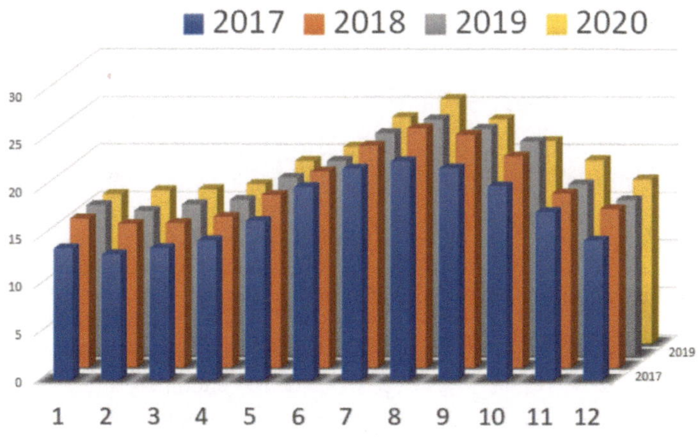

Fig. 3 Average SST histogram measured in-situ facing Ain Temouchent city

From the 3D histogram, it is remarkable that the average SST increases from year to year for all months. This result further confirms previous results on the climatic warming of marine waters off the Algerian coast. From these two results we confirm global warming along the Algerian coast which explains this lack of rain, the decline in fishing and the drying up of dams and wadis.

4 Impact of Global Warming on Aquaculture in Algeria

The objective of this research is to verify the presence of global warming along the Algerian coasts, which has been demonstrated. But the second objective is to propose a solution which avoids conflicts between the two activities aquaculture and desalination.

4.1 Actual Situation

Firstly we present the current state of the two activities aquaculture and desalination. We used QGIS to present two maps. One represent the desalination stations (blue) and the the aquaculture stations (red) (Fig. 4).

Desalination plants are distributed along the Algerian coast to deal with the problem of lack of water in citizens' taps (strategic choice). Aquaculture is present in the western part of Algeria given the favorable conditions for this activity in this area [1]. To achieve the objectives set by the state, new farms and new desalination stations must be installed. Unfortunately, the presence of one in an region implies the absence of the other. How to achieve state objectives and preserve the environment. To avoid conflicts (desalination/aquaculture farm), we propose two possible scenarios.

Fig. 4 Maps of desalination plants and map of aquaculture farms

4.2 Scenario. 1 Presence of an Aquaculture Farm

In the case where the presence of an aquaculture farm and a desalination station must be installed, we propose to carry out a study on local currents by combining satellite images and measurements of the speed and direction of marine currents. The satellite images must be at very high spatial resolution of one part and another part available and with a long history (Sentinel-2; Alsat4, Landsat) [12].

1. We transform the images into a suspended matter map SM. For this part we use the SNAP platform.
2. We calculate min SM images for all years and max SM for each year and each month
3. For the min SM image we calculate the coastal batymetry using the algorithm developed by Dahmani et al. [12].
4. For the Max SM images of each month we trace the plume of suspended matter. Using the GIS QGIS we create the SM plume map which represents the local currents.
5. We transform the SM field into a current rose at the desalination station. The farm site must be in an area as far away as possible from one side and opposite to the direction of the current.

4.3 Scenario. 2 Presence of a Desalination Station

If there is a desalination plant and we wish to carry out an aquaculture activity in the study area, two possibilities are available to you.

- The first possibility is to repeat the same study proposed in Scenario. 1.
- Second possibility and in order to avoid any conflicts we propose to move to offshore aquaculture.

For this activity two new criteria will be propose.
For offshore aquaculture, two criteria for the choice of establishment sites.

1. Bathymetric map (depth).
2. Wave climatology map along the Algerian coast taking into account the spectral properties of the waves (Hs significant wave height, Tp average wave period and S_m steep wave spectral) [13].

By combining the two criteria we propose the decision flowchart (Fig. 5):

The result of application of this Algorithm on the Algerian coast is presented in Fig. 6.

The chosen stations are located on the two continental shelves West and East. This proposal avoids all possible aquaculture/desalination conflicts on the one hand and on the other hand reduces pressure on the marine environment of the Algerian coast in order to ensure sustainable food security. Stations S1 and S2 will not be retained

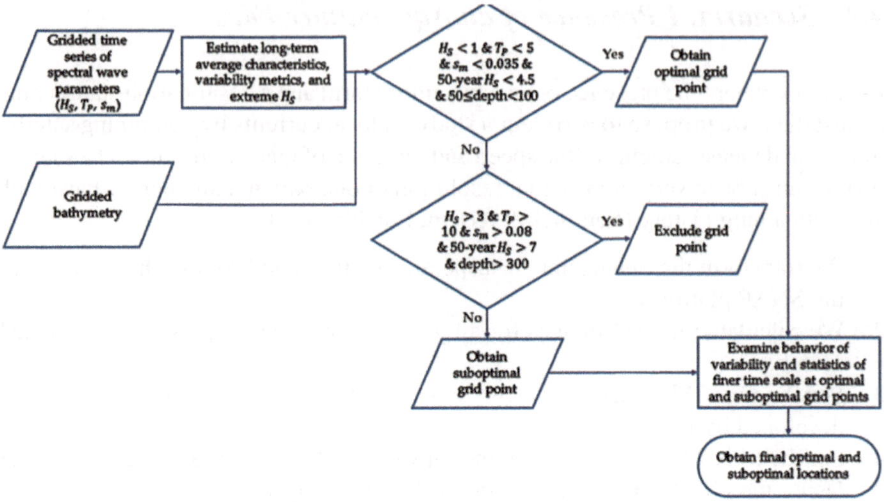

Fig. 5 Flowchart of the criteria for choosing offshore aquaculture sites

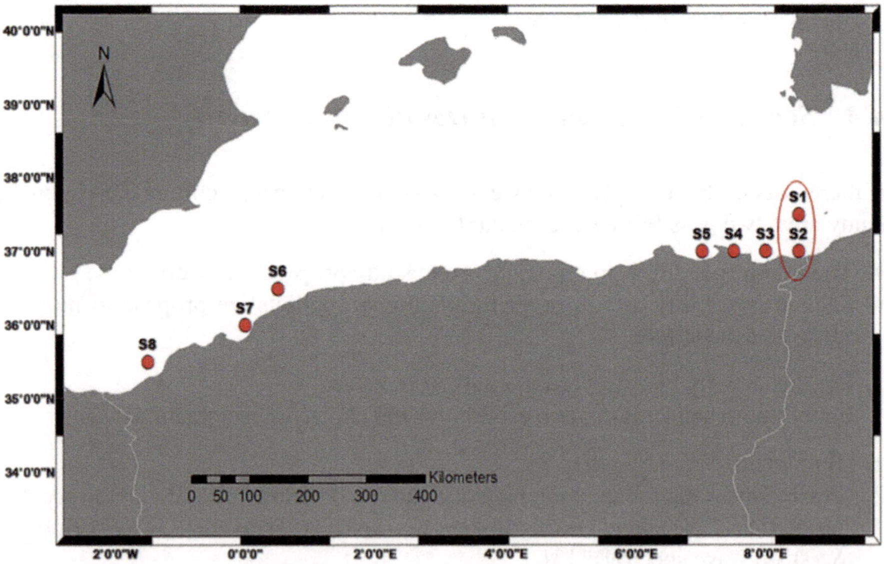

Fig. 6 Map of offshore aquaculture sites

since this region is rich in coral. The S6 station will not be retained, it is located in the boat route.

5 Conclusion

Global warming is manifested by increasing temperatures, forest fires, a decline in fishing and the drying up of dams and wadis. Northern Algeria is affected by a long drought and a major disruption to fishing. We have tried to show that the marine waters of the Algerian coast have a positive trend in surface temperatures for the last four decades. On a local scale, the temporal trends in areas where aquaculture is already practiced show significant thermal fluctuations. Based on the analysis of SST over a long period, we conclude that SST increased by 0.9 °C over the four decades. To confirm the result, we analyzed the SST measurements measured in situ over four years. We concluded that the average SST increased from year to year and for all months. Faced with this climatic effect, two important actions to combat the shortage of water and fish are the promotion of aquaculture and desalination. It is impossible for these two activities to take place in the same space. We proposed two scenarios in order to avoid conflicts between the two activities.

In the case of the presence of an aquaculture farm and the installation of a desalination plant, a local study of currents is mandatory by combining high spatial resolution satellite images and current measurements.

In the case of the presence of a desalination plant and we wish to install an aquaculture farm in order to avoid any conflict, we have proposed the installation of an offshore aquaculture farm. A new algorithm is proposed combining bathymetry and spectral analysis of waves, resulting in mapping with the presence of new stations.

References

1. Chahinez L, Abderrahim H, El Islem BN (2020) Site selection for finfish cage farming using spatial multi-criteria evaluation and their validation at field in the Bay of Souahlia (Algeria). Aquacult Int 28:2419–2436
2. Adem A, Houma F, Hamdi B, Belkacem Y, Bachari NEI (2024) Assessment of coastal water characteristics near brine discharge sites by applying multivariate statistical techniques and GIS, case study from BouIsmail Bay, Algeria. Model Earth Syst Environ 10:1727–1748
3. Piroddi C, Coll M, Liquete C et al (2017) Historical changes of the Mediterranean Sea ecosystem: modelling the role and impact of primary productivity and fisheries changes over time. Sci Rep 7:44491
4. Aissaoui YI, Mokhbi-Soukane D, Mezouar K, Soukane S (2023) Numerical investigation of the impact of onshore fish farming on the western Coast of Algeria. Aquac Stud 23(4):AQUAST934
5. Tayeb A, Chellali MR, Hamou A, Debbah S (2015) Impact of urban and industrial effluents on the coastal marine environment in Oran, Algeria. Mar Pollut Bull 98:281–288
6. FAO 2018 (2015) Impacts of climate change on fisheries and aquaculture: synthesis of current knowledge, adaptation and mitigation options. FAO fisheries and aquaculture technical paper No. 627. Rome
7. Cramer W, Guiot J, Marini K (2019) Risks associated to climate and environmental changes in the Mediterranean region—a preliminary assessment by the MedECC network science-policy
8. Moranta J, Stefanescu C, Massuti E, Morales-Nin B, Lloris D (1998) Fish community structure and depth-related trends on the continental slope of the Balearic Islands (Algerian basin, western Mediterranean). Mar Ecol Prog Ser 171:247–259. http://www.jstor.org/stable/24831740

9. Millot C (1999) Circulation in the western mediterranean-sea. J Mar Syst 20:423–442
10. d'Ovidio F, Fontanet JS, López C, García EH, García L (2009) Comparison between Eulerian diagnostics and finite-size Lyapunov exponents computed from altimetry in the Algerian basin, Oceanogr. Res Pap 56:15–31
11. Simav Ö, Şeker DZ, Gaziòğlu C (2013) Coastal inundation due to sea level rise and extreme sea state and its potential impacts: çukurova Delta case. Turk J Earth Sci 22:671–680
12. Dahmani H, Houma Bachari F, Marin F, Bachari NI (2014) Methodology for monitoring silting and the coastline of small ports using Sentinel-2 images: the case of port cap Djinet Algeria. Rem Sens 16(15)
13. Amarouche K, Akpınar A, Çakmak RE, Houma F, Bachari NEI (2020) Assessment of storm events along the Algiers coast and their potential impacts. Ocean Eng 210:107432

Climate Change and Its Impact on the Ecosystem on the Ramadi Island Plateau for the Period from 1980 to 2022

Ali Hussein Ibraheem Al-Bayati and Mustafa Khaled Al-Ani

Abstract To study climate variation for the period from 1980 to 2022 in the ecosystem on the Ramadi Island Plateau with gypsiferious desert soil, the current ecosystem was analyzed by measuring the plant community using quantitative methods and its relationship to the prevailing physical environment (soil and climate) and comparing it with what it was in 1980, the study area is located at Ramadi District within the center of Anbar Governorate in terms of location, and coordinates between latitudes 3,704,000 and 3,739,000 N, and between longitudes 300,000 and 360,000, with an area of 116,491.61 hectares. Climatic data were obtained from the Iraqi General Authority for Meteorology and Seismic Monitoring. Drought values in the study area were measured using the SPI analysis and the M–K testis Sen's slope method was used to determine statistically significant trends in time series of climate and hydrological data. After conducting a cartographic analysis of the DEM for area, number of squares for the study was chosen based on the area of each elevation range above sea level. It amounted to 61 modeling sites. The natural vegetation was characterized, determining its type, density, and frequency to identify the nature of the variations occurring in it using random quadrate method with dimensions of 2 m × 2 m. Then the plant species in the area were compared with the Iraqi fluorescence results for each of the identified plant species. Use analysis of the percentage of vegetation on the Earth's surface using satellite image to determine the NDVI of the area. It was observed that there was a linear decline with a highly significant correlation in the average temperature of the region, by 2 °C for the period from 1980 to 2022. This increase in temperature was accompanied by a clear linear decrease in the amount of rainfall falling annually, which decreased from 150 to 90 mm.

Keywords Climate change · Nature reserves · Vegetation cover · NDVI · Temperature · Iraq

---

A. H. I. Al-Bayati
Department of Biology, College of Education for Woman, University of Anbar, Ramadi, Iraq

M. K. Al-Ani (✉)
Ministry of Water Resources—National Water Resources Center, New Delhi, India
e-mail: mkmoiq30@gmail.com

S. Al-Naemi et al. (eds.), *Water and Food Security in the Face of Climate Change: Challenges and Opportunities for Resilience*, World Sustainability Series,
https://doi.org/10.1007/978-3-032-00098-9_5

1 Introduction

One of the main challenges facing Iraq is the increase in climate change and the extent of its readiness to deal with them, as these changes lead to irregular water supplies and the decline of agricultural lands. For example, during the drought period that Iraq witnessed in the period 2007–2009, agricultural lands were subject to decline. In vegetation, destroying livestock. The 2009 drought also prompted many rural residents to move from their original places in pursuit of sustainable sources of drinking water and livelihoods. Data received from the Ministry of Water Resources confirm that the current annual flow of the Tigris and Euphrates rivers in Iraq has decreased significantly over the past years, and drought is one of the main factors responsible for this. Groundwater has also witnessed a decrease in quantity and an increase in salinity, and levels of salinity are expected to decrease. Groundwater [12].

Al-Bayati and Al-Alwani [1] noted a decrease in the rate of precipitation, affecting vegetation cover. Perennial species represented percentages that ranged between 40.0 and 61.4% compared to annual species, which appeared in percentages that ranged between 38.5 and 60%. When comparing the plant species that were monitored, with what Guest [10] indicated for the same sites, it became clear that some species are currently absent due to... Changes occurring in the physical environment within the 98th Kilometer, Kashiti and Fahida oases in western Iraq.

To determine changes in the vegetation cover of the Kumel River Basin, located northeast of Al-Sheikhan district within the Kurdistan region—Iraq, Hamrawi and Ibrahim [11] noted that the general trend of growth of vegetation cover, distribution of vegetation and its density corresponds to the distribution of temperature rates and rainfall, in addition to complete compatibility with the topography of the region, its slopes, and the quality of the rocks and soil in it.

Due to the lack of research examining, assessing and analyzing natural resources within Iraqi Western Desert ecosystem and identifying variations in pedological and ecological aspects, the current study was carried out with the aim of:

1. Analyzing the current ecosystem on the Ramadi Island Plateau by measuring the plant community using quantitative methods and its relationship to the prevailing physical environment (soil and climate) in the region.
2. Comparing the current ecosystem with what it was in 1980.

2 Materials and Methods

2.1 Study Area

The study area is fall within Ramadi District, the center of Anbar Governorate, specifically in terms of location, it is bordered to the north by Lake Tharthar and Wadi Ghariba, to the east by Ramadi Tharthar Road, to the west by the Karoush River and

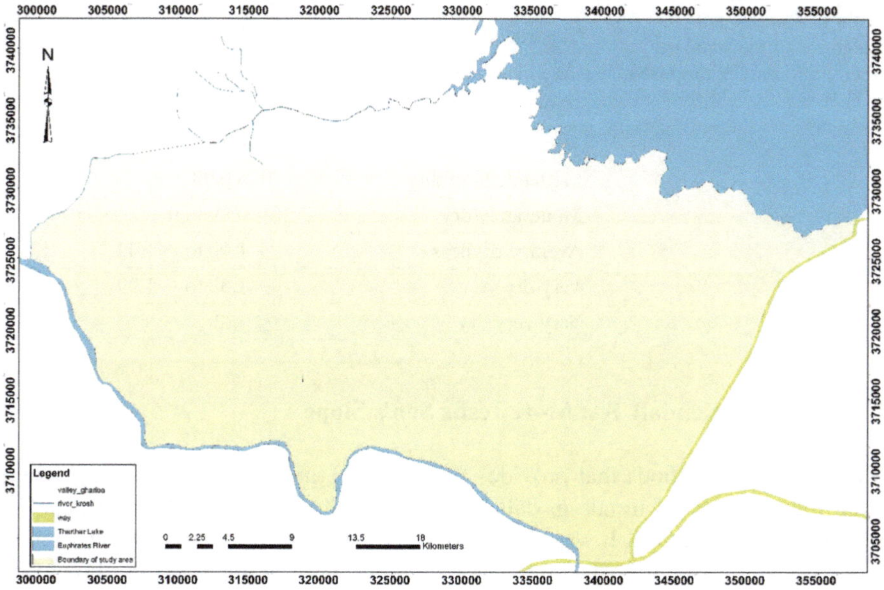

Fig. 1 Map of the administrative location of the study area

the Euphrates River, and to the south by the Euphrates River. As for the astronomical boundaries of the study area, they are limited in coordinates between latitudes 3,704,000 and 3,739,000 north, and between longitudes 300,000 and 360,000, with an area of 116,491.61 hectares. Figurc 1 shows the administrative location of the study area.

2.2 *Climatic Data*

The study relied on official data and government statistics regarding numbers that researchers cannot be diligent about, regarding the vocabulary of climatic elements. Climatic data were obtained from the Iraqi General Authority for Meteorology and Seismic Monitoring, Climate Department. We used rainfall and temperature data recorded during the period 1980–2022 to study trends and differences over 43 years to obtain accurate values in analyzing variables. Drought in the study area was evaluated using Standardized Precipitation Index (SPI) analysis, it is a guide developed by McKee et al. [14]. This index was used over a period of 12 months as an indicator of drought in the medium term to support the analysis and applications of hydrological drought in the study area. Table 1 shows the ranges of SPI values and their types.

Table 1 Ranges and categories of the standard precipitation index proposed by [14]

Classification	SPI values
Very high humidity	≤ 2
High humidity	1.50 to 1.99
Average humidity	1.00 to 1.49
Moderate humidity	0 to 0.99
Moderately dry	− 0.99 to 0
Average dryness	− 1.00 to − 1.49
Very dry	− 1.50 to − 1.99
Very very dry	≤ -2

2.2.1 Mann–Kendall Test M–K Testis Sen's Slope

It is one of the methods that provides trend testing, and a valuable tool for examining the presence of trends in data approved by the Intergovernmental Panel on Climate Change (IPCC). It was also proposed by the World Meteorological Organization (WMO) to identify statistically significant trends in time series of climate and hydrological data [16], as the MK method is used to analyze the trend and obtain trends detected with Sen's Slope values and its statistical significance at different levels [18], based on the trend parameters of Sen, s slope and ß in the M–K methods using the M. Excel [13] and XLSTAT [20] programs.

2.3 Field Procedures

After conducting a cartographic analysis of the digital elevation visualization of the study area (DEM), which was obtained from the ALOS PALSAR satellite (Fig. 2) and as shown in Table 2. The number of squares for the study was chosen based on the area of each elevation range above sea level, amounting to 61 modeling sites (Fig. 2).

The natural vegetation was characterized, and its type, density, and frequency were determined to identify the nature of the variations occurring in it, and the plant species present in the study area during the four seasons (spring, summer, fall and winter) for the year 2023 were identified, by taking plant samples and identifying their types using parts of the Iraqi flora as a reference, as for the unidentified species, they were collected, dried, and later classified using the National Herbarium. Some characteristics of the plant community were analyzed using quantitative methods using the random quadrate method with dimensions of 2 m × 2 m based on what was proposed by Cain and Castro [7], as suggested by Mueller and Ellenbery [15], then the following characteristics of the natural plant were estimated according to the methods presented in Barbour et al. [5].

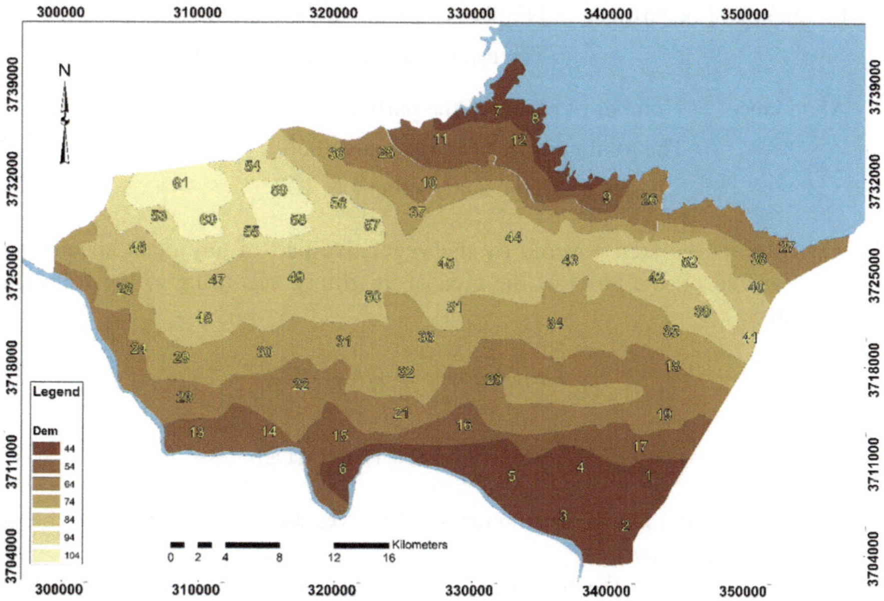

Fig. 2 Digital elevation map of the study area and modeling sites for natural vegetation

Table 2 Cartographic analysis of the digital elevation map of the study area and what it constitutes as a percentage of the area

The elevation range (m)	The area (ha)	The percentage from total Area	Number of squares
44–54	12,786.23	10.98	8
54–64	12,833.43	11.02	9
64–74	24,139.93	20.72	10
74–84	25,777.93	22.13	11
84–94	27,683.89	23.76	13
94–104	8753.88	7.51	6
> 104	4516.31	3.88	4
Total	116,491.60	100	61

1. The number and types of plants present within each square, along with measuring the height and width of the vegetative cluster of each plant.
2. Density according the following equation

$$\text{Density} = \text{The number of individuals of one species} \, / \text{The total number of squares} \quad (1)$$

3. $$\text{Frequency of occurrence} = \text{Number of plants of the particular species} / \text{The total number of squares} \times 100 \quad (2)$$

4. $$\text{Abundance} = \text{Number of plants of the particular species} / \text{The total number of individuals of all species in the study sample} \times 100 \quad (3)$$

5. Coverage is the area covered by the vegetative part of any plant species and is measured on the basis of the area of an ellipse according to the following formulas:

$$\text{Crown cover} = 1/4\pi \text{D1D2} \quad (4)$$

$$\text{Crown volume} = 1/6\pi \text{D1D2h} \quad (5)$$

where D1 and D2 are the diameters of the vegetative part and h is the height of the vegetative part.

6. Biomass represents the weight of the natural plant per unit area, the harvest method was used to estimate it, according to Chapman and Pratt [8].

Then the plant species in the region were compared with the results of [10] in Iraqi flora and the maps prepared by him for each identified plant species.

The satellite image of the study area for the year 2023, which was taken during the month of April, was obtained from the USGS website (https://earthexplorer.usgs.gov) and includes Landsat 8 satellite data, according to system path (Path) 169 and row (Row) 37 and displayed on the area Universal Transverse Mercator (UTM) 38 N on reference WGS84. The video was downloaded in tagged image file format (Geo TIFF), and the study area was cropped from it, using ArcGIS 10.8. Near-infrared vegetation analysis was used the red bands derived from satellite visuals were calculated according to the following equation proposed by Rouse et al. [17]:

$$\text{NDVI} = \frac{(\text{NIR} - \text{R})}{(NIR + R)} \quad (6)$$

whereas:

NIR Near Infrared Band (0.77–0.9 μm).
R Visible red band (0.63–0.69 μm).

The value of NDVI ranges between (− 1, 1), and the closer the value is to 1, this is evidence of the density of vegetation in the area.

3 Results and Discussion

3.1 *The Climate*

Based on the analysis of the moving average of precipitation for the reference period (Fig. 3), it was found that the governorate had begun to suffer from apparent drought in the year 2000 and this continued until the year 2003 and then from 2009 until 2022. It was also found that the total rainfall that fell during 12 years for the period from 2000 until 2011 had decreased significantly, about 25% of the total rainfall witnessed in the same time period preceding the year 2000.

The 5-year moving average of precipitation (Fig. 3) shows that there is a decrease in precipitation values that is less than the average precipitation for the long-term reference period. The current climate conditions are constantly changing, which leads to making the region more vulnerable to a decrease in rainfall, which requires taking all necessary measures.

The use of the M–K testis Sen's slope reached − 0.019*** based on the trend coefficients of Sen's slope and ß in the M–K methods using the [20] and M. Excel [13] programs, indicating the presence of variation in trends, and the Table 3 It shows the severity and frequency of drought in the study area using the SPI index. It is noted that moderate humidity recorded the highest frequency in the region at a rate of 46.5%, while moderate humidity recorded the lowest frequency at a rate of 4.7%, while drought in all its forms recorded the highest frequency during the period from 1980 to 2023.

What confirms the importance of these results is that they were relatively similar to the trends observed in studies of the regional and Mediterranean basin in monitoring the trend, and global in monitoring thc scntiment, including the study of Al-Mashagbah and Al-Farajat [3] in Jordan.

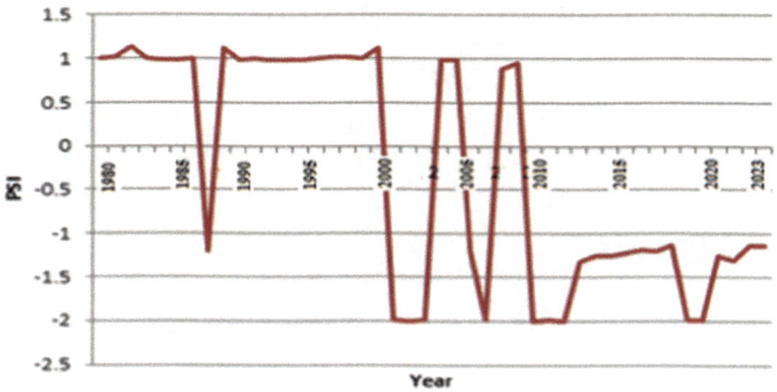

Fig. 3 5-year moving average chart of precipitation for the study area

Table 3 Classifications of drought severity and frequency in determining the frequency of dry and wet years using the SPI index

Classifications of drought severity									
Medium humidity		Moderate humidity		Medium dry		Extremely dry		Very extremely dry	
%	Frequency	%	Frequency	%	Frequency	%	Frequency	%	Frequency
46.5	20	4.7	2	11.6	5	11.6	5	25.6	11

3.2 The Soil

The soil in the study area was classified according to the classification of Buringh [6] into several types, most of them is occupied by gypsum desert soils, at a rate of 70.7% (82,369.1 hectares), followed by valley bottom soils, at a rate of 12.6% (14,759.4 hectares). Then the desert salt marshes with a percentage of 7.3% (8551.1 hectares) and finally the soils of riverbanks and alluvial soils of river basins with a percentage of 5.1 and 4.3% (5936.7 and 5051.2 hectares), respectively.

In terms of soil salinity and its types which identified in the study area, it is noted from the Fig. 7, that the dominance was for class S3, at a rate of 28.49% (33,188.45 hectares), followed by the class S4, at a rate of 22.14% (25,791.24 hectares). While the S1 recorded the lowest area, amounting to 8.21% (9563.96 hectares) (Fig. 4).

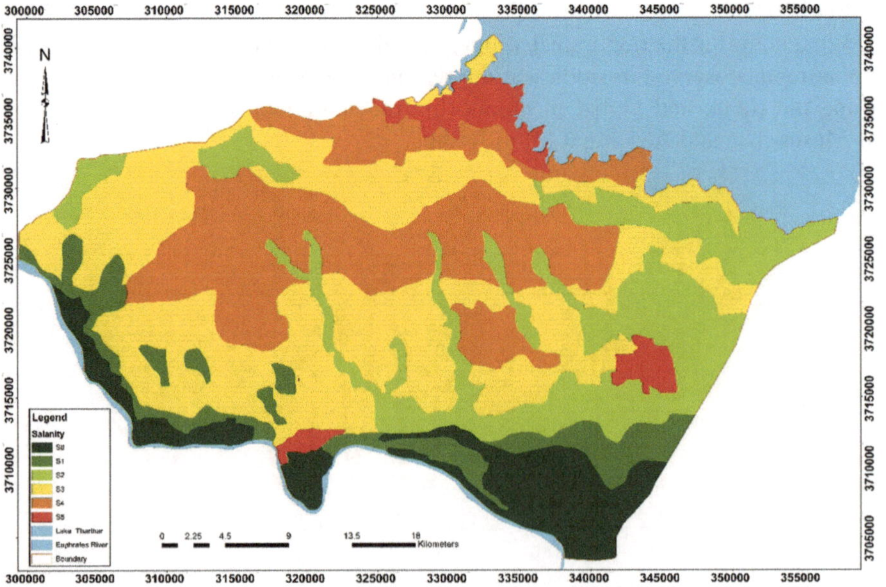

Fig. 4 Spatial distribution of diagnosed soil salinity types in the study area

3.3 *Analysis of vegetation in the study area using the NDVI index*

The purpose of this guide is to study and analyze the proportion of vegetation on the Earth's surface using near-infrared and red bands from satellite images. It is clear from the results of Figs. 5 and 6 that there is variation in vegetation cover according to the NDVI, within the study area for satellite images captured during April of 1980 and 2023, as its lowest and highest values ranged between − 0.1 and 0.8, so values (< 0.1) indicate the presence of little vegetation cover. Values ranging between (0.1–0.4) indicate the presence of low-density vegetation cover representing cultivated lands, while values exceeding 0.4 indicate the presence of dense vegetation cover such as trees and orchards. Based on the method of Al-Shafi'i [4], the study area was divided into three categories:

1. First class (I): Barren Land (< 0.1).
2. Second class (II): Low density vegetation (0.4–0.1).
3. Third class (III): High density vegetation (> 0.4).

The results of the cartographic analysis of the NDVI, showed the presence of variation in the categories of vegetation areas. The area of barren lands (Class I) constituted a percentage of 25.61% (29,833.48 hectares) in 1980, and its percentage increased to 38.87% (45,280.28 hectares) during the year 2023, while lands with low-density vegetation (Class II) constituted the highest percentage, amounting to

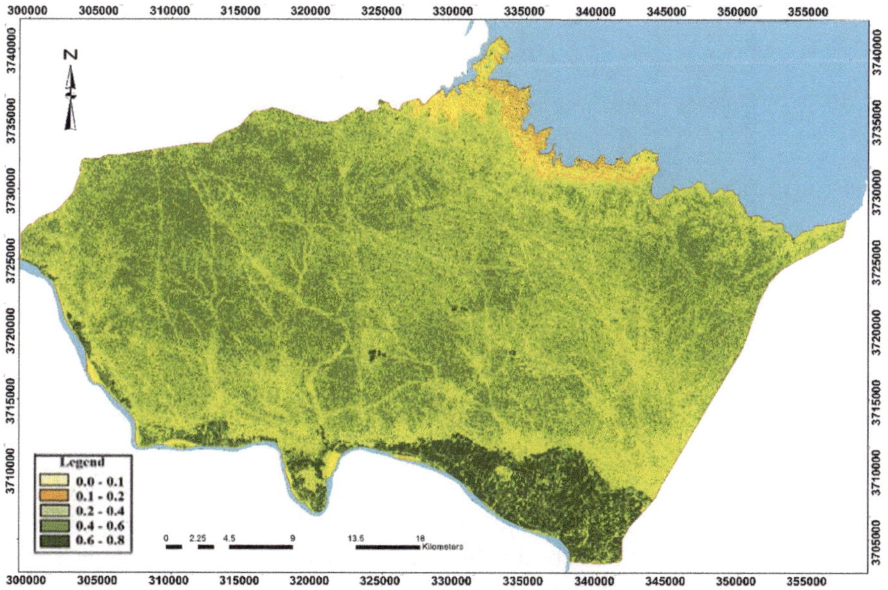

Fig. 5 NDVI analysis of the 1980 satellite image

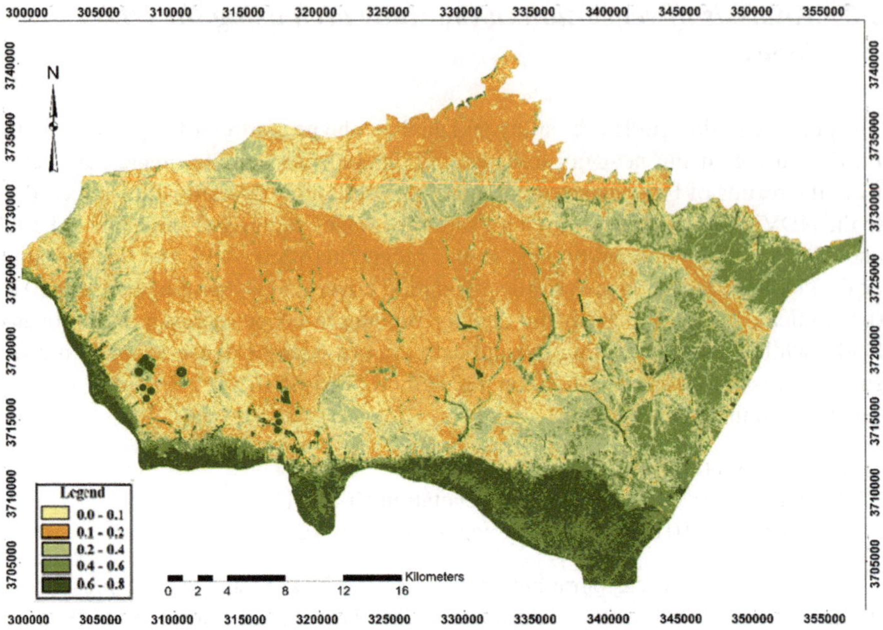

Fig. 6 NDVI analysis of the 2023 satellite image

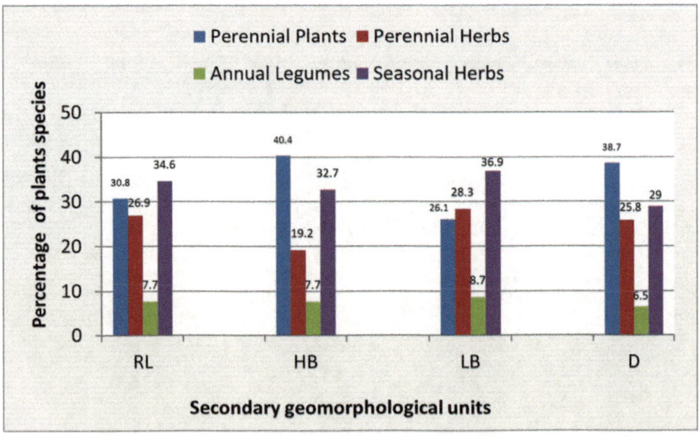

Fig. 7 Percentage of plant species which was identified in secondary physiographic units in the study area

47.21%. (42,111.69 hectares), decreased to 36.15% (54,995.68 hectares) in 2023, while the percentage of lands covered by dense vegetation (Class III) occupied a rate of 38.24% (44,546.43 hectares) in 1980, decreasing to 13.92% (16,215.65) hectares) in 2023. The high percentage of lands class I and class II, may be due to the region being exposed to many security crises that forced people to leave their homes and agricultural lands as a result of the conditions that struck the region, as well as the variation in The climatic factors represented by high temperatures, increased evaporation, and scarcity of rainwater cause an increase in the area of barren lands.

It's observed from results which shows the variation in the types of vegetation over 43 years, is an increase in Class I by 13.26%, compared to a decrease in Classes II and III, 11.06% and 24.38%, respectively. Which indicates that the climate variation recorded in the region had a clear impact on the vegetation in the region.

3.4 Analysis of vegetation in the study area.

It is showed from the Fig. 7 that there is a difference between the secondary physiographic units in terms of the percentage of identified plant species.

The study of plant species and their percentages in the total showed that the perennial species were distributed as shown in the Fig. 8.

As for perennial herbs (Fig. 9).

As for annual legumes (Fig. 10).

As for the nature of the frequency of plant species, it is noticeable from the Fig. 11 that *Achillea fragrantissima* was recorded the highest frequency, reached 100%, while the lowest frequency was for *Camphorosma monspeliaca*, which reached 9.8%. As for perennial herbs, it is noticeable that *Stipa* spp. and Cynddon dactylon L were the dominant at a rate 70%, while *Hirschfeldia incanal* recorded the lowest frequency 12.5%. The annual legumes showed dominance of *Rumex vesicarius* L. with a frequency of 75%, while *Helianthemum lippi* L. had the lowest frequency.

The *Herba-alba* Asso plant recorded the highest abundance, reaching 15.6%, compared to 2.2% for the *Anabasis articulate* plant at physiographic unit D, while the LB unit showed dominance in abundance for the *Atriplex ieucoelada* Boiss plant,

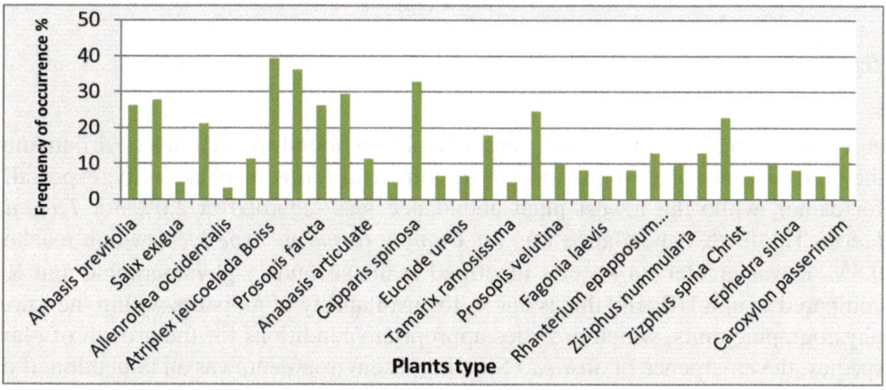

Fig. 8 Total frequency of perennial plants in the study area

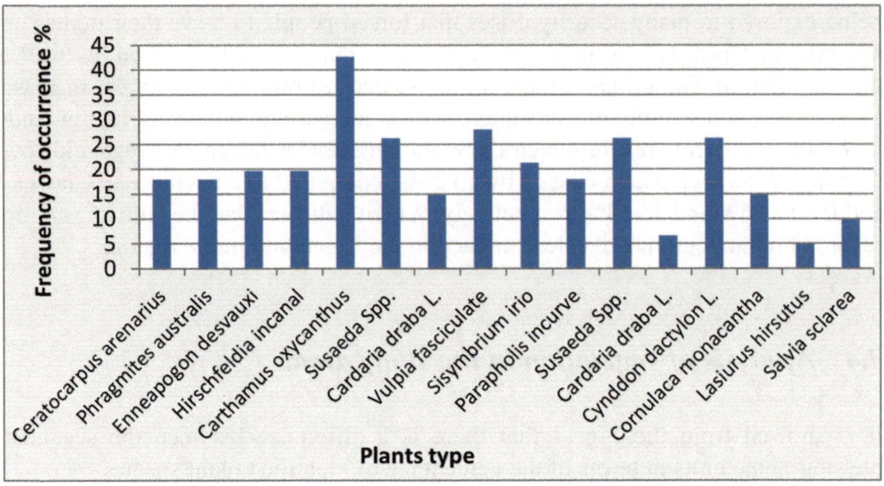

Fig. 9 Relative frequency of perennial herbs plant species in the study area

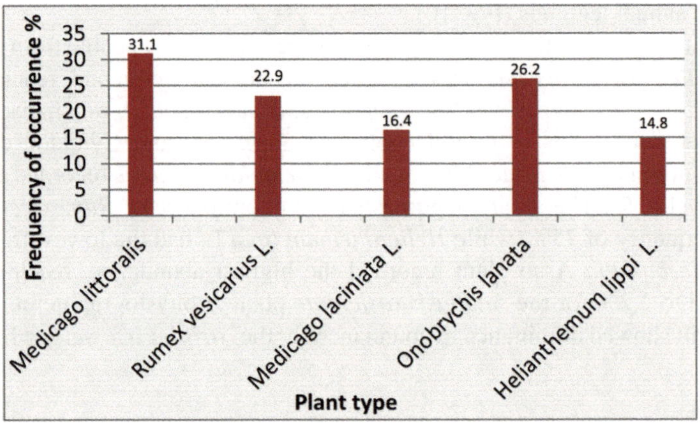

Fig. 10 Relative frequency of annual legumes plant species in the study area

at a rate of 16.9%, which is the perennial shrubs found in such dry environments, they are resistant to drought and are important in terms of palatability, especially for camel, while the lowest plant abundance was recorded at 2.6% for *Fagonia laevis*. The lowest abundance was for *Camphorosma monspeliaca*, which reached 9.8%. It was greater for species identified at the secondary physiographic unit RL compared to unit HB, and this is due to the availability of moisture within the lower physiographic units, which provides appropriate conditions for the growth of plant species, the emergence of *Susaeda* Spp. In this environment, was an indication of its presence in an environment other than its own, as it is one of the plants that spread in

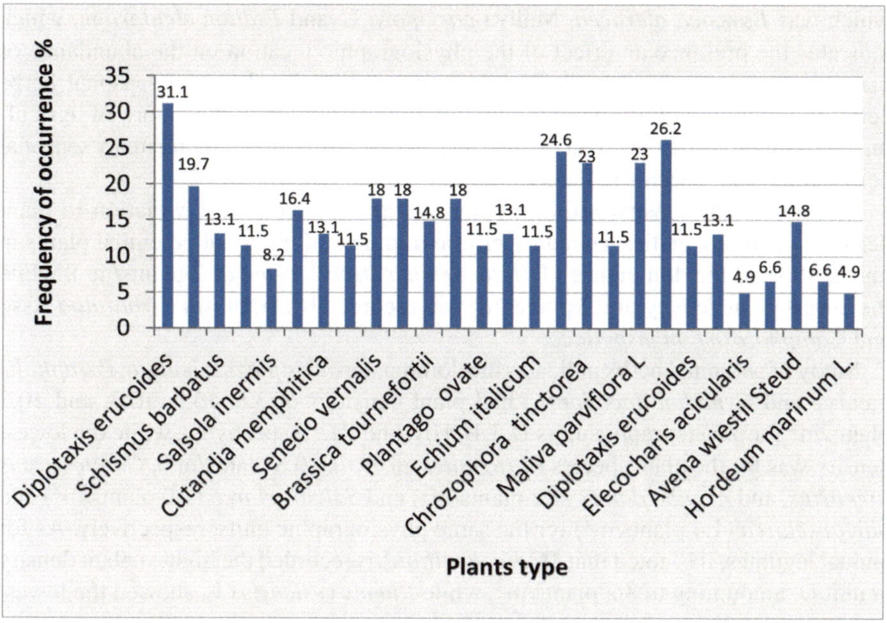

Fig. 11 Relative frequency of seasonal herbs plant species in the study area

most of the Iraqi lands located within the sedimentary areas and close to the banks of rivers and salty lands [10].

Its movement to the region may be due to grazing because it is a plant that is palatable to camels because it contains high levels of water and belongs to the saprophytic family Chenopodiaceae.

The reason for the dominance of *Alhagi mauroram* and *Prosopis farcta* plants in the environment of the island plateau is due to the fact that its soil is of moderate salinity, well-drained, and has good fertility, and the dominant texture type of its soil is moderately soft.

Carthamus oxycanthus recorded the highest abundance, reaching 27.8% at the physiological unit D, followed by the *Cynddon dactylon* L. at 24.1% at the secondary physiological unit RL, while the lowest percentage of abundance in the region was recorded of *Salvia sclarea, Enneapogon desvauxi* and *Hirschfeldia incanal* by 4.3%, 4.5%, and 4.6%, respectively, at the physiographic units HB, LB, and D, respectively. The presence of these plant species in this environment is due to their common presence in fields and farms, as well as in the western desert with mixed soil, as they flower during the last half of the month of February to April and provide good green pastures for animals.

As for the annual legumes identified in the study area, *Rumex vesicarius* L., *Onobrychis lanata, Helianthemum lippi* L., and *Medicago littoralis* were dominant, with abundance reaching 32.7%, 27.2%, 23.9%, and 5.6% for plant species, respectively. Studying the abundance of seasonal plant species present in the study area, which was *Barssica oleracea,* Maliva *parviflora* L. and *Erdium cicutarium*, which indicates the presence an effect of the physiographic location on the abundance of plant species of seasonal herbs The variation in the abundance of seasonal herbs between secondary physiographic units may be attributed to the nature of agricultural exploitation and the effect of agricultural operations in removing many seasonal herbaceous plant species in the long term.

Study of plant density within the study area showed a clear variation in plant density depending on the secondary physiographic unit. Study of perennial plants in unit D showed the dominance of *Halocnemum strobilaceum* (8.2 plants/m^2), while the lowest plant density of 0.3 plants/m^2 was recorded by *Artemisia herba-alba* Asso and *Camphorosma monspeliaca.*

Study of perennial herbs indicates the dominance of *Vulpia fasciculata, Parapholis incurve* and *Cynddon dactylon* L. had plant densities of 3.5, 15.3, 10.3, and 10.2 plants/m^2 for physiographic units D, LB, HB, and RL, respectively, while the lowest density was for the plant species *Phragmites australis* (0.8 plants/m^2), *Ceratocarpus arenarius*, and *Arundo donax* (1.2 plants/m^2) and *Salvia sclarea* (1.0 plants/m^2) and *Salvia sclarea* (1.4 plants/m^2) for the same physiographic units, respectively. As for annual legumes, it is noted that *Medicago littoralis* recorded the highest plant density at unit D, amounting to 8.2 plants/m^2, while *Rumex vesicarius* L. showed the lowest density, amounting to 1.1 plants/m^2.

As for the plant density of seasonal herbs, *Silybum marianum* was dominant and *Plantago ovate, Barssica oleracea*, and *Hordeum marinum* L., with plant densities of 4.2, 13.1, 6.5, and 15.2 plants/m^2 for the physiographic units D, LB, HB, and RL, respectively. Compared to the lowest density of plants: *Salsola inermis, Bromus rubens* L., *Chenopodiastrum murale*, and *Diplotaxis erucoides*, reached 0.2, 0.7, 0.4 and 3.2 plants/m^2.

The observed variation in plant density and species with varying physical location is due to the soil factor, which is one of the factors influencing vegetation cover, what matters to plants is the amount of moisture stored in the soil, and the relationship between the nature of the land scape and vegetation cover is considered crucial for plants, the soil that retains water supports plants [11], in addition to microclimate variation and its impact on meeting the requirements of the plant type in terms of temperature and sunlight and influencing the amount of evaporation [2], as well as the administrative processes accompanying agriculture and what they negatively or positively affect in the density of plant species as well as grazing by animals [1].

A study of the biomass in the study area has shown that it is low, ranging between 842.9 and 3631.8 gm/m^2, at a rate of 2012.9 gm/m^2. The reason for this is due to the shift of the climate system to dry as a result of decreased rainfall and higher temperatures, and the negative effects on the natural plant life. The results of Fig. 12 indicate that there is a clear difference in the biomass with varying physiographic location.

It has been observed from studying the natural flora of the study area, based on previous studies, that some previously identified plant species have disappeared, namely:

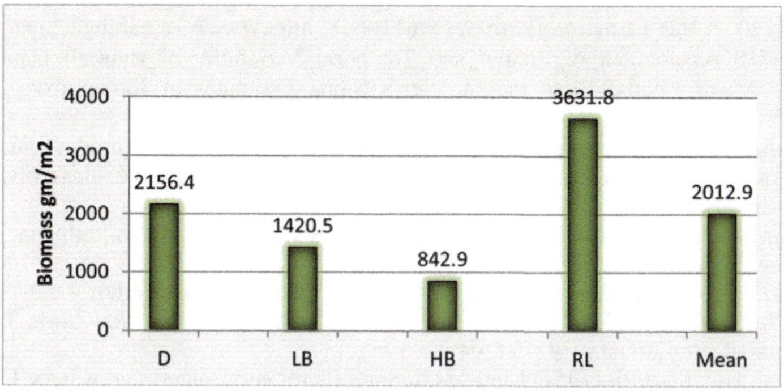

Fig. 12 The biomass of vegetation within the diagnosed physiographic units in the region and its average for all study area

Anastatica hierochuntica L., *Calligonum tetrapterum*, *Leontic leontopetalum*, *Scorzonera papposa* DC, Scorzonera schweinfurthii Boiss, *Gagea reticulate* Schult, *Scoraonera paposa* DC. Iris sisyrinchium L., *Carex stenophylla* WAHLENB, *Papaver hypridum*, Erodium Spp., *Erodium laciniatum*(CAV) WILLD, *Aegilops Crasssa* Bioss Diagan Ser, *Bromous tectorum* Var., *Schismus barbatus* L., *Horrdeum glaucum* steud, *Lolium rigidum* Gaud, *Lophochloa phleoides*, *Rostraria pumila*, *Phalaris minor*, *Schismus arabicus*, *Sisymbrium irio*, *Medicago aschersoniana*, Melilotus Mill, *Plantago lagopus* L., *Plantago boissieri*, *Plantago ciliate*, *Matricaria aurea*(LOEFL), *Poa bulbosa Steudhg*, *Anabsis setifera and Teucerium oliverianum* Ging. Note that Townsend and Guest [19] and Ghazanfar et al. [9] have indicated its presence in the study area, due to human interference in the environment of the study area through random plowing and logging, as well as the plant *Calligonum tetrapterum*, which is one of the small perennial shrubs that the inhabitants of the desert used as the best types of fuel. The current study did not record any presence of it in the study area and overgrazing, in addition to the low amount of rainfall due to harsh climate change, soil deterioration, and sand drift, which are all negative factors that had a prominent role in the decline of plant species and numbers within the study area.

References

1. Al-Bayati AHI, Al-Alwani AKAM (2011) Analytical study of the ecosystem of some desert oases in western Iraq for the period from 1966 to 2009. Al-Araba J Dry Environ 6(2):41–25. https://acsad.org/
2. Al-Esawi KKJ (2020) Analysis of the impact of climatic droughts on vegetation in Al-Anbar Governorate (Al-Jazirah region) using GIS. Iraqi J Des Stud 10(1):13–37. https://ijds.uoanbar.edu.iq
3. Al-Mashagbah A, Al-Farajat M (2013) Assessment of spatial and temporal variability of rainfall data using kriging, Mann Kendall test and the Sen's slope estimates in Jordan from

1980 to 2007. Res J Environ Earth Sci 5:611–618. https://www.researchgate.net/publication/259193718_Assessment_of_Spatial_and_Temporal_Variability_of_Rainfall_Data_Using_Kriging_Mann_Kendall_Test_and_the_Sen's_Slope_Estimates_in_Jordan_from_1980_to_2007

4. Al-Shafi'i WM (2010) Using geographic information systems (GIS) to produce land suitability maps for the Salamiyat Project for some grain crops. M.Sc. thesis, College of Agriculture. University of Baghdad
5. Barbour MG, Burk J, Pitts WD (1980) The nature and properties of soil 6th edn. MacMillan. Co., New York, p 567
6. Buringh P (1960) Soils and soil conditions in Iraq. Ministry of agriculture
7. Cain SA, Castro GM (1959) Manual of vegetation analysis. Harper, New York, 325p. https://catalogue.nla.gov.au/catalog/2194294
8. Chapman HD, Pratt PF (1961) Methods of analysis for soils, plants and waters. University of California, Los Angeles, pp 60–61, 150–179. https://www.scirp.org/reference/ReferencesPapers?ReferenceID=1045783
9. Ghazanfar SA, Edmondson JR (eds) (2014) Flora of Iraq, volume 5 Part 2: Lythraceae to Campanulaceae. https://shop.kew.org/flora-of-iraq-volume-5-part-2-lythraceae-to-campanulaceae
10. Guest ER (1966) Flora of Iraq. Introduction to the Flora. An account of the geology, Soils, Climate and Ecology of Iraq with gazetteer, glossary and biography. Iraq Ministry of Agriculture, p 313. https://www.scirp.org/reference/referencespapers?referenceid=710763
11. Hamrawi MS, Ibrahim IA (2021) Specifying the vegetation cover changes in Komel River using remote sensing techniques.Iraqi National J Earth Sci 21(2):17–26. https://earth.mosuljournals.com
12. International Organization for Migration (IOM) (2010) Iraq Mission, Iraq Displacement Reports Special Focus—Water Scarcity, 3 Sept 2010. https://reliefweb.int/report/iraq/iom-iraq-displacement-reports-special-focus-water-scarcity-september-2010
13. M. Excel (2016) Microsoft Excel. https://microsoft-excel-2016.softonic-ar.com/
14. McKee TB, Doesken NJ, Kleist J (1993) The relationship of drought frequency and duration to time scales. In: 8th Conference on applied climatology, Anaheim, 17–22 Jan 1993, pp 179–184. https://www.scirp.org/reference/ReferencesPapers?ReferenceID=2099290
15. Mueller DD, Ellenbery H (1974) Aims and methods of vegetation ecology. Wiley, New York and London, p 547. https://www.scirp.org/reference/ReferencesPapers?ReferenceID=1651228
16. Rahmat SN, Jayasuriya N, Bhuiyan MA (2012) Trend analysis of drought using standardised precipitation index (SPI) in Victoria, Australia. In: 34th hydrology and water resources symposium, 19–22 Nov 2012, Sydney, Australia, pp 441–448. https://www.researchgate.net/publication/251566189_Trend_analysis_of_drought_usingStandardised_Precipitation_Index_SPI_in_Victoria_Australia
17. Rouse JW, Haas RH, Schell JA, Deering DW (1974) Monitoring vegetation systems in the great plains with ERTS. In: Third ERTS-1 symposium NASA, NASA SP-351, Washington DC, pp 309–317. https://www.scirp.org/reference/referencespapers?referenceid=1411719
18. Taibi S, Meddi M, Mahé G, Assani A (2014) Variability of annual and extreme rainfall over Northern Algeria and relationship with teleconnections patterns. In: Proceedings of the mediterranean meeting on "Monitoring, modelling and early warning of extreme events triggered by heavy rainfalls". PON 01_01503—MED-FRIEND project University of Calabria, Cosenza (Italy), 26th–28th June 2014, p. 3. https://www.researchgate.net/publication/272745426_VariabiltyOf_Annual_And_Extreme_Rainfall_Over_Northern_Algeria_And_RelashioshipWhithTeleconnections_Patterns
19. Townsend CC, Guest E (eds) (1980) Flora of Iraq 4(2):628–1199. Ministry of Agriculture & Agrarian Reform, Baghdad. https://search.worldcat.org/title/Flora-of-Iraq/oclc/12418772
20. XLSTAT (2017) The leading data analysis and statistical solution for Microsoft excel. https://www.xlstat.com/en/

Advancing Sustainable Energy Solutions: Integrating Hybrid Systems and Electric Vehicles for a Greener Future

R. Uthra and Ravi Rangarajan

Abstract Integration of Electric Vehicle (EV) chargers with grid-connected hybrid renewable energy systems to bolster grid stability and sustainability is an essential step in the domain of new energy solutions. In the context of climate change and the global transition to new energy paradigms, the current study emphasizes the role of EV chargers, powered by autonomous photovoltaic (PV) sources, in mitigating voltage sags during grid faults. Through the use of Vehicle to Grid (V2G) technology and a fuzzy-controlled bidirectional converter, the system ensures efficient battery management and rapid response to energy demands. The study findings highlight the significance of EV chargers as potential auxiliary power sources, contributing to sustainable energy solutions, reducing the carbon footprint, and enhancing the resilience of our energy infrastructure in the face of climate challenge.

Keywords Electric vehicle · Sustainable solutions · Green energy · Climate change

1 Introduction

Sustainable Development Goals (SDGs) emphasize building communities and enterprises that can meet current needs while ensuring future generations' ability to use the same resources. Central to this concept is improving the living conditions while ensuring the sustainable preservation of natural resources. With increasing awareness of the depletion of fossil fuels and environmental degradation, the transition to renewable energy has become a critical focus. This transition is vital for addressing climate change, reducing carbon footprints, and promoting sustainable energy consumption. In order to achieve this, combining the most renewable energies, such as wind

R. Uthra
SRM Institute of Science and Technology, Kattankulathur, Tamil Nadu, India

R. Rangarajan (✉)
University of Doha for Science and Technology, Doha, Qatar
e-mail: ravi.rangarajan@udst.edu.qa

S. Al-Naemi et al. (eds.), *Water and Food Security in the Face of Climate Change: Challenges and Opportunities for Resilience*, World Sustainability Series,
https://doi.org/10.1007/978-3-032-00098-9_6

and photovoltaic, as a hybrid system is emerging to be a highly effective strategy for distributing clean energy [1]. These systems can provide clean and sustainable energy, while balancing demand and enhancing the grid's flexibility and adaptability. The integration of hybrid systems into the grid creates an extremely effective mechanism for managing the demand for electricity and the grid itself, reducing peak loads and increasing the grid's flexibility and adaptability. In addition to renewable hybrid systems, Electric Vehicles (EVs) play a transformative role in the energy transition. EVs, when combined with Vehicle-to-Grid (V2G) technology can help in managing renewable energy demand and balance our energy system [2]. V2G allows for two-way flow of electricity between vehicles and the grid. This innovation enables the discharge of energy from EVs back to the grid during periods of high demand contributing to energy efficiency. V2G technology is a key enabler of sustainable energy infrastructure, allowing the grid to effectively manage the increasing share of renewable energy in the system. In addition, this technology aids in reducing the impact of climate change by enabling our energy infrastructure to be more effective in managing an increasing amount of renewable energy for human use.

The adoption of EVs is accelerating globally, with the global market for EV charging stations valued at ~ USD 12 billion in 2022. The EV charging station market is expected to grow exponentially reaching a valuation of ~ USD 177 billion by the year 2030, with a robust Compound Annual Growth Rate (CAGR) of 40%. EVs and the subsequent rise of EV charging stations are revolutionizing the power market and are now being utilized as 'virtual power plants', effectively supplementing the grid [3, 4]. EVs effectively act as virtual power plants, in supporting grid operations, by way of discharging stored energy during peak periods and absorbing excess renewable energy when supply is high. Based on the demand from the grid and the charging station sites, EVs work together to discharge their energy into the grid for a specific period, thereby adjusting the voltage and frequency of the electric power within the grid. The batteries at EV charging stations are powered by an independent renewable energy source, making the entire system sustainable in aspects of decreased reliance on non-renewable energy, reducing carbon footprints associated with Green House Gas (GHG) emissions, and promoting eco-friendly transportation solutions [5, 6].

Furthermore, EV charging stations, powered by independent renewable sources, such as solar panels, are a prime example of sustainable design. Such an integration of renewable energy, hybrid systems, and EVs underpins the global shift toward sustainable development. This closed-loop system not only powers vehicles but also feeds clean energy back into the grid, making the energy cycle more efficient and sustainable. In addition, these technologies directly contribute to achieving multiple SDGs, including SDG 7 (Affordable and Clean Energy), SDG 9 (Industry, Innovation, and Infrastructure), and SDG 13 (Climate Action), by promoting cleaner energy solutions, reducing greenhouse gas emissions, and ensuring energy security for future generations. Hence, understanding and studying the design of V2Gs as an upcoming technology is essential and in here, we present and an optimal design for V2Gs targeting a greener future.

2 EV Charger Fed Grid Integrated Hybrid System

An EV charging station designed to support the electrical grid is shown in Fig. 1. This EV charging station, utilizing V2G technology, serves a dual purpose: it can charge electric vehicles and also feed electricity back into the grid when not in use for charging. This is made possible through a bidirectional power converter, which enables the flow of electricity in both directions—either from the grid to the vehicle or from the vehicle to the grid. The system is regulated by a smart controller that optimizes the timing of charging and discharging operations based on the availability of electric vehicles or any grid faults. It monitors grid conditions and the availability of EVs, ensuring energy is dispatched when it's most needed, such as during peak demand periods or grid disruptions.

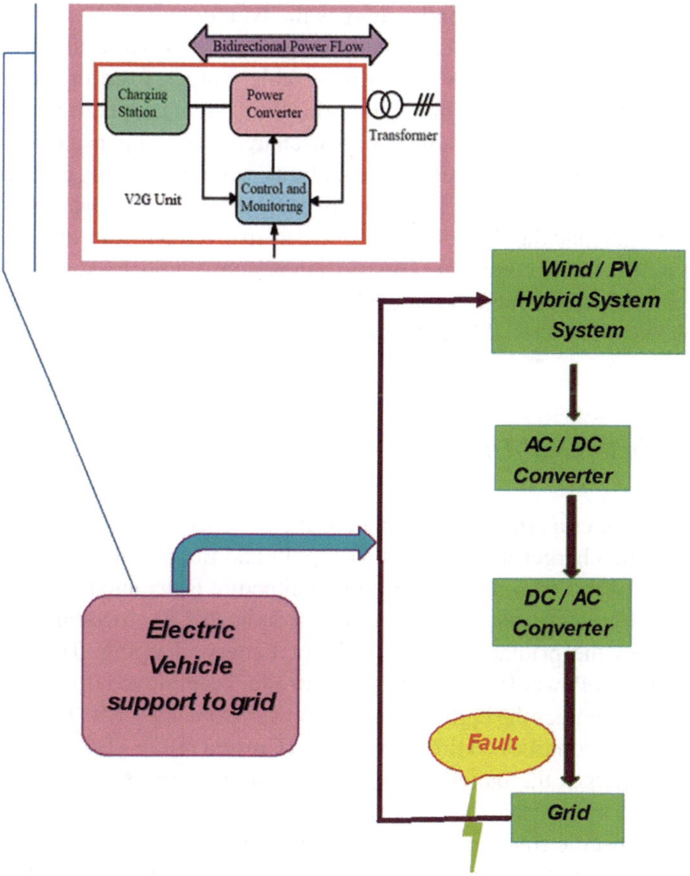

Fig. 1 EV charger integrated with grid

Furthermore, the utilization of renewable energy sources to power these EV charging stations alleviates the load on the grid, thereby enhancing its capacity to support the grid [7, 8]. Additionally, the station can integrate renewable energy sources like solar or wind to power itself, reducing reliance on traditional energy sources and contributing to a more sustainable energy infrastructure. By supporting the grid during high-demand periods and increasing energy flexibility, these stations enhance grid stability while promoting the use of clean energy. This not only alleviates pressure on the energy grid but also advances global efforts toward sustainability goals, reducing the carbon footprints and supporting a transition to renewable energy solutions.

The generated power from the hybrid wind/Photo Voltaic (PV) system is seamlessly integrated with the grid, promoting sustainability and helping mitigate climate change impacts. In this operational mode, both the wind and PV components continue to generate clean energy, which is then fed back into the grid. Simultaneously, an independent PV source consistently charges the battery at the EV charging station, contributing to sustainable and enhanced energy storage. As a result, the battery of the charging station is solely powered by renewable sources, fostering environmental sustainability. When a fault occurs at the grid, the system adapts in itself by modulating the EV charging station's battery to discharge and compensate for the deficit caused by voltage sags. This ensures grid stability while promoting climate-resilient energy infrastructure targeting a future ready energy system. During this process, the battery discharges, and the DC voltage is routed to an inverter of appropriate specifications via a bidirectional converter. This system not only supports sustainable energy management but also enhances the grid's resilience in the face of climate-related disruptions, ensuring a reliable and sustainable energy future.

2.1 Vehicle to Grid Unit

V2G unit constitutes a crucial component of the EV charger, facilitating the connection between the charger and the grid to supply additional voltage. The schematic representation of the V2G unit's charging architecture is depicted in Fig. 2.

The EV charging station is equipped with a dedicated Photovoltaic (PV) system, which serves as the primary input source for battery charging. The Photovoltaic MPPT (Maximum Power Point Tracking) controller is employed in conjunction with the PV source to extract the highest possible power from it [9–12]. The output of the PV system is then conditioned through a Direct Current (DC)—DC converter, which subsequently charges the battery using a bidirectional converter. The bidirectional converter primarily controls the output voltage by adjusting the duty cycle of a pulse generator at a specific frequency and duty ratio. This modulation of pulse generator firing pulses enables the bidirectional converter to both charge and discharge the battery within the EV charging station. A fuzzy controller is utilized to provide the necessary reference voltage for the DC pulse generator. In the discharging mode, the output generated by the bidirectional converter is directed towards an inverter.

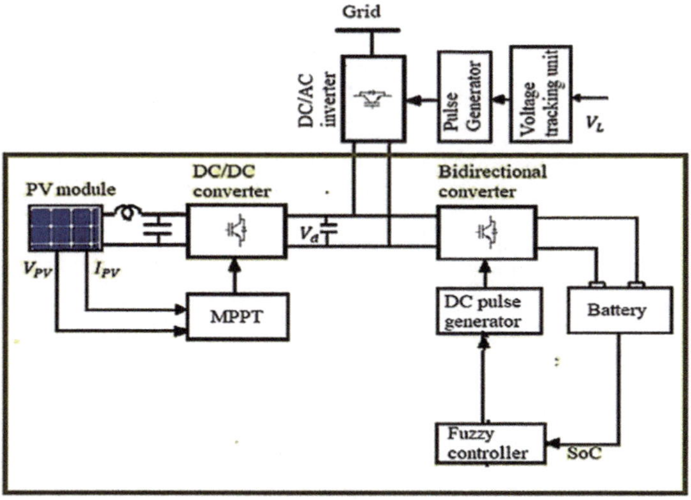

Fig. 2 The design of a vehicle to grid unit

A voltage monitoring instrument regulates the inverter, maintaining synchronization with the grid voltage during a fault condition. This regulation ensures the generation of a reference voltage for the generator, which, in turn, produces inverter pulses. The output of the inverter is linked to the grid via a control circuit, which is activated exclusively during grid faults. Consequently, the inverter output compensates for the deficit voltage experienced during voltage sags.

2.2 *Mode of Operation*

When the EV charger is integrated with a grid-connected hybrid system as shown in Fig. 3, any voltage deficiencies resulting from faults are automatically compensated by the EV charger.

In the event of a grid fault, the battery is discharged, and its DC voltage is redirected to an inverter, which is equipped with appropriate ratings, and regulated by a voltage tracking unit. This unit continuously monitors the voltage of the grid and provides a reference voltage to the Pulse Width Modulation (PWM) controller. This reference voltage is utilized by the PWM controller to initiate firing pulses to the inverter whenever there is a voltage drop due to a fault. Subsequently, the inverter transmits its output to the grid, facilitating efficient and rapid compensation of voltage fluctuations [13, 14].

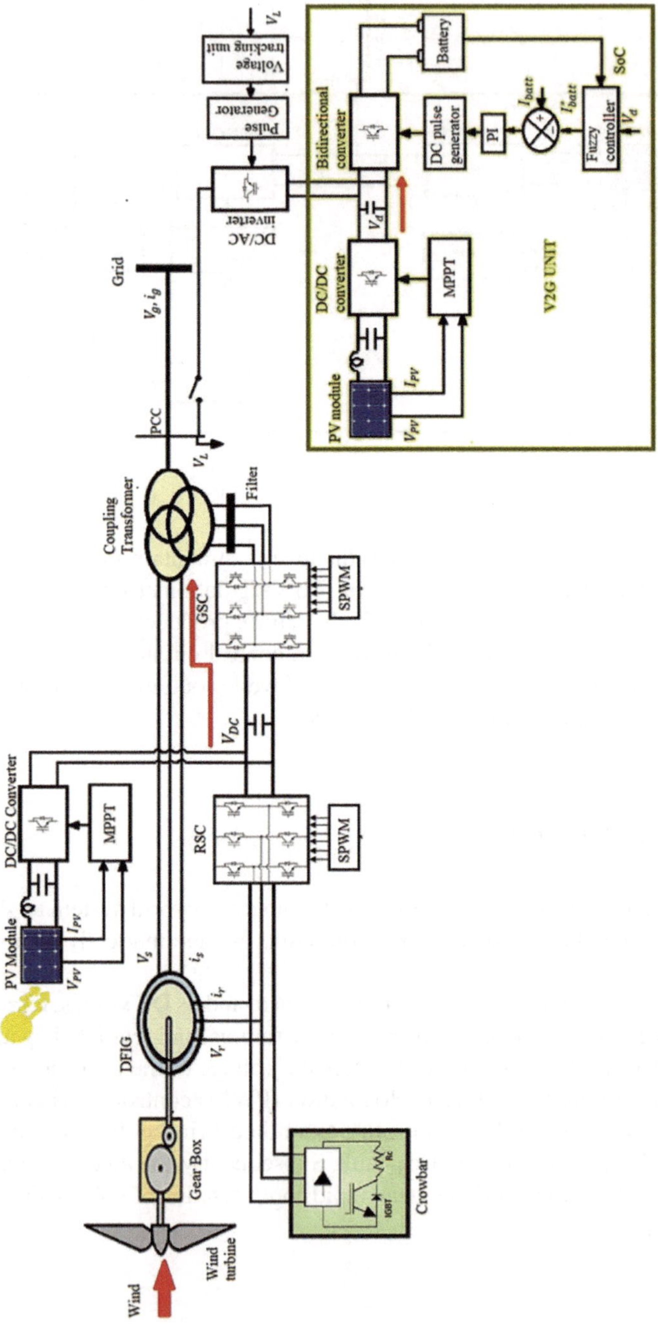

Fig. 3 EV charger with grid integrated hybrid system

3 Results and Discussion

The integration of an EV charger, powered by an autonomous PV source, within a hybrid wind/PV system highlights a crucial development in sustainable energy infrastructure. With a total capacity of 3.5 MW—comprising a 2.5 MW wind turbine and a 1 MW PV system—the hybrid setup efficiently generates renewable energy to meet energy demands while achieving sustainable energy management. The hybrid system has an operational load 0.1 MW, and this ensures that excess energy generated by the hybrid system is available for both charging EV batteries and stabilizing the grid when necessary.

One key advantage of this setup is its capacity to compensate for voltage sags during three-phase faults, demonstrating the resilience and adaptability of such systems. In the event of a three-phase fault causing a 30% voltage sag, the EV charger, using V2G technology, responds by discharging stored energy from the battery into the grid, effectively compensating for the voltage deficit. This process, as depicted in Fig. 4 (Top Panel), shows a symmetrical fault occurring at 3 s, while Fig. 4 (Bottom Panel) illustrates the voltage compensation provided by the V2G unit. The ability of the system to rapidly respond to grid faults ensures grid stability and highlights the crucial role EV chargers can play in modern energy systems as virtual power plants, targeting a future ready energy infrastructure.

The utilization of a fuzzy controller further enhances the system's performance by precisely regulating the bidirectional converter's operations. This control ensures that once the fault is cleared at 3.2 s, the system seamlessly resumes charging the EV battery without any delays. Figure 5 underscores the efficiency of this process, where the fuzzy logic ensures the optimal recharging schedule, enhancing the system's operational efficiency and adaptability.

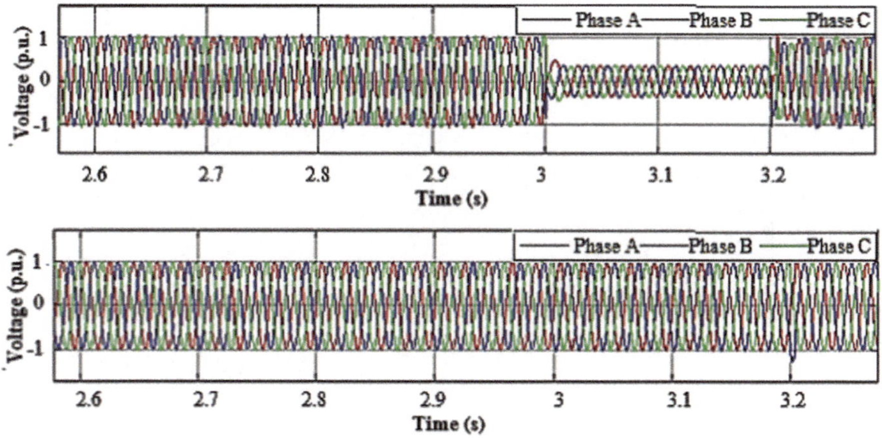

Fig. 4 Three phase with EV charger (top panel) fault voltage (bottom panel) compensated voltage

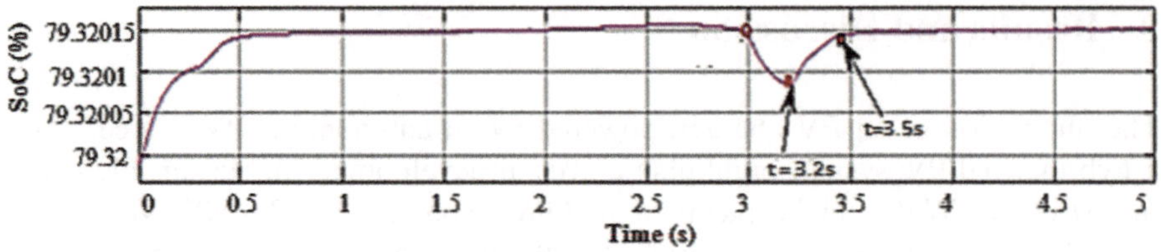

Fig. 5 Battery charging and discharging with fuzzy controller

The active power as well as the DC link voltage experiences a dip at the instant of fault, attributed to the anticipated voltage sag. Through fault ride-through and the subsequent voltage compensation, the active power is maintained at a constant level throughout the fault duration, and this is shown in Figs. 6 and 7. This precise control, in conjunction with V2G technology, supports the broader goals of sustainability and climate resilience by optimizing energy flow, reducing strain on the grid, and facilitating the transition toward renewable energy. Moreover, the system's capability to dynamically discharge and recharge based on real-time grid conditions highlights the efficiency and adaptability of smart energy systems. By leveraging advanced monitoring and analytics, these systems optimize energy flow, balance demand and supply, and enhance grid resilience. This innovation underscores the pivotal role of smart technologies in creating sustainable and responsive energy networks.

In relation to the sustainability aspects of the proposed technological energy solution, voltage compensation during grid fault stabilizes the grid during faults, reducing power outages while maintaining energy reliability, aligning with SDG 7. In addition, the EV charger which is being used as voltage support to the grid during fault, is powered by a renewable source which integrates clean energy into the grid. By preventing grid disruptions, voltage compensation minimizes reliance on backup

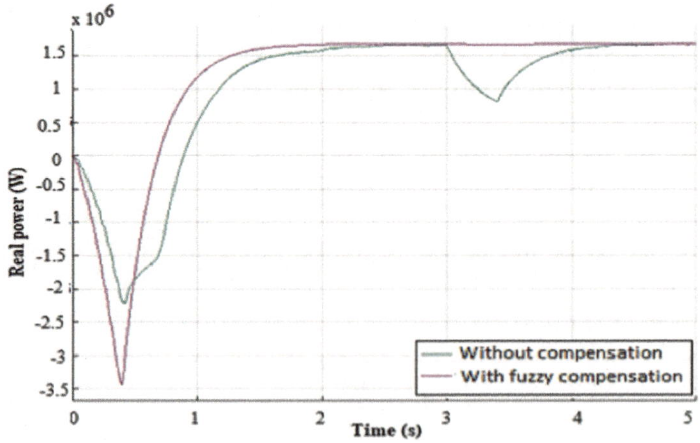

Fig. 6 Active power with wind speed of 8 m/s and irradiance of 1000 W/m^2

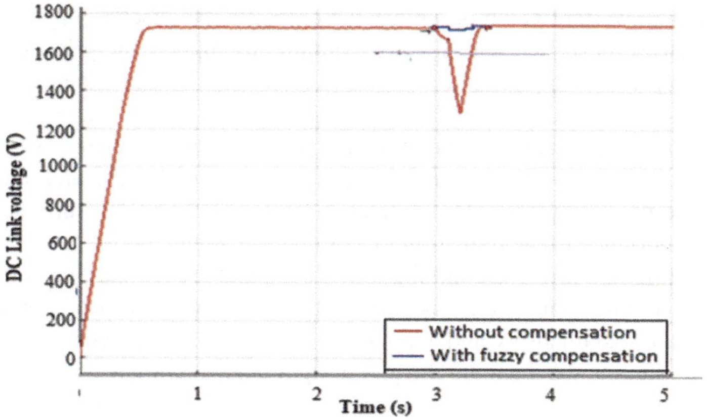

Fig. 7 DC link voltage for a wind speed of 8 m/s and irradiance of 1000 W/m^2

fossil-fuel generators which are often used during outages. This reduces carbon emissions during fault conditions and supports the goal of transitioning to low-carbon energy systems, the overarching goal of most SDGs. Also, voltage compensation plays a crucial role in the development of modern grids, enabling them to accommodate renewable energy sources and withstand the stresses induced by climate-related extreme events. This advancement is a significant step towards constructing infrastructure that is both sustainable and climate-adaptive, in alignment with SDG 13. Thus, the integration of EV charging stations with hybrid renewable energy systems, managed by intelligent controllers, represents a significant step toward achieving key SDGs and offer a climate-resilient and sustainable energy solution that supports the transition to clean energy, enhances grid flexibility, and promotes more holistic environmental sustainability.

4 Conclusion

As the global market for EVs continues to expand, there is a growing emphasis on establishing sustainable EV charging stations to meet the increasing demand. These stations, when powered by renewable energy sources such as wind or solar, significantly reduce the reliance on non-renewable resources, directly contributing to the sustainability and achievement of SDGs by promoting clean energy solutions. The integration of renewable energy into EV charging infrastructure presented here, serves dual purpose of charging the vehicles and acting as distributed power generators, feeding excess energy back into the grid when needed, thereby enhancing grid stability. A critical component of these charging stations is the V2G technology, which allows EVs to return stored energy to the grid during peak demand or grid instability. This capability is managed by an advanced, intelligent fuzzy-controlled bidirectional converter, which optimizes the charging and discharging schedules of the EV batteries, ensuring that they are available to support the grid during faults or voltage sags. This integration of smart technology further enhances the flexibility and resilience of the energy system, aligning with SDG 7 and SDG 13. In the event of a grid fault, such as a voltage sag, these renewable energy-powered EV chargers can provide auxiliary voltage support, stabilizing the grid and reducing the risk of blackouts. The system's ability to quickly respond to grid disturbances underscores its importance in the transition to a climate-resilient energy infrastructure. In here, we highlighted how the combination of sustainable EV charging stations with grid-integrated renewable energy systems not only advances climate action but also strengthens the sustainability and adaptability of our energy networks, supporting a low-carbon future and enhancing the global fight against climate change.

References

1. Marco-Lajara B, Martínez-Falcó J, Sánchez-García E, Millan-Tudela LA (2023) Analyzing the role of renewable energy in meeting the sustainable development goals: a bibliometric analysis. Energies
2. Mojumder MRH, Ahmed Antara F, Hasanuzzaman M, Alamri B, Alsharef M (2022) Electric vehicle-to-grid (V2G) technologies: impact on the power grid and battery. Sustainability
3. Bharathi YH, Dabhi VM, Patel JJ, Prajapati JA, Prayagi SV, Shafeek M (2024) Research and examination of electric vehicles equipped with vehicle to grid (V to G) system. J Electr Syst
4. Hossain S, Rokonuzzaman M, Rahman KS, Habib AKMA, Tan W-S, Mahmud M, Chowdhury S, Channumsin S (2023) Grid-vehicle-grid (G2V2G) efficient power transmission: an overview of concept, operations, benefits, concerns, and future challenges. Sustainability
5. Bibak B, Tekiner-Moğulkoç H (2021) A comprehensive analysis of vehicle to grid (V2G) systems and scholarly literature on the application of such systems. Renew Energy Focus
6. Noori M, Zhao Y, Onat N, Gardner S, Tatari O (2016) Light-duty electric vehicles to improve the integrity of the electricity grid through Vehicle-to-Grid technology: analysis of regional net revenue and emissions savings. Appl Energy
7. Shaaban F, Eajal AA, El-Saadany EF (2014) Coordinated charging of plug-in hybrid electric vehicles in smart hybrid AC/DC distribution systems. Renew Energy 82
8. Mohammad A, Zamora R, Tjing Lie T (2020) Integration of electric vehicles in the distribution network: a review of PV based electric vehicle modelling. Energies 13
9. Arif SM, Lie TT, Seet BC, Ayyadi S, Jensen K (2021) Review of electric vehicle technologies, charging methods, standards and optimization techniques. Electronics
10. Ota Y, Taniguchi H, Nakajima T, Liyanage KM, Baba J, Yokoyama A (2012) Autonomous distributed V2G (vehicle-to-grid) satisfying scheduled charging. IEEE Trans Smart Grid
11. Brenna M, Foiadelli F, Leone C, Longo M (2020) Electric vehicles charging technology review and optimal size estimation. J Electr Eng Technol
12. Dericioglu1 C, Yirik E, Unal E, Cuma MU, Onur B, Tumay M (2018) A review of charging technologies for commercial electric vehicles. Int J Adv Automot Technol
13. Uthra R, Suchitra D (2021) Fault ride through in grid integrated hybrid system using facts device and electric vehicle charging station. Energies
14. Uthra R, Suchitra D, Babu TS, Aljafari B (2022) Fuzzy-based EV charging station and DVR-fed voltage compensation for a DFIG-fed wind energy system during grid faults. Int Trans Electr Energy Syst

Forecasting Temperature Using ARIMA, SARIMA, and the Random Forest Machine Learning Algorithm to Assess Climate Change Impacts in Qatar

Maha A. Dewidar

Abstract Temperature forecasting is deemed one of the most significant elements that contribute in assessing the impacts of climate change. This paper presents a comprehensive analysis of temperature prediction in the state of Qatar using a combination of traditional statistical models and advanced machine learning techniques. The dataset comprises the monthly average of minimum and maximum temperatures in Qatar from January 2011 to February 2024. We employ Autoregressive Integrated Moving Average (ARIMA), Seasonal Autoregressive Integrated Moving Average (SARIMA), and Random Forest (RF) models to forecast minimum and maximum temperatures, aiming to assess the impact of climate change in the region. The study compares the forecasting accuracy of each model, evaluated through Root Mean Square Error (RMSE) and Mean Absolute Error (MAE). The results show that while traditional models like ARIMA and SARIMA provide baseline forecasts, they fail to capture the complex patterns inherent in the temperature data. Random Forest Algorithm demonstrates a much better performance than the previous models in predicting temperature and effectively capturing temperature variations. The results emphasize the importance of applying machine learning approaches in climate modeling to achieve more accurate and reliable forecasts. This work reveals that advanced modeling methodologies is central to the efforts of evaluating the effects of climate change in Qatar.

Keywords Temperature forecasting · ARIMA and SARIMA models · Random forest in machine learning

M. A. Dewidar (✉)
University of Doha for Science and Technology, Doha, Qatar
e-mail: maha.dewidar@udst.edu.qa

S. Al-Naemi et al. (eds.), *Water and Food Security in the Face of Climate Change: Challenges and Opportunities for Resilience*, World Sustainability Series,
https://doi.org/10.1007/978-3-032-00098-9_7

1 Introduction

Climate change is recognized as one of the major challenges humanities will face over the next several hundred years [1, 2] and has become a significant concern for the state of Qatar. The government has acknowledged the need to combating these problems that have added climatic action to the country's agenda [3]. Accurate climate prediction is crucial in mitigating the impacts of climate change. In the last ten years, the purpose for adopting time series in temperature modeling has become most influential towards determining the climatic structures and their patterns. Previous research has shown that time series forecasting is useful for predicting future weather conditions by analyzing past events [4, 5]. Some of the important statistical models used for weather forecasting are the ARIMA and SARIMA models, which are widely applied to predict the temperature. These models' implementation involved linear time series and statistical distributions that have produced improvements in the traditional statistical methods [6, 7]. SARIMA models and machine learning techniques have become essential tools when dealing with climate change outcomes. SARIMA models are especially helpful when analyzing seasonal climates, and result in highlighting trends, fluctuations of temperature, and precipitation in the future [7]. Decision Trees, Random Forest, Support Vector Machines (SVM), and Neural Networks are some of the machine learning algorithms that classify and predict patterns by learning from large datasets [8].

SARIMA together with machine learning can be incorporated to make climate prediction more accurate and help identify the major causes for climate changes. This approach leverages the strengths of both methods, providing a robust foundation for predicting future climate conditions [9]. These combined models can be used as a beneficial and potent tool for decision making and the decomposition of time series to the linear and nonlinear components has a higher accuracy [10]. Random forest time series is a type of ensemble learning specifically developed for time series data. It means a process that includes the use of numerous learned decision trees which are formed based on the bootstrapped samples of the datasets and varied feats or features. They are preferred for problems where the non-sequential nature and interaction of data values are to be captured. Random Forests are advantageous because it is resistant to overfitting problems, can be easily used, and is suitable for using a large amount of data [11–13].

This research aims to assess temperature forecasting in Qatar by comparing traditional time series techniques (ARIMA and SARIMA models) with a machine learning method (Random Forest) to come up with a better understanding of temperature trends as a core process which contributes to enhanced decision making in the adaptation and management of climate change in this region.

2 Methodology

2.1 Data Collection

The dataset applied in this study was extracted from Qatar Open Data Portal (data.gov.qa). The compiled and updated data on this platform are shared in full compliance with the partners' source. In this study, the dataset of temperature is taken for the minimum and maximum average temperature and the time line selected is from January 2011 to February 2024.

2.2 Statistical and Machine Learning Methods

ARIMA and SARIMA Modeling. The ARIMA model is a widely used statistical method for time series forecasting, capturing various types of temporal structures in the data [14]. It combines three components: First, Autoregression (AR) which models the relationship between an observation and a number of lagged observations. It represents the number of previous observations the model uses to predict the current observation (p). Second, differencing (I) which is the integrated part that involves differencing which is used to make the time series stationary. It represents the number of times the data have had past values subtracted (d). Third, Moving Average (MA) which is the size of the moving average window that represents the number of lagged forecast errors in the prediction equation (q) [15]. The ARIMA model is denoted as ARIMA (p, d, q) and it can be mathematically expressed as follow,

Let $\{Y_t\}$ be an arbitrary time series, a sequence of measurements of the same variable collected over time. Let B be the lag operator that shifts $\{Y_t\}$ back k periods, also known as a backshift operator, defined as $B^k Y_t = Y_{t-k}$. Then, in an AR process $\{Y_t\}$ of order p is defined as a linear aggregate of its p predecessors and a stochastic term Z_t and can be written as

$$Y_t = \sum_{j=1}^{p} \beta_j Y_{t-j} + Z_t, \tag{1}$$

where $\{Z_t\} \sim WN\left(0, \sigma^2\right)$ and $WN\left(0, \sigma^2\right)$ is white noise with variance given by σ^2, and the β_j are weights. A moving average process of order q is a linear filter applied to current and q past stochastic terms $\{Z_t\}$ and is defined as

$$Y_t = Z_t + \sum_{l=1}^{q} \alpha_l Z_{t-l}, \tag{2}$$

where $\{Z_t\} \sim WN\left(0, \sigma^2\right)$, and the α_l are weights. In terms of the lag operator B, an AR process of order p, AR(p), and a MA process of order q, MA(q), can be expressed as $Z_t = \left(1 - \sum_{j=1}^{p} \beta_j B^j\right) Y_t = \phi(B) Y_t$ and $Y_t = \left(1 + \sum_{l=1}^{q} \alpha_l B^j\right) Z_t = \theta(B) Z_t$, respectively, where $\phi(B) = 1 - \sum_{j=1}^{p} \beta_j B^j$ and $\theta(B) = 1 + \sum_{l=1}^{q} \alpha_l B^j$ are the polynomials of degree p and q. The integrated component integrated component adjusts for non-stationarity by removing the impact of trends on the time series values over time. This adjustment involves calculating the differences between observations at various lags. The first lag difference of $\{Y_t\}$ is defined by the expression $(1 - B) Y_t = Y_t - Y_{t-1}$. High order differences are defined by repeated application. That is, the second difference can be written as $(1 - B)^2 Y_t = Y_t - 2Y_{t-1} + Y_{t-2}$, and the d difference can be defined by expression $(1 - B)^d Y_t$. Combining the three terms together, a time series $\{Y_t\}$ is an ARIMA process of order p, d, q, with p autoregressive terms, d differences, and q moving average.

$\{Y_t\} \sim ARIMA(p, d, q)$, with $\phi(0) = \theta(0) = 1$ and it can be expressed as

$$\phi(B)(1 - B)^d Y_t = \theta(B) Z_t, \tag{3}$$

where after the dth difference of $\{Y_t\}$, the time series is a stationary ARMA process of order p, q; $\phi(B) = \sum_{j=0}^{p} \beta_j B^j$ and $\theta(B) = \sum_{l=0}^{q} \alpha_l B^j$; B is a backshift operator [16].

Seasonal ARIMA Model (SARIMA) is an extension of the ARIMA model that considers seasonality in the data by incorporating seasonal terms in each of the AR, I, and MA components. The SARIMA model is denoted as ARIMA (p, d, q) (P, D, Q)$_m$, where P, D, and Q represent the seasonal parts of the model, and m is the number of periods in each season [14]. SARIMA models are especially useful for the data with seasonal patterns, as it can model the seasonal fluctuations and improving forecasting accuracy [16]. Akaike Information Criterion (AIC) is used to determine an ARIMA and SARIMA model's parameters, as follow, [17].

K is the number of parameters estimated in the model.

$$\text{ARIMA/SARIMA} : \text{K} = (\text{AR terms} + \text{MA terms} + \text{seasonal AR} + \text{seasonal MA})$$

$$AIC(p) = n\ln(RSS/n) + 2K \tag{4}$$

where RSS is the residual sum of squares and 'n' is the number of data points used. The model with the smallest AIC value is considered the best for forecasting, hence considered better for the study. The initial stage in ARIMA and SARIMA modeling process involves examining the dataset to identify any anomalies or outliers. the data are scaled using a normalization formula t. o ensure the variance is stabilized. The scaled value V_i is computed using the form,

$$V_i = \frac{a_i - \min(a_i)}{\max(a_i) - \min(a_i)} \tag{5}$$

where a_i presents the original data value, and min (a_i) and max (a_i) represent the minimum and the maximum values of the dataset at scan point, respectively. This normalization helps in developing the data for subsequent analysis. In the second step, the rescaled data is used to calculate two very important functions called the Autocorrelation Function [ACF] and Partial Autocorrelation Function [PACF]. These functions are important for identifying the appropriate parameters for the AR and MA components of the model. The ACF analyzes the correlation between observations with a given lag, while the PACF estimates the correlation between observations at different lags, excluding the effect of intermediate lag. This will aid in identifying the potential values for the parameters of the AR(p) and MA(q). This is made useful in the selection of candidate models for time series forecasting. For instance, a clear single peak at lag k in the ACF chart indicates that the data has an MA(k) model, whereas whether there is a single peak at lag k in the PACF chart, it indicates that the data has an AR(k) model. The last of these steps is applying the SARIMA model to forecast the temperature data [17]. Subsequently, and after plotting ACF and PACF, Augmented Dickey Fuller (ADF) test is conducted for hypothesis testing confirming stationary. ADF is a widely used method for checking the stationarity of a time series [15, 18]. The model for the ADF test is as follows:

$$\partial Y_r = \mu + \beta t + \rho Y_{t-1} + \partial Y_{t-1} + \cdots + \partial_p Y_{t-p} + e_{t\ldots} \tag{6}$$

where, μ, ρ and β represent a constant value, autoregressive order and trend, respectively. Moreover, e_t represents a sequence of zero mean and unit variance independent normal random variables. So, the hypotheses for ADF are presented in the following way [11],

$$\begin{array}{ll} \text{Null Hypothesis} & H_0 : |\rho| = 0(\text{Stationary}) \\ \text{Alternative Hypothesis} & H_0 : |\rho| \neq 0(\text{Nonstationary}) \end{array}$$

The SARIMA model is used to estimate the monthly mean temperature over a year, using the following equation [17]:

$$\emptyset(\beta^{12})\Delta_{12}^{D} x_t = \Theta(\beta^{12})\alpha_t \tag{7}$$

where $\Delta_{12}^{D} x_t = x_t - x_{t-12}$, and $\emptyset(\beta^{12})$ and $\Theta(\beta^{12})$ are polynomials in β^{12} of orders p and q, respectively. These polynomials must meet the conditions for stationarity and invertibility. The error component α_t is used to correlate the time series, ensuring the model's accuracy. The value of m is set as 12 because the data used are monthly data with a period of 12.

2.3 Machine Learning Algorithm

The Random Forest (RF) method is an advanced technique that enhances the traditional classification and regression tree (CART) approach. RF, as shown in Fig. 1 [19], utilizes multiple decision trees to improve predictive accuracy and control overfitting [11]. In the case of the RF model, each tree has a root, internal nodes, and leaves, where the internal nodes correspond to test functions according to which the data are split by the corresponding features. The tree in the forest is built with a new bootstrap sample of the original data, and at the node of the tree, the split is performed randomly on a different set of features. This process adds randomness in the selection of trees and helps ensure that the model incorporates different features of the data. The performance of RF is entirely based on the strength of the individual trees and the nature of the correlation between them; the lower the correlation, better the ensemble will be. In regression trees, the splits are made according to a squared residuals minimization algorithm, aiming to minimize the variance in the resulting nodes [18].

In the case of RF for regression, training sets that are obtained through bootstrapping are used and performance of the model and variable levels are tested and assessed by unselected observations, referred to as "out-of-bag" OOB observations. The final model prediction is obtained by averaging the predictions from all trees in the same data. The model is optimized by finding the best number of predictors and trees, depicted by cross-validation (CV) or out of bag error (OOB) error. The CV error is the mean squared error at each variable reduction step obtained from the aggregated predictions from all CV partitions [20].

The mathematical components of the Random Forest method are described as follows [11]: For the data set, $X = \{(x_1, y_1), (x_2, y_2), \ldots, (x_n, y_n)\}$, where x_i represents the input temperature and y_i the output temperature, the Random Forest regression works as follows:

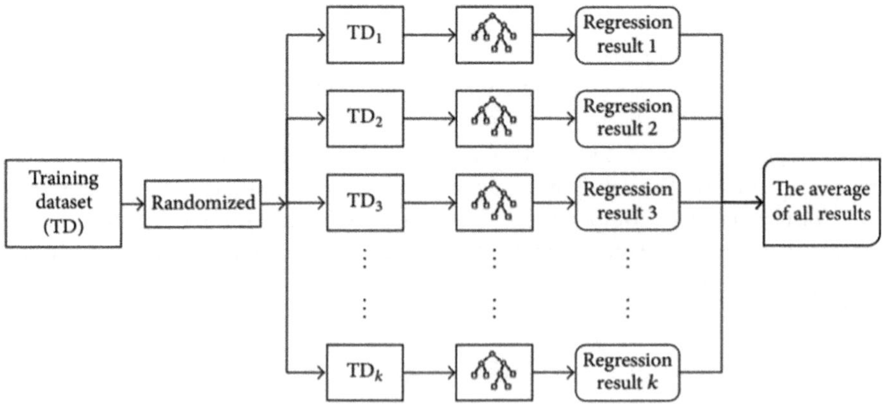

Fig. 1 Structure of a random forest model [19]

1. *Bootstrap Sampling*: Randomly sample B datasets from the original dataset with replacement to create B bootstrap samples.
2. *Growing Decision Trees*: For each bootstrap sample $b = 1,\ldots, B$, grow a decision tree T_b by randomly selecting a subset of features at each node, then Splitting the node using the feature that minimizes the Mean Squared Error (MSE) and continue splitting until the stopping criteria are met (e.g., maximum depth or minimum node size). For a new observation x_i, the Random Forest prediction $\hat{y}$ is obtained by averaging the predictions from all the trees in the forest [21]:

$$\hat{y} = \frac{1}{B}\sum_{b=1}^{B} T_b(x) \tag{8}$$

Here, $T_b(x)$ is the prediction from the bth tree.

The OOB error estimate is the mean squared error of the aggregated prediction errors obtained using the OOB data and the trees grown within the training set and defined as:

$$MSE_{OOB} = \sum_{i=1}^{n}\left(Y_i - \hat{Y}_i^{OOB}\right)^2/n \tag{9}$$

where $\hat{Y}_i^{OOB}$ is the average of the OOB predictions for the ith observation.

All statistical analyses in this study were conducted using "R (version 4.3.0)" with the significance level set at $P < 0.05$.

3 Results and Discussion

A time series decomposition was performed on the temperature historical time series. Using the results from the decomposition, ARIMA, SARIMA and RF models were developed to assess and predict the monthly temperature in state of Qatar. The predictive accuracy of each modeling approach was evaluated using the root mean square error (RMSE) and the normalized root mean square error (NRMSE), based on current data and assumptions [14]. Before applying the time series techniques, it was necessary to investigate the outcome measures for temporal autocorrelation in the residuals using the Durbin-Watson (DW) test. Specifically, under this test there is the null hypothesis that the residuals are serially uncorrelated and an alternative hypothesis that they follow the first-order autoregressive process. The overall DW value equal to 2. As it presented in Fig. 4, the autocorrelation value, 2.0997 (approximately 2), means that there is no autocorrelation in the residuals of the temperature model. The p-value $= 0.7006 > 0.05$ indicates that the null hypothesis (no autocorrelation) cannot be rejected. This means that the residuals do not show a pattern of positive or negative autocorrelation, which supports the validity of the model.

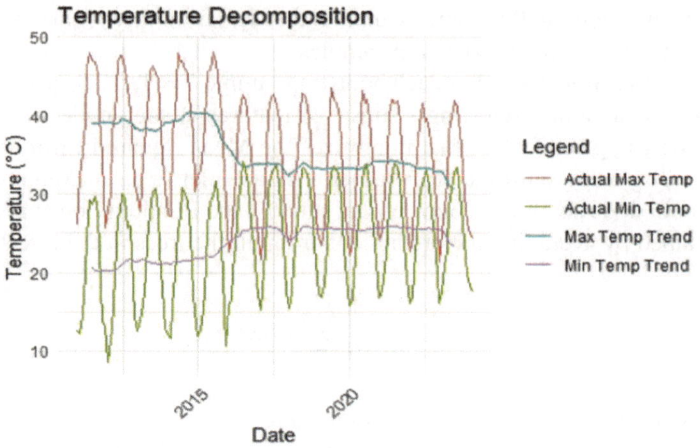

Fig. 2 Monthly maximum and minimum temperature in Qatar from 2011 to 2024

As shown in Fig. 2, the graph illustrates the actual time series data for the monthly average of maximum and minimum temperatures of Qatar. It shows that the highest average maximal temperatures which was 48.5 °C. occurred in July 2015 while the greatest average minimum temperatures, which was roughly 34.2 °C, occurred in July. The decomposition was carried out on the time series and the results revieled a slight upward trend-cycle component in the minimum temperature after 2015, shown in purple, and a decreasing trend in the maximum temperature after 2015 which nearly stabilized by 2024. Although the average maximum temperature in Qatar shows a slight decrease after 2015, the average minimum temperature has increased to levels consistently above 30 °C, never returning to the levels observed before 2015, which were below 30 °C.

Figure 3 shows the Autocorrelation Function (ACF) Plot and the Partial Autocorrelation Function (PACF) Plot. The ACF plot shows that Lag 1 and several other lags have significant correlations (bars exceed the dashed lines).This suggests correlation over multiple time lags, potentially indicating seasonality in the data. Notice that the PACF plot also shows significant correlation only at the first lag, then tails off. This is behavior consistent with a stationary series.

Results from the Augmented Dickey-Fuller (ADF) test in table 1 return p-value $= 0.01 (< 0.05)$, so we reject the null hypothesis that suggests the time series is stationary. Moreover, according to the ACF and PACF plots, there is no evidence for seasonality. Table 2 shows that SARIMA $(2,0,0) \times (2,0,1)_{12}$ has the smallest AIC value observed in the dataset. Therefore, the model is considered the best for forecasting. For Non-Seasonal Components (2,0,0), $p = 2$ indicates the model uses the past two observations to predict the current value; $d = 0$ means no differencing is applied to the series to make it stationary; and $q = 0$ indicates that no past forecast errors are used in the model. For Seasonal Components $(2,0,1)_{12}$; $P = 2$ means that the model uses the past two seasonal periods to predict the current value; $D =$

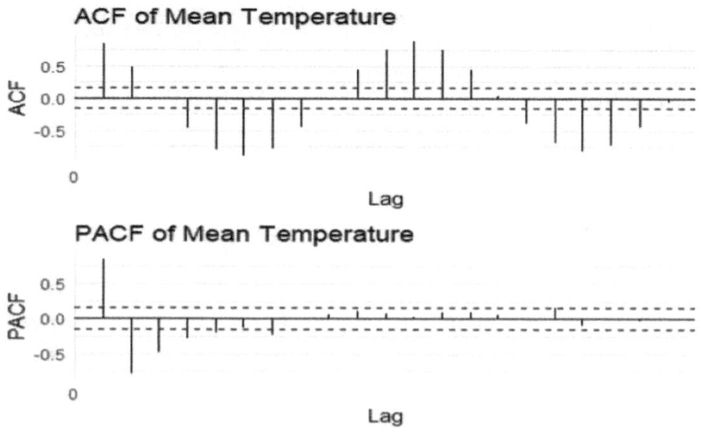

Fig. 3 ACF and PACF of monthly mean temperature during 2011–2024

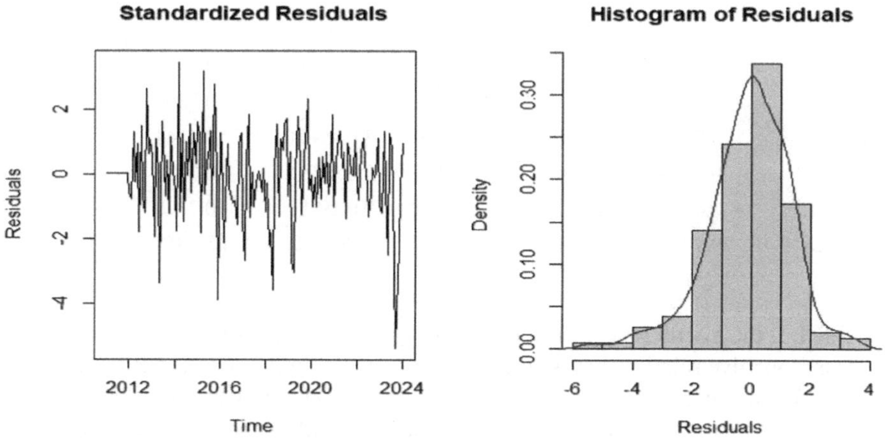

(a) Residuals over time (b) Frequency Distribution Histogram

Fig. 4 **a** Residuals over time **b** Frequency Distribution Histogram

0 means no seasonal differencing is applied, implying the seasonal component is already stationary; Q = 1 indicates that the model uses one past seasonal forecast error to predict the current value, and m = 12 indicates the seasonality occurs on a yearly basis, with 12 months.

The results in Table 3 show the forecast accuracy of the time series based on model validation through diagnostic checks of the data. According to Table 1, the SARIMA $(2,0,0) \times (2,0,1)_{12}$ model has the lowest AIC value, Table 2 also presents findings from the diagnostic dataset generated by the SARIMA model.

Table 1 Augmented Dickey-Fuller test of stationarity for maximum, minimum, and average temperature

Model	ADF test	Lag order	*P*-Value
Max temperature	– 7.1986	5	0.01
Min temperature	– 7.9421	5	0.01
Average temperature	– 9.8934	5	0.01

Table 2 Values of AIC for the SARIMA model

P, d, q	*P, D, Q, m*	AIC value
2, 0, 0	2, 0, 1, 12	596.5979
1, 0, 1	2, 0, 1, 12	601.8898
1, 0, 0	2, 0, 1, 12	607.5365
1, 0, 0	2, 0, 3, 12	611.0414
0, 0, 0	2, 0, 1, 12	706.5393

Table 3 Diagnostic test results for SARIMA (2,0,0) × $(2,0,1)_{12}$

	Coef.	Std err.	z	P > |z|	[0.025	0.975]
AR.L1	0.9807	0.0960	10.2137	< 2e−16	0.7925	1.1689
AR.L2	– 0.3280	0.0911	– 3.6017	0.000316	– 0.506	– 0.1495
AR.S.L1	0.7142	0.1530	4.6689	3.03e−06	0.4144	1.0141
AR.S.L2	0.2450	0.1394	1.7580	0.07875	– 0.0282	0.5182
MA.S.L1	– 0.4336	0.1562	– 2.7754	0.005514	– 0.7398	– 0.1274
Intercept	29.2780	1.9866	14.7374	< 2e−16	25.3843	33.1718

AR.L1, the coefficient for the first autoregressive term is 0.9807. This indicates a strong positive relationship with the previous value in the series. AR.L2, the coefficient for the second autoregressive term is – 0.3280, showing a weaker negative relationship with the value two periods back. AR.S.L1, the first seasonal autoregressive term has a coefficient of 0.7142, suggesting a moderate positive seasonal effect whereas AR.S.L2, the second seasonal autoregressive term has a coefficient of 0.2450, indicating a week positive seasonal effect. MA.S.L1, the first seasonal moving average term has a coefficient of – 0.4336, indicating a moderate negative seasonal effect. The intercept represents the mean level of the series. The Z value is the ratio of the coefficient to its standard error. Higher absolute Z values indicate that the corresponding coefficients are significantly different from zero. AR.L1, AR.L2, AR.S.L1, MA.S.L1, and Intercept have very low *p*-values (< 0.05), indicating that these coefficients are statistically significant whereas AR.S.L2 is not statistically significant (*p*-value = 0.07875).

Fig. 4a Standardized residuals plot, which used to check the goodness of fit of the SARIMA model. The fluctuations of the residuals around zero with no particular pattern indicate that most main features have been captured. However, there are large residuals seen at some periods, like in 2023, which might suggest the presence

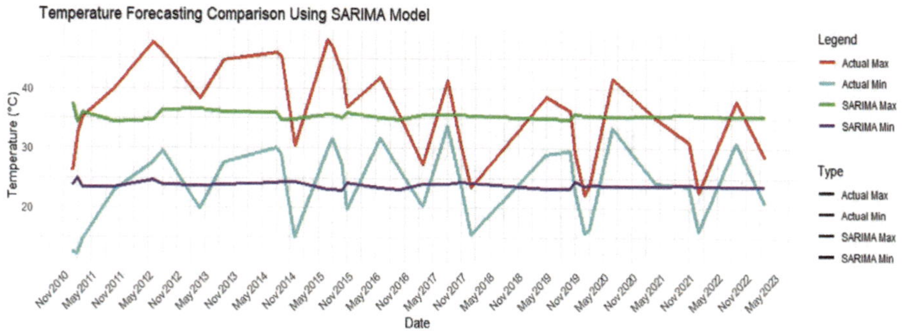

Fig. 5 Max and min temperature forecasting using SARIMA model

of outliers or the inappropriateness of the model for various periods. The residuals do not show a significant trend or seasonal pattern over time indicating that the model has adequately captured the structure of the time series. In particular, we use the histogram of the residuals which appears in Fig. 4b to examine if the residuals appear to come from a normally distributed population, since this is an assumption used in many time-series models including SARIMA. The histogram below suggests that the residuals are approximately normally distributed with a mean around zero

Figure 5 shows that he actual temperature data exhibits more variation and periodicity, while the SARIMA model predictions remain almost constant. This may indicate that the model might not be capturing the underlying seasonal and trend components of the temperature data effectively.

Figure 6 indicates that ARIMA and SARIMA models may not have captured the temperature effectively, leading to predictions with little variation over time. The Random Forest method shows strong performance, capturing the variability in both minimum and maximum temperatures, especially for the minimum temperature. The model effectively captures trends and variations in the data, indicating its strength in handling complex, non-linear relationships.

Table 4 shows that Random Forest outperforms both ARIMA and SARIMA models significantly as it has the lowest values of RMSE (2.514073) and MAE (1.948049) for maximum temperature prediction, as well as the lowest RMSE (1.519) and MAE (1.167) for minimum temperature prediction comparing with RMSE and MAE values for ARIMA and SARIMA predictions. This indicates that Random Forest provides a more accurate and precise prediction for this dataset.

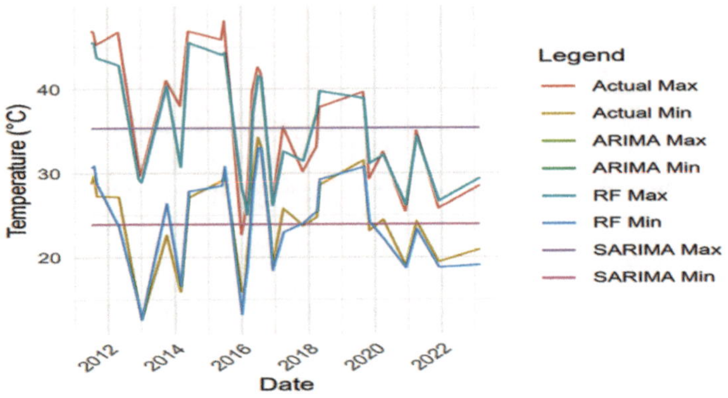

Fig. 6 Forecasting temperature using ARIMA, SARIMA, and random forest (RF)

Table 4 Predictive accuracy evaluated via the root mean square error (RMSE) of ARIMA, SARIMA, and random forest (RF) time series models

Model	Max temperature RMSE	Max temperature MAE	Min temperature RMSE	Min temperature MAE
ARIMA	7.862052	6.935833	12.551	11.232
SARIMA	14.349359	12.097167	5.607	4.597
Random forest	2.514073	1.948049	1.519	1.167

4 Conclusion

This study predicts temperature, which is considered a critical impact of climate change in Qatar. The essential significance of prediction accuracy demonstrates that SARIMA models can be used to predict temperature in the future. However, while SARIMA model produces significant results, it may not be sufficient for capturing complex temperature patterns in this dataset. Random Forest Machine Learning Method offers superior predictive capabilities and performance, capturing the variability in both minimum and maximum temperatures as it had lower values of RMSE and MAE than ARIMA and SARIMA models, which makes it more reliable for forecasting. Therefore, we conclude that the Random Forest Machine Learning Method is more effective for modelling and predicting than time series SARIMA model. We recommend that future studies examine different datasets with longer durations. This will increase the duration of data training set which will have an impact on the dataset forecasting accuracy in SARIMA models. This work underscores the importance of advanced modeling techniques in understanding and predicting climate change's impacts in Qatar.

Competing Interests The authors declare that they have no competing interests.

References

1. Kinney PL et al (2008) Approaches for estimating effects of climate change on heat-related deaths: challenges and opportunities. Environ Sci policy 11(1):87–96
2. Huang C et al (2013) Managing the health effects of temperature in response to climate change: challenges ahead. Environ Health Perspect 121(4):415–419
3. Ministry of Municipality and Environment (2021) Qatar's national climate change action plan. Retrieved from MME Qatar
4. Jones P (2018) Time series forecasting for weather prediction. Int J Forecast 34(2):450–460
5. Smith J, Doe R, Johnson M (2016) Numerical weather prediction and time series analysis. Meteorol Appl 23(1):34–45
6. Khandelwal S, Adhikari R, Verma G (2015) Time series forecasting using ARIMA and SARIMA models. J Stat Comput Simul 85(8):1677–1695
7. Hyndman RJ, Athanasopoulos G (2021) Forecasting: principles and practice. OTexts
8. Hastie T, Tibshirani R, Friedman J (2009) The elements of statistical learning: data mining, inference, and prediction, 2nd edn. Springer, Berlin
9. El-Mallah E, Elsharkawy A (2017) Enhancing performance in time series forecasting models. J Stat Anal Data Mining 10(3):214–228
10. Parviz L (2020) Comparative evaluation of hybrid Sarima and machine learning techniques based on time varying and decomposition of precipitation time series. J Agric Sci Technol 22(2):563–578
11. Breiman L (2001) Random forests. Mach Learn 45(1):5–32
12. Biau G, Scornet E (2016) A random forest guided tour. TEST 25(2):197–227
13. Liaw A, Wiener M (2002) Classification and regression by random forest. R News 2(3):18–22
14. Box GEP, Jenkins GM, Reinsel GC (2008) Time series analysis: forecasting and control. Wiley
15. Hyndman RJ, Athanasopoulos G (2018) Forecasting: principles and practice. OTexts
16. Petukhova T et al (2018) Assessment of autoregressive integrated moving average (ARIMA), generalized linear autoregressive moving average (GLARMA), and random forest (RF) time series regression models for predicting influenza A virus frequency in swine in Ontario, Canada. PloS one 13(6):e0198313
17. Chen P, Niu A, Liu D, Jiang W, Ma B (2018) Time series forecasting of temperatures using SARIMA: an example from Nanjing. In: IOP conference series: materials science and engineering. IOP Publishing 394(2018):052024
18. James G, Witten D, Hastie T, Tibshirani R (2013) An introduction to statistical learning with applications in R. Springer, Berlin
19. Pang B, Yue J, Zhao G, Xu Z (2017) Statistical downscaling of temperature with the random forest model. Adv Meteorol 1:7265178
20. Svetnik V, Liaw A, Tong C, Wang T (2004) Application of Breiman's random forest to modeling structure-activity relationships of pharmaceutical molecules [Internet]. In: International workshop on multiple classifier systems. Springer, Berlin, Heidelberg. https://doi.org/10.1007/978-3-540-25966-4_33
21. Dudek G (2015) Short-term load forecasting using random forests. Intell Syst 2024, 821–828. https://doi.org/10.1007/978-3-319-11310-4_71

References

[illegible]

Status of Marine Intrusion in the Maghreb Countries Bordering the Mediterranean Sea

Larbi Dajbri, Hamza Bouguerra, Fahima Halimi, Azzedine Hani, Nabil Bougherira, Hicham Chaffai, and Salah Eddine Tachi

Abstract The Maghreb countries, particularly Morocco, Algeria, and Tunisia, are bordered to the north by the Mediterranean Sea and are exposed to marine intrusion. This phenomenon is driven by two factors: the adverse effects of climate change, notably the significant decrease in precipitation, which reduces water inflows into dams and aquifers, leading to an imbalance in water budgets; and the second cause stemming from the first. Indeed, the lack of precipitation is exacerbated by the overexploitation of aquifers, resulting in marine intrusion. The compilation of research conducted by scholars from these three countries, as well as to a lesser extent Libya, confirms this trend towards marine intrusion. In Morocco, researchers recently published a study in 2022 on coastal aquifers to assess the sustainability prospects of water resources in these aquifers. In Algeria, studies have focused on the aquifers of the Bay of Algiers, Annaba, Nador (Tipaza), and Bouteldja. The results indicate an almost certain marine intrusion in these aquifers. Tunisian researchers have worked on groundwater monitoring to evaluate artificial recharge in the Mediterranean coastal aquifer of Korba (northeast Tunisia). This work aims to explore the current state and provide a comprehensive review of the extent of saltwater intrusion in the region and the future challenges.

Keywords Maghreb · Salinity · Water · Intrusion

L. Dajbri (✉) · H. Bouguerra · A. Hani · N. Bougherira · H. Chaffai · S. E. Tachi
Laboratoire Ressource en Eau and Développement Durable, Université Badji Mokhtar Annaba, Annaba, Algeria
e-mail: larbi.dajbri@univ-annaba.dz

F. Halimi
Laboratoire de Recherhe en Géologie, Université Badji Mokhtar Annaba, Annaba, Algeria

S. Al-Naemi et al. (eds.), *Water and Food Security in the Face of Climate Change: Challenges and Opportunities for Resilience*, World Sustainability Series,
https://doi.org/10.1007/978-3-032-00098-9_8

1 Introduction

The results on seawater intrusion in the Maghreb, based on published papers, show that significant efforts have been made to understand this phenomenon and develop management strategies in Libya, Tunisia, Algeria, and Morocco. The most commonly used methods involve geochemical data and statistical analysis. In the Maghreb, the coastline extends over 8955 km from Morocco to Libya. These regions are characterized by significant demographic and economic growth. These increases expose the coastal aquifers of the Maghreb to groundwater salinity induced by seawater intrusion, which is caused by the overexploitation of aquifers and climate change, potentially leading to an economic crisis due to the lack of fresh water. Through this work, we aim to provide an overview of marine intrusion in these countries and propose solutions for the development of these regions. We note that research on Libya is almost nonexistent. Therefore, we will focus on the remaining three countries (Morocco, Algeria, and Tunisia).

2 Geographic and Geological Situations

The Maghreb countries, including Algeria, Tunisia, Morocco, and Libya, are bordered to the north by the Mediterranean Sea. Morocco is additionally bordered to the west by the Atlantic Ocean. To the south, these countries are bordered by semi-arid or arid zones (Fig. 1).

3 Geological, Hydrogeological, and Climatic Characteristics

Agoubi [2], in his work published in the journal *Water*, noted that surface water resources in North Africa are limited. The climate is influenced by the Sahara Desert, with rainfall being rare and highly variable in both time and space. Precipitation is higher in the extreme north (200–500 mm/year) and decreases in the southern Saharan zone (50 mm/year) [3]. From a hydrogeological perspective, North Africa features several aquifers containing substantial groundwater reserves. The Jeffara coastal aquifer, covering an area of 200,000 km^2, stretches over 500 km from the northern coast of Gabès to Kohms (Libya), is spanning both countries.

In Algeria, the coastal area, representing 4% of the territory, extends over 1622 km. The main coastal aquifers, which are shallow, are embedded in Quaternary sediments. The high permeability of the aquifer contributes to the vulnerability of groundwater, inducing saltwater intrusion, while also facilitating aquifer recharge through direct precipitation infiltration.

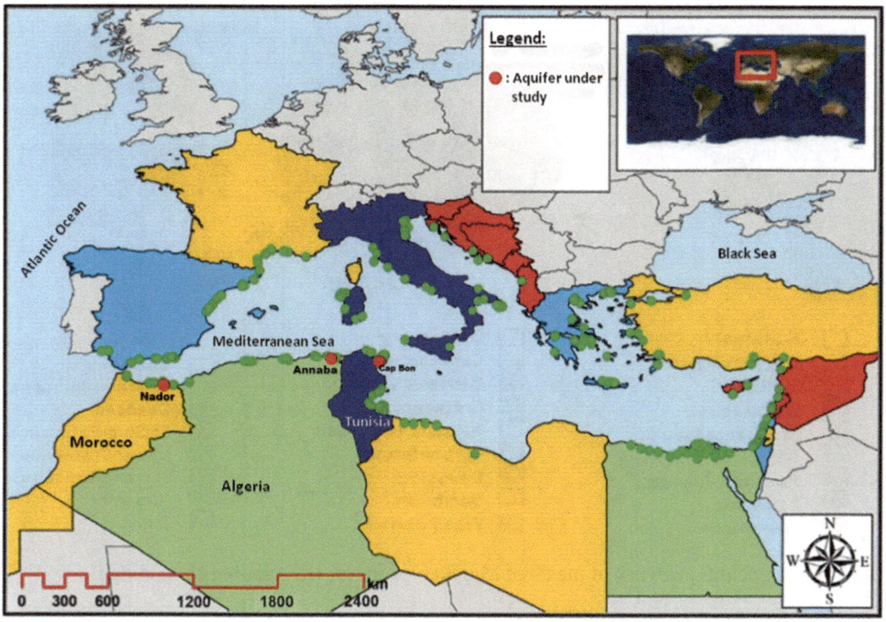

Fig. 1 Geographic situation of the studied area (in Mastrocicco and Colombani [1], modified by us)

Hydrogeologic point of view, Tunisia is characterized by deep aquifers in the south, shared with Algeria and Libya. In the north, surface water (dams and rivers) serves as the primary source of fresh water. Tunisia experiences a semi-arid climate across most of its territory. The main coastal aquifers in Tunisia include the Jeffara aquifer, extending from the Tunisian-Libyan border to south of Sfax, the Sahel aquifer, and the aquifers of Cap Bon, Grombalia, Bizerte, and Jendouba.

The region of Nador, located on the Moroccan Mediterranean coast between the cities of Melilla and Saidia, is bounded to the northwest by Cap des Trois Fourches, to the south by the Gareb plain, to the east by Oued Lakhmis, and to the west by the Gourougou and Beni Bou Ifrour massifs. It consists of a series of reliefs linked to different geodynamic contexts resulting from various tectonic phases: the massifs of Cap des Trois Fourches, Gourougou, and Beni Bou Ifrour, which belong to the Rif mountain range, and the massifs of Kebdana and Gareb-Ziata, part of the Atlas foreland, separated by subsiding Neogene basins such as the Melilla basin and the plains of Gareb and Bou Areg. The Beni Bou Ifrour massif is composed of autochthonous metamorphic terrains of Mesozoic age, intersected by granodiorite laccoliths and dikes, and affected by a network of NE–SW trending faults (Fig. 2).

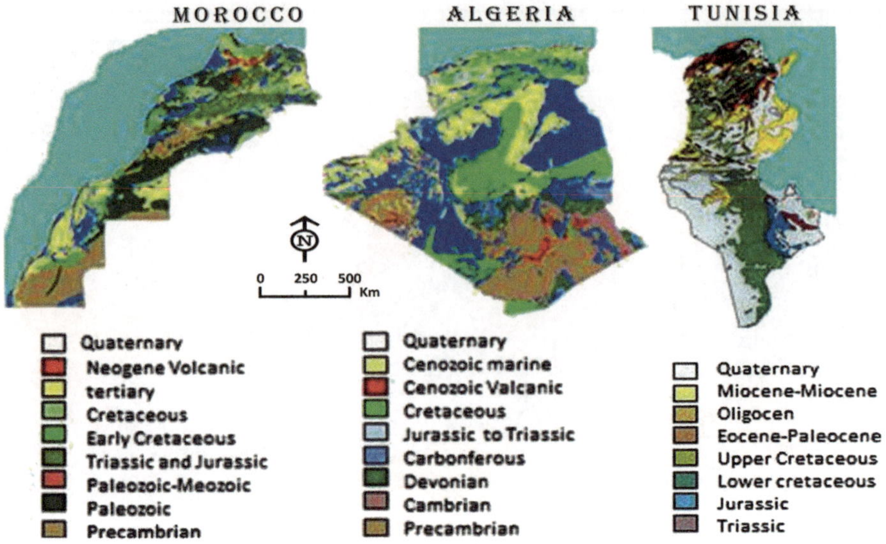

Fig. 2 Geological framework of the three Maghreb countries (in Agoubi [2], modified)

4 Water Resources in the Maghreb

The exploitable groundwater resources of the Maghreb are estimated at 13 km^3/year, with the volumes withdrawn (average values from 2004–2008) estimated at 9.5 km^3/year (Table 1), representing an overall exploitation index (Withdrawal/Exploitable Resources Ratio) of 73%. However, this index rises to 112% when withdrawals are compared to the renewable portion of the resources. This indicator value at national and regional levels conceals significant disparities at several levels. First, within the same country, then between countries, and most importantly, between regions: regions where groundwater is strictly renewable, such as in Morocco, Northern Algeria, and Northern Tunisia, and the Saharan regions of Algeria and Tunisia.

5 Methodology Used

Chemical analysis of available water samples has enabled us to successfully conduct this synthesis on the coastal aquifers of the Maghreb along the Mediterranean coast. Diagrams such as Piper or Durov have been utilized to visualize the distribution of water types based on their chemical composition. It is observed that the waters are relatively saline. The results obtained can assist water sector officials in making decisions and taking actions to preserve coastal aquifers.

Table 1 Groundwater resources in the Maghreb [4]

Groundwater resources in 10^6 m^3/year	Total withdrawals	Renewable resources	Exploitable resources
Northern Algeria	1500	1800	1800
Algerian Sahara	2100	900	5000
Total Algeria	3600	2700	6800
Morocco	3900	4000	4000
Northern and Central Tunisia	1250	1500	1500
Southern Tunisia	780	300	780
Total Tunisia	2030	1800	2280
Total for the Maghreb	9530	8500	13,080

6 Confirmed Water Salinity

Based on studies conducted in the three countries, we demonstrate the presence of seawater salinity.

6.1 In Algeria

In the Annaba region, we utilized results from analyses conducted over time (1996, 2003, 2010, 2014, 2016) from the works of numerous researchers (Djabri, Debieche, Aoun, Zahi, and Aichouri). The compilation of results led us, to create Piper diagram (Fig. 3). Upon observation of Fig. 3, it is noted that the analyzed waters are rich in chloride anions and sodium captions. Observing the diagram, it is observed that some samples are located above the mixing line. The majority of samples fall into the following zones:

- Mixed (mixture), with a maximum number of samples showing various types of water,
- Slight intrusion, characterized by numerous samples,
- Marine intrusion, represented by a significant number of samples, indicating a sea-aquifer relationship,
- Freshwater, comprising about fifteen samples, indicating mineralized waters in the study area,
- Slightly freshwater, characterized by about fifteen samples.

It is observed that waters indicating clear marine intrusion or slight marine intrusion are located above the seawater section of the diagram. The remaining categories fall within the central part of the diamond, representing mixed water types.

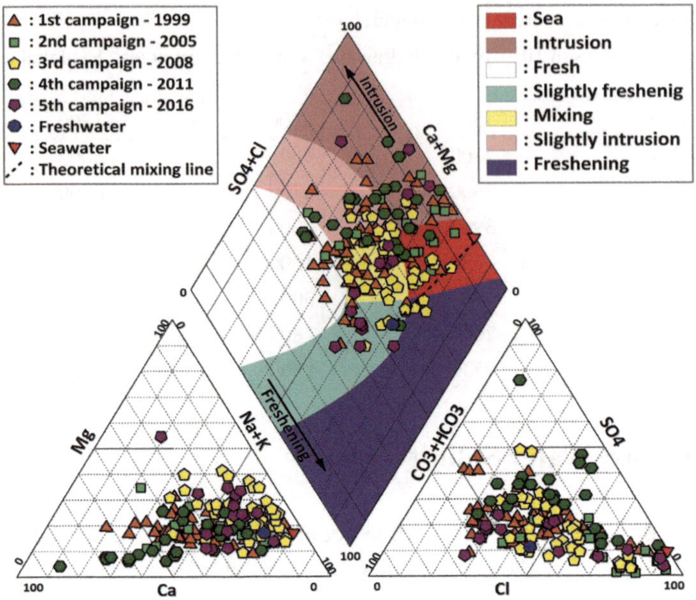

Fig. 3 Piper's trilinear diagram showing the relationship between dissolved ions and water type in Annaba plain

6.2 In Morocco

Research conducted by Yassine Ez-zaouy, Lhoussaine Bouchaou, Aicha Saad, Mohammed Hssaisoune, Youssef Brouziyne, Driss Dhiba, and Abdelghani Chehbouni covers the entire coastal zone of Morocco (both the Atlantic Ocean and the Mediterranean Sea). The diagrams indicate that groundwater from Moroccan coastal aquifers is primarily of Na–Cl type, followed by Ca–Mg–Cl, Ca–Mg–SO_4, Ca–Mg–HCO_3, and Ca–HCO_3–Cl types. Several samples belong to the chloride-sodium facies, indicating a dominance of Na–Cl water type and suggesting interaction of groundwater with saline waters [5]. The anionic triangle shows that several samples tend towards the chloride pole. However, two trends can be noted (Fig. 4). firstly; there is a group between the bicarbonate pole and the chloride pole, and on the other hand, another group between the sulfate pole and the chloride pole. This situation may result from the dissolution of evaporate formations and the effect of seawater intrusion promoting base exchange phenomena [6]. Regarding the cation diagram, it is observed that points tend towards the (Na^{++} K^{+}) domain, with values ranging between 24 and 37%. These values confirm the classification adopted by statistical studies. However, moderately mineralized aquifers have average values ranging from 2 to 4%, and from 1 to 3% for the low mineralization class [7].

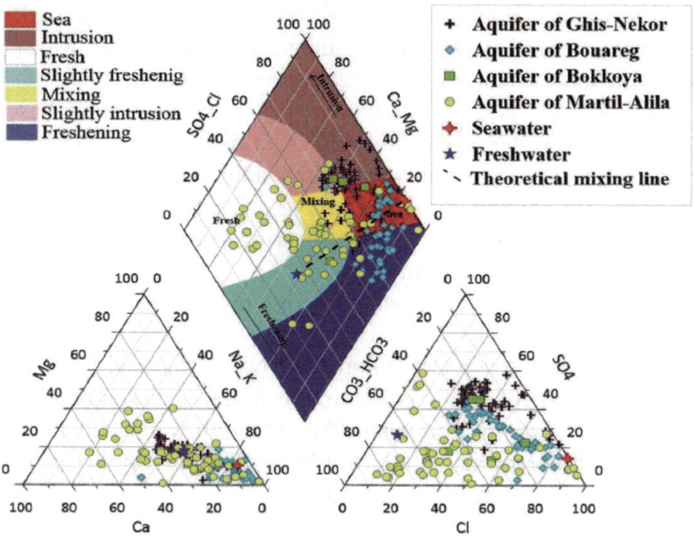

Fig. 4 Piper diagram coastal zones of Morocco [7]

6.3 In Tunisia

Research conducted by Kouzana et al. [8] has shown that the Korba aquifer (located on the coast of Cap Bon, 70 km from Tunis) exhibits high Total Dissolved Solids (TDS) values throughout the region, especially in coastal areas. The high TDS values are likely due to contamination by seawater. This hypothesis is supported by negative piezometric values, high chloride concentrations, and the presence of reverse cation exchange reactions, indicating mixing of seawater with freshwater. It is noteworthy that waters from the coastal zone show low salinity and low chloride concentrations. These low values are attributed to infiltration of meteoric waters through consolidated dune ridges and the presence of clay layers acting as barriers against seawater intrusion.

7 Classification of Waters Using the Durov Diagram

The Durov diagram (Fig. 5) is employed to identify the main geochemical processes and reactions occurring within the aquifer. The position of the samples within fields 1 to 9 forms the basis for this interpretation:

Field 1: HCO_3^- and Ca^{2+} are dominant, which often indicates that the water originates from calcareous, sandy, or similar aquifer formations.

Field 2: The water is dominated by Ca^{2+} and HCO_3^- ions. An association with dolomite is assumed if Mg^{2+} concentrations are significant. However, when Na^+ is also significant, a major ion exchange process is likely.
Field 3: HCO_3^- and Na^+ are dominant, generally suggesting base exchange processes. However, the production of CO_2 at depth may also generate HCO_3^- and result in Na^+ dominance under certain conditions.
Field 4: SO_4^{2-} (or another secondary anion) and Ca^{2+} are dominant. This often indicates water from volcanic or gypsum-bearing formations, or it may result from mixing or simple mineral dissolution.
Field 5: No dominant cation or anion is present, suggesting that water mineralization is due to simple dissolution or mixing.
Field 6: SO_4^{2-} (or a secondary anion) and Na^+ dominate. This water type is uncommon and may reflect a mixing process or a rare dissolution influence.
Field 7: Cl^- and Na^+ are dominant. This mineralization pattern is frequently observed unless the aquifer is affected by pollution. Otherwise, it may result from reverse ion exchange of the Na–Cl type.
Field 8: Cl^- is the dominant anion and Na^+ the dominant cation, indicating that the groundwater is associated with reverse base exchange of the Na–Cl type.
Field 9: Cl^- and Na^+ are dominant, commonly indicating that their origin is related to salt dissolution

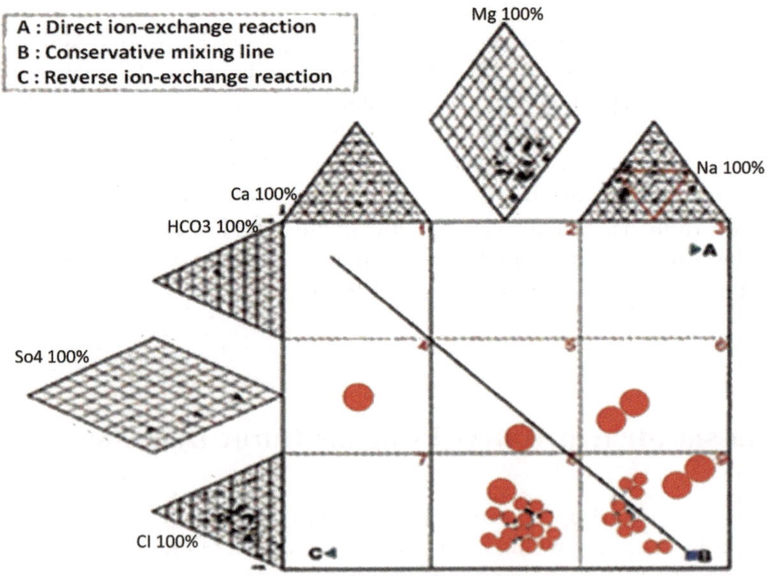

Fig. 5 Extended Durov diagram (Kouzana et al. [8], modified)

- **Case 4**: Shows dissolution of calcium-rich minerals (calcite and gypsum); an example of freshwater is found in this case, used for comparison with other samples.
- **Case 5**: Water composition results from mixing of waters of different origins.
- **Case 6**: Involves both mixing and ion exchange processes. Only one sample is found in this case, belonging to the inner zone of the aquifer.
- **Case 9**: Indicates mixing with seawater.

In this study, the majority of samples are located in fields 8 and 9 of the Durov diagram. Field 8 is characterized by Cl^- as the dominant anion and Na^+ as the dominant cation, while in field 9 both Cl^- and Na^+ are dominant. These samples plot below the conservative mixing line, indicating the occurrence of reverse ion exchange and mixing processes. In this mechanism, calcium from the clay fraction of the aquifer is exchanged with sodium from the water, resulting in higher calcium and lower sodium concentrations in the groundwater.

8 Conclusion

This study was conducted on three different countries: Morocco, Algeria, and Tunisia. Hydrochemical tools were used to understand the origins of water mineralization in these regions. Moroccan researchers monitored the spatial evolution of chemical characteristics of coastal waters (both Atlantic and Mediterranean zones), while Algerian research focused on temporal monitoring of water quality in the Annaba region. The analyses conducted facilitated the creation of Piper diagrams in Morocco and highlighted the mixing of freshwater and seawater using the Durov diagram in Tunisia.

By using different methods such as Piper and Durov diagrams, the authors identified water salinity associated with proximity to the sea. Factors influencing water salinity include overexploitation of aquifers, water–rock interactions leading to base exchange, and decreased precipitation, which accelerates seawater intrusion processes.

8.1 Implementation of Management

The Mediterranean coastline stretches over 26,000 km and represents a rich biodiversity hotspot, with over half of the 25,000 plant species endemic to the region. It is an important area for migratory birds traveling between Africa and Europe. However, this region suffers from uncontrolled development, unplanned urbanization, national and international tourism pressures, terrestrial and marine pollution, and overexploitation of natural resources, particularly freshwater. This degraded situation has prompted the European Union to develop programs for development and conservation in partnership with countries on the southern Mediterranean coast.

These programs advocate for promoting national policies that integrate biodiversity conservation of these ecosystems, including the creation, classification, and implementation of management plans. It's worth noting that water resource authorities in the Maghreb countries are aware of the general water issue and specifically seawater intrusion. Tunisia and Morocco have taken proactive measures by implementing artificial recharge of aquifers, which has somewhat mitigated marine pollution. Algeria has developed a new water strategy focusing on unconventional water sources, including seawater desalination and reuse of treated water. Unfortunately, achieving these goals fully has been hindered by financial constraints, exacerbated by the COVID-19 pandemic, which disrupted markets and funding.

8.2 General Conclusion

Through this research work, it is evident that water resources management in the Maghreb countries is not fully mastered. This situation is largely influenced by changes in leadership responsible for water resources. Furthermore, research conducted by scientists is often undervalued by decision-makers who may not prioritize collaboration with universities. This study has demonstrated that similar causes lead to similar effects across the three countries consider.

References

1. Mastrocicco M, Colombani N (2021) The issue of groundwater salinization in coastal areas of the mediterranean region: a review. Water 13(1):90
2. Agoubi B (2021) A review: saltwater intrusion in North Africa's coastal areas—current state and future challenges. Environ Sci Pollut Res 28:17029–17043
3. Hamed Y, Hadji R, Rehaounia B, Zighmi K, Baali F, El Gayar A (1918) Eaux de surface et souterraines en Afrique du Nord: Synthèse globale des resultats et recommendations. Revue Euro-Mediterranéene pour l'integration environementale 3:25
4. Besbes M (2011) Gestion des eaux souterraines au Maghreb—Académie d'Agriculture de France. 250ème Anniversaire. UNESCO, 11p
5. Mondal NC, Singh VP, Singh VS, Saxena VK (2010) Determining the interaction between groundwater and saline water through groundwater major ions chemistry. J Hydrol 388(1–2):100–111
6. Bahir M, Ouazar D, Goumih A, Ouhamdouch S (2019) Evolution of the chemical and isotopic composition of groundwater under a semi-arid climate; the case of the Cenomano-Turonian aquifer within the Essaouira basin (Morocco), Environmental Earth Sciences. https://doi.org/10.1007/s12665-019-8349-2
7. Ez-Zaouy Y, Bouchaou L, Saad A, Hssaisoune M, Brouziyne Y, Dhiba D, Chehbouni A (2022) Morocco's coastal aquifers: recent observations, evolution and perspectives towards sustainability. Environ Pollut 293:118498

8. Kouzana L, Ben Mammou A, Gaaloul N (2007) Intrusion marine et salinisation des eaux d'une nappe phréatique côtière (Korba, Cap-Bon, Tunisie). Geo-Eco-Trop 31:57–70

Machine Learning Screening Tool for the Prediction of Removal of Water Hardness Performance in Desalination Pretreatment Process

Rokia Youcef, Nassila Sabba, Amel Benhadji, and Mourad Taleb Ahmed

Abstract Membrane technology performance can be predicted, analyzed, and designed using machine learning (ML). This involves figuring out the best operating parameters to increase the pretreatment process's effectiveness and stop membrane clogging forming on the membrane layer. Pretreatment was used both before and after reverse osmosis to remove water hardness at the Palm Beach desalination facility on an industrial scale. A laboratory-based experimental methodology was created to assess a number of parameters, including pH, conductivity, salinity, TAC, and TH. Artificial neural networks (ANNs) are the most widely used algorithm for simulating membrane separation-based desalination of saltwater for the removal of water hardness. By contrasting it with the testing dataset, the trained model will be verified. The most practical way to assess the trained model which will be validated by comparing it to the testing dataset. the most usable to evaluate the performance of the developed ANN model as Associated metrics include Mean Absolute Error (MAE = 0.0104), a Root Mean Squar Error (RMSE = 0.028), (prediction average = 0.742) and (R^2 = 0.991). The prediction of water hardness after pretreatment process as " sand filter' and "Cartridge filters" has an effect on membrane performance as permeate flux variation, treatment cost and successfully analyzed complex large scale water quality evaluation.

Keyword Water hardness · Desalination · Membrane layer · Pretreatment process

R. Youcef (✉) · N. Sabba · M. Taleb Ahmed
Laboratory of Recovery and Recycling of Matter for Sustainable Development, USTHB, University of Science and Technology Houari Boumediene, Algiers, Algeria
e-mail: rokia.youcef@usthb.edu.dz

A. Benhadji
Laboratory of Reaction Engineering, University of Science and Technology Houari Boumediene, Algiers, Algeria

S. Al-Naemi et al. (eds.), *Water and Food Security in the Face of Climate Change: Challenges and Opportunities for Resilience*, World Sustainability Series,
https://doi.org/10.1007/978-3-032-00098-9_9

1 Introduction

The development of science and technology is reaching new horizons include environmental disorder and a major pollution problem to have clean drinking water production needs. One of the solutions that meets the requirements of sustainable development is 'Desalination production of sea water' [1].

In Algeria, the large operational desalination plants spread over a 1200 km coastline, with a total capacity of about 2.17 million m^3 of desalinated water per day which represents about 18% of drinking water consumption. The membrane process technology as Reverse Osmosis (RO) is used in 11/12 plants, with the exception of the "Kahrama plant" in Oran use a distillation technique as Multistage Flash (MSF).

The optimization of pretreatment prior to membrane process at the Palm Beach desalination plant aim to reduce the clogging potential of the water to provide a satisfactory quality of pretreated water. This step is necessary in order to successfully implement seawater desalination processes. The water treatment and management systems need the applications of different machine learning algorithms in different step of treatment. Machine learning-based desalination is a new sector integrates analyze real-time data to solve performance investigation employing for the world's water shortage problem. Due to the potential benefits of better data management and the development of prediction models for broad applications, big data and data analysis are gaining more attention [2–4].

The aim of our work is to study the pretreatment prior to reverse osmosis at the Palm Beach desalination plant. This step is necessary in order to successfully evaluate seawater desalination processes treatment with analysis in operating conditions in the composition of the water, such as the temperature of the water, carbonate hardness, total hardness, pH, conductivity and salinity of the treated water. Moreover, One common form of data-driven modeling approach is the use of artificial neural networks (ANNs), which begin with training algorithms to explain data phenomena and how they affect the performance of membrane processes.

2 Materials and Methods

2.1 Prediction of Water Hardness

The evolution of pretreatment performance with deep learning model as an application, we have implemented the ANNs method. Therefore, We have used this method as an application. In order to assess the resulting model on validation, the implementation proceeds via the training phase utilizing the training dataset. The algorithm has been implemented using the phyton programming language.

2.2 The Dataset

The 570 data points in the dataset used for the machine learning study each represented water hardness readings following membrane desalination treatment (Fig. 1). In the regression analysis, water hardness served as the dependent variable, while the independent variables were pH, conductivity (mS/cm), Turbidity (NTU), and TDS (mg/l) obtained after the pretreatment process with the "sand filter." These independent variables were chosen due to their potential impact on the membrane desalination process. This analysis considered water sourced from the Palm Beach desalination plant in Staouali, Algiers, Algeria.

Data visualization. The numerical variables were subjected to descriptive statistics. In order to determine possible multicollinearity problems and to comprehend the direction and intensity of the relationship between the dependent and independent variables, a correlation matrix was created. **Regression analysis**. This study employed regression analysis with the analysis parameters following the pretreatment process as input to estimate the overall hardness of water and calcium hardness after membrane desalination. The investigation included three regression model artificial neural networks (ANN). The ANN, a strong and adaptable algorithm, was essential to our investigation. Given a related input, it can potentially approximate any output. The Artificial Neural Network are versatile and powerful algorithms which have the potential to approximate any output given a related input. The ANN is based in non-linear combinations of inputs in multiple layers which are fed via a series of weights. These weights are learned during training through backpropagation, which computes the specific contribution of each weight to the global error defined in the loss function. To prevent overspecification of neurons and overfitting, regularization techniques were used during the training, i.e. Batch Normalization and Dropout [1].

3 Results and Discussions

3.1 Sand Filter Process

As seen in Table 1. Descriptive statistics of the dataset for sand filter process and the influence of treatment on membrane process desalination.

After treatment by membrane process, the dataset's total hardness of water ranged from 51 to 65 (PPM of $CaCO_3$), the calcium harness varied from 35 to 62 (PPM of $CaCO_3$), the pH ranged from 7.72 to 8.01, the TDS ranged from 34,203.3 to 34,351.65 mg/L, the turbidity ranged from 0.05 to 0.2 NTU, and the conductivity ranged from 49.57 to 50.08 mS/cm. Regarding how the variables interact with one another, The entire calcium and water harness values. The ideal outcome of the membrane desalination process can be achieved by minimizing the conductivity

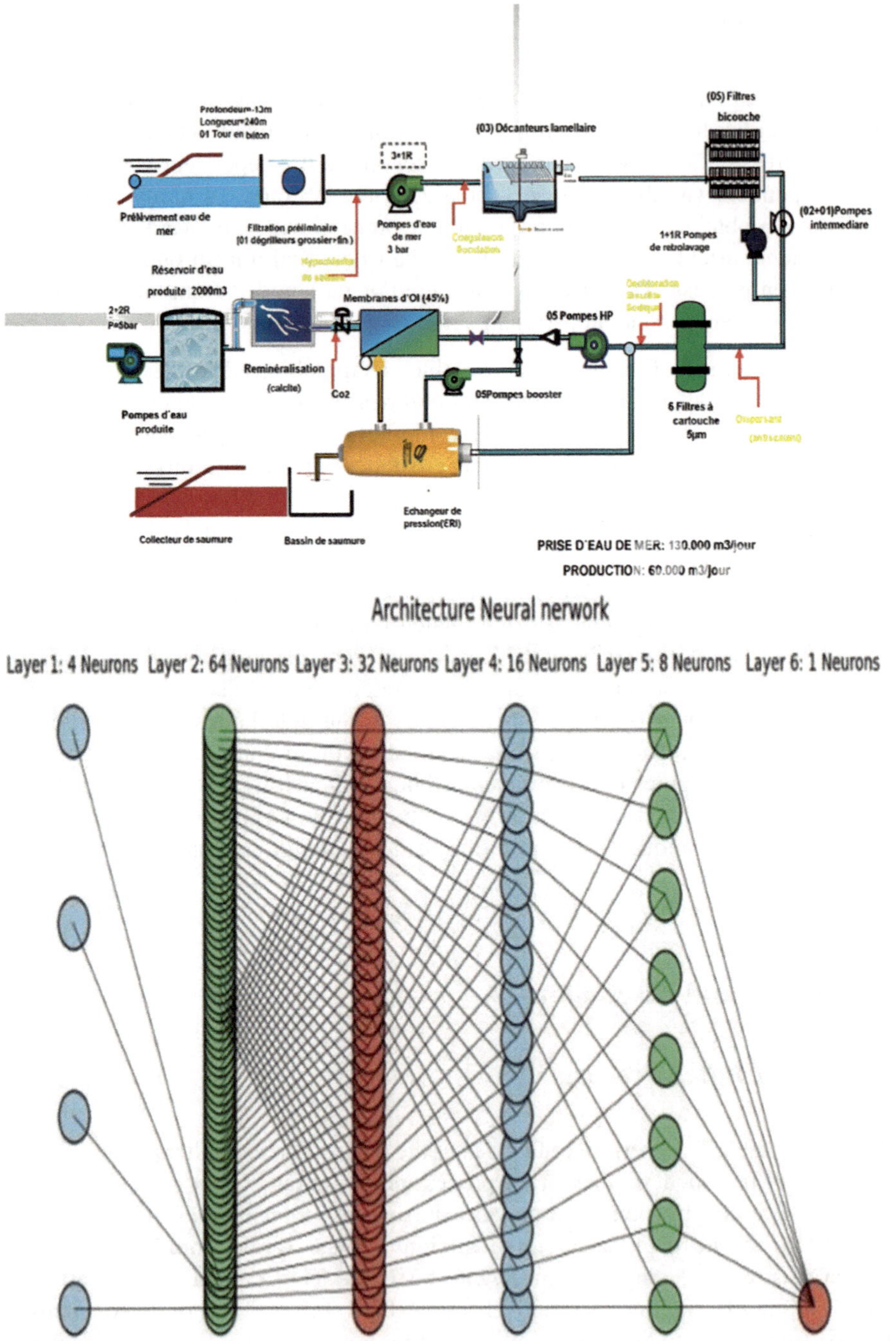

Fig. 1 Machine learning for water desalination process management and schematic ANN, artificial neurol network

Table 1 Descriptive statistics of sans filter dataset on the performance of water hardness removal

	Temperature (°C)	pH	Conductivity (mS/cm)	Turbidity (NTU)	TDS (mg/l)	Calcium hardness (PPM de $CaCO_3$)	Total hardness (PPM de $CaCO_3$)
Count	30.0	30.0	30.0	30.0	30.0	30.0	30.0
Mean	18.64	7.89	49.79	0.115	34,358.29	48.96	61.5
Std	0.61	0.065	0.12	0.030	86.86	7.146	3.42
Min	17.4	7.72	49.57	0.05	34,203.3	35.0	51.0
25%	18.22	7.85	49.72	0.09	34,306.8	45.25	60.0
50%	18.6	7.9	49.78	0.12	34,351.65	48.5	63.0
75%	18.975	7.94	49.87	0.13	34,351.65	53.75	64.0
Max	20.1	8.01	50.08	0.2	34,351.65	62.0	65.0

distance to the nearest 49.6–49.7 mS/cm. Regarding TDS variation, it ranges from 34,200 to 34,400 mg/L. A overview of all the descriptive statistics and visualizations is provided in Table 1 and Fig. 2.

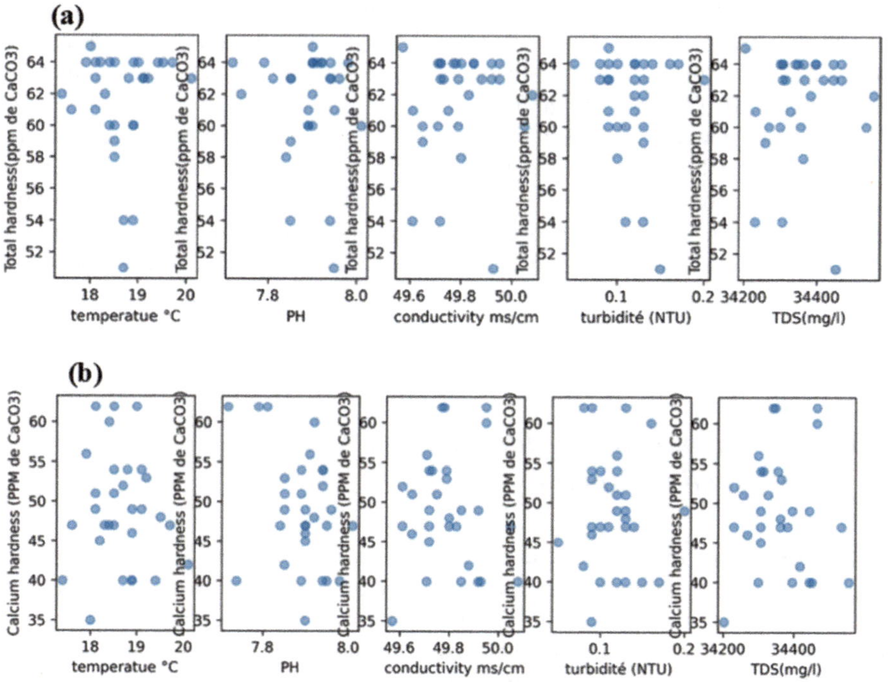

Fig. 2 Sand filter process parameters and their influence on desalination process **a** The total harness of water **b** The calcium harness

Table 2 Descriptive statistics of Cartridge filter dataset on the performance of water hardness removal

	Temperature (°C)	pH	Conductivity (mS/cm)	Turbidity (NTU)	TDS (mg/l)	Calcium hardness (PPM de $CaCO_3$)	Total hardness (PPM de $CaCO_3$)
Count	30.0	30.0	30.0	30.0	30.0	30.0	30.0
Mean	18.44	7.88	49.83	0.067	43,706.4	48.96	61.5
Std	0.587	0.06	0.147	0.014	57,133.4	7.146	3.421
Min	16.8	7.73	49.57	0.04	3422.4	35.0	51.0
25%	18.125	7.83	49.8	0.06	34,362.0	45.25	60.0
50%	18.35	7.91	49.82	0.07	34,380.7	48.5	63.0
75%	18.7	7.93	49.94	0.08	34,458.6	53.75	64.0
Max	19.9	7.98	50.26	0.09	344,724	62.0	65.0

3.2 Cartridge Filters

The range of values in the dataset for the total hardness of water after desalination was 51–65 (PPM of $CaCO_3$), the calcium hardness was between 35 and 62 (PPM of $CaCO_3$), the pH was between 7.73 and 7.98, the TDS was between 3422.4 and 3,447,224 mg/L, the turbidity was between 0.04 and 0.09 NTU, and the conductivity was between 49.57 and 50.26 mS/cm.

Regarding how the variables interact with one another, The entire calcium and water harness values. The ideal outcome for the membrane desalination process is obtained when the conductivity distance is as close to 49.6–50 mS/cm as possible. Table 2 and Fig. 3 provide an overview of all the visualizations and Descriptive statistics. The descriptive statistics of the dataset for the Cartridge filters process and the effect of treatment on the membrane process desalination are shown in Table 2 [2].

3.3 Prediction of Water Harness After Membrane Desalination Process

The ANN's error metrics for the test set following Bayesian optimization-based hyperparameter adjustment. Water hardness and the analytical parameters following the pretreatment process were the features employed in this regression model. The best results, after 750 trials, were achieved with 5 layers, 64, 32, 16, 8, and 1 neurons, using ReLU activation functions in all non-output layers and batch normalization only in the final deep layer. The ANN model produced the best results for an Adam optimizer with a batch size of 50 and a validation split of 0.05, resulting in an R^2 Score of 0.991. indicates that the model explain about 90% of the variance in the

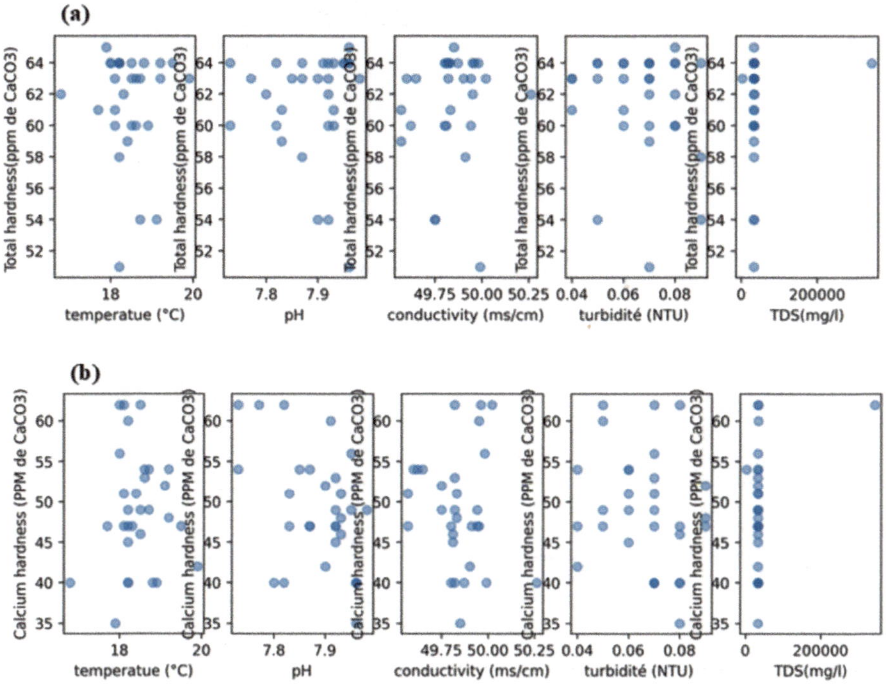

Fig. 3 Cartridge filter process parameters and their influence on desalination process **a** The total harness of water **b** The calcium harness

data, Prevision of 0.742 seems to be the average of forecasts, which is consistent with a good R^2 score, Mean Absolute Error (MAE) of 0.0104 is relatively low indicates a low average error and Root Mean Square Error (RMSE) of 0.028 higher than MAE suggest a presence of outliers. Therefore, the Artificial neuron networks (ANN), was shown as an excellent tool for the results of saline water analysis after sand filter process pretreatment screening that the change of conductivity have an influence on membrane desalination especially on variation of water hardness [4–7].

According to Fig. 4. We noticed that the variation of loss decreases during the training of the model, which indicates that the model is learning and improving its performance on the training data.

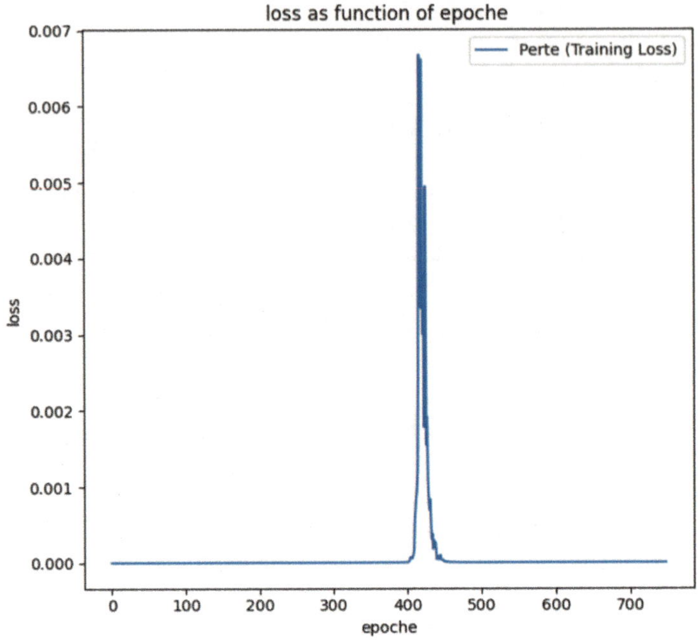

Fig. 4 The Variation of loss during training of the model

4 Conclusion

ML approaches provide water quality assessment and an improvement on desalination performance for optimization, prediction and operation management. The professional background knowledge apply in engineering practice require the implementation of ML in practical application. The study showed that ANN model is suitable for prediction of water hardness elimination after Reverse Osmosis (RO) by evaluate the variation of operating parameters in pretreatment process such as sand filter and cartridge filte and their influence on efficiency of desalination. The study could be enhanced by comparing between model on prediction of water hardness to give to engineering the best conditions for desalination plant process design.

Acknowledgements The funding agencies (ATRST) have provided funding for this study. The University of Science and Technology Houari Boumediene provides financial support for its execution. The Palm Beach desalination and El Marsa station team's insightful recommendations are appreciated by the authors.

CRediT Authorship Contribution Statement **Rokia Youcef**—Investigation, Methodology, Software, Conceptualization, Funding acquisition, Writing—original draft; **Nadia Belkhir and Faiçal Houari-** Resources, Methodology; **Nassila Sabba** review and editing, original draft, Project administration, Funding acquisition. **Amel Benhadji**—Reviewing and Editing **Mourad Taleb Ahmed**—Conceptualization, Supervision, original draft, Reviewing and Editing.

Declaration of Competing Interest The authors state that none of the work described in this study could have been influenced by any known competing financial interests or personal relationships.

References

1. Dansawad P, Li Y, Li Y, Zhang J, You S, Li W, Yi S (2023) Machine learning toward improving the performance of membrane-based wastewater treatment: A review. Adv Membr 3:100072
2. Ministère de l'ÉnergielAlgérie
3. Ashraf WM, Jamil MA, Uddin GM, Shboul B, Ishfaq K, Ng KC, et al (2024) Machine learning assisted improved desalination pilot system design and experimentation for the circular economy. J Water Process Eng 63:105535
4. Abd Elaziz M, Zayed ME, Abdelfattah H, Aseeri AO, Tag-eldin EM, Fujii M, Elsheikh AH (2024) Machine learning-aided modeling for predicting freshwater production of a membrane desalination system: a long-short-term memory coupled with election-based optimizer. Alex Eng J 86:690–703
5. Medina-Collana JT, Reyna-Mendoza GE, Montaño-Pisfil JA, Rosales-Huamani JA, Franco-Gonzales EJ, Córdova García X (2022) Evaluation of the performance of the electrocoagulation process for the removal of water hardness. Sustainability 15(1):590
6. Casas A, Rodríguez-Llorente D, Rodríguez-Llorente G, García J, Larriba M (2024) Machine learning screening tools for the prediction of extraction yields of pharmaceutical compounds from wastewaters. J Water Process Eng 62:105379
7. Zhu Y, Lian B, Wang Y, Miller C, Bales C, Fletcher J, Yao L, Waite TD (2022) Machine learning modelling of a membrane capacitive deionization (MCDI) system for prediction of long-term system performance and optimization of process control parameters in remote brackish water desalination. Water Res 227:119349

Global Warming, Agriculture and Food Security: A Review of Impact of Rising Temperature on Crop Yields and Food Security in Northeastern, Nigeria

K. M. Malabe and S. B. Mustapha

Abstract This article reviews the impact of rising temperatures on agriculture and food security in Northern Nigeria, a region heavily reliant on agriculture due to its high population dependence on subsistence farming. Agriculture in Northern Nigeria is critical for the economy but faces significant challenges due to its hot, semi-arid climate. Climate change, characterized by increasing global temperatures, exacerbates these challenges by affecting crop yields, which in turn threatens food availability and security. Rising temperatures in this region adversely affect key crops such as maize, millet, sorghum, wheat, and rice. Studies reviewed indicate a significant negative impact of high temperatures on these crops, leading to reduced yields and increased food prices, which worsen food insecurity. Research have shown a direct correlation between higher temperatures and decreased maize, rice, millet, wheat and sorghum yields. This resulted in economic pressures on farmers and higher food prices for consumers, thereby impacting local economies and food security. To address these issues, the article emphasizes the need for comprehensive adaptation strategies. Key recommendations include the development and adoption of heat-resistant crop varieties, implementation of advanced water management technologies, and enhancement of agricultural support services to help farmers cope with temperature fluctuations. These measures are crucial for improving crop resilience and ensuring food security in the face of rising temperatures.

Keywords Rising temperature · Food security · Crop yields · Global warming

1 Introduction

In Nigeria, agriculture is a crucial sector that significantly contributes to economic growth and development. It employs over 70% of the population, thereby playing a substantial role in enhancing the Gross Domestic Product (GDP) [13]. This sector

K. M. Malabe (✉) · S. B. Mustapha
Department of Agricultural Extension Services, University of Maiduguri, Maiduguri, Nigeria
e-mail: kmalabemohammed@gmail.com

S. Al-Naemi et al. (eds.), *Water and Food Security in the Face of Climate Change: Challenges and Opportunities for Resilience*, World Sustainability Series,
https://doi.org/10.1007/978-3-032-00098-9_10

is vital for providing employment and sustaining livelihoods, particularly within the farming communities of Northern Nigeria [24]. However, agriculture is highly susceptible to climatic factors, especially extreme temperatures. Temperature fluctuations can have both positive and negative effects on crop yields [16]. In the Sahelian region of Northern Nigeria, characterized by a hot and semi-arid climate, crops require specific temperature ranges at various growth stages [1]. Extreme temperatures can adversely affect crop yields, leading to reduced food availability and challenges in food access, utilization, and stability [20]. Climate change, characterized by rising global temperatures, has emerged as one of the most pressing challenges of the twenty-first century [17]. Its effects penetrate various sectors, but none are as directly impacted as agriculture, which is sensitive to shifts in climatic conditions [16]. In Northern Nigeria, where agriculture is the primary livelihood for many and of great importance of the economy, the consequences of increasing temperatures are particularly critical [9]. This region, predominantly defined by its sahelian hot and semi-arid climates, faces unique challenges as global warming continues to intensify [8].

Rising temperatures are expected to exacerbate existing challenges, such as reduced crop yields and altered growing seasons. These changes threaten not only agricultural productivity but also food security [22]. Considering these dynamics is essential for policymakers, researchers, and practitioners working in the fields of agriculture and food security. This review will provide a comprehensive overview of the effects of climate change on Northern Nigeria's agricultural systems, offering insights into the challenges and effective response for enhancing food security against warming climate.

1.1 Conceptual Definition of Terms

Global warming: refers to a prolonged gradual increase in the overall temperature of the earth's atmosphere as a result of absorption of heat by greenhouse gases from sun's light, a process known as the greenhouse effect. This caused by increase in the levels of greenhouse gases such as carbon dioxide (CO_2), methane (CH_4), and nitrous oxide (N_2O). These gases in the earth's atmosphere, traps heat and prevent it from escaping into space. Consequently, global warming leads to a various climatic changes, such as excessive high temperatures, drought and floods.

Food Security: Food security refers to having a reliable supply of sufficient food quantities so that individuals can obtain and use it effectively to meet their nutritional requirements, along with ensuring consistent food availability. Adeyemo and Bakare [3] food security refers to the condition in which all individuals have reliable and consistent access to sufficient, safe, and nutritious food to meet their dietary needs and food preferences for an active and healthy life. It encompasses four key dimensions: availability, access, utilization, and stability.

Food crop Yields: refers to the quantity of harvested plants grown for human consumption, produced from a given area of land. This area can be quantified in multiple scales such as hectares, acres, or square meters, subject to the region and the measurement standard. According to FAO [11] crop yields refer to the amount of a particular crop produced per unit area of land. Conceptually, it measures the efficiency and productivity of agricultural practices in converting resources, such as land, water, and nutrients, into harvestable crops. High crop yields indicate successful farming practices and can result from factors like improved seed varieties, effective pest management, optimal soil conditions, and advanced agricultural techniques [2]. According to [18] low crop yields might signal issues such as poor soil health, inadequate irrigation, or adverse weather conditions. As stated by [19] crop yields are a key indicator of agricultural efficiency and sustainability.

Rising Temperatures: refers to the increase in average temperature extents of a particular area throughout a timeframe. Conceptually, rising temperature is often associated with global warming and climate change, where higher temperatures are observed due to various factors including increased concentrations of greenhouse gases such as carbon dioxide and methane in the atmosphere [6]. Rising temperatures can lead to a range of environmental impacts, including more frequent and severe heatwaves, altered weather patterns, melting ice caps, and rising sea levels [17]. It is a key indicator of the broader impacts of human activities on the climate system and has significant implications for ecosystems, agriculture, and human societies.

1.2 Empirical Review of Rising Temperature on Crop Yields and Food Security in Northeastern Nigeria

Northeastern Nigeria, characterized by its semi-arid to arid climate, is particularly vulnerable to the impacts of rising temperatures. As global temperatures continue to rise due to climate change, the region faces significant challenges related to agricultural productivity and food security. The region relies intensively on crops such as Maize, Millet, Sorghum, and Rice which are vulnerable to temperature changes. Rising temperatures have reduced yields by causing premature grain development and lowering overall crop quality [4]. Declines in crop yields due to rising temperatures directly impact food availability in many parts of northeastern region part of Nigeria. More than 70% of the population highly dependent on subsistence farming, and the reduced agricultural output led to food shortages and higher food prices, worsening food insecurity in the region currently [11]. The declining crop yields resulted in reduced income for farmers and higher food prices for consumers. These economic pressures forced local economies and contribute to wider economic uncertainty [23].

Haruna et al [12] small holder maize farmers' perceptions and adaptations to climate change in Bauchi state, Nigeria. Utilizing historical temperature data and maize yields records, the researchers employed statistical analysis techniques to identify correlations between temperature variability and maize yields. The findings from the study revealed a significant negative relationship between high temperature and maize yield which threatens food availability. This study stressed the need for mitigation and adaptation strategies to the impacts of temperature fluctuations in northeastern Nigeria.

Ezra et al. [10] examined the impact of rising temperature variability on rice yield in Taraba State, Nigeria. The statistical models and temperature data analysis techniques used in the study showed the relationship between high temperature and rice yields over a long period of time period. The results showed significant adverse effects of high temperatures on rice yields with adverse effect on food security socio-economic stability. The study emphasized the necessity of carrying out temperature-resilient agricultural practices and water management strategies to enhance rice production-control against extreme temperature conditions in the region.

Azare et al. [7] analyzed the effects of climate change on pearl millet production in Nigeria. Through integration of field surveys and temperature data analysis in the study, indicated the relationship between extreme temperature and millet yields. Results revealed that rising temperatures have negative significant impact millet yield with detrimental consequences for food security. The study highlighted the need for implementing climate-smart agricultural practices and investing in adaptive strategies to reduce adverse effect of high temperatures and enhance millet yield.

Lawrence and Elisha [15] investigate sorghum yield in Patiskum local government area of Yobe State, Nigeria. By means of field experiments and temperature modeling method, the research assessed the sensitivity of sorghum yields to high temperatures. Results revealed significant negative impact on sorghum yields with negative influence on food access. The study illustrated the imperative of actualizing mitigation and adaptation strategies and soil management practices to enhance sorghum resilience to high temperature impacts.

Akaangee et al. [5] investigated yield of wheat varieties in northern guinea savanna ecological zone of Taraba State, Nigeria. Using remote sensing data and crop modeling techniques, the research analyzed the effects of temperature changes on wheat yields over time. Findings revealed that rising temperatures negatively impacted wheat production, particularly in the northeastern regions of the country. The study highlighted the sensitivity of wheat farming to high temperature risks and this emphasized the need for putting adaptation measures in to practice such as heat-tolerant wheat varieties.

1.3 Dynamic Response Shifting Temperatures Paradigms

Heat Resistant Crop Varieties: Actively supporting and encouraging the idea of crop varieties that could adapt mechanism to tolerate heat stress is critical. Crops with as deep root system to access water deeper in the soil and heat-shock proteins to protect cellular functions. Growing temperature-resistant crop varieties is a key strategy to mitigate the effects of high temperature [25].

Water Optimization Technologies: putting water management into practices, such as enhanced water conservation methods and water treatment technologies, can protect ecosystem, enhance food security, ensure water quality, and build resilience. These water management methods improve crop yield in response to of rising temperatures [14].

Agricultural Service and Farmer Support: enhancing farmers' capacity to respond to temperature shifts through targeted training and information dissemination. Helping farmers navigate temperature fluctuations through expert guidance and training makes a substantial impact on increasing crop yields [21].

2 Conclusion

It is concluded that rising temperatures advisedly affected crop yields and food security in Northeastern part of Nigeria. Addressing these challenges requires a mixed approach that includes the development of heat-tolerant crop varieties, water management technologies, and enhance socioeconomic status of the farmers. Putting mitigation and adaptation strategies into practice simultaneously will be crucial in ensuring the region's agricultural productivity and food security against rising temperature.

References

1. Abdullahi Y, Idris M (2023) Temperature requirements for crop growth in the Sahelian region of Northern Nigeria. J Nigerian Agricu Sci 20(2):78–92
2. Adebayo AA, Ojo O (2021) Impact of improved seed varieties and pest management on crop yields in Nigeria. J Agric Res Dev 15(3):45–59
3. Adeyemo A, Bakare O (2021) Understanding food security: definitions and implications for Nigeria. J Nigerian Food Secur Stud 12(2):77–89
4. Agada LE, Yakubu MS (2023) Heat wave and its impact in Northeastern Nigeria: a case study of Yobe State, Nigeria. FUDMA J Sci 6(1):112–122
5. Akaangee MM, Gungula DT, Jakusko BB, Meer BB (2023) Evaluation of three varieties of wheat (*Triticum aestivum L.*) on yield and yield components in Jalingo Metropolis, Northern Guinea Savanna ecological zone of Taraba State, Nigeria. Nigerian Agric J 54(2):234–321
6. Akinyemi AR, Okoro EJ (2021) The role of carbon dioxide and methane in global warming: evidence from Nigerian climate data. Nigerian J Environ Sci Technol 23(3):234–250
7. Azare IM, Dantata IJ, Abdullahi MS, Adebayo AA, Aliyu M (2020) Effects of climate change on pearl millet (*Pennisetum glaucum* [L. R. Br.]) production in Nigeria. J Appl Sci Environ Manag 24(1):157–162
8. Bakare YA, Ogunleye AS, Kehinde AD (2019) Impact of microcredit access on climate change adaptation strategies adoption strategies and rice yield in Kwara State, Nigeria. Retrieved from https://doi.org/10.1016/j.wds.2023.100047
9. Bello AS, Ibrahim AA, Yakubu S (2023) Analysis on adaptation to climate change among rice farmers in Western Zone of Bauchi State, Nigeria. J Entrepreneurship Sustain Dev 6(1):20–23
10. Ezra A, Adebayo AA, Umar AS, Martins IK (2020) Effects of agro-climatic variables on yield of selected varieties of rice in Taraba State, Nigeria. Bayero J Pure Appl Sci 13(1):32–39

11. FAO (2021) The state of food and agriculture 2021: Making agri-food systems more resilient to shocks. Food and agriculture organization of the United Nations. https://openknowledge.fao.org/handle/20.500.14283/cb4476en
12. Haruna Y, Abubakar AI, Gambo M, Shugaba YA (2023) Small holder maize farmers' perceptions and adaptations to climate change Bauchi LGA, Bauchi State, Nigeria. J Agric Environ Sci Res 2(2):234–237
13. Ibrahim M, Olayemi JA (2023) The role of agriculture in Nigeria's economic growth and employment. Nigerian J Econ Dev 18(2):102–117
14. Jones RT, Lee MA (2023) Adaptive management strategies for enhancing crop yields under climate change. Agric Syst. https://doi.org/10.1016/j.agsy.2023.103123
15. Lawrence UE, Elisha I (2022) Relationship between rainfall variability and sorghum yield in Patiskum Local Government Area of Yobe State, Nigeria. J Geogr Reg Plann 15(1):10–17
16. Mohammed A, Bello S (2023) The effects of temperature fluctuations on crop yields in Nigeria. J Nigerian Clim Agric Stud 12(1):55–68
17. NASA (2023) Global climate change: vital signs of the planet. National Aeronautics and Space Administration. Retrieved from https://climate.nasa.gov
18. Nwankwo CP, Oke OA (2021) Assessing soil degradation and its impact on crop yields in Northern Nigeria. J Soil Environ Res 16(1):65–79
19. Ogundele TO, Aliu O (2023) Crop yields as indicators of agricultural efficiency and sustainability in Nigeria. Nigerian J Agric Econ 20(1):88–102
20. Okonkwo C, Nwosu A (2022) Climate change and its impact on agricultural productivity in Nigeria: a case study of extreme temperature effects. J Nigerian Agric Sci 58(3):45–58
21. Smith JA, Brown LK (2022) Improving agricultural resilience: training and information for climate adaptation. J Agric Sci Technol 15(3):45–58
22. Tahir S (2018) Flood destroys over 120,000 hectares of farmlands in Jigawa. Independent Newspaper. http://www.independent.ng/floodsdestroys-over-120000-hectares-of-farmlands-inJigaw. 26th Nov 2018
23. Tunde A (2022) Economic impacts of declining crop yields on Nigerian farmers. J Nigerian Agric Econ 15(3):45–59
24. Usman S, Ahmed R (2022) The impact of agriculture on employment and livelihoods in Northern Nigeria. J Nigerian Rural Dev Stud 10(1):30–45
25. Zhang Y, Patel SR (2022) Breeding temperature-resistant crops: strategies and impacts on agricultural resilience. J Clim Agric 18(4):789–805

Combining Multisource Data to Understand the Decline in Small Pelagic Fishing Along the Algerian Coast

Fouzia Houma, Lamia Ali, Harid Roumaissa, and Nour El Islam Bachari

Abstract Sardines and other small pelagic were part of the daily meal of Algerian citizens in coastal towns. These quantities fished are low and therefore the price of sardines is no longer accessible to thousands of families. Each year explanations are given, for example the unsuitability of fishing techniques, the effect of the climate, the construction of several dams and the effects of global warming. Global warming is felt in Algeria by a long drought, forest fires, lack of rain, and increase in temperatures. It is interesting to understand the mechanism that links the production of small pelagic with environmental factors measured by satellite Sea Surface Temperature (SST) and Chlorophyl (Chl-α) and ecological index. Fishing statistics (FS) are considered the control parameter of sardine abundance in the Algerian coast. We conclude that the spatial variability of FS agrees with the spatial variability of diatom/ dinoflagells ecological index and inversely with surface temperature. We correlated the statistics of small pelagic (FS) with environmental factors Sea Surface Temperature (SST) and chlorophyll (Chl-α). FS (Sardina pilchardus), is strong related to SST ($r^2 = 0.84$) and moderate related to chlorophyll ($r^2 = 0.43$) and slightly stronger with the two parameters together (r2 = 0.72).

Keywords Fishing statistics · Surface temperature · Chlorophylle · Ecological index

F. Houma (✉) · H. Roumaissa
Marine and Coastal Ecosystems Laboratory (ECOSYSMarL), National Higher School of Marine Sciences and Coastal Management (ENSSMAL), Algiers, Algeria
e-mail: fouzia.houma.bachari@enssmal.edu.dz

L. Ali · N. E. I. Bachari
Laboratory of Biological Oceanography and the Marine Environment (LOBEM), University of Science and Technology of Houari Boumedien (USTHB), Algiers, Algeria

S. Al-Naemi et al. (eds.), *Water and Food Security in the Face of Climate Change: Challenges and Opportunities for Resilience*, World Sustainability Series,
https://doi.org/10.1007/978-3-032-00098-9_11

1 Introduction

The functioning of the marine environment is a physicochemical process combined with geological, climatic and dynamic factors that allow ecosystems to evolve. The nutrient-phytoplankton-zooplankton-fish trophic chain is a natural process operating at all spatiotemporal scales (Fig. 1).

Phytoplankton plays an important role in this chain and any dysfunction of these particles has consequences on the distribution of fish [1]. Satellite observations currently constitute an important contribution in view of the large space oceanography observed a repeated manner over time. Satellites do not observe species directly and even fish but allow the observation of the color and transparency of marine waters. The radiation on its sun-sea-satellite optical path depends on the charge of marine waters by particles suspended in the water. The concentration of chlorophyll in the water influences the color of sea water. Since the 1980s, researchers have implemented algorithms to calculate Chl-α concentrations depending on the sensor used [2–4]. The MODIS AQUA series given the spectra used, the repeatability of observation and the spatial resolution has allowed researchers and oceanographers to carry out almost continuous monitoring of Chl-α [2–4]. Among the satellites that provide synoptic for the entire Algerian coast with two observations per day that the MODIS satellite. The problem with optical observations is the non-observation in the presence of clouds. Harid et al. were able to put an algorithm to assimilate the missing data for the Mediterranean [5, 6]. By using adequate corrections and calibration were able to provide an important time series of Chl-α distribution along the Algerian coast. Using official fishing statistics, it is remarkable the presence of a negative West–East gradient of small pelagic along the Algerian coast on the one hand and a significant decline over the last five years. The last five years have been marked by a drop in rainfall which has a direct consequence of the drying of the wadis and therefore no supply of terrestrial nutrients to coastal waters.

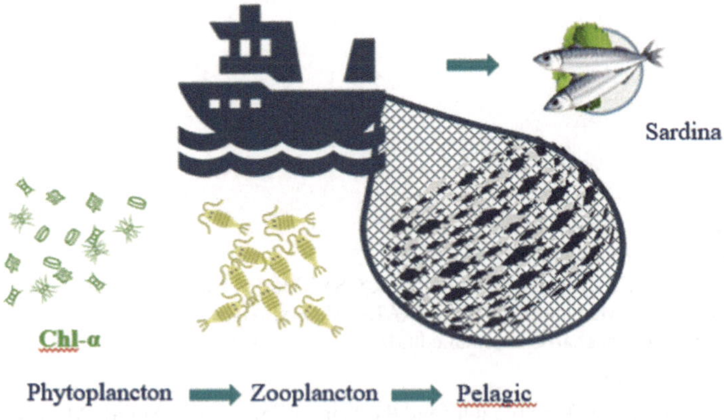

Fig. 1 Trophic chain

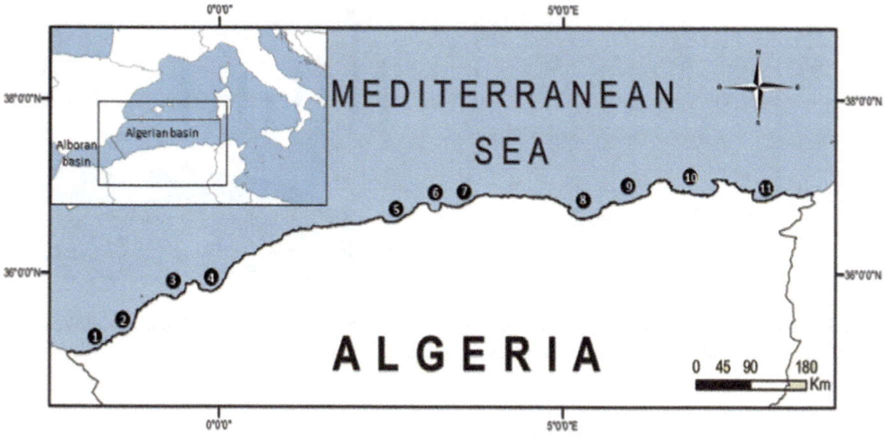

Fig. 2 Localization measure station

In this work we are interested in answering the following questions:

1. Can we explain the West–East abundance gradient of small pelagic fish?
2. Can variations in Chl distribution be explained by variations in environmental factors (SST; O_2; current)?
3. What are the correlations between fishing statistics and environmental factors?

2 Materials and Method

2.1 Study Area

Located in the north of the African continent, Algeria, has a maritime coastline of more than 1822 km, equivalent to approximately 3.96% of the 46,000 km of the Mediterranean coast line (Fig. 2). According to the division of the Mediterranean basins, the Algerian coast has been subdivided into two sub-basins, the South-East (S.E.) of the Alboran sub-basin from 2°W to 0°5′E (Algerian-Moroccan borders) and the Algerian sub-basin from 0°5′E to 8°40′E (Algerian borders). The Algerian coast is characterized by its latitudinal shape and consequently the Coriolis forces are constant on the one hand and on the other hand by a very weak tide even negligible compared to the effect of barometric forces.

2.2 Data Set

To make this research we use multi-source data taken from different dates with different spatial and temporal resolutions.

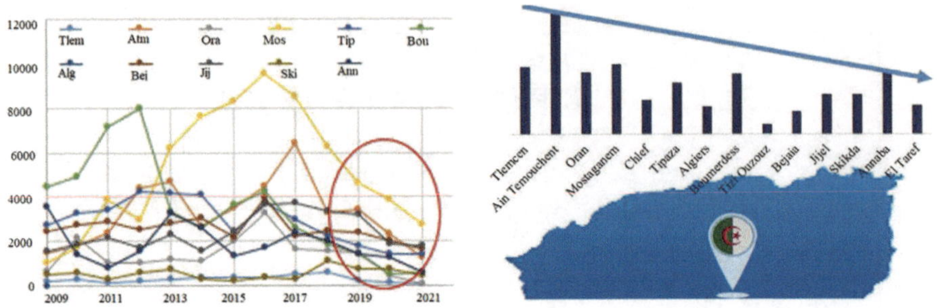

Fig. 3 Spatiotemporal variation of fishing statistics along the Algerian coast. We note the West–East fishing gradient and the decline of fisheries for the last five years

2.2.1 Fishing Statistical Data (Fs)

The fishing statistics are official data from the Ministry of Fisheries given in our case for 14 official stations. Small pelagic constitute 75% of fishing along the Algerian coast. In Fig. 3 we present a histogram of the cumulative fishing of small pelagic on the one hand and on the other hand the curves of fishing statistics according to the year.

2.2.2 Campaign Data

Since 2014, oceanographic surveys have been carried out each year by the national research and aquaculture center (CNRDPA). We analyzed the samples in the laboratory in order to calculate the Chl concentrations and classify the species into two large classes diatoms and dinoflagellates. We present the results in three classes West, Center and East. Physicochemical parameters: temperature, salinity, pH, dissolved oxygen, chlorophyll-a and MES were sampled also and measured [7].

2.2.3 Satellite Data

In this work we used the MODIS sensor (Moderate Resolution Imaging Spectroradiometer, NASA) onboard the AQUA satellite AQUA: Launched in May 2002 at 705 km of altitude, with a spatial resolution of 1 km. The MODIS satellite allows two observations of the Algerian coast per day. It is most suitable for spatio-temporal mapping of chlorophyll (Chl-α) concentrations and surface temperature (SST) (Fig. 4).

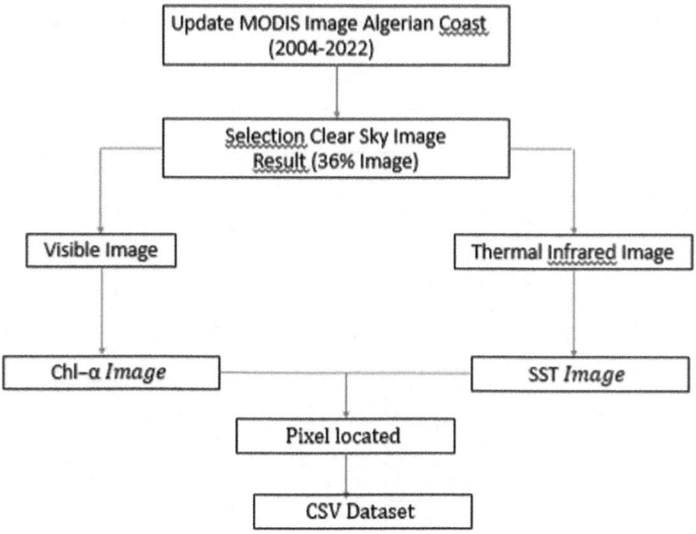

Fig. 4 Transformation of MODIS images into Chl and SST data base

2.3 Method

For the statistical analysis we used XLSTAT and for the mapping the QGIS. SeaDAS (SeaWiFS Data Analysis System): is the software used to the processing and analysis satellite images in this study. It is an open-source software, which has been developed by NASA to view, process and control the quality of all products from SeaWiFS. It also allows viewing of ocean products MODIS.

3 Results and Discussion

In our analysis instead of working for each station we will group the stations into three classes West-Centre-East. The western zone brings together the stations (1–4), the central zone the stations (5–7) and the eastern zone the stations (8–11) in Fig. 1

3.1 Chl-α and SST Variation Along the Algerian Coast

Table 1 show the Chl-α concentrations (mg.m^{-3}) and the SST measurements (°C) by basin 5 km from the coast.

Chlorophyll concentrations follow seasonal variations in all three sectors. In the West sector, Chl-α concentrations do not exceed 0.9 mg.m^{-3} from November to

Table 1 Average values of the Chl-α concentrations ($mg.m^{-3}$) and the SST measurements (°C) by basin at 5 km

Chl-α			
Sector/Month	West	Centre	East
Jan	0.626	0.63	0.61
Feb	0.677	0.687	0.784
Mar	0.863	0.768	0.649
Apr	0.425	0.518	0.392
May	0.303	0.342	0.22
Jun	0.215	0.246	0.149
July	0.156	0.314	0.121
Aug	0.139	0.22	0.127
Sept	0.193	0.342	0.16
Oct	0.19	0.21	0.176
Nov	0.385	0.57	0.433
Dec	0.591	0.67	0.607
Annual mean	0.397	0.46	0.369

SST (°C)			
Sector/Month	West	Centre	East
Jan	14.51	14.74	14.27
Feb	13.44	14.18	13.87
Mar	13.97	14.29	13.76
Apr	15.11	15.27	16.5
May	16.9	17.45	17.76
Jun	19.76	20.06	20.78
July	21.82	22.25	23.01
Aug	23.02	23.97	25.13
Sept	22.34	23.18	24.38
Oct	20.52	21.6	21.93
Nov	17.51	18.21	19.22
Dec	15.23	15.31	16.11
Annual mean	17.84	18.37	18.89

April and are less than 0.4 mg.m^{-3} from May to October. In the Centre sector, Chl-α concentrations vary from 1 mg between 0.5 and 0.8.m^{-3} from November to April and less than 0.4 mg.m^{-3} from May to October. Chl-α concentrations in the Eastern sector vary between 0.4 and 0.8 mg.m^{-3} from November to March and less than 0.4 mg.m^{-3} from April to October. At 15 km from the coast, we notice that all three sectors have similar variations in Chl-α concentrations [8].

The SST also follows seasonal variations in the three sectors, with warmer and warmer periods moving from west to east. Indeed, in the western sector, the SST does not exceed 15.2 °C from December to April, and it varies between 16.9 and 23 °C from May to November. In the central sector, the SST does not exceed 15.3 °C from December to April and varies between 17.5 and 24 °C from May to November. The SST varies between 13.8 and 16 °C from December to March in the Eastern sector and between 16.5 and 25 °C from April to November. We note a decreasing West–East gradient of Chl-α with a peak in the central sector, and there is an increasing West to East gradient in SST values, with a difference of 1 °C between the western and eastern sectors.

The gradient in fishing statistics is similar to the temperature gradient. This study allowed us to explain that the abundance of sardines is more influenced by temperature than by chlorophyll.

3.2 Observation and Identification of Phytoplankton

Observations of the phytoplankton settling ponds were made with an inverted OPTKA type microscope. After counting and grouping the stations by region we calculate the diatom/dinoflagells index (dia/dino) [9]. Changes in the dia/dino ratio can be of great importance for the nutrition of zooplankton and subsequent trophic levels. They even influence the benthos, as diatom blooms descend quickly and provide more food for the zoobenthos than dinoflagellates, which remain in the water column longer. Diatom dominance over dinoflagellate dominance is typical in the spring. Therefore, diatom dominance, i.e., standard Dia/Dino index > 0.5, is typical for a non-impacted ecosystem and is considered a good status (Wasmund 2017). The dia/dino index results are put in Table 2

In the Western sector, the index is higher than 0.5, indicating the dominance of diatoms (Dia/Dino index = 0.65). In the central sector, the index is below the threshold (0.4), indicating the dominance of dinoflagellates. On the other hand, the index is much lower than the threshold (0.15) in the Eastern sector, indicating the exclusive dominance of diatoms. The index study shows that the ecological status in the Western sector is good, and it is less good in the Central sector and much less

Table 2 Ecological index

Sector	West	Centre	East
Dia/Dino index	0.65	0.40	0.15

in the Eastern sector. The fishing statistics gradient is similar to the Dia/Dino index gradient. Consequently, the ecological quality of marine waters plays an important role in the abundance of small pelagic.

3.3 Correlative Analysis

The study of the interaction between small pelagic and the abiotic parameters of their environment is a non-negligible factor. Indeed, chlorophyll pigments (Chl-α) concentration is a proxy to indicate phytoplankton biomass, while Sea Surface Temperature (SST) explains the environment conducive to marine productivity. Studying these parameters helps to understand and explain the local marine environment and the availability of favorable conditions in an ecosystem to establish a good strategy for exploiting fishery resources. It could analyze ecological health based on indicators using chlorophyll and sea surface temperature and fish catch data. Several studies have shown the importance of these parameters in the distribution, growth, and abundance of small pelagic species [10, 11]. We summarize the correlative analysis between fishing statistics and environmental factors in Table 3

The Sardine is very strongly related to SST ($r^2 = 0.84$) and moderately related to Chl-α ($r^2 = 0.43$). There is no doubt that abiotic factors play a major role in the distribution of sardines which live in the water column with a high concentration between 35 and 60 m deptht Azmi et al. [11]. Furthermore, Sabates et al. [12] demonstrated the close relationship between abundance and distribution of round sardinella and water temperature in the western Mediterranean. Fach [13] obtained the same result in the eastern Mediterranean for the same species.

The main factor is temperature and therefore any variation in temperature directly influences the abundance of sardines. The last five years have seen an impressive increase in SST in the Mediterranean, which has been noted in the quantity fished during this period. This study allowed us to explain the impact of global warming and marine pollution on the abundance of small pelagic. A deregulation of SST, a lack of rain and pollution by wadis contribute negatively to this resource rich in vitamin omega. This result is expected given that the abundance of FS is significant in summer and autumn. In perspective we try to integrate other parameters such as salinity, oxygen and sea current [14–16].

Table 3 Regression analysis

Parameters	Equation	R^2
T°C	$SF_S = 0.0002 \times Exp(0.576 \times SST)$	0.84
Chl	$SF_S = 601.23 \times Chl^{-2.045}$	0.43
SST;Chl	$SF_S = -31368 - 1215 \times Chl + 1982.6 \times SST$	0.72

4 Conclusion

- The study of biotic/abiotic interactions is complex and requires the integration of all physicochemical and hydrographic parameters. Based on fishing statistics for small pelagics (sardines), we were interested in explaining its spatial variability and decline over the last five years.
- The West–East spatial gradient of FS is similar to the West–East gradient of the Diatom/Dinoflagellate ecological index. The abundance of small pelagics depends on the ecology of the water. The abundance of small pelagics is influenced by marine pollution.
- The West–East spatial gradient of FS is inverse with the West–East gradient of water surface temperature (SST). The warmer the water, the lower the abundance. This result explains the decline in FS for the last five years which presents overheated sea water.
- The correlative analysis between the FS and the abiotic factors gave us a strong correlation between the FS and the two parameters temperature and chlorophyll. The dependence is more remarkable with temperature.

References

1. FAO (2018) Impacts of climate change on fisheries and aquaculture: synthesis of current knowledge, adaptation and mitigation options. FAO Fisheries and Aquaculture Technical Paper No. 627. Rome
2. Gilbert NP, Humphreys TJ, Kattawar, George W (1978) Color of the ocean. Appl Opt **17**:1432–1446
3. O'Reilly JE, Maritorena S, Mitchell BG, Siegel DA, Carder KL, Garver SA, Kahru M, McClain C (2015) Ocean color chlorophyll algorithms for SeaWiFS. J Geophys Res Ocean 103(C11):24937–24953
4. Piroddi C, Coll M, Liquete C et al (2017) Historical changes of the Mediterranean Sea ecosystem: modelling the role and impact of primary productivity and fisheries changes over time. Sci Rep 7:44491
5. Harid R, Demarcq H, Keraghel M-A, Ait-Kaci M, Zerrouki M, Bachari N-E-I, Houma F (2022) Spatio-temporal variability of a chlorophyll-a based biomass index and influence of coastal sources of enrichment in the Algerian Basin. Cont Shelf Res 232:104629
6. Abdallah KW, Harid R, Demarcq H et al (2024) Retrieval of micro-phytoplankton density using Sentinel-3 and MODIS satellite sensors on the eastern Algerian coast. Thalassas 40:285–297
7. Aminot A, Kérouel R (2004) Hydrologie des écosystèmes marins: Paramètres et analyses. Ifremer, Brest
8. Wasmund N (2017) Recruitment of bloom-forming cyanobacteria from winter/spring populations in the Baltic Sea verified by a mesocosm approach. Boreal Environ Res 22:445–455
9. Godhe A, Asplund ME, Härnström K, Saravanan V, Tyagi A, Karunasagar I (2008) Quantification of diatom and dinoflagellate biomasses in coastal marine seawater samples by real-time PCR. Appl Environ Microbiol 74(23):7174–7182
10. Suhartono N, Muzzneena M, Tukimat L (2013) The relationship between sea surface temperature and chlorophyll-a concentration in fisheries aggregation area in the archipelagic waters of Spermonde using satellite images. In: AIP conference proceedings, vol 1571, No 1. American Institute of Physics, pp 466–472

11. Azmi S, Agarwadkar Y, Bhattacharya M, Apte M, Inamdar A (2015) Indicator based ecological health analysis using chlorophyll and sea surface temperature along with fish catch data off Mumbai coast. Turk J Fish Aquat Sci 15(4):923–930
12. Sabatés ANA, Martín P, Lloret J, Raya V (2006) Sea warming and fish distribution: the case of the small pelagic fish, Sardinella aurita, in the western Mediterranean. Glob Change Biol 12(11):2209–2219
13. Fach BA (2014) Modeling the influence of hydrodynamic processes on anchovy distribution and connectivity in the black sea. Turk J Fish Aquat Sc 14
14. Firad B, Bachari NEI, Amarouche K, Ghoulam AK, Salim L (2021) Temporal analysis of the distribution of thynnus thunnus in relation to environmental factors along the Algerian coast. Egypt J Aquat Biol Fish 25(5):553–571
15. Ali L, Bachari NEI (2021) Spatio-temporal variation of sea surface temperature (SST) and marine chlorophyll concentrations along the algerian coast (se alboran sub-basin and algerian sub-basin). Analele Universităţii din Oradea, Fascicula Biologie Tom 28(2):184–193
16. Moranta J, Stefanescu C, Massuti E, Morales-Nin B, Lloris D (1998) Fish community structure and depth-related trends on the continental slope of the Balearic Islands (Algerian basin, western Mediterranean). Mar Ecol Prog Ser 171:247–259

Securing West Africa Water Supply for Food Security: Strategic Actions Against Climate Change

Peter Sanjo Adewale, Hakeem A. Ajonbadi, Anthony Kola-Olusanya, Florence Adeoti Yusuf, Yusuf Shogbesan, and Oluwaseun Oyebisi Odetola

Abstract Climate change (CC) is a global crisis increasingly threatening the existence of humans and rapidly changing West Africa's water landscape. Therefore, this study investigated the impact of CC on water supply and food security in the West African region. The phenomenology research method, which involves in-depth interviews, focus group discussion, observation, narrative and content analysis, was used to collect data from purposively selected farmers, traditional leaders, community leaders, water managers and agricultural extension officers in West Africa. Data were analysed using thematic analysis, hermeneutic analysis, coding and categorisation. The result shows that the CC effect on the water supply directly endangers the region's food security and livelihood. Unpredictable rainfall, drought, soaring temperatures, and increasingly severe weather events are depleting vital water resources essential for agriculture and domestic purposes. These disruptions threaten crop yields, degrade soil quality, cause pest invasion and disease spread and reduce water availability for irrigation, escalating the risk of widespread food shortages. To combat these threats, it is imperative to implement adaptive water management, boost irrigation efficiency, and accelerate the adoption of drought-resistant crops. Policymakers and farmers must prioritise climate-resistant crop cultivation in West Africa. By taking decisive action now, West Africa can fortify its food systems against the mounting challenges of CC.

Keywords Climate change · Water resources · Food security · Drought · Water supply · West Africa

---

P. S. Adewale (✉) · A. Kola-Olusanya · F. A. Yusuf · Y. Shogbesan · O. O. Odetola
Osun State University, Osogbo, Nigeria
e-mail: peter.adewale@uniosun.edu.ng

H. A. Ajonbadi
University of Science and Technology, Doha, Qatar

S. Al-Naemi et al. (eds.), *Water and Food Security in the Face of Climate Change: Challenges and Opportunities for Resilience*, World Sustainability Series,
https://doi.org/10.1007/978-3-032-00098-9_12

1 Introduction

West Africa is home to about 444 million people, accounting for 5.4% of the world population and about 6.5 million square kilometres [1]. However, this region has only 2% of the world's renewable water resources, which is also unevenly distributed across the region, with some areas experiencing severe water scarcity and drought simultaneously [2]. This has spread diseases and food insecurity across the region, making life difficult. Stressing the extreme poverty level of West African countries [3] affirmed that the average annual per capita income of countries in the West Africa region was less than $ 1000 US dollars in 2007, with about three-fifths of the people living on less than one dollar a day. The per capita water availability in the region has decreased by more than 40% since 1980 [4] explained the contributing factors for the decline in food production in the region. He argued that the structural adjustment programme staged in the region some years back was responsible for low food production. However, [5] believed that environmental changes such as rising temperatures, rising sea levels, changing precipitation patterns, and increased evaporation reduce water availability.

Scholars have stressed that climate change (CC) accounts for over 20% of global water scarcity, significantly impacting food production and security. Rising temperatures are warming faster than the global average, making lakes and rivers dry [7]. It increases the evaporation rate, reducing water availability and growing demand for farm irrigation. The reoccurring droughts and floods reduced the quantity and quality of available water, respectively [8]. Bhaga et al. [9] concluded that sea-level rise could cause coastal erosion into freshwater sources, limiting the availability of potable water, loss of arable land, damage to buildings, roads and bridges and increased flooding. Similarly, [10] stressed that sea-level rise has also negatively affected saltwater intrusion into fresh water.

Water scarcity in Africa causes the spread of diseases and affects agricultural productivity, industry, and domestic use. This has increased poverty and food insecurity for the people. Arfasa et al. [6] predicted that by 2025, more than 1.8 billion people will face water scarcity globally. The resultant effects may include contamination of freshwater sources, harm to agriculture and aquaculture, increased risk of waterborne diseases and economic loss from industries that rely on freshwater for their production [11]. The poverty level of developing countries in Africa could be responsible for their increased vulnerability to CC impact. These consequences include food insecurity, loss of livelihood, water scarcity, increased poverty, migration and displacement, especially among the youth.

About 40% of the West African population lacks access to safe water, while not less than 60% of arable land is rain-fed and vulnerable to climate variability. Water scarcity and issues in West Africa cost less than 5% of the region's GDP annually. Mitigating CC's impact on water in West Africa will enhance agricultural resilience, thereby achieving United Nations Sustainable Development Goals 2, 6 and 13 of promoting food security, improved livelihoods and climate resilience, respectively [12]. Dembélé et al. [13] opined that mitigating CC through digital agriculture could increase food production, ensure food security, reduce greenhouse gas emissions, and improve soil health and water efficiency.

There is a need to develop regional strategic action against CC impact and secure the region's water supply for improved food security. Therefore, this study explored strategies for securing West Africa's water supply for food security in the face of CC. The specific objective of this study is to assess the impact of CC on West Africa's water resources and food security. Identify adaptive strategies for water management and irrigation efficiency and investigate the potential of drought-resistant crops in enhancing food security in West Africa. The following research questions guided the study. How does CC affect water resources and food supply? What adaptive strategies can West Africa adopt to enhance water management and irrigation efficiency? What policy and institutional reforms can mitigate CC in agriculture in West Africa?

2 Methodology

The phenomenology research method was adopted to investigate strategies for securing West Africa's water supply for food security in the context of CC. The technique helps explore human experiences and perspectives [14]. This method is helpful in social and health sciences because it captures live experiences in the first person, allowing researchers to examine how individuals make meaning of the experiences of respondents' realities.

2.1 Study Area

This study was conducted in the West Africa region of Africa. West Africa is made up of countries. Following the United Nations definition, West Africa is a region of African that is made up of the following countries Nigeria, Burkina Faso, Benin, Cape Verde, The Gambia, Togo, Ivory Coast, Guinea, Niger, Guinea-Bissau, Liberia, Mali, Mauritania, Sierra Leone, Senegal, and Ghana. Nigeria is the most populous country out of all these countries.

The West African region is geographically located in the west of Africa continent. The region spans about 5.1 km^2 and is made up of 16 counties (Benin, Burkina Faso, Cape Verde, The Gambia, Ghana, Guinea, Guinea-Bissau, Ivory Coast, Liberia, Mali, Mauritania, Niger, Nigeria, Senegal, Sierra Leone, and Togo) recognised by United Nations. The region has diverse landscapes comprising savannas, deserts, forests and coastal areas. There were about four significant rivers in the region, namely: River Volta, River Gambia, River Niger and River Senegal. Adamawa Plateau, Guinea Highlands and Fouta Djallon are the major mountain ranges in the region. The region is highly vulnerable to CC impacts due to the poor economy and people's standard of living [15]. Rising temperatures, unpredictable rainfall patterns, sea-level rise and increased drought frequency, especially in the Sahel region, are the significant climate variables that threaten the livelihood of the citizens of the countries in the region.

Moreover, more than 40% of the people in West Africa lack access to safe and quality water [16]. The decreasing water availability in the region is primarily due to CC, over-extraction of groundwater, and pollution of water bodies from agricultural runoff and industrial activities [17]. Food security is an important issue that threatens the existence of more than 30% of the region's population. There has been a continuous decrease in crop yield primarily due to CC-related issues. Major countries that are mainly vulnerable to CC impact in the region include Mali, Burkina Faso, Niger, and Mauritania are highly susceptible to drought. At the same time, the coastal parts of Nigeria, Ghana, Cape Verde and Senegal are vulnerable to flooding and erosion.

2.2 Sampling and Sampling Techniques

The purposive sampling method was used to select farmers, traditional leaders, community leaders, water managers and agricultural extension officers across the region. These people represent diverse stakeholders and possess specialised knowledge and experience relevant to the area's CC, water management, and agriculture. Purposive sampling allows researchers to identify and access hard-to-reach populations such as rural farmers and traditional leaders in this region. The method also enables the selection of information-rich cases that provide in-depth insights into CC's impact on the area. The participants in this study were selected based on their roles and expertise in farming, water management, CC impacts and roles in their communities.

2.3 Research Instrument

The instruments used for this study include in-depth interviews, focus group discussions, observation, narrative, and content analysis. In-depth interviews were conducted with policymakers, local farmers, local community leaders, traditional leaders, and water managers in the region. Focus group discussions were conducted with farmers and local community members. Moreover, key informant interviews were conducted with CC experts as well as experts in water management. In addition, observational studies of water management practices and agricultural activities in Nigeria were conducted.

2.4 Data Analysis

Data collected were analysed using thematic analysis of patterns and themes in interviews and focus group discussions conducted with farmers, community leaders, traditional leaders, and water managers in the region. Hermeneutic analysis was used to interpret and understand the meaning of text responses and the respondents' experiences. The responses were coded and categorised.

3 Results

3.1 Impact of CC on West Africa's Water Resources and Food Security

Climate variables in the West Africa region continuously threaten water availability and quality. As reported by most respondents, some variables include changes in precipitation patterns and reduced rainfall, which affect plant growth and agricultural productivity. It was observed that CC alters precipitation patterns and reduces rainfall, increasing water evaporation from reservoirs and lakes. It increases the risk of drought floods and decreases river flows and lake levels in most West African countries. Specifically, it reduced the recharge rates of groundwater. It increases salinisation due to sea-level rise and contamination of groundwater due to flooding. CC also increases water pollution from agricultural runoff and industrial activities. It changes water pH levels and temperature, reducing the quality of water. Moreover, across West Africa, CC decreased water availability for irrigation and drinking, increasing competition for water resources.

Most respondents believed that rising temperatures threaten their countries' water supply. They reported that rising temperature increases evaporation, producing insufficient water for plant growth. Most respondents reported the impact of the threat of CC on the water supply, which included reduced crop yields and productivity, increased pest and disease pressure, and shifts in growing seasons and planting dates. Others mentioned an increased risk of water-borne diseases, droughts, and floods experienced in Ghana and the Benin Republic due to the impact of CC on the water supply. Most respondents agreed that there had been a decrease in livestock productivity and increased food prices, as well as food insecurity observed in their countries due to CC's impact on water supply.

Most respondents from Nigeria agreed that reduced rainfall affected agriculture in the northern part of the country, with increased flood risk in the south. For respondents from Ghana, droughts impact maize and cocoa production, while respondents from Senegal claimed that rising sea levels threaten their country's coastal agriculture and fishing communities. Increased temperature and reduced rainfall affect grain and legume production in Mali, while droughts and floods impact cotton and maize production in Burkina Faso. Farmers from Cote d'Ivoire agreed that CC affected their cocoa production. Landslides and floods were the significant impacts of CC observed in Guinea, while severe floods and droughts affected rice and cassava production in Sierra Leone. Rubber and cocoa production were impacted in Liberia, and it was observed that CC threatens coastal agriculture and fishing communities in The Gambia.

The majority of respondents from the northern part of Nigeria reported that CC had badly hit Lake Chad. The same story was told by respondents from Ghana, who expressed concern about the impact of CC on Volta Lake and its river systems. The coastal aquifers threatened by sea-level rise in Senegal were one of the major concerns of the residents of the country. Mali has experienced reduced rainfall in the past few years, which has affected their inland deltas. As reported by respondents, the consequences of the reduced water supply due to CC include food insecurity, increased risk of water-borne diseases, and economic losses in agriculture, energy and industry. Others reported social tensions and conflict over water resources. Addressing CC's impacts on water supply in the region requires concerted efforts from government agencies, civil society, local communities and assistance from international organisations.

3.2 Adaptive Strategies for Water Management and Irrigation Efficiency

Water supply issues in West Africa were mainly due to CC, agricultural expansion and rapid population growth in the region. Respondents who were farmers agreed that mulching reduces soil evaporation and retains moisture. Others decided that choosing drought-tolerant and water-efficient crops could reduce water challenges on their farms, as such crops used less water. Specifically, over the years, Ghana, Senegal and Burkina Faso have developed a vital institution for irrigation of their farms.

Residents of countries in West Africa use traditional strategies to ensure water availability for domestic and agricultural purposes. Some techniques they put in place include collecting rainwater and storage for potable and non-potable uses. Most respondents dug boreholes around their homes and farms to access groundwater for domestic and farm uses. Clay pots, concrete tanks, plastic tanks, and other containers were used to store water during rainy and dry seasons. It was observed that in some farms, flood water and wastewater were used for irrigation. Others used water from canals, dams, and waterwheels to irrigation vegetable farms in Nigeria and Ghana. According to most respondents, residents tried to resolve water issues in their areas by protecting and restoring watersheds to maintain their quality. Others treat and reuse wastewater for irrigation and industrial purposes, while some use flood water to irrigate their farmland.

Moreover, some farms use modern water management strategies such as drip irrigation and solar-powered water pumps as renewable energy sources for pumping water. Modern water treatment plants were found in Lagos, Nigeria, and other urban centres in most countries under study. This strategy was adopted to improve water quality for urban and rural communities. In addition, efforts were made to implement conservation agricultural practices in some of these countries, such as Ghana and Nigeria. The Benin Republic established an organic farm at Porto-Novo, where various conservation agricultural practices were implemented. Like some of the other countries in West Africa, Nigeria has adopted innovative water management technology. From field study, precision agriculture was observed in some parts of the region. At the trial stage, modern farmers now use drones and sensors in water, herbicides and pesticide applications.

The majority of the countries under the study faced many challenges that limited their efforts in mitigating CC's impact on water availability and quality. Some of the difficulties observed include financial constraints. The current inflation in Nigeria and other West African countries has badly affected their economies. The purchasing power of many households was meagre, and they had limited financial resources. Climate variability includes increasing temperature, rising sea levels and flooding, and compounded water availability and quality in most regions. While most of the farmers in the developing countries of West Africa had the limited technical capacity required to run and manage modern water management facilities, others were faced with inadequate infrastructure for irrigation.

4 Discussion

Increasing temperature, sea-level rise, increased evaporation, flooding and droughts continuously jeopardize water availability and quality across West Africa. CC has led to decreased precipitation, especially in the Sahel region, while increasing evaporation rates due to increased temperature diminishes the quantity of water in reservoirs and lakes across the region [18]. Rising temperatures and reduced rainfall also decreased groundwater recharge rates, while sea-level rise contributes to the salinization of freshwater resources in coastal areas [19, 20]. The impact of CC is also felt in water quality degradation, as agricultural runoff during heavy rainfall alters water pH levels. The increasing demand for irrigation due to low water supply and availability has heightened competition for water resources in both rural and urban areas in West Africa. This result aligned with the findings of [18].

Respondents identified various impacts of CC on water supply and agricultural productivity, including reduced crop yields, heightened pest pressures, and evaporation rates. Rising temperatures increase evaporation and reduce water availability for irrigation and crop growth, especially in the Sahel region, where irrigation demand has tripled [18]. The decline in precipitation and higher temperatures decreased surface runoff and groundwater recharge [21]. Warmer temperature is also expected to increase pest survival and reproduction rates, increasing crop pressures and decreased agricultural output [18]. Changes in rainfall patterns and rising temperatures are increasingly shifting planting and growing seasons, making planning for the farming complex more difficult [22]. Climate variables also affect livestock productivity, as increased temperature kills poultry [23].

West Africa, particularly Ghana, Nigeria and Mali, faced socio-economic and water scarcity challenges occasioned by CC. The result of this study underscores the urgent need for collaborative efforts to address these issues. Lake Chad in Nigeria was reportedly severely impacted by CC, with the result showing that there is acute water scarcity in the region, with metrics indicating that some areas had less than a total of 100 m^3/capital/year, leading to social tensions and conflicts [24]. The case in Ghana around the Volta Lake reflects a similar trend of water resources crisis coupled with rapid population growth leading to food insecurity, increased water-borne diseases and economic losses. This result aligns with the findings of [24, 25].

To ameliorate the impacts of CC on agriculture and food supply, farmers recognised and adopted practices like mulching and selection of drought-resistant crops on their farms [26]. Most farmers in the region rely heavily on rain to get water for their crops [27]. However, Burkina Faso, Senegal and Ghana farmers have established robust irrigation systems for farming purposes. With the success of irrigation systems in some of the countries in the region, especially in Ghana, irrigation has positively impacted farm revenue and food security [28]. However, the irrigation coverage remains low in the region compared to available arable land [28]. Mulching and using drought-resistant crops to conserve moisture and other sustainable practices adopted by farmers have encouraged community resilience in the region [28]. Although some success has been recorded, there are still pockets of water supply crisis in the region occasioned by ineffective government policies [29].

Respondents also highlighted that residents engaged in traditional water harvesting, borehole digging, and water storage using various containers to ameliorate CC's impact on water supply [18, 30, 31]. Other farmers used floodwater and wastewater as adaptative strategies to cope with water scarcity and irrigation of their vegetable farms [30, 32]. Watershed and wetland protection was an essential strategy for water supply in the region practised by communities [18].

Despite successes recorded in the efforts of communities, farmers and other stakeholders in mitigating the water crisis in West Africa, there remain multifaceted challenges facing West African countries in mitigating CC's impact on water supply in the region. Inadequate infrastructure, climate variability and poor economy were the significant challenges facing the region's water supply. This result resonates with the findings of other researchers like [31, 32]. Many countries in West Africa could not invest in water supply due to poor economy and inflation. Other constraints include household and farmers' low purchasing power, lack of public awareness about water management, and market failures, which hinder agricultural growth [33, 34]. Drought, floods, and farmers' poor technical skills in utilising available modern water management technologies like drones have also been identified as significant challenges facing water management issues in the West Africa region [35–38]

5 Recommendations

The findings of this study call for a comprehensive and coordinated approach to address the challenges posed by climate change on water supply and food security in West Africa. West African farmers are urged to adopt adaptive water management practices, enhance irrigation efficiency, and accelerate the adoption of drought-resistant crops. [1] indicates that such measures are critical to mitigating the effects of rising temperatures and erratic rainfall patterns on both water availability and crop yields, as evidenced by studies highlighting increased evaporation and decreased groundwater recharge in the region. However, it is essential to recognise that while these practices hold significant potential, their broader implementation is often hindered by economic constraints and limited technical capacity among smallholder farmers, a challenge also noted in previous research [4].

Integrating climate literacy into the region's strategic framework is not merely an educational exercise but a fundamental pillar for building resilience against the existential threat of climate change. This study underscores the severe impacts of erratic rainfall, drought, and flooding on water resources and food security in West Africa. Enhancing climate literacy among policymakers, community leaders, and farmers can empower these stakeholders to understand the complex dynamics of climate change, its far-reaching effects on water quality and agricultural productivity, and the necessity of adaptive strategies. Such education would facilitate more informed decision-making and promote the widespread adoption of innovative practices, ranging from adaptive water management to cultivating drought-resistant crops [27]. A climate-literate society not only comprehends the science behind climate variability but also appreciates the socio-economic implications of environmental degradation. By incorporating climate literacy programmes into national curricula, community outreach initiatives, and professional training workshops, governments can create a knowledgeable populace better equipped to engage with and implement adaptive strategies. This approach could also help mitigate the fragmented and delayed responses noted in the study, fostering a culture of proactive engagement rather than reactive crisis management. In essence, bolstering climate literacy is a critical investment in the region's future. This investment enhances community resilience, promotes sustainable water and food security practices, and ultimately strengthens the region's capacity to withstand and adapt to the escalating challenges of climate change.

Policymakers must, therefore, prioritise cultivating climate-resistant crops and support the development of climate-resilient agricultural practices. Evidence from Ghana, Nigeria, and Mali suggests that aligning governmental policies with scientific recommendations can foster sustainable water use and improve food security. Beyond policy development, governments need to provide the necessary financial and technical support to implement modern irrigation technologies and other adaptive strategies. This support is crucial for overcoming the infrastructural and economic barriers that currently limit the scalability of effective water management systems, as documented in the literature [16].

The urgency of the climate crisis in West Africa necessitates immediate and collaborative action at both national and regional levels. Regional cooperation especially among the 16 West African nations is vital for sharing knowledge, pooling resources, and developing cohesive strategies to manage transboundary water challenges [19]. Adaptation programs should specifically target vulnerable communities, ensuring that those most affected by climate change receive prioritised support. Moreover, the role of researchers is paramount; further studies assessing the impacts of climate change on specific crops and agricultural systems are needed to refine climate-resilient water management models and to inform evidence-based policy and practice [28].

In summary, a multi-pronged strategy that integrates adaptive water management, supportive government policies, regional cooperation, and continued research is essential to fortify West Africa's food systems against the mounting threats of climate change. While the existing literature supports these recommendations, their successful implementation will depend on addressing the underlying economic and infrastructural challenges that currently impede progress.

6 Conclusion

The intertwined challenges of climate change, water scarcity, and food insecurity in West Africa present an urgent existential crisis that demands immediate, uncompromising action. The evidence is clear: unpredictable rainfall, pervasive drought, and destructive flooding are not isolated events but integral elements of a rapidly evolving climate system undermining water quality and agricultural productivity across the region. Despite some localised efforts in adaptive water management and adopting drought-resistant crops, these measures are grossly inadequate in the face of such overwhelming challenges.

This crisis is compounded by the fragmented and delayed responses from national governments and regional bodies, whose policies have failed to keep pace with the accelerating impacts of climate change. The situation in countries like Nigeria, Ghana, Cape Verde, Mali, and others is critical, as their water resources—already stretched thin—are being further depleted and degraded. West African nations must abandon incremental, piecemeal strategies in favour of bold, coordinated regional initiatives. Such initiatives must encompass significant investment in water infrastructure, integrating advanced irrigation technologies, and rapidly scaling up successful adaptive practices.

Without a concerted and assertive response, the dual threats to water supply and food security will not only erode the livelihoods of millions. Still, they will also destabilise the region's broader socio-economic and ecological balance. West African leaders and policymakers must prioritise regional cooperation and innovative solutions to safeguard their communities and secure a sustainable future. The time for cautious optimism has long passed; what is needed now is decisive, radical reform to combat the existential threat of climate change.

References

1. Aboah M (2024) WASH levels and associated human health risks in war-prone West African countries: a global indicators study. Environ Health Insights 18. https://doi.org/10.1177/11786302241228427
2. Ackah I, Graham E (2020) Meeting the targets of the Paris agreement: an analysis of Renewable energy (RE) governance systems in West Africa (WA). Clean Technol Environ Policy 23(2):501–507. https://doi.org/10.1007/s10098-020-01960-6
3. Akpoti K, Groen T, Dossou-Yovo E, Kabo-Bah AT, Zwart SJ (2022) Climate change-induced reduction in agricultural land suitability of West Africa's inland valley landscapes. Agric Syst 200:103429. https://doi.org/10.1016/j.agsy.2022.103429(2022)
4. Akudugu MA, Millar KK, Akuriba MA (2021) The livelihoods impacts of irrigation in Western Africa: the Ghana experience. Sustainability 13(10):5677. https://doi.org/10.3390/su13105677
5. Antognazza MR (2024) Phenomenology. In: Oxford University Press eBooks, pp 111–125. https://doi.org/10.1093/oso/9780198831518.003.0007
6. Arfasa GF, Owusu-Sekyere E, Doke DA, Ampofo JA (2024) Impacts of climate and land use on the sustainability of irrigation water in West Africa: a systematic review of evidence. All Earth 36(1):1–13. https://doi.org/10.1080/27669645.2024.2308371
7. Baah-Kumi B, Ward FA (2020) Poverty mitigation through optimised water development and use: insights from the Volta Basin. J Hydrol 582:124548. https://doi.org/10.1016/j.jhydrol.2020.124548
8. Balasundram SK, Shamshiri RR, Sridhara S, Rizan N (2023) The role of digital agriculture in mitigating climate change and ensuring food security: an overview. Sustainability 15(6):5325. https://doi.org/10.3390/su15065325
9. Bhaga TD, Dube T, Shekede MD, Shoko C (2020) Impacts of climate variability and drought on surface water resources in sub-Saharan Africa using remote sensing: a review. Remote Sens 12(24):4184. https://doi.org/10.3390/rs12244184
10. Billa SA, Fogwe ZN (2023) Stakeholder's domestic water harvesting infrastructure: resultant conflicts and management strategies in the Upper Noun Valley, North West region of Cameroon. J Geogr Environ Earth Sci Int 27(7):57–68. https://doi.org/10.9734/jgeesi/2023/v27i7696
11. Büyükcangaz H, Alhassan M, Harris JN (2017) Modernized irrigation technologies in West Africa. Turk J Agric Food Sci Technol 5(12):1524–1527. https://doi.org/10.24925/turjaf.v5i12.1524-1527.1429
12. Cantelon JA, Guimond JA, Robinson CE, Michael HA, Kurylyk BL (2022) A review: vertical Saltwater intrusion in coastal aquifers driven by episodic flooding. Water Resour Res 58(11). https://doi.org/10.1029/2022wr032614
13. Dembélé M, Zwart S, Ceperley N, Mariéthoz G, Schaefli B (2020) Multivariate and spatially calibrated hydrological model for assessing climate change impacts on hydrological processes in West Africa. EGU General Assembly 2020. https://doi.org/10.5194/egusphere-egu2020-9143
14. Fabian PS, Kwon H, Vithanage M, Lee J (2023) Modelling, challenges, and strategies for understanding impacts of climate extremes (droughts and floods) on water quality in Asia: a review. Environ Res 225:115617. https://doi.org/10.1016/j.envres.2023.115617
15. Griggs G, Reguero BG (2021) Coastal adaptation to climate change and Sea-Level rise. Water 13(16):2151. https://doi.org/10.3390/w13162151
16. Henchiri M, Zhang J, Li S, Essifi B, Wilson K (2024) Comprehensive assessment of drought vulnerability and resilience over North and West Africa during 1980–2100. Agric Water Manag 296:108804. https://doi.org/10.1016/j.agwat.2024.108804
17. Henrietta UU, Fani DCR, Ngo NV, Emmanuel ON, Odularu G (2020) Crop and livestock production responses to rainfall and temperature variation in West Africa. In: Springer eBooks, pp 13–28. https://doi.org/10.1007/978-3-030-47875-9_2
18. Hudgens J, Trillo R, Calonnec N (2003) The rough guide to West Africa. Rough Guides

19. Hussain S, Sharif U, Hadeed MZ, Raza A, Aman MK, Safdar M, Habib N, Munir M (2024) Recent approaches for mitigating climate change in agriculture. In: Advances in environmental engineering and green technologies book series, pp 114–132. https://doi.org/10.4018/979-8-3693-3807-0.ch010
20. Iizumi T, Iseki K, Ikazaki K, Sakai T, Shiogama H, Imada Y, Batieno BJ (2024) Increasing heavy rainfall events and associated excessive soil water threaten a protein-source legume in dry environments of West Africa. Agric For Meteorol 344:109783. https://doi.org/10.1016/j.agrformet.2023.109783
21. Johnson J, Becker M, Dossou-Yovo ER, Saito K (2023) Farmers' perception and management of water scarcity in irrigated rice-based systems in dry climatic zones of West Africa. Agron Sustain Dev 43(2). https://doi.org/10.1007/s13593-023-00878-9
22. Kargbo JM (2005) Impacts of monetary and macroeconomic factors on food prices in West Africa. Agrekon 44(2):205–224. https://doi.org/10.1080/03031853.2005.9523710
23. Katic PG, Namara RE, Hope L, Owusu E, Fujii H (2013) Rice and irrigation in West Africa: achieving food security with agricultural water management strategies. Water Resour Econ 1:75–92. https://doi.org/10.1016/j.wre.2013.03.001
24. Mahato A, Upadhyay S, Sharma D (2022) Global water scarcity due to climate change and its conservation strategies with special reference to India: a review. Plant Arch 64–69. https://doi.org/10.51470/plantarchives.2022.v22.no1.009
25. Merem EC, Twumasi Y, Wesley J, Isokpehi P, Fageir S, Crisler M, Romorno C, Hines A, Ochai GS, Leggett S, Nwagboso E (2021) Appraising variations in climate change parameters along the lower West African region. J Saf Eng 7(1):1–22. https://doi.org/10.5923/j.safety.20180701.01
26. Merem EC, Twumasi Y, Wesley J, Olagbegi D, Crisler M, Romorno C, Alsarari M, Isokpehi P, Hines A, Hirse G, Ochai GS, Nwagboso E, Fageir S, Leggett S, Offiah J, Emeakpor S (2022) Assessing issues in water resources use within countries in West Africa. Int J Ecosyst 12(1):1–19. https://doi.org/10.5923/j.ije.20221201.01
27. Muccione V, Ewen T, Vaghcfi SA (2025) A scoping review on climate change education. PLOS Clim 4(1):e0000356. https://doi.org/10.1371/journal.pclm.0000356
28. Nkiaka E, Bryant RG, Kom Z (2024) Understanding links between water scarcity in the Sahel and Lake Chad Basin using the Water Footprint and carbon concept. Earth S Futur 12(2). https://doi.org/10.1029/2023ef004013
29. Ojo S, Mensah H, Albrecht E, Ibrahim B (2020) Adaptation to climate change Effects on water resources: understanding institutional barriers in Nigeria. Climate 8(11):134. https://doi.org/10.3390/cli8110134
30. Pienaah CKA, Batung E, Saaka SA, Mohammed K, Luginaah I (1944) Early warnings and perceived climate change preparedness among smallholder farmers in the Upper West region of Ghana. Land 12(10):2023. https://doi.org/10.3390/land12101944
31. Sedegah DD, Agyekum MW, Kyeremeh B Local-level water conservation and management practices in rural communities: a case study in Dormaa municipality, Bono region, Ghana. Irrig Drain 72(4):1095–1108. https://doi.org/10.1002/ird.2849(2023)
32. Semenza JC (2020) Cascading risks of waterborne diseases from climate change. Nat Immunol 21(5):484–487. https://doi.org/10.1038/s41590-020-0631-7
33. Sesay A, Omotosho M (2011) The politics of regional integration in West Africa
34. Shvili J (2017) West African countries. WorldAtlas. https://www.worldatlas.com/geography/west-african-countries.html
35. Song F, Zhang GJ, Ramanathan V, Leung LR (2022) Trends in surface equivalent potential temperature: a more comprehensive metric for global warming and weather extremes. Proc Natl Acad Sci 119(6):e2117832119
36. Tilahun ZA, Bizuneh YK, Mekonnen AG (2023) The impacts of climate change on hydrological processes of Gilgel Gibe catchment, southwest Ethiopia. PLoS ONE 18(6):e0287314. https://doi.org/10.1371/journal.pone.0287314
37. Veettil AV, Mishra AK (2018) Potential influence of climate and anthropogenic variables on water security using blue and green water scarcity, Falkenmark index, and freshwater provision indicator. J Environ Manage 228:346–362. https://doi.org/10.1016/j.jenvman.2018.09.012

38. Yira Y, Diekkrüger B, Steup G, Bossa AY (2017) Impact of climate change on hydrological conditions in a tropical West African catchment using an ensemble of climate simulations. Hydrol Earth Syst Sci 21(4):2143–2161. https://doi.org/10.5194/hess-21-2143-2017

Modeling and Simulation of Rainfall Prediction in Climate Data Using LSTM, GRU, and N-BEATS: A Comparative Analysis

Murad A. Yaghi, Huthaifa Al-Omari, and Shahed AlNawaiseh

Abstract Accurate climate modeling and simulation become essential for predicting and mitigating extreme climate events. Rainfall prediction holds significant importance due to its direct effect on droughts and floods. In this research, we have applied a recent deep learning model called (N-BEATS) in order to predict rainfall in time series data, then we compared the results with state-of-the-art traditional methods, namely LSTM and GRU. Our analytical study compared these algorithms using many performance metrics for multiple window sizes (5, 10 & 20). Our analytical study proved the superiority of the N-BEATS architecture in predicting future rainfall with much faster training time, and more accuracy.

Keywords Climate change · N-BEATS · LSTM · GRU · Rainfall prediction · Deep learning · Neural networks · Modeling and simulation

1 Introduction

Artificial intelligence (AI) is being used by a growing number of scientists to handle massive amounts of data, especially when analyzing climate data that requires a large amount of processing capacity, this due to the huge quantity and complexity of data. Many sensors around the world sending data simultaneously and these big and enormous data needs very high computing power to analyze, the data collected from sensors and other sources provides a wealth of knowledge essential to understanding climate dynamics, predictions patterns, and guiding decision-making. However, climate modeling and simulation are considered among the most computationally demanding tasks that usually needs supercomputers for accurate prediction. Therefore, many scientists around the world started to use artificial intelligence especially deep learning and neural networks techniques to gain accurate predictions while minimizing the computation power needed by such models. The

M. A. Yaghi (✉) · H. Al-Omari · S. AlNawaiseh
School of Computing and Informatics, Al Hussein Technical University, Amman, Jordan
e-mail: Murad.Yaghi@htu.edu.jo

S. Al-Naemi et al. (eds.), *Water and Food Security in the Face of Climate Change: Challenges and Opportunities for Resilience*, World Sustainability Series,
https://doi.org/10.1007/978-3-032-00098-9_13

type of data associated with climate modeling is usually time series data [1]. This research contributes to the ongoing efforts in climate and environmental modeling by providing insights into the strengths and limitations of these models in the context of rainfall prediction. The findings underscore the importance of selecting appropriate modeling techniques and window sizes based on the specific requirements of climate models, ultimately aiding in the development of more accurate and reliable environmental predictions. Recent advances in the application of machine learning (ML) for climate modeling have significantly enhanced the ability to predict climate variability and extreme weather events. The research in [2] highlighted the effectiveness of supervised learning techniques, such as Random Forests and neural networks, in predicting severe weather and droughts. They also emphasized the role of unsupervised learning approaches, including deep neural networks and Gaussian processes, in uncovering unknown patterns in climate data, which contributes to improving global climate predictions. In terms of model-specific developments, the study in [3] demonstrated that LSTM (Long Short-Term Memory) and Temporal Convolutional Networks (TCN) significantly outperform traditional models like ARIMA, WRF, and SVR in weather forecasting. These models excel in capturing long-term dependencies in sequential data, making them superior for both short-term and long-term predictions. These advancements are particularly critical in forecasting extreme weather events, where traditional numerical models are often less accurate. In [4], the scientists utilized LSTM models to predict temperature data. Their application of unsupervised learning methods to extract latent climate patterns contributed to the model achieving high accuracy in climate forecasts, demonstrating the potential of deep learning in managing large and complex climate datasets. Moreover, the study in [5] introduced hybrid models combining machine learning with traditional climate models to address systemic biases and improve the overall accuracy of climate predictions. Rainfall forecasting has also seen significant improvements with the application of neural network models. The analytical study in [6] compared NN-TensorFlow and NN-MPCA models with conventional methods, showing that the neural network models were more reliable for seasonal rainfall predictions, particularly in critical seasons like summer and winter. Additionally, the research in [7] demonstrated that LSTM models are more effective in long-term rainfall predictions compared to the North Atlantic Oscillation (NAO) index, further showcasing the versatility of deep learning models in climate-related tasks. These studies underscore the growing impact of machine learning techniques in enhancing climate forecasting capabilities, making them indispensable in the current era of climate science.

The rest of paper is organized as follows: Sect. 2 investigates the architecture of the three models used for this analysis. Section 3 includes experimental results and analysis for applying these methods on real data and shows the superiority of applying N-BEATS model over traditional methods. Finally, Sect. 4 summarizes the results and outcomes from this analytical study.

2 Methodology

2.1 The Dataset

The dataset utilized in this study known as "weatherAUS," comprises daily weather observations from various Australian locations, including Sydney [8]. It encompassing daily meteorological data over an extensive period from 2008 to 2017. This dataset consists of 145,460 entries, each capturing essential weather parameters that can significantly impact weather forecasting and climate analysis. This dataset is sourced from the Australian Commonwealth Bureau of Meteorology (BOM) and is available through the R package 'rattle'. The data includes parameters such as temperature, rainfall, wind speed, humidity, and atmospheric pressure.

2.2 N-BEATS Model

The N-BEATS (Neural Basis Expansion Analysis Time Series) architecture is an advanced approach to time series forecasting that reformats the problem as a nonlinear multiple regression. In the core of N-BEATS model, there is a fully connected nonlinear regressor, which processes time series history in order to predict multiple points simultaneously [9]. N-BEATS architecture relies on basic building blocks that are stacked together using residual connections. This stacking architecture allows the model to handle deep models with up to 100 layers. This deep stacking technique, often referred to as "backcasting," is very useful for handling long time series data [10]. Each layer consists of a linear regression followed by ReLU nonlinear activation function, this approach simplifies the model conceptually and in terms of implementation [11]. Each BasicBlock in the N-BEATS model generates two outputs: forecast and backcast. These blocks contain fully connected layers, followed by two linear layers, which compute forward and backward coefficients for prediction. The output of one block becomes the input for the next, allowing the model to refine its predictions as more layers are added [12]. The forecast (y_t) and backcast (x_t) values are calculated by summing the contributions from each BasicBlock, combined via base layers as demonstrated by the following equations:

$$y_t = \sum_{i=1}^{dim(k_f)} k_{f,i} y_{f,i} \tag{1}$$

$$x_t = \sum_{i=1}^{dim(k_b)} k_{b,i} v_{b,i} \tag{2}$$

The residual connections between blocks ensure that each block concentrates on unexplained aspects of the input, which is supposed to enhance prediction accuracy

[13]. Before applying these computations in the N-BEATS model, it is preferred that input data is normalized to have a consistent scaling for all features in the dataset. However, it is important that after having the final predictions in the N-BEATS model to rescale the data back to its original form.

2.3 LSTM (Long Short-Term Memory)

LSTM is another type of neural network that is designed to handle time series data. It uses some type of gates to control the flow of information between the cells. It has mainly three types of gates, and each gate has specific role in deciding which information to keep or pass to the other cells. It is specifically useful for time series data where there are time dependencies between the samples presented in the dataset. The input gate is responsible for deciding which information to keep. Output gate is deciding which information to pass to other cells. The forget gate decides which information to discard. A Recurrent Neural Network (RNN) adds a looping structure to a regular neural network. In regular neural networks, all layers are fully connected, but the nodes within the same layer don't connect. This setup doesn't work well for sequence data because it can't capture the order of information. In RNNs, however, nodes in the same layer are connected, so the output considers both the current input and the previous state, allowing RNNs to remember information. However, RNNs often struggle to learn long-term patterns due to issues like exploding or vanishing gradients during training. To fix this, Hochreiter and Schmidhuber (1997) introduced the Long Short-Term Memory (LSTM) network. LSTM uses special units with three "gates" that control the flow of information, deciding what to keep or discard. These gates use a sigmoid function, which outputs values between 0 and 1, determining how much information passes through [14]. The calculations inside each LSTM cell can be depicted with the following equations:

(a) Candidate cell state:

$$\widetilde{C}_t = \tanh\left(W_C \cdot \left[h_{t-1}, x_t\right] + b_C\right) \quad (3)$$

(b) Cell state:

$$C_t = f_t * C_{t-1} + i_t * \widetilde{C}_t \quad (4)$$

(c) Final output:

$$h_t = o_t * \tanh(C_t) \quad (5)$$

where (i_t) represent the input term, (f_t) represent the forget term, and (o_t) represent the output term [15].

2.4 GRU (Gated Recurrent Unit)

GRU is another form of neural networks that is specialized for handling time series data. It has two main gates, reset and update. By that, it is considered simpler than LSTM model, but proved to have comparable performance, and in some situations, it could outperform LSTM models [16]. The GRU operations performed in each cell can be depicted by the following equations:

(a) Reset gate:

$$r_t = \sigma\left(W_r \cdot \left[h_{\{t-1\}}, x_t\right] + b_r\right) \tag{6}$$

(b) Update gate:

$$z_t = \sigma\left(W_z \cdot \left[h_{t-1}, x_t\right] + b_z\right) \tag{7}$$

(c) Current memory:

$$\widetilde{h_t} = \tanh\left(W_h \cdot \left[r_t * h_{t-1}, x_t\right] + b_h\right) \tag{8}$$

(d) Final output:

$$h_t = (1 - z_t) * h_{t-1} + z_t * \widetilde{h_t} \tag{9}$$

GRU's simpler structure makes it more efficient than LSTM for many time series tasks while still being more effective [17].

2.5 Evaluation Metrics

In this study, the performance of the three predictive models have been evaluated using three key metrics: Mean Absolute Error (MAE), Root Mean Squared Error (RMSE) and Mean Absolute Percentage Error (MAPE). MAE is used to calculate the average of the absolute differences between predicted and actual values, which aid in providing a value of the overall error magnitude without severely penalizing large errors. This metric is preferred when all errors are having the same importance. RMSE takes the square root of the average of squared differences between predicted and actual values, so larger errors contain more significant value for RMSE than MAE. This particularly can be useful when smaller errors are less significant than larger errors. Lastly, MAPE calculates the average percentage difference between the predicted and actual values, which provides some insight into the model's performance in relative terms. This metric is especially useful when evaluating predictions across different scales, where it presents error values as percentages, making it easier to interpret in practical terms [18–20].

3 Experimental Results

This section presents a detailed analysis for each model. Firstly, the data set has been normalized to scale each feature equally, and then divided into training and testing parts. After that, we have trained each of the three model on the training set.

Figure 1 shows how the loss function is related to the epoch number during training for LSTM model.

As shown in the chart above, the model's loss has decreased during training for all window sizes. However, the loss value for all the three windows didn't get below 30 during the training process. This indicates the limitations of LSTM model in detecting long time dependences in time series data (Fig. 1).

For GRU model, as shown in the chart above, the model's loss decreases slowly. Generally, higher window sizes achieved better accuracy. However, the model couldn't achieve a loss value less than 20 for all window sizes during the training process (Fig. 2).

Finally, the N-BEATS model, as depicted in the chart above, proved to have excellent performance. The loss has been decreased for all window sizes much faster than GRU and LSTM especially for higher window sizes (Fig. 3).

After training the models, each one of them has been tested using the test portion of the dataset. Table 1 Shows the testing results for each model using MAE, RMSE and MAPE as performance metrics.

Testing results shown that generally LSTM and GRU achieved comparable accuracy on training data, while N-BEATS outperformed them in each aspect. Surprisingly N-BEATS achieved the best result for a window size of 20, indicating the superiority of this model in handling long time dependencies in the data.

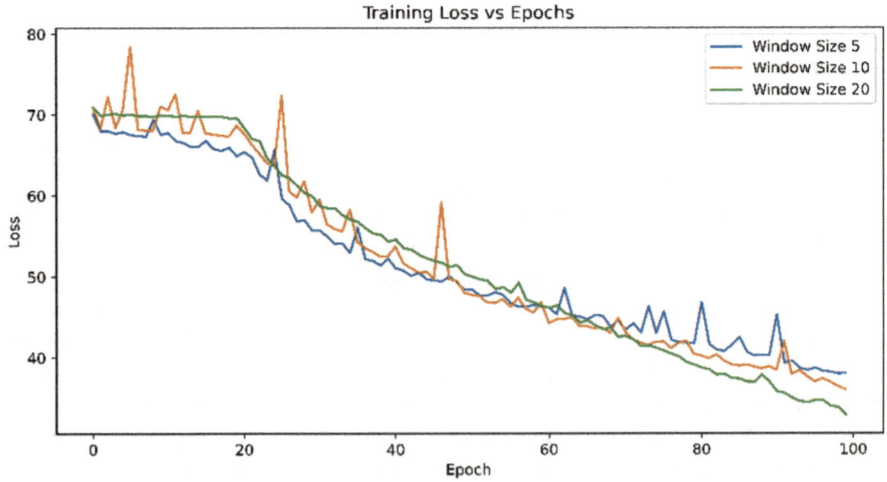

Fig. 1 Epochs versus loss for LSTM model

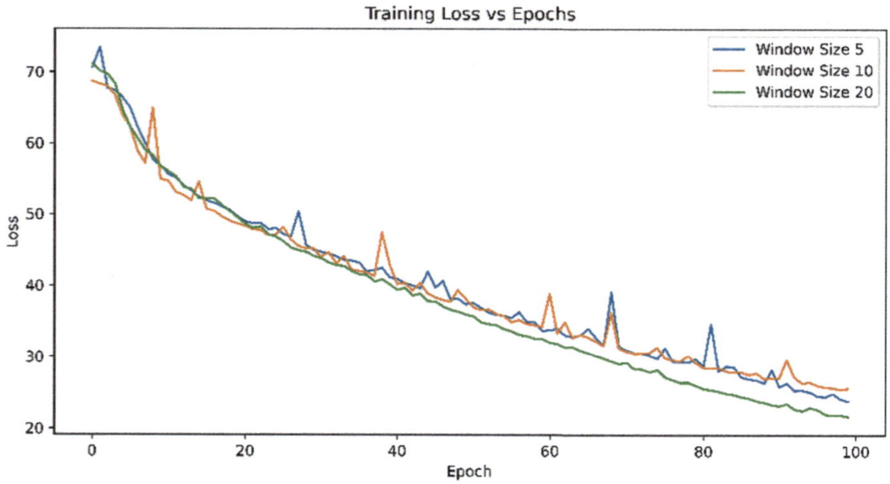

Fig. 2 Epochs versus loss for GRU model

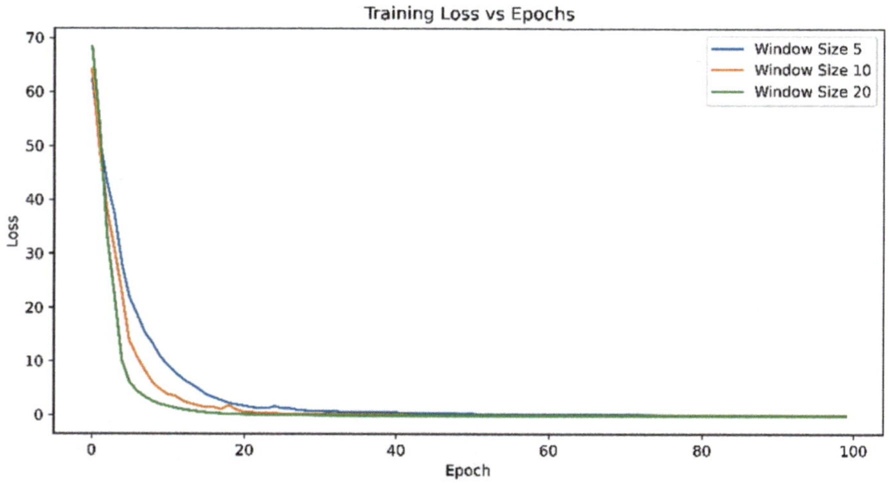

Fig. 3 Epochs versus loss for N-BEATS model

4 Discussion

The comparison plots of RMSE, MAE, and MAPE for different models (LSTM, GRU, and N-BEATS) across various window sizes (5, 10, and 20) provide significant insights into model performance and their suitability for time series forecasting tasks. The RMSE plot in Fig. 4 indicates that the N-BEATS model has better accuracy than both LSTM and GRU. This can be justified by the special architecture of the N-BEATS model that has fully connected layers at each cell, while featuring backward

Table 1 Performance results for LSTM, GRU and N-BEATS

Model	window size	RMSE	MAE	MAPE
LSTM	5	7.0228	3.2849	4.4734
LSTM	10	7.4188	3.4380	4.6918
LSTM	20	7.6811	3.3012	4.0419
GRU	5	7.8405	3.3018	4.4193
GRU	10	8.3951	3.5701	3.7802
GRU	20	8.6148	3.6377	4.2807
N-BEATS	5	7.0605	2.9249	3.1618
N-BEATS	10	7.2906	3.1353	3.0507
N-BEATS	20	6.3617	2.7786	2.9131

residual links. This complex architecture proved to have better accuracy in capturing hidden pattern and dependencies than the simple recurrent architectures of LSTM and GRU. The difference in accuracy between these models becomes more notable when increasing the window size, where the gap between N-BEATS and the other models increases.

When analyzing the MAE plot in Fig. 5, it has been noticed that N-BEATS has also achieved the lowest MAE, proving its reliability in producing more accurate forecasts

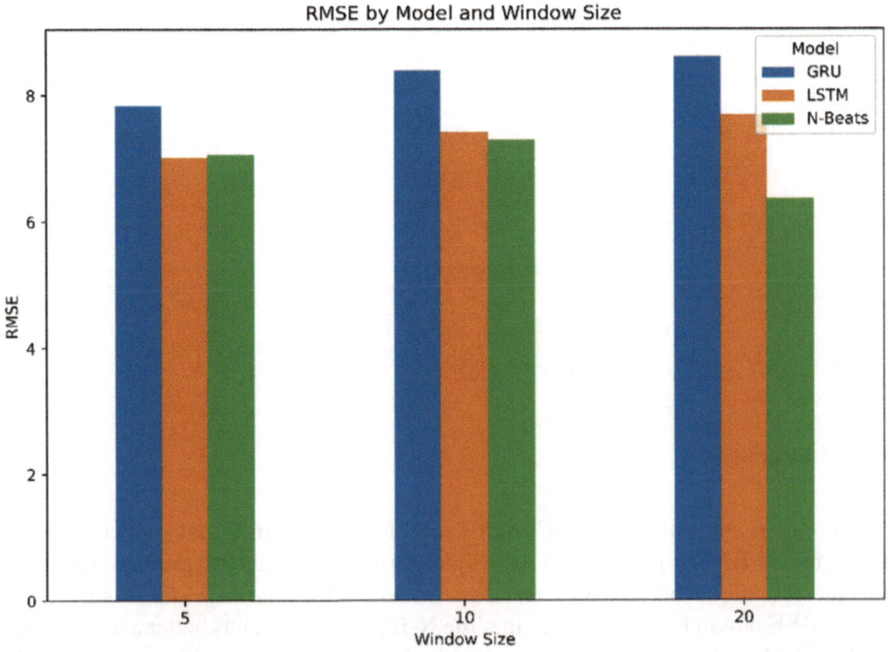

Fig. 4 RMSE comparisons for GRU, LSTM and N-BEATS

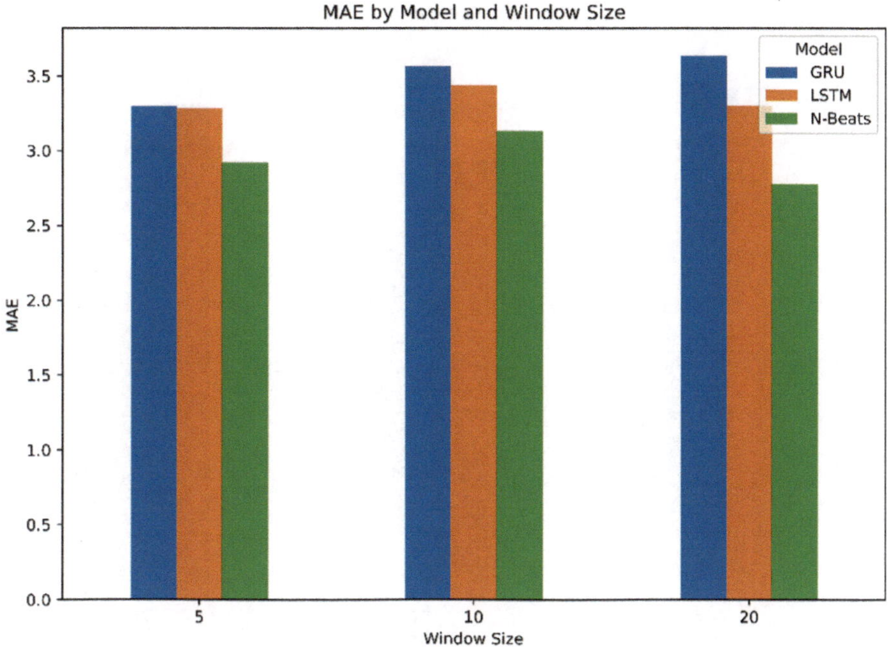

Fig. 5 MAE comparisons for GRU, LSTM and N-BEATS

with minimal error at different window sizes allowing it to capture complex patterns more effectively than GRU and LSTM. This also proved that N-BEATS has better generalization capabilities than the other models. On the other hand, LSTM showed higher MAE value than N-BEATS but generally lower than GRU. This shows that LSTM can also be reliable especially for small window sizes, but this reliability decreases compared to N-BEATS when window sizes becomes greater.

MAPE is particularly useful for assessing variability in the data. It expresses errors as percentages, reflecting the relative performance across different scales. As shown in Fig. 6 N-BEATS had the lowest MAPE value that indicates better accuracy than LSTM and GRU models in handling variability in the data set. LSTM and GRU have approximately similar MAPE values for small window sizes. While the difference between them varies for larger window sizes. GRU is better for window size of 10, while LSTM performed better when the window size increased to 20. N-BEATS performed better than all of the other architectures for all window sizes.

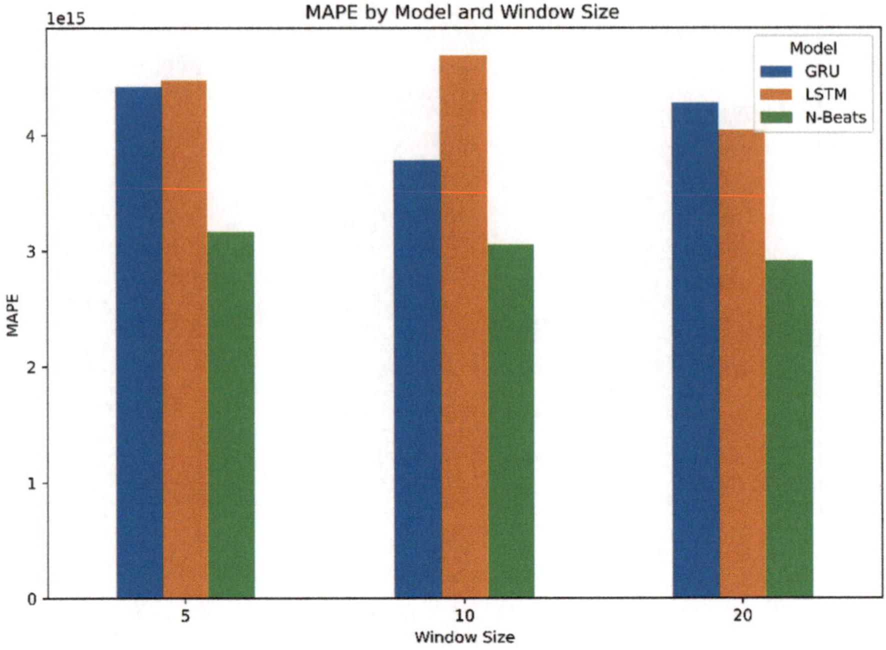

Fig. 6 MAPE Comparisons for GRU, LSTM and N-BEATS

5 Conclusion and Future Research

This study presents a comparative analysis of three advanced deep learning models LSTM, GRU, and N-BEATS for rainfall prediction across different temporal window sizes (5, 10, and 20 days). This study is important for having accurate rainfall prediction which is essential for climate modeling and simulation. The performance of these models was evaluated using a comprehensive dataset, considering key metrics such as mean absolute error (MAE) and root mean square error (RMSE). The results indicate that while all three models exhibit strong predictive capabilities, the N-BEATS model consistently outperforms LSTM and GRU, especially with larger window sizes, proving its potential superiority in capturing long-term dependencies in capturing rainfall patterns. On the other hand, LSTM and GRU show comparable performance in shorter window sizes, making them suitable for short-term forecasting applications. However when the window sizes increases, the gap in performance between N-BEATS and the other models increases significantly. This proved the superiority of N-BEATS model in handling long term dependencies compared to other recurrent based models.

As future research, the N-BEATS model can be compared to other models for different types of data sets related to climate change to prove its consistent superiority for handling long dependencies. It can also be applied with other architectures such as CNN to handle spatial dependencies in the data set, which is essential to forecast weather at different locations.

References

1. Hewamalage H, Bergmeir C, Bandara K (2021) Recurrent neural networks for time series forecasting: current status and future directions. Int J Forecast 37(1):388–427
2. Hill AJ, Herman GR, Schumacher RS (2020) Forecasting severe weather with random forests. Mon Weather Rev 148(5):2135–2161
3. Fente S, Singh G (2022) Deep learning-based effective fine-grained weather forecasting model using LSTM and TCN. Pattern Anal Appl 25(4):789–805
4. Ham YG, Kim JH, Luo JJ (2019) Deep learning for multi-year ENSO forecasts. Nature 573(7775):568–572
5. Slater LJ, Arnal L, Boucher M-A, Chang AY-Y, Moulds S, Murphy C, Nearing G, Shalev G, Shen C, Speight L, Villarini G, Wilby RL, Wood A, Zappa M (2023) Hybrid forecasting: blending climate predictions with AI models. Hydrol Earth Syst Sci 27:1865–1889
6. Araujo JA et al (2021) machine learning for climate precipitation prediction modeling. Remote Sens
7. Poornima S, Pushpalatha M (2019) Prediction of rainfall using intensified LSTM based recurrent neural network with weighted linear units. Atmosphere 10(11):668
8. Australian Commonwealth Bureau of Meteorology (n.d) WeatherAUS dataset. Rattle package, version 5.5.1. [Online]. Available: https://www.rdocumentation.org/packages/rattle/versions/5.5.1/topics/weatherAUS
9. Olivares KG, Bandara K, Seeger M (2023) Neural basis expansion analysis with exogenous variables: forecasting electricity prices with NBEATSx. Energy and AI 9(3):101030
10. Jeon M, Moreno-Pino F, Seeger M (2023) Feature-aligned N-BEATS with Sinkhorn divergence for time-series forecasting. J Time Ser Anal 44(2):217–235
11. Bandara K et al (2023) N-BEATS model for financial time series forecasting. J Forecast 42(1):15–28
12. Oreshkin BN, Carpov D, Bengio Y (2023) N-BEATS: Neural basis expansion analysis for interpretable time series forecasting. J Mach Learn Res 24:48–65
13. Puszkarski B, Hryniów K, Sarwas G (2022) Comparison of neural basis expansion analysis for interpretable time series (N-BEATS) and recurrent neural networks for heart dysfunction classification. Physiol Meas 43(6):064006
14. Greff K, Srivastava RK, Koutnik J, Steunebrink BR, Schmidhuber J (2023) LSTM: a search space odyssey. IEEE Trans Neural Netw Learn Syst 34(2):2222–2232
15. Smagulova K, James AP (2023) A review on the long short-term memory model for time series forecasting. Artif Intell Rev 56(4):104–122
16. Jianwei Y et al (2023) Unemployment rate forecasting: LSTM-GRU hybrid approach. J Labour Mark Res 57:12–28
17. Lee X et al (2023) CNN-GRU model for large-scale energy storage optimization in smart grid systems. Front Energy Res 11:123–135
18. Lenton TM et al (2024) Remotely sensing potential climate change tipping points across scales. Nat Commun
19. Grabar V et al (2024) Long-term drought prediction using deep neural networks based on geospatial weather data. Environ Model Softw
20. Ma Y et al (2022) Corn yield prediction and uncertainty analysis using Bayesian neural network. Remote Sens Environ

Nature-Based Solutions to Restore the Wetlands Along the Critical Zones of River Nyong, Cameroon

Norbert F. Tchouaffe-Tchiadje and Michel Tchotsoua

Abstract The study examines possible solutions to restore the River Nyong's wetlands, Mbalmayo, Cameroon through Nature-based solutions (NbS) using Tchouaffé's Theory of change (TToC). TToC is a new educational model to assess performance of network of actors co-working to overcome socio-environmental challenges. Many study from the same authors, to find innovative approaches and solutions to challenging wetlands stated that the major challenges or causes of the degradation of the river Nyong were mostly: Deforestation; recurrent pollutions and sedimentation of the river bed and the associated consequences are: The growth of floating macrophytes or eutrophication, which in turn induces a gradual reduction in the flow of the river and unprecedented damage to its ecosystem as well. So the main objective of this study consists of diagnosing the root causes of the eutrophication and to prescribe solutions to remediate them. To reach this aim we carried out geo-referencing of one of the critical zones of river Nyong and we also carried out focus group discussions to collect information from 50 riperian populations. As key findings 48 out of 50 interviewed respond and 60% of them were favorable for Nature-based solution while for the rest 40% were favorable for Ecological Engineering solution (EES.) and Ecological based adaptation (EbA). Women compared to men are more self-engaged for wetlands restoration with self-engaged performance of 32.5% against 25% for men. In nutshell, both men and women working together enables to reach a self-engaged performance of 57.5% which is an effective performance according to Tchouaffe's theory of change model and while combining two or three proposed solutions, the trophic status of the river Nyong could be optimized and improved for a long term basis.

N. F. Tchouaffe-Tchiadje (✉)
Forestry, Water and Environment, Higher Institute for Agriculture, University of Ebolowa, Ebolowa, Cameroon
e-mail: ntchoua@kesmonds-edu.ac

M. Tchotsoua
Department of Geography and Geomatic, University of Ngaoundéré, Ngaoundéré, Cameroon

N. F. Tchouaffe-Tchiadje
Kesmonds International University, Ngaoundéré, Cameroon

S. Al-Naemi et al. (eds.), *Water and Food Security in the Face of Climate Change: Challenges and Opportunities for Resilience*, World Sustainability Series,
https://doi.org/10.1007/978-3-032-00098-9_14

Keywords Wetlands · Deforestation · Theory of change · Pollution · Buffer-zone · Renaturation

1 Introduction

This scoping study has been developed to enable better understanding the river Nyong's local context in term of policy and legal framework and reveal key challenges for the application of Nature-nased solutions (NbS).

It investigates the main causes of problem associated to the degradation of the river Nyong, which are pollution, deforestation, and sedimentation of the river bed. The Consequences associated to these challenges are: The growth of floating macrophytes which in turn causes a gradual reduction in the flow of the river and unprecedented damage to its ecosystem as well. To overcome these challenges, the natural-based solutions which entail: action to protect, sustainably manage and restore wetlands, seams the solutions to be addressed after geo-referencing of one critical zone and analysis of the perception of riparian populations. The main objective of this study consists of diagnosing the root causes of the degradation of wetlands and to prescribe solutions to tackle them. It raises awareness of the role and contributions of respondents to preserve the river Nyong wetlands.

2 Review of Literature

2.1 Context of Study

Actions to restore wetlands are gathering momentum around the world, according to the Partnership for Environment and Disaster Risk Reduction [21]. They are home to 40% of all biodiversity. Specefically, peatlands store twice as much carbon as the world's forests. Wetlands play a key role in helping reduce flood damage and purifying polluted water. However, the rate of wetland loss has accelerated since 2000.

Today wetlands which cover about 6% of the Earth's land surface are recognized to be far more important than what we have believed. We now realise that they are a very important ecosystem; they are vital for human health, food supply, tourism, and jobs. They are among the most valuable and productive ecosystems we have, rich in biodiversity, a breeding ground for fauna, and flora.

Wetland conservation, restoration and watershed protection are becoming more prominent features of sustainable agriculture land-use planning [17]. The same reference also points out that "extreme caution is necessary about what is advocated for inclusion in the agreement as additional sink categories, given the absence of a complete understanding of the net GHG (Green House Gas) impact of wetlands".

Methane is a potent GHG estimated to warm the Earth's atmosphere 81 to 83 times as much as carbon dioxide over a 20 year period [30]. The river Nyong wetlands as many others around the world can sequest emissions of methane; the second contributor of climate change. Methane accounts for around 20% of global greenhouse emissions, behind only carbon dioxide (NAS, 2023); According to Daniel Jacob [18] about one third of global methane emissions occur naturally, mostly from wetlands and it has a most higher global-warming potential than carbon dioxide does-up to 80 times as much and is responsible for about a third of current global warming. The same author said 'If we are trying to address climate change over the next decade or two, methane is a very powerful lever.' Therefore there is a need for Cameroonians to be thinking about these as a means to generate biogas from wetlands and to secure water security at the wake of climate change.

2.2 Definition of Wetlands

Wetlands are defined as: "... areas of marsh, fen, peatland or water, whether natural or artificial, permanent or temporary, with water that is static or flowing, fresh, brackish or salt, including areas of marine water the depth of which at low tide does not exceed six metres" (Ramsar Convention Article 1.1).

According to article 2.1 of the Ramsar Convention,"Wetlands may incorporate riparian and coastal zones adjacent to the wetlands, and islands or bodies of marine water deeper than six metres at low tide lying within the wetlands."

Wetlands are considered the most diverse of all ecosystems. Plant life found in wetlands includes mangrove, water lilies, cattails, sedges, tamarack, black spruce, cypress, gum, and many others. Animal life includes many different amphibians, reptiles, birds, and furbearers. Most of all wetland is a haven for water for all uses.

2.3 Importance of Wetlands

Wetlands provide us with clean water, they protect us from storms, floods and fires, they sustain biodiversity; 40% of the world's species live or breed in wetlands. Many wetlands also store large carbon stocks, for example, peatlands alone store twice the carbon of all the forests in the world. Yet 35% of wetlands have disappeared in the last 50 years.

Wetlands are an important store of water for local people, as well as regulating flows downstream. A natural, well functioning wetland needs little management, and can withstand natural fluctuations in the quantity and timing of the water it receives. But problems arise when degradation affects the natural functioning of the system, for example through drainage, excessive water extraction, or diversion of water upstream. Water management is necessary to avoid or mitigate these problems

([28], P.3). If this initiative is maintained in Cameroon, the issue of water security will be guaranteed.

According to Wetlands International ([28], P.5), many conservationists have a special sympathy for wetlands. They see it as:

Unique watery habitats filled with birds, wild animals and plants, with an indefinable mystique. That magic calls for special protection—not only of wetland wildlife but because wetlands have their own intrinsic values. Yet this romantic picture is far removed from the reality faced by wetland communities: farmers and fishers who live with seasonal floods and drought, the constant threat of waterborne diseases, and an unremitting depletion of the wetland foods and products which are mainstay of their livelihoods and cultural traditions.

Wetlands are also helpful in controlling floods, replenishing groundwater, protecting biodiversity and providing livelihood to local population [15]. Thus the observance of wetlands Day is meant to improve public awareness about the importance of wetlands. Wetlands restoration aims to bring back shrub and tree cover for multiple purposes, including carbon and methane sequestration, water regulation, biodiversity restoration, or improving the livelihoods of rural residents [16].

Wetlands play an important role in fresh water cycle. They are the link between water and land. They act as filters, thereby protecting sources of drinking water [15] thus guaranteeing water security. Wetlands hold rainwater and sediments and purify water. The shrinking wetlands prove to be disastrous to the fresh water supply [12, 15] and in this case they need to be protected at all so as to maintain the continuous supply of water for consumption.

Wetlands also serve as a vital environmental sanctuary and a source of livelihood for the—often poor—people who live in and around them. They are also important as important as water reservoirs and purifiers for everyone [28].

Wetlands are also very important in that they can sequester carbon, provide habitat and purify water [29] which is indispensable to the poor.

Wetlands are an important store of water for local people, as well as regulating flows downstream [28]. A natural, well functioning wetland needs little management, and can withstand natural fluctuations in the quantity and timing of the water it receives. But problems arise when degradation affects the natural functioning of the system, for example through drainage, excessive water extraction or diversion of water upstream. ([28], P.3).

2.4 Wetlands in Cameroon

Cameroon ratified the Ramsar Convention that was adopted in 1971, amended in 1982, in 1987, and in 2006. Joining this treaty late meant that Cameroon knew the importance of these natural resources just late. That is about three years ago. In March 2006, Cameroon deposited with the Director-General its instrument of accession to the convention on wetlands of international importance. The main wetlands of Cameroon are the Ndop Flood plain, the Lake Chad basin, the Ndian Creeks, the

Coast of Limbe, Tiko, Kribi, Douala, Edea, Lagdo, Songloulou, Maga. Many of them were degraded by human abuses.

2.5 Degradation of Wetlands in Cameroon

Many administrative authorities in Cameroon see wetlands as underexploited resources or hindrance to development: That is water that can be extracted, land that can be converted to farming, timber to be felled, a source of mosquitoes and diseases and a barrier to transport. As a result the wetland areas have been wantonly and indiscriminately destroyed. Conflicts have often arisen between groups of people over the use of shrinking wetlands and their fertile lands, fish stocks, or fresh water resources, [25, 28].

2.6 Wetlands and Nature-Based Solutions

Nature-based Solutions are ecosystem-based ways to increase carbon storage in vegetation and soils, such as extensive reforestation and soil restoration.

- Each year, a mature tree cleans the air by absorbing 22 kg of carbon dioxide (CO_2) and releasing oxygen (O_2) in exchange.
- Peatlands are the world's largest organic carbon stock; the area covered by near natural peatland worldwide (> 3 million km^2) sequesters 0.37 gigatonnes of CO_2 a year.
- Blue carbon is the carbon stored in coastal and marine ecosystems such as mangroves, seagrasses and salt marshes.
- These ecosystems sequester more carbon per unit than terrestrial forests.
- Destroying these ecosystems releases CO_2 into the atmosphere. Ecosystem-based ways to reduce the impacts from disasters and climate change include: green and infrastructure, forest restoration, wetland restoration, climate smart agriculture/ agroforestry and urban greening. These can be combined with early warning systems or grey infrastructure such as sea walls, if not harmful to the environment.

2.7 Types of Infrastructure:

- **Green infrastructure** refers to trees, lawns, hedgerows, parks, fields, forests, etc. Blue infrastructure refers to water elements, like rivers, canals, ponds, wetlands, floodplains, water treatment facilities, etc. These terms come from urban planning and land-use planning.

  Grey infrastructure refers to built or engineered infrastructure.

- Hybrid infrastructure encompasses both green and grey approaches, as well as "blue" infrastructure, which mimics natural systems using artificial materials or through combining natural and nonnatural structures.
- Ecological restoration is the process of assisting the recovery of an ecosystem that has been degraded, damaged, or destroyed. At the heart of both are environmental management approaches, which can be combined with measures that explicitly reduce disaster and climate change impacts.

3 Methodology

The methodology of study consists of geo-refencing the critical zone of the of research and of carrying out focus group discussion with reparians populations and of using the TToC model to assess their self-engaged performance:

3.1 Geo-Referencing of One of the Critical Zone

We carried out an observational study along the Japan Market (Fig. 1) using geo-referencing to identify six critical zones which affect the waterway of river Nyong during dry season as you may see in Figs. 2 and 3.

When NT/PT > 7.2 then the limiting factor is the phosphorus otherwise when NT/PT < 7.2 the limiting factor is nitrogen. In short, for this case study all NT/PT ratios analysed in this context are superior to 7.2, this means phosphorus is the limiting targeted factor on which the trophic status of the river Nyong should be considered. Figures 1 and 2 and Table 1 show that the trophic status of river within the critical zone shifts from Oligo-mesotrophy, up-stream to Hyper-eutrophy due to culmulative pollutions and sedimentation of the river bed.

3.2 Focus Group Discussion

We carried out focus group discussion with 50 riparian populations around the Japan's market, Mbalmayo to seek their viewpoints related to the degradation of river Nyong from 7 to 8th July 2022.

On our request, we explore the current degradation situation of the river Nyong with our respondents focusing on critical points and ask them to choose the appropriate approach which fit to remediate to the problem:48 out of 50 responded to our request which the results are present in the following Table 1 matrix.

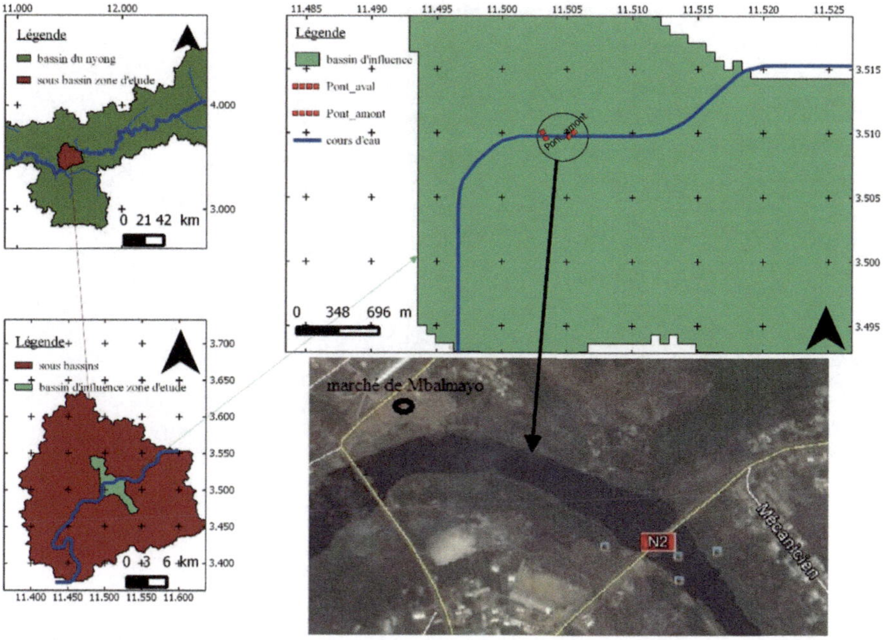

Fig. 1 Study area or Japan market

Fig. 2 Geo-referencing the critical zones. *Source* (Geomatic Laboratory of the Department of forestory, University of Dschang, 2021)

Fig. 3 Critical zones of river Nyong

3.3 Theoretical Contributions

This section introduces TToC (Tchouaffé's Theory of change: Coded TToC) model as an alternative approach which brings together transdisciplinary change makers to co-create transformative solutions to solve a particular challenge. It helps to assess multiple-actors innovation platforms.

Self-assessment using TToC worksheet: Considering the above respondents of Table 2 as potential change-makers, we use TToC model to analyse data and to assess the performance of women, men and both to solve the problem related to the river Nyong degradation. So the below Table 3 which recognizes the importance of inclusion and gender balance helps us achieve that goal and make a real difference with the existing management. It is conceived to assess the respondents readiness to adopt green techniques to preserve River Nyong wetland. The table compiled the scores of the five indicators of TToC model that are not only introduced and clarified some of the most basic and important formulas and concepts of TToC model but provided a how to manage on addressing these issues in practice. The value of involving people with lived experience in the reseach process (ie, actors engagement) is increasingly being recognised in various research domains.

Scoring Criteria

For each criterion, you will need to provide a score of 0–4

0 = Not Addressed

1 = Limited attempt to address

2 = Some success in addressing the various elements

Table 1 Redfield [22] NT/ PT ratio

Critical zones	Z0	Z1	Z2	Z3	Z4	Z5	Z6
Samples	× 0	× 1	× 2	× 3	× 4	× 5	× 6
Total Phosphorus (mg/l)	0.03	0.02	0.03	0.04	0.03	0.12	1.09
Total nitrogen (mg/l)	15	16.58	16.38	14.30	14.00	16.33	25.53
NT/PT	500	829	546	357.50	466.66	136.08	23.42
Trophic Status per zone		Oligo-mesotrophy	Meso-eutrophy	Eutrophy	Meso-eutrophy	Hyper-eutrophy	Hyper-eutrophy
Limiting factor	NT/PT > 7.2 means phosphorus is the limiting factor and NT/PT < 7.2 means nitrogen is the limiting factor; According to Redfield (22)						

Source [26]

Here × 0 is a limiting variable accoding to OCDE [20].

And the **Redfield** ratio NT/PT determines the limiting factor. According to Redfield [22],

Table 2 Interview matrix

Number of participants	Number of respondents	Favorable for Nature approach/ renaturation	Favorable for Mixte approaches		Favorable for any solution
		NbS	EES	EbA	
24 Women	23	11	4	8	1
26 Men	25	18	2	5	1
Total:50	48	29	6	13	2
Pourcentage	100%	60%	12.5%	27%	0/4%
Terminology		NbS: Nature -based Solution	EES: Ecological-engineering Solution [19]	EbS: Ecosystem-based Adaptation	

3 = Good attempt to address the element

4 = Addressed very effectively

The scoring system can be used to assess each principle individually indicating what action has been taken by an innovator to implement it.

TToC Formulation and self-engaged performance (Ps): The self-engaged performance Ps is calculated in Table 2 using Tchouaffé's formula:

$Ps = \left[1/n^2 \sum \left(A_{agenti}\right)\right] \times 100$; **Formula Authored by**[26]

Where n > 1 stands for total scores; $\mathbf{A_{agenti}}$ stands for the average score per person and solution chosen.

If the Global self-engaged performance is Ps < 50%, the chosen network of actors is ineffective;

If the Global self-engaged performance is Ps > 50%, the due network is effective [26].

Ps is based on existing framework to co-develop a Decision Support System (DSS) with potential stakeholders to determine their self-engagement, it is also called effectiveness.

Here the framework is the focus group made of riparian populations, which have to collaborate more effectively to perform change along the river Nyong. In short Ps allows to know how they have planned to guide restoration efforts on the ground and how their participation and engagement being effectively achieved. In addition, successful wetlands restoration requires a concerted effort ftom governments, civil society and the private sector.

4 Key Findings

4.1 Focus Group Discussion Analysis

As key findings 48 out of 50 interviewed respond and, which to be more effective 60% of respondents were favorable for Nature-based solution (NbS) while for rest 40% were favorable for other solutions (EES and EbS). To preserve the river Nyong wetlands, the majority of respondents (80%) demanded a strict policy on the usage and management of water bodies, the forests and wood industries along the river nyong saying that these had sparked serious effects on their life because hosting malaria vectors. They also recommended the government to take a robust decision and serious law enforcement especially on wetland degradation. They suggested that the government should build up a plateform associating them to sensitizing and distribute trees that are environmental friendly to be planted along the rivers. Some (27%) fovorable for EbA, also cited the need for adoption of sustainable agriculture and land management practices like agroforestry, conservation agriculture which will serve food and increased carbon sinks role. Others demanded eviction of communities living along river banks and wetlands and to adopt green technique or EES to preserve the wetlands.

It was also noted that there was a need to sensitize farmers on agro-ecology and organic farming practices and dangers of excessive deforestation and usage of chemicals, with adoption of ecofriendly energy saving like eco-stoves, biogas and chatcoal briquettes being some of the alternatives to reducing the rate of deforestation. Adopting more sustainable farming practices may not only improve the wetlands of river Nyong but protect them for the future generations; more sustainable and climate-friendly behaviours to counter growing threats like eutrophication and climate change.

4.2 Local Population Self-Engagement for Green Techniques

Almost the riparians populations are willing to confront the degradation of the river Nyong wetlands.

Women with a self-engaged performance of 32.5% are likely more than men 25% to bring a solution the River Nyong degradation. They are convinced that the restoration of this wetlands will reduce the hosts of maleria in this area. With a global self-engaged performance of 52.5% the contribution of respondents are effective according to the Tchouaffe's theory of change [26]. Most of them are willing to adopt green techniques to improve their livelihoods and ecosystems.

4.3 Solutions to Restore Wetlands of River Nyong

Wetlands are very indispensable resources along the river Nyong in guaranteeing water security if sustainably managed. They act as watersheds, water catchments and reservoirs to many water resources in many communities especially reparian communities. They do not only serve as sources of freshwater for home use, they also serve as source of water for irrigation and water for livestock. Among the managing responses to restore the wetlands, we have Nature-based Solutions (NbS), Ecological Engineering Solutions (EES) and Ecological-based Adaptation (EbA).

Nature-based solutions (NbS) are actions to protect, sustainably manage, or restore natural ecosystems, that address societal challenges such as climate change, human health, food and water security, and disaster risk reduction effectively and adaptively, simultaneously providing human well-being and biodiversity benefits. Nature-based solutions such as green roofs, rain gardens, or constructed wetlands can minimize damaging runoff by absorbing stormwater, reducing flood risks and safeguarding freshwater ecosystems. NbS can contribute to disaster risk reduction, build resilient cities, improve water management and contribute to long-term food security.

Sustainable restoration of wetlands at the coastal regions requires planning for a more extreme future climate by returning critical water resources in the coastal landscape to levels that existed before human began modifying the region [14, 27]. In this light stakeholders have to rehabilitate human settlements away from these natural resources so that they can be restored for the rejuvenation of the ecosystem and water in particular for its sustainable and equitable use. If this is not done, cities like Balmayo, Douala, Edea and Kribi will keep on suffering from freshwater for home consumption.

Wetland soils and vegetation naturally store water, filter sediments and pollutants from fresh water supplies, and help stabilise shorelines by reducing erosion and storm surges associated with rising sea levels [14, 26, 27]. From this perspective the wetlands along the River Nyong need to be restored so that the water from the tributary So'o, specifically from Mbalmayo needs to be filtered in order to deliver freshwater to Yaounde's population and its neibourghoods whose water supply is not the best. This is because these wetlands serve as water sheds to these areas of the Centre Region.

For wetlands to be restored and sustainably managed as a means to guarantee water security is for enough awareness to be raised to the community especially in the areas where these wetlands are found. If this is done, it is going to demystify the concept and believe that people have about wetlands. Instead the public should be made aware that wetlands are vital resources that need to be catered for because of their potentials and water being the most important.

The government also has to lay down a firm law on wetlands as it has done to protect other natural resources. If this is done and stakeholders join efforts to restore wetlands Cameroon will gradually move out of the water insecurity it is presently facing as a result of loss of wetlands. They will serve as a sanctuary for the ecosystem and an ecotourism destination. In that same light, managing the wetlands will pose particular challenges because they will reconcile the different interest of conservation and development dealing with conflicts over water and other resources and major seasonal fluctuations in wetlands.

Ecological Engineering solutions are hybrid solutions which consist of combining green and gray infrastructures, to bring together policy makers, planners, engineers and communities to find the best solutions for managing Wetlands degadation.

Applying ecological engineering to resolve the eutrophication, the findings of this study clearly show that eutrophication of the water body of river Nyong could be resorbed through retention of nutrients, pre-dam creation, treatment of tributary water or pipe and restoration of river functions. The explanation for each typology and limits are:

- **Retention of nutrients**: It's difficult to transpose the experience acquired on extensive systems such as natural lagooning or filters planted with reeds to ZRVs 'Zones de Rejet Végétalisées: vegetated discharge areas' because processing systems operate different from those of the ZRVs. A slight reduction in nitrogen fluxes from surface water can be expected if the passage times are not too short. In the case of permeable or highly permeable soils, large hydraulic transfers will lead to a deep leaching of the nitrates formed, the possible denitrification being considered negligible
- **Pre-dam creation**: The primary objective of pre-dams is to prevent rapid siltation of the main reservoir by simply reducing the flow velocity and sedimentation of particles including those related to phosphorus. Intakes thus accumulate at the bottom of the basin and remain trapped there as long as the oxygen content is sufficient [23, 26].

Treatment of tributary water or pipe: The physico-chemical treatment of tributary waters upstream of the water can be done by setting up a dephosphatation station using Fe^{3+} or Al^{3+} ferric ions as precipitating agent, followed by decantation [23, 24, 26].

Ecological-based Adaptation (EbA) consists here to restore the smooth functioning of the River Nyong, while protecting the wetlands.

At the local level, the basic principle is to use buffer zones, whose main function is to filter the waters before their arrival in the aquatic environment. These buffer zones can be dry or wet (Grassed strips, vegetated ditches, Forest buffer zones (foresters), Artificial wet buffer zones (ZTHA) which is in an agricultural environment, a retention pond, an existing pond of variable depth and height of water, vegetated or not [25, 26].

From Theory to Practice: It is advisable to treat tributary water up-stream or pipes at the household level to mitigate impact of pollution down-stream of the river, to create Pre-dam on waterways up-stream to mitigate overflood in rainy season; and down-stream, the flow diversion could be made for the bioretention of nutrients (Phosphorus, Nitrogen), through the creation of a buffer zone, which could possibly generate Bio-fuel from Biomass coming from macrophytes. So, Coupling NbS and

Table 4 Ratio of biomass and water used to produce bio-gas

Biomass used	Weight of biomass	Liters of water used
Microphytes	10 kg	10 L
Cow dung	10 kg	10 L
Food waste	10 kg	10 L
Co-digestion		
Microphytes and cow dung	5 kg each	10 L
Microphytes, cow dung, and food waste	3.3 kg each	10 L

(*Source* authors)

EES it is possible to reprofile the waterways in dryness and to make a flow diversion as well, by-passing part of water resources at the critical points towards a buffer zone to return sediments. In nutshell combining NbS, EES and AbS allows to restore the smooth functioning of the River Nyong (Renaturation or creation of buffer) as shown in Figure 5. All the three options are potential solutions to mitigate methane; emissions second major greenhouse gas and to mitigate climate change as well. The methane is shorter-lived than carbon dioxide so near-term methane reduction efforts could make a big impact on climate change.

The sediments recovered from the buffer zone could be use to produce compost and the biomass generated could be used to produce bio-fuel and bio-gas as well. In term bio-fuel the due Biomass generated could be combined to cow dung or any organic waste to produce briquettes of ecological charcoal. Sieving, adding sediments binder, compacting the mixture are among other processes to be followed before obtaining the eco-friendly charcoal. The bio-gas could also be produced from degradation of macrophytes biomass (Table 4), which is not just renewable energy source but also an appropriate way of managing waste, having potential to replace fossil fuel.

5 Conclusion

As key findings of this study, 48 out of 50 interviewed respond and 60% of respondents were favorable for Nature-based Solution (NbS) while for rest 40% were favorable for EES and EbA. Women compared to men are more self-engaged for wetlands restoration with self-engaged performance or effectiveness of 32.5% against 25% for men. In nutshell, both men and women working together enables to reach a self-engaged performance of 57.5%, which is effective according to Tchouaffe's theory of change model.

In short, Nature is the most resource-efficient solution to help adapting to climate change and mitigating disasters as well; all the solutions proposed conccur to the rehabilitation and restoration of the critical ecosystems like the deplition of wetlands of the river Nyong and while integrating two ore three proposed solutions, we could optimize and improve the trophic status of the due river for a long term basis.

We have to learn to balance today's needs with future environmental needs. Despite the biological richness and abundance of water in wetlands along the river Nyong, Cameroon, people who live in and around it keep on edging it away until a point it is at their own detriment. The destruction of these wetlands might have been because of poverty of the local populations or lack of awareness, from government. Untill Cameroonians know that these resources are a cornerstone that they rejected, little will be done to come out of the water crisis which is at the threshold of most Cameroonian. In this effort the Cameroonian government and stakeholders have to build up a networking platform to restore these lost resources and acknowledge their importance. This has to be done in a participatory approach where the government, non-governmental organisations and local communities are involved. From this dimension there will be hope that these neglected resources will be restored in order to contribute to the SDGs and ensure the betterment of the local populations. As way forward, we do pay tribute to the local people for whom indigenous principles and values are based on a deep and harmonic relationship with nature. The three solutions proposed here, show how natural features can outperform man-made technical solutions. This recognition and appreciation of Nature as our ally and protector demonstrates how intrinsic and beautiful risk reduction can look like.in sum, the biomass generated from macrophytes could be an alternative potential to reduce wood cutting and green gases emissions through the production of ecological charcoal briquettes.

References

1. http://aquapedia.waterdiplomacy.org/wiki/index.php?title=Water_Governance_in_Cameroon_%E2%80%93_Opportunities_for_New_Approaches_and_Capacity_Building
2. http://nbs-africa.org
3. http://www.fao.org/3/I9553EN/i9553en.pdf
4. https://alterneteurope.eu/wp-content/uploads/2021/02/Alternet_Day_PWPT_Presentation_Norbert.pptx.pdf
5. https://cran.r-project.org/web/packages/MuMIn/MuMIn.pdf
6. https://naturalsciences.ch
7. https://wetland.org
8. https://worldbank.org
9. https://www.SSRN.com/
10. https://www.eawag.ch/en/department/ess/main-focus/environmental-and-health-psychology-ehpsy/
11. https://www.riob.org/sites/default/files/documents/INBO%2027%20Web.pdf
12. Atwood TB et al (2017) Global patterns in mangrove soil carbon stocks and losses. Nat Clim Chang 7:523–528
13. Barton K (2018) MuMIn: multi-model inference. R package version 1.40.4
14. Daily GC, Polasky S, Goldstein J, Kareiva PM, Mooney HA, Pejchar L, Shallenberger R (2007) Ecosystem services in decision making: time to deliver. Front Ecol Environ 7(1):21–28. https://doi.org/10.1890/080025
15. Deccan Herald (2003, August 15) Mangroves play key role in coastal protection
16. Food and Agriculture Organization of the United Nations (FAO) (2023) The state of the world's forests 2023: forests and health. FAO. https://doi.org/10.4060/cc4073en

17. Government of India (1998) National carbon sinks table. Ministry Environ Forests
18. Harvard University (2023) Climate resilience through ecosystem-based adaptation: research highlights. Cent Environ Harvard University
19. Mitsch WJ (2012) What is ecological engineering?. Ecol Eng 45 :5–12
20. Organisation for Economic Co-operation and Development (OECD) (1982) The costs and benefits of environmental regulations. Paris: OECD Publishing
21. Partnership for Environment and Disaster Risk Reduction (PEDRR) (2023) Nature-based solutions for disaster risk reduction and climate resilience: policy brief. PEDRR Secretariat
22. Redfield AC (1958) The biological control of chemical factors in the environment. Am Sci 46(3):205–221
23. Tchouaffé TNF (2018) Desastres Embientals Pelo Mundo. FILHO; Edson Oleveira Braga; Santana Stella Emery-Belo Horizonte, Editora Vorto, p 191–222
24. Tchouaffé TNF, Tchotsoua M (2018) Gender-equitable pathways to achieve fish marketing in Cameroon: a case study of Mbalmayo markets. Nat Resour Conserv Res 1:6. https://doi.org/10.24294/nrcr.v1i2.855(2018b)
25. Tchouaffé TNF (2019) Diagnostic analysis of current Nyong river management in the Mbalmayo floodplain. INBO Newsletter, vol 27, p 20
26. Tchouaffé TNF, Fonteh TM, Tchamba M (2021) Ecological engineering to mitigate eutrophication in the flooding zone of the river Nyong, Cameroon. In: Leal Filho W et al (eds) Crown. https://link.springer.com/referenceworkentry/https://doi.org/10.1007/978-3-030-57281-5_8
27. Twilley RR (2007) Coastal wetlands and global climate change: processes, impacts, and policy implications. Ecol dimensions clim change 81–107. Island Press
28. Wetlands International (2009) Planting trees to protect the coast and its people: learning from mangrove projects in Asia. Wetlands Int
29. World Bank (2019) The changing wealth of nations 2018: building a sustainable future. World Bank, Washington, DC. https://doi.org/10.1596/978-1-4648-1046-6
30. Zhang Y (2023) The role of blue carbon ecosystems in climate mitigation. J Environ Manage 325:116541

Construction of Photobioreactor for Algal Cultivation: An Approach Towards Eco-Friendly Textile Wastewater Treatment

Sidra Pervez, Aqeela Bibi, Muhammad Asif Nawaz, Hadia Qureshi, and Summayya Asghar

Abstract This research provides an overview of photobioreactor (PBR) technology and its applications in algae cultivation, optimization and use for textile wastewater treatment. PBR design, operation, and optimization are covered, with an emphasis on important variables that affect the development of algae and the effectiveness of wastewater treatment. Microalgae was isolated from local sources. Using microscopy, two types of algae were identified. The best growth medium for these microalgae was subsequently identified by culturing them in three distinct nutritional media (BBM, Fog's, and M4N) with differing contents. The growth in photobioreactor involved many parameters, including temperature, pH, and culture length. Chlorella and Spirogyra were identified as algal strains. The highest concentrations of carbohydrates (173.2 mg/ml), protein (133 mg/ml), and sugar (59.6 mg/ml) were observed at 40 °C in Fog's medium for sample A, and at 25 °C in BBM for sample B (carbohydrates: 208 mg/ml, protein: 162.2 mg/ml, sugar: 62.3 mg/ml). Spirogyra exhibited its highest growth and concentration at pH 5 in Fog's medium (carbohydrates: 185.7 mg/ml, protein: 188.5 mg/ml, sugar: 94.97 mg/ml), and at pH 8 in BBM (carbohydrates: 208 mg/ml, protein: 192.9 mg/ml, sugar: 100 mg/ml). By using extracellular enzymes and photosynthetic assimilation of organic compounds, algae can effectively lower the COD levels (41%) from 2875 mg/L upto 1189 mg/L and BOD levels (20%) from 7521 mg/L upto 1510 mg/L in wastewater. Overall, monitoring both COD and BOD in algae-based wastewater treatment allows for a comprehensive assessment of the organic pollutant removal efficiency and the overall treatment performance.

Keywords Phycoremediation · Biodegradability · COD and BOD analysis

S. Pervez (✉) · A. Bibi · H. Qureshi · S. Asghar
Department of Biochemistry, Shaheed Benazir Bhutto Women University (SBBWU), Peshawar, Pakistan
e-mail: drsidrapervez@sbbwu.edu.pk

M. A. Nawaz
Department of Biotechnology, Shaheed Benazir Bhutto University (SBBU), Sheringal, Upper Dir, Pakistan

S. Al-Naemi et al. (eds.), *Water and Food Security in the Face of Climate Change: Challenges and Opportunities for Resilience*, World Sustainability Series,
https://doi.org/10.1007/978-3-032-00098-9_15

1 Introduction

The rise in global population raises the need for more food, energy, technology and fresh water. However, both domestic and mechanized origins create further wastewater as an outcome. The release of untreated organic and inorganic waste into the ecosystem resulted in affecting the environment. So far, may treatments have been used for the removal of waste materials from water before their release into the environment. Traditional treatment approaches are often costly, energy-rigorous, frequently incompetent and incapable of directing all challenges caused and initiated by the generated wastewater [1]. In the past decade, the microalgae-based wastewater treatment systems have gained much attention because of their ability to grow in wastewater without any expensive medium components and sequestration of industrial carbon dioxide. Phycoremediation removes heavy metals as well as hazardous organic chemicals such as toxic dyes without causing secondary pollution [2]. Microalgae contain many important macromolecules like carbohydrates, proteins, and lipids along with micromolecules like carotenoids, vitamins and chlorophylls [3]. Producing microalgae heterotrophically shows great promise as a method for both microalgae cultivation and environmental remediation, particularly in treating wastewater from the food sector [4].

A photobioreactor (PBR) is any culture technique established and constructed for photoautotrophic organisms using sun or synthetic light to foster and progress photosynthesis. PBRs are often exploited to increase and grow certain mosses, microalgae, and cyanobacteria [5]. Microalgae may grow under photoheterotrophic, heterotrophic, autotrophic, and mixotrophic circumstances, demonstrating their adaptability in metabolic environments. Examples that may flourish in photoautotrophic, heterotrophic, and mixotrophic environments include *Haematococcus pluvialis*, *Arthrospira platensis*, and *Chlorella vulgaris*. But other strains, including *Scenedesmus acutus* and *Selenastrum capricornutum*, may survive in photoautotrophic, heterotrophic, or photoheterotrophic environments [6]. Selecting the appropriate species is the pivotal initial step in microalgae cultivation, significantly influencing the sustainability and economic viability of the system. Robust microalgae species, capable of thriving in challenging environments, are identified during this stage. Once species selection is complete, microalgae are cultivated using either batch or continuous systems. While open systems typically offer higher durability and biomass productivity compared to closed systems, closed reactors excel in flexibility, contrasting with the more rigid nature of open reactors [7]. de Oliveira and Bragotto, [8] reported that Food and Agriculture Organization of the United Nations (FAO) have approved and registered around 87 tons of algal cultivation in 11 countries. Another main reason for using microalgae as treatment method is that this is sustainable and renewable method for recovery of purified water. Algal strains use organic material and CO_2 present in wastewater as their food for their growth and as a result produce variety of valuable by products including biomaterials, biolpolymers and bioenergy [1]. Therefore, this study was designed with the aim to design photobioreactor and cultivate indigenous microalgae in this reactor using optimized parameters. Moreover, this study also focuses on algal based wastewater treatment and the efficiency of process will be investigated using Chemical Oxygen Demand (COD) and Biological Oxygen Demand (BOD) analysis of wastewater along with dye degradation.

2 Materials and Methods

2.1 Collection of Microalgae Sample

Two samples of microalgae were collected from courtyard garden and from stagnant water bodies at Shaheed Benazir Bhutto Women University, Peshawar, Pakistan. After collecting, the microalgae samples were kept at room temperature (28 °C) and exposed to constant light until they were processed further.

2.2 Recognition of Microalgae Based on Appearance

For the initial assessment, the term "microscopic characteristics" was examined using a microscope that had 200X and 1000X magnifying lenses attached. Various features of the algae cells, including their color, basal body, structure, shape, and arrangement, were observed and documented.

2.3 Cultivation of Microalgae

Inoculum Preparation

Algae were cultivated in Bold Basal Medium (BBM). Successive dilutions (ranging from 10^{-1} to 10^{-5}) were prepared. During the initial eight days of the growth experiment, each specimen tube held 1 ml of microalgal sample mixed with 9 ml of Bold basal media.

2.4 Design of a Photobioreactor.

The microalgae cultivation system consisted of a 10-L glass photobioreactor (PBR) illuminated by an external LED light mounted on the top of the vessel. Prior to inoculation, the microalgal strains underwent pre-cultivation. Subsequently, they were inoculated into the PBR containing 7 L of Bold's Basal Medium (BBM) along with a 1-L inoculum. The PBR operated at a temperature of 28 °C and maintained a pH level of 8 throughout the cultivation process. Different components of PBR were designed accordingly.

Optimization of PBR Conditions for Algal Cultivation

Different physical and chemical conditions were optimized including inoculum size, cultivation temperature, cultivation pH and cultivation time to increase the yield of microalgae and to increase different components in algal cells. Furthermore, various reported media including BBM, M4N and Fog's media were also used and among them one was selected and modified according to growth of isolated organisms.

Determination of Microalgae Cell Concentration

A UV/VIS spectrophotometer calibrated to a wavelength of 688 nm was used every other day to measure the concentration of algae cells within the photobioreactor culture.

2.5 *Characterization of Microalgae*

After optimizing various parameters, the microalgae were separated through filtration and the cells were used for determination of total carbohydrate, total protein and total sugar and glucose concentration for its characterization using Anthrone, Lawry, DNS and GOD-PAP method respectively.

2.6 *Wastewater Treatment and Dye Degradation*

Synthetic textile wastewater

Textile wastewater is characterized by elevated pH levels, suspended solids, chlorides, nitrates, and metals such as manganese, sodium, copper, and iron. Additionally, it typically exhibits high levels of both Biochemical Oxygen Demand (BOD) and Chemical Oxygen Demand (COD). Its distinctive dark- brown color is indicative of contaminants derived from textile manufacturing processes. The composition of synthetic textile wastewater, as outlined in Table 1.

Table 1 Composition of the synthetic textile wastewater

Component	Concentration (mg L^{-1})
Glucose	1000
Urea	62.5
Cobalt chloride	0.04
Sodium bicarbonate	1000
Dye	300

Wastewater Treatment

The synthetic wastewater was treated by incubating it with microalgae. COD and BOD were analyzed before and after treatment using standard protocol and treatment efficiency was calculated.

Qualitative and Quantitative Analysis of Dye Degradation

For qualitative analysis, different concentrations of bromophenol blue, methyl orange and congo red were mixed with agar. Wells were formed and algae solution was poured into wells and zone of hydrolysis was monitored. For quantitative analysis, dyes were mixed with algal solutions in different concentrations and allowed to stand for different time intervals at room temperature. After a specific time interval, the solution was centrifuged at 10,000 revolutions per minute for 15 min, the decolorization of dyes was calculated spectrophotometrically using following formula:

$$\text{Decolorization}\% = (\text{initial absorbance} - \text{final absorbance}) / \text{initial absorbance} \times 100.$$

3 Results and Discussion

3.1 Microalgae Isolation and Identification

The samples were selected according to their texture, color, and shape. Following selection, these algal samples were grown in Bold's Basal Medium (BBM). After an incubation period of 8 days, the chosen algae samples were transferred to larger fermenter bottles.

3.2 Identification of Microalgae Based on Morphology

Microscopic characterization was used to determine the microalgae's morphological identity. From microscopic identification it was observed that algae were *Chlorella*, which had a thin membrane concealing its spherical, flagella-free shape and *Spirogyra*, which had long, stringy, hair-like strands that were usually colored green or yellow and formed filamentous sequences that resembled threads. Specimens of Spirogyra ranged in diameter between 40 and 100 μm (Fig. 1).

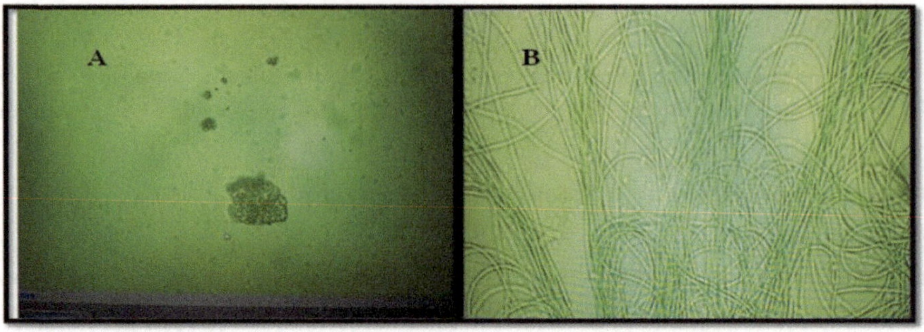

Fig. 1 Microscopic observation of microalgae **a** (Chlorella) **b** (Spirogyra)

Fig. 2 PBR before (left) and after (right) cultivation of algae

3.3 Cultivation of Microalgae

The PBR enabled robust algae growth was determined by closely monitoring and controlling environmental factors including temperature, light intensity, and nutrient availability. This was evidenced by the increase in algae biomass over time, as quantified through measurements of cell density or biomass concentration. the effectiveness of the PBR design in providing a controlled environment conducive to algae growth was emphasized. The efficient utilization of light energy for photosynthesis, coupled with appropriate aeration to ensure adequate oxygen supply for respiration, played crucial roles in maximizing algae productivity (Fig. 2).

3.4 Optimization of Conditions for PBR

Optimization of inoculum size

Different amounts of algal inoculum (5, 10, 15, and 20 ml) was analyzed but all other parameters remained the same. Based on the fresh weight gain, it was discovered that the inoculum size significantly affected the filamentous algae growth rate. The lowest measured growth rate, measuring 23.53 mg/ml, was obtained when the BBM. The growth rate increased to 30.7 mg/ml when the inoculum size was increased from 5 to 10 ml. The inoculum size was increased to 15 ml, which resulted in a quicker development rate of 60.5 mg/ml. Nevertheless, the growth rate dropped to 34.87 mg/ml when the size of the inoculum was raised over this limit, especially to 20 ml (Fig. 3a). These results show that the inoculum's size had a considerable impact on the algae's growth rate/

Optimization of temperature

The algal growth in BBM Media at three different temperatures: 25 °C, 30 °C, and 40 °C. Spirogyra sp., an algal species, was discovered to be affected by temperature. At 25 °C, Spirogyra sp. exhibited the highest growth rate in BBM media, with carbohydrate concentrations of 207 mg/ml, protein concentrations of 162.2 mg/ml, and sugar concentrations of 62.3 mg/ml. However, when the temperature was raised to 30 °C, the growth of Spirogyra sp. slowed down, resulting in reduced carbohydrate concentrations of 163 mg/ml, protein concentrations of 128.4 mg/ml, and sugar concentrations of 38.3 mg/ml. Spirogyra sp. growth was significantly inhibited by raising the temperature to 40 °C further (Fig. 3b). This decrease in growth at higher temperatures aligns with findings from Converti et al. [9], who noted that at 35 °C, algal growth decreased by 17%, and at 38 °C, cell death occurred.

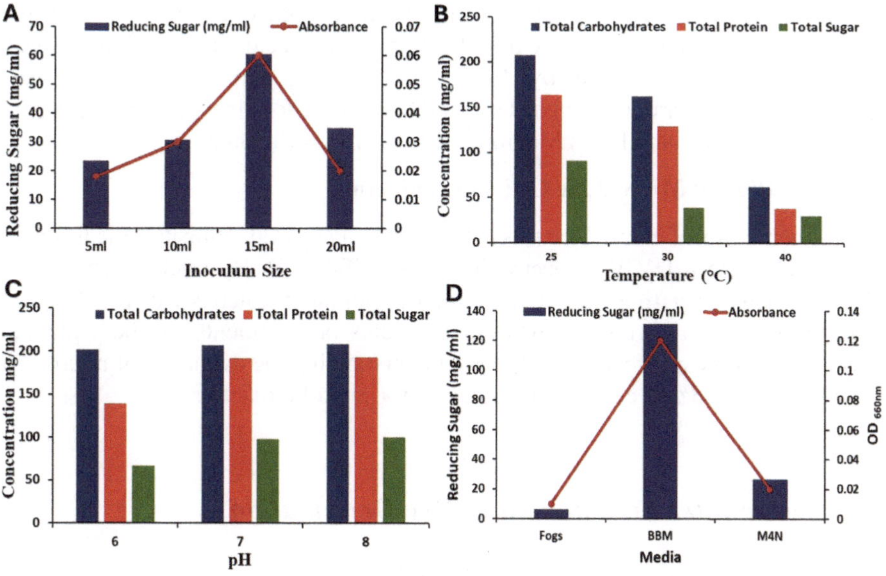

Fig. 3 **a** Effect of inoculum size, **b** effect of temperature on algal growth, **c** effect of pH on algal growth, **d** effect of medium

Table 2 Direct OD of microalgae cell concentration with time

Media's	5th day	10th day	15th day
Bold basal media	0.016	0.060	0.088
Fog's media	0.134	0.336	0.356

Effect of pH on Algal Growth using BBM

Spirogyra was cultivated at various pH values, and the algae showed substantial growth responses. When algal growth was evaluated at pH values of 6, 7, and 8, significant differences were found. Spirogyra sp. showed slower development at pH 6, with concentrations of 66.67 mg/ml sugar, 138.5 mg/ml protein, and 201.6 mg/ml carbs. Nonetheless, the algal growth reached its peak and most active range at pH 7 and 8. Similarly, Spirogyra sp. showed more growth at pH 8, exhibiting quantities of 100 mg/ml sugar content, 192.8 mg/ml protein, and 209 mg/ml carbohydrates (Fig. 3c). These findings are consistent with those of Zhao et al., [10], who found that growth was best in settings near pH 8, and that algae grew fastest in medium with an initial pH of about 7.10.

Optimization of medium components including carbon and nitrogen sources

Microalgae was cultivated using three different media: BBM, Fog's Media, and M4N media, to assess the performance of their growth. It was found that microalgae exhibited the best growth in BBM followed by Fog's Medium and then M4N Medium (Fig. 3d). This aligns with previous research indicating adverse effects of elevated nitrogen concentrations on algal growth [11]. BBM and Fog's medium was then selected for additional microalgae modification and characterization in light of these findings. Additionally, studies by Henderson et al. [12], and Grobbelaar [13] have highlighted the favorable outcomes of synthetic media for algal growth, supporting our choice of BBM and Fog's Medium for further experimentation.

Determination of Microalgae Cell Concentration

The concentration of microalgae was monitored daily by evaluating its optical density at 688 nm using a UV/Vis Spectrophotometer (Table 2). By tracking changes in optical density over time, the microalgae development kinetics can be observed, and the efficacy of various growth environments or treatments can be evaluated. This approach provides valuable data for understanding the dynamics of microalgal growth and optimizing cultivation strategies for photobioreactor.

3.5 Wastewater Treatment and Dye Degradation

Chemical Oxygen Demand (COD) analysis

The results showed that Chemical Oxygen Demand (COD) analysis for algae-based wastewater treatment revealed significant reductions in organic pollutant levels.

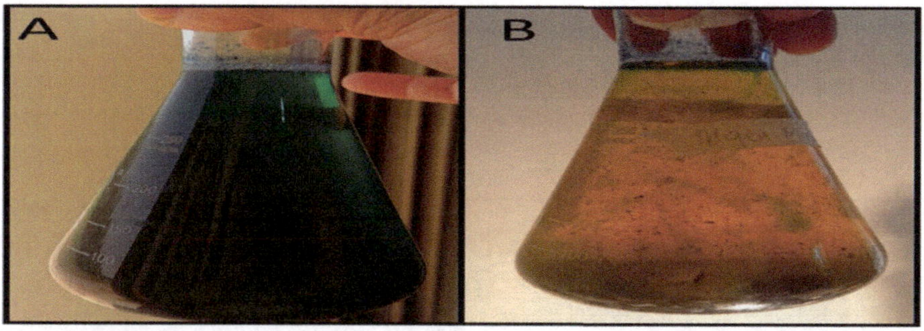

Fig. 4 **a** Mixture of algae and synthetic textile wastewater before treatment, **b** degrading property after 2 days of incubation after treatment

Table 3 COD values showing the effectiveness of algae-based wastewater treatment

COD	Initial (mg/L)	Final
Untreated wastewater	2875	Unchanged
Treated wastewater with Algae	2875	1189 mg/L

Through the metabolic activities of algae, organic contaminants present in the wastewater were effectively broken down, leading to a decrease in COD concentrations over time (Fig. 4 and Table 3). It is well known that algae use organic waste as a source of carbon to thrive, which lowers the number of contaminants in the water. This process, known as bioremediation, can be highly efficient, especially when coupled with appropriate nutrient management and environmental control. Additionally, the potential for algae to remove nutrients such as nitrogen and phosphorus through assimilation into biomass was considered, further enhancing the treatment efficiency [14].

Biochemical Oxygen Demand (BOD) analysis

Biochemical Oxygen Demand (BOD) analysis showed tha algae-based wastewater treatment often demonstrates substantial reductions in organic pollutant levels. The quantity of dissolved oxygen used by microbes during the breakdown of organic waste in the water is measured by the BOD assay. As algae metabolizes organic compounds, they utilize oxygen from the water, resulting in a decrease in BOD values (Table 4). The discussion surrounding BOD analysis in algae-based wastewater treatment typically emphasizes the efficiency and effectiveness of the treatment process. Algae's ability to remove organic pollutants from wastewater by utilizing them as a carbon source for growth is a key mechanism driving reductions in BOD levels. Additionally, the role of algae in oxygen production during photosynthesis helps to replenish dissolved oxygen levels in the water, which is beneficial for aquatic organisms and overall water quality improvement [15].

Table 4 BOD values showing the effectiveness of algae-based wastewater treatment

BOD	Initial (mg/L)	Final
Untreated wastewater	7521	Unchanged
Treated wastewater with Algae	7521	1510 mg/L

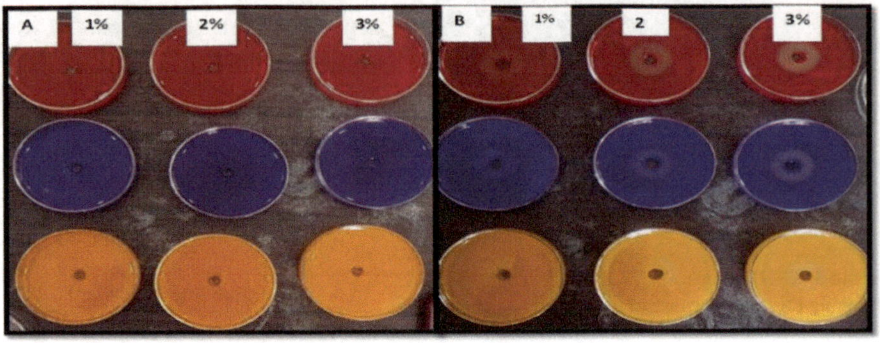

Fig. 5 **a** Dyes before algal treatment, **b** dyes after algal treatment

3.6 Qualitative Analysis (Decolorization on Petri Plates)

Azo dye degradation was conducted to assess the capability of microalgae in decolorizing the dyes. Following an overnight incubation, clear zone inhibition was observed (Fig. 5). The zones formed were less pronounced at 1% dye concentration, moderately visible at 2% concentration, and most prominently visible at 2.5% concentration for all dyes, including Congo red, methylene orange, and bromophenol blue. This phenomenon is largely attributed to bioconversion, with dye elimination showing a strong dependence on concentration [16]. According to Daneshvar et al., [17] there was a rapid fading of the Congo red dye within five days, which was followed by a gradual decolorization rate over time.

3.7 Quantitative Analysis (Spectroscopic Analysis)

The quantitative analysis results, conducted via spectroscopic methods, offered valuable insights into the efficiency of algae in degrading dyes (Table 5). Spectral measurements revealed notable degradation of dye compounds by the algae throughout the experiment. The discussion primarily focused on elucidating the mechanisms through which algae facilitated dye degradation, potentially involving enzymatic reactions or adsorption onto algal surfaces. Additionally, the implications of these findings for wastewater treatment and environmental remediation were explored, underscoring the potential of algae-based approaches in mitigating dye pollution [18, 19].

Table 5 OD of each dyes measured after every hour at different wavelengths

Dye degradation	1st hour	2nd hour	3rd hour
Bromophenol blue	0.255	0.126	0.080
Congo red	0.325	0.180	0.073
Methylene orange	0.286	0.152	0.087

Algal and microbial wastewater treatment processes have been widely employed for nutrient removal and organic matter degradation. Conventional methods include activated sludge systems, trickling filters, and anaerobic digesters. Algal-based systems, such as high-rate algal ponds (HRAPs) and photobioreactors, utilize algae for nutrient uptake and oxygen production. Emerging technologies integrate algae and microbes for symbiotic wastewater treatment, enhancing efficiency and sustainability. These processes are critical for reducing pollutants and supporting resource recovery.

4 Conclusion

This research study underscores the significance of photobioreactor (PBR) technology in cultivating indigenous microalgae for various applications, particularly in textile wastewater treatment. The study clarifies PBR design, operation, and optimization while highlighting important variables that affect algae growth and wastewater treatment effectiveness, including temperature, pH, media selection, light intensity, and nutrient availability. Through microscopy and cultivation in different nutrient media, two distinct types of microalgae, Chlorella (sample A) and Spirogyra (sample B), were identified and characterized. The study highlights the effectiveness of Fog's medium for Chlorella growth at 40 °C and BBM for Spirogyra growth at 25 °C, with specific pH conditions favoring optimal growth and concentration. Furthermore, the characterization of microalgae samples using various biochemical assays revealed valuable insights into their carbohydrate, protein, and sugar content. The fifteenth day of culture was identified as the peak activity period, demonstrating the potential for biomass and valuable compound production. Algae-based wastewater treatment is a promising method for reducing values of COD upto 41% and BOD upto 20%, providing a sustainable and cost-effective approach to water remediation.

Acknowledgements Authors would like to extend heartfelt gratitude to the Directorate General of Science and Technology (DGST), Pakistan (DGST/BSTIET/MS/2022-2023/8773) for their invaluable support and funding and also like to acknowledge Higher Education Commission (HEC) of Pakistan (Ref# HEC/R&D/TG/213024) for providing travel assistance.

References

1. Plöhn M, Spain O, Sirin S, Silva M, Escudero-Oñate C, Ferrando-Climent L, Funk, Song Y, Wang L, Qiang X, Gu W, Ma Z, Wang, G (2022) The promising way to treat wastewater by microalgae: Approaches, mechanisms, applications and challenges. J Water Process Eng 49:103012
2. Mustafa HM, Hayder G (2021) Recent studies on applications of aquatic weed plants in phytoremediation of wastewater: a review article. Ain Shams Eng J 12(1):355–365
3. Tang DYY, Khoo KS, Chew KW, Tao Y, Ho SH, Show PL (2020) Potential utilization of bioproducts from microalgae for the quality enhancement of natural products. Biores Technol 304:122997
4. Mata TM, Martins AA, Caetano NS (2010) Microalgae for biodiesel production and other applications: a review. Renew Sustain Energy Rev 14(1):217–232
5. Suh IS, Lee C (2003) Photobioreactor engineering: Design and performance. Biotechnol Bioprocess Eng 8(6):313–321
6. Chojnacka K, Márquez-Rocha FJ (2003) Kinetic and stoichiometric relationships of the energy and carbon metabolism in the culture of microalgae. Biotechnology 3(1):21–34
7. Richmond A (2013) Biological principles of mass cultivation of photoautotrophic microalgae. Wiley, Ltd eBooks, pp 169–204
8. de Oliveira APF, Bragotto APA (2022) Microalgae-based products: food and public health. Futur Foods 6:100157
9. Converti A, Casazza AA, Ortiz EY et al (2009) Effect of temperature and nitrogen concentration on the growth and lipid content of Nannochloropsis oculata and Chlorella vulgaris for biodiesel production. Chem Eng Process 48:1146–1151
10. Zhao L, Peng C, Zhang J, Tang Z (2021) Synergistic effect of microbubble flow and light fields on a bionic tree-like photobioreactor. Chem Eng Sci 229:116092
11. Farooq A, Khan AU, Yasir A (2013) Transesterification of oil extracted from different species of algae for biodiesel production. Afr J Environ Sci Technol 7(6):358–364
12. Henderson R, Parsons SA, Jefferson B (2008) The impact of algal properties and pre-oxidation on solid/liquid separation of algae. Water Res 42:1827–1845
13. Grobbelaar JU (2000) Physiological and technological considerations for optimizing mass algal cultures. J Appl Phycol 12:201–206
14. Pandey BC, Gupta S (2022) Effectiveness of algae in wastewater treatment systems. Int J Res Appl Sci Eng Technol 10(8):520–526
15. Goswami RK, Agrawal K, Verma P (2022) Microalgae-Based technologies for removal of textile wastewater. In: Clean energy production technologies, pp 83–101
16. Aydin H, Baysal G (2006) Adsorption of acid dyes in aqueous solutions by shells of bottom (Pistacia khinjuk stocks). Desalination 196:248–259
17. Daneshvar N, Ayazloo M, Khataee AR, Pourhassas M (2007) Biological decolorization of dye solution containing Malachite green by microalgae Cosmarium sp. Biores Technol 98:1176–1182
18. Premaratne M, Nishshanka GKSH, Liyanaarachchi VC, Nimarshana P, Ariyadasa TU (2021) Bioremediation of textile dye wastewater using microalgae: current trends and future perspectives. J Chem Technol Biotechnol 96(12):3249–3258
19. Mehariya S, Plöhn M, Jablonski P, Stagge S, Jönsson LJ, Funk C (2023) Biopolymer production from biomass produced by Nordic microalgae grown in wastewater. Biores Technol 376:128901

Climate Change Effects on Pesticide Use: Farm Workers' Knowledge and Attitudes Toward Exposure Safety

Norr Hassan, Ravi Rangarajan, and Haruna Musa Moda

Abstract Pesticides have been in use for decades to help improve crop production, in terms of quality and quantity, by controlling pests and their diseases. However, their efficiency and use are shaped by environmental factors including increase temperature and elevated CO_2 concentrations, which alter photosynthetic activity, plant growth and at the same time enabling dilution pesticide concentration absorbed by plants. Structured pretested questionnaire was used to assess farmers' personal and environmental safety awareness during pesticide application on farmland. Twenty nine percent of the farmers interviewed worked with pesticide application between 6 to10 years, while 26.7% have handled pesticide during planting season from 11–15 years. Farmers' knowledge on the impact of pesticide on human health significantly correlate with years of practice experience and highest educational level (<0.001 and < 0.001) while both years handling pesticide on farms and educational attainment related with awareness of the increased impact of pesticides on the environment (0.002 and < 0.001, respectively). Related ill health experienced among the respondents include headache (27.3%), itchy eyes (14.3%), dizziness (12.5%), and skin irritation (10%). These findings highlight the multifaceted nature of pesticide-related health risks among farmers, and raise awareness on the importance of developing and implementing comprehensive training programs that teach climate adaptation strategies, including safety measures, such as the proper and complete use appropriate personal protective equipment (PPE), also information on different types of pesticides, integrated pesticide management system, safe handling, safe disposing, safe storing of pesticides. Furthermore, there is need to strengthen policy to regulate pesticide use, which will go along a long way to limit farmers access to banned products.

Keywords Climate change · Farmers · Pesticide use · Pesticide exposure · Knowledge · Attitude

N. Hassan (✉) · R. Rangarajan · H. Musa Moda
Department of Environmental Health and Safety, University of Doha for Science and Technology, Doha, Qatar
e-mail: norr.hassan@udst.edu.qa

S. Al-Naemi et al. (eds.), *Water and Food Security in the Face of Climate Change: Challenges and Opportunities for Resilience*, World Sustainability Series,
https://doi.org/10.1007/978-3-032-00098-9_16

1 Introduction

Pesticides have been popular and extensively used to prevent and control pests and their diseases, thereby improving the agricultural crop quality and quantity [16]. These compounds are categorized according to intended target organism use, as insecticides, herbicides, fungicides, and others [56]. Additionally, their mode of action, chemical classes, toxicity level, application [28], and efficiency, and influence by environmental conditions (Delcour, 2015) have all been well studied in the literature. The Food and Agriculture Organization (FAO) reported that globally, around 3.70 million tons of active ingredients (Mt) of pesticides are used in agriculture yearly [21]. However, this number is projected to increase with the increased number and type of pests and the diminished pesticide residue on crops because of climate change. Thus, as an adaptation mechanism, farmers may have to spray more frequently and in larger volumes during the growing season [34, 56], resulting in long-term health and environmental concerns [56].

Climate change, specifically temperature and precipitation extremes, directly impact crops and indirectly effect soil nutrients [46], also influence pest infestation [17], development and survival, and reproductive rates [56]. Pest species flourish in warm and humid environments, affecting crop–pest interactions [56]. Elevated temperatures and CO_2 concentrations, correlate with a significant change in photosynthetic activity, promoting plant growth [26], and causing a dilution of concentration of pesticide absorbed by plants. This in-turn decreases the pesticide residue and effect, and triggers a probable increase in types and quantities of pesticides used [36]. The drastic change in precipitation levels can create a flourishing environment for pathogen attacks [44], causing changes in interspecific relationships between organisms, rising microbial contaminants in food, and decreasing food supply. Pesticide intake in plants is influenced by precipitation and will be restricted in case of reduced transpiration in dry conditions [56]. In addition to that, climate change leads to the creation of ecological niches presenting potential advantage for insect pests to shift from one area to another or even disperse in new geographic regions [56], eventually changing in pest distribution and abundance, and elevating pest outbreaks [16].

As Fig. 1 depicts, insect physiology is vulnerable to temperature, CO_2 and precipitation changes. Elevated temperature has significant influence on agricultural pests, including an increase in the number of generations, doubling metabolic rates, extending their geographic areas, and boosting overwintering survival. The increase in temperature can also result in outbreaks of plant diseases transmitted by insects and cause spatial desynchronization, leading to loss of synchrony between pests and their host plants [56]. Furthermore, rise in CO_2 levels can increase pest population [56] and insects feeding on plants [50]. Moreover, changes in precipitation patterns affect insect dynamics, where rise in precipitation promotes expedited growth in some populations, simultaneously, decreased rainfall results in more insect attacks due to heightened plant susceptibility [49, 56]. All of this means that farmers will

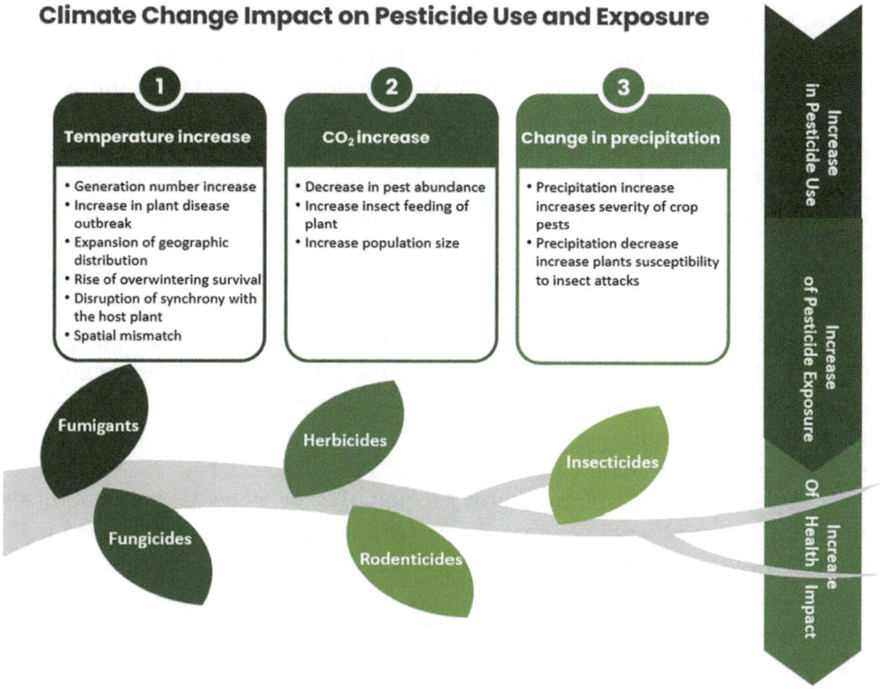

Fig. 1 Impact of climate change (temperature increase, CO_2 increase, precipitation changes) on pesticide exposure

deal with more pests, thus, obliging them to use more pesticides including fumigants, fungicides, herbicides rodenticides and insecticides. This will raise pesticide exposure by farmers resulting in more negative health illnesses.

Various research indicated the connection between perception of farmers' and their adaptation actions to climate change [6, 20, 56]. Pesticide manufacture, transportation, preparation, and application at the farm all leads to occupational pesticide exposures [30]. Pesticides are linked to a group of adverse human health effect, both acute and chronic, that undermine quality of life [35, 37]. Acute health effects caused by pesticide short-term exposure are usually experienced directly after exposure, such as headache, nausea, vomiting, dizziness, skin and eye irritation and others [14]. Chronic effects, on the other hand, are associated with long-term pesticide exposure and lead to severe health consequences, including neurological disorders such as Alzheimer's disease [29], Parkinson's disease [41], neurotoxic effect [23], and cognitive effects on children's cognitive capacities [11]. Cancer is another significant health issue from direct pesticide exposure [15], in which several studies showed increased occurrence of breast cancer [43], bladder and colon cancer [4], brain cancer [2], lung cancer [19], Liver cancer [8] and Leukemia [42]. The effect on reproduction has also been researched [18], in which pesticide exposure by parents was consistently associated with higher risk of any birth abnormality [22, 24]. Fetal death is also pesticide-exposed effects [5] similarly, respiratory disorders like asthma [31].

Highly dangerous pesticides that are banned in developed countries, are unfortunately, frequently used and easily accessible in Low- and Middle- Income Countries (LMICs) [13]. Studies found robust correlation between farmers' practices and their knowledge, attitudes, and views regarding the usage of pesticides in LMICs [12, 33]. However, the data on safe pesticide use showed that farmers' knowledge is very inadequate [39] and farmers frequently failed to have the fundamentally required training in this field [7, 10]. This led to the inappropriate use of pesticides in many situations [1, 10, 25]. Other research revealed that even though farmers had positive attitudes and strong understanding about pesticides, these did not translate into standard farming methods, which resulted in the abuse of pesticides [48]. Studies in the Philippines, China, India, and Thailand showed a significant number of farmers did not follow protective measures when handling with pesticides [3].

A study on farmers in the Çukurova region, Turkey, showed that none of the farmers had received pesticide use training [39]. A study conducted in Iran on 311 farmers showed that despite dealing with compounds ranging from high to low hazard compounds, most farmers have not received formal training on pesticide use. The farmers also showed lack of knowledge on pesticide use, risks associated, and proper handling methods of pesticides and associated waste [51]. Even with this extensive experience, farmers in Iran still report poisoning effects including nausea and headaches while applying pesticides [3].

Within the African region, in a study in Southwest Ethiopia, the detrimental effects of pesticides on human health were acknowledged positively by farmers, despite their subpar practices. Approximately 42% of farmers never utilized Personal Protective Equipment (PPE) to guard against and protect themselves from pesticide exposure. Farmers complained of symptoms like headaches, nausea, vomiting, skin irritation, and abdominal pain; these were perceived as complications associated with pesticide exposure [25]. Thus, in LMICs, the widespread use of highly dangerous pesticides, combined with inadequate knowledge and lack of proper training, poses significant health risks to farmers. Despite awareness of pesticide dangers, unsafe practices persist, leading to improper handling and frequent pesticide poisoning. Addressing these gaps and understanding the knowledge, awareness, and perceptions on proper use of pesticides and training on pesticide safety is crucial to mitigate these risks. Therefore, this research focuses on evaluating farmers knowledge in Northern Nigeria on pesticide use, symptoms associated with it, pesticide use in relation to climate change, and to analyze the factors that effect this knowledge, with an aim to identify gaps and determine areas for training. This study also aims to portray ramifications of climate change on farmers occupational exposure to pesticides and endorse evidence-based and more-informed adaptation strategies.

2 Materials and Methods

2.1 *Study Design and Participants*

Farmers across the northern Nigeria participated in the cross-sectional study during the period 15 February–30 June 2021, to measure their personal and environmental safety awareness regards pesticide application on crops. The ecological zones in the northern region are defined by three distinct vegetation of namely Guinea, Sudan and Sahel Savannah respectively. These support both crop production and animal husbandry employment to around 35% of the inhabitants of the area. Participants are considered eligible for inclusion in the study, where they are above 18 years of age, they consider as part of the work farm routine as handling and application of pesticides on farm. In addition, they are involved with purchase and storage of pesticides for farming purposes. Local farmers' association and farm extension officers were approached to facilitate farmers' participation in the study where convenient snowball sampling technique applied in selecting participants.

2.2 *Survey Questionnaire Development and Variables*

Structured pretested questionnaire was used to collect data among target population which comprise of their socio-demographic characteristics, pesticide storage and disposal, environmental fate, attitudes regard hazardous effect of pesticides, source of pesticide knowledge and health issues. Farmer's safety knowledge factors considered during handling of pesticide on the farm and possible health and environmental effect include common safety practices adopted during and after pesticide use on farm was also considered. Variables measured in the study include farmers' attitudes regards pesticides application, storage and associated health and environmental impact. These were measured using a 5-point Likert scale consisting of eight items. Response was set as; strongly disagree = 1, disagree = 2, undecided = 3, agree = 4 and strongly agree = 5 respectively. At the end of the survey period, 531 based on the quality of response provided were adopted in the study.

2.3 *Statistical Methods*

Statistical package for social sciences (SPSS) 28.0 software package for Windows was used to analyze the data that informed the study outcome. Prior to the analysis, the collected data were inspected and cleaned for any irregularity and responses with significant percentage of missing answers were expunged. Descriptive results were expressed as frequencies and percentages to evaluate farmers response regards each variable. Frequency and central tendencies were applied to measure participants categorical data related to their knowledge and attitude towards pesticide application and its associated impact. The Chi-square test ($\chi 2$) was used to measure the possible association between nominal variables. We used $\alpha \leq 0.05$ as a criterion for statistical significance.

2.4 *Ethical Considerations*

Informed consent of participants was obtained prior to each taking part in the study. The study was conducted in accordance with Helsinki declaration and its associated protocol approved by the Ethics Committee of the Department of Agricultural Technology, Forestry Research Institute of Nigeria, Jos (FCFJ/MMU/001/02/2021), on 9 February 2021.

3 Results

Table 1 represents the demographics of farmers who participated in this study. The majority of them were men (79.3%) between the age of 31–40 years old (28.7%). Almost 4.3% of them were still practicing farming although they were above 60 years old. None of them had went to universities, and only 5.6% had no formal educational background, and 29.2% had secondary level education. Relatedly, 29.1% of the farmers had 6–10 years of experience in pesticide use while 26.7% had experience of 11–15 years. Thus, most of them are well-versed. In terms of health symptoms, a total of twelve health symptoms seems to be the most common occurrence, however, the most reported out of the twelve were headaches (27.3%), dizziness (12.5%), itchy eyes (14.3%) and skin irritation (10%).

Table 2 presented an associated link between the participants' pesticide knowledge with education level attained and years spent working with pesticides. Overall, the table highlights important areas where education and experience have a major impact on understanding of pesticide safety.

Farmers' knowledge on the impact of pesticide on human health was significantly correlated by years of farm practice experience and highest educational level (< 0.001 and < 0.001, respectively). Similarly, years of experience and educational attainment were substantially linked (0.002 and < 0.001, respectively) with awareness of the effects of pesticides on the environment. The knowledge on the importance of reading the safety data sheet/label on the pesticide pack prior to use (< 0.001 and < 0.001 respectively) is also directly correlated to both, the number of years working with pesticide on farm and highest level of education. The importance of properly storing pesticides before and after use on farm and safe way to dispose pesticides residue/expired products were correlated with the number of years working with pesticide on farm (< 0.001 and < 0.001 respectively) but not highest educational level. Knowledge of how to use PPE was significantly correlated with education (0.041), but not

Table 1 Farmer's demographic characteristics (N = 531)

Variables	Frequency (n)	Percentage (%)
Age in Years		
> 20	17	3.3
21–31	80	15.5
31–40	148	28.7
41–50	170	33
51–60	78	15.1
> 60	22	4.3
Gender		
Male	407	79.3
Female	98	19.1
Education		
No formal education	29	5.6
Primary school	33	6.4
Secondary school	150	29.2
Tertiary school	302	58.8
Years working in the Pesticide farm		
0–5	123	23.8
6–10	150	29.1
11–15	138	26.7
16–20	69	13.4
> 20	36	7
Commonly experienced health symptoms associated with pesticide use)		
Headache	137	27.3
Dizziness	63	12.5
Skin irritation	50	10
Vomiting	15	3
Nausea	47	9.4
Itchy eyes	72	14.3
Coughing	21	4.2
Stomachache	26	5.2
Poor vision	8	1.6
Shortness of breath	8	1.6
Excessive sweating	15	3
Fatigue	26	5.2
Other form of health effect	14	2.8

Note All variables (categorical) are presented as frequency and percentage

[a] Multiple responses allowed

Table 2 Knowledge level regarding pesticide use and impact in relation to level of education and years of working with pesticides

Knowledge questions	Years working handling pesticide	Highest Level of Education
	2-sided significance*	
Pesticide does present health effect in humans?	< 0.001*	< 0.001*
Pesticides affect health of livestock?	0.261	< 0.001*
Pesticide has effects on the environment?	0.002*	< 0.001*
Pesticides are considered essential for high crop yield?	0.506	0.616
I am aware of how to store if not in use?	< 0.001*	0.377
I am aware of safe way to dispose pesticides residue/expired products?	< 0.001*	0.454
I am aware of how-to use of personal protective equipment when handling pesticide?	0.385	0.041*
I am aware of the safe means to apply or spray pesticides on the farm?	0.923	0.035*
I read safety data sheet/label on the pesticide pack prior to use?	< 0.001*	< 0.001*

p-value < 0.05 indicates significance
* Denotes Chi-square test

with experience ($p = 0.385$). Similarly for knowledge of pesticide effects on livestock health and application techniques, which were substantially correlated with educational attainment (< 0.001 and 0.035, respectively), but not years of experience (0.261 and 0.923, respectively). Finally, the need of pesticide use for higher agricultural production was not shown to be significantly correlated with neither year of experience or educational level (0.506 and 0.616), indicating a potential subject for further research.

4 Discussion

The need for increased farm produce yield has led to frequent unsafe handling, storage of pesticide alongside unsafe disposal of used containers among farm workers, potentially leading to adverse environmental consequences and both acute and chronic health effects on humans [27]. This could be associated to a lack of education, safety awareness and attitude regarding the risk from exposure to these harmful products. In this study, 79.3% of the farmers were males, and most of them (94.4%) have been to schools but never to universities, and 5.6% have no formal education of any kind (Table 1). This could mean that those with lower educational level take the road of farming and agriculture, maybe due to accessibility, economic necessities and lower entry barriers. In addition, the present study showed many farmers experienced ill-health and acute pesticide poisoning symptoms when they encounter these products thereby expanding on the earlier known health effect among these group of farm workers. Headache is the most common type of symptom reported (27.3%), followed by itchy eyes (14.3%), dizziness (12.5%), and skin irritation (10%). Shortness of breath and poor vision, which are more serious health issues, were reported by only 1.6% of the farmers. Similar findings have been reported in many countries, such as southwestern Nigeria, Morrocco, Ethiopia and Pakistan [37]. One of the reasons of ill-health outcomes described in our study, despite adequate pesticide safety knowledge identified among the participants, may have been attributed with inadequate regulatory systems that allows farmers have access to internationally regulated pesticides due to high toxicity and exposure potentials [5632, 45]. These findings require urgent action, given the potential for climate change to decrease crop resilience towards Rising temperatures, heat stress and altered rain patterns, resulting in greater reliance on pesticides to protect the production of crops. Increased pesticide uses for higher crop yields, combined with inadequate education and insufficient regulatory oversight increase health and environmental risks, particularly in the phase of climate change [16, 40].

In LMICs, the unsafe pesticides exposure among farmers is correlated to the insufficient and inadequate understanding and awareness of the pesticide products and proper way of handling them (Nwadike et al., 2021), however this was not the case in our study. Most farmers claimed they are knowledgeable and aware that pesticides affect human health (94.6%), livestock health (89.7%), and the environment (85.6%), yet farmers' knowledge regarding pesticide utilization were driven by high crop yield, as 89.8% stated that pesticide use is unavoidable for high crop yield. Also, the majority confirmed that storage knowledge pre and post usage (86.9%), safe way to dispose residues and expired products (81.9%), safe use of PPE while handling pesticides (94.5%), safe farm application of pesticides (91.2%), and reading Safety Data Sheet/Label of pesticides (87.9%) are all important to protect self from health impact of pesticide use. It is evident that despite farmers' awareness of the health and environmental risks associated with pesticide use, their practices remain largely driven only by high crop yields, highlighting a critical gap between knowledge and application of safety measures. Hence the need for further study that evaluate farmers, knowledge attitude and practice regarding safe pesticide and associated climate change impact [27].

Farmers' safety knowledge was positively influenced by educational level and years of farm practice experience. The data showed that the knowledge of farmers on the effect of pesticide on human health and environment is both coming from experience in the agricultural land (< 0.001 and 0.002 respectively) and dependent on education level (< 0.001 and < 0.001 respectively). Also, the knowledge on the importance of reading the safety data sheet/label on the pesticide pack prior to use (< 0.001 and < 0.001 respectively) is also directly correlated to both years working with pesticide on farm and highest level of education. These findings suggest that farmers with higher education levels have more scientific information on pesticide impacts on human health, livestock, and the environment which leads them to take more informed decisions, such as use protective equipment, safely apply pesticides, and read the safety data sheets or labels when handling pesticides. In addition to that, the hands-on practical experience in farming makes farmers learn from experience, such as what happens if they spray a specific kind of banned pesticide or if they don't wear their PPEs. However, it is imperative that farmers with lower educational levels and less experience may lack critical knowledge of pesticide safety and climate change impacts, leading to improper handling and increased health and environmental risks.

The importance of properly storing pesticides before and after use on farm and safe way to dispose pesticides residue/expired products were correlated with the number of years working with pesticide on farm (< 0.001) but not with highest educational level. It could be that farmers form habits on pesticide handling from experiences, especially that they don't learn this in formal school education which is more linked to theoretical knowledge, and learnings from various media sources. Thus, there should be tailored education and training programs around safe pesticide management and disposal practices to reduce health and environmental risks, prevention of the release of pesticide residues into the surrounding environmental media and reduce farmers' exposure [47].

Ensuring proper use of personal protective equipment when handling pesticide and safely applying pesticides were directly correlated with the highest education level of farmers (< 0.041 and 0.035 respectively) but not years of experience. This highlights the need for better use of PPE and continuous monitoring for farmers to make sure they are property and continuously using them to reduce pesticide exposure risks. Also, there is a need for further assessment regarding the effectiveness of PPE usage, especially under extreme working conditions. Earlier research has demonstrated that at elevated temperatures, class 1B "highly hazardous" pesticide like Dichlorvos has a substantial ability to permeate gloves and absorbed human skin (Thredgold et al., 2019). Considering the climate change impacts and expected increase in pesticide use coupled with exposure among farmers, tailored educational and training programs to help change their perceptions and modify behavior towards safe handling of pesticides is essential to mitigate the increasing health and environmental burden arising from it. Another action to be considered and implemented is the close monitoring of farmers by public agencies like government and municipalities or non-governmental organizations to prevent the misuse or reduce the overuse of pesticides. In addition, to mitigate pesticide exposure risks, it is crucial to ensure proper PPE usage, provide continuous monitoring, and implement tailored educational programs, especially given the rising impacts of climate change and the coupled increase in pesticide use. Furthermore, there is need to strengthen policy to regulate pesticide use, which will go along a long way to limit farmers access to banned products [9].

5 Limitations

There are some limitations within the study including the use of convenience sampling, which may limit the generalizability of the findings to the wider population of farmers. Additionally, no information was collected on actual practices and behaviors of farmers when it comes to the use and handling of pesticides, thus, preventing the researchers from linking knowledge to practice. Besides, some responses were self-reported, which may have introduced biases such as social desirability bias.

To address these limitations, it is recommended that future research adopt random or stratified sampling methods to improve representativeness and increase both geographic and demographic coverage. It is also suggested to incorporate triangulation by including qualitative questions and analyses to strengthen and enhance validy of the findings.

6 Conclusion

With the rise in temperature and frequent use of pesticide for farming activity, many farmers are at risk of pesticide exposure with long-term health impact and reduced quality of life. On this premise, there is an urgent need to ensure tailored interventions are factored into farm extension workers service delivery to farmers to help raise their safety awareness and improve their behavior to guarantee productivity and resilience among the farmers while navigating through existing environmental challenges. Other forms of pest control strategy should be encouraged as an alternative to frequent pesticide application, which is mostly persistent.

References

1. Abadi B (2018) The determinants of cucumber farmers' pesticide use behavior in central Iran: Implications for the pesticide use management. J Clean Prod 205:1069–1081
2. Ahmad MF, Ahmad FA, Alsayegh AA, Zeyaullah M, AlShahrani AM, Muzammil K, Saati AA, Wahab S, Elbendary EY, Kambal N, Abdelrahman MH, Hussain S (2024) Pesticides impacts on human health and the environment with their mechanisms of action and possible countermeasures. Heliyon
3. Akter M, Fan L, Rahman MM, Geissen V, Ritsema CJ (2018) Vegetable farmers' behaviour and knowledge related to pesticide use and related health problems: a case study from Bangladesh. J Clean Prod 200:122–133
4. Amr S, Dawson R, Saleh DAA, Magder LS, St. George DM, El-Daly M, Squibb K, Mikhail NN, Abdel-Hamid M, Khaled H, Loffredo CA (2015) Pesticides, gene polymorphisms, and bladder cancer among Egyptian agricultural workers. Arch Environ Occup Health 70(1):19–26
5. Arbuckle TE, Sever LE (1998) Pesticide exposures and fetal death: a review of the epidemiologic literature. Crit Rev Toxicol 28(3):229–270
6. Asrat P, Simane B (2018) Farmers' perception of climate change and adaptation strategies in the Dabus watershed North-West Ethiopia. Ecol Process 7(1):1–13

7. Bagheri A, Emami N, Allahyari MS, Damalas CA (2018) Pesticide handling practices, health risks, and determinants of safety behavior among Iranian apple farmers. Hum Ecol Risk Assess Int J 24(8):2209–2223
8. Barsouk A, Thandra KC, Saginala K, Rawla P, Barsouk A (2021) Chemical risk factors of primary liver cancer: an update. Hepatic Med Evid Res 179–188
9. Blanco-Muñoz J, Lacasaña M (2011) Practices in pesticide handling and the use of personal protective equipment in Mexican agricultural workers. J Agromedicine 16(2):117–126
10. Bondori A, Bagheri A, Sookhtanlou M, Allahyari MS, Damalas CA (2018) Pesticide use in cereal production in Moghan plain, Iran: risk knowledge and farmers' attitudes. Crop Prot
11. Bouchard MF, Chevrier J, Harley KG, Kogut K, Vedar M, Calderon N, Eskenazi B (2011) Prenatal exposure to organophosphate pesticides and IQ in 7-year-old children. Environ Health Perspect 119(8):1189–1195
12. Chen R, Huang J, Qiao F (2013) Farmers' knowledge on pest management and pesticide uses in Bt cotton production in china. China Econ Rev 27:15–24
13. Damalas CA (2011). Pesticide exposure, safety issues, and risk assessment indicators. Int J Environ Res Public Health 8(5):1402–1419. https://doi.org/10.3390/ijerph8051402
14. Damalas CA, Koutroubas SD (2016) Farmers' exposure to pesticides: toxicity types and ways of prevention. Toxics 4(1):1
15. de Graaf L, Boulanger M, Bureau M, Bouvier G, Meryet-Figuiere M, Tual S, Baldi I (2022) Occupational pesticide exposure, cancer and chronic neurological disorders: a systematic review of epidemiological studies in greenspace workers. Environ Res 203:111822
16. Delcour I, Spanoghe P, Uyttendaele M (2015) Literature review: Impact of climate change on pesticide use. Food Res Int 68:7–15
17. Deutsch CA, Tewksbury JJ, Huey RB, Sheldon KS, Ghalambor CK, Haak DC, Martin PR (2008) Impacts of climate warming on terrestrial ectotherms across latitude. Proc Natl Acad Sci 105(18):6668–6672
18. Fang L, Liao X, Jia B, Shi L, Kang L, Zhou L, Kong W (2020) Recent progress in immunosensors for pesticides. Biosens Bioelectron 164:112255
19. Faria NMX, Fassa AG, Meucci RD, Fiori NS, Miranda VI (2014) Occupational exposure to pesticides, nicotine and minor psychiatric disorders among tobacco farmers in southern Brazil. Neurotoxicology 45:347–354
20. Fierros-González I, López-Feldman A (2021) Farmers' perception of climate change: a review of the literature for Latin America. Front Environ Sci 9:672399
21. Food and Agriculture Organization of the United Nations (2022) Pesticides use and trade. 1990–2022: New FAO STAT data release. Retrieved on September 10 from https://www.fao.org/statistics/highlights-archive/highlights-detail/pesticides-use-and-trade-1990-2022/en#:~:text=Main%20findings%3A,and%20a%20doubling%20since%201990
22. García J, Ventura MI, Requena M, Hernández AF, Parrón T, Alarcón R (2017) Association of reproductive disorders and male congenital anomalies with environmental exposure to endocrine active pesticides. Reprod Toxicol 71:95–100
23. Gargouri B, Yousif NM, Attaai A, Bouchard M, Chtourou Y, Fiebich BL, Fetoui H (2018) Pyrethroid bifenthrin induces oxidative stress, neuroinflammation, and neuronal damage, associated with cognitive and memory impairment in murine hippocampus. Neurochem Int 120:121–133
24. Garry VF, Harkins ME, Erickson LL, Long-Simpson LK, Holland SE, Burroughs BL (2002) Birth defects, season of conception, and sex of children born to pesticide applicators living in the Red river valley of Minnesota, USA. Environ Health Perspect 110(3):441–449
25. Gesesew HA, Woldemichael K, Massa D, Mwanri L (2016) Farmers knowledge, attitudes, practices and health problems associated with pesticide use in rural irrigation villages Southwest Ethiopia. PLoS ONE 11(9):e0162527
26. Gutierrez AP, Ponti L, d'Oultremont T, Ellis CK (2008) Climate change effects on poikilotherm tritrophic interactions. Clim Change 87(1):167–192
27. Jallow MF, Awadh DG, Albaho MS, Devi VY, Thomas BM (2017) Pesticide knowledge and safety practices among farm workers in Kuwait: results of a survey. Int J Environ Res Public Health 14(4):340

28. Khan BA, Nadeem MA, Nawaz H, Amin MM, Abbasi GH, Nadeem M, Ayub MA (2023) Pesticides: impacts on agriculture productivity, environment, and management strategies. Emerging contaminants and plants: Interactions, adaptations and remediation technologies. Springer International Publishing, Cham, pp 109–134
29. Liu H, Reynolds GP, Wei X (2023) The influence of agricultural work and plasma uric acid on hospital admission for alzheimer's disease. J Alzheimers Dis 92(4):1283–1287
30. Maroni M, Fanetti AC, Metruccio F (2006) Risk assessment and management of occupational exposure to pesticides in ag-riculture. MedLav 97:430–437
31. Mattila T, Santonen T, Andersen HR, Katsonouri A, Szigeti T, Uhl M, Wąsowicz W, Lange R, Bocca B, Ruggieri F, Kolossa-Gehring M, Sarigiannis DA, Tolonen H (2021) Scoping review—the association between asthma and environmental chemicals. Int J Environ Res Public Health 18(3):1323
32. Moda HM, Anang DM, Moses N, Manjo FM, Joshua VI, Christopher N, Doka P, Danjin M (2022) Pesticide safety awareness among rural farmers in Dadinkowa, Gombe State, Nigeria. Int J Environ Res Public Health 19(21):13728
33. Mohanty MK, Behera BK, Jena SK, Srikanth S, Mogane C, Samal S, Behera AA (2013) Knowledge attitude and practice of pesticide use among agricultural workers in Puducherry, South India. J Forensic Leg Med 20(8):1028–1031
34. Müller R, Bandow C, Seeland A, Fennel D, Coors A, Ebke PK, Förster B, Martinez-Arbizu P, Moser T, Oetken M, Renz J, Römbke J, Schulz N, Oehlmann J (2010) BiK F AdaMus: a novel research project studying the response and adaptive potential of single species and communities to climate change in combination with other stressors. J Soils Sediments 10:718–721
35. Muñoz-Quezada MT, Lucero BA, Iglesias VP, Muñoz MP, Cornejo CA, Achu E, Baumert B, Hanchey A, Concha C, Brito AM, Villalobos M (2016) Chronic exposure to organophosphate (OP) pesticides and neuropsychological functioning in farm workers: a review. Int J Occup Environ Health 22(1):68–79
36. Noyes PD, McElwee MK, Miller HD, Clark BW, Van Tiem LA, Walcott KC, Erwin KN, Levin ED (2009) The toxicology of climate change: environmental contaminants in a warming world. Environ Int 35(6):971–986
37. Nwadike C, Joshua VI, Doka PJ, Ajaj R, Abubakar Hashidu U, Gwary-Moda S, Danjin M, Moda HM (2021) Occupational safety knowledge, attitude, and practice among farmers in Northern Nigeria during pesticide application—a case study. Sustainability 13(18):10107
38. Öztaş D, Kurt B, Koç A, Akbaba M, İlter H (2018) Knowledge level, attitude, and behaviors of farmers in Çukurova region regarding the use of pesticides. Biomed Res Int 2018(1):6146509
39. Pesticide Action Network (n.d.) Pesticide action network. Retrieved on 27 Oct 2024 from https://www.panna.org/
40. Priyadarshi A, Khuder SA, Schaub EA, Shrivastava S (2000) A meta-analysis of Parkinson's disease and exposure to pesticides. Neurotoxicology 21(4):435–440
41. Rafeeinia A, Asadikaram G, Moazed V, Darabi MK (2023) Organochlorine pesticides may induce leukemia by methylation of CDKN2B and MGMT promoters and histone modifications. Gene 851:146976
42. Rani L, Thapa K, Kanojia N, Sharma N, Singh S, Grewal AS, Kaushal J (2021) An extensive review on the consequences of chemical pesticides on human health and environment. J Clean Prod 283:124657
43. Roos J, Hopkins R, Kvarnheden A, Dixelius C (2011) The impact of global warming on plant diseases and insect vectors in Sweden. Eur J Plant Pathol 129:9–19
44. Röösli M, Fuhrimann S, Atuhaire A, Rother HA, Dabrowski J, Eskenazi B, Dalvie MA (2022) Interventions to reduce pesticide exposure from the agricultural sector in Africa: a workshop report. Int J Environ Res Public Health 19(15):8973
45. Rosenzweig C, Iglesius A, Yang XB, Epstein PR, Chivian E (2001) Climate change and extreme weather events-Implications for food production, plant diseases, and pests
46. Sam KG, Andrade HH, Pradhan L et al (2008) Effectiveness of an educational program to promote pesticide safety among pesticide handlers of South India. Int Arch Occup Environ Health 81:787–795. https://doi.org/10.1007/s00420-007-0263-3

47. Sandoghi A, Yousefi A, Amini AM (2016) Farmers' knowledge, attitude and practice of healthy greenhouse cucumber and tomato production in Isfahan County. J Soil Plant Interact Isfahan Univ Technol 7(3):155–167
48. Schneider L, Rebetez M, Rasmann S (2022) The effect of climate change on invasive crop pests across biomes. Curr Opin Insect Sci 50:100895
49. Senthil-Nathan S (2021) Effects of elevated CO2 on resistant and susceptible rice cultivar and its primary host, brown planthopper (BPH), Nilaparvata lugens (Stål). Sci Rep 11(1):8905
50. Sharafi K, Pirsaheb M, Maleki S, Arfaeinia H, Karimyan K, Moradi M, Safari Y (2018) Knowledge, attitude and practices of farmers about pesticide use, risks, and wastes; a cross-sectional study (Kermanshah, Iran). Sci Total Environ 645:509–517
51. Skendžić S, Zovko M, Živković IP, Lešić V, Lemić D (2021) The impact of climate change on agricultural insect pests. Insects 12(5):440
52. Thredgold L, Gaskin S, Quy C, Pisaniello D (2019) Exposure of agriculture workers to pesticides: the effect of heat on protective glove performance and skin exposure to dichlorvos. International J Environ Res Public Health 16(23):4798
53. Tudi M, Daniel Ruan H, Wang L, Lyu J, Sadler R, Connell D, Chu C, Phung DT (2021) Agriculture development, pesticide application and its impact on the environment. Int J Environ Res Public Health 18(3):1112
54. Usmail AJ, Maja MM, Lakew AA (2023) Farmers' perceptions of climate variability and adaptation strategies in the rural areas of Dire Dawa administration, eastern Ethiopia. Heliyon 9(5)
55. Xin F, Susiarjo M, Bartolomei MS (2015) Multigenerational and transgenerational effects of endocrine disrupting chemicals: a role for altered epigenetic regulation. In: Seminars in cell & developmental biology, vol 43. Academic Press, pp 66–75
56. Zinyemba C, Archer E, Rother HA (2021) Climate change, pesticides and health: considering the risks and opportunities of adaptation for Zimbabwean smallholder cotton growers. Int J Environ Res Public Health 18(1):121

Climate Change, Hunger, and Extreme Poverty in Africa: A Case Study of Egypt

Khaled El-Sayed Hassan

Abstract The global crises of poverty and hunger, exacerbated by climate change, affect millions worldwide. This study investigates climate change's impact on achieving SDGs 1 and 2, focusing on hunger and poverty in Africa, with a specific focus on Egypt. Findings reveal that climate variability harms livestock, fisheries, and food security, while rising sea levels and melting ice increase land salinity, reducing agricultural productivity. In Egypt, rising sea levels threaten the Nile Delta, jeopardizing agricultural output. The study recommends infrastructure enhancements, international cooperation, and promoting heat-tolerant, water-efficient crops to address these challenges.

Keywords Climate change · Hunger · Extreme poverty · SDGs

1 Introduction

Climate change intensifies hunger and extreme poverty in Africa. In 2022, over 282 million Africans faced severe food insecurity due to climate-related factors [16]. Climate disruptions lower agricultural yields, increase food prices, and deepen poverty [27]. Sub-Saharan Africa, with 433 million people living in extreme poverty in 2021, is highly affected [52]. In Egypt, 5.8 million people were food insecure in 2022, with climate impacts diminishing water resources and agricultural productivity [16].

K. E.-S. Hassan (✉)
Former Demographic Consultant at UNDP-Kuwait, Kuwait City, Kuwait
e-mail: kehassan@gmail.com

Vice-president of the Egyptian Society for Migration Studies (EGYMIG), Cairo, Egypt

S. Al-Naemi et al. (eds.), *Water and Food Security in the Face of Climate Change: Challenges and Opportunities for Resilience*, World Sustainability Series,
https://doi.org/10.1007/978-3-032-00098-9_17

1.1 Study Objectives

This study explores the relationship between climate change, hunger, and poverty in Africa and Egypt, assessing current and future impacts, and proposing mitigation strategies.

1.2 Study Methodology

A systematic review of 85 studies was conducted, with 63 selected based on relevance, reliability, quality, and recent publication [32, 34].

1.3 Literature Review

Climate Change and Agriculture in Africa. Climate change reduces agricultural yields and food security [22]. Vulnerable communities are at risk due to climate shocks [40].

Food Security in Africa. The FAO [15] shows how climate change worsens food insecurity by impacting agricultural output and livestock health.

Poverty and Vulnerability. [42] states that climate change increases poverty risks by affecting food prices and productivity.

Climate Change in Egypt. Climate-related challenges in Egypt, including reduced water resources and rising temperatures, threaten food security [2, 35].

SDGs and Climate Change. Climate change impedes the achievement of SDGs 1 and 2 in Africa [23]. Adapting development policies to climate change is crucial.

Future Scenarios. Without significant adaptation, food insecurity and poverty will worsen in Africa [6, 27].

2 Climate Change, Hunger, and Extreme Poverty

The link between climate change, hunger, and extreme poverty is evident, as climate change worsens vulnerabilities in food systems and economies.

2.1 Definitions

Climate change refers to long-term shifts in temperature, precipitation, and extreme weather due to human activities like fossil fuel burning [25, 26]. These shifts affect agriculture and livelihoods.

Hunger is the lack of sufficient food, while extreme poverty is living on less than $1.90 per day, leading to severe deprivation. Climate change worsens both by increasing extreme weather events and disrupting food production.

2.2 Manifestations

Climate Change Impact. Rising temperatures, sea levels, and erratic rainfall reduce yields. In Egypt, salinity from rising sea levels threatens the Nile Delta [20]. Droughts and unpredictable rainfall impact irrigation and food security [47].

Hunger Consequences. Food insecurity and resource shortages lead to conflicts. Climate change intensifies hunger, as seen in Egypt's declining agricultural productivity [10, 18].

Extreme Poverty Effects. Agriculture-dependent communities suffer as climate change restricts resources. Water scarcity and dam projects worsen poverty in Egypt [33], with gender disparities amplifying economic hardships [21].

2.3 Theoretical Frameworks

Climate Change and Hunger. Rising temperatures, erratic rainfall, and extreme weather impact African food production. In sub-Saharan Africa, where 60% rely on farming, droughts have left millions food-insecure [13, 14].

Staple crops like maize face threats, with yields in the Arab African region projected to drop by 40% by 2050 [41]. In Egypt, rising sea levels may reduce arable land by 15% by 2050, harming rice and wheat production [39].

Framework Mechanisms

Declining Productivity. Climate shifts lower yields.

Rising Food Prices. Lower supply increases costs, limiting access.

Vulnerability of the Poor. Food costs consume a significant portion of low-income household budgets.

Regional Disparities. Sub-Saharan Africa, reliant on rain-fed agriculture, faces heightened risks.

Key Measures

Adaptation. Climate-resilient crops, improved irrigation, and heat-resistant agriculture.

Social Protection. Strengthening safety nets to protect vulnerable populations.

Climate change intensifies hunger and poverty, particularly in Africa, where agriculture and water resources are under strain. Addressing these challenges requires global cooperation, sustainable farming investments, and policy reforms.

Table 1 and Fig. 1 illustrate undernourishment rates across continents and African regions (2005–2022). Africa accounts for 38.4% of the global undernourished population, second only to Asia (54.7%). Central Africa has the highest undernourishment rate (29.1%) due to conflicts affecting farming, particularly in the DRC. East Africa's rate (28.5%) stems from drought, conflict, and displacement in Somalia, South Sudan, and parts of Ethiopia and Kenya. Sub-Saharan Africa's rate was 22.5% in 2022, worsened by climate change, conflict, and weak agricultural infrastructure.

Table 1 Prevalence of undernourished by regions, 2005–2022

Regions	Years					
	2005	2010	2015	2020	2021	2022
World	12.1	8.6	7.9	8.9	9.3	9.2
Africa	19.2	15.1	15.8	18.7	19.4	19.7
Northern Africa	6.2	4.7	5.4	6.0	6.9	7.5
Sub-Saharan Africa	22.5	17.6	18.2	21.6	22.2	22.5
Eastern Africa	31.7	23.8	24.6	28.1	28.4	28.5
Central Africa	31.9	22.5	23.3	27.6	28.5	29.1
Southern Africa	5.1	7.2	9.3	9.5	10.0	11.1
Western Africa	12.2	10.8	10.6	13.7	14.5	14.6

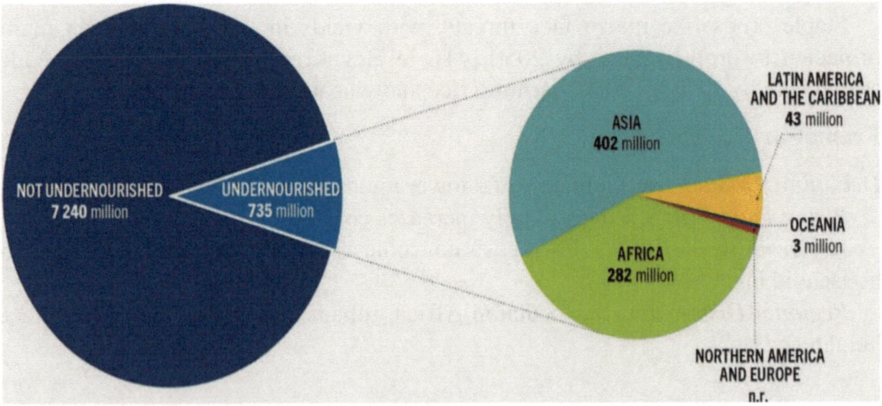

Fig. 1 The global distribution of undernourished by continents 2022. Sources [25, 26]

Climate Change and Poverty. Climate change worsens poverty in Africa, particularly due to agriculture's sensitivity to climate shifts. Over 60% of sub-Saharan Africa's population relies on agriculture [53, 54]. Droughts, erratic rainfall, and extreme temperatures reduce crop and livestock productivity, heightening food insecurity in nations like Mozambique, Ethiopia, and Nigeria. In the Sahel, desertification shrinks arable land, triggering migration and conflict [38]. Egypt faces declining Nile water flow due to reduced Ethiopian Highlands rainfall, threatening agriculture and water security [7]. Rising sea levels further endanger Egypt's northern coast and Nile Delta, intensifying food shortages.

Framework Mechanisms

Higher vulnerability. Poor communities are most exposed and least equipped to adapt.

Economic instability. Climate events disrupt agriculture and infrastructure, weakening economies.

Resource competition. Scarcity fuels migration and conflict, exacerbating instability.

Displacement. Shrinking farmland forces migration to overburdened areas, deepening poverty.

Mitigation Measures

Resilience-building. Climate-smart agriculture, water management, and safety nets.

Global action. Support for adaptation policies to aid vulnerable populations.

Figures 2 and 3 illustrate Africa's extreme poverty trends versus global patterns and its share of global poverty from 1990 to 2020.

Figure 2 highlights a slower decline in Extreme poverty in Africa than in the rest of the world. Between 1994 and 2019, Africa's poverty rate fell from 51 to 34%—a 33% drop. Meanwhile, global poverty fell from 35% in 1990 to just 3% in 2019, a 90% reduction. This contrast highlights Africa's persistent economic challenges despite global progress.

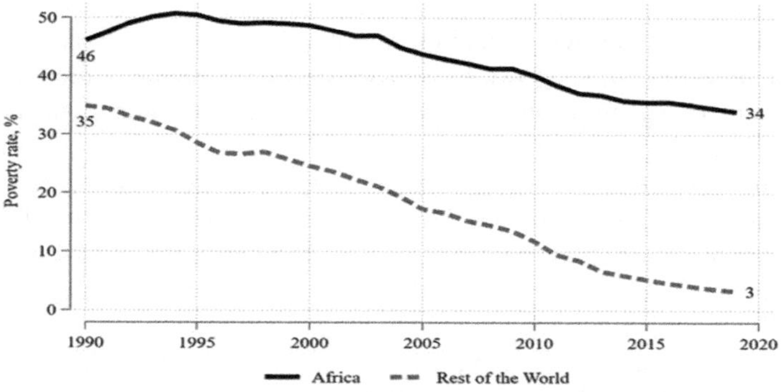

Fig. 2 Trends in extreme poverty rate in Africa, 1990–2020. *Source* [1]

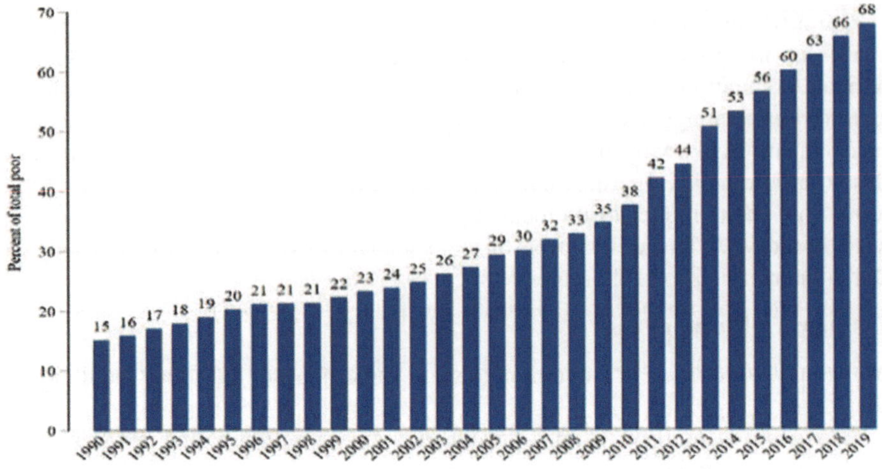

Fig. 3 Share of Africa in the global extreme poverty, 1990–2019. *Source* [1]

Figure 3 shows Africa's increasing share of global poverty, rising from 15% in 1990 to 68% in 2019, largely due to stronger economic growth in other regions. East Asia, the Pacific, and South Asia, which had high poverty rates in 1990, experienced substantial income growth, with per capita income rising by 3.1% and 3.8% annually, respectively. In contrast, Sub-Saharan Africa saw only 1.5% growth, while the Middle East and North Africa recorded just 0.1% [53, 54].

3 Impact of Climate Change on Hunger and Extreme Poverty in Africa

3.1 Current Impacts

Africa's reliance on rain-fed agriculture, weak adaptive capacity, and widespread poverty increase its vulnerability to climate change. Irregular rainfall, droughts, heatwaves, and rising temperatures disrupt food production, intensifying hunger and poverty.

In the Horn of Africa, recurring droughts have devastated farming and livestock. The 2017 drought in Somalia left 6.7 million people in need of food aid, triggering displacement and worsening poverty [13, 14]. Similarly, the 2015–2016 El Niño drought slashed Zimbabwe's maize yield by 53%, affecting Malawi and Mozambique as well [37].

In the Sahel, desertification and unpredictable rainfall are shrinking agricultural output. Lake Chad has diminished by 90% since the 1960s, impacting over 10 million people in the region [44]. Climate-related food shortages are aggravated by poor infrastructure, governance issues, and conflict. In South Sudan and the Democratic Republic of Congo, climate and political instability have disrupted food supply chains, worsening hunger and poverty [].

Table 2 highlights key climate change impacts on agriculture and food security in Africa.

It shows that climate change severely impacts African agriculture, reducing yields and worsening food insecurity. In the Horn of Africa, droughts have affected 6.7 million people in Somalia [13, 14]. The 2015–2016 El Niño drought cut Zimbabwe's maize production by 53%, exacerbating food shortages [37]. In Southern Africa and the Sahel, desertification and erratic rainfall could lower yields by 20–50% by 2050 [28]. Ethiopia experienced a 90% crop loss due to severe droughts [12], while heatwaves in Kenya slashed maize and wheat output by 40%, inflating food prices [51]. Rising sea levels in Kenya and Nigeria damaged rice farms, intensifying food insecurity [46]. Shrinking Lake Chad has affected 10 million people in the region [44].

Table 2 Current impacts of climate change on agricultural production and food security in Africa

Region/country	Climate impact	Effect on agriculture	Impact on food security	Source
Horn of Africa	Recurrent droughts	Severe reduction in crop yield, loss of livestock	6.7 million people in need of food assistance (Somalia, 2017)	[13]
Southern Africa	El Niño-induced drought (2015–2016)	53% decline in maize production (Zimbabwe)	Widespread hunger, increased food imports, rising national debt	[37]
West Africa (Sahel)	Desertification, erratic rainfall	20–50% decline in crop yields by 2050	Increased food shortages, rural livelihoods threatened	[28]
Ethiopia	Droughts, irregular rainfall	90% loss of crops during severe drought years	Displacement and food insecurity for millions	[12]
Kenya	Increased heatwaves and drought	40% reduction in maize and wheat production	Increased food prices, widespread hunger	[51]
Nigeria	Sea level rise, floods	Rice farming areas flooded, leading to crop losses	Displacement and heightened food insecurity	[46]
lake chad basin	Shrinking water resources	Reduced agricultural land, depletion of fish stock	Over 10 million people facing food shortages	[44]

3.2 Future Impacts

Without urgent action, climate change will further worsen hunger and poverty in Africa. Temperatures are rising faster than the global average, with staple crop yields projected to decline by 50% by 2050 [25, 26]. This will disproportionately harm the poorest, increasing food prices and malnutrition. In East Africa, shifting weather patterns could cut crop yields by 40% in some Kenyan regions by 2030 [19], worsening hunger and mortality.

Water shortages will intensify, with reduced rainfall and rising evaporation in the Nile Basin threatening Egyptian agriculture [48]. In West Africa, sea-level rise will flood farmland, increase soil salinity, and drive migration [38]. Climate change could push 100 million people into extreme poverty by 2030, with Sub-Saharan Africa hit hardest [53, 54]. Given that agriculture employs over 60% of Africans, urgent measures—such as climate-resilient farming, improved water management, and social protection—are essential to reduce risks (Table 3).

Table 3 Future impacts of climate change on agriculture, hunger, and extreme poverty in Africa by 2050

Region/country	Projected impact	Details	Source
Sub-Saharan Africa	Decrease in crop production	Up to 50% reduction in staple crops like maize, millet, and sorghum	[25, 26]
East Africa (Ethiopia, Kenya, Uganda)	Reduced growing seasons and crop yields	Maize and bean yields could decrease by up to 40% by 2030	[19]
Nile river basin (Egypt, Sudan, Ethiopia)	Reduced water availability	Decreased rainfall and increased evaporation rates could lower water availability	[48]
West Africa (Nigeria, Senegal)	Rising sea levels and soil salinization	Inundation of farmland and increased soil salinization leading to reduced agricultural viability	[38]
Poverty levels (Sub-Saharan Africa)	Increase in extreme poverty	An additional 100 million people could be pushed into extreme poverty by 2030	[53, 54]
Agricultural sector (Sub-Saharan Africa)	Economic impact on smallholder farmers	Climate-related shocks could reduce income opportunities and exacerbate poverty	
Food prices (Sub-Saharan Africa)	Increased food prices	Decreased production will drive up food prices, reducing access to nutritious food for many households	

4 Impact of Climate Change on Hunger and Extreme Poverty in Egypt

4.1 Current Impacts

Climate change has worsened hunger and poverty in Egypt, largely due to its dependence on the Nile River for agriculture. Reduced water availability, higher temperatures, and erratic rainfall have led to water shortages, lowering wheat and rice production in the Nile Delta and deepening rural poverty [20]. Rising temperatures also affect livestock, reducing milk and meat yields, increasing food prices, and exacerbating poverty, particularly in Upper Egypt [3]. Egypt's reliance on food imports makes it vulnerable to price shocks, as seen in 2020–2021 when wheat prices surged, straining food subsidies [10, 18].

The Nile Delta faces flooding and saltwater intrusion due to rising sea levels, displacing farmers and fishers, worsening poverty [47]. Extreme weather events like droughts and heatwaves have damaged crops, increasing poverty in Upper Egypt [33]. Food shortages and higher prices have raised malnutrition, especially among rural children [31]. Overall, climate change is disrupting agriculture, water resources, and food imports, deepening economic hardship.

4.2 Future Impacts

Future climate change effects on hunger and poverty in Egypt will intensify:

Water Scarcity & Agricultural Decline. Egypt's water gap could reach 20–30% by 2050, reducing crop yields by 30% and worsening rural poverty [8].

Food Price Inflation. Declining domestic production will increase reliance on imports, driving food prices up by 20–30%, disproportionately affecting the poor [11].

Sea Level Rise & Displacement. A 1-m rise in sea levels by 2100 could submerge parts of the Nile Delta, displacing millions and reducing food production [20].

Health & Malnutrition. Child malnutrition may rise by 20–30% by 2050 due to declining food supply and higher prices [4].

Extreme Weather Events. More frequent heatwaves, droughts, and floods will disrupt farming and infrastructure, raising food insecurity. By 2050, heatwaves up to 50°C could further reduce yields [47].

Coastal Vulnerability & Migration. Up to 10 million people in the Nile Delta may be displaced by 2050 due to rising sea levels and soil salinity, straining urban areas and increasing poverty [57].

Without urgent adaptation measures, climate change will escalate hunger and poverty in Egypt, worsening water scarcity, agricultural losses, and economic instability (Table 4).

Table 4 Current and projected impacts of climate change on hunger and extreme poverty in Egypt (2023–2050)

Indicator	Years				Source
	Current impact	Projected impact			
	2023	2030	2040	2050	
Prevalence of undernourishment (%)	8	10.5	12	15	[16]
Extreme poverty rate (%)	29.7	30.5	33	36.5	[49]
Crop yield loss (%)	5	5	10	15	[27]
Increase in food prices (%)	15	20	30	50	[29]
Vulnerable population (millions)	30	35	40	45	[45]

Projection Classification, Assumptions, and Results

Classification. 2023 Baseline: Represents current hunger and poverty levels as a reference point.

2030 Projection: Short-term outlook with partial adaptation, leading to moderate increases in food insecurity.

2040 Projection: Medium-term scenario showing worsening conditions due to limited adaptation.

2050 Projection: Long-term outlook predicting severe agricultural disruptions, escalating hunger and poverty.

Assumptions. Undernourishment: Declining agricultural productivity increases food insecurity (FAO).

Extreme Poverty: Agricultural losses worsen poverty, especially in low-income groups (World Bank).

Crop Yield Loss: Rising extreme weather events reduce yields (IPCC).

Food Prices: Declining production and higher demand drive price hikes (IFPRI).

Vulnerable Population: Population growth and climate stress increase risks (UNDP).

Main Findings. Undernourishment: Expected to rise from 8% in 2023 to 15% by 2050.

Extreme Poverty: Projected to increase from 29.7% to 36.5%.

Crop Yield Loss: Expected to grow from 5% in 2023 to 15%.

Food Prices: Predicted to rise by 50%.

Vulnerable Population: Estimated to increase from 30 to 45 million by 2050.

5 Policies, Recommendations, and Successful Experiments

To address climate change threats to food security, urgent policy actions are required. This study recommends strategies backed by successful examples:

Climate-Resilient Crops: Promote flood-tolerant crops like those developed by IRRI [30].

Infrastructure and Land Use: Enhance infrastructure to counter climate disruptions, as seen in the Netherlands' water-absorbing urban planning [36].

Heat-Tolerant Plants: Support research on heat-resistant crops, following Jordan's example [43].

Water-Saving Crops: Encourage water-efficient crops like stevia in Egypt [9, 17].

Saline Crops: Grow salt-tolerant plants, as the UAE does using seawater [56].

Targeted Food Policies: Implement food security programs for vulnerable groups, similar to Brazil's Zero Hunger Program [10, 18].

Integrated Social Policies: Combine social and climate policies, as in Ethiopia's Productive Safety Net Program [50].

Carbon Taxes: Use carbon tax revenues to support low-income communities, as in British Columbia [51].

Fossil Fuel Subsidy Reform: Redirect fossil fuel subsidy savings to social programs, as done in Indonesia [24].

Regional Cooperation: Strengthen collaboration, following the African Union's CAADP model [5].

Postponing adaptation and social protection measures heightens risks. Immediate action is essential to secure food security and protect vulnerable populations.

References

1. Abay KA, Yonzan N, Kurdi S, Tafere K (2023) Revisiting poverty trends and the role of social protection systems in Africa during the COVID-19 pandemic. J Afr Econ 32(Suppl 2), ii44–ii68. https://doi.org/10.1093/jae/ejac041
2. Abdelaal HS, Fathy S, Abdelaal HM, El-Ramady H (2019) Grains production prospects and long-run food security in Egypt. Sustainability 11(16):4457
3. Abdel-Fattah M, Galal T, Zahran M (2019) Impact of climate change on livestock production in Egypt: a case study. Afr J Environ Sci Technol 13(2):47–56. Available from https://doi.org/10.5897/AJEST2018.2639
4. Abdelhady D, Fawzy E, Ghazal F (2021) Climate change and malnutrition: a study of future impacts on Egypt's vulnerable populations. Int J Environ Res Public Health 18(8):3955–3967. https://doi.org/10.3390/ijerph18083955
5. AU (2021) CAADP: Comprehensive Africa agriculture development programme. African Union. https://www.au.int
6. Béné C, Newsham A, Davies M et al (2021) Resilience and food security: a review of the evidence. Food Secur 13(2):425–439. https://doi.org/10.1007/s12571-021-01140-2
7. Conway D, Hulme M (2021) Recent fluctuations in the flow of the Nile river and their climatic implications: a review. Prog Phys Geogr Earth Environ. 45(2):155–173. https://doi.org/10.1177/0309133321100622
8. Elsharkawy H, Abdelatif R, Eid H (2021) Projected climate change impacts on crop yields in Egypt by 2050. Clim Chang 169(1–2):117–130. https://doi.org/10.1007/s10584-021-03256-2
9. FAO (2021) Water-efficient crops and climate adaptation in Egypt. Food and Agriculture Organization. https://www.fao.org
10. FAO (2019) Zero hunger program: a pathway to ending hunger and poverty in Brazil. Food and Agriculture Organization. https://www.fao.org

11. FAO (2020) The state of food security and nutrition in the world 2020: Transforming food systems for affordable healthy diets. FAO, Rome. https://doi.org/10.4060/ca9692en. Last accessed 23 Sept 2024
12. Food and Agriculture Organization of the United Nations (2016) The impact of climate change on agriculture and food security in Ethiopia: an analysis of vulnerability and livelihood. FAO, Rome. Available from: http://www.fao.org/3/I6653E/I6653E.pdf
13. Food and Agriculture Organization (2018) Somalia: drought response plan 2017. FAO, Rome. Available from: http://www.fao.org/3/I8424EN/i8424en.pdf
14. Food and Agriculture Organization (2018) The impact of disasters and crises on agriculture and food security 2017. FAO, Rome
15. Food and Agriculture Organization (2015) The state of food insecurity in the world 2015. FAO, Rome
16. Food and Agriculture Organization (2023) The state of food security and nutrition in the World 2023. FAO, Rome. Available from: https://www.fao.org/publications/sofi/2023/en/
17. Food and Agriculture Organization (2021) The state of food security and nutrition in the world 2021: transforming food systems for food security, improved nutrition, and affordable healthy diets for all. FAO, Rome. Available from: https://www.fao.org/documents/card/en/c/cb4474en
18. Food and Agriculture Organization (2019) The state of food security and nutrition in the world 2019: safeguarding against economic slowdowns and downturns. FAO, Rome
19. Funk C, Shukla S, Klotz S (2019) Climate change and agricultural yields in East Africa: a review. Nat Clim Chang 9(1):1–10. Available from: https://www.nature.com/articles/s41558-018-0291-6
20. Ghoneim E (2020) Rising sea level and its implications for Egypt's Nile delta. Clim Chang 162(3):437–449. Available from: https://doi.org/10.1007/s10584-020-02858-z
21. Glazebrook T, Noll S, Opoku E (2020) Gender matters: climate change, gender bias, and women's farming in the global south and north. Preprints. Available from: https://doi.org/10.20944/preprints202006.0302.v1
22. Hallegatte S, Bangalore M, Bonzanigo L, Fay M, Kane T, Narloch U, Rozenberg J, Treguer D, Vogt-Schilb A (2016) Shock waves: managing the impacts of climate change on poverty. The World Bank, Washington, D.C.
23. Hassan K (2022) Climate change and the global sustainable development goals. LAP LAMBERT Academic Publishing GmbH & Co., Germany
24. IEA (2015) Indonesia's fuel subsidy reforms and their impact on climate and social protection. International Energy Agency. https://www.iea.org
25. Intergovernmental Panel on Climate Change (2021) Climate change 2021: impacts, adaptation, and vulnerability. IPCC, Geneva. Available from: https://www.ipcc.ch/report/ar6/wg2/
26. Intergovernmental Panel on Climate Change (2021) Climate change 2021: the physical science basis. Cambridge University Press, Cambridge. Available from: https://www.ipcc.ch/report/ar6/wg1/
27. Intergovernmental Panel on Climate Change (2022) Climate change 2022: impacts, adaptation, and vulnerability. IPCC, Geneva. Available from: https://www.ipcc.ch/report/ar6/wg2/
28. Intergovernmental Panel on Climate Change (2019) Climate change and land: special report on climate change, desertification, land degradation, sustainable land management, food security, and greenhouse gas fluxes in terrestrial ecosystems. IPCC, Geneva. Available from: https://www.ipcc.ch/srccl/
29. International Food Policy Research Institute (IFPRI) (2023) Global food policy report. IFPRI, Washington, DC
30. IRRI (2019) Flood-tolerant rice: Innovations in climate resilience. International Rice Research Institute. https://www.irri.org
31. Janssens C, Havlík P, Krisztin T et al (2020) Global hunger and climate change adaptation through international trade. Nat Clim Chang. 10(9):829–835. Available from: https://doi.org/10.1038/s41558-020-0847-4
32. Liberati A, Altman DG, Tetzlaff J, Mulrow C, Gøtzsche PC, Ioannidis JP et al (2009) The PRISMA statement for reporting systematic reviews and meta-analyses of studies that evaluate

health care interventions: explanation and elaboration. PLoS Med 6(7). https://doi.org/10.1371/journal.pmed.1000100
33. Mason-D'Croz D, Sulser TB, Wiebe K, et al (2019) Agricultural investments and hunger in Africa: Modeling potential contributions to SDG2—zero hunger. World Dev 116, 38–53. Available from: https://doi.org/10.1016/j.worlddev.2018.12.006
34. Moher D, Liberati A, Tetzlaff J, Altman DG (2015) PRISMA group: preferred reporting items for systematic reviews and meta-analyses: the PRISMA Statement. PLoS Med 6(7). https://doi.org/10.1371/journal.pmed.1000097
35. Nassr S (2021) Analysis of climate change effects on food security in Egypt using IMPACT model. Egypt J Agric Econ 31(3):743–755
36. OECD (2020) Climate adaptation in land use planning: the dutch experience. Organisation for Economic Co-operation and Development. https://www.oecd.org
37. ReliefWeb (2017) Southern Africa: humanitarian impact of El Niño, 2015–2016. ReliefWeb. Available from: https://reliefweb.int/report/southern-africa/southern-africa-humanitarian-impact-el-ni241o-2015-2016
38. Rigaud K, de Sherbinin A, Jones B, Bergmann J, Clement V, Ober K, Midgley A (2018) Groundswell: preparing for internal climate migration. The World Bank Group, Washington, DC. Available from: https://openknowledge.worldbank.org/handle/10986/29461
39. Sayed M, Mansour R (2020) Climate change and agricultural productivity in Egypt: impacts and adaptation. Environ Sci Policy. Available from: https://www.sciencedirect.com/science/article/pii/S1462901120302176
40. Schmidhuber J, Tubiello FN (2007) Global food security under climate change. Proc Natl Acad Sci U S A 104(50):19703–19708
41. Sultan B, Gitani S (2016) Agricultural impacts of climate change in West Africa: a review. Clim J. Available from: https://www.climatejournal.org/article/1012736
42. The World Bank (2015) Climate change complicates efforts to end poverty. World Bank, Washington, D.C
43. UNDP (2018) Advances in heat-tolerant crops in the Middle East. United Nations Development Programme. https://www.undp.org
44. UNICEF (2020) Lake Chad basin crisis: UNICEF response plan 2020. United Nations Children's Fund. Available from: https://www.unicef.org/reports/lake-chad-basin-crisis-unicef-response-plan-2020
45. United Nations Development Programme (UNDP) (2023) Human development report 2023. UNDP, New York
46. United Nations Environment Programme (2020) Climate change and floods in Nigeria: risk and adaptation strategies. UNEP, Nairobi. Available from: https://www.unep.org/resources/report/climate-change-and-floods-nigeria-risk-and-adaptation-strategies
47. Waha K, Krummenauer L, Adams S et al (2017) Climate change impacts in the Middle East and Northern Africa (MENA) region and their implications for vulnerable population groups. RegNal Environ Chang 17:1623–1638. Available from: https://doi.org/10.1007/s10113-017-1144-2
48. Wheeler T, von Braun J (2020) The impact of climate change on water resources in the Nile basin. Nature 578(7794), pp 500–503. Available from: https://www.nature.com/articles/s41586-020-2606-1
49. World Bank (2023) Poverty and shared prosperity 2023: reversals of fortune. World Bank, Washington, DC
50. World Bank (2017) Productive safety net program in Ethiopia: a climate resilience success. World Bank. https://www.worldbank.org
51. World Bank (2019) Kenya climate smart agriculture strategy 2017–2026. World Bank, Washington, DC. Available from: https://www.worldbank.org/en/country/kenya/publication/kenya-climate-smart-agriculture-strategy-2017-2026
52. World Bank (2021) Poverty and inequality platform (PIP). World Bank, Washington, DC. Available from: https://pip.worldbank.org/

53. World Bank (2020) Poverty and shared prosperity 2020: reversals of fortune. The World Bank Group, Washington, DC. Available from: https://www.worldbank.org/en/publication/poverty-and-shared-prosperity
54. World Bank (2020) Poverty and shared prosperity: reversals of fortune. World Bank, Washington, DC
55. World Food Programme (2023) Hunger map 2023. WFP, Rome. Available from: https://hungermap.wfp.org/
56. WRI (2020) Salt-tolerant crops and water scarcity solutions in the UAE. World Resources Institute. https://www.wri.org
57. Zohry A (2020) Climate change-induced migration in Egypt: implications for the Nile delta. J Environ Stud 18(4):101–112

Progress of Adapting VarMix® Technology for Ensuring Food Security in Unfavorable Areas in the Philippines

Imeldalyn G. Pacada, Loida M. Perez, Roel R. Suralta, Jesusa C. Beltran, Jonathan M. Niones, Evelyn H. Bandonill, Nenita V. Desamero, and Juliet P. Rillon

Abstract Food security in the face of climate change is a fundamental challenge, particularly for developing countries like Philippines, where rice is also the major staple food. The threat of multiple biotic and abiotic stresses causes low productivity particularly for rice farmers in rainfed lowland areas. Varietal mix or VarMix technology is currently being explored in the Philippines to combat environmental stresses particularly in the unfavorable lowland areas. The exploration of this technology to address both biotic and abiotic stresses has far-reaching implication, not only for increasing profitability of rice farmers but also an eye-opener for rice research community and industry to further explore the released varieties with multi-traits and can be used for generating next-generation of VarMix combination to address future challenges in water scarcity, pest and diseases. It is crucial for rice research and rice breeders to develop a stopgap strategy to address these both abiotic and biotic stresses. Specifically until now, no single multi-trait local variety is well adapted to unfavorable rice areas in the Philippines, thus this paper.

I. G. Pacada (✉) · L. M. Perez · N. V. Desamero
Plant Breeding and Biotechnology Division, Science City of Munoz, Philippines
e-mail: imeldalyn.pacada@gmail.com; igpacada@exchange.philrice.gov.ph

R. R. Suralta
Crop Biotech Center, Santo Domingo, Philippines

J. C. Beltran
Socio Economics Division, Science City of Munoz, Philippines

J. M. Niones
Genetic Resource Division, Science City of Munoz, Philippines

E. H. Bandonill
Rice Chemistry and Food Science Division, Science City of Munoz, Philippines

J. P. Rillon
Crop Protection Division, Science City of Munoz, Philippines

I. G. Pacada · L. M. Perez · R. R. Suralta · J. C. Beltran · J. M. Niones · E. H. Bandonill · N. V. Desamero · J. P. Rillon
Phillipine Rice Research Institute, Science City of Munoz, Nueva Ecija, Philippines

S. Al-Naemi et al. (eds.), *Water and Food Security in the Face of Climate Change: Challenges and Opportunities for Resilience*, World Sustainability Series,
https://doi.org/10.1007/978-3-032-00098-9_18

Keywords VarMix · Climate change mitigation · Water scarcity · Pest · Diseases

1 Introduction

In the Philippines, a total of 2.8 million hectares are planted with rice, and 30% of these are rainfed lowland areas. Rice farmers in these areas grow rice only once a year, with low productivity is common due to multiple abiotic and biotic stresses [1].

Varietal Mix (VarMix) is a rice technology currently being explored in the Philippines to address environmental stresses, particularly in lowland unfavorable areas. VarMix is a mixture of selected Philippine released variety with similar agronomic traits but have genetic dissimilarities [2–4]. The selected varieties have documented reactions to pests and diseases and have validated for their diverse functional traits. This functional diversity enhances the expression of adaptive characteristics, contributing to the mitigation of both biotic stresses (i.e. pest and diseases) and abiotic stresses (i.e. water-related constraints) [2–4].

Contemporary research on VarMix mainly focused in disease management and disease resistance and assessing VarMix on productivity under water scarcity is limited. Thus, this study demonstrates the exploration journey [5] of VarMix in addressing both biotic and abiotic stresses during experimental stage (2013–2018) to adoption of technologies (2019–2024).

2 Methodology

First exploration.

The first field experiment was conducted in Umiray, Dingalan, Aurora Philippines (15°14′2.97″N, 121°24′24.56″E) where rice diseases (i.e. bacterial leaf blight rice blast, etc.) is perennial.

Primary selection for candidate varieties to be combined was based on the Top 20 varieties used by Region III rice farmers, surveyed data by the Socio-Economic Division (SED) of Philippine Rice Research Institute (PhilRice). To identify the genetic dissimilarities among 20 varieties, the application of molecular markers (simple sequence repeats or SSRs) were applied. From these, four varieties were selected based on different group/subgrouping. To evaluate the reaction to diseases in the area, two checks (resistant and susceptible) were included and, and a combination formula was applied to generate 15 and 20 VarMix combinations with 1:1 and 1:1:1 ratio, respectively. The initial optimization of candidate varieties for creating VarMix combinations and mode of planting (inter-row) was not successful. The agronomic characteristics and grain quality of the two varieties were not similar, and the labor-intensive planting method (separate planting of every single varieties) was not acceptable to farmers. As a result, the approach was modified by eliminating two varieties, and mixing the varieties before sowing. In 2014, VarMix combination with 1:1 and 1:1:1 ratio was selected having high yield stability amid environmental stresses.

Discovering the full potential of VarMix

To further explore the potential of VarMix and evaluate capabilities to mitigate both biotic and abiotic stresses, selected combinations were established in selected areas across three major islands in the Philippines namely Luzon, Visayas and Mindanao covering four climatic zones in 2015 (Fig. 1). Detailed information of establishing experimental set up and the formula used to determine the mixing effects and yield performance of VarMix is discussed in the paper by Pacada and Francia [2]. In the VarMix combinations, their single variety, and the ratio of each combination are presented in Table 1 and actual image of six single varieties used in creating VarMix combination is shown in Fig. 2.

Aside from VarMix evaluation for yield and yield stability, validation and confirmation of genetic similarities of six identified varieties in comparison to 31 PSB/

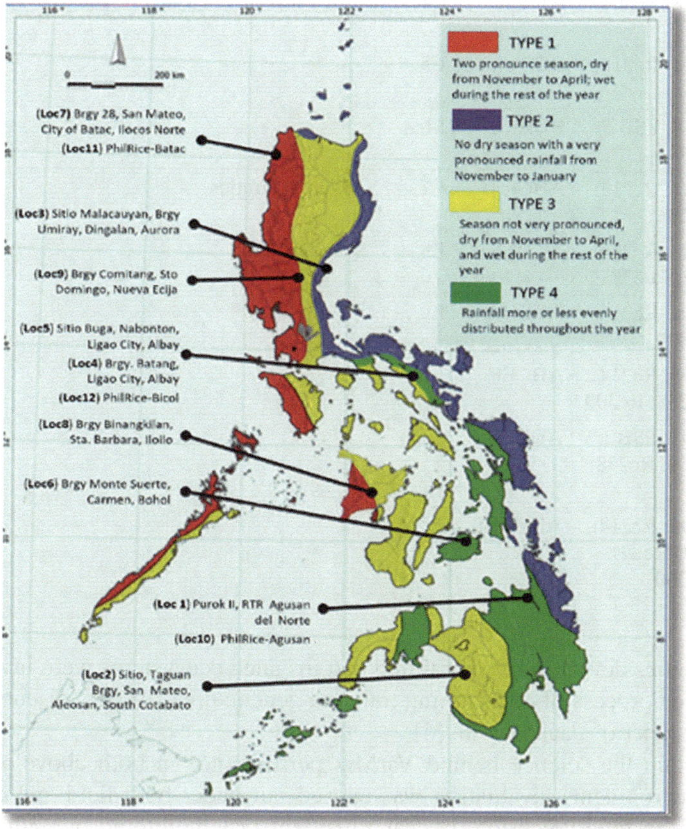

Fig. 1 Climatic zones based on modified coronas classification (MCC) and the actual location of 12 experimental set up [6]

Table 1 Ratio of single varieties to create VarMix combination and their codes

Single variety/VarMix combination	Analysis code	Simlify VarMix code	Remarks
NSIC Rc298	A		Single variety source used by Pacada et al. [3], for establishing VarMix combinations
NSIC Rc214	B		
NSIC Rc216	C		
PSB Rc82	D		
NSIC Rc238	E		
NSIC Rc300	F		
NSIC Rc298: PSB Rc82	AD	L1	Tested ratios of Pacada et al. [3]
NSIC Rc298: Rc238	AE	L2	
NSIC Rc298: Rc214: Rc216	ABC	L3	
NSIC Rc298: Rc300: Rc238	AFE	L4	
NSIC Rc214: Rc216: PSB Rc82	BCD	L5	
NSIC Rc214: PSB Rc82: Rc238	BDE	L6	
NSIC Rc298: Rc214: Rc216: Rc238	ABCE	L7	Untested ratios
NSIC Rc298: Rc216: PSB Rc282: Rc238	ACDE	L8	
NSIC Rc298: Rc216: Rc238: Rc300	ACEF	L9	
NSIC Rc298: Rc214: Rc216: Rc238: Rc300	ABCEF	L10	
NSIC Rc298: PSB Rc82: Rc216: Rc238: Rc300	ADCEF	L11	
NSIC Rc298: Rc214: Rc216: PSB Rc82: Rc238: Rc300	ABCDEF	L12	

NSIC varieties developed from rainfed and irrigated ecosystems were investigated. The detailed process of determining trait and genetic diversity was discussed in the published paper of Pacada et al. [4].

To explain the science behind VarMix performance in both above and below ground environments, evaluation was carried out under both field and controlled conditions. In field condition, pest and diseases was monitored by gathering data during reproductive and maturity stage using standard evaluation system for rice or SES [7] was used. Below-ground environment investigation was carried out using

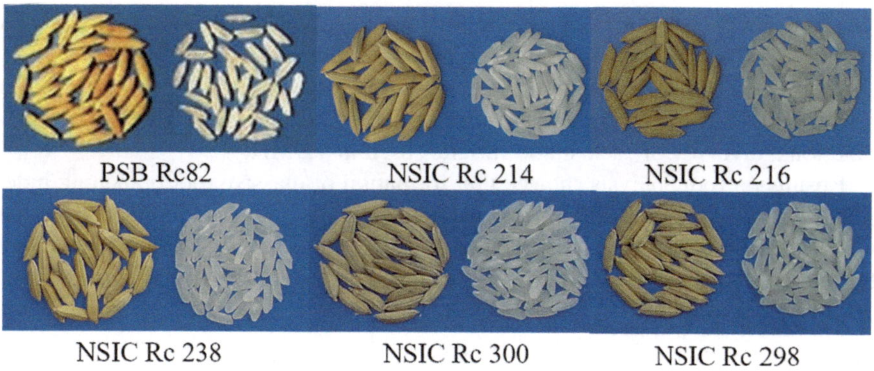

Fig. 2 Actual images of six Philippine rice variety used to create VarMix combination [2]

controlled condition. For root analysis, the growth performance and yield of VarMix under soil moisture fluctuation stress during reproductive stage was used and detailed protocol was demonstrated in the published article of Pascua et al. [8]. Initial evaluation for adverse settings such as drought and salinity screening were also implemented using the screening protocol of IRRI knowledge bank [6] and Gregorio et al. [9].

Market acceptability was also assessed to determine how acceptable the harvested VarMix seeds were to maket playes, namely rice agents, millers and traders. This is important since VarMix seeds are not currently popular in the rice market. Comprehensive survey process in Region III was described in Pacada et al. research [10, 11].

Progress of technology adaptation from 2019 to 2024 was also discussed, and this consist of the technology demonstration of selected VarMix combination adapted to selected areas in Region VI and Luzon.

3 Results and Discussion

Yield performance of and their corresponding single varieties across the 12 locations using basic statistics. Four VarMix combinations including BCD, ABCE, ADCEF, ADCEF (red arrow, Fig. 3) and single variety B (NSIC Rc214) have mean of 3.7t/ha. However, VarMix having five single varieties ADCEF yielded the highest at 7.7t/ha (red box with red arrow, Fig. 3) followed by single variety B with 7.5 t/ha (Fig. 3). Despite the presence of biotic and abiotic stresses, some VarMix combination showed its abilities to lessen the effect of actual environmental condition. Profile of evaluated biotic stress, water scarcity profile, detailed results in mixing effect, inter-varietal diversity, and yield stability were presented in published work of Pacada and Francia [2]. Table 2 displays the location and specific VarMix combination that showed positive relative mixing effect ($MErel_{ml}$). The list on this table will be the basis of location specific VarMix combination for future deployment in similar unfavorable areas in the Philippines that need stop-gap solution due to climate change.

The validated and confirmed genetic dissimilarities of six selected varieties can be attributed to separate breeding institution (Table 3), thus contribute to express functional diversity of each single varieties used in VarMix.

Results on root analysis showed that formation of massive root branching index (RBI) was developed during drought conditions. The RBI may increase the grain weight and collective reaction of single varieties, thus enhance VarMix mechanism for tolerance to adverse conditions. This conclusion could explain why some VarMix

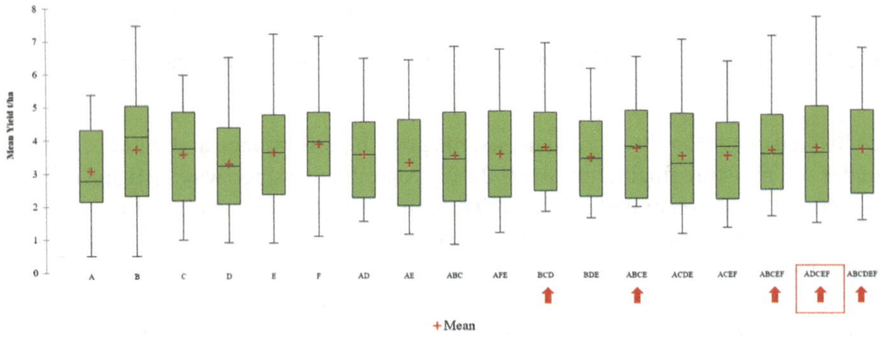

Fig. 3 Statistics summary of yield performance of VarMix combination and its single varieties

Table 2 VarMix combination having positive $MErel_{ml}$ in selected location in the Philippines

VarMix combination	Location with positive $MErel_{ml}$	Location	VarMix combination with positive $MErel_{ml}$
L7	3, 7, 8, 12, 1, 5, 6, 9, 11	Loc9	L1, L2, L3, L4, L5, L6, L7, L8, L10, L11, L12
L10	3, 7, 8, 4, 6, 9, 10, 11	Loc7	L1, L2, L3, L4, L5, L7, L8, L10, L11, L12
L1	7, 8, 12, 1, 4, 5, 6, 9, 10	Loc8	L1, L2, L4, L7, L8, L9, L10, L11, L12
L6	3, 12, 1, 4, 6, 9	Loc5	L1, L2 L3, L4, L5, L7, L9, L11, L12
L2	7, 8, 4, 5, 6, 9, 10	Loc6	L1, L2, L3, L5, L6, L7, L8, L10, L12
L5	3, 7, 12, 1, 4, 5, 6, 9, 10, 11	Loc10	L1, L2, L3, L4, L5, L8, L10, L12
L3	3, 7, 5, 6, 9, 10	Loc3	L3, L4, L5, L6, L7, ACEF, L10, L11, L12
L12	3, 7, 8, 1, 4, 5, 6, 9, 10	Loc4	L1, L2, L5, L6, L8, L9, L9, L12
L8	7, 8, 1, 4, 6, 9, 10	Loc1	L1, L5, L6, L7, L8, L11, L12
L11	2, 7, 8, 1, 5, 9, 10	Loc12	L1, L5, L6, L7
L4	3, 7, 8, 5, 9, 10	Loc11	L5, L7, L9, L10
L9	3, 8, 4, 5, 11	Loc2	L11

Table 3 Single varieties used in generating VarMix combination and their year released, breeding institution, and the identified clustering based on 42 SSRs markers [4]

Varieties	Year released	Breeding Institution	Genetic similarities groupings [4]
PSB Rc 82	2000	IRRI	I
NSIC Rc 214	2009	IRRI	II, 2.1a
NSIC Rc 216	2009	PhilRice	II, 2.1b.1.2
NSIC Rc 238	2011	IRRI	II, 2.1b.1.2
NSIC Rc 298	2012	PhilRice	II, 2.1b.1.2
NSIC Rc 300	2012	PhilRice	II, 2.1b.2.2

combination reach its full potential under field condition [8]. Similar result was observed during the initial evaluation of VarMix and its single varieties under salinity and drought condition.

The VarMix market acceptability study showed that millers quoted higher prices for VarMix ABC (NSIC Rc 298:214:216) and AFE (NSIC Rc 298:300:238) compared with known premium prices such as NSIC Rc 160 and Rc 216 (Fig. 4). Results indicates that VarMix has a potential niche market in the rice industry, particularly in improving the grain quality of various released varieties [10, 11]. Mixing of variety is a known approach and already adapted in rice growing country like Thailand where they improved the market potential of their released varieties by marketing Jasmine rice, a known branded name for quality rice, with a product of varietal mixture of KDML 105 and RD15 [5].

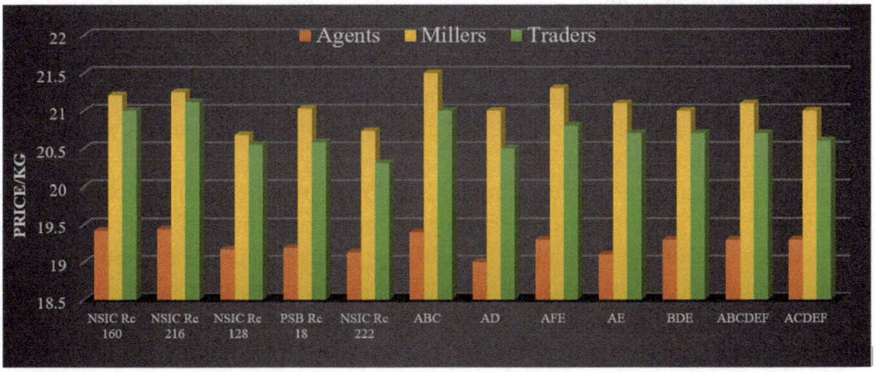

Fig. 4 NSIC Rc 160, Rc 216, Rc 128, Rc 18 are varieties having premium milled rice and NSIC Rc222, a known variety that known for agents, millers, and traders having low quality or having less price in the market due to its grain quality [10, 11]

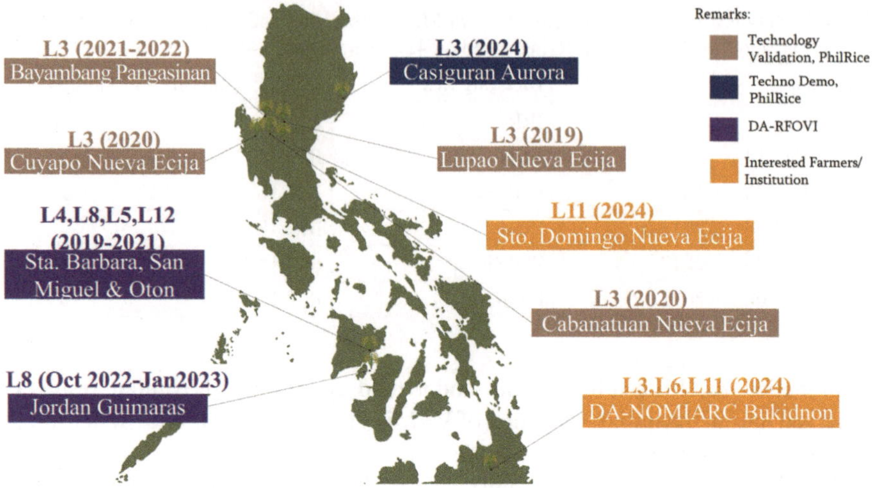

Fig. 5 VarMix technology demonstration assessment area and location of some interested farmers and institution in the Philippines

3.1 Adoption of Technologies

VarMix yield performance during experimental stage piqued interest to the rice farmers of Brgy Binangkilan, Sta Barbara Iloilo (a known Province of Region 6 having > 200ha of unfavorable areas), and neighboring Barangay, thus they are the early takers of technology.

Technology demonstration of selected VarMix combination assessed in Ilocos, Central Luzon, Western Visayas and Northern Mindanao Region is shown in Fig. 5. These area was identified due to perennial problem of water scarcity, pest and diseases. Positive feedback was received currently from rice farmers as well as from partner agency like DA-RFO VI.

In June 2024, a memorandum circular No. 25 was released and this pertain to the guidelines for the scaling of rice technologies under the National Rice Program. One of the objectives of the Masagana Rice Industry Development Program (MRIDP) are increasing rice productivity by optimizing yields, reducing risk, and increasing resilience through climate change adaptation measures. Thus, VarMix will play an important role in the future particularly for ensuring food security in unfavorable areas in the Philippines.

3.2 Conclusion and Future Outlook

It has been validated that the application of genetic diversity is not only effective in controlling diseases but also serves as an emerging approach to mitigate pest infestations. Moreover, the selection process of single varieties for generating VarMix combination is novel not just for biotic but also in abiotic stresses particularly for addressing water scarcity.

The exploration journey of VarMix technology has come a long way from proof-of-concept confirmation (science behind), ensuring its capability for promoting yield stability and to market acceptability validation. The discovery of VarMix combinations with good to excellent grain quality is also exceptional. This can be another plus factor of the VarMix technology where it can also enhance the productivity of farmers.

Unlocking the potential of VarMix is exhilarating, specifically the emerging techniques of 'omics that can facilitate the discovery of hidden capabilities of this technology and it can also enhance technical precision of selecting other released Philippine varieties that can be used to address challenges in adapting the changing climate.

References

1. Corales AM, Santos RC, Banayo NMC, Bueno CS, Johnson DE, Kato Y (2019) Dissemination pathways for drought-tolerant rice cultivars: a farmer participatory evaluation in the Philippines. World Dev Perspect 15:1000131
2. Pacada IG, Francia FJA (2023) Mixing effects and yield performance of rice varietal mixture (VarMix) in selected areas in the Philippines. Philipp Am Acad Sci Eng 16(1):8–18
3. Pacada IG, Cabanting RMF, Pocsedio AE, Dalusong VG, Perez LM (2014) Varietal mixtures as potential approach for increasing rice yield in rainfed lowland condition. In: Abstract of papers presented during 44th crop science scientific conference of the federation of crop science societies of the Philippines, vol 39, no 1, p 108
4. Pacada IG, Celestino MTF, Miranda MA (2022) Genetic and traits diversity of selected rice varieties for Varietal Mixture (VarMix). Rice Based Biosyst J 7(1):49–58
5. Pacada IG, Suralta RR, Beltran JC, Niones JM, Desamero NV, Bandonill EH, Rillon JP (2024) Are there good reason to push for VarMix in unfavorable rice-growing areas?. Rice Sci Decis Makers. https://www.philrice.gov.ph/wp-content/uploads/2021/09/RS4DM_VarMix.pdf. last accessed October 31, 2024
6. IRRI Knowledge Bank (2017) http://www.knowledgebank.irri.org/ricebreedingcourse/Breeding_for_drought_resistance.htm 21 Jun 2017
7. IRRI (International Rice Research Institute) (2014) Standard evaluation system for rice (SES) 5th Edition International Rice Research Institute
8. Pascua AG, Pacada IG, Suralta RR (2020) Growth performance and yield of rice variety mixture (VarMix) under soil moisture fluctuation stress during reproductive stage. Rice Based Biosyst J 7:37–47
9. Gregorio GB, Senadhira S, Mensoza RD (1997) Screening rice for salinity tolerance. International Rice Research Institute
10. Pacada IG, Martin MFC, Beltran JC (2020) Assessing the potential market for rice varietal mixture. In Abstract of Papers presented during 42nd national academy of science and technology annual scientific meeting. Manila Hotel Philippines 42(1):169
11. Pacada IG, Beltran JC, Bandonil EH, Martin MFC (n.d.) Assessing the acceptability of market players on VarMix: the case of Central Luzon. Philippine-American Academy of Science and Engineering (unpublished)

Tackling Climate Change in Yemen: Environmental Policy Gaps and Organizational Solutions

Mohamed Saleh Al-Haj

Abstract Yemen is grappling with two of the most pressing humanitarian challenges in its history: climate change and armed conflict. Since the onset of the war in September 2014, the country has faced recurrent environmental disasters, particularly in mountainous regions and coastal areas bordering the Red Sea, Gulf of Aden, the Arabian Sea, and Socotra Island. These regions remain highly vulnerable to severe cyclones and extreme weather events. From 2021 to 2024, the compounded effects of conflict and climate-related disasters have significantly worsened Yemen's humanitarian crisis, underscoring the urgent need for effective disaster management policies. Incorporating historical techniques from Ancient Arabia Felix, including advanced water storage methods and terrace farming, offers essential lessons for developing sustainable strategies to address future climate challenge. This paper examines the challenges facing Yemen in managing natural disasters amidst political instability and climate change. It evaluates the roles of key environmental organizations, highlights gaps in disaster management, and proposes actionable recommendations inspired by successful international models. Empirical evidence is drawn from qualitative research, including semi-structured interviews with government authorities, meteorological reports, international news sources, and the author's personal observations. The findings reveal that Yemen's response to tropical storms has been catastrophic, despite the availability of advanced early-warning technologies. The lack of a cohesive disaster management strategy and a clear vision for climate resilience has exacerbated the crisis. This study recommends structural reforms, including improving coordination between government agencies, NGOs, and international organizations, as well as adopting proven international frameworks to enhance Yemen's capacity to mitigate natural disasters in the context of its political instability.

Keywords Climate change · Disaster management · Arabian tropical storms · Environmental policy · Cultural heritage preservation · Social media · Early warning schemes · Ancient water management systems integration

M. S. Al-Haj (✉)
Faculty of Humanities, Sana'a University, Sanaa, Republic of Yemen
e-mail: ibbcity@gmail.com

S. Al-Naemi et al. (eds.), *Water and Food Security in the Face of Climate Change: Challenges and Opportunities for Resilience*, World Sustainability Series,
https://doi.org/10.1007/978-3-032-00098-9_19

1 Introduction

Yemen is confronting some of the most severe humanitarian challenges in its history, driven by the dual crises of extreme weather phenomena and ongoing armed conflict. Vulnerable groups within the Yemeni population are disproportionately affected, as these forces continue to devastate lives, infrastructure, and cultural heritage. A report from the International Committee of the Red Cross, titled "When Rain Turns Dust," highlights the profound humanitarian impact of these intertwined crises, which demand urgent scholarly and policy attention. The frequency and intensity of tropical storms over the past decade underscore the escalating risks Yemen faces [1]. In August 2015, two tropical cyclones struck the country within a single week, an extraordinary event in recent history. Similar devastation occurred in May 2018, when MEKUNU cyclones hit the same regions in rapid succession. The years 2019–2020 saw an unprecedented four cyclones within a single year—a phenomenon unseen since 1902. Although the number of tropical storms slightly declined from 2020 to 2023, the severity of tropical depressions and tidal wave events significantly increased, causing widespread damage. By July–August 2024, heavy rainstorms and flooding destroyed homes, displaced thousands of families, and severely damaged critical infrastructure, including schools, roads, and agricultural terraces. This devastation has left countless Yemenis without shelter, clean water, or access to essential services. Simultaneously, a cholera outbreak—aggravated by limited healthcare access and rising food insecurity has further strained the nation. Yemen's vulnerability to natural disasters, exacerbated by political instability and climate change, poses significant threats to its population, infrastructure, and cultural heritage. Despite repeated disasters, the country lacks an effective disaster management framework. This paper examines Yemen's existing institutions, their limitations, and opportunities for improvement. It highlights the growing role of social media and smartphone technology in international development as potential tools for disaster preparedness and response. Additionally, these traditional approaches exemplify the ingenuity of Yemen's ancestors in managing environmental challenges and could serve as a foundation for modern disaster management strategies. However, leveraging these historical practices requires integrating them with contemporary technologies and policy frameworks, emphasizing coordination between local communities, governmental agencies, and international partners.

2 Method

This study employs a combination of qualitative and observational research methods, integrating diverse data sources spanning from 2005 to 2024. The data utilized include both published and unpublished materials, incorporating satellite imagery, climatology data from Tropical Storm Risk (TSR), and reports from organizations such as OCHA and Relief-Web. These sources are complemented by imagery maps and photographic documentation, offering a comprehensive understanding of Yemen's environmental challenges. Additionally, the author conducted semi-structured interviews with government officials responsible for managing environmental disasters. These interviews provided critical insights and allowed for reflection on key questions surrounding Yemen's disaster management practices, institutional limitations, and potential strategies for improvement. The inclusion of the author's first-hand observations further enriches the study, adding a grounded perspective to the analysis. By synthesizing these diverse data sources and methodologies, the study aims to offer a robust evaluation of Yemen's vulnerability to climate-related disasters, the effectiveness of its institutional responses, and actionable references for building resilience against future challenges.

3 Climate Change and Tropical Storms in Yemen

Yemen faces significant vulnerabilities to climate change impacts, including droughts, extreme flooding, pest infestations, sudden disease outbreaks, and changes in rainfall patterns which demand urgent attention. Though, research by Evan and Camargo [2] highlights that the Arabian Sea experiences 1–2 tropical cyclones annually. This is due to dry air from the Arabian Peninsula and monsoon winds that intensify storm formation. Warm ocean waters further amplify convection, fueling stronger cyclones. Similarly, Murakami et al. [3] used numerical climate models to explain the recent surge in post-monsoon cyclones, attributing it to global warming. Other studies [4, 5] connect Yemen's unpredictable storms to climate change trends. However, Thompson [6] cautions that attributing specific storms to long-term processes remains complex. Table 1 summarizes major tropical storms affecting Yemen and their impacts on neighboring countries.

4 Climate Change and Floods in Yemen: Patterns and Impacts

Between June and August, Yemen has increasingly experienced heavy rainfall and widespread flooding, exacerbating destruction to property and livelihoods. The floods have damaged roads, bridges, and infrastructure, with loss of lives reported. In recent years (2021–2024), waves of intense rains and destructive floods have affected even traditionally arid areas like Eastern and Northern areas bordering the Empty Quarter Desert. This section examines major flooding events and their socio-economic impacts.

Table 1 Major tropical storms affecting Yemen, and its neighboring countries

Tropical storm names	Impacts on Yemen	Impacts on neighboring countries
2019: VAYU, HIKKA, KYARR, MAHA (Oct 24—Nov 3)	Yemen's Southern coast and Socotra experienced 4 cyclones. KYARR intensified into a "Super Cyclonic Storm" (250 km/h), damaging lives, livelihoods, and infrastructure [7, 8]	Oman faced property damage but no human casualties
2018: MEKUNU & LUBAN (LUBAN: Oct 14)	**LUBAN**: Heavy damage in Al-Mahra; 17 dead, 100 injured, and 2,000 families displaced. **MEKUNU**: Swept through Socotra and Gulf of Aden; 40 missing, 7 boats capsized, 1 million displaced [9, 10]	**LUBAN**: Flooding killed 1 in Oman. **MEKUNU**: 7 dead, $400 M damage in Oman
2015: CHAPALA & MEGH (Oct 28 & Nov 4	1.1 million people affected, 40,000 displaced, 38 dead, 800 + houses, roads, and public infrastructure damaged. Fisheries, farms, and livestock devastated [11–13]	Minimal regional impact outside Yemen
2023: TEJ (Oct 21)	Severe Cyclonic Storm TEJ affected 18,000 + households, displacing thousands and damaging public infrastructure Relief [14]	Minimal effects reported in neighboring countries
2008: Major tropical storm	Worst disaster since 1990s; 180 deaths, $1.6 billion in damages, 6% of Yemen's GDP [15]	Heavy rainfall affected Saudi Arabia and Oman

4.1 August 2013 Floods

On August 14, 2013, storms and heavy rainfall triggered a week-long deluge that devastated mountainous regions, especially in Hajjah Governorate near the Saudi border. Flash floods and lightning resulted in 50 deaths, displaced over 8000 camp residents, and submerged vast agricultural lands [16].

4.2 The 2020–2021 Floods

During May–August 2020, torrential rains ravaged Yemen, killing 172 people and displacing 300,000. Refugee camps housing 862,000 internally displaced persons (IDPs) were particularly affected. In April 2020, southern coastal regions saw 150,000 people impacted across 13 governorates, with significant damage reported in Aden and Lahj [17, 18]. In September 2021, widespread flooding damaged homes, farms, and infrastructure across 18 governorates, displacing 34,383 families [19].

4.3 The 2022–2023 Floods

In August 2022, Yemen recorded cumulative rainfall of 2500 mm—45% above prior years. Approximately 72 people lost their lives, and 73,854 families were affected [20]. In 2023, intensified rain events caused tidal waves that devastated irrigation infrastructure and triggered hazardous rockfalls. April floods in Ibb Governorate killed 31 people and injured 37. By June, over 165,000 individuals were affected in 70 districts [21].

4.4 The 2024 Rainstorms

In August 2024 brought record-breaking rainfall, with 8 million people affected across 20 out of 22 governorates. The floods caused 240 deaths, 635 injuries, and displacement of thousands, severely damaging shelters, roads, and agricultural terraces. Mareb Governorate saw rainstorms destroy 73 displacement sites, affecting over 21,000 individuals [22, 23]. In Al-Mahwit Governorate, a catastrophic flood on August 28 destroyed a water barrier, burying families in mud and rubble, with over 50 deaths reported and dozens missing [24].

4.5 Notable Observations

1. **2008 Storm**: The October 24, 2008 storm caused momentous damage despite early warnings from the author and Prof. Towfick Sufian, Vice President of Sana'a University, four days before the storm hit Yemen. The early warning data available on line by the Tropical Storm Risk (TSR) included a probability of hazard, magnitude, frequency, etc. But the government were unable to utilize it or mobilize assets and alert citizens in the wake of a disaster.
2. **Back-To-Back Cyclones (2015)**: CHAPALA and MEGH struck yemen within one week, marking the first recorded instance of such consecutive severe storms in the region.

4.6 *Discussion*

The recent flood events highlight Yemen's vulnerability to extreme weather conditions exacerbated by climate change. The destruction of critical infrastructure, displacement of populations, and loss of agricultural productivity underline the need for adaptive measures. Historical examples of Yemen's ancient disaster management practices, including water storage systems and terrace farming, offer insights into potential solutions for mitigating future climate impacts.

5 Environmental Policy for Natural Disasters in Yemen: Organizational Challenges and Strategic Recommendations

Effective disaster management requires robust institutional frameworks and coordination among relevant agencies. However, Yemen faces significant structural challenges, with many of its environmental institutions either ill-equipped or misaligned to handle the increasing frequency and intensity of natural disasters. This section explores the roles of Yemen's key environmental organizations, their historical contexts, and the gaps in their operational capacities.

5.1 *Yemen's Environmental Organizations*

Disaster management encompasses organizing resources and assigning responsibilities to mitigate emergencies and reduce casualties of human lives and natural resources [25]. Despite Yemen's long history of climatological observation, its institutions are weak, hindered by political instability and outdated infrastructure.

(a) **Civil Aviation and Metrology Authority (CAMA)**

Yemen's meteorological legacy dates back to 1839, during the British colonial administration in Aden [26]. Aden Port served as a crucial hub for maritime trade routes between Europe, India, East Asia, and East Africa. An official meteorological office was established in 1870, followed by a surface meteorological station at Aden International Airport in 1947 to monitor atmospheric conditions. Today, CAMA operates under the Civil Aviation Authority, which limits its operational independence and hinders its disaster response capabilities.

(b) **Ministry of Water and Environment**

Founded in 2003 by Presidential Decree No. 105, the Ministry of Water and Environment prioritizes water resource management, including urban and rural drinking water supply, wastewater treatment, and planning for limited resources. However, the ministry lacks a dedicated department for addressing environmental disasters. Its reliance on international donors further highlights its operational inefficiencies.

(c) **Environmental Protection Council (EPC)**

Established in 1990 following Yemen's unification, the EPC is tasked with managing environmental policies, including climate change. Despite its broad mandate, the council has no defined role in natural disaster management. Yemen has not ratified the Paris Agreement, although it has submitted a National Adaptation Program of Action (NAPA) to the UN Framework Convention on Climate Change. Without an actionable disaster management strategy, the EPC remains ineffectual in addressing Yemen's growing environmental crises.

6 Challenges in Institutional Coordination

6.1 Challenges in Institutional Coordination

Conflict of Interest

The roles of Yemen's institutions in disaster management often overlap, resulting in inefficiencies and a lack of accountability. Institutions like CAMA report marginalization and limited financial support, which impedes their operations.

Absence of Early Warning Systems: Despite recurring disasters, Yemen lacks an integrated early warning system. This gap has led to significant losses, as evidenced by the devastating 2008 floods, where early warnings went unheeded.

Centralized Decision-Making: Regional governments, such as Aden and Hadramout, have independently implemented early warning systems, demonstrating the potential of decentralized approaches. However, the central government's inefficacy undermines such localized efforts.

7 Failed States, Humanitarian Crises, and Public Challenges

Yemen is exceptionally vulnerable to climate-change-associated impacts within its political, social, and economic sectors. The country's public challenges and widespread poverty unfold against the backdrop of a complete absence of the rule of law, especially since Yemen's designation as a failed state [27]. Currently, Yemen's fragile political framework is divided into two competing proxy governments: the Houthis in the north and another governing entity in the south and east. Despite this political fragmentation, Yemen is united by a shared vulnerability to its most pressing environmental issue: endemic water scarcity. Ranked among the world's most water-stressed nations [28], Yemen frequently experiences regional droughts exacerbated by its naturally arid climate. Climate change contributes to higher mortality rates from diseases, particularly in light of extreme rainstorms that have destroyed irrigation systems and agricultural infrastructures, further aggravating food insecurity and economic hardship. This combination of climate risks and conflict undermines economic and food security, exacerbates disparities in well-being, limits access to essential services, and weakens government capacity to mobilize resources. Healthcare access is a critical concern as Yemenis face outbreaks of vector-borne diseases like malaria and cholera [29]. The humanitarian crisis, as highlighted by UN 2019 reports, has claimed tens of thousands of lives, with at least 17,700 civilian deaths, 80% of the population starving or malnourished, and over 3.3 million vulnerable groups (women, children, and minorities) deeply affected. The August 2024 floods destroyed homes, displaced thousands, and severely damaged critical infrastructure, leaving many without shelter, clean water, or access to healthcare in the face of cholera outbreaks and worsening food insecurity.

8 Cultura Heritage and Climate Change

In addition to its humanitarian crises, Yemen's rich cultural heritage faces severe threats due to climate change. Torrential rains and flooding have damaged numerous historical and archaeological sites across Yemen, particularly in Marib, the ancient seat of the Sabaean civilization, and other areas like Al-Jawf, Shabwa, Hadramout, and Tarim. Flooding not only destroyed many monuments but also unearthed hidden artifacts, which remain undocumented due to limited international attention. Coastal and highland regions have also seen substantial destruction in settlements such as Ibb, Hajjah, Dhamar, and Rimah. While UNESCO responded to damage in key World Heritage sites like Zabid, Shibam, and Sana'a, its efforts remain insufficient, leaving countless cultural landmarks unprotected. The destruction of Yemen's cultural heritage not only deprives communities of their history but also worsens living conditions in historic settlements already strained by conflict and climate disasters.

9 The Role of Social Media for Reducing Environmental Disasters

Social media is emerging as a vital tool for disaster response, enabling the exchange of information and fostering interaction during emergencies. In Western nations, platforms like Twitter, Facebook, and YouTube have been effectively used to issue emergency warnings and facilitate knowledge-sharing between disaster management agencies Barnett and Adger [30] and Stromberg [31] and Faust 1994. These tools promote awareness and collaboration, reducing redundancy in disaster response measures. However, in Yemen, the use of social media for disaster mitigation is irregular and largely limited to networks of individuals, communities, and unofficial media outlets. Since 2015, Yemeni social media groups have utilized social media and smart phones to document and share real-time updates during tropical storms, with maps, photos, and videos. These platforms have helped disseminate critical information, often embedding global weather forecasts and news to alert communities to imminent threats.

Despite these advances, Yemen's digital infrastructure faces significant challenges. Internet shutdowns and restrictions on free expression, stemming from both historical governance under former President Ali Abdullah Saleh and ongoing conflict, impede the effectiveness of social media in disaster response [32]. In contrast, countries like India have adopted community-driven approaches to disaster risk reduction. Indian networks integrate government agencies, NGOs, and local communities to oversee disaster management, enhancing preparedness and response [33].

10 Discussion

Flood events have starkly demonstrated Yemen's vulnerability to extreme weather conditions, which are increasingly aggravated by climate change. The widespread destruction of critical infrastructure, the displacement of thousands, and the loss of agricultural productivity underscore the urgent need for adaptive measures to enhance resilience.

11 General Recommendations

Yemen's vulnerability to natural disasters is heightened by the inefficiency of its environmental institutions and the absence of a cohesive disaster management strategy. To address these challenges, structural reforms and international collaboration are crucial. By adopting proven frameworks and prioritizing early warning systems, Yemen can mitigate the impacts of future disasters, thereby protecting its population and resources.

(a) **Institutional Independence**: One of the first steps toward enhancing Yemen's disaster management capabilities is restructuring the Civil Aviation and Meteorology Authority (CAMA) as an independent entity. This change would empower CAMA to operate with greater autonomy, improving its ability to coordinate with other agencies and respond effectively to emergencies. An independent and specialized institution is essential for fostering accountability and ensuring that disaster management efforts are both comprehensive and timely.

(b) **Adopting Ancient Arabia Felix Models**: Yemen's historical practices, such as sophisticated water storage systems and terrace farming, provide valuable insights into sustainable solutions for mitigating future climate impacts. These traditional approaches exemplify the ingenuity of Yemen's ancestors in managing environmental challenges and could serve as a foundation for modern disaster management strategies. However, leveraging these historical practices requires integrating them with contemporary technologies and policy frameworks, emphasizing coordination between local communities, governmental agencies, and international partners.
(c) **Adopting International Models**: Yemen can draw valuable lessons from disaster management frameworks implemented in other countries. For instance, India's Disaster Management Authority (DMA) provides a successful example of integrating national and regional response mechanisms., allowing for swift and efficient responses. Adopting a similar model in Yemen would enable better coordination between national and local agencies, ultimately improving disaster preparedness and response outcomes.
(d) **Early Warning Systems**: Establishing robust early warning systems is vital for reducing the devastating effects of natural disasters. By leveraging technology and social media platforms, Yemen can disseminate timely alerts to vulnerable communities, enabling them to take precautionary measures. These systems should be supported by reliable meteorological data and communication networks to ensure their effectiveness. Early warning systems have proven to significantly reduce casualties and property loss in many disaster-prone regions, making them a critical component of Yemen's disaster management strategy.
(e) **Capacity Building**: Capacity building initiatives are essential for strengthening Yemen's disaster response framework. International organizations and NGOs should focus on training programs for local authorities and community leaders, equipping them with the skills and knowledge needed to manage disasters effectively. Community-based disaster management initiatives can empower local populations to take proactive measures, fostering resilience at the grassroots level. Building local expertise ensures that disaster management efforts are sustainable and tailored to Yemen's unique needs.

By implementing these reforms and fostering international cooperation, Yemen can transform its disaster management approach, mitigating risks and safeguarding its communities from the impacts of natural disasters.

12 Conclusion

Yemen's rising exposure to tropical storms and extreme weather highlights the critical need for improved disaster preparedness and effective climate change mitigation strategies. While global warming is a key driver of increasing cyclone intensity and frequency, the lack of robust early warning systems and coordinated government action exacerbates the human and environmental toll of such events. The ongoing conflict further complicates progress, but proactive measures—such as adopting international best practices, fostering community engagement, and utilizing social media for real-time disaster communication—can significantly mitigate the impacts of natural disasters. Additionally, preserving Yemen's rich cultural heritage must be prioritized. These historical assets are not only integral to the nation's identity but also hold untapped potential for contributing to economic recovery and sustainable development. A holistic approach that combines resilience-building, cultural preservation, and strategic international cooperation is essential for safeguarding Yemen's future.

Acknowledgements The author would like to express his sincere gratitude to Dr. Daniel Varisco, President of the American Yemeni Studies Centre, for his invaluable insights and guidance. Special thanks are also extended to Dr. Lily Filson, a lecturer of Liberal Arts at Tulane University, USA, for her academic support and encouragement throughout the research process. Furthermore, the authors deeply appreciate the assistance and motivation provided by Dr. Salem Al-Naemi, President of the University of Doha for Science and Technology. Their collective expertise and encouragement have been instrumental in shaping this study.

References

1. Worland J (2015) Why Yemen was hit by a rare tropical cyclone. Time, 3 Nov 2015. http://time.com/4097413/yemen-cyclone-chapala
2. Amato E, Camargo S (2010) A climatology of Arabian sea cyclonic storms. J Clim 24(1):140–158. https://doi.org/10.1175/2010JCLI3611.1.
3. Murakami H, Vecchi GA, Underwood S (2017) Increasing frequency of extremely severe cyclonic storms over the Arabian. Nature Springer Nov. https://doi.org/10.1038/s41558-017-0008-6
4. Emanuel K (2003) Tropical cyclones. Annu Rev Earth Planet Sci 31(1):75–104. https://doi.org/10.1146/annurev.earth.31.100901.141259
5. Woodward A (2017) Climate change blamed for Arabian Sea's unexpected hurricanes. DAILY NEWS, 13 Nov 2017. https://www.newscientist.com/
6. Thompson A (2015) Yemen sees unprecedented tropical cyclone double-whammy. Climate Signals beta, 10 Nov 2015
7. IMD (2019) Cyclone Vayu likely to weaken into 'well-marked low pressure area. India Meteorological Department (IMD), 18 June 2019. https://www.freepressjournal.in/india/cyclone-vayu-likely-to-weaken-into-well-marked-low-pressure-area-imd
8. NASA (2019) Cyclone Hkkia hits Oman. Earthobservatory.NASA.gov, September 24, 2019. https://earthobservatory.nasa.gov/images/145645/cyclone-hikaa-hits-oman
9. CNN (2018) Cyclone pounds Yemeni island ahead of landfall on Yemen, Oman coast. CNN, 25 May 2018. https://edition.cnn.com/2018/05/25/middleeast/yemen-oman-cyclone-mekunu-wxc-intl/index.html
10. Relief Web (2018) Yemen: cyclone LUBAN. UN office for the coordination of humanitarian affairs, 17 Oct 2018. https://reliefweb.int/sites/reliefweb.int/files/resources/
11. Relief Web (2015) Tropical Cyclones Chapala and Megh-15 (UNOCHA, NOAA, Local Media, Relief web 2015 in OCHA services, 12 Nov 2015. https://reliefweb.int/report/yemen/yemen-tropical-cyclones-chapala-and-megh-15-unocha-noaa-local-media

12. UN News (2015) Yemen: as cyclone weakens, UN warns more than 1million people could be impacted by flooding 4 Nov 2015. https://news.un.org/en/story/2015/11/514552-
13. Weather Channel (2015) Cyclone Megh makes direct hit on Socotra; second final landfall in Yemen (RECAP). Weather Channel 10 Nov 2015. https://weather.com/storms/hurricane/news/cyclone-megh-five-yemen-socotra--arabian-peninsula
14. Relief Web (2018) Yemen: cyclone Mekunu. Relief Web, 27 May 2018. https://reliefweb.int/report/yemen/yemen-cyclone-mekunu-flash-update-2-27-may-2018-enar
15. World Bank (2009) To the republic of Yemen for a flood protection and emergency reconstruction project, WB, March 6, 2009. http://documents1.worldbank.org/curated/en/767681468170962811/pdf/471960PJPR0P11101Official0Use0Only1.pdf
16. The Guardian (2013) Yemen flash floods destroy camps for displaced people. theguardian.com, 29 August 2013. www.theguardian.com/global-development/2013/aug/29/
17. Aljazeera (2020) At least 172 killed in Yemen flash floods this month. Aljazeera Com, 12 August 2020. https://www.aljazeera.com/news/2020/08/172-killed-yemen-flash-floods-month-200812112022171.html
18. Relief Web (2020) Yemen operational update. Relief web, 20 August 2020. https://reliefweb.int/disaster/ff-2020-000123-yem
19. OCHA (2021) Yemen: floods—Jul 2021. In Relief Web, 5 Sep 2021. https://reliefweb.int/disaster/fl-2021-000110-yem
20. OCHA (2022) Yemen: flood snapshot (as of 30 Sept 2022). In Relief Web, 10 Oct 2022. https://reliefweb.int/report/yemen/yemen-flood-snapshot-30-september-2022
21. OCHA (2023) Cyclone Tej, 5 Nov 2023. Format Situation Report Source OCHA. https://reliefweb.int/report/yemen/yemen-situation-update-cyclone-tej-5-november-2023
22. Climate Centre (2024) Extreme weather in Yemen offers glimpse of country's climate future. Climate Centre, 11 Sept 2024. https://www.climatecentre.org/14478/extreme-weather-in-yemen-offers-glimpse-of-countrys-climate-
23. IFRC (2024) Operation update: Yemen, flood operation. International Federation of Red Cross and Red Crescent, Nov. 13, 2024. Glide №: FL-2024–000147-YEM
24. Mareb Press (2024) More than 50 people die in Malhan, Mahwit, and the local authority launches an urgent distress call. Mareb Press, 28 August 2024. https://marebpress.net/news_details.php?sid=205858
25. Comfort L (2005) Risk, security, and disaster management. Annu Rev Polit Sci 8(1):335–356. https://https://doi.org/10.1146/annurev.polisci.8.081404.075608
26. CAMA (2023) The emergence and development of meteorology in Yemen. Civil Aviation and Metrology Authority, 13 Jun 2023. https://cama.gov.ye/ar
27. Coppi G (2018) The humanitarian crisis in Yemen: beyond the man-made disaster. The International Peace Institute (IPI), Jan 2018. https://css.ethz.ch/con
28. Wiebelt M, Breisinger C, Ecker O, Al-Riffai P, Robertson R, Thiele R (2011) Climate change and floods in Yemen: impacts on food security Research Institute, pp 9–27. http://cdm15738.contentdm.oclc.org/utils/getfile/collec
29. ICRC (2020) When rain turns to dust. International Committee of Red Cross, July, 2020. https://www.icrc.org/sites/default/files/topic/file_plus_list/rain_turns_to_dust_climate_change_conflict.pdf.cama
30. Barnett J, Adger WN (2018) Mobile worlds: choice at the intersection of demographic and environmental change. Annu Rev Environ Resour 43(1):245–265. https://doi.org/10.1146/annurev-environ-102016-060952
31. Stromberg D (2015) Media and politics. Annu Rev Econ 7(1):173–205. https://doi.org/10.1146/annurev-economics-080213-041101
32. Sayadi E, Taye B (2020) As Yemen's war goes online, internet shutdowns and censorship are hurting Yemenis. #KeepItOn, Internet shutdowns and censorship, 25 June 2020. https://www.accessnow.org/keepiton-as-yemens-war-goes-online-internet

33. Walch C (2019) Adaptive governance in the developing world: disaster risk reduction in the state of Odisha India. Clim Dev 11(3):238–252. https://doi.org/10.1080/17565529.2018.1442794
34. UN (2019) Humanitarian crisis in Yemen remains the worst in the world, warns UN. UN News Global perspective Human stories 14 Feb 2019

Unlocking Desert Plants' Potential for Climate Change Adaptation and Food Security

Alshayma N. Al-Thani, Asma Ghafoor Jan, Madhurima Chatterjee, Yasmin A. Hageer, Farzin Shabani, and Ahmad M. Alqudah

Abstract Climate change's current and future scenarios strongly influence plant growth and yield. Desert lands already have the most severe environmental conditions, such as heat, salinity, and drought on the earth and recent abrupt climatic disasters lead to increased global desertification. Deserts with harsh conditions of arid terrain forced the plants to develop novel adaptive mechanisms to survive. Such conditions provide an exceptional opportunity to study the adaptation mechanisms, and its genetic basis that can help understanding plants' secrets to withstand climate change. Genome sequencing of desert plants can strengthen our interpretation on physiological and biochemical mechanisms, helping to identify genetic sequences crucial for climate change adaptation. Also, transcriptome profiling and metabolomics reveal actively expressed genes and secondary metabolites under various stress conditions. These molecular insights contribute to accumulate valuable information about the plants' resilience to environmental stresses and assist in illustrating the genetic architecture and regulatory networks, mostly relevant for evolution driven by climate change. Further, these steps assure interventions of new strategics to support programs for enhancing food security. In this review chapter, we describe the morphological, anatomical, physiological, biochemical, and genetic features of desert plants and discuss the basic genetic mechanism behind their systemic adjustments to harsh climatic conditions. The chapter highlighted how desert plants can be a good source of help for plants affected by climate change. This knowledge can be transferred to crop plants, enhancing their ability to withstand harsh environments and reducing crop loss.

Keywords Desert plants · Climate change · Food security · Stress-tolerant · Adaptation mechanism · Genomics

A. N. Al-Thani · A. G. Jan · M. Chatterjee · F. Shabani · A. M. Alqudah (✉)
Department of Biological and Environmental Sciences, College of Art and Science, Qatar University, Doha, Qatar
e-mail: aalqudah@qu.edu.qa

Y. A. Hageer
Department of International Affairs, College of Arts and Sciences, Qatar University, Doha, Qatar

S. Al-Naemi et al. (eds.), *Water and Food Security in the Face of Climate Change: Challenges and Opportunities for Resilience*, World Sustainability Series,
https://doi.org/10.1007/978-3-032-00098-9_20

1 Introduction: Background of the Review

Dryland ecosystems as arid, semi-arid, and hyper-arid region extend over 40% of the Earth's surface and are found on every continent due to their climatic diversity [1]. Naturally, plants are constantly exposed to additional harsh climatic scenarios in deserts and strive to maintain their vigor, overcoming destructive challenges, including water scarcity, nutrient deficiency, highly saline soil, radiation, extreme heat, and heavy metals. Environmental factors such as climate, seasonal timing and quantity of rainfall, temperature, variation in soil texture, and their physical and chemical properties are the root causes of the atypical diversity of plant biota found in deserts [2]. Despite extreme desert conditions, many plant species persist across generations due to unique adaptation strategies, modifying their root, stem, leaf, pollen, flower, and fruit structures [3]. Therefore, desert-inhabitant plant species must withstand intense selection pressure to accomplish their physiological processes, including photosynthesis, transpiration, water retention, etc. [4]. Moreover, they have the unique adaptive capability and specialized anatomical, physiological, chemical, and morphological features, unlike regular crops to store water and maximize photosynthetic rate, minimizing water loss and simultaneous regulation of energy metabolism [5]. Desert plants use different mechanisms to fix CO_2 to cope with high temperatures and reduce the amount of water loss during photosynthesis, such as C_4 plants separate carbon fixation spatially (between mesophyll and bundle sheath cells), while Crassulacean Acid Metabolism (CAM) plants follows temporal separation of carbon fixation according to the time (night/day cycle) [6]. Additionally, heterogeneity of the plant habitat in deserts is one of the consequences of surface and sub-surface bioprocesses led by microorganisms. However, the soil microbial environment governing nitrogen cycles gets negatively influenced by harsh temperatures and shortage of water, leading to nutrient-depleted soils and deteriorated soil function and conservation [7].

In addition to their physiological and biochemical traits, understanding the genetic mechanisms underlying the unique adaptations of desert plants is essential. Reports revealed that the phytohormone abscisic acid-insensitive genes control stomata opening and closing in response to water availability and protein storage to ensure adaptation to various abiotic stress stimuli in the desert [8]. Furthermore, the other crucial phytohormone auxin-related genes are primarily allied with the refinement of root architectural parameters, including length, depth, and number of lateral roots also regulate metabolic homeostasis [9].

The desert plants' physiological and biochemical aspects and comprehensive genetic studies remain limited. As climate changes become more common globally, improving the resistance mechanisms of plants to combat unexpected climatic conditions, e.g., heat-drought resistance, becomes necessary. On the other hand, as desert plants have unique mechanisms and features, it is crucial to explore the most recent scientific proof of concept behind plant adaptation to overcome the mounting danger of climate change and address global food security. Moreover, changing climatic conditions and anthropogenic factors are directly responsible for losing biodiversity and genetic resources of desert plants' wild relatives, as prolonged genetic evolution causes irreversible loss of genetic information and combinations [10]. Thus, the conservation of genetic resources through retrieval of detailed genetic information of desert plants utilizing modern biotechnological tools not only contributes in deciphering the molecular and genetic base of their adaptation strategies but also helps to obtain information about the genetic diversity and evolutionary history of those plants. Further, with the survey of resourceful scientific literature, the unexplored special features and the genetic background of desert plants could be disseminated to the international scientific community to preserve the botanical heritage of deserts. In this direction, efficient production of food crops with improved adaptive qualities can be achieved by genetically altering target crops and acquiring the information on genetic and molecular mechanisms of stress-tolerant desert plants to ensure a more long-term and secure future in farming. Although success rates remain low, researchers continue to develop crop varieties that withstand extreme weather and utilize drylands for sustainable agriculture [5]. Therefore, through this review article, we attempted to represent the crux of existing information on physiological and genetic adaptations of desert plants combating severe climatic conditions. Hopefully, this will support the reconsideration and design of constructive experimental procedures for improving and sustaining desert plants and food crops. We anticipate that these insightful steps will be capable enough to broaden the scope of restoring the fragile balance of desert ecosystems by unlocking the genetic codes contained in desert plants and eradicating global food insecurity for the benefit of our society.

2 Desert Ecosystem and Climate Change

The climate of a region is influenced by its geographical location and atmospheric circulation patterns, which determine precipitation levels [11]. Thus, the arid terrestrial ecosystem of deserts results from irregular or very low rainfall and lack of evapotranspiration. Moreover, in recent years, rapid alterations in climatic conditions due to extensive urbanization and increased global population have gradually induced desertification of vegetative lands, one of the most hazardous conditions for the existing ecosystem of that particular location. In this condition, arid and semi-arid regions face tremendous loss of vegetation and biodiversity. Additionally, human activities, such as excessive grazing, deforestation, and unsuitable agricultural practices, significantly contribute to this process. Simultaneously, rapid modernization and drastic lifestyle changes have also further deteriorated the climatic scenario of naturally occurring deserted areas. Thus, in the current situation, deserts and dry lands face strong disruption of ecosystems and loss of precious habitats due to harsh climatic conditions such as extreme heat, unpredictable and limited rainfall, drought, high salinity, heavy metals contamination in soil, and nutritional deficiencies.

Moreover, desertification challenges not only the ecosystem and biodiversity but also the livelihoods of millions reliant on these areas for survival. Furthermore, the depletion of desert vegetation can worsen global environmental issues like soil erosion, diminished carbon storage, and altered local weather conditions [12]. Scientists seek to understand the genetic features that allow these hardy species to survive in severe environments which could transform agriculture, benefiting arid regions and mitigating climate change impacts on global food security.

3 The Physiological and Biochemical Properties of Desert Plants

Desert plants evolved with valuable traits that simultaneously increase water uptake capacity and minimize water loss to cope with variable weather conditions and water scarcity in arid lands. Based on the morphological characteristics of their above-ground organs, desert plants are classified into three types: (a) Annual desert plants, having shorter life cycles and rapid growth; (b) Perennial plants, mostly survive during the dry season; and (c) Succulents comprising of shallow root system, having excellent water storage capacity in their stems [13]. Most annual plants only thrive in favorable climatic conditions and adapt to extreme conditions while completing their lifecycle, so they fall under the drought-avoiding category [13]. On the other hand, perennials and succulents are counted as drought resisters and have significantly reduced transpiration activity and extensive water storage capacity [5].

3.1 Leaf Architecture of Desert Plants

Maintenance of a balance between transpiration rate and heat dissipation by latent heat transfer is the ultimate base for leaf modification as a part of adaptive strategies acquired by dessert plants [14]. Smaller leaves with reduced surface area are key adaptations that minimize transpiration, which is critical during low moisture and water availability [15]. The smaller leaves absorb less heat; thus, water loss is much reduced through minimal transpiration, which helps maintain a cooler temperature across the plant body. Further, reduced temperature minimizes photorespiration and helps achieve the optimum temperature for efficient photosynthesis [5]. Some other interesting features, such as thick layers of reflective wax, trichomes, and leaf hairs on the leaf surface, work as competent photoreflectors of sunlight and help decrease the leaf temperature by efficiently reducing solar radiation [14, 15]. Furthermore, in earlier studies, the impact of diaheliotrophism (leaf angle perpendicular to the sunrays) and paraheliotrophism (leaf angle parallel to the sunrays) were exhibited as two other beneficial adaptive features corresponding to potent heat absorption and dissipation of desert plants depending on their growing period and lifecycle [5]. Interstingly, in some desert plants, like *Lupinus arizonicus*, shifting between diaheliptrophism and paraheliotrophism is noticeable at different times of the day, depending on the direction of sun rays [16]. The leaves of these plants have encrypted sunken stomata either in crypts or grooves, which protects the stomata from the damaging effects of external elements [17].

3.2 *Physiological Properties of the Root System of Desert Plants*

The role of below-ground organs is highly crucial, while the difficulties for survival are faced by plants grown in drought-prone desert lands. Naturally, water distribution in desert soil is quite unstable and dependent on soil depth. The desert soils consist of two water reservoirs, one at the surface and the other in deeper soil layers [18]. Although, in summer, precipitation water can't penetrate deep soil layers because of high heat-induced water evaporation, it can be possible in winter and spring [19]. Specifically, desert plants have acquired deep-growing tap root systems to attain the required water level from below the ground [13]. Root architecture and distribution generally determine whether the plant will absorb water from water reservoirs of the soil surface, deep soil layers, or both [13, 18]. This is one of the reasons why desert plants prefer to adopt more extended root systems than usual to access deeper water reservoirs of soil layers. In the Chihuahuan desert, some bushes are prone to absorb the maximum available water derived from rainfall as they have horizontal root distribution a few meters deep into the shallow soil. However, very few roots are up to five meters deep in the soil [19]. Further, some evidence assured the formation of rhizosheath around the roots of desert plants, which is a promising adaptive feature of desert plants [20]. Furthermore, Cacti, being succulents, have ~ 2.5 m horizontal root distribution from the plant stem, and they spread ~ 1.5 m at the upper part of soil layers and also exhibit an adaptive trait of adventitious root formation from shoots to boost water uptake capacity and anchorage with soil [13, 21].

3.3 *Strategies for Photosynthesis in Stem and Other Non-Foliar Tissues*

Desert plants preferably use their stem and other non-foliar parts as photosynthetic organs due to their smaller and narrower leaves. Photosynthetic stems positively impact maintaining the equilibrium of carbon content in desert plants because they have two primary carbon sources [5]. Alongside small, thick leaves and trichomes, developing deep embedded chloroembryo inside the supporting tissues of desert plants is highly beneficial to avoid transpirational water loss during their life cycle. Usually, chloroembryo participates in photosynthesis using a low chlorophyll a/b ratio, and it also has a lesser number of Calvin cycles than leaves [22]. The occurrence of patterns and pigmentation in chloroembryo depends on the species' characteristics, and also, the accumulation of chlorophyll can be found in particular parts/organs of a plant body [22]. Cactus from deserts are capable of accomplishing photosynthesis in their stem. A recent interesting study using *Clusia minor* L. proved that blocking light access to the photosynthetic stem can interfere with root growth and regulation of carbon distribution in the plant body, which deciphers the significance of adaptive photosynthesis in non-foliar tissues [23]. Firstly, Ribulose-1,5-bisphosphate carboxylase (Rubisco) catalyzes the assimilation of atmospheric CO_2, and the absorbed CO_2 diffuses into the cells through stomatal pores [24]. Furthermore, pre-respired carbon can be recycled by stems and leaves to limit carbon diffusion during respiration [25]. After examining 11 plant species, an interesting study proved strong coordination between photosynthetic and hydraulic characteristics in photosynthetic stems showing same functionalities as leaves with support of certain adaptive traits such as carbon gain, water absorption, and loss [25]. The stomata present in the stem are more minor than leaves, resulting in less water loss and carbon gain. Therefore, the photosynthetic rate is higher in stems during the dry season and improved drought-tolerance of desert plants [24]. Additionally, many studies demonstrate the occurrence of efficient Photosystem II photochemistry in non-foliar tissues, which assures functional electron transport in the same way as leaf tissue [24].

3.4 Characteristics of Succulents

Succulent plant types have consistently increased the curiosity and interest of the plant research community due to their unusual morphology and structure. Succulent plants are named such because they have a unique yet adaptive characteristic of water storage to combat drought stress, resulting in the swollen appearance of leaves and stems, roots, flowers, and other parts [26]. This plant consists of achlorophyllous parenchyma cells, also known as hydrenchyma cells, with ample water storage capacity [27]. They have various unique characteristics, such as some plants are leafless, and photosynthesis occurs in their stem. Some plants are also categorized as geophytic as they have primarily underground storage organs [26]. In crassulacean plants, these hydrenchyma cells contribute to photosynthesis, while in regular plants, those cells are not involved in photosynthesis. The succulents can store up to 95% water content in their cells [28]. Perennial desert plants mostly have this specialty of storing water and using it to overcome drought stress, buffer transpiration and refilling the hydrenchyma cells [29]. With these incredible traits, desert plants can maintain ecophysiological balance in the environment. Interestingly, most succulents accomplish photosynthesis with a CAM- associated pathway [30]. Further, several comparative studies have reported that the CO_2 fixation rate is almost similar in C_3, C_4, and CAM plants. However, CAM succulents have ~ eight times higher water storing and usage capacity than C_3 and C_4 plants [31]. Thus, the efficiency of the succulents makes them exceptional in preserving their growth and development in severe weather conditions of dry lands.

3.5 *Biochemical Characteristics Adapted by Desert Plants for Photosynthesis*

Conducting any physiological activities in desert climates is stressful for plant species despite having multiple traits adapted through evolution. Fewer stomata are a limiting factor for CO_2 uptake from the environment and therefore C_3, C_4, and CAM photosynthetic pathways help desert plants to overcome those hindrances. Desert plants prefer to possess C_4 and CAM photosynthesis with open stomata at night and reduce photorespiration, while the C_3 pathway results in excessive carbon gain [5]. In the C_4 pathway taking place in mesophyll cells, phosphoenol pyruvate carboxylase (PEP) works as a CO_2 pump and catalyzes the reduction of the absorbed atmospheric CO_2 from stomata and converts it to oxaloacetate (Four-Carbon; C_4), therefore the cycle named C_4 [32]. Consequently, through plasmodesmata, the reduced oxaloacetate to malate gets transferred into the bundle sheath, possesses decarboxylation, and forms a CO_2-saturated environment, resulting in reduced photorespiration [33]. They induce efficient photosynthesis with heavy light intensity and high temperature. Moreover, C_4 plants are more water-use efficient than C_3 plants, making them ideal for growing and surviving in desert conditions [32]. The C_4 and CAMP plants function almost the same regarding photosynthesis. Again, the CAM cycle occurs in the same cell instead of two separate cellular locations in the C_4 cycle [5]. The CAM plants open their stomata at night, uptake the atmospheric CO_2, convert to malate following the same C_4 cycle, and get stored in the vacuole and CO_2 release occurs due to decarboxylation of malate [34, 35]. Due to closed stomata at daytime, the Rubisco conducts the Calvin cycle and maintains a high CO_2:O_2 ratio inside the plant cell, which leads to the avoidance of photorespiration [28]. In another way, C_3 and C_4 plants help reduce water loss due to transpiration by closing the stomata during the day.

4 The Genetic Mechanism of Adaptation in Desert Plants

Physiological responses to environmental stimuli rely on cellular signaling, mediated by interconnected genetic networks within the plant system. Thus, comprehending the genetic base of desert plants through functional genomics approaches is the utmost prerequisite to support the physiological adaptive phenomena of the plant system. Further, an in-depth genomic study allows us to learn about the responsible genes/gene networks correlated with the traits. Exploration of genomic strategies acquired by desert plants is a newly emerging research area that is genuinely beneficial for protecting desert plant diversity. Studying an entire desert ecosystem can open multiple avenues to identify common or specific genes and their adaptive traits evolved through parallel/convergent evolution [36]. Hence, to understand the population diversity across the desert plant species, genome-wide sequencing to identify Single Nucleotide Polymorphisms (SNP) can contribute to detecting variation of allelic frequency between populations [6]. Recently, many accomplished research reports have employed omics-based approaches that revealed the existence of gene regulatory networks associated with the plant defense mechanism towards environmental stimuli and stressors [37, 38]. Further, the advancement of high throughput sequencing technologies accelerated the usage of molecular markers such as Expressed Sequence Tag (EST)-based Simple Sequence Repeats (SSR) markers for critical gene mapping and diversity analyses, which successfully contributed to the process of identifying the genes related to stress management in desert plants [39]. Figure 1 summarizes the key genetic mechanisms contributing to understand desert plant adaptation and their implications for crop improvement. Also, QTL mapping, along with transcriptomic analyses, has successfully discovered the specific loci responsible for addressing abiotic stress factors [40]. Moreover, these findings can be utilized in several Marker Assisted Selection-based (MAS) breeding programs to improve the crops.

A recent interesting study uncovered key genes and their contribution to shared adaptive strategies of various plant species of the Atacama desert, Chile, one of the most hyperarid non-polar regions globally [36]. This study identified several

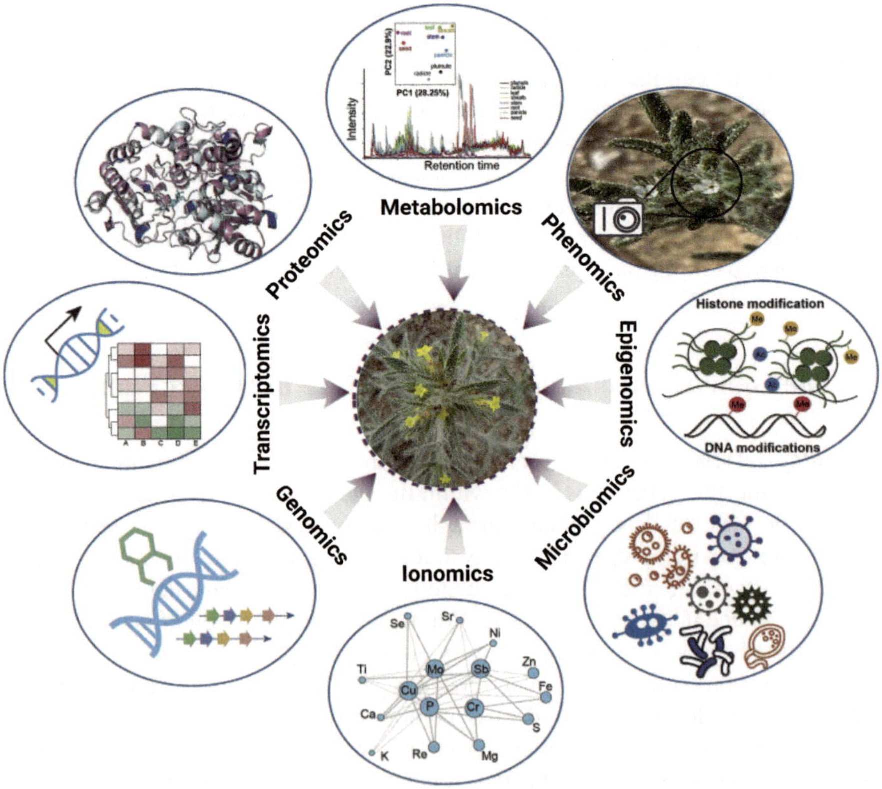

Fig. 1 Schematic representation of the scientific contribution for understanding the genetic adaptation of desert plants

genes associated with the response to DNA damage, light stimulus, photosynthetic light reaction, and plastid organization to confront extreme high-light radiation in Atacama. Again, several genes were involved in efficient stress tolerance, balancing poor nutrient uptake by regulating multiple abiotic stress, metal ions, detoxification, energy, N, P, and S metabolism, etc. [36].

4.1 Adaption to Drought Stress

Drought stress is one of the significant constraints responsible for the slow, unhealthy growth of desert plants. Although desert plants evolve with different physiological, cellular, and molecular mechanisms to combat drought stress, the stress tolerance capability vastly depends on the species' characteristics and developmental stages [41]. In recent years, associative studies on transcriptome, proteome, and metabolomics have successfully revealed the involvement of various regulatory and functional proteins in the complex molecular signaling mechanism of plant's response to water deficiency, which results in perturbation of several drought-responsive genes at the transcript level [42]. Therefore, investigating the role of basal-performing genes is always advantageous in response to stress; regulatory genes/gene encoding proteins, namely, protein kinases, protein phosphatases, phospholipid metabolism-related genes, and stress-induced transcription factors, do participate in signal transduction mechanism and regulation of several effector molecules present in the signal cascade [42]. The contribution of essential genes and transcription factors (TFs) in abiotic stress management, proven by available literature, is summarized in Table 1.

Various functional genes, namely, pyrroline-5-carboxylate synthetase (P5CS), Δ-pyrroline-5-carboxylate reductase (P5CR), ornithine-δ-aminotransferase (δ-OAT) enzymes are generally responsible for proline accumulation, one of the most prominent mechamism for regulation of osmotic balance during drought-stress condition [43]. A study on the characterization of P5CS in one of the drought-tolerant wild medicinal desert plants, *Calotropis procera*, revealed the over-expression of the P5CS gene in response to severe environmental stress factors like nutritional deficiency, increased drought and salinity, which is also common in other wild desert plant species [43]. Further, elevated proline level was recorded in *P5CS*-transgenic tobacco lines under water stress conditions [44]. Similar observations on P5CS in transgenic rice [45], potato [46], sugarcane [47], soybean [48], etc., confirmed the enhancement of osmotolerance ability under drought stress. Further, in tobacco and rice, overexpression of ornithine-δ-aminotransferase (δ-OAT) increased proline content under stress conditions [49, 50]. Also, biosynthesis of another important functional gene product, betaine, was significantly increased due to low temperature, drought, and salinity stress through NAD^+ dependent oxidation reaction catalyzed by betaine aldehyde dehydrogenase (BADH). Plants under stress showed a rise in BADH synthesis, while in normal conditions, no spike was determined [51]. Interestingly, this theory was implied by improved drought tolerance depicted in transgenic

Table 1 Summarization of the roles of significant stress-responsive genes and transcription factors combating abiotic stress factors

Name of the genes/TFs	Role in stress adaption	References
Drought stress tolerance		
Δ-pyrroline-5-carboxylate synthetase (*P5CS*)	Proline biosynthesis	[43, 44]
Ornithine-δ-aminotransferase (*δ-OAT*)	Proline biosynthesis	[49, 50]
Betaine aldehyde dehydrogenase (*BADH*)	Betaine biosynthesis:	[51]
Sucrose phosphate synthase 1 (*SPS1*)	Sucrose biosynthesis	[54]
Plasma membrane intrinsic proteins 2;3 (*PIP2;3*)	Aquaporin; maintenance of water homeostasis	[55]
Plasma membrane intrinsic proteins 2;7 (*PIP2;7*)	Aquaporin; improvement of seeds	[56]
Tonoplast Intrinsic Proteins 1;3 (TIPs) (*TIP1;3*)	Aquaporin; improvement of seeds	[57]
9-cis-epoxy carotenoid dioxygenase (*NCED*)	ABA biosynthesis	[58]
Calcium-dependent protein kinase 1; 2 (*CDPK1; CDPK2*)	Water and osmotic stress tolerance	[59]
Calcium-dependent protein kinase 10; 30 (*CDPK10; CDPK30*)	Regulation of drought tolerant signaling pathway	[60]
MAP kinase 3; 4; 6 (*MPK3; MPK4; MPK6*)	Involved in MAP kinase signaling pathway	[61]
MAP Kinase kinase 1 (*MEK1*)	Involved in MAP kinase signaling pathway	[61]
MAP Kinase kinase kinase 1 (*MEKK1*)	Involved in MAP kinase signaling pathway	[61]
Phytoene synthase gene (*PSY*)	Abscisic acid induction	[62]
Detoxification efflux carriers (DTX) / multidrug and toxic compound extrusion (DTX/MATE)	Translocation of abscisic acid	[63]
NAC with Transmembrane Motif 1- like Transcription Factor 9 (NTL9)	Sensing abiotic stress factors	[64]
Nuclear factor YA10 (NFYA10)	Regulation of drought tolerance	[65]
Myeloblastosis oncogene transcription factor (MYB52)	ABA hypersensitivity and drought tolerance	[66]
Salinity stress tolerance		
Hyperosmolarity gated Calcium ion channel (*OSCA1*)	Osmosensing and Ca^{2+} signaling	[70]
Potassium (K^{+}) exchange antiporters 1; 2; 3 (*KEA1, KEA2, KEA3*)	Elevation of Ca^{2+} osmosensors	[71]
Monocation-induced $[Ca^{2+]}_i$ increases 1 (*MOCA1*)	Activation of Ca^{2+}ion channels	[72]
Hydrogen-peroxide-induced Ca^{2+} increases 1 (*HPCA1*)	Mediate the influx of Ca^{2+}ions	[73]

(continued)

Table 1 (continued)

Name of the genes/TFs	Role in stress adaption	References
Heat/cold stress tolerance		
Heat stress transcription factor family 1; 2; 7 (HSFA1, HSFA2, HSFA7)	Heat stress-responsive mechanism	[81, 82]
Heat shock transcription factor B (HSFB)	Heat stress-responsive mechanism	[81, 82]
Inducer of CBF expression 1; 2 (ICE1, ICE2)	Cold stress responsive mechanism	[86]

tobacco harboring barley-specific *BADH* gene [51]. Also, in transgenic sweet potato [52] and tomato [53] containing the *SoBADH* gene from spinach, higher BADH activity increased betaine production, and simultaneously improved stress tolerance was observed. Again, several studies on transgenic tobacco containing *A. thaliana* specific sucrose phosphate synthase (SPS), *AtSPS1*, directly linked with the sucrose synthesis pathway, exhibited enhanced sucrose content with increased plant height, thickening and lengthening of stem diameter and fibers, respectively [54].

Furthermore, plant aquaporin (AQP) is a membrane channel transporting water, gases, and small molecules across biofilm. It maintains the intracellular water transport and maintenance pathways while plants are under stress, especially drought stress [55]. A research report on transgenic *A. thaliana*, over-expressing rose PIP2 gene (plasma membrane intrinsic proteins) (*CrPIP2;3*) exhibited enhanced plant growth and survival rate and also stress defense under drought stress by maintaining water homeostasis across plant body [56]. Similarly, over-expression of drought-resistant plant Jatropha specific aquaporins *JcPIP2;7* and *JcTIP1;3* (Tonoplast Intrinsic Proteins) improved the germination rate, quality, yield, and vigor of seeds and also the root length of *A. thaliana* [57].

In association with abscisic acid (ABA)- dependent and independent regulatory mechanisms, multiple signal transduction cascades such as mitogen-activated protein kinases (MAPKs), calcium-dependent protein kinases/calmodulin-like domain protein kinases (CDPKs), reactive oxygen species (ROS) get triggered through the activation of TFs responsible for defense response upon sensing water loss, drought, and other environmental stressors [40]. Research on *A. thaliana* depicted the over-expression of AtNCED3 leads to a significant increase in drought tolerance with a decrease in transpiration rate [58]. On the other hand, over-expression of AtCDPK1 and AtCDPK2, AtCDPK10, and *AtCDPK30* exhibited induction in drought tolerance and proved their involvement in osmotic and water stress balance in *A. thaliana* [59, 60]. Similarly, in *A. thaliana*, MAP kinase encoding genes such as *AtMPK3, AtMPK4, AtMPK6, AtMEK1*, and *AtMEKK1 were* identified to be immensely involved in MAP kinase signaling pathway to induce drought, salinity, cold stress tolerant mechanism [61].

Furthermore, the phytoene synthase gene (*PSY*) was reported to be responsible for the abscisic acid induction process during the elevation of salinity and drought in rice and maize [62]. Also, enhancement of drought, salt, and cold stress tolerance was observed in transgenic *A. thaliana* by overexpression of a candidate gene of detoxification efflux carriers (DTX) / multidrug and toxic compound extrusion (MATE) DTX/MATE from cotton (*Glossypium hirsutum*) [63]. Further, NAC with Transmembrane Motif 1- like Transcription Factor 9 (NTL9), a membrane-associated transcription factor induced by osmotic stress, plays a significant role in promoting leaf senescence, which is the last stage of leaf development and one of the adaptive features for nutrient cycling across the different parts of plants. upregulating the senescence-associated genes [64]. Again, overexpression of wheat and soybean nuclear factor YA10 (NFYA10) positively regulated the drought resistance in transgenic model plant *A. thaliana* [65]. Further, in *A. thaliana*, Myeloblastosis oncogene (MYB) protein group transcription factor MYB52 exhibited ABA hypersensitivity and drought tolerance [66].

4.2 Adaptation to Salinity Stress

Like drought stress, high salt concentration in soil and the environment is another significant abiotic stress factor faced by the plant kingdom, especially in arid/desert areas. Naturally occurring salt ions in soil and water get settled on the soil surface and around the root growth, and it's primarily responsible for restricting the growth of desert plants or any other crops of naturally saline areas like coastal [40]. Halophytes and glycophytes are two plant groups classified according to their efficiency in salt tolerance [67]. While halophytes can survive in saline environments, glycophytes are not capable of thriving and to mention, and most of the plant species fall under the glycophytes category [67].

According to some studies, sensing Ca^{2+} influx proved highly beneficial in supporting plant salt stress tolerance [68].

Further, it was reported upon the detection of salt stress by Ca^{2+} sensors, cytosolic Ca^{2+} concentration gets aggravated and the ion channel gets activated [69]. In *A. thaliana*, *OSCA1* was identified in the plasma membrane as a calcium-permeable ion channel essential for osmosensing and Ca^{2+} signaling [70]. Further, plastidial potassium (K^+) exchange antiporters (KEA) 1, 2, and 3 were discovered to be involved in the elevation of Ca^{2+} osmosensors in *A. thaliana* [71]. In another study, *MOCA1* was identified as encoding glucuronosyltransferase in the plasma membrane of *A. thaliana* for glycosyl inositol phosphorylceramide (GIPC) sphingolipids to activate Ca^{2+} ion channels and upregulate the Na^+/H^+ antiporters [72]. Furthermore, in *A. thaliana*, an increment in major ROS, H_2O_2, due to stressors was detected by a leucine-rich-repeat receptor (LRR) kinase, and the hydrogen-peroxide-induced Ca^{2+} increases 1 (*HPCA1*) which was found to mediate the influx of Ca^{2+} ions through the induction of H_2O_2 and is also responsible for stomatal closure [73]. Annexin 1 was also found to be involved in ROS activation and triggers calcium signaling due to sensing the extracellular presence of NaCl [74]. Additionally, calcium-dependent protein kinases (CDPKs) and calcineurin B-like proteins (CBLs) are involved in downstream transcription and protein activation as a result of a calcium signaling pathway [75, 76].

According to the recent RNA-seq analyses, some significant TFs, WRKY, bZIP, C2H2 zinc finger genes, AP2, NAC, and DREB families are associated with the activation of cellular signaling process through binding to their corresponding promoter and regulation of expression pattern of several target genes on sensing high salt concentration in plant environment [40]. In a study, overexpression of basic leucine zipper (bZIP) genes was observed in salt-sensitive wheat cultivars while exposed to high salt concentrations. However, decreased expression was observed in salt-tolerant wheat cultivars [77]. Again, in *Arabidopsis*, upregulation of WRKY8 due to salt stress was observed [95]. Increased NAC expression was linked to enhanced salt tolerance in rice and wheat [78]. Further, TFs viz. DREB1/CBF, DREB2, AREB/ABF, OsNAC5, and ZFP 19 showed increased expression under salinity stress conditions in rice [40].

4.3 Adaptation to Heat and Cold Stress

Deterioration in plant growth and development and a very low percentage of germination rate are the common challenges plants face in desert/dry regions due to high heat intensity in the summer season and chilling temperatures in winter. Apart from those, sunburn, curling and burning of leaves, leaf senescence, and discoloration of leaves and fruits are the hazardous morphological deformities of desert plants caused by heat stress [79]. In a study on the desert plant *Alhagi sparsifolia* S., disruption of photosynthesis was observed due to oxidative damage of photosynthetic proteins and thylakoid membranes due to high temperature [80].

The involvement of various TFs, notably MAPKs, CDPKs, and ROS-producing NADPH oxidase, is crucial. At the same time, the signal transduction process gets initiated against heat stress and promotes stress-responsive genes in plants [81]. Additionally, when a plant undergoes heat stress, several TFs from the heat stress transcription factor family (HSFs) and Heat Shock Proteins (HSPs) such as HSFA1, HSFA2, HSFA7, HSFBs, and DREB2A get activated for interaction with HSP70 and HSP90 to accelerate thermotolerance in plants [81, 82]. Further, Similar to drought and salinity stress responses, other important TFs, WRKY, bZIP, NAC, and MYB, are also associated with regulating heat-stress response [83]. Besides heat stress, cold stress is another highly devastating factor for desert plants to withstand. In both chilling (when the environmental temperature is 0–20°C) and freezing stress (when the environmental temperature is < 0°C), plants are capable of tolerating extremely low temperatures despite changes in their physiological, biochemical, and cellular functionalities [84]. ICE-CBF-COR is reportedly activated in response to cold stress in most plant species, which later triggers osmoregulation by expressing downstream genes [85]. Further, the involvement of ICE1 and ICE2 genes in the regulation of CBF expression during cold stress was identified [86]. Further, through comprehensive sequencing, 109 genes were identified associated with the activation of cold-stress-responsive pathways in plants, confirming the presence of potent regulatory mechanisms in response to environmental cold stress [40]. Consequently, the heat/cold tolerance in native desert plants could be improved by utilizing the present genetic information, which contributes to enhancing the stress-responsive mechanism of other plants through designing biotechnological research procedures.

Apart from stress-responsive genes toward drought, salinity, and high/low temperature, the function of several genes was also characterized. LARP1 (La ribonucleoprotein 1) - associated with adaptive response to environmental stimuli and abiotic stress-responsive gene, involved in ribosome biogenesis, hormonal stimulation [87, 88]; APX2 (ascorbate peroxidase 2) – responsive towards high light (HL) and drought stress [89]; LSD1 (lysine-specific demethylase 1) – regulation of low temperature related cell death [90]; PHR2 (photolyase receptor 2) – involved in blue light photoreception were deciphered [91, 92]. Further, several genes were identified to be responsible for the improvement of photosynthesis, viz. JAC1 (J-type co-chaperone 1) – involved in reducing the size of stomatal aperture, efficient water usage, regulation of H_2O_2, and controlling cell death [93], CP24 (chlorophyll a-b binding protein 24) – associated with the maintenance of grana membrane structure, electron transport [94, 95]. Again, RHL1 (root hairless 1) [96], and ZFP1 (zinc finger protein 1) [97] were identified to be involved in the maintenance of root growth, development of root hairs in *A. thaliana.* In this direction, understanding the fundamental genetic regulatory mechanisms of desert plants' evolutionary adaptive features could help ensure improved growth and development of desert plants with an enhanced tolerance response towards environmental extremities in dry, rainfed, nutrient-deficient desert lands.

Functional characterization of newly identified genes through population genomics study is mainly done using the sequence database of the model plant *A. thaliana,* which provides foolproof of most experimental evidence [98]. However, as *A. thaliana* is a stress-sensitive plant, in recent days, the native desert plants/extremophiles have been the preferable choice for the investigation of novel stress-responsive genes and their mechanism [98]. Likewise, a recent research reported an Arabidopsis desert relative, *Anastatica hierochuntica,* native of the Negev Desert, exhibited much greater tolerance than *A. thaliana* from heat stress and low nitrogen (N) stress [98]. We presume that the comprehensive application of available information on genetic interference with adaptive stress-response of desert plants will be more significant if a detailed comparative study could be performed using model plants of versatile characteristics mainly originating from extreme environments. This will enable the vision of the entire agri-scientific community to be more focused on making our world greener, vegetative, and flourishing with food crops in the near future.

5 Omics-Based Studies on Desert Plants

Characterization of desert plants with high-quality genome sequencing is critical to understand their ability to survive and reproduce under harsh conditions. Precise and extensive genome sequencing, along with demanding annotation techniques, exposes the genetic sequences and the roles of genes, regulatory elements, and intricate pathways. This broad understanding helps to explore the complex mechanisms underlying these extraordinary adaptation strategies. This information can be the foundation for using these valuable features for agricultural improvement programs.

In recent years, researchers have tried to make considerable advances using advanced omics approaches aiming to unravel the genetic composition of many species, mainly native desert plants naturally adapted to severe weather conditions. Desert plant genome sequencing is an essential field of research with far-reaching implications for understanding plant adaptability, ecological resilience, and future agricultural and conservation uses. One such example is the sequencing of the *Welwitschia mirabilis* [99]. The plant's genome sequencing has successfully become a focus of genomic research due to its newly explored remarkable adaptation to harsh desert conditions. The study also has implications for biodiversity conservation, as the genomic data identified evolutionarily conserved characteristics of this unique species. In another study, a draft genome sequence of the desert tree *Populus pruinosa* unveiled genes associated with coping mechanisms for drought, heat stress, and oxidative stress, shedding light on its ability to thrive in arid conditions [100]. Similarly, chromosome level genome assembly of xerophytic plant, *Haloxylon ammodendron* provided new insights in understanding the genetic network of desert plants [101].

On the other hand, Next-generation sequencing (NGS) has revolutionized the study of genomes, transcriptomes, and metabolomes by providing rapid, cost-effective methods for analyzing DNA, RNA, and small molecules. NGS technologies enabled comprehensive genome sequencing, including whole genome sequencing (WGS) and de novo assembly, facilitating the detection of genetic variations and structural alterations. Transcriptome sequencing with NGS allows for precise measurement of gene expression levels, identification of alternative splicing events, and discovery of non-coding RNAs, shedding light on transcriptional regulation and cellular processes. Additionally, NGS-based metabolomics offers insights into metabolic pathways, biomarker discovery, and environmental metabolite profiling, enhancing our understanding of metabolism and its role in health and disease [102–104]. Overall, NGS has transformed biological research, empowering scientists to unravel the complexities of living systems at the genomic, transcriptomic, and metabolomic levels. Gene editing and genetic modification ensure the addition of these resilience-enhancing features into various crop kinds, boosting their capacity to adapt and yield even in adverse conditions. This genetic knowledge transmission offers a bright opportunity for global food security, revolutionizing crop breeding practices to maintain a consistent food supply in the face of climate change.

Further, a recent study demonstrated applications of CRISPR/Cas9 genome editing for improving date palm production based on the whole genome sequence of date palm [104]. Another survey of *Arabidopsis thaliana* investigated the possibility of improvement of drought stress tolerance using a CRISPR system targeting ABA-responsive signal binding protein 1 (AREB1) [105].

6 Challenges and Future Prospects of Desert Plant Studies to Transfer Genetic Knowledge to Crop Plants

For many years, the irreversible loss of genes and genetic combinations in wild desert plants due to environmental and anthropogenic pressures has been a significant concern for the scientific community [38]. Research studies on desert plants stand very resourceful as they can provide a provision to understand the plants' physiological and genetic adaptation strategies to abiotic stress in plants and save the plant diversity across different dry/arid regions. Further, these genetic resources assist in the production of varieties of food or commercially important crops with increased productivity and stress tolerance efficiency to cope with climatic difficulties [38]. Recently, scientists have developed tremendous interest in applying this genetic information to practical use by transferring specific genes or regulatory components responsible for competent drought tolerance or water management from desert plants to diverse crop species [13]. However, the entire genome of important native desert plants is still unexplored. Although the creation of databases for different desert plants has been started to address ecological and metabolic importance, there is still a significant insufficiency of present genetic information on native desert plants due to complex genetic gaps scattered among thousands of species due to evolutionary adaptations. Desert plants are unique, with physiological structures different from regular plants, and comprise various biochemical features. Interestingly, most physical and biochemical properties are adapted based on climatic variations in different geographical locations and long-term evolutionary changes. Therefore, maintaining plant genetic diversity is crucial to avoid biodiversity loss and contributes to deciphering a plant's potential value in food and agricultural sectors. Nevertheless, while the prospect of transferring genetic information from desert plants to crops is significant, ethical, environmental, and regulatory concerns must be addressed. With proper implementation and assessment, transferring genetic traits from desert plants to crops could enable resilient agricultural systems to address climate change challenges and ensure food security for future generations.

Many wild desert plant species possess valuable biochemical and nutritional traits; however, factors such as soil salinity, heat stress, and nutrient depletion can compromise their growth and metabolite production. Thus, choosing and raising the appropriate genotype of the target plant is highly essential and challenging at the same time. To overcome these challenges, geneticists and plant specialists need to validate the genetic and nutrition profile of the target plant at the molecular level. The revolutionary research reports provide valuable conceptions on genetic adaptation and the involvement of several transcription factors in crop plants and desert plants towards environmental stress stimuli, which are discussed in this review. They assure the affluent design and implementation of genetic strategies along with progressive sequencing of native plants. This genetic information will lead the plant scientist community to affirm the selection of target genes and eliminate the detrimental effect of stressors in plants of desert/rainfed locations. In this manner, the global population can achieve climate-adaptive yet nutrition-rich foods and pave the way to overcome the hurdles caused by hunger and malnutrition.

7 Conclusion

This study provided a cumulative insight into the core genetic mechanism adapted by desert plants and the revelation of advances in molecular and bioinformatic tools such as RNA-Seq and genome sequencing to uncover the resilience of those plants. This information assures a roadmap for strengthening crops against environmental adversity. The transfer of genetic information from these desert-adapted species to crops indicates belief, offering essential resilience traits crucial for thriving in evolving climates. The present attempt to accumulate interesting theories on the regulatory mechanisms of desert plants will surely assist the plant geneticist in protecting the plant diversity of desert lands and promoting sustainable global food production with creative agricultural practices.

Conflict of Interest The authors declare no conflict of interest.

References

1. Faheem Z, Kazmi JH, Shaikh S, Arshad S, Mohammed S (2024) Random forest-based analysis of land cover/land use LCLU dynamics associated with meteorological droughts in the desert ecosystem of Pakistan. Ecol Ind 159:111670
2. Wang M, Dong Z, Luo W, Lu J, Li J (2015) Spatial variability of vegetation characteristics, soil properties and their relationships in and around China's Badain Jaran desert. Environ Earth Sci 74:6847–6858
3. Carvajal DE, Loayza AP, Rios RS, Gianoli E, Squeo FA (2017) Population variation in drought-resistance strategies in a desert shrub along an aridity gradient: Interplay between phenotypic plasticity and ecotypic differentiation. Perspect Plant Ecol, Evol Syst 29:12–19
4. Ward D (2016) The biology of deserts. Oxford University Press
5. Mohanta TK, Mohanta YK, Kaushik P, Kumar J (2024) Physiology, genomics, and evolutionary aspects of desert plants. J Adv Res 58:63–78
6. Cushman JC (2001) Crassulacean acid metabolism. A plastic photosynthetic adaptation to arid environments. Plant Physiol 127(4):1439–1448
7. Benitez-Alfonso Y, Soanes BK, Zimba S, Sinanaj B, German L, Sharma V, Foyer CH (2023) Enhancing climate change resilience in crops. Curr Biol 33(23):R1246–R1261

8. Sah SK, Reddy KR, Li J (2016) Abscisic acid and abiotic stress tolerance in crop plants. Front Plant Sci 7:571
9. Mal S, Panchal S (2024) Drought and salt stress mitigation in crop plants using stress-tolerant auxin-producing endophytic bacteria: a futuristic approach towards sustainable agriculture. Front Plant Sci 15:1422504
10. Sandquist DR (2014) Plants in deserts. In: Monson R (eds) Ecology and the environment. The plant sciences, vol 8. Springer, New York, NY
11. Kulshrestha S (2023) An introduction to the ecology of desert systems. Int J Innov Sci Res Technol 8(3):556–583
12. Bombi P, Salvi D, Shuuya T, Vignoli L, Wassenaar T (2021) Climate change effects on desert ecosystems: a case study on the keystone species of the Namib desert *Welwitschia Mirabilis*. Plos One 16(11)
13. Kirschner GK, Xiao TT, Blilou I (2021) Rooting in the desert: a developmental overview on desert plants. Genes 12(5):709. Kooyers NJ (2015) The evolution of drought escape and avoidance in natural herbaceous populations. Plant Sci 234:155–162
14. Liu LB, Bai WP, Li HJ, Tian Y, Yuan HJ, Garant TM, Liu HS, Zhang J, Bao AK, Rowland O, Wang SM (2021) ZxABCG11 from the xerophyte *Zygophyllum xanthoxylum* enhances drought tolerance in *Arabidopsis thaliana* through modulating cuticular wax accumulation. Environ Exp Bot 190:104570
15. Serna L, Martin C (2006) Trichomes: different regulatory networks lead to convergent structures. Trends Plant Sci 11(6):274–280
16. Wainwright CM (1977) Sun-tracking and related leaf movements in a desert lupine (*Lupinus arizonicus*). Am J Bot 64(8):1032–1041
17. Hassiotou F, Evans JR, Ludwig M, Veneklaas EJ (2009) Stomatal crypts may facilitate diffusion of CO_2 to adaxial mesophyll cells in thick sclerophylls. Plant, Cell Environ 32(11):1596–1611
18. Ehleringer JR, Phillips SL, Schuster WSF et al (1991) Differential utilization of summer rains by desert plants. Oecologia 88:430–434
19. Gibbens RP, Lenz JM (2001) Root systems of some Chihuahuan desert plants. J Arid Environ 49(2):221–263
20. Ndour PMS, Heulin T, Achouak W, Laplaze L, Cournac L (2020) The rhizosheath: from desert plants adaptation to crop breeding. Plant Soil 456:1–13
21. Shishkova S, Las Peñas ML, Napsucialy-Mendivil S, Matvienko M, Kozik A, Montiel J, Patiño A, Dubrovsky JG (2013) Determinate primary root growth as an adaptation to aridity in Cactaceae: towards an understanding of the evolution and genetic control of the trait. Ann Bot 112(2):239–252
22. Puthur JT, Shackira AM, Saradhi PP, Bartels D (2013) Chloroembryos: a unique photosynthesis system. J Plant Physiol 170(13):1131–1138
23. Kocurek M, Kornas A, Wierzchnicki R et al (2020) Importance of stem photosynthesis in *Clusia minor* plant carbon allocation. Trees 34:1009–1020
24. Simkin AJ, Faralli M, Ramamoorthy S, Lawson T (2020) Photosynthesis in non-foliar tissues: implications for yield. Plant J 101(4):1001–1015
25. Ávila E, Herrera A, Tezara W (2014) Contribution of stem CO_2 fixation to whole-plant carbon balance in nonsucculent species. Photosynthetica 52:3–15
26. Ogburn RM, Edwards EJ (2010) The ecological water-use strategies of succulent plants. Advances in botanical research, vol 55. Academic Press, pp 179–225
27. Cabrita PJV (2024) Water exchange between the Chlorenchyma and the Hydrenchyma and its physiological role in leaves with Crassulacean acid metabolism. Physiol Plant 176(2):e14221
28. Griffiths H, Males J (2017) Succulent plants. Curr Biol 27(17):R890–R896
29. Males J (2017) Secrets of succulence. J Exp Bot 68(9):2121–2134
30. Sayed OH (2001) Aridity and plant survival in desert environments. In: Prakash I (ed) Ecology of desert environments. Scientific Publishers, Jodhpur, pp 87–103
31. Kluge M, Ting IP (2012) Crassulacean acid metabolism: analysis of an ecological adaptation vol. 30. Springer Science & Business Media

32. Ehleringer JR, Monson RK (1993) Evolutionary and ecological aspects of photosynthetic pathway variation. Annu Rev Ecol Syst 2(1):411–439
33. Su P (2010) Photosynthesis of C_4 desert plants. In: Ramawat K (eds) Desert plants. Springer, Berlin, Heidelberg
34. Martinoia E, Rentsch D (1994) Malate compartmentation-responses to a complex metabolism. Annu Rev Plant Physiol Plant Mol Biolology 45:447–467
35. Maier A, Zell MB, Maurino VG (2011) Malate decarboxylases: evolution and roles of NAD (P)-ME isoforms in species performing C4 and C3 photosynthesis. J Exp Bot 62(9):3061–3069
36. Eshel G, Araus V, Undurraga S, Soto DC, Moraga C, Montecinos A, Moyano T, Maldonado J, Díaz FP, Varala K, Nelson CW, Contreras-López O, Pal-Gabor H, Kraiser T, Carrasco-Puga G, Nilo-Poyanco R, Zegar CM, Orellana A, Montecino M, Maass A, Allende ML, DeSalle R, Stevenson DW, González M, Latorre C, Coruzzi GM, Gutiérrez RA (2021) Plant ecological genomics at the limits of life in the Atacama desert. Proc Natl Acad Sci 118(46):e2101177118
37. Baeshen MN, Ahmed F, Moussa TAA, Abulfaraj AA, Jalal RS, Noor SO, Baeshen NA, Huelsenbeck JP (2021) A comparative analysis of de novo transcriptome assembly to understand the abiotic stress adaptation of desert plants in Saudi Arabia. Appl Ecol Environ Res 19(3):1753–1782
38. Gairola S, Al Shaer KI, Al Harthi EK, Mosa KA (2018) Strengthening desert plant biotechnology research in the United Arab Emirates: a viewpoint. Physiol Mol Biol Plants 24:521–533
39. Long Y, Zhang J, Tian X, Wu S, Zhang Q, Zhang J, Dang D, Pei XW (2014) De novo assembly of the desert tree *Haloxylon ammodendron* (CA Mey.) based on RNA-Seq data provides insight into drought response, gene discovery and marker identification. BMC Genomics 15:1–11
40. Lovio-Fragoso JP, De Jesús-Campos D, Razo-Mendivil FG, García-Coronado H, Domínguez-Rosas E, Trillo-Hernández EA, Hayano-Kanashiro C, Hernández-Oñate MÁ (2023) Desert plant transcriptomics and adaptation to abiotic stress. In: Transcriptome profiling. Academic Press, pp 199–256
41. Farooq M, Wahid A, Lee DJ, Ito O, Siddique KHM (2009) Advances in drought resistance of rice. Crit Rev Plant Sci 28(4):199–217
42. Bashir SS, Hussain A, Hussain SJ, Wani OA, Zahid Nabi S, Dar NA, Baloch FS, Mansoor S (2021) Plant drought stress tolerance: Understanding its physiological, biochemical and molecular mechanisms. Biotechnol Biotechnol Equipment 35(1):1912–1925
43. Ramadan AM, Hassanein SE (2014) Characterization of P5CS gene in *Calotropis procera* plant from the de novo assembled transcriptome contigs of the high-throughput sequencing dataset. Comptes Rendus Biol 337(12):683–690
44. Kishor PK, Hong Z, Miao GH, Hu CAA, Verma DPS (1995) Overexpression of [delta]-pyrroline-5-carboxylate synthetase increases proline production and confers osmotolerance in transgenic plants. Plant Physiol 108(4):1387–1394
45. Zhu B, Su J, Chang M, Verma DPS, Fan YL, Wu R (1998) Overexpression of a Δ1-pyrroline-5-carboxylate synthetase gene and analysis of tolerance to water-and salt-stress in transgenic rice. Plant Sci 139(1):41–48
46. Hmida-Sayari A, Gargouri-Bouzid R, Bidani A, Jaoua L, Savouré A, Jaoua S (2005) Overexpression of Δ1-pyrroline-5-carboxylate synthetase increases proline production and confers salt tolerance in transgenic potato plants. Plant Sci 169(4):746–752
47. Guerzoni JTS, Belintani NG, Moreira RMP, Hoshino AA, Domingues DS, Filho JCB, Vieira LGE (2014) Stress-induced Δ1-pyrroline-5-carboxylate synthetase (P5CS) gene confers tolerance to salt stress in transgenic sugarcane. Acta Physiol Plant 36:2309–2319
48. Zhang GC, Zhu WL, Gai JY, Zhu YL, Yang LF (2015) Enhanced salt tolerance of transgenic vegetable soybeans resulting from overexpression of a novel Δ 1-pyrroline-5-carboxylate synthetase gene from *Solanum torvum* Swartz. Hortic Environ Biotechnol 56:94–104
49. Roosens NH, Bitar FA, Loenders K, Angenon G, Jacobs M (2002) Overexpression of ornithine-δ-aminotransferase increases proline biosynthesis and confers osmotolerance in transgenic plants. Mol Breeding 9:73–80

50. Wu L, Fan Z, Guo L, Li Y, Zhang W, Qu LJ, Chen Z (2003) Over-expression of an Arabidopsis δ-OAT gene enhances salt and drought tolerance in transgenic rice. Chin Sci Bull 48:2594–2600
51. Ishitani M, Nakamura T, Han SY, Takabe T (1995) Expression of the betaine aldehyde dehydrogenase gene in barley in response to osmotic stress and abscisic acid. Plant Mol Biol 27:307–315
52. Fan W, Zhang M, Zhang H, Zhang P (2012) Improved tolerance to various abiotic stresses in transgenic sweet potato (*Ipomoea batatas*) expressing spinach betaine aldehyde dehydrogenase. PLoS ONE 7(5):e37344
53. Li M, Li Z, Li S, Guo S, Meng Q, Li G, Yang X (2014) Genetic engineering of glycine betaine biosynthesis reduces heat-enhanced photoinhibition by enhancing antioxidative defense and alleviating lipid peroxidation in tomato. Plant Mol Biol Report 32:42–51
54. Huber SC, Huber JL (1996) Role and regulation of sucrose-phosphate synthase in higher plants. Annu Rev Plant Biol 47(1):431–444
55. Kaldenhoff R, Fischer M (2006) Aquaporins in plants. Acta Physiol 187(1–2):169–176
56. Zheng J, Lin R, Pu L, Wang Z, Mei Q, Zhang M, Jian S (2021) Ectopic expression of CrPIP2; 3, a plasma membrane intrinsic protein gene from the halophyte *Canavalia rosea*, enhances drought and salt-alkali stress tolerance in Arabidopsis. Int J Mol Sci 22(2):565
57. Khan K, Agarwal P, Shanware A, Sane VA (2015) Heterologous expression of two Jatropha aquaporins imparts drought and salt tolerance and improves seed viability in transgenic *Arabidopsis thaliana*. PLoS ONE 10(6):e0128866
58. Iuchi S, Kobayashi M, Taji T, Naramoto M, Seki M, Kato T, Tabata S, Kakubari Y, Yamaguchi-Shinozaki K, Shinozaki K (2001) Regulation of drought tolerance by gene manipulation of 9-cis-epoxycarotenoid dioxygenase, a key enzyme in abscisic acid biosynthesis in Arabidopsis. Plant J 27(4):325–333
59. Seo M (2002) Complex regulation of ABA biosynthesis in plants. Trends Plant Sci 7:41–48
60. Sheen J (1996) Ca2+-dependent protein kinases and stress signal transduction in plants. Science 274(5294):1900–1902
61. Mizoguchi T, Irie K, Hirayama T, Hayashida N, Yamaguchi-Shinozaki K, Matsumoto K, Shinozaki K (1996) A gene encoding a mitogen-activated protein kinase kinase kinase is induced simultaneously with genes for a mitogen-activated protein kinase and an S6 ribosomal protein kinase by touch, cold, and water stress in *Arabidopsis thaliana*. Proc Natl Acad Sci 93(2):765–769
62. Welsch R, Wust F, Bar C, Al-Babili S, Beyer P (2008) A third phytoene synthase is devoted to abiotic stress-induced abscisic acid formation in rice and defines functional diversification of phytoene synthase genes. Plant Physiol 147(1):367–380
63. Lu P, Magwanga RO, Kirungu JN, Hu Y, Dong Q, Cai X, Liu F (2019) Overexpression of cotton a DTX/MATE gene enhances drought, salt, and cold stress tolerance in transgenic Arabidopsis. Front Plant Sci 10:299
64. Yoon HK, Kim SG, Kim SY, Park CM (2008) Regulation of leaf senescence by NTL9-mediated osmotic stress signaling in Arabidopsis. Mol Cells 25(3):438–445
65. Yu Y, Bai Y, Wang P, Wang Y, Wan H, Liu C, Ni Z (2020) Soybean nuclear factor YA10 positively regulates drought resistance in transgenic *Arabidopsis thaliana*. Environ Exp Bot 180:104249
66. Park MY, Kang JY, Kim SY (2011) Overexpression of AtMYB52 confers ABA hypersensitivity and drought tolerance. Mol Cells 31:447–454
67. Munns R, Gilliham M (2015) Salinity tolerance of crops—what is the cost? New Phytol 208(3):668–673
68. Zhao C, Zhang H, Song C, Zhu J-K, Shabala S (2020) Mechanisms of plant responses and adaptation to soil salinity. In: The innovation, vol 1, no 1. New York, N.Y, p 100017
69. Dodd AN, Kudla J, Sanders D (2010) The language of calcium signaling. Annu Rev Plant Biol 61:593–620
70. Yuan F, Yang H, Xue Y, Kong D, Ye R, Li C, Zhang J, Theprungsirikul L, Shrift T, Krichilsky B, Johnson DM, Swift GB, He Y, Siedow JN, Pei ZM (2014) OSCA1 mediates osmotic-stress-evoked Ca^{2+} increases vital for osmosensing in Arabidopsis. Nature 514(7522):367–371

71. Stephan AB, Kunz HH, Yang E, Schroeder JI (2016) Rapid hyperosmotic-induced Ca^{2+} responses in *Arabidopsis thaliana* exhibit sensory potentiation and involvement of plastidial KEA transporters. Proc Natl Acad Sci USA 113(35):E5242–E5249
72. Jiang Z, Zhou X, Tao M, Yuan F, Liu L, Wu F, Wu X, Xiang Y, Niu Y, Liu F, Li C, Ye R, Byeon B, Xue Y, Zhao H, Wang HN, Crawford BM, Johnson DM, Hu C, Pei C, Zhou W, Swift GB, Zhang H, Vo-Dinh T, Hu Z, Siedow JN, Pei ZM (2019) Plant cell surface GIPC sphingolipids sense salt to trigger Ca2+ influx. Nature 572(7769):341–346
73. Wu F, Chi Y, Jiang Z, Xu Y, Xie L, Huang F, Wan D, Ni J, Yuan F, Wu X, Zhang Y, Wang L, Ye R, Byeon B, Wang W, Zhang S, Sima M, Chen S, Zhu M, Pei J, Johnson DM, Zhu S, Cao X, Pei C, Zai Z, Liu Y, Liu T, Swift GB, Zhang W, Yu M, Hu Z, Siedow JN, Chen X, Pei ZM (2020) Hydrogen peroxide sensor HPCA1 is an LRR receptor kinase in Arabidopsis. Nature 578 (7796), 577–581
74. Laohavisit A, Richards SL, Shabala L, Chen C, Colac¸o RDDR, Swarbreck SM, Shaw E, Dark A, Shabala S, Shang Z, Davies JM (2013) Salinity-induced calcium signaling and root adaptation in Arabidopsis require the calcium regulatory protein Annexin1. Plant Physiol 163(1):253–262
75. Boudsocq M, Sheen J (2013) CDPKs in immune and stress signaling. Trends Plant Sci 18(1):30–40
76. Weinl S, Kudla J (2009) The CBL-CIPK Ca (2+)-decoding signaling network: function and perspectives. New Phytol 184(3):517–528
77. Johnson RR, Wagner RL, Verhey SD, Walker-Simmons MK (2002) The abscisic acid-responsive kinase PKABA1 interacts with a seed-specific abscisic acid response element-binding factor TaABF, and phosphorylates TaABF peptide sequences. Plant Physiol 130(2):837–846
78. Hu, Y., Chen, L., Wang, H., Zhang, L., Wang, F., & Yu, D (2013) Arabidopsis transcription factor WRKY8 functions antagonistically with its interacting partner VQ9 to modulate salinity stress tolerance. Plant J 74(5):730–745
79. Hasanuzzaman M, Nahar K, Alam MM, Roychowdhury R, Fujita M (2013) Physiological, biochemical, and molecular mechanisms of plant heat stress tolerance. Int J Mol Sci 14(5):9643–9684
80. Xue W, Li XY, Lin LS, Wang YJ, Li L (2011) Effects of elevated temperature on photosynthesis in desert plant *Alhagi sparsifolia* S. Photosynthetica 49:435–447
81. Kaushal N, Bhandari K, Siddique KHM, Nayyar H (2016) Food crops face rising temperatures: An overview of responses, adaptive mechanisms, and approaches to improve heat tolerance. Cogent Food Agric 2(1)
82. Ohama N, Sato H, Shinozaki K, Yamaguchi-Shinozaki K (2017) Transcriptional regulatory network of plant heat stress response. Trends Plant Sci 22(1):53–65
83. Ren S, Ma K, Lu Z (2019) Transcriptomic and metabolomic analysis of the heat-stress response of *Populus tomentosa* Carr. Forests 10(5):383
84. Chinnusamy V, Zhu J., Sunkar R (2010) Gene regulation during cold stress acclimation in plants. In: Methods in molecular biology, vol 639. Springer, Clifton NJ, pp 39–55
85. Ritonga FN, Chen S (2020) Physiological and molecular mechanism involved in cold stress tolerance in plants. Plants 9(5):560
86. Guo X, Liu D, Chong K (2018) Cold signaling in plants: insights into mechanisms and regulation. J Integr Plant Biol 60(9):745–756
87. Zhang B, Jia J, Yang M, Yan C, Han Y (2012) Overexpression of a LAM domain containing RNA-binding protein LARPIc induces precocious leaf senescence in Arabidopsis. Mol Cells 34(4):367–374
88. Zhang W, Fang D, Ye Z, Hu F, Cheng X, Cao J (2023) Identification and molecular evolution of the La and LARP genes in 16 plant species: a focus on the *Gossypium hirsutum.* Int J Biol Macromol 224:1101–1117
89. Rossel JB, Walter PB, Hendrickson L, Chow WS, Poole A, Mullineaux PM, Pogson BJ (2006) A mutation affecting ASCORBATE PEROXIDASE 2 gene expression reveals a link between responses to high light and drought tolerance. Plant Cell Environ 29(2):269–281

90. Dietrich RA, Richberg MH, Schmidt R, Dean C, Dangl JL (1997) A novel zinc finger protein is encoded by the Arabidopsis LSD1 gene and functions as a negative regulator of plant cell death. Cell 88(5):685–694
91. Petersen JL, Lang DW, Small GD (1999) Cloning and characterization of a class II DNA photolyase from Chlamydomonas. Plant Mol Biol 40:1063–1071
92. Liu Q, Su T, He W, Ren H, Liu S, Chen Y, Gao L, Hu X, Lu H, Cao S, Huang Y, Wang X, Wang Q, Lin C (2020) Photooligomerization determines photosensitivity and photoreactivity of plant cryptochromes. Mol Plant 13(3):398–413
93. Czarnocka W, Rusaczonek A, Willems P, Sujkowska-Rybkowska M, Van Breusegem F, Karpiński S (2020) Novel role of JAC1 in influencing photosynthesis, stomatal conductance, and Photooxidative stress Signalling pathway in *Arabidopsis thaliana*. Front Plant Sci 11:1124
94. Kovacs L, Damkjær J, Kereïche S, Ilioaia C, Ruban AV, Boekema EJ, Jansson S, Horton P (2006) Lack of the light-harvesting complex CP24 affects the structure and function of the grana membranes of higher plant chloroplasts. Plant Cell 18(11):3106–3120
95. De Bianchi S, Dall'Osto L, Tognon G, Morosinotto T, Bassi R (2008) Minor antenna proteins CP24 and CP26 affect the interactions between photosystem II subunits and the electron transport rate in grana membranes of Arabidopsis. Plant Cell 20(4):1012–1028
96. Schneider K, Mathur J, Boudonck K, Wells B, Dolan L, Roberts K (1998) The ROOT HAIRLESS 1 gene encodes a nuclear protein required for root hair initiation in Arabidopsis. Genes Dev 12(13):2013–2021
97. Han G, Wei X, Dong X, Wang C, Sui N, Guo J, Yuan F, Gong Z, Li X, Zhang Y, Meng Z (2020) Arabidopsis ZINC FINGER PROTEIN1 acts downstream of GL2 to repress root hair initiation and elongation by directly suppressing bHLH genes. Plant Cell 32(1):206–225
98. Eshel G, Shaked R, Kazachkova Y, Khan A, Eppel A, Cisneros A, Acuna T, Gutterman Y, Tel-Zur N, Rachmilevitch S, Fait A (2017) *Anastatica hierochuntica*, an Arabidopsis desert relative, is tolerant to multiple abiotic stresses and exhibits species-specific and common stress tolerance strategies with its halophytic relative, *Eutrema* (Thellungiella) *salsugineum*. Front Plant Sci 7:1992
99. Wan T, Liu Z, Leitch IJ, Xin H, Maggs-Kölling G, Gong Y, Li Z, Marais E, Liao Y, Dai C, Liu F, Wu Q, Song C, Zhou Y, Huang W, Jiang K, Wang Q, Yang Y, Zhong Z, Yang M, Yan X, Hu G, Hou C, Su Y, Feng S, Yang J, Yan J, Chu J, Chen F, Ran J, Wang X, Van de Peer Y, Leitch AR, Wang Q (2021) The Welwitschia genome reveals a unique biology underpinning extreme longevity in deserts. Nat Commun 12(1):4247
100. Yang W, Wang K, Zhang J, Ma J, Liu J, Ma T (2017) The draft genome sequence of a desert tree *Populus pruinosa*. Gigascience 6(9):gix075
101. Wang M, Zhang L, Tong S, Jiang D, Fu Z (2022) Chromosome-level genome assembly of a xerophytic plant, *Haloxylon ammodendron*. DNA Res 29(2):dsac006
102. He C, Jin M, Liu W, Shan L, Li Y, Zhang Z, Li Q, Li M, Jin N, Hu, F (2024) The metabolomics and transcriptomics of *Atraphaxis bracteata* reveal age-related molecular alterations under naturally high temperature and drought conditions. Physiol Plant 176(1):e14129
103. Wang J, Jiang X, Zhao C, Fang Z, Jiao P (2020) Transcriptomic and metabolomic analysis reveals the role of CoA in the salt tolerance of *Zygophyllum* spp. BMC Plant Biol 20:1–14
104. Sattar MN, Iqbal Z, Tahir MN, Shahid MS, Khurshid M, Al-Khateeb AA, Al-Khateeb SA (2017) CRISPR/Cas9: a practical approach in date palm genome editing. Front Plant Sci 8:1469
105. Roca Paixão JF, Gillet FX, Ribeiro TP, Bournaud C, Lourenço-Tessutti IT, Noriega D. D, de Melo BP, de Almeida-Engler J, Grossi-de-Sa MF (2019) Improved drought stress tolerance in Arabidopsis by CRISPR/dCas9 fusion with a Histone AcetylTransferase. Sci Rep 9(1):8080

Valorization of Desalination Reject Brine for Calcium Recovery

Aiste Zukaityte, Roneta Chaliulina, Bryant Montenegro, and Yousef Alhorr

Abstract With the growing need for sustainable solutions to manage desalination by-products and reduce carbon emissions, mineralization offers a promising approach converting CO_2 into valuable and stable metal carbonates. This study investigated the effects of pH and reaction time on calcium carbonate precipitation from a Qatari desalination reject brine. The experiments involved a two-step synthesis, where (1) CO_2 reacted with NaOH, producing Na_2CO_3, (2) the produced solution was mixed with reject desalination brine at equimolar molar ratio of Ca^{2+} to $CO_3{}^{2-}$ 1:1. The initial synthesis pH was adjusted by adding concentrated NaOH solution. Reactions were carried out at pH levels from 9.25 to 10.5 and the influence of pH on precipitation speed, phase formation, and magnesium interaction as evaluated. The study found that the 9.5 pH optimal for calcium carbonate precipitation with the highest Ca^{2+} recovery. The presence of magnesium and its inhibitory effect on crystallization was a significant factor in the process, especially at higher pH levels.

Keywords Calcite · Calcium carbonate · Mineralization · CCU · Carbon capture and utilization. · Desalination reject brine

1 Introduction

Economic growth, driven by the change in energy consumption and industrial activity, has led to increased CO_2 emissions [1]. Since 1990, global CO_2 emissions have risen by more than 60%. In 2023, these emissions grew by 1.1%, reaching a new record of 37.4 billion metric tons ($GtCO_2$), according to the International Energy Agency (2024) [2]. CO_2 is a significant contributor to global warming, and it is released into the atmosphere through the combustion of fossil fuels and various industrial activities [3].

A. Zukaityte · R. Chaliulina (✉) · B. Montenegro · Y. Alhorr
Gulf Organisation for Research and Development, Ar-Rayyan, Qatar
e-mail: r.chaliulina@gord.qa

S. Al-Naemi et al. (eds.), *Water and Food Security in the Face of Climate Change: Challenges and Opportunities for Resilience*, World Sustainability Series,
https://doi.org/10.1007/978-3-032-00098-9_21

A surge in population and rapid economic expansion have increased pollution and environmental degradation [4]. Many countries have faced severe water scarcity due to a combination of factors such as over-extraction of groundwater and pollution of the freshwater sources [5]. To address the growing demand, seawater desalination has emerged as a dominant source of freshwater [6]. Nowadays, desalination is used in over 150 countries, particularly in the Middle East-North Africa (MENA) region [7]. However, alongside freshwater, desalination generates reject brine as a by-product, which is often discharged into the environment with adverse ecological consequences [8].

Research suggests a potential to address the environmental impacts of CO_2 and reject brine simultaneously [5, 9, 10]. Mineral carbonation is a promising technique for CO_2 storage and utilization due to its ability to transform CO_2 into thermodynamically stable metal carbonates [11–16]. Notably, mineral carbonation presents reduced leakage risks, distinguishing it from other CCU technologies [17, 18]. Reject brine, which is rich in calcium, can potentially be used to form calcium carbonate ($CaCO_3$), a material widely used in industries such as paper, paint, and construction [19].

One of the most explored methods of brine carbonation involves increasing the pH of the brine and injecting CO_2. This process leads to the formation of hydroxides from Mg and Ca^{2+}, which then react with CO_2 producing carbonates. However, the resulting solids are not phase pure, as they contain mixed crystalline phases, such as $MgCO_3$ and $CaCO_3$ [9, 20]. To selectively precipitate $CaCO_3$, other techniques have been explored. One such method is the pH swing method, where the brine pH is increased to around 10.5, causing Mg to precipitate as $Mg(OH)_2$. The magnesium-free brine is then further alkalinized to a pH above 12 to precipitate $Ca(OH)_2$, which reacts with CO_2 to form $CaCO_3$ [21, 22]. While this method yields a high amount of phase pure $CaCO_3$, it is complex and involves multiple steps, making it challenging to implement on a large scale.

Another investigated method for selective $CaCO_3$ precipitation in reject brine involves the addition of reagents such as sodium carbonate (Na_2CO_3) or sodium bicarbonate ($NaHCO_3$) [23, 24]. Unlike NaOH, the addition of these weaker bases has a relatively mild effect on the pH of the feed water: $NaHCO_3$ typically raises the pH to around 8–9, whereas Na_2CO_3 to 10–11. Since magnesium ions precipitate as $Mg(OH)_2$ at pH above 10.5, at this lower pH they would remain dissolved in solution, while calcium ions in the presence of carbonate and bicarbonate ions would preferentially precipitate. The preferential precipitation occurs due to difference in calcium and magnesium carbonates' solubility products, influenced by pH. Therefore, under typical conditions, calcium carbonate precipitates first, and magnesium carbonate may only precipitate if the carbonate ion concentration is sufficiently high or the pH is increased further, allowing for selective calcium removal as calcium carbonate, e.g. calcite [21]. However, the presence of $Mg2^+$ in brine can adversely impact the rate of calcite precipitation [25, 26]. Mg^{2+} can be incorporated into the calcite crystal lattice at Ca^{2+} sites, altering crystal morphology and slowing its growth. It has been therefore suggested that bicarbonate is used as a buffer to promote the formation of calcium carbonates and keep the pH within the preferred range 8.5–9.5 for selective calcium precipitation. While research has focused on optimizing parameters such as temperature, time, and pH to efficiently remove $Ca2^+$ while minimizing $Mg2^+$ loss, the characteristics of the obtained $CaCO_3$ have not been extensively studied [27].

Calcium carbonate ($CaCO_3$) exists in three anhydrous crystalline polymorphs: calcite, aragonite, and vaterite, as well as hydrated forms: calcium carbonate monohydrate (MHC), calcium carbonate hemihydrate and calcium carbonate hexahydrate (ikaite) [23, 25, 28]. These forms originate from the transformation of amorphous calcium carbonate (AC) [23], pH and additives are two dominating factors controlling polymorphism during calcium carbonate precipitation [25, 29, 30].

Literature studies often derive optimal pH values (like 9.5 for calcite) under controlled conditions with pure solutions or well-defined ionic compositions. In contrast, the presented experiments use desalination brine as the feed introduce a complex matrix of ions that can significantly alter the reaction pathways and kinetics. This study investigates how pH and reaction time impact calcium carbonate precipitation in the presence of magnesium in desalination reject brine, aiming to determine the most efficient synthesis parameters. This research addresses a critical gap in the field of carbon capture and mineralization by focusing on the optimal conditions for calcium carbonate precipitation from actual feedstocks, offering practical insights for improving the efficiency and scalability of mineralization processes.

2 Methodology

The experimental process, as shown in Fig. 1, consisted of two main stages, CO_2 absorption and $CaCO_3$ precipitation. The syntheses were conducted in two 30 L jacketed reactors at ambient pressure and 25 °C.

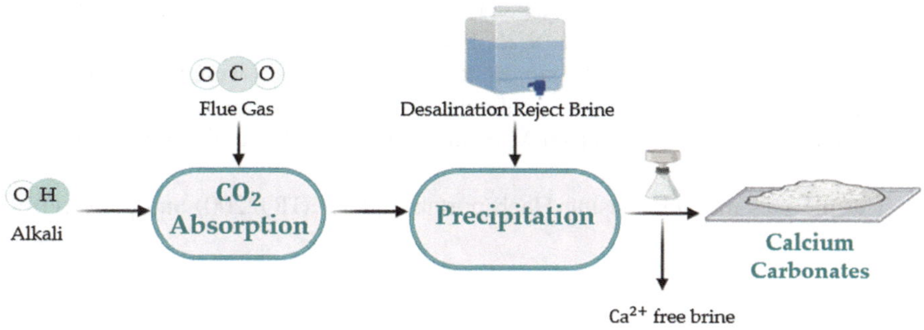

Fig. 1 Experimental process flow for calcium carbonate precipitation from brine via NaOH and CO_2 bubbling

2.1 CO_2 Absorption

Sodium carbonate solution was produced via CO_2 absorption in alkali solution. The procedure began with sodium hydroxide (assay 48–50%, Elite Trading, Qatar), diluting the solution to 3 M molar concentration. Then CO_2 gas (assay 99.99%, Buzwair Scientific and Technical Gases, Qatar) was slowly bubbled for approximately 5 h, until the pH reached 10.67 ± 0.05. The pH was measured using Hach portable meter (HQ2200) and calibrated pH electrode (PHC10101). The produced carbonate concentration was 1.5 M ± 0.01.

$$2NaOH_{(aq)} + CO_{(2g)} \rightarrow Na_2CO_{(3aq)} + H_2O \quad (1)$$

2.2 Reject Brine Preparation and Analysis

The desalination reject brine used for this study, was provided by Umm Al Houl SWRO, Qatar. The brine was kept at 25 °C temperature and mixed for at least 30 min upon each collection, to ensure the representative data. ICP-MS analyses were carried out to identify the initial concentrations of metals. Automatic titrator was used for CO_3^{2-}/HCO_3^- concentration identification. The total dissolved solids (TDS) were measured using Yinmik multifunctional water tester (BLE-C600).

2.3 Syntheses

All syntheses were performed at 25 °C temperature, ambient pressure. Precipitation was investigated at different pH (9.25, 9.5, 9.75, 10, 10.25 and 10.5), adjusted by adding NaOH (19 M). The pH changes during the syntheses were recorded with pH meter. Every 15 min, a sample was collected for filtration, using a vacuum pump and filter paper (Grade 601, Ahlstrom Munksjö, Finland). The total syntheses time—3 h.

The pH was measured using Hach portable meter (HQ2200) and pH electrode (PHC10101).

2.4 Characterization

Yields were analyzed by x-ray diffraction (XRD) and scanning electron microscope (SEM) for morphological and crystalline phase identifications. The supernatant obtained from filtration was analyzed with automatic titrator for CO_3^{2-}/HCO_3^-

concentrations identification and inductively coupled plasma mass spectrometry (ICP-MS) for Ca^{2+} concentration identification.

X-ray diffraction analyses (XRD)

XRD were carried out using a PANalytical Aeris XRD in Bragg—Brentano geometry using CuKα radiation. The angular scan range was from 5° to 60° 2θ with a step size of 0.0108° 2θ and 39.27 s per step.

Scanning electron microscope (SEM)

Morphological analyses were carried out using Phenom ProX SEM, secondary electron detector (SED). Prior to the analyses, the samples were mixed, ensuring the proper distribution and representation of the sample. A small amount of the powder was placed on Adhesive Carbon Disc.

CO_3^{2-}/HCO_3^- concentration

The concentrations of carbonate (CO_3^{2-}) and bicarbonate (HCO_3^-) ions were determined using a Mettler Toledo G10S automatic titrator with 0.1 M HCl (Sigma-Aldrich, USA) as the titrant. The system was calibrated with buffer solutions (Inorganic Ventures, USA), and the measurements were performed in triplicate.

Ca^{2+} concentration quantification

The concentrations of Ca^{2+} in the reject brine and syntheses supernatants were quantified using Inductively Coupled Plasma Mass Spectrometry (ICP-MS). Samples were diluted 200 times using 1% nitric acid (69–71%, NICE, India). Calibration standards for Ca were prepared by diluting Supelco IC multi-element standard VI (Sigma-Aldrich, Germany) and used to create a calibration curve the nitric acid used for both the preparation of standards and the dilution of samples remained consistent to avoid discrepancies in Ca concentration. The results were validated by including quality control samples and replicates in the analysis.

2.5 *Statistical Analysis*

The statistical analysis was performed using SPSS software to evaluate the effect of pH and time on the yield of the reaction and calcium precipitation across different pH ranges. Simple linear regression was conducted to identify the relationship between time, pH, and yield/Ca precipitation. The results were interpreted based on the R value, R^2, Adjusted R^2, and p-values for the coefficients. The significance level was set at $p < 0.05$. Analysis of variance (ANOVA) was used to assess whether the overall regression models were statistically significant. F-values and associated p -values were used to interpret the significance of the models. A pairwise comparison analysis was carried out to evaluate differences in yield across different pH levels. The significance of these differences was determined using $p < 0.05$.

Table 1 Desalination reject brine composition

Ca^{2+}	0.96 g/L
Mg^{2+}	2.9 g/L
TDS	g/L
pH	7.98 pH

3 Results and Discussion

The experimental study conducted across different pH levels and reaction times revealed significant insights into the calcium carbonate ($CaCO_3$) precipitation process from desalination reject brine. The results highlight the complex relationship between pH, reaction kinetics, and the yield and composition of the precipitates, which are essential for optimizing conditions in industrial applications.

3.1 Desalination Reject Brine

The total dissolved solids, pH and ICP-MS results of Ca^{2+} and Mg^{2+} concentrations are listed in Table 1.

3.2 Precipitation Yield

The yield of precipitate (total powder obtained, in g/L) was measured at 15 min intervals, revealing a dependency on both the initial pH and the duration of the reaction (Fig. 2).

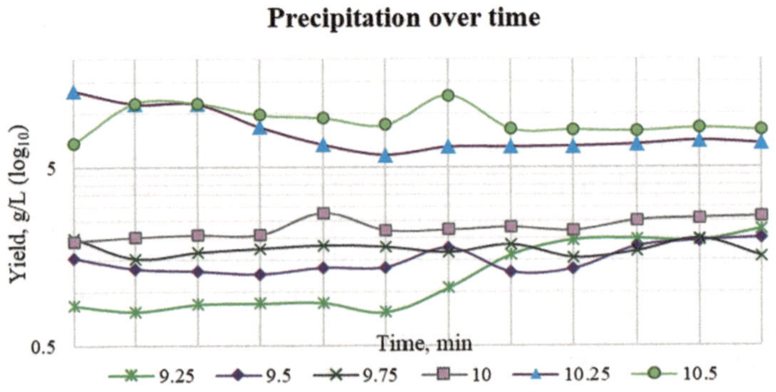

Fig. 2 Yield evolution over time at different initial pH

The experiments with the lowest initial pH value (9.25) had the lowest yields, which began to increase at 90 min of reaction and stabilized at 135 min. Syntheses with initial pH of 9.5–10 were relatively low and stable, showing little variation throughout the reaction. The initial high yields at pH 10.25 and 10.5 suggest a rapid initial reaction, leading to less stable precipitate forms that may have dissolved over time, as indicated by reduced yields over time, e.g. higher pH can be attributed to formation of $Mg(OH)_2$, as its solubility decreases with the increased pH [31].

The experiments with the lowest initial pH value (9.25) had the lowest yields, which began to increase at 90 min of reaction and stabilized at 135 min. Syntheses with initial pH of 9.5–10 were relatively low and stable, showing little variation throughout the reaction. The initial high yields at pH 10.25 and 10.5 suggest a rapid initial reaction, leading to less stable precipitate forms that may have dissolved over time, as indicated by reduced yields over time, e.g. higher pH can be attributed to formation of $Mg(OH)_2$, as its solubility decreases with the increased pH [31].

The findings contrast some of the previous trends discussed in the literature, e.g. Wang et al. [21] reported that the yield increases by increasing reaction time and stabilizes at 40 min [15]. However, it can be seen in Fig. 2, that reaction rate kept increasing over time, especially, at lower pH (9.25–9.5). This might be attributed to lower buffering capacity of seawater and the influence of salt species which increase the solubility of calcium carbonate, prolong the time required for calcium carbonate crystals formation to approximately 1.5–4.0 h [31]. The authors suggest maintaining reaction pH between 9 and 10 to obtain a significant CO_3^{2-} concentration and to avoid/reduce $Mg(OH)_2$ precipitation [21]. To translate this into applications concerning carbonation of desalination reject brines, it is possible to reduce the pH by introducing more CO_2, or by adding small amounts of acid, or chelating agents to bind with Mg^{2+}.

The aforementioned discrepancies between the preferred pH values reported in the literature and the experimental data can be explained by several factors, particularly the composition of our feed solution—desalination reject brine. Further studies should include the quantification of ionic strength of the brine, interactions of other ions, as well as the effect of descaling agents on the inhibition of nucleation and growth of calcium carbonates.

3.3 Statistical Analysis

The experimental results are summarised through statistical analyses to assess the impact of different pH levels on yield, as illustrated below. The results highlight the critical influence of pH on the yield of the reaction. Figure 3 displays a graphical representation of pairwise comparisons of yield rates across different pH levels, with connections between pH points indicating the significance of the differences in yield rates.

The graph supports the interpretation that the yield is highly sensitive to changes in pH. The graph shows that higher pH levels (10.00, 10.25, and 10.50) have generally

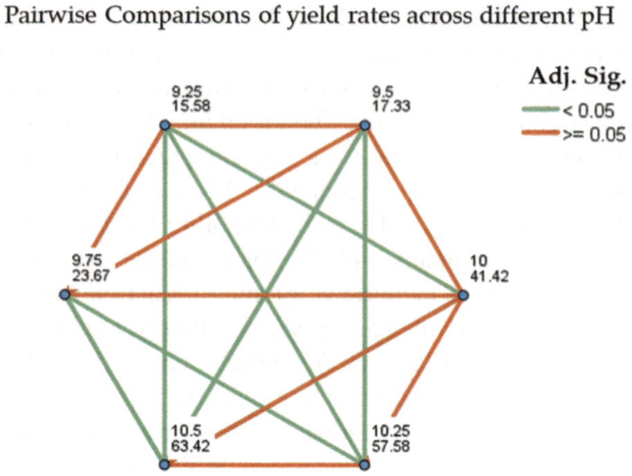

Fig. 3 Pairwise Comparisons of yield across the pH. Green lines indicate pairwise comparisons where the difference in yield rates is statistically significant ($p < 0.05$). Orange lines indicate pairwise comparisons where the difference in yield rates is not statistically significant ($p \geq 0.05$)

higher average ranks compared to lower pH levels (9.25–9.75), albeit with higher variability. Non-significant lines at lower pH conditions indicate that the yield rates are statistically similar. In practical terms, it indicates flexibility in process conditions, which are essential while working with desalination reject brines, or similar industrial by-products.

The regression analysis conducted for the pH range of 9.25–10.5 revealed several findings shown in Table 2. The R-value of 0.833 indicates a strong positive correlation between the predictors (pH and Time) and the dependent variable (Yield). The R^2 value of 0.694 suggests that approximately 69.4% of the variance in yield can be explained by the predictors (pH and Time). Additionally, the Adjusted R^2 value of 0.685 confirms that about 68.5% of the variance in yield is accounted for by the model, adjusting for the number of predictors. The standard error of the estimate is 1.985, indicating the average distance that the observed values fall from the regression line, with a lower value suggesting a better fit of the model.

The ANOVA results, with an F-value of 78.097 and a *p*-value of less than 0.001, demonstrate that the overall regression model is statistically significant. The time coefficient is $-$ 0.005, but it is not statistically significant ($p = 0.269$). In contrast, the pH coefficient is 6.823, and it is statistically significant ($p < 0.001$), indicating that for each unit increase in pH, the yield increases by approximately 6.823 units.

A more focused analysis within the pH range of 9.25–9.75 reveals different results as shown in Table 3. In this narrower pH range, both pH and time play a significant role in increasing yield.

The key findings for this specific range show an R-value of 0.693, indicating a moderate positive correlation between the predictors (pH and Time) and yield. The R^2 value of 0.48 suggests that approximately 48% of the variance in yield can be explained by the predictors. The model summary also indicates that time has a significant positive effect on yield, with a coefficient of 0.004 ($p < 0.001$). This means that within the pH range of 9.25–9.75, each unit increase in time results in an approximate 0.004 increase in yield. Therefore, it is evident that while time does

Table 2 Summary of regression and ANOVA analysis for yield prediction across all studied pH conditions

Regression	
R-value	0.833
R^2	0.694
R^2 adjusted	0.685
Standard error	1.985
ANOVA	
F-value	78.097
p-value	0.001
Coefficients	
pH coefficient	6.823
p-value	< 0.001
Time coefficient	– 0.005
p-value	0.269

Table 3 Statistical analysis results for pH range 9.25–9.75 influence on yield

Regression	
R-value	0.693
R^2	0.480
R^2 adjusted	0.449
Standard error	0.299
ANOVA	
F-value	15.249
p-value	< 0.001
Coefficients	
Time coefficient	0.004
p-value	< 0.001
pH coefficient	0.832
p-value	0.002

not significantly affect yield across the entire pH range, it does contribute positively within the more constrained pH range of 9.25–9.75.

3.4 Calcium Precipitation Over Time

Calcium carbonate formation is dependent on the availability of carbonate ions, which combine with calcium ions to precipitate as $CaCO_3$. The availability of carbonate ions in brines is governed by the carbonate-bicarbonate equilibrium which is highly pH dependent. Literature studies typically determine optimal pH values,

8.5–9.5 for calcite formation, under controlled conditions with pure solutions or well-defined ionic compositions. In contrast, our experiments using desalination brine as feed involve a complex mixture of ions, which can significantly alter the reaction pathways and kinetics. We have hence experimented at values that are both around and above optimum. Overall, the results coincide with the consensus that in order to enhance phase-purity it is essential to shift the equilibrium below 10.5 towards the formation of $CaCO_3$ instead of $Mg(OH)_2$.

In Appendix A, the detailed experimental data (Table A2) presents the unreacted calcium (Ca%), and carbonate (CO_3%) across a range of pH levels (9.25–10.5) and different reaction times (15–180 min). As time progresses, the percentage of unreacted calcium generally decreases, indicating that calcium is reacting and possibly precipitating out of solution. The percentage of unreacted carbonate also decreases over time, indicating that CO_3^{2-} is being consumed, likely in the formation of the precipitate (such as $CaCO_3$).

As illustrated in Fig. 4, at higher initial pH levels, specifically between 10.0 and 10.5, the precipitation of $CaCO_3$ occurred more rapidly due to the reduced solubility of calcium carbonate under alkaline conditions, leading to a significant reduction in unreacted calcium concentrations during the initial sampling period (0–60 min). By the end of the 3-h synthesis, the levels of unreacted calcium were lower, indicating that a substantial proportion of calcium had precipitated as $CaCO_3$.

Throughout the synthesis, it was observed that the precipitation dynamics differed significantly based on the initial pH. The first 60 min showed a rapid decline in unreacted calcium levels at higher pH, while lower pH conditions exhibited a slower decrease. Between 60 and 120 min, precipitation rates stabilized at higher pH levels, whereas lower pH levels continued to show relatively high concentrations of unreacted calcium. By the final sampling period (120–180 min), unreacted calcium concentrations approached the lowest levels at lower pH, indicating effective precipitation.

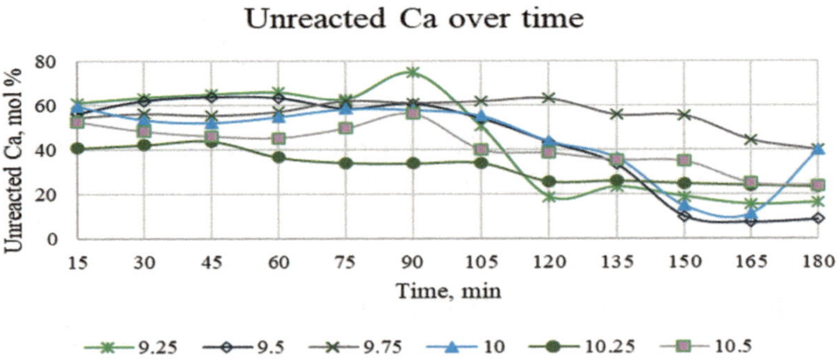

Fig. 4 Unreacted Ca content (mol %) over time at different pH conditions

Table 4 Regression and ANOVA statistical analyses on pH and time influence on Ca precipitation

Regression	
R-value	0.706
R^2	0.499
R^2 adjusted	0.484
Standard error	12.255
ANOVA	
F-value	34.359
p-value	< 0.001
Coefficients	
Time coefficient	– 0.685
p-value	< 0.001
pH coefficient	– 0.174
p-value	0.045

Statistical analysis on Ca precipitation

Regression and ANOVA statistical analyses were performed to tests pH and time influence on Calcium precipitation. The results are listed in Table 4.

The regression analysis reveals a strong relationship between time and pH on calcium concentrations, with an R-value of 0.706, indicating a robust correlation between the predictors and the dependent variable. The model explains approximately 49.9% of the variance in calcium levels ($R^2 = 0.499$), with an adjusted R^2 of 0.484, confirming its validity. The ANOVA results further support the model's significance, with an F-value of 34.359 and a p-value of < 0.001. The coefficients indicate that for each unit increase in time, calcium concentrations decrease by 0.224 units, while each unit increase in pH results in a 6.887 unit decrease in calcium concentration, both of which are statistically significant ($p < 0.001$ for time and $p = 0.045$ for pH). These findings underscore the importance of both time and pH in influencing calcium levels in your experimental conditions.

3.5 *Crystallinity and Polymorphism*

Table A1 with XRD figures in the Appendix A section and Table 5 display the results of XRD analysis conducted across a range of pH levels (9.25–10.50) over various time intervals (15–180 min).

The degree of crystallinity, as evidenced by the sharpness and intensity of the XRD peaks, was strongly influenced by pH. Lower pH levels (9.25–9.75) resulted in slower and less efficient crystallization. At pH levels of 9.25 and 9.5, the XRD spectra indicated slow gradual crystallization process. The early stages (15–75 min) were characterized by broad and poorly defined peaks, suggesting the dominance of

Table 5 Crystalline phases based on XRD results

Time (min)	9.25 pH	9.5 pH	9.75 pH	10 pH	10.25 pH	10.5 pH
15	AC	AC	AC + C	AC + C	CC + H + M	CC + A + M
45	AC	AC	AC + C	AC + C	CC + M	CC + A + M
75	AC + C	AC + C	AC + C	AC + C + H	CC + A + M	CC + A + M
105	AC + C	AC + C	AC + C	AC + C	CC + A + M	CC + A + M
135	C	C	AC + C	C	CC + A + M	CC + A + M
165	C	C	AC + C	C	CC + A + M	CC + A + M

AC amorphous calcium carbonate, *C* calcium carbonate monohydrate, *CC* calcite, *A* aragonite, *M* magnesium hydroxide, *H* halite

amorphous or poorly crystalline phases. It was only after 75 min that the peaks began to sharpen, indicating the onset of crystallization. At pH 9.75, Calcium carbonate monohydrate peaks could be identified earlier; however, based on the background, it can be stated that some of the precipitates remained partially amorphous even after the full 3-h synthesis time.

The XRD patterns at pH 10.00 showed a significant acceleration in the crystallisation process, where precipitation was rapid and crystallisation was highly efficient, highlighting the strong influence of pH on the kinetics of crystal formation. By 75 min, the peaks were much sharper and more intense compared to lower pH levels, indicating rapid nucleation and growth of crystalline phases. At the highest pH levels 10.25 and 10.5, the material powders exhibited complex mixture of phases, including magnesium hydroxide, aragonite and calcite. The crystallisation process was highly efficient, with sharp and intense peaks observed as early as 15 min.

The trends seen in this study generally align with previously published research. Song et al. (2022) observed that lower pH levels (around pH 9.0) led to slower crystallization, with the formation of amorphous particles initially, gradually evolving into crystalline forms over time. At higher pH levels (around pH 10.25 and above), faster nucleation and growth rates were observed, leading to more immediate formation of crystalline structures, including vaterite and calcite [32]. While some authors observed $Mg(OH)_2$ formation at 10.5 and higher pH [21], in this study, $Mg(OH)_2$ was detected at 10.25 pH.

Despite the observed increasing yields and crystallinity at elevated pH values, the equilibrium conditions favour the nucleation and growth of magnesium hydroxides, which can act as inhibitors for calcium carbonate crystallisation. These magnesium compounds may passivate the surface of growing calcium carbonate crystals, reducing their growth rate, altering their morphology, and causing polymorphism [33].

The observed polymorphism, with phases such as aragonite accompanying calcite at higher pH levels further highlighted the complexity of the desalination reject brine valorisation. Evidently, faster kinetics at higher pH will lead to disordered amorphous or metastable calcium carbonate phase formation, alongside Mg containing phases, which may be an effective way to strip divalent ions from desalination brine, but

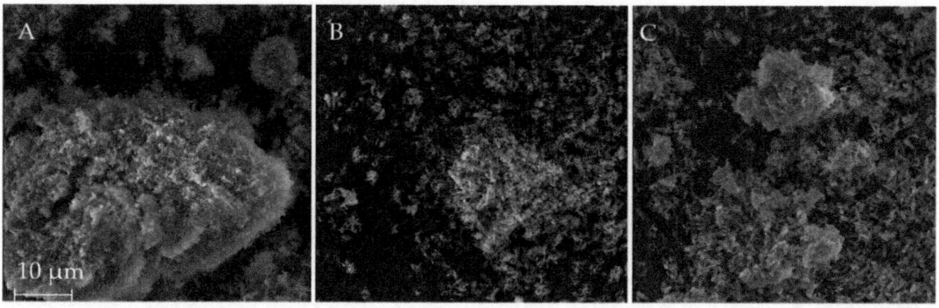

Fig. 5 Morphological evolution over time in lower (9.25–10.0) pH

at the same time, potentially compromising the value of the recovered product. On the other hand, reducing the pH to enhance phase-purity of calcites may result in dramatic reduction of total yields, as seen in Fig. 2.

3.6 Morphological Characterization

Scanning Electron Microscopy (SEM) was used to characterise and explore how pH and time affected the shape and texture of synthesised calcium carbonate. These factors showed how the mineral formed and changed over time under different pH conditions.

At the early stages of the reaction under pH conditions of 9.25–10.0, the obtained precipitate appeared poorly crystalline and with nano size crystallites (See A in Fig. 5). The crystal aggregates were observed to transform into rice-like structures (B) at 90 min for pH 9.25. Similar transformations occurred at 105 min for pH 9.5 and 9.75 and 90 min for pH 10 at 105 min. By 165 min, these structures became more defined (C), indicating progression in the morphological evolution of the precipitate.

At higher pH (10.25–10.5) the morphological transition differed from the one observed at lower pH. At higher pH (Fig. 6) the particles were observed to form larger cluster with smooth surfaces and uneven edges (A) at the early stage of reaction. This shape transforms to vesicular structures (B), where spherical particles are attached to a porous surface, notably observed at the 30-min mark under pH 10.25 conditions. At pH 10.5 within 15 min, large particles of said texture morphology immediately form, marking a distinctive early stage in the morphological evolution.

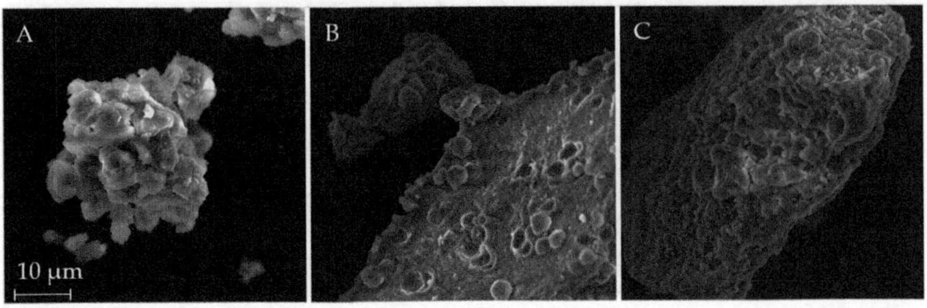

Fig. 6 Morphological evolution over time in higher (10.25–10.5) pH

4 Conclusion

The study demonstrated the significant influence of both pH and reaction time on the precipitation of calcium carbonate ($CaCO_3$) from Qatari desalination reject brine. While the highest yields of $CaCO_3$ were observed at elevated pH levels (10.25–10.5), these conditions also resulted in the formation of additional phases, compromising product purity. In contrast, lower pH levels (9.25–10) led to slower nucleation rates and reduced overall yields but allowed calcium recovery without contamination from secondary phases. The study concluded the optimum pH for phase-pure calcium carbonates from Qatari desalination brine to be at 9.5.

Future studies should further investigate the influence of brine composition, ion interactions, and the potential effects of descaling agents to optimize calcium carbonate crystallization and recovery in real-world desalination processes.

These findings have practical implications for carbon capture and mineralization research, suggesting that optimizing pH and reaction time can enhance the recovery of valuable $CaCO_3$, mitigate environmental impacts of brine discharge, and improve operational efficiency by transforming waste into a resource. This optimized process supports scalability and integration into existing desalination operations, providing environmental benefits, cost savings, and material recovery opportunities.

References

1. Acheampong AO (2018) Economic growth, CO_2 emissions and energy consumption: what causes what and where? https://doi.org/10.1016/j.eneco.2018.07.022.
2. CO_2 Emissions in 2023-Analysis—IEA. Accessed: 3 Aug 2024 [Online]. Available: https://www.iea.org/reports/co2-emissions-in-2023
3. El-Naas MH, Mohammad AF, Suleiman MI, Al Musharfy M, Al-Marzouqi AH (2017) A new process for the capture of CO_2 and reduction of water salinity. Desalination 411:69–75. https://doi.org/10.1016/j.desal.2017.02.005
4. Abdul Rahman F et al (2017) Pollution to solution: capture and sequestration of carbon dioxide (CO2) and its utilization as a renewable energy source for a sustainable future. https://doi.org/10.1016/j.rser.2017.01.011

5. Mustafa J, Mourad AAHI, Al-Marzouqi AH, El-Naas MH (2020) Simultaneous treatment of reject brine and capture of carbon dioxide: a comprehensive review. Desalination 483:114386. https://doi.org/10.1016/J.DESAL.2020.114386
6. Missimer TM, Maliva RG, Whitaker UA (2017) Environmental issues in seawater reverse osmosis desalination: intakes and outfalls. https://doi.org/10.1016/j.desal.2017.07.012
7. Panagopoulos A, Haralambous K-J (2020) Environmental impacts of desalination and brine treatment—challenges and mitigation measures. Mar Pollut Bull 161:111773. https://doi.org/10.1016/j.marpolbul.2020.111773
8. Panagopoulos A, Haralambous KJ, Loizidou M (2019) Desalination brine disposal methods and treatment technologies—a review. Sci Total Environ 693:133545. https://doi.org/10.1016/J.SCITOTENV.2019.07.351
9. A-HI Mourad A, Mohammad AF, Al-Marzouqi AH, El-Naas MH, Al-Marzouqi MH, Altarawneh M (2022) CO_2 capture and ions removal through reaction with potassium hydroxide in desalination reject brine: statistical optimization. Chem Eng Process: Process Intensif 170:108722. https://doi.org/10.1016/j.cep.2021.108722
10. Ruan H et al (2022) Capturing CO_2 with NaOH solution from reject brine via an integrated technology based on bipolar membrane electrodialysis and hollow fiber membrane contactor. Chem Eng J 450:138095. https://doi.org/10.1016/J.CEJ.2022.138095
11. Chaliulina R (2023) On sustainable production of $CaCO_3$ via monohydrocalcite—a carbon capture and mineralisation product from waste brines. Green Sustain Chem 13(01):34–61. https://doi.org/10.4236/gsc.2023.131004
12. Chaliulina R (2022) Utilisation of carbon capture and mineralisation products: considering monohydrocalcite as a potential functional filler. Open Access Library J 9(12):1–19. https://doi.org/10.4236/OALIB.1109436
13. Chaliulina R (2019) Precipitated calcium carbonate: recycling carbon dioxide and industrial waste brines. University of Aberdeen, Aberdeen
14. Gálvez-Martos Y, Chaliulina JL, Medina-Martos R, Elhoweris E, Hakki A, Mwanda A, Al-Horr J (2021) Eco-efficiency of a novel construction material produced by carbon capture and utilization. Accessed: 30 Sep 2021 [Online]. Available: https://www.sciencedirect.com/science/article/abs/pii/S2212982021001128
15. Martos J-LG, Elhoweris A, Chaliulina R, Mwanda JA, Bannerman C, AlHorr Y (2022) On the eco-efficiency and life cycle performance of decarbonization approaches of cementitious materials. SSRN Electron J. https://doi.org/10.2139/SSRN.4282839
16. Chaliulina R, Galvez-Martos JL, Hakki A, Elhoweris A, Mwanda J, Al-Horr Y (2021) Eco materials from CO_2 capture: compressive strengths of a plasterboard alternative. Constr Build Mater 312. https://doi.org/10.1016/J.CONBUILDMAT.2021.125276
17. Hepburn C et al (2019) The technological and economic prospects for CO_2 utilization and removal. Nature 575(7781):87–97. https://doi.org/10.1038/s41586-019-1681-6
18. Ho HJ, Iizuka A, Shibata E (2019) Carbon capture and utilization technology without carbon dioxide purification and pressurization: a review on its necessity and available technologies. Ind Eng Chem Res 58(21):8941–8954. https://doi.org/10.1021/ACS.IECR.9B01213/ASSET/IMAGES/MEDIUM/IE-2019-01213A_0005.GIF
19. Jimoh OA, Ariffin KS, Bin Hussin H, Temitope AE (2018) Synthesis of precipitated calcium carbonate: a review. Carbonates Evaporites 33(2):331–346. https://doi.org/10.1007/s13146-017-0341-x
20. Bang J-H, Yoo Y, Lee S-W, Song K, Chae S (2017) CO_2 mineralization using brine discharged from a seawater desalination plant. Minerals 7(11):207. https://doi.org/10.3390/min7110207
21. Wang Y, Qin Y, Wang B, Jin J, Wang B, Cui D (2020) Selective removal of calcium ions from seawater or desalination brine using a modified sodium carbonate method. https://doi.org/10.5004/dwt.2020.24828
22. Bang J-H, Chae S-C, Song K, Lee S-W (2021) Optimizing experimental parameters in sequential CO_2 mineralization using seawater desalination brine. Desalination 519:115309. https://doi.org/10.1016/j.desal.2021.115309

23. Stashans A, Chamba G (2011) A new insight on the role of Mg in calcite. Int J Quantum Chem 111(10):2436–2443. https://doi.org/10.1002/QUA.22538
24. Molinari R et al (2021) Selective precipitation of calcium ion from seawater desalination reverse osmosis brine. J Clean Prod 328:129645. https://doi.org/10.1016/j.jclepro.2021.129645
25. Bin Hu Y, Wolthers M, Wolf-Gladrow DA, Nehrke G (2015) Effect of pH and phosphate on calcium carbonate polymorphs precipitated at near-freezing temperature. Cryst Growth Des 15(4):1596–1601. https://doi.org/10.1021/CG500829P
26. House WA, Howson MR, Pethybridge AD (1988) Crystallisation kinetics of calcite in the presence of magnesium ions. J Chem Soc Faraday Trans 1: Phys Chem Condens Phases 84(8):2723–2734. https://doi.org/10.1039/F19888402723
27. Morse JW, Mackenzie FT (1990) Geochemistry of sedimentary carbonates. Accessed: 3 Aug 2024 [Online]. Available: https://books.google.com.qa/books?hl=lt&lr=&id=3Cvk2lkukgEC&oi=fnd&pg=PP1&ots=bSZgKfZPny&sig=k2ZsgnZQ_Eti4WHWAzwi9-a06bs&redir_esc=y#v=onepage&q&f=false
28. Zou Z et al (2019) Solid-state chemistry: a hydrated crystalline calcium carbonate phase: calcium carbonate hemihydrate. Science (1979) 363(6425):396–400. https://doi.org/10.1126/SCIENCE.AAV0210/SUPPL_FILE/AAV0210_ZOU_SM.PDF
29. De Yoreo JJ, Waychunas GA, Jun Y-S, Fernandez-Martinez A (2013) In situ investigations of carbonate nucleation on mineral and organic surfaces. Rev Mineral Geochem 77(1):229–257. https://doi.org/10.2138/rmg.2013.77.7
30. Blue CR, Giuffre A, Mergelsberg S, Han N, De Yoreo JJ, Dove PM (2017) Chemical and physical controls on the transformation of amorphous calcium carbonate into crystalline CaCO3 polymorphs. Geochim Cosmochim Acta 196:179–196. https://doi.org/10.1016/j.gca.2016.09.004
31. Qu D, Wang J, Wang L, Hou D, Luan Z, Wang B (2009) Integration of accelerated precipitation softening with membrane distillation for high-recovery desalination of primary reverse osmosis concentrate. Sep Purif Technol 67(1):21–25. https://doi.org/10.1016/J.SEPPUR.2009.02.021
32. Song K, Bang J-H, Chae S-C, Kim J, Lee S-W (2022) Phase and morphology of calcium carbonate precipitated by rapid mixing in the absence of additives. RSC Adv 12(30):19340–19349. https://doi.org/10.1039/D2RA03507C
33. Bearat HA (2004) Mineral sequestration of carbon dioxide: enhancing carbonation reactivity of Brucite and Forsterite

Investigating the Role of Ethnicity and Perceived Inequality in Shaping Resilience in Malaysia: Insights for Advancing SDG No. 11 in the Face of Climate Change

Mohammad Javad Maghsoodi Tilaki and Massoomeh Hedayati Marzbali

Abstract This study explores the resilience of infrastructure and communities in the face of climate change, focusing on the vulnerability of urban infrastructure and its impact on societal tensions. The resilience and interdependence of these systems are crucial, especially for cities in the Global South striving to achieve SDG No.11. Despite the importance of social resilience and community attitudes in disaster management, there is a significant gap in understanding how vulnerable communities behave when faced with unequal access to infrastructure during disruptions. This research examines how perceived inequality affects individual resilience in a Penang neighborhood experiencing a water supply disruption, particularly when nearby areas have resolved their issues. The study utilizes water supply disruption data and employs structural equation modeling (SEM) to quantitatively analyze individual resilience levels in a resettled, marginalized, and vulnerable community in Penang, Malaysia. As one of the first studies to investigate the intersection of ethnicity, perceived inequality, and resilience in Malaysia, the findings reveal that ethnicity plays a significant role in shaping individual resilience, with varying levels observed across different ethnic groups. Furthermore, residents who perceive higher levels of inequality experience greater stress and reduced access to resources, leading to lower resilience, regardless of ethnicity. The research makes three key contributions: (i) validating a resilience framework for assessing the impacts of infrastructure disruptions on achieving safe and resilient cities, (ii) analyzing the roles of adaptability, perceived inequality, and individual resilience in response to climate change,

M. J. Maghsoodi Tilaki (✉)
Department of Geography, School of Humanities, Universiti Sains Malaysia, Penang 11800, Malaysia
e-mail: maghsoodi@usm.my

M. Hedayati Marzbali
Department of Urban Planning, School of Housing, Building and Planning, Universiti Sains Malaysia, Penang 11800, Malaysia

S. Al-Naemi et al. (eds.), *Water and Food Security in the Face of Climate Change: Challenges and Opportunities for Resilience*, World Sustainability Series,
https://doi.org/10.1007/978-3-032-00098-9_22

and (iii) comparing resilience among different ethnic groups within a vulnerable community.

Keywords Social resilience · Perceived inequality · Individual resilience · Multi-ethnic communities · Climate change mitigation

1 Introduction

Climate change, one of the twenty-first century's most critical threats, impacts both the environment and essential infrastructure systems. Community resilience depends on robust infrastructure, adaptable social systems, and equitable resource distribution. Water supply systems, crucial for survival and societal functioning, are particularly vulnerable to climate change, which can intensify inequalities and social tensions, especially in multi-ethnic communities with pronounced disparities in resource access [12].

Malaysia, with annual rainfall of 2000–5000 mm, has prioritized water resource management for decades. Initiatives like the Tenth and Eleventh Malaysia Plans (2011–2020) emphasized efficient water management, infrastructure expansion, and improved resource reliability to meet growing demands [8]. The Twelfth Malaysia Plan (2021–2025) advances these efforts through the Water Sector Transformation 2040 (WST 2040) Agenda, aiming to ensure water security and establish Malaysia as a Regional Water Hub by 2040.

However, Penang's rapid industrial growth and expanding urbanization have intensified water shortages, with rising temperatures further increasing demand. Rapid development, industrial activities, and the water-intensive digital sector further strain its water resources, as the digital industry requires substantial water for cooling and operational processes [14].

Penang's water supply is sourced from two dams serving the island's northern regions, while a pipeline from mainland Malaysia provides water to the central, southeastern, southern, and southwestern areas. This pipeline is crucial for high-density residential zones, the airport, the main public university, and key industrial areas. Freshwater is sourced from natural bodies such as rivers and undergoes a series of processes—pre-treatment, coagulation, sedimentation, filtration, disinfection, and fluoridation—before distribution.

The last water supply disruptions in Batu Maung, Penang, starting in December 2023, highlighted vulnerabilities in the infrastructure. The two-stage repair process, initially planned for 21 pipeline locations, caused extended maintenance and significant hardship for residents and businesses. Households in low-cost housing and landed property, particularly vulnerable groups, were severely affected, lacking pre-installed water storage facilities [14]

In January 2024, a main pipeline failure during repairs caused a 48–96-h disruption, affecting nearly 600,000 consumers in Penang, with significant impacts on daily life, businesses, and overall resilience. Batu Maung, a region in the Southwest district of Penang, was one of the area's most severely impacted, providing a compelling case for examining how residents perceive and respond to such infrastructure failures. This study develops a theoretical framework to examine the impact of perceived inequality on residents' resilience during water supply disruptions. To achieve this aim, the study focuses on the following objectives: (1) validating the measurement of individual resilience by evaluating eight key dimensions of RABIT framework, (2) examining the impact of perceived inequality across different racial groups on individual resilience.

By focusing on the experiences of individuals in Malaysian communities, this research contributes to a deeper understanding of how infrastructure inequities influence resilience to enhance social sustainability and cohesion in the face of climate-induced disruptions.

2 Literature Review

2.1 Perceived Inequality and Individual Resilience

Resilience is the ability of individuals, communities, systems, or environments to withstand, adapt to, and recover from disruptions or stressors. It spans social, economic, environmental, and psychological dimensions. On an individual level, psychological resilience involves coping mechanisms, social support, problem-solving, and emotional balance during crises. Achieving resilience requires cross-sector collaboration focused on sustainability and equity, enabling communities and systems to adapt, recover, and thrive amid challenges. Literature reveals that poor infrastructure management worsens social inequalities, leaving underprivileged communities more vulnerable to disruptions like power outages and water shortages. Participatory approaches, equitable resource allocation, and innovative technologies are vital for addressing these challenges [12].

Perceived inequality undermines resilience by eroding social cohesion, trust, and support networks, essential for coping with adversity. This diminishes individuals' sense of control, self-worth, and well-being, increasing vulnerability to mental health challenges [13]. Perceived inequality can cause chronic stress, negatively affecting mental and physical health, including impaired cognition, reduced immunity, and increased risk of depression and heart disease [4].

Constant feelings of disadvantage create ongoing stress, hindering recovery from setbacks. Research links perceived inequality to reduced self-esteem (sense of self-worth) and self-efficacy (belief in one's ability to succeed), both crucial for resilience. Feeling undervalued or incapable diminishes motivation and problem-solving abilities, creating a cycle that further weakens resilience [6].

In conclusion, perceived inequality undermines resilience by affecting social cohesion, increasing stress, lowering self-esteem and self-efficacy, and limiting resource access.

2.2 Measuring Community Resilience Using the RABIT Framework

Measuring resilience after crises is vital for understanding recovery and adaptation in social, economic, environmental, and psychological dimensions. Social resilience focuses on community functionality, cohesion, trust, and support networks, though subjective factors like relationships and trust complicate evaluation [13]. Quantitative tools like social network analysis are useful but often lack qualitative depth. Mixed-method approaches, combining surveys and participatory assessments, offer a more comprehensive understanding of post-crisis social dynamics [6].

A holistic understanding of post-crisis recovery requires integrating multiple resilience dimensions. However, measurements often remain fragmented, focusing on individual aspects rather than systemic perspectives [10]. Since resilience evolves over time, timely assessments are critical, as immediate metrics may not capture long-term adaptability [6].

Recent studies highlight the effectiveness of the RABIT (Resilience Assessment Benchmarking and Impact Toolkit) Framework in measuring and enhancing resilience across various contexts. RABIT evaluates absorptive, adaptive, and transformative capacities while incorporating governance and sustainability aspects. It provides tools for assessing resilience at social, economic, and institutional levels, proving useful for diverse settings, from urban communities to agricultural systems [7]. The framework emphasizes core attributes vital for community stability and adaptability, offering a versatile approach to resilience assessment and improvement, as shown in Table 1.

3 Theoretical Framework

This study integrates Resilience Theory and Social Inequality Theory to examine the interplay between local authorities' crisis management and residents' behaviors during the 96-h water disruption in Batu Maung from January 10–13, 2024. It assesses residents' sensitivity, behavioral changes, and resilience in response to this infrastructural failure.

Resilience Theory highlights the importance of preparation, prevention, immediate response, and post-crisis recovery to mitigate crisis impacts like water scarcity (Coombs 2015). In Batu Maung, pre-crisis measures may have included water rationing plans, communication strategies, and resource protocols. During the crisis, emergency water distribution and usage restrictions were key responses, followed by post-crisis evaluations to refine future strategies.

Behavioral Response Theory examines how individuals and communities react to crises, emphasizing risk perception as a critical factor shaping responses (Slovic 2000). This theory sheds light on how residents' perceptions of the water shortage influenced their actions during the disruption.

Table 1 The RABIT components, definition and relevant indicators

Resilience attribute	Indicators
Robustness	– Facilities and infrastructure; – Collaboration among individuals, the community, and local authorities in the region
Self-organization	– Trust levels among community members; – Collaborative networks; – Trust of individuals in the community and local authorities, including community leaders in the region
Learning	– Recognition of current risks; – Availability of drills and training programs; – Exchange of knowledge among members
Redundancy	– Backup plans; – Varied income sources; – Outside assistance
Rapidity	– Availability of early warning systems; – Prompt action in emergency situations; – Rapid support from external networks during crises
Scale	– Engagement between individuals, families and community; – Partnerships between the community and both the private and public sectors; – Interconnected relationships across different levels
Diversity	– Range of options accessible to the community; – Adoption of innovative approaches; – Viewing change as an opportunity rather than a threat
Equality	– Involvement and increased skills; – Inclusiveness and openness

This study adopts a dual-perspective approach by integrating Social Inequality Theory with Resilience Theory to evaluate the impact of inequality on resilience and explore strategies to enhance community resilience in Batu Maung. The framework assesses the community's ability to cope with crises and adapt to future disruptions, examining the effectiveness of current resilience strategies while identifying opportunities for improvement.

By combining institutional (top-down) and individual (bottom-up) perspectives, this approach provides a comprehensive understanding of water scarcity management. The findings aim to inform improved resilience strategies that address immediate challenges, mitigate long-term psychological impacts, and promote sustainable crisis management practices in Batu Maung and similar climate-affected communities. Based on the conceptual model in Fig. 1, this study aims to test the following two hypotheses:

H1. The eight dimensions (robustness, self-organization, redundancy, rapidity, scale, diversity and flexibility, equality, perceived inequality) and their respective indicators provide a valid measurement of the individual resilience construct.

H2. Perceived inequality of different races has a significant effect on resilience.

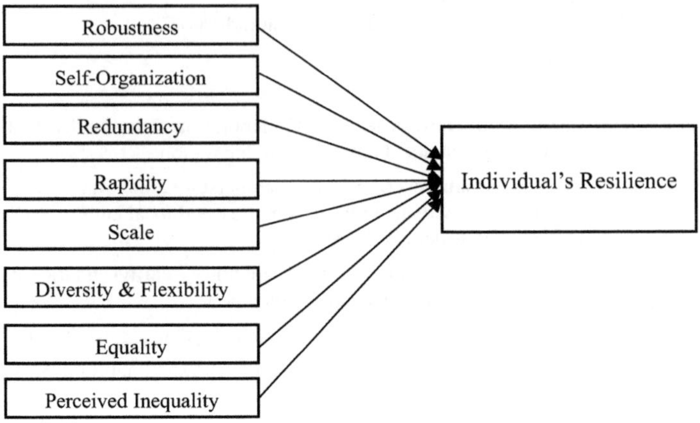

Fig. 1 The study conceptual model

4 Methods

This study investigates the impact of water supply disruption on perceived inequality, behavior, and resilience in Batu Maung, Penang. Using a quantitative empirical approach, the research comprises two parts: validating a social resilience framework through SEM-PLS. Additionally, correlation analysis will explore links between social resilience and behavioral changes, examining whether delayed recovery amplifies perceptions of social inequality compared to other affected areas.

4.1 Data Collection

The study employed a quantitative approach, collecting data in Batu Maung in late February 2024. Six trained student teams administered questionnaires covering demographics and 42 resilience-related attributes, including robustness, self-organization, equality, perceived inequality, and happiness. After excluding 28 incomplete responses, 386 valid entries were analyzed.

4.2 Study Location

This study focuses on Batu Maung in Mukim 12, Southwest Penang-a northern Malaysian state—an area that has rapidly developed following the completion of Penang's Second Bridge, attracting real estate investments. Batu Maung hosts tourist attractions like the War Museum and Batu Maung Aquarium and is near the Free

Industrial Zone. This has transformed the area into a multi-industry hub. Historically a fishing village, Batu Maung's demographics now include Malays (53.08%), Chinese (36.15%), Indians (9.50%), and others (1.27%). Land reclamation and industrial expansion attracted multinational companies, increasing the population due to relatively affordable rents compared to Bayan.

Lepas and Bandar Bayan Baru. These challenges highlight residents' sensitivity, behavioral adaptations, and resilience to infrastructure failures.

5 Analysis

The socio-demographic profile of participants reveals that 55.45% were female (n = 214) and 44.55% male (n = 172). The largest age group was 50–60 years (21.1%, n = 81), followed by 40–50 years (18.8%, n = 72). Most respondents (63.73%, n = 246) were married, and 54.66% (n = 211) had a high school education. Employment data showed 35.4% (n = 137) were employed, 24.3% (n = 94) self-employed, and 18.6% (n = 72) housewives. Nearly half fell within the B40 income group, with 44.81% (n = 173) earning below RM 2500/month.

Ethnically, Malays comprised 53.36% (n = 206), followed by Chinese (21.50%, n = 83) and Indians (17.61%, n = 68). Education levels varied, with 31.35% (n = 121) holding university degrees and 9.34% (n = 36) having only primary education, while 2.85% (n = 11) did not disclose. This demographic profile underscores the diversity of the study participants, highlighting significant variations across gender, age, race, marital status, employment, education, and income levels. These factors provide valuable context for interpreting the study's findings.

5.1 Measurement Model Results

Hypothesis testing was conducted using partial least squares (PLS) analysis with SmartPLS 3 software [9]. The measurement model was evaluated based on outer loadings, convergent validity, composite reliability, and discriminant validity (Table 2). Following [3], a cut-off value of 0.6 was applied for outer loadings to ensure items effectively represented their latent variables. All standardized loadings were significant and exceeded the 0.6 threshold.

Composite reliability, evaluating internal consistency, exceeded the 0.7 threshold [1] for all constructs, confirming good reliability. Convergent validity, assessed via average variance extracted (AVE) with a 0.5 cut-off [2], was also achieved for all constructs.

Recommended by [3], discriminant validity was assessed using the heterotrait-monotrait (HTMT) ratio, a robust measure where values below 0.85 indicate distinct constructs. In this study, HTMT ratios ranged from 0.0185 to 0.2153, well below the threshold, confirming good discriminant validity. Constructs such as Robustness,

Table 2 Measurement model results for the latent constructs

Construct	Items	Composite reliability (CR)	Average variance extracted (AVE)	Mean (SD)
Robustness	6	0.871	0.548	2.56 (0.96)
Self-organization	6	0.867	0.791	3.45 (1.12)
Redundancy	4	0.872	0.681	3.02 (0.92)
Rapidity	5	0.883	0.593	2.93 (0.99)
Scale	4	0.822	0.722	2.89 (0.98)
Diversity and flexibility	4	0.831	0.692	3.04 (1.07)
Equality	4	0.842	0.648	3.00 (0.89)
Perceived inequality	4	0.814	0.621	2.63 (1.03)
Individual's resilience	5	0.874	0.659	2.80 (0.92)

Self-Organization, Redundancy, Rapidity, Scale, Diversity and Flexibility, Equality, Perceived Inequality, and Individual Resilience were found to be distinct within the model. The narrow confidence intervals further validate the precision of these estimates, reinforcing the constructs' reliability and uniqueness.

5.2 *Assessment of the Hierarchical Individual Resilience Construct*

This study defines individual resilience as a second-order construct comprising eight first-order reflective constructs: robustness, self-organization, redundancy, rapidity, scale, diversity and flexibility, equality, and perceived inequality, measured using 42 items. The explained variance (R^2) of this hierarchical construct is reflected in its individual components: robustness ($R^2 = 52\%$), self-organization ($R^2 = 54.6\%$), redundancy ($R^2 = 52.3\%$), rapidity ($R^2 = 50.8\%$), scale ($R^2 = 51.2\%$), diversity and flexibility ($R^2 = 49.1\%$), equality ($R^2 = 56.2\%$), and perceived inequality ($R^2 = 55.9\%$). Each path coefficient linking individual resilience to its respective dimensions is statistically significant at $p < 0.01$, thereby supporting Hypothesis 1 (H1).

5.3 *Relationship Between Respondents' Race and Individual's Resilience*

Analysis of Variance (ANOVA) was performed to examine differences in composite resilience scores among the racial groups. Additionally, an Analysis of Covariance (ANCOVA) was conducted to include perceived inequality as a covariate. As shown in Table 3, the ANOVA analysis revealed significant differences in composite resilience

Table 3 ANOVA results for the effects of race and perceived inequality on resilience

Source	Sum of squares	df	Mean square	F value	Sig.
C (race)	26.726362	3	8.908787	5.216582	0.001535
Residual	652.372913	382	1.707768		

Table 4 ANCOVA summary of analysis of covariance for the effects of race and perceived inequality on resilience

Source	Sum of squares	df	Mean square	F value	Sig.
C (race)	15.701862	3	5.233954	3.627487	0.013201
PercIneq1	0.538593	1	0.538593	0.373282	0.541588
PercIneq2	12.984286	1	12.984286	8.998997	0.002880
PercIneq3	18.130392	1	18.130392	12.565599	0.000442
PercIneq4	19.670810	1	19.670810	13.633214	0.000255
Residual	545.400839	378	1.443913		

scores among the four racial groups. The *p*-value (0.001535) indicates that there are significant differences in resilience scores among the four racial groups.

Furthermore, the ANCOVA analysis, with perceived inequality included as a covariate, reveals that significant differences in resilience scores persist even after accounting for perceived inequality (see Table 4). Moreover, three of four perceived inequalities significantly influence resilience scores, as detailed in Table 4 (H2).

6 Discussion

Baes on the obtained results, the average perceived inequality scores were: (1) Malay-Race (Malaysian): 5.2, (2) Chinese-Race (Malaysian): 4.8, (3) Indian-Race (Malaysian): 5.5, and (4) Foreign: 5.0. These differences highlight varying experiences and vulnerabilities during disruptions to essential services like water supply. In addition, the relationship between respondents' race and their resilience was examined. The average resilience scores by race were: Indian (6.3), Malay (6.1), Chinese (5.9), and Foreign (5.8). These differences underscore the complex interplay between perceived inequality and resilience, suggesting that racial background may influence how individuals cope with and adapt to disruptions in essential services.

Disparities in perceived inequality and resilience among racial groups reveal significant socio-cultural dynamics. Indian respondents reported the highest perceived inequality (5.5) and resilience (6.3), possibly reflecting a coping response to systemic challenges, consistent with research linking socio-economic pressures to heightened resilience. Similarly, Malay respondents showed relatively high inequality (5.2) and resilience (6.1), indicating notable adaptive capacity. In contrast, Chinese respondents reported the lowest inequality (4.8) and moderate resilience

(5.9), suggesting differing social or economic influences on their perceptions and adaptability.

In alignment with the current research findings, [11] and by [5] highlight that such disruptions often heighten perceptions of inequality, especially in multi-ethnic urban areas where neighboring communities may recover faster.

Water supply disruptions, exacerbated by climate change, intensify resource scarcity and socio-economic inequalities, disproportionately impacting marginalized communities. The paradox of higher resilience among groups facing greater inequality, such as Indian and Malay respondents, underscores the role of social dynamics in adaptation. Strong community networks, resource-sharing practices, and cultural cohesion likely bolster their resilience, as highlighted by studies emphasizing the importance of solidarity and alternative resources in overcoming systemic challenges. As climate change worsens, disruptions to essential services like water supply will increasingly threaten vulnerable populations. This study underscores the link between social dynamics and resilience, emphasizing the need for policies that reduce inequalities, enhance social cohesion, and ensure equitable resource access. Leveraging the adaptive strengths of marginalized communities through targeted interventions and collective action is key to building infrastructure resilience and mitigating vulnerabilities in diverse urban contexts.

Funding This research was funded by the Malaysian Ministry of Higher Education through a Fundamental Research Grant Scheme (FRGS) Grant No. (FRGS/1/2021/SSI02/USM/02/5).

References

1. Bagozzi RP, Yi Y (1988) On the evaluation of structural equation models. J Acad Mark Sci 16:74–94
2. Fornell C, Larcker DF (1981) Evaluating structural equation models with unobservable variables and measurement error. J Mark Res 18(1):39–50
3. Hair JF, Ringle CM, Sarstedt M (2011) PLS-SEM: indeed a silver bullet. J Mark Theor Pract 19(2):139–152
4. Hart A, Gagnon E, Eryigit-Madzwamuse S, Cameron J, Aranda K, Rathbone A, Heaver B (2016) Uniting resilience research and practice with an inequalities approach. SAGE Open 6(4):2158244016682477
5. Li T, Fung HH (2013) Age differences in trust: an investigation across 38 countries. J Gerontol B Psychol Sci Soc Sci 68(3):347–355
6. Lynne F (2009) Mental health, resilience and inequalities. W. H. Organization. https://apps.who.int/iris/bitstream/handle/10665/107925/E92227.pdf
7. Ospina AV, Heeks R (2011) ICTs and climate change adaptation: enabling innovative strategies. Strategy Brief, Centre for Development Informatics, UK
8. Prime Minister of Malaysia (2012) Tenth Malaysia Plan: 2011–2015
9. Ringle C, Wende S, Becker J (2015) SmartPLS 3. Boenningstedt: SmartPLS GmbH. Accessed 5 Jan 2018
10. Sharifi A, Yamagata Y (2015) A conceptual framework for assessment of urban energy resilience. Energy Procedia 75:2904–2909
11. Smith LG, Thomas EF, McGarty C (2015) "We must be the change we want to see in the world": integrating norms and identities through social interaction. Polit Psychol 36(5):543–557

12. Subramanyam N, Education V (2024) Untapped transformative potential in decentralised groundwater systems that improve households' resilience in India's water-scarce urban environments. Habitat Int 151:103140
13. Tseliou F, Ashfield-Watt P (2022) The association between resilience resources, contextual factors and mental health status: a national population-based study. BMC Public Health 22(1):602
14. Zainudin N, Ahmad NHC, Jamaluddin A, Sabri B, Nordin N, Nordin N (2022) Understanding water saving behavior among sub-urban households in Penang. Jurnal Pengguna Malaysia, Malaysia

Temporal Patterns and Intra-seasonal Isotopic Characteristics of Indian Summer Monsoon Precipitation in Lucknow, India

K. Prasanna, Anurag Kumar, Anupam Sharma, Supriyo Chakraborty, and Ravi Rangarajan

Abstract The study of the connection between isotopic depletion in precipitation and stratiform rains is limited. This study reports the relationship between the temporal characterization of stable isotopes in precipitation in Lucknow of Central India that has been lacking up to the present. Stable isotope analyses were carried out on collected precipitation samples from three locations in the city of Lucknow in northern India, from July 2018 to September 2018 is reported here. A least-squares regression of the isotope data was used to obtain the local meteoric water line (LMWL), and the obtained LMWL was $\delta D = 7.75\ \delta^{18}O + 7.2$ ($R^2 = 0.99$). The average $\delta^{18}O$ values of the samples ranged from − 16.27 to 0.45‰ while δD values varied from − 112.42 to 10.51‰ respectively.

Keywords Stable isotopes · d-excess · Precipitation · Cloud · Climate change

1 Introduction

Global warming in recent times has led to frequent occurrences of extreme climate events and has become a concerning factor impacting human life [1–3]. Climate change research has attained global importance owing to its impact on modifying the regional temperature and frequent occurrences of extreme events endangering human life and settlements [1]. The change in climate, especially in the global temperature,

K. Prasanna · A. Kumar · A. Sharma
Birbal Sahni Institute of Palaeosciences, Lucknow, India
e-mail: prasanna@bsip.res.in

S. Chakraborty
Ministry of Earth Sciences, Indian Institute of Tropical Meteorology, Pune, India

R. Rangarajan (✉)
University of Doha for Science and Technology, Doha, Qatar
e-mail: ravi.rangarajan@udst.edu.qa

S. Al-Naemi et al. (eds.), *Water and Food Security in the Face of Climate Change: Challenges and Opportunities for Resilience*, World Sustainability Series,
https://doi.org/10.1007/978-3-032-00098-9_23

has significant impact on modifying the regional precipitation patterns [4, 5]. Indian sub-continental landmass receives rain during summer and winter due to reversals of wind patterns in different seasons [6]. Indian Summer Monsoon (ISM) Rainfall constitutes the main supplier of fresh water, contributing to 90% of agricultural and other water demands of ~ 20% of the world's population occupying the region [5, 7]. These changes in monsoonal rainfall patterns have strong implications on the economic growth and the productivity of the region [5].

Isotopes in precipitation are reliable tracers in the hydrological cycle [8–10]. Peninsular Indian region receives rainfall during summer and winter time, associated with Southwest monsoon (ISM) and Northeast monsoon (NEM) winds [11, 12]. Precipitation isotopic compositions at the inland stations are affected by factors like Rayleigh's fractionation, surface air temperature, latitudinal position, elevation from the sea level, humidity [8, 13]. In addition, factors like recycling of evaporated water [14–16], wind trajectory changes and source moisture variability at seasonal time scales [17, 18] also affects $\delta^{18}O$ in precipitation. Among all these factors listed, studies on isotopic composition of precipitation from the Indian region have suggested that the role of source moisture is the most dominant factor controlling the isotopic ratio observed in the precipitation [19, 20].

Stratiform rains, which are typically characterized by extended periods of light to moderate rainfall, play a significant role in influencing the isotopic composition of precipitation [21]. The study of the relationship between isotopic depletion in precipitation and stratiform rains remains underexplored, particularly in the context of global warming related climatic changes and regional meteorological patterns within the Indian sub-continent [22, 23]. Consequently, the processes driving isotopic depletion in these rain events are not fully understood.

In this paper, we address this research gap by providing a comprehensive analysis of the temporal behavior of stable isotopes in precipitation in Lucknow, located in Northern India, where such data have been sparse until now. The present study focuses on the isotopic characterization of rainwater collected from three strategically selected locations across the city of Lucknow, situated in the state of Uttar Pradesh, during the ISM months of July to September 2018. The objective was to examine the temporal trends and isotopic variations in precipitation across these locations, with a specific emphasis on understanding the underlying factors driving these changes. This research is especially relevant in light of growing concerns over climate change and its potential impacts on regional weather patterns, including changes in precipitation isotopic composition.

2 Methods

2.1 Sample Collection

Daily precipitation samples were collected from the three selected sites on precipitation days (Fig. 1). These three sites were selected near Lucknow, Uttar Pradesh, India, and the collected samples were used for measurements of stable isotope ratios. The three sites were: Site 1—Jalaudi Nagar (26.76 N, 81.15 E) on the eastern side of Lucknow; Site 2—Para Manek Nagar (26.48 N, 80.53 E) on the western side of Lucknow and near to bank of Ganga River; Site 3—Chabil Chowki- Barabanki (26.94 N, 81.16 E) in the northern side of the Lucknow city. A recent study has documented precipitation isotopic records from these sites and investigated the influence of low-pressure systems on precipitation isotopic variability [24].

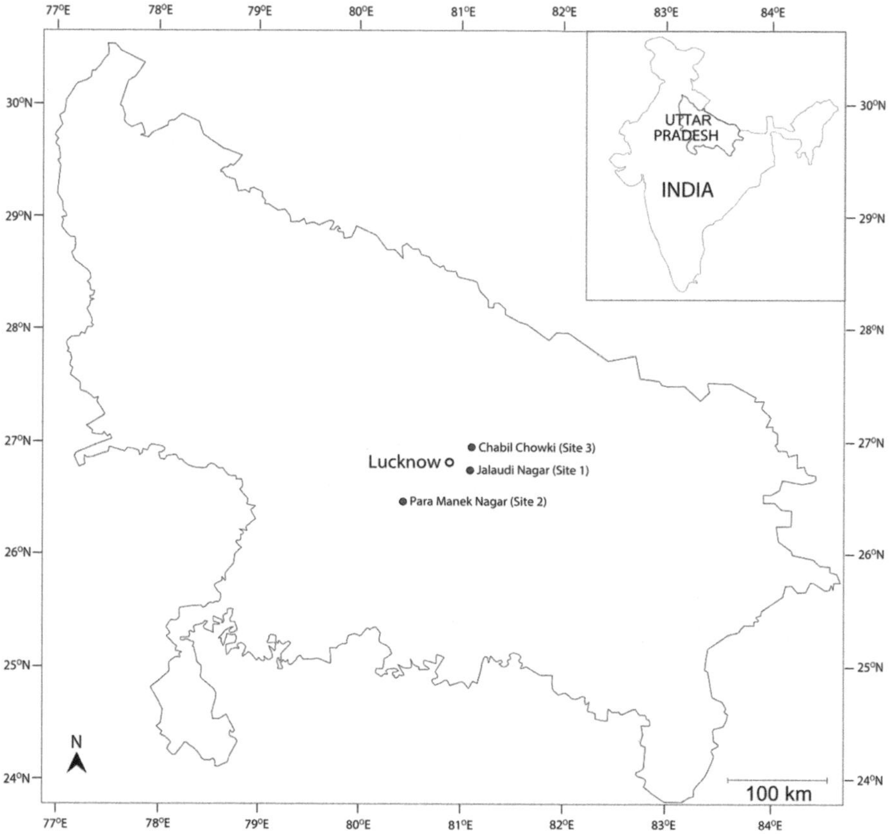

Fig. 1 Sampling locations from 3 sites around Lucknow city. Site 1—Jalaudi Nagar (26.76 N, 81.15 E), Site 2—Para Manek Nagar (26.48 N, 80.53 E), Site 3—Chabil Chowki- Barabanki (26.94 N, 81.16 E)

The rainfall data were obtained from the India Meteorological Department (IMD) under the Ministry of Earth Sciences, Government of India. The temperature and relative humidity data were obtained from ERA 5 (Fifth generation ECMWF reanalysis for the global climate and weather for the past 4–7 decades) [25]. Samples were collected using a funnel of 250 mm diameter, and a total of 110 daily rainwater samples were collected in 60 ml airtight bottles and were stored at low temperatures to prevent evaporative enrichment of stable isotopes. In addition, cloud cover in the form of cloud percentage/fraction on all the sampling days were obtained from TRMM observations (https://gpm.nasa.gov/missions/trmm) for comparison with the precipitation isotopes.

2.2 Back Trajectories and Source Identification

The 5-day back trajectory for specific sampling dates was calculated using MeteoInfo software (Version 3.0.4) and is presented in Fig. 2. The termination points of these trajectories indicate the moisture source for the precipitation. This source region, spanning approximately 855 × 920 km^2, is generally considered as the primary moisture source for all sampling days during the ISM. A height of 1500 m, approximately corresponding to the 850 mb pressure level, was selected for these trajectories, as this level represents the core of air transport through the low-level jet stream, a key pathway for moisture transport contributing to ISM precipitation [26].

2.3 Stable Isotopic Analyses

The isotopic analyses ($\delta^{18}O$ and δD) of rainwater samples were done using the LGR's Liquid Water Isotope Analyzer [27]. The overall analytical precision of the measurements obtained for $\delta^{18}O$ (δD) was about 0.1‰ (< 1‰). All the samples were analyzed at the Indian Institute of Tropical Meteorology (IITM), Pune, India.

2.4 Rain Water Isotope Model

The comparison of observed and modeled $\delta 1^{8}O$ values allows for model validation and refinement for better hydrological and climate studies. The isotopic composition of the atmospheric water vapor reaching Lucknow that causes precipitation can be conceptualized through a model delineating four distinct stages: The first stage is the initial evaporation occurring in the source region. In this study it is assumed as a box in Arabian Sea (shown in Fig. 2) for the period of observations. The second stage is the transportation of moisture and the resultant rainout during its transit across marine and terrestrial environments. The third stage is the recycling of moisture facilitated

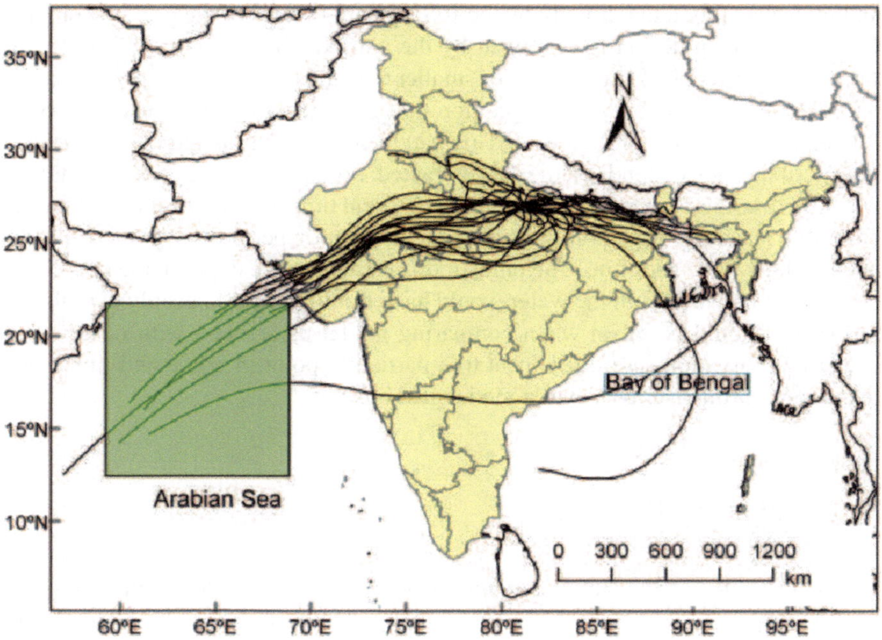

Fig. 2 Back trajectories of the expected moisture parcels arriving at the study station during the ISM period

by evapotranspiration processes originating from the land. The fourth and final stage is the Rayleigh rainout at the precipitation collection point in Lucknow. The model calculations are simplified by assuming that the whole box over the AS acts as a single source for generating the water vapor whose isotopic composition and was obtained from published literature [28]. The assumed Arabian Sea water composition was 0.79‰. The Total precipitable water W" refers to the total amount of water vapor present in a vertical column of the atmosphere, calculated as if all that water vapor were to condense and fall as precipitation, essentially representing the depth of water in millimeters if it were to rain down at once; it is often used in weather forecasting and climate studies to assess potential precipitation levels based on the atmospheric moisture content, and is commonly calculated using the relationship between surface dew point temperature and the logarithm of precipitable water, as described by both Reitan and Smith in their research. The fraction of remaining moisture after rainout is calculated as $f =$ total overhead precipitable water/total rainfall. Total overhead precipitable water here is derived from TRMM, TP that is the accumulated liquid and frozen water, comprising rain and snow, that falls to the Earth's surface. It is the sum of large-scale precipitation and convective precipitation. Large-scale precipitation is generated by the cloud scheme in the ECMWF Integrated Forecasting System (IFS). The cloud scheme represents the formation and dissipation of clouds and large-scale precipitation due to changes in atmospheric quantities (such as pressure, temperature and moisture) predicted directly by the IFS at spatial scales of the grid box or larger. Convective precipitation is generated by the convection scheme in the IFS, which represents convection at spatial scales smaller than the grid box. This parameter does not include fog, dew or the precipitation that evaporates in the atmosphere before it lands at the surface of the Earth. This parameter is accumulated over a particular time period which depends on the data extracted. For the reanalysis, the accumulation period is over the 1 h ending at the validity date and time. For the ensemble members, ensemble mean and ensemble spread, the accumulation period is over the 3 h ending at the validity date and time. The units of this parameter are depth in meters of water equivalent. It is the depth the water would have if it were spread evenly over the grid box. Care should be taken when comparing model parameters with observations, because observations are often local to a particular point in space and time, rather than representing averages over a model grid box.

3 Results and Discussion

The $\delta^{18}O$ values of the samples ranged from − 16.27 to 0.45‰ while δD values varied from − 112.42 to 10.51‰ respectively, demonstrating substantial isotopic depletion during the sampling period. The temporal variation plot (Fig. 3) shows an overall pattern is like a saw tooth change. These sawtooth fluctuations are characteristic of isotopic enrichment during dry periods, followed by depletion during wetter periods, indicating variability in rainout processes and evaporation dynamics within the region. The δD values showed similar pattern to $\delta^{18}O$ as expected. d-excess values were calculated using the equation: d-excess = $\delta D - 8 * \delta^{18}O$ [13]. d-excess helps understand the factors controlling precipitation isotopic variability when analyzed alongside other environmental parameters, such as temperature and humidity. The d-excess values in our samples range from − 0.76 to 17.73‰, exhibiting a similar trend that oscillates around the mean with an amplitude of approximately 4–5‰ (Fig. 3). The overall mean d-excess value observed was 12.5 ± 4.5‰, which is close to the expected equilibrium value of 10‰ typically observed in meteoric waters. These d-excess values primarily depend on relative humidity and sea surface temperature at the moisture source, and they are largely independent of rainout fractionation processes.

Lower d-excess values in precipitation are generally observed when relative humidity is high over the primary moisture source, while higher values are associated with lower humidity conditions. While discussing the rainfall isotope variability, it is of interest to explore the correlation between rainwater isotope ratios and parameters like relative humidity and temperature. Figure 3 illustrates the temporal variability between measured $\delta^{18}O$ values in rainwater and rainfall amount, relative humidity, and temperature. Notably, no significant interdependence was found between $\delta^{18}O$ values and either temperature, relative humidity or amount of rainfall was observed.

We obtained the Local Meteoric water line equation with slope and intercept as: $\delta D = 7.75 * \delta^{18}O + 7.2$ ($R^2 = 0.99$). This equation was similar to the regional ($\delta D =$

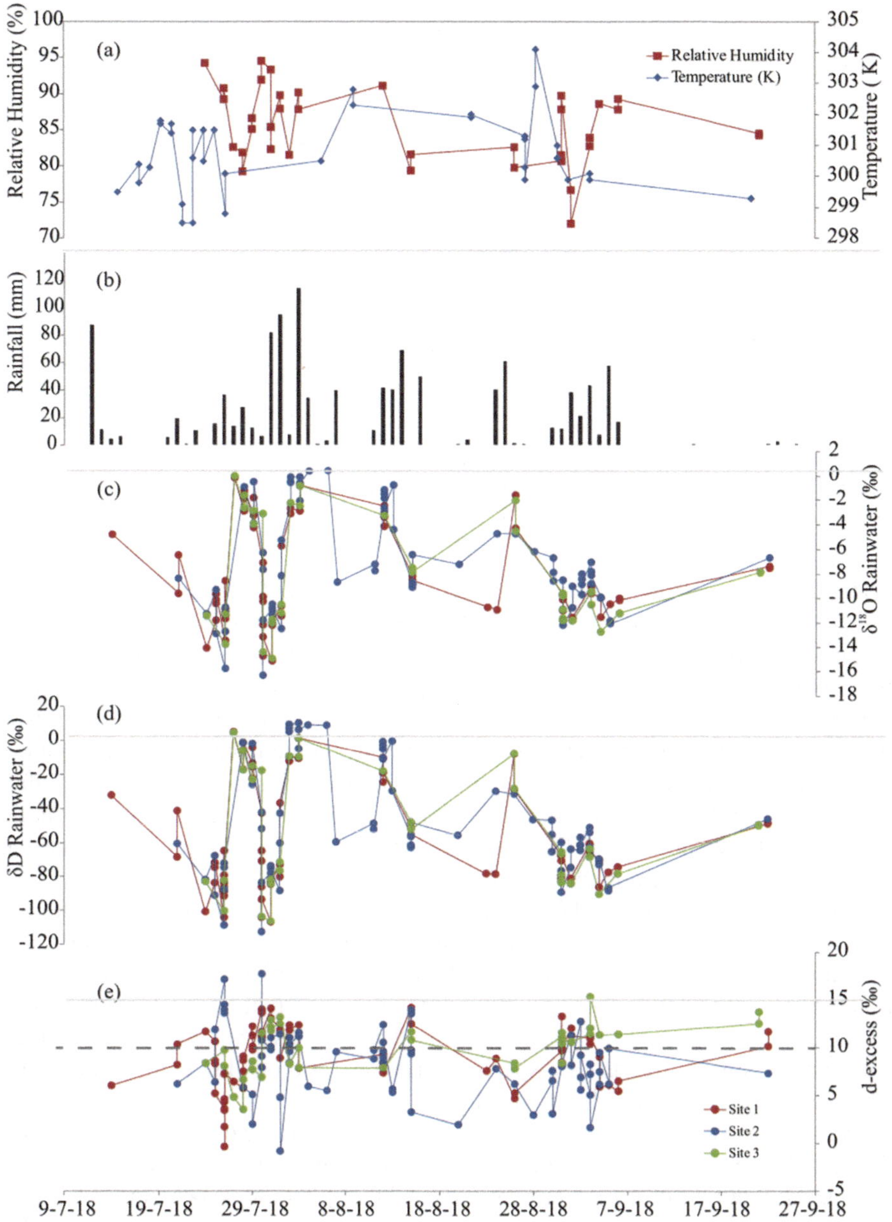

Fig. 3 **a** Intra-seasonal variability of relative humidity and temperature profiles during the sampling period. **b** Amount of Rainfall received at the study location. **c**–**e** Measured $\delta^{18}O$, δD and d-excess of the collected water samples

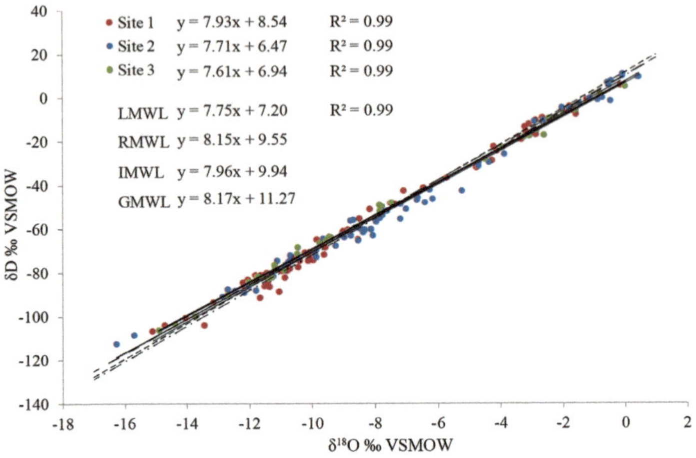

Fig. 4 LMWL of the study area in comparison with the regional, Indian and global meteoric water lines

8.15 * δ^{18}O + 9.55), Indian (δD = 7.96 * δ^{18}O + 9.94) and GMWL (δD = 8.2 * δ^{18}O + 11.27) and is shown in Fig. 4 [8, 29, 30]. The strong correspondence between the Local Meteoric Water Line (LMWL) derived in this study and other regional/global lines indicates that Lucknow's rainfall isotopic characteristics are largely consistent with broader regional and global patterns. Local climatic and environmental factors influencing rainfall isotopes are not significantly deviating from regional and global trends. It is commonly thought that rainwater isotope ratios may depend on rainfall amount or local temperature and relative humidity. However, several authors in the recent past have shown that for interdependency of factors controlling monsoon rainfall is often more complex [20, 24, 30].

The observed and modeled δ^{18}O values in precipitation across the three different sites are shown in Fig. 5. The modelled temporal variability in δ^{18}O values fluctuate with positive and negative excursions. The model predictions generally follow the trends of observed values, however there are regions where the model either overestimates or underestimates δ^{18}O compared to the field observations. While all three sites exhibit similar trends, site-specific variations in modelled δ^{18}O values were observed compared to the observed values.

We then investigated the effect of cloud forms on isotopic depletion in precipitation, building on evidence that cloud microphysics significantly influences rain-water isotope ratios during stratiform rain formation [31, 33]. The cloud percentage/fraction during all the rain events versus the δ^{18}O ratios of the collected precipitation samples across all the three sites is shown in Fig. 6. In all the sites, we observed a negative correlation with respect to the stratiform cloud percentage and the isotopic ratios. The isotopic composition of stratiform rains is expected to be relatively more depleted possibly due to the transport of vapor to lower levels by the process of downdraft, and this vapor is reused for precipitation [32]. Based on a global compilation, it has

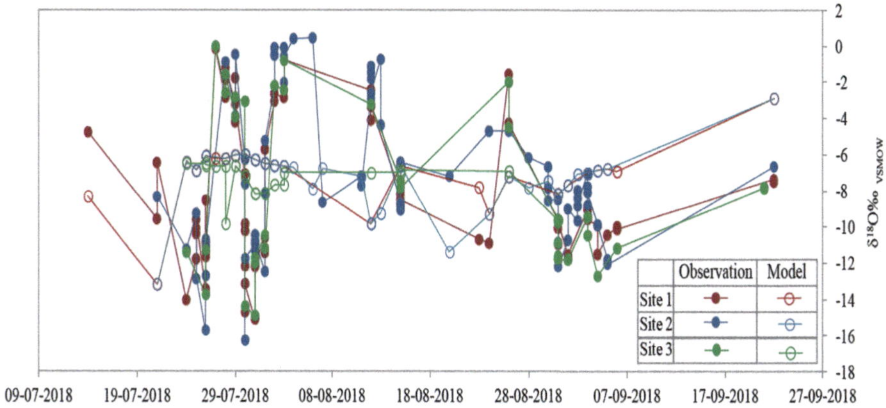

Fig. 5 Comparison of the measured Lucknow rainwater isotopic composition ($\delta^{18}O$ values expressed in ‰ relative to VSMOW) with that modeled. Filled and open circles represent observed and modelled values respectively

been shown that the precipitation $\delta^{18}O$ value decreases by about 2‰ for an increase of 10% in the stratiform rains [31, 33].

This isotopic recycling, influenced by cloud dynamics, is particularly relevant in the context of climate change. As global temperatures rise, changes in atmospheric circulation, cloud formation, and precipitation patterns are expected. With global warming, cloud formation during the Indian Summer Monsoon is expected to undergo significant changes. Rising sea surface temperatures (SSTs) in the Indian Ocean will enhance evaporation, increasing moisture and fueling deeper convection, leading to more intense cumulonimbus clouds and heavy rainfall events. Stratiform rain events may increase in frequency, and the recycling of moisture could become more pronounced, potentially amplifying the isotopic depletion observed in precipitation. On the contrary, with increase in surface temperature, more evaporation is expected resulting in more convective activities. This uncertainty in sensitivity associated with cloud dynamics suggests that changes in rainfall patterns due to climate change could have measurable impacts on the isotopic composition of precipitation, with implications for regional water cycles and hydrological models.

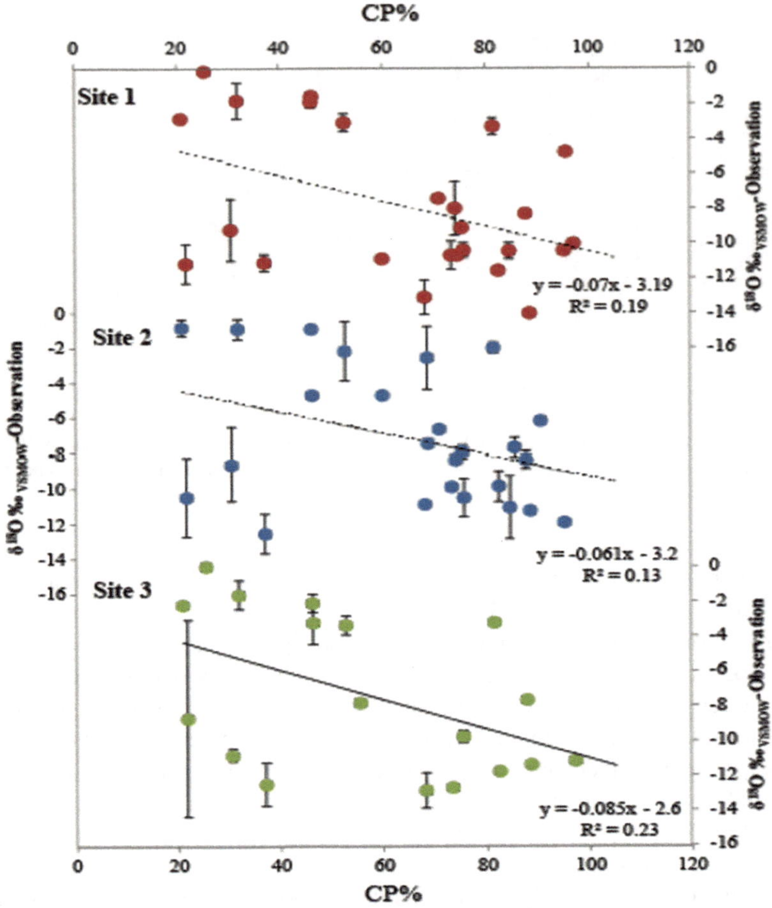

Fig. 6 Daily cloud percentage/fraction versus precipitation $\delta^{18}O$ in the collected samples from all the three sites

4 Conclusion

The isotopic composition of the precipitation samples exhibited a broad range of values. Specifically, $\delta^{18}O$ values ranged from − 16.27 to 0.45‰, while δD values varied from − 112.42 to 10.51‰. The isotopic depletion observed in the precipitation samples is likely influenced by factors such as temperature and humidity at the time of precipitation, which are all sensitive to climate change. These findings provide critical insights into the regional hydrological cycle and highlight the need for further studies to better understand the implications of isotopic variations in precipitation under changing climatic conditions. The current study represents a significant contribution to the understanding of isotopic depletion in precipitation in northern India, offering a foundation for future research on the impacts of climate change on regional water cycles. By establishing baseline isotopic data for Lucknow, India, this research also sets the stage for long-term monitoring efforts that could help detect shifts in precipitation patterns linked to broader climatic trends.

Acknowledgments The authors thank The Director, Birbal Sahni Institute of Palaeosciences, Lucknow for supporting this study. We are also thankful to all the help obtained during the sampling and analysis throughout the project. The authors are thankful for the research support provided to RR through the grant CCEC01-1108-230167 from Qatar National Research Fund.

References

1. IPCC (2007) Climate change 2007: impacts, adaptation and vulnerability. In: Parry ML, Canziani OF, Palutikof JP, van der Linden PJ, Hanson CE (eds) Contribution of working group II to the fourth assessment report of the intergovernmental panel on climate change. Cambridge University Press, Cambridge, pp 976
2. Rahat SH, Saki S, Khaira U et al (2024) Bracing for impact: how shifting precipitation extremes may influence physical climate risks in an uncertain future. Sci Rep 14:17398. https://doi.org/10.1038/s41598-024-65618-9
3. Tabari H (2020) Climate change impact on flood and extreme precipitation increases with water availability. Sci Rep 10:13768
4. Philander SG (2004) Our affair with El Nino. Princeton University Press, Princeton
5. Krishna KK, Rupa KK, Ashrit RG et al (2004) Climate impacts on Indian agriculture. Int J Climatol 24:1375
6. Gadgil S (2003) The Indian monsoon and its variability. Ann Rev Earth Planet Sci 31:429
7. Rangarajan R, Ghosh P (2011) Rainwater management and harvesting strategies for Human needs. Environ Sci Technol 45:9469
8. Rozanski K, Araguas-Araguas L, Gonfiantini R (1993) Isotopic patterns in modern global precipitation. In Swart PK et al (eds), Climate change in continental isotopic records. American Geophy. Union Geophys. Monograph, vol 78, pp 1–36
9. Gat JR, Matsui E (1991) Atmospheric water balance in the Amazon basin: an isotopic evapotranspiration model. J Geophys Res 96:13179
10. Clark I, Fritz P (1997) Environmental isotopes in hydrogeology. Lewis Publishers, Boca Raton, pp 1–328
11. Gupta SK, Deshpande RD, Bhattacharya SK et al (2005) Groundwater δ18O and δD from central Indian Peninsula: influence of the Arabian Sea and the Bay of Bengal branches of the summer monsoon. J Hydrol 303:38
12. Gupta SK, Deshpande RD (2005) Groundwater isotopic investigations in India: what has been learned? Curr Sci 89:825
13. Dansgaard W (1964) Stable isotopes in precipitation. Tellus 16:436
14. Chakraborty S, Burman PKD, Sarma D et al (2022) Linkage between precipitation isotopes and biosphere-atmosphere interaction observed in northeast India. Clim Atmos Sci 5:10. https://doi.org/10.1038/s41612-022-00231-z
15. Jeelani G, Deshpande RD, Galkowski M et al (2018) Isotopic composition of daily precipitation along the foothills of the Himalayas: impact of marine and continental sources of atmospheric moisture. Atmos Chem Phys 18:8789–8805
16. Rangarajan R, Laskar AH, Bhattacharya SK et al (2017) An insight into the western Pacific wintertime moisture sources using dual water vapor isotopes. J Hydrol 547:111–123
17. Simpkins WW (1995) Isotopic composition of precipitation in central Iowa. J Hydrol 172:185

18. Amundson RG, Chadwick OA, Kendall C et al (1996) Isotopic evidence for shifts in atmospheric circulation patterns during the late quaternary in mid North America. Geology 24:23
19. Sengupta S, Sarkar A (2007) Stable isotope evidence of dual (Arabian Sea and Bay of Bengal) vapour sources in monsoonal precipitation over north India. Earth Planet Sci Lett 250:511
20. Fousiya AA, Aravind GH, Achyutan H et al (2022) Modulation of the precipitation isotopes by the dynamic and thermodynamic variables of the atmosphere in southern parts of India. Water Resour Res 58:e2021WR030855
21. Landshuter N, Aemisegger F, Mölg T (2024) Stable water isotope signals and their relation to stratiform and convective precipitation in the tropical Andes. J Geophys Res Atmosph 129:e2023JD040630
22. Lekshmy PR, Midhun M, Ramesh R (2018) Influence of stratiform clouds on δD and δ18O of monsoon water vapour and rain at two tropical coastal stations. J Hydrol 563:354–362
23. Munksgaard NC, Kurita N, Sánchez-Murillo R et al (2019) Data Descriptor: Daily observations of stable isotope ratios of rainfall in the tropics. Sci Rep 9:14419
24. Chakraborty S, Sarkar A, Datye A et al (2025) Precipitation isotopes and monsoon dynamics in the core monsoon zone of India. Sci Rep 15:6761
25. Hersbach H, Bell B, Berrisford P et al (2020) The ERA5 global reanalysis. Q J R Meteorol Soc 146:1999–2049
26. Rahul P, Prasanna K, Ghosh P et al (2018) Stable isotopes in water vapor and rainwater over Indian sector of Southern Ocean and estimation of fraction of recycled moisture. Sci Rep 8:7552
27. Chakraborty S, Sinha N, Chattopadhyay R et al (2016) Atmospheric controls on the precipitation isotopes over the Andaman Islands. Sci Rep 6:19555
28. Rahul P, Ghosh P, Bhattacharya SK (2016) Rainouts over the Arabian Sea and Western Ghats during moisture advection and recycling explain the isotopic composition of Bangalore summer rains. J Geophys Res Atmos 121:6148–6163
29. Gat JR (2003) Isotope composition of air moisture over the Mediterranean Sea: an index of the air-sea interaction pattern. Tellus Ser B-Chem Phys Meteorol 55(5):953–965
30. Kumar B, Rai SP, Saravana Kumar U et al (2010) Isotopic characteristics of Indian precipitation. Water Resour Res 46:W12548
31. Lekshmy P, Midhun M, Ramesh R et al (2014) (2014) 18O depletion in monsoon rain relates to large scale organized convection rather than the amount of rainfall. Sci Rep 4:5661. https://doi.org/10.1038/srep05661
32. Kurita N (2013) Water isotopic variability in response to mesoscale convective system. J Geophys Res-Atmos 118:10376–10390
33. Risi C, Bony S, Vimeux F (2008) Influence of convective processes on the isotopic composition (δ18O and δD) of precipitation and water vapor in the tropics: 2. Physical interpretation of the amount effect. J Geophys Res 113:D19306

Nano-laccase: Silver Nanoparticle-Immobilized Laccase for Efficient and Sustainable Wastewater Treatment

Muhammad Asif Nawaz, Sidra Pervez, and Latif-ur-Rahman

Abstract Rapid urbanization and industrialization in past few decades have led to substantial increase of manmade pollutants in water reservoirs that are difficult to biodegradation due to their synthetic origin. These pollutants are not only toxic to humans but also to the ecosystem. Several traditional wastewater treatment methods are available, but they are not effective against removal of heavy metals, synthetic dyes and other pollutants. Among enzymes, microbial laccases belong to a group of enzymes that have been used extensively for water purification and waste removal. However, soluble laccase faces several drawbacks that can be overcome by immobilization of the enzyme using appropriate support. Therefore, in this study, laccase was produced from indigenous bacterial strain. The enzyme production was also optimized to improve the yield of laccase. After that it was immobilized using silver nanoparticles. These nanoparticles were prepared and characterized using scanning electron microscopy (SEM), energy-dispersive electron microscopy (EDEX), and X-ray diffraction (XRD). Different immobilization parameters were also optimized. The thermal stability data showed that immobilized enzymes showed 80% residual activity at 50 °C after 4 h of incubation. Similarly, the storage stability profiling also exhibited that enzyme is stable for 1 month at room temperature. The reusability of laccase is also very higher, and it showed 96% activity even after 10 cycles. Immobilized laccases effectively lower the COD levels and BOD levels in wastewater. So, it can be concluded that silver nanoparticles immobilized laccase can be efficiently used in various bioreactors for continuous and sustainable purification of wastewater.

Keywords Nano-silver immobilization · Bioremediation · Dye degradation

M. A. Nawaz
Department of Biotechnology, Shaheed Benazir Bhutto University (SBBU), Sheringal, Dir Upper, Pakistan

S. Pervez (✉)
Department of Biochemistry, Shaheed Benazir Bhutto Women University (SBBWU), Peshawar, Pakistan
e-mail: drsidrapervez@sbbwu.edu.pk

Latif-ur-Rahman
Institute of Chemical Sciences, University of Peshawar, Peshawar, Pakistan

S. Al-Naemi et al. (eds.), *Water and Food Security in the Face of Climate Change: Challenges and Opportunities for Resilience*, World Sustainability Series,
https://doi.org/10.1007/978-3-032-00098-9_24

1 Introduction

The increasing global population has resulted in significant environmental challenges, impacting human health and exacerbating global pollution. In response, researchers are devising innovative and eco-friendly solutions to safeguard the ecosystem from further harm. Water body contamination has emerged as a pressing global, social, and environmental issue, driven by ongoing urbanization, industrialization, and inadequate wastewater treatment and management strategies [1, 2]. The use of laccase as a treatment approach for removal of pollutants from industrial water has gained much attention in the past few years as ecofriendly and effective tools for treating and improving water quality. Laccases (E.C. 1.10.3.2) are group of copper containing oxidases that are responsible for catalyzing oxidation of various phenolic and non-phenolic substrates while reducing molecular oxygen to water. Laccase has been produced using various sources including fungi, bacteria and plants due to its versatile and eco- friendly applications in different industrial processes [3]. Besides, this *Aspergillus, Bacillus and Rhizopus* have also been explored due to their limited requirements and presence in all habitats. This large number of laccase sources have resulted in its promising applications in different fields such as bioremediation, textile processing, biomedical properties, food, pulp and paper industries, and wastewater treatment due to their ability to degrade a wide range of pollutants, including synthetic dyes, phenolic compounds, and other toxic substances [4]. Therefore, this study explores the nanoimmobilization of laccase for the treatment of textile wastewater. The innovation of this research lies in designing a highly stable and reusable laccase-nanoparticle system, capable of effectively degrading synthetic dyes and other pollutants under industrial conditions. This work is noteworthy as it overcomes the limitations of soluble laccase, offering a sustainable and cost-efficient approach to wastewater treatment while advancing green and sustainable chemistry [4].

2 Material and Methods

2.1 Production of Laccase

Laccase was produced by using previously isolated strain of *Bacillus subtilis* [5].

2.2 Silver Nanoparticles Synthesis

Silver nanoparticles were produced by chemical method and characterized based on SEM, XRD and EDX.

2.3 Immobilization of Laccase on Silver Nanoparticles

For preparation of nano-immobilized laccase, enzyme solution was mixed with silver nanoparticles along with crosslinking agent i.e. glutaraldehyde. The immobilization yield of silver nanoparticles immobilized laccase was calculated using the following formula mentioned.

$$\text{Immobilization Yield} = \frac{\text{Total activity of immobilized enzyme}}{\text{Total activity of soluble enzyme}} \times 100. \quad (1)$$

2.4 Optimization of Conditions for Nano-laccase Synthesis

The enzyme immobilization was optimized by considering various parameters including glutaraldehyde concentration (1–5%), enzyme load (100–500 U/mg), crosslinking time (1–4 h) as well as crosslinking temperature (20–50 °C).

2.5 Characterization of Laccase

The soluble and immobilized laccase was characterized based on different parameters including reaction time, temperature, pH and ionic strength of buffer.

2.6 Analysis of Kinetic Parameters

The kinetic parameters (Km and Vmax) of laccase immobilized on silver nanocarriers and soluble enzyme were determined by measuring the enzyme assay using different substrate concentrations. The values of Km and Vmax were determined by Lineweaver Burk plot [6].

2.7 Reusability Analysis

The importance of immobilization was the reusability of nano-laccase for subsequent cycles. For this purpose, enzyme was incubated with substrate and after product determination, the substrate was removed, nanoparticles were recovered, washed and then again activity was performed. This process remains continuous for 10 cycles.

2.8 Wastewater Treatment Using Soluble and Nano-laccase

Synthetic wastewater was prepared and immobilized and soluble enzyme was mixed with it for different time intervals. COD and BOD was calculated as per standard procedure.

2.9 Dye Degradation by Laccase

Three different dyes including bromophenol blue, methyl orange and congo red were used in this study. For qualitative analysis, zone of hydrolysis was noted while for quantitative analysis, dyes were mixed with enzyme solutions in different concentrations and allowed to stand for different time intervals at room temperature. The decolorization of dyes was calculated spectrophotometrically using following formula:

$$\text{Decolorization}\,\% = (\text{initial absorbance} - \text{final absorbance})/\text{initial absorbance} \times 100.$$

3 Results and Discussion

3.1 Characterization of Silver Nanoparticles

Nanoparticles of silver has enhanced surface area which improves its applicability as compared to normal and micro sized silver. The synthesized nanoparticles were characterized by SEM, XRD and EDX (Figs. 1 and 2). The size and appearance of nanoparticles was determined by SEM.

3.2 Optimization of Immobilization Parameters

Glutaraldehyde concentration was varied for immobilization of laccase on silver nanoparticles. Table 1 showed that 4% glutaraldehyde was found to be the optimum concentration for cross-linking of laccase with silver nanoparticles. Similarly, the effect of laccase concentration was also monitored, and it was observed that 300 U/mg of laccase were found to be optimum concentration for highest immobilization yield (Table 1). Moreover, the effect of cross-linking time for immobilization of laccase on silver nanoparticles was also investigated while fixing glutaraldehyde and enzyme concentration. As time increases, the immobilization yield increases, and it reaches its maximum after 03 h of incubation. Further increase in incubation time will

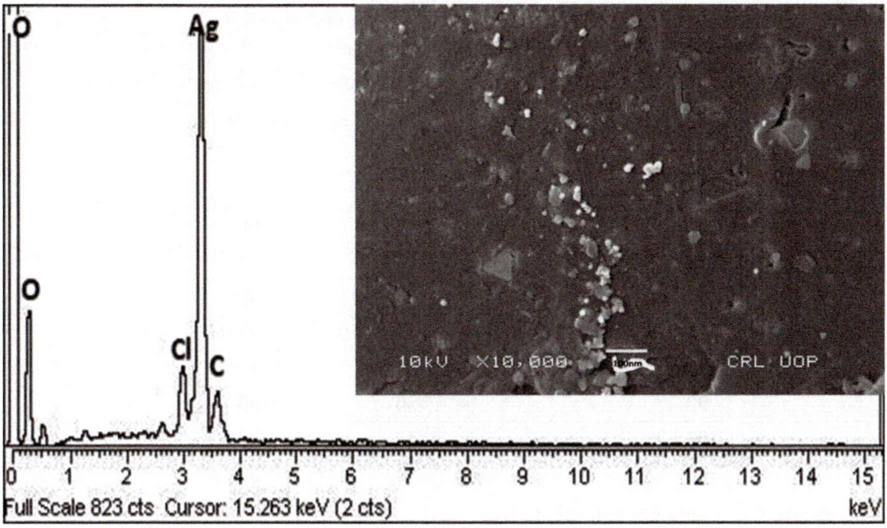

Fig. 1 SEM and XRD analysis of silver nanoparticles

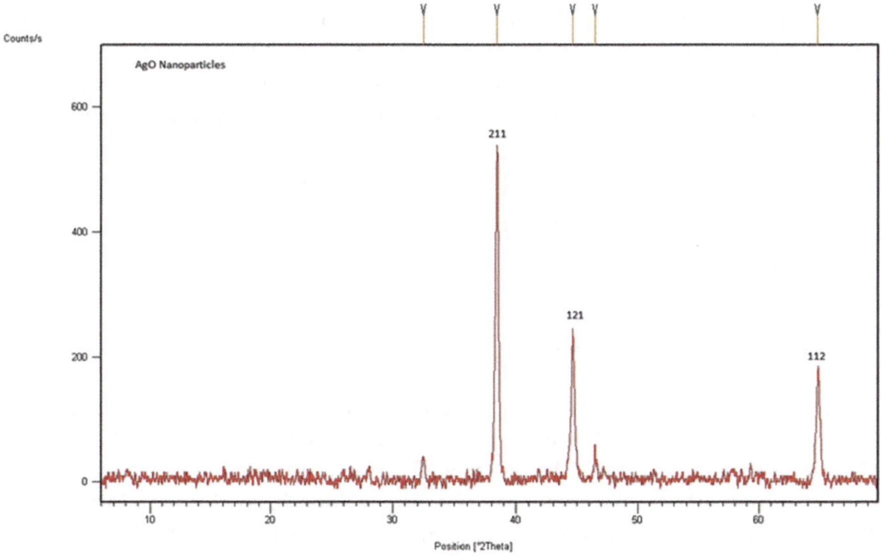

Fig. 2 EDX analysis of silver nanoparticles

Table 1 Effect of glutaraldehyde concentration, laccase concentration, cross-linking time and cross-linking temperature on immobilization yield of laccase immobilized silver nanoparticles

Parameters	Immobilization yield (%)	Parameters	Immobilization yield (%)
Glutaraldehyde concentration (%)		*Laccase concentration (U/mg)*	
1	45	100	47
2	76	200	79
3	82	300	**100**
4	**100**	400	91
5	79	500	90
Cross-linking time (minutes)		*Cross-linking temperature (°C)*	
1	81	20	56
2	92	30	**100**
3	**100**	40	82
4	86	50	67

Bold means maximum immobilization yield

significantly affect the immobilization yield or the catalytic activity of the nanocatalyst. The nanoparticles and laccase immobilization were also studied at different cross-linking temperatures ranging from 20 to 50 °C. At 30 °C, a 100% immobilization yield was achieved which gradually declined afterwards. Similar findings have been previously reported by immobilizing pectinase using silver nanoparticles [7].

3.3 *Characterization of Soluble and Nano-immobilized -Laccase*

3.3.1 Effect of Reaction Time

The enzyme substrate reaction time (10–60 min) was evaluated for soluble and silver nanoparticles immobilized laccase and it was found that reaction time has changed after immobilization. Table 2 showed that after immobilization, the reaction time was increased from 10 (soluble enzyme) to 20 min (immobilized enzyme). Furthermore, the relative activity of immobilized enzyme is also greater at longer time period as compared to soluble form. Umar and Ahmed [8] also observed optimum reaction time of 10 min for soluble laccase produced from *Ganoderma leucocontextum.* The pectinase immobilized using silver nanocatlyst also exhibited same catalysis time profile for pectinase [7].

Table 2 Effect of reaction time on the activity of soluble and immobilized laccase

Reaction time (minutes)	Relative activity (%)	
	Nano-silver immobilized enzyme activity	Soluble enzyme activity
10	83	**100**
20	**100**	92
30	95	87
40	90	83
50	87	75
60	82	60

Bold means maximum relative activity at given conditions

3.3.2 Effect of Reaction Temperature

The effect of temperature on laccase activity was also analyzed to determine the optimal temperature for both forms. As shown in Table 3, soluble laccase exhibited an optimum temperature of 40 °C, while immobilized laccase had a higher optimum temperature of 55 °C. Additionally, the nano-silver immobilization also increases the stability of laccase at elevated temperatures, retaining 81% and 65% relative activity at 65 and 70 °C, respectively, whereas the soluble enzyme became completely inactive at 65 °C. Similarly, laccase from *Bacillus subtilis* MTCC 2414 has been reported to exhibit higher activity at 70 °C [9].

Table 3 Effect of reaction temperature on the activity of soluble and immobilized laccase

Reaction temperature (°C)	Relative activity (%)	
	Soluble enzyme activity	Nano-silver immobilized enzyme activity
30	68	45
35	89	60
40	**100**	75
45	71	86
50	45	93
55	30	**100**
60	10	90
65	0	81
70	0	65

Bold means maximum relative activity at given conditions

Table 4 Effect of Reaction pH on the activity of soluble and immobilized laccase

Reaction pH	Relative activity (%)	
	Nano-silver immobilized enzyme activity	Soluble enzyme activity
6	68	48
6.5	81	63
7	**100**	**100**
7.5	92	72
8	86	54

Bold means maximum relative activity at given conditions

3.3.3 Effect of Reaction pH

The variation of pH is also an important parameter for oxidation of guaiacol [10]. Zou et al. [11] reported that most of the laccases are active in acidic and neutral pH and become inactive in alkaline environment. However, both forms of laccases in this study also exhibited oxidation of guaiacol in wide range of pH with relative activity of 70 and 69% at pH 3 and 9 (Table 4). The soluble and immobilized both enzymes showed maximum activity at pH 7 while the stability of immobilized enzyme is more as compared to soluble one.

3.3.4 Kinetic Analysis

The catalytic efficiency of soluble and immobilized laccase was observed by determining its Km and Vmax. The affinity and maximum velocity of laccase was not affected by silver nano-immobilization, and it was observed that no significant difference in Km and Vmax was observed. Soluble laccase has Km and Vmax of 1.36 mM and 2498 μM/min while immobilized has 1.57 mM and 2389 μM.

3.4 Reusability Analysis

The key advantage of immobilization is the separation of nanoparticles from substrate mixture and then its continuous reuse. In this study, the reusability of silver immobilized laccase was determined for 10 cycles. Figure 3 demonstrated that nano-immobilized laccase has retained almost 96% of its activity after 10 cycles. This indicates that the crosslinking of enzymes with silver nanoparticles using glutaraldehyde facilitated their reusability in continuous industrial applications.

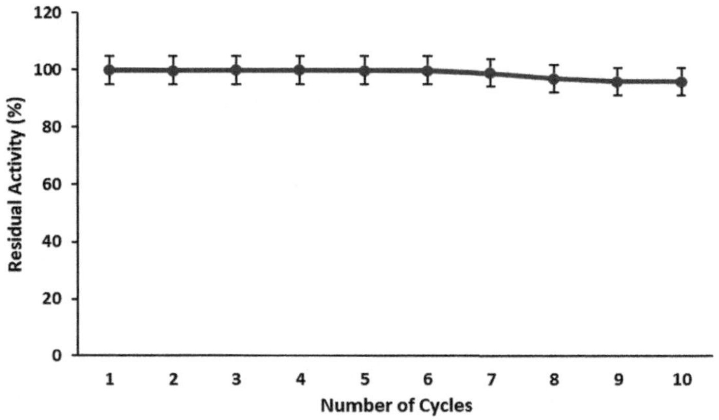

Fig. 3 Reusability profile of silver nanoparticles immobilized laccase

3.5 *Wastewater Treatment and Dye Degradation*

The wastewater treatment efficiency was evaluated by COD and BOD analysis. It was observed that the COD and BOD degradation efficiency of immobilized laccase is more i.e. 58 and 79% as compared to soluble enzyme i.e. 41 and 52%. These findings suggest that immobilized laccase can be repeatedly reused and can increase the efficiency of treatment process by stabilizing enzyme molecules.

3.6 *Dye Degradation Efficiency*

The soluble and immobilized laccase was also studied to check their dye degradation efficiency. For this purpose, enzyme solution was mixed with 03 dyes (congo red, bromophenol blue and methyl orange). The qualitative analysis showed colorless zones around wells that suggested the potential of laccase for dye degradation. After this quantitative analysis was performed to check the % efficiency of dye degradation by laccases. The OD of dyes decreases with passage of time and % degradation increases. Deng et al. [12] used soluble and immobilized forms of LAC-4 for degradation and detoxification of different dyes. They found that soluble and immobilized enzyme degrade and detoxify different dyes and their mixture with an efficiency of 80%.

Zhaou et al. [13] explained various methods of laccase immobilization and their usage as wastewater treatment. They have demonstrated different supports for immobilization of laccase and their advantages and drawbacks along with pollutants removal. Moreover, various studies are also available in which macro and micro supports are also used for immobilization of laccase using entrapment, adsorption and encapsulation methods [14–17].

4 Conclusion

In this study, laccase was successfully produced from Bacillus subtilis and immobilized onto silver nanoparticles to enhance its bioremediation capabilities. The silver nanoparticle-immobilized laccase demonstrated higher relative activity across a broad range of conditions, highlighting its improved versatility. Furthermore, the immobilized enzyme exhibited excellent reusability, enhancing product oxidation efficiency while reducing time and effort. Both soluble and immobilized laccase were effectively applied in wastewater treatment and dye degradation, with the immobilized form showing superior efficiency. These findings underscore the potential of silver nanoparticle-immobilized laccases as a powerful bioremediation agent for treating polluted water in textile industries, facilitating its conversion into clean water.

Acknowledgements Authors would like to extend heartfelt gratitude to Higher Education Commission (HEC) of Pakistan (Ref# HEC/R&D/TG/213324).

References

1. Bilal M, Asgher M, Parra-Saldivar R, Hu H, Wang W, Zhang X, Iqbal HMN (2017) Immobilized ligninolytic enzymes: an innovative and environmental responsive technology to tackle dye-based industrial pollutants—a review. Sci Total Environ 576:646–659
2. Spina F, Cordero C, Schilirò T, Sgorbini B, Pignata C, Gilli G, Bicci C, Varese GC (2015) Removal of micropollutants by fungal laccases in model solution and municipal wastewater: evaluation of estrogenic activity and ecotoxicity. J Clean Prod 100:185–194
3. Patel SK et al (2020) Laccase production and applications: a comprehensive review. Biotechnol Adv 38:107–131
4. Duran N et al (2015) Potential applications of laccase in the degradation of contaminants. Appl Microbiol Biotechnol 99(3):1025–1038
5. Behram T, Pervez S, Nawaz MA et al (2024) Synthesis and analysis of silver nanocarriers for pectinase immobilization: Enhancing enzymatic stability for continuous industrial applications. Heliyon 10:e23549
6. Lineweaver H, Burk D (1934) The determination of enzyme dissociation constants. J Am Chem Soc 56:658–666
7. Umer A, Ahmed S (2022) Optimization, purification and characterization of laccase from Ganoderma leucocontextum along with its phylogenetic relationship. Sci Rep 12:2416
8. Muthukumarasamy NP, Jackson B, Raj AJ, Sevanan M (2015) Production of extracellular laccase from Bacillus subtilis MTCC 2414 using agroresidues as a potential substrate. Biochem Res Int 2015:765190
9. Wang H, Tang C, Yu G, Li P (2013) A novel membrane-surface liquid co-culture to improve the production of laccase from Ganoderma lucidum. Biochem Eng J 80:27–36
10. Morozova OV, Shumakovich GP, Shleev SV, Yaropolov AI (2007) Laccase-mediator systems and their applications: a review. Appl Biochem Microbiol 43:523–535
11. Bonomo RP, Cennamo G, Purrello R, Santoro AM, Zappala R (2001) Comparison of three fungal laccases from Rigidoporus lignosus and Pleurotus ostreatus: correlation between conformation changes and catalytic activity. J Inorg Biochem 83(1):67–75
12. Deng X et al (2024) Efficient degradation and detoxification of structurally different dyes and mixed dyes by LAC-4 laccase purified from white-rot fungi Ganoderma lucidum. Ecotoxicol Environ Saf 279:116450

13. Zhou W, Zhang W, Cai Y (2021) Laccase immobilization for water purification: a comprehensive review. Chem Eng J 403:12627
14. Rodríguez-Couto S (2023) Immobilized-laccase bioreactors for wastewater treatment. Biotechnol J 19:2300354
15. Majeau JA et al (2010) Laccases for removal of recalcitrant and emerging pollutants. Biores Technol 101(7):2331–2350
16. Sharma KK, Shrivastava B, Sastry VRB, Sehgal N, Kuhad RC (2013) Middle-redox potential laccase from Ganoderma sp.: its application in improvement of feed for monogastric animals. Sci Rep 3:1299
17. Zou YJ, Wang HX, Ng TB, Huang CY, Zhang JX (2012) Purification and characterization of a novel laccase from the edible mushroom Hericium coralloides. J Microbiol 50:72–78

Future Water Scarcity Appearance in the Arabian Peninsula

Zahrah Abdulghafour and Masoud Irannezhad

Abstract Water is crucial for achieving the United Nations' 2030 Sustainable Development Goals (SDGs), particularly in arid climate regions like the Arabian Peninsula (AP). Terrestrial water storage (TWS) generally serves as the fundamental indicator of water resources availability, which is a key component of water security. Although previous studies have investigated historical variability in TWS, its future changes have received less attention. Employing GRACE satellite data combined with global hydrological (GHMs) and land surface models (LSMs), this study analyzed alterations in TWS throughout the AP under low (RCP2.6) and medium–high (RCP6.0) greenhouse gas concentration scenarios during the mid (2030–2059) and late (2070–2099) future periods. Significant decreases in TWS are predicted for southeastern Saudi Arabia and eastern Yemen, while northern areas, including Kuwait, are expected to experience increases. The rate of TWS decline is projected to be more apparent in the late twenty-first century and under RCP6.0. These insights play a significant role in achieving water security in the AP, thereby contributing to global economic, social, and environmental sustainability.

Keywords Kuwait · GRACE · RCP · Saudi Arabia · Sustainability · Yemen

1 Introduction

The Sixth Assessment Report (AR6) of the Intergovernmental Panel on Climate Change (IPCC) confirmed that the global surface air temperature (SAT) was about 1.1 °C warmer during the period 2011–2020 compared to 1850–1900 [1]. The rate

Z. Abdulghafour · M. Irannezhad (✉)
Department of Civil Engineering, College of Engineering, Australian University, West Mishref, Safat, Kuwait
e-mail: m.irannezhad@au.edu.kw

M. Irannezhad
Water, Energy and Environmental Engineering Research Unit, Faculty of Technology, University of Oulu, Oulu, Finland

S. Al-Naemi et al. (eds.), *Water and Food Security in the Face of Climate Change: Challenges and Opportunities for Resilience*, World Sustainability Series,
https://doi.org/10.1007/978-3-032-00098-9_25

of warming SAT was notably higher over land surface (1.6 °C) than over oceans (0.9 °C). Global mean sea surface temperature (SST) has also risen in recent decades [2]. These increases in SAT and SST were primarily attributed to substantial anthropogenic greenhouse gas (GHG) emissions into the Earth's atmosphere [1, 2]. Based on the Clausius-Clapeyron (C–C) equation, such warming trends in SAT and SST principally intensify atmospheric water content [3], change the global climate system, influence various components of the hydrological cycle, and consequently alter water resources worldwide [4].

Water is not only essential for all forms of life but also for key human activities such as agriculture, forestry, recreation, and energy production. Improving human well-being while protecting nature, water plays a crucial role in achieving the United Nations's 2030 Sustainable Development Goals (SDGs) [5, 6]. Specifically, "Clean Water and Sanitation" (SDG6) is the fundamental prerequisite for advancing 63 out of 162 targets associated with other SDGs [5]. In the era of global warming and climate change, hence, alterations in water resources pose serious economic, social, and environmental challenges to humanity. This is particularly true for developing countries, where high population density, vulnerable infrastructures, and poor land use planning and management exacerbate such challenges [6].

Assessing water security is crucial for mitigating sustainability risks, particularly in severely arid environments. Terrestrial water storage (TWS)—the sum of continental water stored in rivers, lakes, reservoirs, canopies, snow, ice, wetlands, groundwater, and soil—serves as a fundamental indicator of water resources availability, which is one the key components of water security [7]. While previous studies have explored historical variations in local, regional, and global TWS, future changes have only been addressed in a limited number of recent publications. Therefore, estimating TWS in the hyper-arid climate of the Arabian Peninsula (AP) over this century is imperative for developing adaptation and mitigation strategies, which can be essential to achieving global sustainability.

The overall aim of this study was to assess future changes in water resources availability within the severely arid environment of the AP. To achieve this, we analyzed alterations in TWS across the AP, comparing projections for the mid (2030–2059) and late (2070–2099) twenty-first century to the historical baseline period (1976–2005) under low (Representative Concentration Pathway 2.6 or RCP2.6) and medium–high (RCP6.0) greenhouse gas concentration scenarios. Our findings provide a foundation for rapidly attaining water security throughout the AP, thereby contributing economic, social, and environmental sustainability around the world.

2 Material and Methods

As the largest peninsula in the world, the AP is in the southwestern part of Asia (Fig. 1a), between 10°–35° N latitude and 35°–60° E longitude. It is bordered by the Arabian Sea to the south, the Red Sea to the west and the Persian Gulf to the northeast (Fig. 1b). Covering an area of about 3.2 million km^2, the AP includes seven countries of Bahrain, Kuwait, Oman, Qatar, Saudi Arabia, the United Arab Emirates (UAE), and Yemen (Fig. 1b). The region is dominated by vast deserts, including the "Rub' Al Khali" (Empty Quarter), the world's largest continuous sand desert (Fig. 1b). The Ap is also home to various mountain ranges, such as the Hajar and Sarawat Mountains in Oman and Saudi Arabia, respectively. Its climate is predominantly severe to hyper- arid, with extremely high SAT in summer and minimal precipitation (less than 100 mm) throughout the year [8, 9]. Such unique geography and climate make water scarcity a critical issue in the AP, impacting nature and humanity, especially in the context of global warming [10].

Meteorological forcing data were obtained from climate simulations conducted by four General Circulation Models (GCMs), which are part of the Coupled Model Intercomparison Project Phase 5 (CMIP5) and included in the IPCC's Fifth Assessment Report (AR5). These GCMs were GFDL-ESM2M, HadGEM2-ES, IPSL-CM5A-LR, and MIROC5. The climate variables incorporated into the forcing data include precipitation, SAT, both shortwave and longwave solar radiations, wind speed, specific humidity, and surface pressure [7]. These variables were bias-adjusted and downscaled to a $0.5° \times 0.5°$ spatial resolution to be employed by five global hydrological models (GHMs) of CWatM, H08, MPI-HM, PCR-GLOBWB, and WaterGAP2 [11]; one global land surface model (LSM) of CLM4.5 [11]; and one dynamic global vegetation model of LPJmL [12], to simulate critical terrestrial hydrological processes. To validate such TWS simulations within these models, finally, the TWS data provided by the Gravity Recovery and Climate Experiment (GRACE) satellite mission was used [13]. For this, the multi-model mean was computed by applying weights to each ensemble member, based on two factors: their skill, measured by the root mean squared error (RMSE) of the area-weighted seasonal cycle of TWS compared to GRACE data, and their independence, which reflects how distinct the model results are from one another [14].

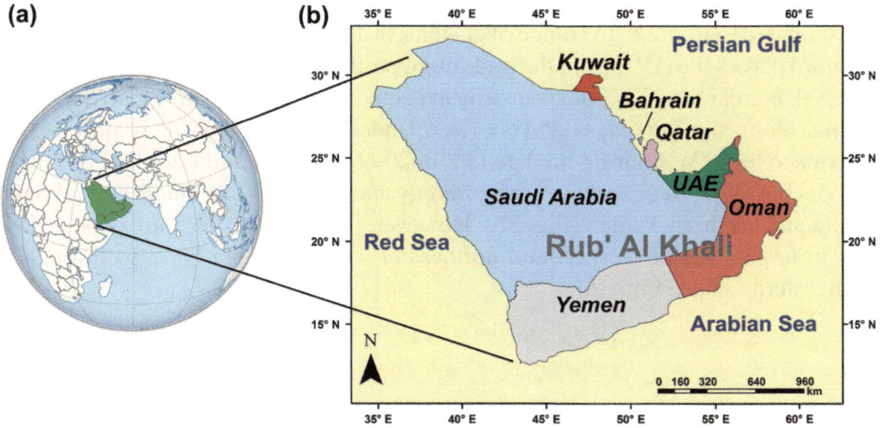

Fig. 1 **a** the location of the Arabian Peninsula (AP), and **b** its seven countries (Bahrain, Kuwait, Oman, Qatar, Saudi Arabia, the UAE, and Yemen) considered in this study

For each GCM, four radiative forcing scenarios were evaluated over different periods: the pre-industrial control (1861–2099), the historical climate scenario (1861–2005), RCP2.6 (2006–2099), and RCP6.0 (2006–2099). These simulations followed the standard protocol of the Group-2 simulation scenario design under ISIMIP2b (https://www.isimip.org/). The RCP2.6 and RCP6.0 are the only scenarios for which TWS data were available from all models in the ISIMIP2b simulations. For each combination of GCM and radiative forcing scenario, the hydrological models were run with time-varying land and water management activities and socioeconomic conditions for the historical period (HIST). However, for future projections (2006–2099 under RCP2.6 and RCP6.0), these factors were fixed at their 2005 levels [7]. Changes in the climatological mean seasonal cycle of TWS for both mid (2030–2059) and late (2070–2099) future periods under RCP2.6 and RCP6.0 scenarios were finally calculated relative to the average for the historical baseline period (1976–2005).

3 Results

The AP is projected to experience changes in the climatological mean seasonal cycle of TWS (hereafter TWS), ranging from − 50 to 50 mm, under RCP2.6 during the mid-future period (2030–2059) (Fig. 2a). By 2050, substantial water scarcity, indicated by decreases in such TWS, is expected across much of Saudi Arabia and Qatar (Fig. 2a). In contrast, TWS will increase in Oman, Yemen, the UAE, and northern Saudi Arabia (Fig. 2a). Similarly, during the late future period (2070–2099), TWS changes under the RCP2.6 scenario are expected to range between -50 and 50 mm (Fig. 2b). Compared to the mid future, larger areas in Oman, Yemen, the UAE, Kuwait, and Saudi Arabia are projected to experience increases in TWS (Fig. 2b). However, water scarcity is still anticipated over the inland regions of Saudi Arabia (Fig. 2b).

Compared to RCP2.6, the range of changes in TWS was generally larger (− 100 to 100 mm) across the AP during the mid future period (2030–2059) under the RCP6.0 (Fig. 3a). Increases in TWS are primarily expected in Kuwait and northeastern Saudia Arabia, along the Persian Gulf (Fig. 3a). Under RCP6.0, the AP is projected to experience less TWS during the late (2070–2099) than the mid (2030–2059) future periods (Fig. 3b). Water scarcity is accordingly anticipated in the inland areas of Saudi Arabia and northern Yemen (Fig. 3b). However, increases in TWS are expected not only in large parts of western and northeastern Saudi Arabia but also throughout southeastern Oman (Fig. 3b).

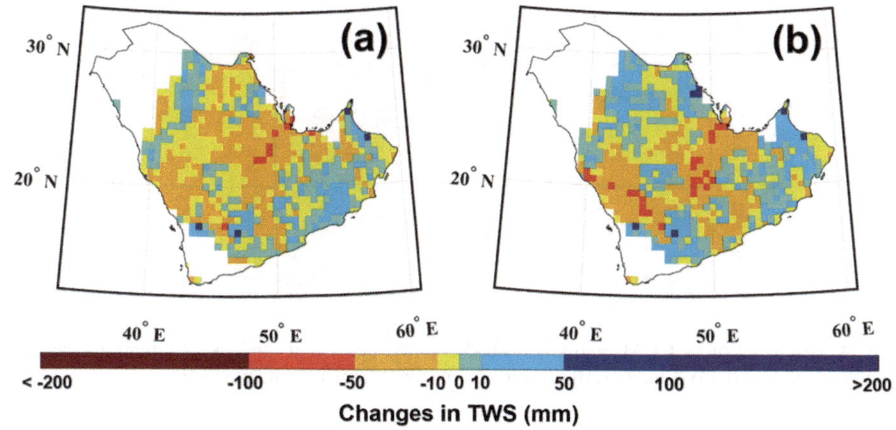

Fig. 2 Changes in the climatological mean seasonal cycle of terrestrial water storage (TWS) (mm) in the Arabian Peninsula (AP), averaged during both the **a** mid (2030–2059) and **b** late (2070–2099) future periods, under a low greenhouse gas concentration scenario (RCP2.6), relative to the average for the historical baseline period (1976–2005)

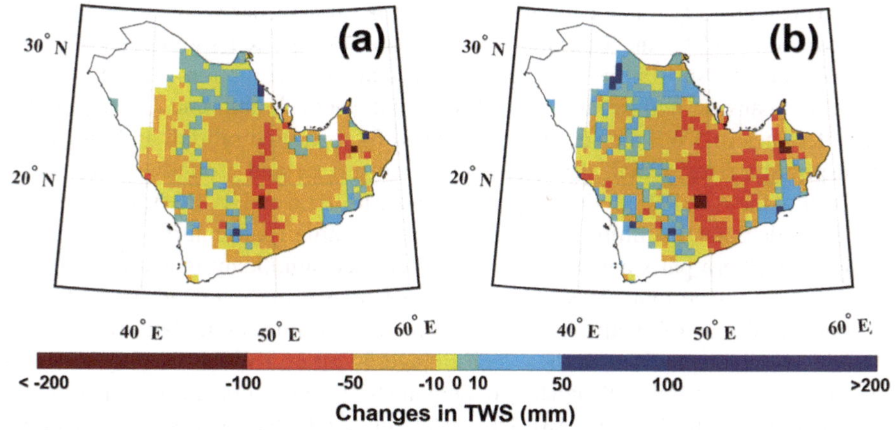

Fig. 3 Changes in the climatological mean seasonal cycle of terrestrial water storage (TWS) (mm) in the Arabian Peninsula (AP), averaged during both the **a** mid (2030–2059) and **b** late (2070–2099) future periods, under a low greenhouse gas concentration scenario (RCP6.0), relative to the average for the historical baseline period (1976–2005)

4 Discussion

The variation in TWS simulations among the ensemble members is primarily influenced by two key sources of uncertainty: climate forcing from GCMs and the parameterizations used in GHMs and LSMs. For each RCP scenario, GCM-related uncertainty tends to be greater than that from GHM/LSM parameterizations across the AP

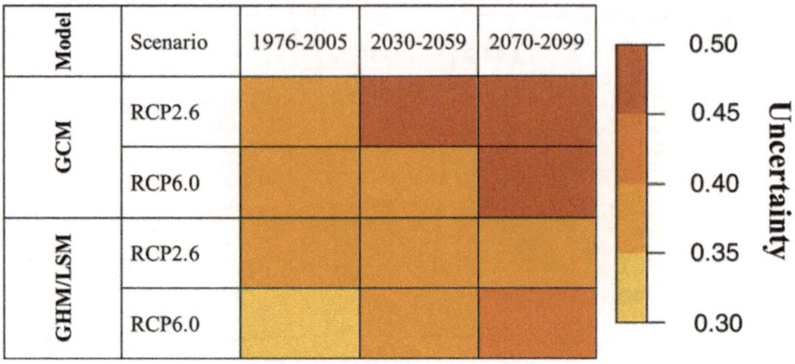

Fig. 4 Contribution of GCMs and GHMs/LSMs to the uncertainty in TWS simulations based on the quantile-based TWS index [20], averaged over the Arabian Peninsula (AP)

during historical and future periods, particularly in 2070–2099 (Fig. 4). During 1976–2005, the GCM and GHM/LSM uncertainty ranges remain relatively small (Fig. 4), likely due to the strong agreement in seasonal amplitude and temporal variability of TWS when compared with GRACE data [7]. This consistency likely reflects the advantages of bias correction using observational data from the same period.

The present study found declines in future TWS mostly across the inland areas of Saudi Arabia and Qatar. Similarly, previous studies reported significant decreases in TWS due to substantial increases in groundwater depletion, driven by agricultural expansion and urbanization throughout Saudi Arabia and Qatar [15, 16]. Such declines in TWS are further exacerbated by restricted rainfall and high evapotranspiration rates, which strain already scarce surface and groundwater resources throughout the AP [17]. Model projections under RCP2.6 and RCP6.0 also suggested moderate declines in future TWS across Saudi Arabia (particularly inland areas), while its northern parts and Kuwait may experience significant increases [10]. Our results also showed that TWS will decrease (increase) in the inland parts of Saudi Arabia and Yemen (Kuwait and northeastern Saudi Arabia) under both RCP2.6 and RCP6.0. based on previous studies, such changes in future TWS are largely attributed to warmer SAT, less rainfall, and higher demand for water resources due to population growth and industrialization [18]. Hence, effective adaptation strategies, including the development of alternative water resources, will be crucial not only for mitigating the impacts of future TWS reductions across the AP but also for supporting sustainable development in its respective countries [19].

5 Conclusions

This study explored distinct spatio-temporal variability in projected TWS throughout the AP under global warming and climate change. Significant declines in future TWS are primarily expected over the southeast of Saudi Arabia and the eastern parts of Yemen. In contrast, the northern areas of the AP will generally experience substantial increases in TWS during 2030–2099. The decline rate of TWS is anticipated to not only be more evident in 2070–2099 than in 2030–2059, but also higher under the RCP6.0 scenario compared to RCP2.6. Interestingly, however, the rate of TWS increase remains fairly consistent between the mid (2030–2059) and late (2070–2099) future periods. Hence, the AP is vulnerable to intensified water stress under future climatic change, particularly in medium–high emission scenarios (e.g., RCP6.0). Such insights about future alterations in TWS urge the development of adaptation and mitigation strategies for acting towards achieving water security, thereby contributing to economic, social, and environmental sustainability across the AP.

Acknowledgements This study was funded by the Kuwait Foundation for the Advancement of Sciences (KFAS), Grant No. CN23-44SE-2126.

References

1. IPCC, Masson-Delmotte V, Zhai P, Pirani A, Connors SL, Péan C, Berger S, Caud N, Chen Y, Goldfarb L et al (2021) Climate Change 2021: the physical science basis. Contribution of Working Group I to the sixth assessment report of the intergovernmental panel on climate change
2. Iz HB (2018) Is the global sea surface temperature rise accelerating? Geod Geodyn 9:432–438. https://doi.org/10.1016/j.geog.2018.04.002
3. Ali H, Mishra V (2017) Contrasting response of rainfall extremes to increase in surface air and dewpoint temperatures at urban locations in India. Sci Rep 7:1228. https://doi.org/10.1038/s41598-017-01306-1
4. Gudmundsson L, Boulange J, Do HX, Gosling SN, Grillakis MG, Koutroulis AG, Leonard M, Liu J, Schmied HM, Papadimitriou L et al (2021) Globally observed trends in mean and extreme river flow attributed to climate change. Science (80-) 371:1159–1162. https://doi.org/10.1126/science.aba3996
5. UN About the Sustainable Development Goals (2015) United Nations sustainable development. Sustainable development goals
6. Irannezhad M, Ahmadi B, Liu J, Chen D, Matthews JH (2022) Global water security: a shining star in the dark sky of achieving the sustainable development goals. Sustain Horizons 1:100005. https://doi.org/10.1016/j.horiz.2021.100005
7. Pokhrel Y, Felfelani F, Satoh Y, Boulange J, Burek P, Gädeke A, Gerten D, Gosling SN, Grillakis M, Gudmundsson L et al (2021) Global terrestrial water storage and drought severity under climate change. Nat Clim Chang 11:226–233. https://doi.org/10.1038/s41558-020-00972-w
8. Bahaj AS, Mahdy M, Alghamdi AS, Richards DJ (2020) New approach to determine the importance index for developing offshore wind energy potential sites: supported by UK and Arabian Peninsula case studies. Renew Energy 152:441–457. https://doi.org/10.1016/j.renene.2019.12.070
9. Wehbe Y, Temimi M (2021) A remote sensing-based assessment of water resources in the Arabian Peninsula. Remote Sens 13
10. Odhiambo GO (2017) Water scarcity in the Arabian Peninsula and socio-economic implications. Appl Water Sci 7:2479–2492. https://doi.org/10.1007/s13201-016-0440-1
11. Haddeland I, Clark DB, Franssen W, Ludwig F, Voß F, Arnell NW, Bertrand N, Best M, Folwell S, Gerten D et al (2011) Multimodel estimate of the global terrestrial water balance: setup and first results. J Hydrometeorol 12:869–884. https://doi.org/10.1175/2011JHM1324.1

12. Bondeau A, Smith PC, Zaehle S, Schaphoff S, Lucht W, Cramer W, Gerten D, Lotze-Campen H, Müller C, Reichstein M et al (2007) Modelling the role of agriculture for the 20th century global terrestrial carbon balance. Glob Chang Biol 13:679–706. https://doi.org/10.1111/j.1365-2486.2006.01305.x
13. Zhao M, Velicogna I, Kimball JS et al (2017) Satellite observations of regional drought severity in the continental United States using GRACE-based terrestrial water storage changes. J Clim 30:6297–6308. https://doi.org/10.1175/JCLI-D-16-0458.1
14. Eyring V, Cox PM, Flato GM, Gleckler PJ, Abramowitz G, Caldwell P, Collins WD, Gier BK, Hall AD, Hoffman FM et al (2019) Taking climate model evaluation to the next level. Nat Clim Chang 9:102–110. https://doi.org/10.1038/s41558-018-0355-y
15. Wada Y, Gleeson T, Esnault L (2014) Wedge approach to water stress. Nat Geosci 7:615–617. https://doi.org/10.1038/ngeo2241
16. Rodell M, Famiglietti JS, Wiese DN, Reager JT, Beaudoing HK, Landerer FW, Lo M-H (2018) Emerging trends in global freshwater availability. Nature 557:651–659. https://doi.org/10.1038/s41586-018-0123-1
17. Aloui S, Zghibi A, Mazzoni A, Elomri A, Triki C (2023) Groundwater resources in Qatar: a comprehensive review and informative recommendations for research, governance, and management in support of sustainability. J Hydrol Reg Stud 50:101564. https://doi.org/10.1016/j.ejrh.2023.101564
18. Amin Mir M, Waqar Ashraf M (2023) The challenges and potential strategies of Saudi Arabia's water resources: a review in analytical way. Environ Nanotechnol Monit Manag 20:100855. https://doi.org/10.1016/j.enmm.2023.100855
19. Sherif M, Liaqat MU, Baig F, Al-Rashed M (2023) Water resources availability, sustainability and challenges in the GCC countries: an overview. Heliyon 9:e20543. https://doi.org/10.1016/j.heliyon.2023.e20543
20. Samaniego L, Kumar R, Zink M (2013) Implications of parameter uncertainty on soil moisture drought analysis in Germany. J Hydrometeorol 14:47–68. https://doi.org/10.1175/JHM-D-12-075.1

Waste to Wealth: A Review on Sustainable Electromagnetic, Thermal and Acoustic Shielding Material from Waste Face Mask

Aparna Krishna, Mohammed Ouda, and Jamal Kurdi

Abstract Plastic pollution has emerged as a significant environmental challenge, intensified by the pandemic and the increased use of disposable face masks. This pollution poses significant risks to both marine and terrestrial ecosystems, impacting wildlife and human health. Addressing this problem requires robust waste management strategies, especially those centered on creative recycling methods. This review examines the environmental effects of discarded face masks, emphasizing their growing accumulation in landfills and waterways. It provides an extensive overview of recent developments in converting face masks into value-added products, such as thermal insulation, sound absorption, and electromagnetic shielding materials, highlighting the concept of "waste to wealth". The review also explores various processing techniques, types of masks, and composites incorporating metals, oxides, carbon nanotubes, and polymers. By assessing the functional properties of face mask-derived materials, this review supports sustainable approaches to managing mask waste and offers insights into developing environmentally friendly shielding and insulation solutions to help reduce ecological impacts.

Keywords Acoustic shielding material · Circular economy · Electromagnetic (EM) shielding · EM absorption · Plastic waste · Sustainable shielding material · Thermal shielding material

A. Krishna (✉)
Applied Research, Innovation and Economic Development Directorate, University of Doha for Science and Technology, Doha, Qatar
e-mail: aparna.krishna@udst.edu.qa

M. Ouda
Electrical Engineering, University of Doha for Science and Technology, Doha, Qatar

J. Kurdi
Chemical Engineering, University of Doha for Science and Technology, Doha, Qatar

S. Al-Naemi et al. (eds.), *Water and Food Security in the Face of Climate Change: Challenges and Opportunities for Resilience*, World Sustainability Series,
https://doi.org/10.1007/978-3-032-00098-9_26

1 Introduction

Plastic waste has become a significant part of municipal solid waste, posing a serious environmental challenge intensified by rapid population growth, urbanization, and industrial development. The extensive use of synthetic polymers since the mid-twentieth century has entrenched plastics in everyday life, with applications in packaging, construction, healthcare, automotive, and consumer goods [1, 2]. Their versatility, durability, and cost-effectiveness have rendered them indispensable; however, this widespread utilization has led to alarming production rates. The challenges associated with plastic waste are complex and multifaceted. Many plastics are non-biodegradable, remaining in the environment for centuries and creating long-term pollution issues. The COVID-19 pandemic has intensified this problem, resulting in a surge in the use of single-use plastics, especially personal protective equipment (PPE) like face masks (FMs) and gloves, leading to a marked increase in waste disposal. Disposable surgical masks are primarily composed of materials such as polyethylene (PE), polyurethane, and polycarbonate, with polypropylene (PP) being a key component [3]. The hydrophobic properties of this plastic make it notably resistant to degradation, causing a substantial portion to persist in the environment as microplastics that have infiltrated nearly every ecosystem, often making their way into oceans and other bodies of water [4]. In marine ecosystems, FM waste presents serious risks to various species through ingestion and entanglement, comprising a significant percentage of marine debris and endangering vulnerable species [4]. A comprehensive understanding of the environmental impact of plastic pollution, including microplastics (FM), is crucial. This should be coupled with the exploration of sustainable solutions that convert discarded plastics into valuable products, ultimately reducing plastic waste and mitigating its harmful effects on ecosystems.

The concept of a circular economy emphasizes the importance of reusing and recycling materials to reduce waste and promote sustainability. Recently, the incorporation of plastic fibers from PPE into concrete has attracted interest from both the construction industry and academia due to its sustainability benefits compared to traditional steel reinforcement [5–7]. Research has demonstrated that using PPE waste as a binder in construction materials can replace cement and enhance the strength and durability of the resulting composites. These composites have shown promising results, exhibiting superior strength and durability compared to conventional construction materials such as mud blocks and traditional cement-based products [5]. In a notable study, researchers investigated the integration of surgical FMs into mortar mixtures without any pretreatment, adding up to 5% by weight of cement [7]. They analyzed the mechanical and microstructural properties of the composite to assess its performance. Even though the strength and durability were lower than those of the reference material, they remained within acceptable limits for practical applications [7].

Although studies have examined waste PPE in the civil construction industry [6], research on using polypropylene-based FMs for shielding materials with thermal, acoustic, and EM properties remains limited. This review evaluates existing research on composite materials derived from waste FMs for these applications, focusing on the mechanisms of EM shielding, thermal and acoustic insulation, and methods for converting waste FMs into effective shielding materials.

2 Fundamentals of EM, Thermal and Acoustic Shielding Mechanisms

EM, thermal, and acoustic shielding materials play vital roles in various industries, from electronics and automotive to construction and telecommunications [8, 9].

2.1 Electromagnetic Shielding

The rapid advancements in electronic and communication technologies have resulted in a substantial increase in electronic devices across various sectors, including scientific, military, and commercial domains. This technological growth has contributed to the rise of electromagnetic pollution, commonly referred to as electromagnetic interference (EMI) [9]. EMI can significantly affect the performance and reliability of electronic devices, particularly in critical sectors such as defense and healthcare. Additionally, prolonged exposure to electromagnetic waves (EMWs) has been associated with several health concerns, including anxiety, fatigue, brain tumors, insomnia, nervousness, and leukemia [9, 10]. As a result, high-performance shielding materials are vital for effective EMI control, ensuring both the protection of electronic devices and the preservation of human health. These shielding materials are generally classified into two categories: EM shielding materials by reflection and EM shielding materials by absorption (EM absorbers). The ability of a material in shielding against EMW is measured by its shielding effectiveness (SE) which can be defined as the ratio of the strength of the incident field to the strength of the transmitted field, expressed in decibels (dB) which depends on three main mechanisms: reflection, absorption, and multiple reflections [11].

When the EMW interacts with EMI shielding material, a fraction of the incoming wave is reflected back from the rear interface of the shielded material due to the impedance mismatch between the material and the surrounding free space. A few parts of the incoming waves penetrate the shielding material, where they are further attenuated through dielectric and magnetic losses, as well as additional attenuation caused by multiple reflections within the material's internal interfaces. These processes cause the EMWs to be absorbed and reflected multiple times, leading to the dissipation of the attenuated energy as heat. Any remaining EMWs are then transmitted through the material's rear surface [11].

2.2 Acoustic Shielding

Acoustic shielding is essential for reducing noise pollution, which negatively affects human health by contributing to stress, sleep disturbances, and long-term health issues such as cardiovascular diseases [12]. Acoustic shielding operates through two primary mechanisms: sound absorption and sound insulation. Sound absorption refers to a material's ability to reduce sound reflections by absorbing sound energy, which helps minimize echoes in a space [9]. In contrast, sound insulation focuses on blocking sound transmission, effectively separating areas, and maintaining acoustic privacy. The effectiveness of acoustic materials is measured using the sound absorption coefficient (SAC), which represents the ratio of sound energy absorbed to the total incident sound energy. SAC values range from 0, indicating perfect reflection, to 1, indicating perfect absorption. A higher SAC signifies improved performance; however, it is important to note that this coefficient varies with frequency, as some materials may effectively absorb sound at high frequencies but be less efficient at lower frequencies [13].

2.3 Thermal Insulation (Shielding)

Thermal shielding is essentially important in insulating systems against unwanted heat transfer, ensuring thermal stability, and protecting sensitive components from excessive heat or cold. The effectiveness of thermal shielding is determined by thermal conductivity (λ or Keff), specific heat capacity, and the thickness of the insulating layer [14]. Thermal conductivity measures how well a material can conduct heat and the materials with low thermal conductivity resist the transfer of heat, while those with high conductivity allow heat to pass through more easily. The unit of thermal conductivity is expressed in watts per meter-kelvin (W/mK) [15]. Materials with a thermal conductivity value of less than 0.07 W/mK are typically regarded as effective thermal insulators. Conventional thermal insulators are usually produced from petrochemicals (mainly polystyrene) or highly processed natural materials such as glass and mineral wool [15].

3 State of the Art of Sustainable Thermal, Acoustic Insulation and EM Shielding Materials from Waste FM

As previously explained, the growing production and disposal of FMs present significant environmental challenges. Due to their non-biodegradable nature and the release of harmful substances when improperly discarded, waste FMs pose serious risks to both health and the environment. Therefore, recycling discarded masks is crucial to recover valuable materials that can be used as fillers in composite materials. Additionally, researchers have identified the considerable commercial potential of waste FM in developing shielding materials for electromagnetic, thermal, and acoustic applications [16, 17].

3.1 EM Shielding Material

Yu et al. developed a green and high-yield conversion procedure to upcycle waste FMs into carbon nanotube (CNT)/Ni hybrids through catalytic carbonization for microwave absorption application [18]. Initially, the waste masks were prepared by removing ear straps and nose clips, leaving only the PP fabric. The fabric was soaked in a NaOH aqueous solution to eliminate potential viruses as shown in Fig. 1. After the initial treatment, a NiCl2 solution is added to promote the formation of $Ni(OH)_2$ nanoparticles on the surfaces of PP fabrics. This is followed by the carbonization of the resulting PP-$Ni(OH)_2$-$NiCl_2$ composites at high temperatures which enhances their structural integrity and functional characteristics. To do the microwave absorption testing, the obtained CNT/Ni hybrids were dispersed into paraffin wax at a weight content of 30 wt% and pressed into coaxial rings. Scattering parameters were measured using a vector network analyzer across a frequency range of 2–18 GHz [18]. The CNT/Ni hybrids produced from 50 wt% $Ni(OH)_2$ exhibited impressive microwave absorption performance, achieving a minimum reflection loss (RLmin) of − 56.3 dB at 12.5 GHz and an effective absorption bandwidth of 4.3 GHz, ranging from 10.8 to 15.1 GHz. This method highlights a valuable approach for recycling waste masks into high-value shielding material [18].

In another study, Xiang et al. focused on the innovative use of FMs, polyethylene terephthalate microparticles (PET MPs), and MXene to create EMI shielding composites characterized by isolated conductive networks [19]. The fabrication process involved cutting FMs into pieces, which were then treated with a polydopamine (PDA) adhesive to facilitate the deposition of MXene on their surfaces. The resulting FMs-PDA-MXene composites demonstrated a remarkable EMI SE of

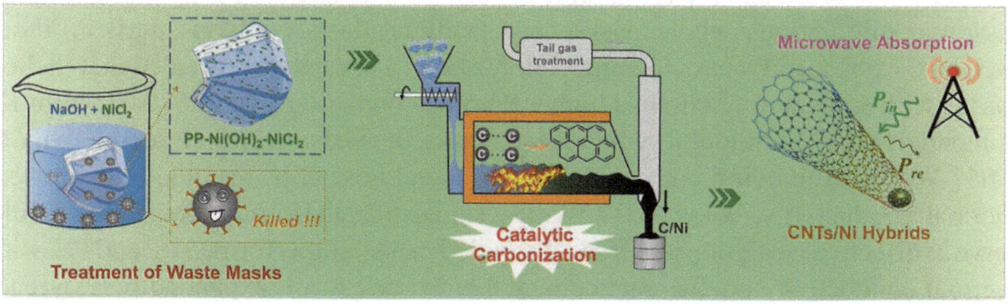

Fig. 1 Schematic representation of the catalytic carbonization of waste masks into high-value CNTs/Ni hybrids [18]

30.5 dB at 10 GHz, achieved with only 1.5 vol.% MXene content. This enhancement in EMI SE is largely attributed to the effective interfacial polarization provided by MXene, which interacts with incoming electromagnetic waves [19].

Liu et al. proposed a highly innovative approach to analyze the environmental impact of waste masks by converting them into valuable microwave absorbers [20]. This study introduced an efficient and simple method for recycling waste masks into high-value magnetic metal-doped carbon microparticles with a columnar structure. The process begins by cutting the masks into blocks, sterilizing them through ultrasonic washing with deionized water and ethanol, and then drying them. The sterilized masks are treated with concentrated sulfuric acid at 150∘C for 10 h, resulting in sulfonated mask (SM) samples. After thorough washing with DI water to achieve a near-neutral pH, the SMs are immersed in solutions of iron, cobalt, and nickel nitrates for 4 h to allow the absorption of magnetic metal ions. Following this, the original and metal absorbed SMs are carbonized in a tube furnace under an argon atmosphere at 800∘C for 2 h, yielding carbonized sulfonated masks (CSM), iron-doped CSM (CSM-Fe), cobalt-doped CSM (CSM-Co), and nickel-doped CSM (CSM-Ni) microparticles. The cobalt-doped carbon micro-particles achieved a minimum reflection loss (RLmin) of 52.8 dB at 7.0 GHz with a thickness of 3 mm, and the nickel-doped carbon materials showed an RLmin of 26.2 dB at 8.8 GHz, along with a wide effective absorption bandwidth (EAB) of 5.3 GHz at 2.4 mm. These absorption capabilities exceed those of many existing materials, making the carbonized waste masks a strong candidate for EMI shielding and other microwave absorption applications [20].

While these innovative shielding materials from waste FM support waste management and provide eco-friendly EMI shielding, challenges remain regarding the scalability of the fabrication process, long-term stability under various environmental conditions, and the need for further studies on durability and mechanical properties.

3.2 Thermal and Acoustic Shielding Material

Ali et al. [21] investigated the potential of using discarded FMs as sustainable materials for thermal insulation and sound absorption in building applications. They prepared both loose and melted samples of FMs to evaluate their properties, creating various configurations: whole masks, masks without elastic ear loops, masks without metal nose clips, and masks lacking both features. The analysis revealed that melted masks with a metal nose clip had higher thermal conductivity (0.086–0.088 W/mK) compared to those without it, while loose masks showed even lower conductivity at 0.057 W/mK. Removing both ear loops and metal nose clips resulted in the lowest thermal conductivity, below 0.04 W/mK, suggesting these modified masks could serve as effective insulation materials. Recycled masks outperformed PP and PE, with thermal conductivity coefficients comparable to rock wool, expanded polystyrene, phenol formaldehyde foam, and polyurethane foam. Sound absorption was evaluated using an impedance tube, with the loose mask sample without a metal nose clip demonstrating the highest absorption coefficient, exceeding 0.6 for frequencies above 580 Hz. In contrast, the melted samples showed lower absorption capabilities, performing better only at higher frequencies due to the loss of their porous structure [21].

In [22], the acoustic properties of PP fibers sourced from used surgical FMs were analyzed. Eighty masks were disinfected and shredded into fibers, resulting in nine samples with varying bulk densities, porosities, and thicknesses. Acoustic characterization was conducted using an impedance tube, revealing that sound absorption performance improved with increased thickness, demonstrating higher Noise Reduction Coefficient (NRC) values (0.45–0.85) and SAA values 0.43–0.83. Samples exhibited effective sound absorption, particularly at medium to high frequencies, and comparisons with commercial rock wool materials showed that the recycled fibers performed comparably or even better in some instances. These findings suggest that recycled surgical masks can serve as effective sound absorbers in building construction, presenting a sustainable alternative for applications like acoustic ceiling panels and noise barriers [22]. In another study, Fabiani et al. assessed the thermal and acoustic performance of recycled FMs for insulation, including new white, new blue, used blue, and shredded masks [23]. Panels were made by attaching masks to cardboard and securing them to a plasterboard. Thermal tests using the hot disk method showed conductivity values from 0.0425 W/mK for shredded masks to 0.0539 W/mK for new white masks, all classified as thermally insulating. Shredded masks had higher thermal diffusivity due to increased voids. Acoustic performance was evaluated using an impedance tube across various thicknesses, ranging from 0.5 to 3 cm. The blue masks demonstrated the highest absorption coefficients at frequencies above 2300 Hz, while shredded masks excelled at frequencies over 4000 Hz at a thickness of 3 cm. Notably, the absorption coefficients for the 0.5 cm thickness exceeded 0.6 at frequencies above 2300 Hz [23].

Recently, in [16], environmentally friendly sound-absorbing, and thermal insulating panels were developed using discarded waste polypropylene FMs. The discarded polypropylene FMs were shredded into fibrous mass using a Garnet machine. These fibers were then mixed with a bio-based binder called Serishoom, and different panel samples with varying thickness and diameter were fabricated using response surface methodology (RSM) to test their thermal and acoustic properties as shown in Fig. 2. These prepared samples were used for SAC measurements and thermal conductivity testing. The sound absorption properties were tested using an impedance tube, and the sound absorption average (SAA) was calculated for frequencies between 200 and 2500 Hz. The guarded hot-plate method was used to measure the thermal conductivity of the panels. The panels' SAA ranged from 0.28 to 0.53, and their Keff was measured between 0.031 and 0.056 W/mK, demonstrating high performance in both acoustic and thermal insulation [16].

From this review, it is clear that utilizing discarded FMs in the development of innovative shielding and insulation materials not only addresses the pollution problem associated with mask waste but also contributes to reducing environmental impact, making this approach a sustainable alternative in material development.

Fig. 2 Procedure of synthesis of Waste FM as thermal and acoustic insulation material [16]

4 Conclusion

This review provides a comprehensive analysis of composites developed from discarded FMs, highlighting their potential for applications in electromagnetic shielding and thermal and acoustic insulation. The use of FMs during the pandemic has significantly reduced virus transmission but has also contributed to substantial plastic pollution. Transforming these waste masks into new materials can mitigate environmental impact and address the pressing issue of mask waste. The review details essential information on the collection of waste masks, various fabrication techniques, and the resulting properties of the composites, which exhibit promising thermal conductivity, sound absorption capabilities, and effective EM shielding properties that are essential for protecting sensitive electronics. Further research is necessary to evaluate their performance under varying environmental conditions, such as temperature fluctuations, humidity, and repeated electromagnetic exposure, to ensure their long-term effectiveness in real-world applications.

Acknowledgements The Research reported in this publication, was supported by the Qatar Research Development and Innovation Council [ARG01-0504-230068]. The content is solely the responsibility of the authors and does not necessarily represent the official views of QRDI.

References

1. Cabrera FC (2021) Eco-friendly polymer composites: a review of suitable methods for waste management. Polym Compos 42(6):2653–2677
2. Williams AT, Rangel-Buitrago N (2022) The past, present, and future of plastic pollution. Mar Pollut Bull 176:113429
3. Abbasi SA, Khalil AB, Arslan M (2020) Extensive use of face masks during covid-19 pandemic: (micro-)plastic pollution and potential health concerns in the Arabian Peninsula. Saudi J Biol Sci 27(12):3181–3186
4. Dharmaraj S, Ashokkumar V, Hariharan S, Manibharathi A, Show PL, Chong CT, Ngamcharussrivichai C (2021) The covid-19 pandemic face mask waste: a blooming threat to the marine environment. Chemosphere 272:129601
5. Mohan HT, Jayanarayanan K, Mini KM (2022) A sustainable approach for the utilization of PPE biomedical waste in the construction sector. Eng Sci Technol Int J 32:101060
6. Kilmartin-Lynch S, Saberian M, Li J, Roychand R, Zhang G (2021) Preliminary evaluation of the feasibility of using polypropylene fibres from covid-19 single-use face masks to improve the mechanical properties of concrete. J Clean Prod 296:126460
7. Castellote M, Jiménez-Relinque E, Grande M, Rubiano FJ, Castillo A (2022) Face mask wastes as cementitious materials: a possible solution to a big concern. Materials 15(4)
8. Kondawar SB, Modak PR (2020) Theory of EMI shielding. In: Materials for potential EMI shielding applications. Elsevier, pp 9–25
9. Mago J, Sunali, Negi A, Stuart Bolton J, Fatima S (2024) Sound insulation: key concepts and technologies. Springer Nature Singapore, Singapore, pp 1–44
10. Redlarski G, Lewczuk B, Żak A, Koncicki A, Krawczuk M, Piechocki J, Jakubiuk K, Tojza P, Jaworski J, Ambroziak D et al (2015) The influence of electromagnetic pollution on living organisms: historical trends and forecasting changes. Biomed Res Int 2015(1):234098
11. Lakshmanna B, Nagalaksmi K, Madakka M (2023) Mobile phone and base stations radiation and its effects on human health and environment: a review. Sustain Technol Entrepr 2(2):100031
12. Cheng J, Li C, Xiong Y, Zhang H, Raza H, Ullah S, Wu J, Zheng G, Cao Q, Zhang D, Zheng Q, Che R (2022) Recent advances in design strategies and multifunctionality of flexible electromagnetic interference shielding materials. Nano-Micro Lett 14(1)
13. Tao Y, Ren M, Zhang H, Peijs T (2021) Recent progress in acoustic materials and noise control strategies—a review. Appl Mater Today 24:101141
14. Massoudinejad M, Amanidaz N, Santos RM, Bakhshoodeh R (2019) Use of municipal, agricultural, industrial, construction and demolition waste in thermal and sound building insulation materials: a review article. J Environ Health Sci Eng 17:1227–1242
15. Halashi K, Taban E, Soltani P, Amininasab S, Samaei E, Moghadam DN, Khavanin A (2024) Acoustic and thermal performance of luffa fiber panels for sustainable building applications. Build Environ 247:111051
16. Asdrubali F, D'Alessandro F, Schiavoni S (2015) A review of unconventional sustainable building insulation materials. Sustain Mater Technol 4:1–17
17. Dehdashti Z, Soltani P, Taban E (2024) Utilizing discarded face masks to fabricate sustainable high-performance panels for enhanced building thermal and acoustic comfort. J Clean Prod 446:141304
18. Neri M, Cuerva E, Levi E, Pujadas P, Müller E, Guardo A (2023) Thermal, acoustic, and fire performance characterization of textile face mask waste for use as low-cost building insulation material. Dev Built Enviro 14:100164

19. Yu R, Wen X, Liu J, Wang Y, Chen X, Wenelska K, Mijowska E, Tang T (2021) A green and high-yield route to recycle waste masks into CNTS/NI hybrids via catalytic carbonization and their application for superior microwave absorption. Appl Catal B 298:120544
20. Xiang M, Yang R, Zhuang H, Jing W, Liu C, Yang Z, Dong S (2023) Electromagnetic interference shielding composites obtained from waste face masks, mxene and polyethylene terephthalate micro plastics. J Hazard Mater Adv 12:100365
21. Liu J-T, Zheng Y-C, Hou X, Feng X-R, Jiang K, Wang M (2024) Constructing magnetic metal doping carbon micro-particles with columnar structure via carbonization of waste masks for microwave absorption. Polymer 294:126739
22. Ali M, Almuzaiqer R, Al-Salem K, Alabdulkarem A, Nuhait A (2021) New novel thermal insulation and sound-absorbing materials from discarded facemasks of covid-19 pandemic. Sci Rep 11(1):23240
23. Maderuelo-Sanz R, Acedo-Fuentes P, García-Cobos FJ, Sánchez-Delgado FJ, Mota-López MI, Meneses-Rodríguez JM (2021) The recycling of surgical face masks as sound porous absorbers: preliminary evaluation. Sci Total Environ 786:147461
24. Fabiani C, Cavagnoli S, Chiatti C, Pisello AL (2022) Management of disposable surgical masks for tackling pandemic-generated pollution: thermo-acoustic investigations and life cycle assessment of novel recycled building panels. Resour Conserv Recycl 186:106509

Food Security and Sustainability

A Deep Neural Network for Tomato Yield Prediction in Controlled Environment Agriculture

Hassan Ali, Rachid Benlamri, Aitazaz A. Farooque, Raziq Yaqub, Fahim Ullah Khan, and Ahmed Badawi

Abstract To optimize agricultural operations and guarantee food security, it is imperative to accurately predict drop yields. This paper introduces a deep neural network (DNN) model for predicting tomato yield in a controlled environment agriculture (CEA) system. The model capitalizes on a high-dimensional dataset that encompasses a diverse array of environmental and agronomic variables. With a test root mean square error (RMSE) of 0.07, mean absolute error (MAE) of 0.05, and R-squared (R^2) of 0.96, the DNN model exhibited high predictive accuracy. The cross-validated RMSE of 0.07 (95% Confidence Interval (CI): 0.03–0.12), MAE of 0.06 (95% CI: 0.01–0.12), and R^2 of 0.82 (95% CI: 0.56–1.08) further validated the model's performance. These results indicate the potential of the proposed DNN model to significantly enhance precision agriculture practices.

Keywords Tomato yield prediction · Deep Neural Network (DNN) · Controlled Environment Agriculture (CEA) · Precision agriculture · Predictive modeling

---

H. Ali (✉) · R. Benlamri
Centre of Excellence for Sustainability and Food Security, University of Doha for Science and Technology, Doha, Qatar
e-mail: hassan.ali@udst.edu.qa

H. Ali · A. Badawi
Department of Electrical Engineering, University of Doha for Science and Technology, Doha, Qatar

A. A. Farooque
Canadian Centre for Climate Change and Adaptation, University of Prince Edward Island, Charlottetown, PE, Canada

R. Yaqub
Department of Electrical Engineering and Computer Science, A&M University, Huntsville, AL, USA

F. U. Khan
Department of Agriculture, Hazara University Dodhial Mansehra, Khyber Pakhtunkhwa, Pakistan

S. Al-Naemi et al. (eds.), *Water and Food Security in the Face of Climate Change: Challenges and Opportunities for Resilience*, World Sustainability Series,
https://doi.org/10.1007/978-3-032-00098-9_27

1 Introduction

The global agriculture sector faces increasing pressure to boost food production through sustainable methods. Machine learning (ML) has emerged as a powerful tool for precision agriculture, particularly in crop yield prediction and disease detection. ML algorithms can capture complex, non-linear relationships in agricultural data, outperforming traditional statistical models [1]. Various ML techniques, including Random Forest, XGBoost, and Support Vector Machines, have demonstrated high accuracy in predicting crop yields [2, 3]. These models utilize diverse input parameters such as weather conditions, soil fertility, and pesticide usage [4, 5].

Recent advancements in deep learning (DL) have created new opportunities for predicting crop yields in controlled environment agriculture (CEA), particularly for tomatoes [6]. Deep neural networks (DNNs), particularly convolutional neural networks (CNNs) and long short-term memory (LSTM) models, have demonstrated potential in capturing the intricate relationships between environmental variables and agricultural productivity [7, 8]. These models can incorporate a wide range of predictors, including climate parameters, soil conditions, and real-time sensor data, to enhance prediction accuracy [9, 10]. However, challenges remain, such as the risk of overfitting [9]. Additionally, data imbalance—where yield outcomes are available in the last phase of the growing cycle—can badly impact the model's predictions and reduce its generalizability.

To address the above challenges, this study focuses on employing DNNs with dropout regularization [11], cross-validation [12, 13], and data stratification [14, 15] techniques. Dropout layers help mitigate overfitting by randomly deactivating a fraction of neurons during training, ensuring the model does not become overly dependent on specific patterns. Cross-validation provides a reliable measure of model performance across different data subsets, thereby enhancing the model's robustness. Stratification ensures that the training data accurately represents the full range of yield outcomes, particularly important in cases of imbalanced datasets where yield outcomes are missing.

A DNN model with the proposed strategies is developed and evaluated for predicting tomato yield. For a CEA tomato crop, the model incorporates fifty-seven average weekly predictor variables for 24 weeks (a small training dataset). A number of measures, including as root mean square error (RMSE), mean absolute error (MAE), and R-squared (R^2), are used to evaluate the model's performance.

Rest of the paper is organized as follows: Sect. 2 describes the methodology, detailing the data collection, preprocessing, DNN model architecture and training and evaluation. Section 3 covers the results, presenting the performance metrics and 95% confidence intervals (CI) for the cross-validated performance measures of the DNN model. Finally, Sect. 4 concludes the paper by summarizing the key findings.

2 Methodology

2.1 Data Collection

The dataset was generated during the second edition of Wageningen University & Research (WUR) Autonomous Greenhouse Challenge by the team AICU [16]. The dataset includes information on the following: the status of actuators, climate setpoints, resource consumption, harvest, crop-related parameters, tomato quality, analysis of irrigation and drain samples, and root-zone/slab information. Additionally, it includes data on the outdoor and interior greenhouse climate. This data was collected during a six-month cherry tomato production (cv. Axiany) in a high-tech glasshouse compartment at the Wageningen Research Centre in Bleiswijk (The Netherlands). Raw and processed data are included in the dataset. Raw data were collected through manual registrations, climate and irrigation process computers, weather stations, and climate measuring devices and sensors. The authors transformed this data set into the average weekly data for the 24 weeks. In kilograms, the tomato yield per square meter is the dependent variable, which is assessed on a weekly basis. For the work in this paper, the AICU dataset was converted to average weekly data for the 24 weeks. The dependent variable is the tomato Class A yield (ProdA) in Kg per square meter, assessed on a weekly basis. Please note that Class A means first quality trusses that can be commercially traded.

2.2 Data Preprocessing

The dataset underwent preprocessing to address missing values, outliers, and inconsistencies in the data. The StandardScaler from scikit-learn was used to standardize the continuous variables. This was done to ensure that all the features were on the same scale, which is important for the convergence of the DNN model. Subsequently, the data was divided into training and testing sets with stratified sampling to maintain the distribution of the target variable.

2.3 Model Architecture

The DNN model (shown in Fig. 1) was implemented via TensorFlow's Keras API. The model starts with an input layer that receives the predictors of the dataset. The input layer, which comprises 58 nodes, is coupled to the first hidden layer, which comprises 32 neurons. Every neuron in this layer employs the ReLU (Rectified Linear Unit) activation function, which provides non-linearity into the model, allowing it to acquire intricate patterns and relationships within the data.

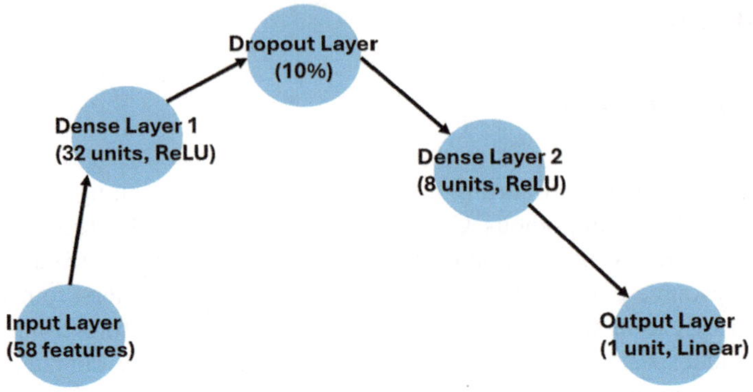

Fig. 1 DNN model architecture

Following the first hidden layer, a dropout layer is added, with a dropout rate of 0.1. This layer functions as a regularization approach by randomly setting 10% of the neurons to zero during the training process. This procedure mitigates overfitting by ensuring that the model does not excessively depend on particular neurons and instead acquires more generalized properties. The second hidden layer consists of 8 neurons, which also utilize the ReLU activation function. The second hidden layer enhances the knowledge obtained from the previous hidden layer, enabling the model to improve its comprehension of the data. Finally, the network is completed by an output layer that comprises a solitary neuron. This neuron is accountable for generating the ultimate prediction, which in this instance is the anticipated tomato production. The model is assembled with the Adam optimizer, a widely favoured option in deep learning because to its efficacy and capacity to dynamically adjust the learning rate during training. The utilized loss function is mean squared error (MSE), which calculates the average of the squared discrepancies between the expected and actual values. Through the process of minimizing this loss function, the model strives to attain the highest level of precision in its predictions. In summary, this DNN structure is specifically created to effectively record and analyze intricate connections within the data. Additionally, the incorporation of dropout layers serves to improve the model's resilience and capacity to apply its findings to new situations.

2.4 Model Training and Evaluation

The dataset was split into training and testing sets. The model was trained on the training dataset using 100 epochs and a batch size of 2. Cross-validation was performed using a fivefold method to assess the model's robustness. RMSE, MAE, and R^2 metrics were used for comprehensive evaluation of DNN model's performance, capturing both the accuracy of its predictions and its explanatory power. The three-performance metrics are described below.

1. RMSE is a commonly employed measure for assessing the precision of a prediction model. RMSE is a statistical measure that calculates the square root of the average of the squared differences between the observed and predicted values. This metric is highly responsive to significant errors, as it calculates the square of the differences before taking the average, so assigning greater importance to larger errors. Mathematically, the RMSE is precisely defined as:

$$RMSE = \sqrt{\frac{1}{n}\sum_{i=1}^{n}\left(y_i - \hat{y}_i\right)^2}, \tag{1}$$

where y_i represents the actual values, $\hat{y}_i$ represents the predicted values, and n is the total number of observations.

2. MAE is a metric commonly employed to assess the precision of a model's predictions. MAE, in contrast to RMSE, prioritizes the average size of the mistakes, irrespective of their orientation. The term "average of the absolute differences" refers to the calculation of the mean value obtained by subtracting the projected values from the actual values and taking the absolute value of each difference. The mathematical expression for MAE is:

$$MAE = \frac{1}{n}\sum_{i=1}^{n}\left|y_i - \hat{y}_i\right|. \tag{2}$$

This metric provides a straightforward interpretation of the model's prediction accuracy, as it directly measures the average error.

3. R^2, often known as the coefficient of determination, is a statistical metric that quantifies the proportion of the variability in the dependent variable that can be explained by the independent variables. It primarily evaluates the model's fit by comparing it to a baseline model that predicts the average value of the target variable for all observations. The equation for the coefficient of determination, R^2, is as follows:

$$R^2 = 1 - \frac{\sum_{i=1}^{n}\left(y_i - \hat{y}_i\right)^2}{\sum_{i=1}^{n}\left(y_i - \overline{y_i}\right)^2}, \tag{3}$$

where $\overline{y}$ is the mean of the actual values. An R^2 value of 1 indicates that the model explains all the variance in the target variable, while a value of 0 indicates that it explains none.

Furthermore, CIs were calculated for these metrics to comprehend the variability of the model's performance across various data splits. The CIs offer further understanding on the dependability of the model's predictions. The low confidence ranges for the RMSE and MAE indicate that the model consistently performs well. However, the wider range for the R^2, indicates areas where the model might potentially be enhanced, such as accurately capturing the variability in the data.

3 Results

The DNN model exhibited strong predictive performance, as shown in Fig. 2.

The outcomes derived from the DNN regression model exhibit a significant degree of precision in forecasting tomato yield. The model's predictions exhibit a high level of accuracy, as evidenced by a Test RMSE of 0.07 and Test MAE of 0.05, indicating a close proximity to the actual observed yields. The R-squared value of 0.96 for the test implies that the model is able to account for 96% of the variation in tomato yield, demonstrating its high predictive ability.

Figure 3 displays the 95% CI for the cross-validated performance measures of the DNN model used for predicting tomato yield. The CI of each metric offers valuable information about the variation and dependability of the model's predictions across many folds in the cross-validation procedure. The dots represent the mean values, while the error bars indicate the upper and lower confidence intervals.

The mean value of the RMSE is 0.07, with a 95% confidence interval (CI) of 0.03–0.12. The model's error in predicting tomato yield remains consistent across various subsets of the data, as indicated by the relatively limited range of the RMSE CI. The model has the potential to produce exceptionally accurate predictions in certain scenarios at a minimum threshold of 0.03, while the maximum threshold of 0.12 shows that the prediction error remains within an acceptable range even in the worst-case scenario. This serves as an illustration of the model's ability to maintain minimal prediction errors across a variety of data distributions.

The average value of the MAE is 0.06, with a 95% CI ranging from 0.01 to 0.12. The CI for MAE likewise has a limited range, resembling that of RMSE, with a minimum value as small as 0.01. This implies that the model could provide highly precise predictions with minimal absolute errors in certain situations. The maximum value of 0.12 corresponds to the RMSE, which confirms that the model consistently

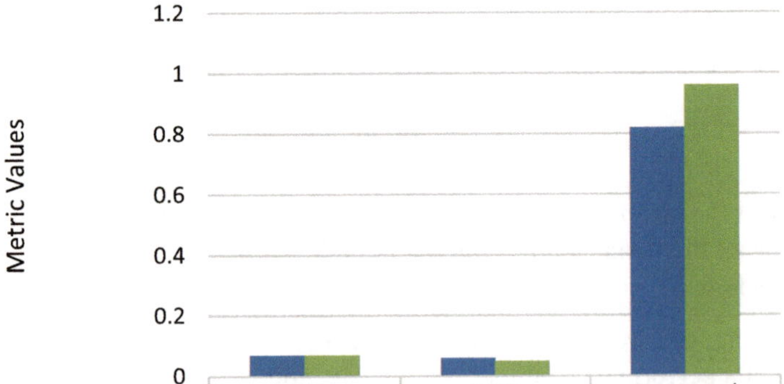

	RMSE	MAE	R-squared
Cross-validation	0.07	0.06	0.82
Test	0.07	0.05	0.96

Fig. 2 DNN Regression performance metrics

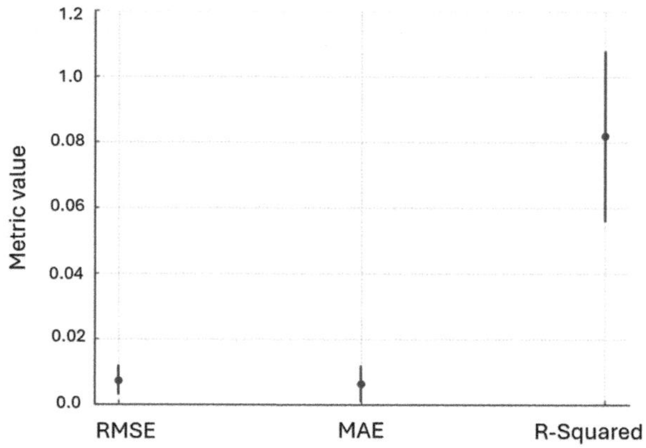

Fig. 3 Cross-validated metrics with 95% CIs for the DNN model

maintains low prediction errors. The low MAE values highlight the model's capacity to accurately forecast the actual tomato output with minimum variation.

The average R^2 value is 0.82, with a 95% CI ranging from 0.56 to 1.08. The CI for R^2, which quantifies the amount of variation accounted for by the model, spans from 0.56 to 1.08. The minimum value of 0.56 suggests that, even in the most unfavorable situation, the model can account for more than half of the variability in the tomato production data, which is considered a robust outcome. The upper limit barely surpasses 1, which is a typical when interpreting R^2, however it could be due to small overfitting in certain folds during cross-validation. In general, the confidence interval indicates that the model effectively captures the fundamental patterns in the data, however there may be some variation based on the subset of data.

4 Conclusion

Utilizing weekly environmental and agronomic predictors, we created and assessed a DNN model for tomato yield prediction. To capture intricate patterns in the data and prevent overfitting, the model was meticulously constructed with multiple layers, including dense and dropout layers. The final model exhibited a robust predictive capability, as evidenced by a test RMSE of 0.07, a test MAE of 0.05, and a test R^2 value of 0.96. As indicated by these findings, the model is highly accurate, capable of accounting for 96% of the variance in the tomato yield data.

The model's reliability was further bolstered by the cross-validation procedure, as the cross-validated RMSE and MAE values closely matched those observed in the test set. The cross-validated R^2 value of 0.82, together with its 95% confidence interval of 0.56–1.08, suggests a robust model performance. However, the R^2 metric's variability suggests that the model's explanatory power may differ depending on the specific training subset used.

Above findings emphasize the importance of employing sophisticated ML methods, including DNNs, to predict agricultural yields. The model's potential utility for farmers and agronomists in the planning and optimization of agricultural production is underscored by its high accuracy and generalizability. Additionally, the consistent reliability of the model's predictions across various data divisions is indicated by the low CIs for RMSE and MAE.

This study is subject to numerous constraints, despite the model's exceptional performance. The model was trained and validated using historical data, which may not completely capture future environmental variations or unforeseen challenges in tomato production. Furthermore, although the model's accuracy is high, the minor variability in the R^2 CI indicates that the model could benefit from additional data or further fine-tuning.

Acknowledgements This research was supported by the Qatar Research, Development and Innovation (QRDI) Council through the Grant No. MME03-1121-210025.

References

1. Geetha K, Vidhya D, Kiran A (2023) An extensive study on precision farming based on crop yield using integrated approaches to learning. In: Proceedings of the 2023 international conference on research methodologies in knowledge management, artificial intelligence and telecommunication engineering (RMKMATE). IEEE, pp 1–8. https://doi.org/10.1109/RMKMATE59243.2023.10369738
2. Shawon SM, Barua Ema F, Mahi AK, Mohsin Sarker Raihan M (2023) Crop yield prediction: robust machine learning approaches for precision agriculture. In: Proceedings of the 2023 26th international conference on computer and information technology (ICCIT). IEEE, pp 1–6. https://doi.org/10.1109/ICCIT.2023.10369738
3. Gupta S, Geetha A, Sankaran KS, Zamani AS, Ritonga M, Raj R, Ray S, Mohammed HS (2022) Machine learning- and feature selection-enabled framework for accurate crop yield prediction. J Food Qual. https://doi.org/10.1002/jfq.12345
4. Dureen T, Rayen V, Pushpalatha D, Kavitha D (2023) Effective crop yield prediction using gradient boosting to improve agricultural outcomes. In: Proceedings of the 2023 international conference on networking and communications (ICNWC). IEEE, pp 1–6. https://doi.org/10.1109/ICNWC.2023.12345678
5. Salokhe MR (2023) Machine learning: applications in agriculture (crop yield prediction, disease and pest detection). Int J Adv Res Sci Commun Technol. https://doi.org/10.1234/ijarsct.2023.56789
6. Ojo MO, Zahid A (2022) Deep learning in controlled environment agriculture: a review of recent advancements, challenges and prospects. Sensors (Basel, Switzerland) 22. https://doi.org/10.3390/s22020678
7. Alhnaity B, Pearson S, Leontidis G, Kollias SD (2019) Using deep learning to predict plant growth and yield in greenhouse environments. ArXiv, abs/1907.00624
8. Gong L, Yu M, Jiang S, Cutsuridis V, Pearson S (2021) Deep learning based prediction on greenhouse crop yield combined TCN and RNN. Sensors (Basel, Switzerland) 21. https://doi.org/10.3390/s21072654
9. Oikonomidis A, Catal C, Kassahun A (2022) Deep learning for crop yield prediction: a systematic literature review. N Z J Crop Hortic Sci 51:1–26. https://doi.org/10.1080/01140671.2022.2087534

10. Jeevan Nagendra Kuma Y, Chandan R, Harsh Somanini S, Vadtya S, Ram Lohit Pranay Y, Mohammed KA, Chandrashekar R, Kansal L, Kalra R (2024) Predictive modeling for enhanced plant cultivation in greenhouse environment. In: E3S web of conferences. https://doi.org/10.1051/e3sconf/2024.123456
11. Kim N, Ha K-J, Park N-W, Cho J, Hong S, Lee Y-W (2019) A comparison between major artificial intelligence models for crop yield prediction: case study of the Midwestern United States, 2006–2015. ISPRS Int J Geo Inf 8(5):240. https://doi.org/10.3390/ijgi8050240
12. Peryanto YA, Umar R (2020) Klasifikasi Citra Menggunakan convolutional neural network dan K fold cross validation. J Appl Inf Comput 4(1):45–51. https://doi.org/10.30871/jaic.v4i1.2017
13. Mojarad SA et al (2011) Cross validation evaluation for breast cancer prediction using multilayer perceptron neural networks. Am J Eng Appl Sci 4(4):576–585. https://doi.org/10.3844/ajeassp.2011.576.585
14. Sadaiyandi J, Arumugam P, Sangaiah AK, Zhang C (2023) Stratified sampling-based deep learning approach to increase prediction accuracy of unbalanced dataset. Electronics 12(21):4423. https://doi.org/10.3390/electronics12214423
15. Wikipedia Contributors (2024) Stratified sampling. Wikipedia. https://en.wikipedia.org/wiki/Stratified_sampling
16. Hemming S, de Zwart F, Elings A, Petropoulou A, Righini I (2020) Autonomous greenhouse challenge, 2nd edn. Wageningen University & Research. https://doi.org/10.4121/uuid:88d22c60-21b3-4ea8-90db-20249a5be2a7

Challenges and Opportunities in AI-Driven Supply Chain Sustainability: A Review Article

Ghoson Abdulaziz AL-Obaidly

Abstract This tries to review challenges and opportunities in AI-driven supply chain sustainability comprehensively. Thus, qualitative research aims to study the implications and possibilities of Artificial Intelligence (AI) technologies associated with supply chain sustainability (SCS). Sustainability is one of the most crucial components of modern ethical and social practices due to environmental problems, such as global warming and resource depletion. At the same time, SCS is applied to attain economic sustainability and maintain competitive business models with environmental and social responsibility. The AI technologies can project demands, record optimization, predictive maintenance, and supplier risk management to facilitate supply chain operations. So, AI significantly optimizes supply chain operations by facilitating decision-making processes and saving suppliers' and customers' time and money. However, the trained AI model could be unfair or unethical and may lack transparency and accountability, thereby raising the privacy issues of both suppliers and customers. The integration of AI technologies into SCS also presents several opportunities for organizations. For instance, demand forecasting may be enhanced with AI by using historical data to predict trends in business, which will affect inventory handling. It will allow the avoidance of overstock situations and stockout shelves, subsequently saving costs and inventory wastage for organizations. Another important aspect is that the production demand fluctuations for a particular product could be predicted, and adequate inventory levels could be maintained. It is also helpful for better management of sourcing and supplier relations. It will help organizations to pursue 'green' sourcing practices, rely on more ethical sourcing solutions, and optimize their supply chains.

Keywords Artificial intelligence · Environmental challenges · Opportunities · Supply chain · Sustainability

G. A. AL-Obaidly (✉)
University of Doha, Doha, Qatar
e-mail: ghoson.alobaidly@udst.edu.qa

S. Al-Naemi et al. (eds.), *Water and Food Security in the Face of Climate Change: Challenges and Opportunities for Resilience*, World Sustainability Series,
https://doi.org/10.1007/978-3-032-00098-9_28

1 Introduction

1.1 Background and Significance

The notion of SCS is increasingly emphasized as crucial for businesses in the twenty-first century. The fact that the world's population is becoming aware of environmental problems, such as climate change, depletion of resources, pollution, and new legislative actions that expect businessmen to be more socially responsible and ethical, drove a huge number of organizations to reconsider their supply chain strategies. The SCS is mainly targeted at cost reduction, efficiency, and service improvement. It involves environmental and social dimensions, implying that a business should implement activities that prevent environmental impacts and positively affect the well-being of the population [1]. This is a passionate update to the perception of sustainability of the supply chain, which encounters many issues associated with environmental harm and disregard for human labor. At the same time, AI and Machine Learning (ML) are extraordinary technologies that affect the functioning of a huge number of industries, including SCS [2]. AI provides numerous opportunities for SCS to become more sustainable through better data processing, predictive analytics, and automation. Thus, the interplay of SCS and AI is highly relevant and provides fertile grounds for future development.

1.2 Problem Statement

The introduction of AI in the management of the supply chain marks a period of transformation in which the traditional boundaries are transcended by technology, providing unprecedented chances for optimization, as well as efficiency. Such evolution is not just a transformation of technology but an elementary change in how supply chains are conceived, designed, and operated. A change from the conventional procedures to the solutions driven by AI in SCM reflects a wide narrative of technological advancement and the effect that it has on global commerce [6]. However, the AI integration in supply chain management can be characterized by AI's ability to analyze large datasets, predict trends, and automate critical decision-making procedures. According to [13], there is a paradigm transition brought by AI in the fast world of business, with the effect of AI extending beyond the operational efficiency towards strategic business planning, as well as competitive advantage. However, AI's significance is underscored by [13], in improving the strategic, as well as, creative potentials of the human resources, thus augmenting the human intelligence instead of replacing it.

Supply chain management's landscape is going through a metamorphic change that is navigated by the quick advancement, as well as integration of the AI technologies. Such evolution is improving the present operation along with paving the path for upcoming trends that promise to reevaluate the efficiency, flexibility, and sustainability of the supply chains. But, implementing AI also has its challenges like gaps in skill, ethical issues, as well as complexities related to integration of data. Such issues necessitate strategic navigation as well as organizational preparedness [6]. However, despite the challenges posed by AI, there are bright prospects for AI in the supply chain with the advancements in the technology expected to offer new chances for innovation, flexibility, and competitiveness. For the practitioners, the adoption of a strategic approach towards the integration of AI is critical. It requires investment in AI technologies that align with the objectives of the business, fostering an innovative culture, and creating digital competencies within the workforce. Moreover, the policymakers would need to focus on establishing a conducive regulatory setting that would provide support for the adoption of AI, along with ensuring the protection of data and ethical standards [14]. There is a need for collaborative efforts among the industry, academics, and the government to drive innovation in the supply chain and address the challenges associated with the integration of AI.

According to Esan [5], the integration of AI in the supply chain poses various challenges and barriers. These challenges can be divided into the following categories: technical challenges, organizational challenges, and data challenges. Technical challenges are related to the complexity of AI integration with the present IT infrastructure, issues regarding the quality of data, and the need for specific skills for developing, implementing, and handling AI applications. According to Orikpete and Ewim [11], the organizational challenges focus on aligning the AI initiatives with the goals of business, handling change, and guaranteeing stakeholder's buy-in. Data challenges are related to privacy of data, security, as well as compliance with the regulatory demands. To overcome these challenges, there is a need for a strategic approach to the integration of technology, investment in the development of skills, and focus on the practices related to the management of data. Thus, integrating AI into the supply chain presents a transformative change towards a more intelligent, flexible and efficient management of the supply chain. Existing research underscores AI's potential in revolutionizing the operations of the supply chain, making them adaptable towards the changing conditions of the market, as well as the needs of the customer [12]. Since such technologies are continuously evolving, growth is expected in their adoption in the supply chain, providing opportunities for competitive advantage, as well as innovation [5]. Exploring how AI may enhance supply chain sustainability is a major research avenue, aligning with the global efforts for achieving environmental, as well as social sustainability.

1.3 Objectives of the Research

The key purpose of this research is to discover potential challenges and opportunities regarding the bigger use of AI in supply chains. Therefore, by synthesizing the findings of multiple studies, the present work aims to provide an extensive analysis of the current situation and prospects. The primary objectives of the research are as follows:

1. This study examines the key principles of sustainable supply chain practices, and the challenges organizations experience during implementation. The research stresses the need for balanced economic, environmental, and social sustainability approaches.
2. Is to examines how AI may improve supply chain efficiency, responsiveness, and sustainability. This includes assessing present uses, advantages, and supply chain optimization prospects.
3. Is to investigate the data management, ethics, and talent shortages in AI-integrated sustainable supply chains. It also discusses AI's potential in demand forecasting, inventory management, and logistics, as well as future trends, technologies, and research paths.

1.4 The Rationale of the Study

The field of AI is rapidly evolving, with the potential to transform people's lives and a variety of industries. At the same time, the application of AI in SCS is especially appealing. One of the reasons why this study is relevant is because it primarily observes pressing environmental issues that have put sustainability at the core of modern ethical and social practice. Thus, sustainability represents a critical factor, primarily in supply chains. However, ensuring supply chain sustainability is challenging and associated with specific barriers. In particular, fostering collaboration and managing all supply chain networks with diverse stakeholders is a complex task [7, 17]. Whereas the second rationale is the importance of AI and trusting that AI technologies represent frequent openings for promoting SCS. Indeed, a variety of AI technologies used for demand prediction, catalogue optimization, contractor risk organization, and predictive maintenance offer enormous opportunities for improving efficiency and lowering costs [17]. On the contrary, deploying AI in supply chains is challenging due to a variety of reasons. The first set of potential challenges includes crucial issues of fairness, transparency, and accountability. The other is associated with the potential implications of using AI on customer privacy. Thus, the core rationale is the dual nature of AI integration, and the study aims to investigate how its positive effects can be enhanced to promote SCS [3]. The implications of AI on SCS are both challenging and opportune. While one type of application can lead to better demand forecasting and optimized inventory and sourcing, the other can result in substantial equity and privacy issues. The study will aim to investigate available AI solutions to these issues by suggesting scenarios of their use and ways to mitigate the adverse effects of deploying AI in supply chains [7].

2 Theoretical Background

2.1 Defining Supply Chain Sustainable Management

The SCS is understood as the management of material, data, and wealth movements and assistance between the corporations along the network while taking goals from all three dimensions of sustainable growth, specifically monetary, conservational, and societal, into account [4]. The purpose of it is to generate, guard, and produce long-term eco-friendly, communal, and financial value for all shareholders who are associated with transporting goods and amenities to the marketplace [1].

2.2 AI Application in Supply Chain Sustainability

The main applications of AI in SCS include demand forecasting, inventory management, supplier risk assessment, and maintenance planning [2]. AI and ML leverage the patterns in data to provide more accurate forecasts, optimize supply chain activities, and even automate several processes [5]. The deployment of technologies to enhance the efficiency of SCS applications allows for better planning and a level of satisfaction with all components of the supply chain [6]. A detailed explanation of these applications is provided as follows: In this application, AI and ML algorithms analyze historical data on sales or other relevant performance indicators and make the most accurate prediction of future demand based on the patterns they recognize. As a result, organizations can optimize the required inventory levels, prevent stockouts and overstock when excess products must be sold off in retail chains at lower prices, and make sure that a product is available [3, 7]. AI systems can monitor the current levels of inventory, recognize demand trends, and even place orders for new goods automatically. This way, organizations can reduce holding costs and minimize their waste. When a product is available in the supply chain on time, there is no risk of it turning to waste because of expiration [7, 8].

AI can also be implemented in the evaluation of supply chain risks by involving in the assessment of progress, compliance with ethics or environmental standards, or by recognition of some specific predefined patterns. In environments with increased risks, AI can help to select the appropriate strategy and tactics of procurement to minimize SCS disruption and follow schedule [5]. AI-driven predictive maintenance helps to monitor the performance of equipment and determine when it may likely fail. Organizations can replace the facilities or their parts in advance which can save time and money on emergency replacements and keep the equipment working for a longer period [9]. AI can optimize transportation routes, delivery schedules, and overall logistics efficiency. This will result in the reduction of fuel consumption and transportation costs, as well as, lower environmental pollution from logistics operations [7].

2.3 Challenges in AI Integration with SCS

Although the use of AI with SCS is promising, there are several substantial challenges related to its application in SCS. These challenges include data complexity, ethical issues, and personnel shortages [7].

Data Complexity: The first challenge is related to data complexity. According to Gautam et al. [7], it is critical to have proper and verified data when AI is used. A major issue regarding supply chains is the scattered data or unequal quality of data which makes analytical insights not precise sufficiently. SCS generates vast quantities of data from suppliers and manufacturers to distributors and customers. Moreover, data is often disorganized, unstructured, and heterogeneous as many sources are used that support diverse systems. As a result, substantial investment in maintaining data quality and proper storage is necessary since the volume and complexity of vast data make it big data and big data management issues may ensue [5]. AI can facilitate the process through the application of NLP and ML algorithms to organize and analyze the data. However, it is necessary to ensure proper big data management strategies to benefit from AI applications and avoid issues related to data utility, integration, and security [8].

Excessive costs of Implementation: There can be expensive upfront investments involved in AI technology implementation like the hardware, software, as well as expertise related to AI. Such costs can be expensive for several small and medium-sized firms [7]. There is a disagreement among the workforce concerning the skills underlying the data, as well as AI. There are some challenges related to workforce constraints for the AI regarding sustainability projects which include that it can be difficult for organizations to source for, as well as progress talents which can be helpful to deploy sustainable programs of AI [7].

Ethical Issues: AI application in SCS is also associated with several critical ethical issues. Specifically, AI algorithms tend to reflect the prejudices present in the training data, and this feature may result in unfair or unethical outcomes. Therefore, it is critical to ensure that decision-making processes developed with the use of AI are fully transparent and explainable. Moreover, AI application poses severe risks to supply chain participants' privacy when sensitive information is involved. This issue related to SCS and AI integration must be approached thoughtfully [6]. The protection of data along with its confidentiality seems to be a critical element of any firm since there is an increase in cyber threats [7].

Personnel Shortages: There is a growing demand for specialists who can work with AI systems and implement them. Organizations will have to invest in personnel training and development programs [7]. Implementation of AI and sustainability can be difficult since they are frequently part of an organization's culture. This is because the majority of the employees and stakeholders tend to be resistant to adopting new technologies or processes [7]. Engagement with distinct environmental legislations as well as implementation of them may be an intimidating activity. It is essential to understand that AI solutions must cope with the switching among distinct compliance levels.

2.4 Opportunities in Integrating AI with SCS

Although the implementation of AI technologies in SCS is associated with a series of evident challenges, it is also endowed with a range of opportunities.

Enhanced demand forecasting: An AI-driven demand forecasting model analyzes historical data as well as the trends present in the market and other external factors to determine the demand for the target product or services more accurately. The adoption of AI technologies in forecasting generates enormous benefits concerning inventory management, as it eliminates the cases of stockouts and overstock, leading to a considerable level of customer satisfaction [7, 17].

Optimized inventory management: AI-enabled systems are capable of constant monitoring of inventory levels, predicting supply and demand fluctuations, and automatically reordering. The application of AI in inventory management significantly decreases holding costs, minimizes waste, and ensures the provision of decent product amounts at the desired time of purchase [3].

More sustainable sourcing: The new technologies can be used to assess the alternatives, based on which the most appropriate suppliers will be selected. Further, AI allows us to investigate the level of each supplier's compliance with ethical and environmental standards and discover thc areas where they can be maximally effective. As a result, more ethical and respectful to the environment supply chains will be employed. In addition, the understanding of their possibilities and benefits will inevitably lead to the improvement of the suppliers' motivation [7].

2.5 The Need for a Holistic Approach in SCS

Despite the evident challenges and risks, the implementation of AI technology in SCS continues to be associated with a range of unique opportunities. At the same time, to realize those advantages, it is crucial to understand that the solution, which in this case, can be found only via the fundamental approach, which considers all possible technological and organizational aspects [10]. Thus, SCS managers need to develop a clear strategy that determines how recently adopted AI applications can help them reach the established goals of sustainability. It will also include the appropriate choice of technologies, training of staff, and building appropriate capacities (Habib, 2024). It is also worth noting that to take full advantage of the AI for SCS and overcome the challenges, collaboration between various stakeholders is necessary. Such stakeholders include technology providers, suppliers, customers regulators, and industry organizations. As such, these stakeholders can work together to develop standards, share best practices, and drive the emergence of AI solutions that are responsible and sustainable [7].

3 Literature Survey

The literature on SCS and AI represents a large and multi-dimensional field of research. SCS accounts for the balance between monetary, conservational, and community magnitudes. The transmission to AI implies the introduction of a new advanced approach that promises better efficiency and responsiveness in supply chain operations. Only a few well-consulted examples have been covered in this literature survey to investigate further challenges, scope, and opportunities of implementing AI in SCS [3, 10].

3.1 Supply Chain Sustainable Management

SCS has been noticed and taken into action increasingly as the new perception of sustainability in supply chain procedures has become a need for businesses. Supporting the TBL concept of bringing two other dimensions besides profit, people, and planet, found the foundation of SCS. Agyemang et al. [1] stress the importance of SCS as it serves as an approach to solving global environmental issues, ensures ethical sourcing, and stands for social equity. Moreover, it is impossible to avoid negative economic consequences in the short term since SCS is offering long-lasting sustainability for coming generations [3]. The findings of their study suggest the drivers of the SCS phenomenon, which are regulatory and market pressures, and corporate social responsibility initiatives. Davenport and Ronanki [4] bring light to the challenges of SCS, which include the complexity of supply chain networks, the requirement for full transparency, and the demand for an appropriate engagement of all stakeholders involved. Carter and Rogers [2] reveal that the role of SCS is designed to highlight the importance of managing and organizing an ethical sourcing system, accompanied by the appropriate supplier selection, monitoring, and evaluation. Other SCS challenges are connected to difficulties in adding environmental and social criteria to the suppliers' assessments due to the lack of common standards and the limited amount of available data [7].

3.2 AI in Supply Chain Management

AI technologies, including ML, NLP, and computer vision, provide ample opportunities to improve supply chain management. The existing literature on AI applications in supply chains covers forecasting the demands, management of inventory, risk management of the supplier, and predictive maintenance. Particularly, Esan et al. [5] explored the use of AI technologies in demand forecasting. The study highlights how AI-driven algorithms can be used to improve sales forecasting. In these terms, the evaluation indicated the ability of AI methods to identify historical sales, marketplace trends, and other exterior factors contributing to precise demand estimations in comparison with traditional methods. It led to the improved performance of inventory management systems, such as timely response to changing demand, fewer stock-outs, fewer overstock situations, and better customer satisfaction [3]. Meanwhile, their study on inventory management optimization maintained that an AI system can be used to optimize the inventory workload. In these regards, the AI-run inventory systems are used to continuously monitor inventory levels in real-time, forecast changes in demand, and re-order automatically, reducing inventory holding costs, and waste, and ensuring that products are available on time. Lastly, in their research exploring AI adverse use, Eyo-Udo [6] described how AI-driven algorithms can help to assess supplier performance, remotely monitor suppliers' compliance with their plans to maintain high ethical and environmental standards and avoid further abuse or failure. At the same time, the authors focus on the challenges associated with the employment of AI tools in SCS, like algorithm bias, secrecy infringement, and lack of informed consent.

3.3 Integration of AI and SCS

Esan et al. [5] discuss the complexity of data in AI-driven SCS. They state that "supply chains are considered to be data-intensive systems thus generating huge amounts of data from multiple sources, which is mostly unstructured and heterogeneous". As a result, data quality, integration, and security are serious challenges. AI technologies can be applied to manage and analyze this data, but proper data management strategies should be put in place. Habib et al. [9] focus on the ethical dilemmas posed by AI in supply chains. The authors claim that "if AI algorithms are predicated on patterns, and are used to make decisions, then they will perpetuate whatever patterns were in their training sets", meaning that unfair and unethical outcomes will be created. This underscores the importance of guaranteeing that the administrative decision procedures are transparent and can be accounted for. It is also important to mention that AI may pose "substantial new privacy challenges". Sensitive data concerning suppliers, customers, the organization's reputation, and operations may be at risk [3].

3.4 Opportunities in AI-Driven SCS

According to Sikander et al. [17], personnel should also be considered. The researchers state that "most importantly, the challenges are related to people: it is chiefly human factor driving expectations from managers, facilitating the development of AI technology, and implementing it". This means that there is a shortage of employees who can apply AI in practice. Individuals possessing expertise in AI are required to design and implement the requisite systems. Workforce training and development should be considered. Habib et al. [9] have researched the prospects of AI-driven demand forecasting. Their research demonstrates that AI models can use historical data, market analysis, and consideration of other external factors to predict demand. As a result, it makes inventory management more effective, significantly reducing stockouts and overstock and improving customer satisfaction. Gautam [7] has researched AI-driven inventory optimization. They have found that AI systems can monitor the level of goods in stock in real time, predict changes in demand, and automatically order more goods before it happens. This minimizes holding costs and waste and ensures that there are products to sell. The challenges of these systems are discussed [7]. They include the need to encourage adequate data quality to ensure that the AI has relevant and correct data to rely on, as well as security. Gupta and George [8] have researched the applications of AI in sustainable sourcing. They have found that AI tools can help to evaluate the work of a supplier, ensure that the warehouse follows the required ethical and environmental standards, and suggest possible applications. The deployment of AI allows sourcing goods more ethically, reducing environmental impact, and improving relations with suppliers. Howells [10] has researched AI-driven predictive maintenance. Their review has found that AI systems can evaluate equipment status and predict possible malfunctions in the future, allowing maintenance to act before it happens. Challenges include ensuring the quality and security of data.

4 Methodology

4.1 Research Design, Data Collection and Analysis

This research uses a qualitative research design to determine the challenges and opportunities posed by AI-driven SCS. Secondary data is used to provide an overview of the present context and determine the crucial determinants of AI for SCS implementation and the effect of AI technologies on SCS. The relevant literature examined in the study was retrieved from academic journals, and databases including Web of Science, Google Scholar, and Scopus. The sources were explored in pursuit of journal papers, conference papers, and industry reports. Additionally, various company case studies documenting the successful application of AI in SCS and studies compiled from business journals and company reports were analyzed [3, 4]. The challenges and opportunities regarding the adoption of AI in SCS found through secondary sources are highlighted in the results section.

5 Results and Discussion

On the one hand, the integration of AI technologies in SCS is an opportunity for companies to improve their performance [3]. On the other hand, it presents several issues that must be analyzed. Based on the knowledge obtained from the articles, the following vital aspects of AI in SCS may be highlighted.

5.1 Addressing Data Complexities

The implementation of AI significantly relies on the quality and management of data. However, it also has a direct relation to some SCS issues. According to [3] as a rule, AI may be confused by a large amount of data. Therefore, unless properly preprocessed, companies will not derive any insights. So, data should be properly handled in terms of accuracy and security and the ways data preprocessing can improve the quality of decision-making.

5.2 Navigating Ethical Dilemmas

There are many moral apprehensions regarding the implementation of AI in SCS. First, a great risk of algorithmic bias exists, and an inability to solve this issue may result in biased decisions causing losses of customers and profits. AI technologies generate existing correlations of demand, for example, high demand for certain products, using historical data. The latter might be related to inequalities in incomes and living standards. To avoid this issue, AI implementations should be regularly audited. Second, there are significant data privacy concerns and the need to impose privacy measures ruling the ways of these technology's implementation [6]. According to [1], there are too few AI experts compared to an existing tremendous and rapidly growing demand. As a result, the SCS industry may face issues while attracting specialists to properly implement AI.

5.3 Capitalizing on Fine-Tuned Demand Forecasting.

AI allows us to predict future demand using predictive analytics analyzing historical data and the existing market context. Because it may help reduce stocks and stock-outs, which will also lead the companies to reduce waste, it is very helpful [3]. A summary of challenges and opportunities found from the secondary sources is highlighted in Table 1, reflecting the challenges and opportunities presented by using AI technologies in the sustainability of the supply chain. The data demonstrating

Table 1 Summary of challenges and opportunities in AI-driven supply chain sustainability

Aspect	Challenges	Opportunities
Data class	Erratic and uneven information bases	Improved analytics and policymaking
Practical intricacy	Increased operation expenses and skill requirements	Regularization of supply chain jobs
Moral deliberations	AI unfairness, work shift, and secrecy fears	Better risk management and sustainability profiles
Governing agreement	Variable procedures across sections	Good gain and modernization
Expertise breaches	Scarcity of expert specialists	Expansion of custom-made AI results
Alternate administration	Managerial struggle to alteration	Collective networks and corporations

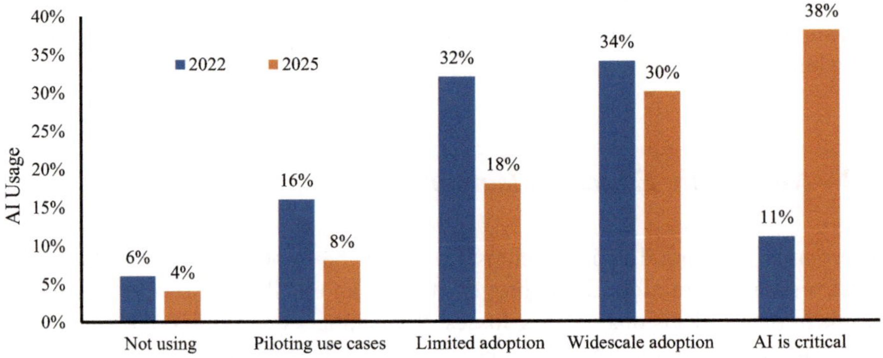

Fig. 1 Worldwide AI adoption rate in the supply chain in 2022 and projected for 2025. *Source* Habib [9]. https://www.veraqor.io/ai-in-supply-chain-management/

the steps in the usage of AI technologies in 2022 and its projected score in 2025 is given in Fig. 1. Figure 1 displays a vibrant surge in the global implementation of AI in the supply chain, with corporations and trades considering AI critical for their companies to stay viable and not become outdated [9].

5.4 Advancements in AI Algorithms

The first critical perspective in the future of AI-driven SCS refers to the development of more sophisticated AI algorithms to support demand forecasting, inventory management, and composite assessment of sustainable purchasing. Specifically, a greater focus on the development of algorithms that can be better equipped to manage the complexities associated with the supply chain data and provide more accurate and better-tailored answers to these challenges will become a part of future research in the field. As a result, more accurate predictions of demand, more efficient inventory management, and the implementation of evaluation tools for sustainable purchasing will be implemented. In addition, an increased focus on AI algorithms will result in a decreased risk of bias through the implementation of less predictable decision-making processes [9].

5.5 Transparency and Interpretability of AI Models

With the use of AI in SCS processes, it becomes increasingly important to ensure that it is viewed as an understandable and interpretable technology. The stakeholders involved in the deployment of AI in their SCS processes need to comprehend the decision-making factors leading to the most optimal decisions. Therefore, an emphasis on the development of algorithms that are not only accurate in decisions but also interpretable will be an essential priority for future research. Lastly, tools and methods to render AI decisions interpretable for people outside the field can be developed [6].

5.6 Development of Ethical Guidelines

The ethical implications of AI use in SCS are significant and should not be overlooked. Future research should focus on the development of ethical guidelines that regulators can use to ensure that upon adoption, the technology complies with the set standards about algorithm precision and accuracy, and data privacy. The necessary guidelines should take a broader approach and include for example, how AI has contributed to the economy, the cons of AI, about bad AI and its beginning, algorithmic bias, how it may ruin human life, how computers make mistakes, and data privacy. The guidelines will then have to involve industry stakeholders, academics, and regulators in reaching a conclusion that reflects and upholds the standards to which SCS technology is developed. Requiring organizations to abide by these guidelines will ensure that AI tools are designed responsibly and ethically, with fairness and social equity in mind [17].

5.7 Social and Environmental Impacts of AI Integration

Another factor that could determine the uptake and acceptance of AI technology for SCS is the social implications and how the adoption of the technology impacts the environment. Future research should seek to find out how this technology could affect the various stakeholders directly involved or who are in proximity to the SCS. The other implication that also has to be evaluated is how the technology may impact the SCS process itself, more specifically, the automated tasks, and its impact on factors, such as, working conditions for machine operators, and the loss of jobs, such as, cleaners due to the deployment of autonomous vehicles [9]. Technology developers have always boasted that AI technology ensures accuracy and efficiency, and the findings will have to be tested against the greater benefit of the environmental goal, for example, a reduction in vehicle stationary times by 10% can result in a 100% reduction of carbon emissions and a reduction in both, fuel and time wastage. All the theorized implications should be put to the test to determine their validity [9]. Hence, it can be concluded that the future of AI-driven SCS management is quite promising. By continuing to develop AI algorithms, ensuring transparency, establishing ethical guidelines, and assessing the social and environmental impacts of such supply chains, organizations can benefit from the capabilities of AI to create responsible, efficient, and sustainable supply chains. However, significant research and collaboration are required to overcome the associated challenges and explore the possibilities.

6 Conclusions

The present study arrives at several key conclusions, based on the analysis of published data and the discussions provided in this review. Firstly, one of the significant challenges highlighted is the necessity for effective data management. The capability of AI to generate meaningful insights is heavily reliant on the quality and accessibility of input data. However, the vast amounts of unstructured data often obscure critical information, posing a barrier to optimal utilization. The reviewed studies underscore the importance of employing advanced data management and analytical tools to ensure data accuracy, reliability, and security, which are essential for leveraging AI's full potential. One more challenge is the shortage of skilled specialists. The development and implementation of AI technologies require profound and specialized knowledge, which is currently in short supply and high demand. Organizations thus need to invest in developing and training their staff to effectively handle AI, as well as, creating the appropriate culture and mindset in the company. Whereas a, along with the challenges, there are significant opportunities provided for or positively affecting SCS by the integration with AI. First, it should be noted that it becomes crucial for SCS to mitigate the impact of improved demand forecasting. By analyzing the enormous amounts of historical data and current trends, predictive analytics delivered by AI provide more accurate forecasting. This guarantees that the supply is well-prepared for the expected demand and thus the danger of stockouts or surplus lists is minimized, and the proficiency of the whole stock chain is optimized. It can be regarded as one of the fundamental benefits of AI integration into SCS.

Whereas another advantage is inventory management optimization. AI algorithms can identify the intricacies, patterns, or trends in data that the human ability to observe and analyze would not allow identifying. Thus, the process is more precise and allows reducing holding costs. In these terms, the efficiency of operations is increased, while the sustainability goal is reached by facilitating the reduction of waste and resource consumption. Finally, it should be noted that the necessity of ensuring and promoting ethical sourcing is related to the outstanding role of AI. Not being limited in the ability to collect and analyze data on suppliers' standards and practices, AI can help to evaluate suppliers according to environmental and social criteria. In this way, the system helps the organization reach its sustainability goals and monitor whether its supply chain meets ethical and environmental standards. Additionally, this study reveals that, the introduction of AI into SCS is a potentially transformative development. Although the issues faced, particularly in the fields of data complexities, increased costs of implementation, skill gaps, ethical dilemmas, resistance to change, regulatory and compliance problems, and increasing need for the organization of skilled personnel working with the technology, present obstacles to the usage of AI in supply chain processes, the benefits recognized underscore the potential of AI applications in forecasting, inventory management, and sustainable purchasing.

Therefore, the focus on further research and innovation is critical for the acceleration of AI's benefits in supply chain processes and the pervasion of technology as the most efficient and sustainable decision-making tool in the field of SCS. Moreover, following the results obtained, the future of AI-driven SCS can be predicted by a combination of several emerging trends and evident technological advancements. Among the most critical aspects mentioned, it is possible to refer to the increased focus on the development of AI algorithms.

List of Abbreviation

AI	Artificial Intelligence
ML	Machine Learning
SCS	Supply Chain Sustainability (SCS)
SCSM	Supply Chain Sustainable Management
TBL	Technology based Learning

References

1. Agyemang M, Zhu Q, Adzanyo M, Antarciuc E, Zhao S (2018) Evaluating barriers to green supply chain redesign and implementation of related practices in the West Africa cashew industry. Resour Conserv Recycl 136:209–222. https://doi.org/10.1016/j.resconrec.2018.04.011
2. Carter CR, Rogers DS (2008) A framework of sustainable supply chain management: moving toward new theory. Int J Phys Distrib Logist Manag 38(5):360–387
3. Choi TM, Wallace SW, Wang Y (2018) Big data analytics in operations management. Prod Oper Manage 27(10):1868–1883. https://doi.org/10.1111/poms.12838
4. Davenport TH, Ronanki R (2018) AI for the real world. Harv Bus Rev 96(1):108–116

5. Esan O, Ajayi FA, Olawale O (2024) Supply chain integrating sustainability and ethics: strategies for modern supply chain management. World J Adv Res Rev 22(1):1930–1953. https://doi.org/10.30574/wjarr.2024.22.1.1259
6. Eyo-Udo N (2024) Leveraging artificial intelligence for enhanced supply chain optimization. Open Access Res J Multidiscipl Stud 7(2):001–015. https://doi.org/10.53022/oarjms.2024.7.2.0044
7. Gautam H (2024) The impact of AI and sustainability in supply chains. https://appvintech.com/the-impact-of-ai-and-sustainability-in-supply-chains/
8. Gupta M, George JF (2016) Toward the development of a big data analytics capability. Inform Manage 53(8):1049–1064. https://doi.org/10.1016/j.im.2016.07.004
9. Habib MB (2024) The impact of AI in supply chain management and the way forward. https://www.veraqor.io/ai-in-supply-chain-management/
10. Howells R (2024) Supply chain predictions 2024: AI, sustainability top of mind. https://www.forbes.com/sites/sap/2024/01/02/supply-chain-predictions-2024-ai-sustainability-top-of-mind/
11. Onyiriuka EJ, Ewim DRE, Abolarin SM (2023) An optimization technique to identify simulation assumptions for various nanofluids using machine learning. International Heat Transfer Conference Digital Library
12. Orikpete OF, Ewim DRE (2023) A review of noise management practice in Nigeria. Environ Sci Sustain Dev 8(1):31–44. https://doi.org/10.21625/essd.v8i1.931
13. Pattnaik M, Vijayalakshmi NS, Sharma M, Kumar A, Sharaschandra KS (2022, July) A novel paradigm to artificial intelligence in transforming supply chain management in the agile business world. In: 2022 international conference on innovative computing, intelligent communication and smart electrical systems (ICSES). IEEE, pp 1–6. https://doi.org/10.1109/ICSES55317.2022.9914144
14. Reis O, Eneh NE, Ehimuan B, Anyanwu A, Olorunsogo T, Abrahams TO (2024) Privacy law challenges in the digital age: a global review of legislation and enforcement. Int J Appl Res Soc Sci 6(1):73–88. https://doi.org/10.51594/ijarss.v6i1.733
15. Sakala CM, Bwalya SM (2023) The role of AI in optimizing supply chain performance. J Procur Supply Chain Manage 2(1):1–14
16. Seuring S, Muller M (2008) From a literature review to a conceptual framework for sustainable supply chain management. J Clean Prod 16(15):1699–1710. https://doi.org/10.1016/j.jclepro.2008.04.020
17. Sikander U (2024) Top 4 global supply chain challenges and opportunities. Citation. https://ecomexp.com/top-4-global-supply-chain-challenges-and-opportunities-in-2024/
18. Sodhi MS, Tang CS (2019) Research opportunities in supply chain transparency. Prod Oper Manage 28(12):2946–2959. https://doi.org/10.1111/poms.13115
19. Wang Y, Han JH, Beynon-Davies P (2019) Understanding blockchain technology for future supply chains: a systematic literature review and research agenda. Supply Chain Manage Int J 24(1):62–84

Food Security Between Norms and Net-Zero Climate Future

Mai Magdy Abdelrazeq, Taghreed El-Kasaby, and Shaimaa R. Dalal

Abstract Food system is a highly complex topic; food scientists, local farmers, food manufacturers, private sector and governments regularly gather for decisive decisions to be dictated in terms of food security and achieving sustainable diet goals. However, food system is interconnected, here is the major conundrum of Sustainable Diet within Global food system. This Chapter deals with Sustainable Diet Notions, tackling food waste and agro-waste, understanding different supply chains in the food system and how circular economy should be effectively applied. All are introduced with evidence-based innovative idea from multi-disciplinary and circular-economy approaches, to upcycle agro waste and household waste.

Keywords Sustainable diet · Climate action · International cooperation

1 Introduction

Sustainable Diet as per UN SDG2; achieving zero hunger, is mainly about achieving healthy and nutritious diet which meets sociocultural norms, economically affordable, available throughout the year, and finally with minimal environmental impacts.

Other Sustainable Development Goals are incorporated in the Sustainable Diet planning including: SDGs1, 3, 12, 13 and 14; No Poverty, Good Health, Responsible Consumption and Production, Climate Action and Life Below Water.

Sustainable Diet has a pivotal role on reducing Greenhouse Gas emissions (air footprint), Soil footprint (protection of topsoil, low erosion, low intensive farming,

M. M. Abdelrazeq (✉)
National Research Centre, Dokki, Cairo, Egypt
e-mail: M.magdy.23@abdn.ac.uk

The Rowett Institute, University of Aberdeen, Aberdeen, UK

T. El-Kasaby · S. R. Dalal
Mansoura University, Mansoura, Egypt

S. Al-Naemi et al. (eds.), *Water and Food Security in the Face of Climate Change: Challenges and Opportunities for Resilience*, World Sustainability Series,
https://doi.org/10.1007/978-3-032-00098-9_29

reforestation) and water footprint with (no physical and chemical pollutants). As well as, no overexploitation to rangeland, sustainable fishing methods, adopting regenerative traditional agriculture method with new innovative technologies and avoiding industrialized conventional agricultural methods; i.e. monoculture (planting one crop in the field such as rice, maize or wheat) with intensive use of insecticides, pesticides and inorganic fertilizers with heavy machinery and low labor in the field [1].

In theory food security seems applicable, but in practice, it is not easily possible, multiple drivers must be considered in achieving Global Sustainable Diet System.

International Multiple-Stakeholder's collaboration of sustainable diet usually make it a very hard controversial choice, it is quite difficult to gather manufacturers, local farmers, private sectors, food banks, governmental and non-governmental institutions towards final legislations for achieving a sustainable diet system [2]. Civil society and consumers are a major element in achieving Sustainable Diet goals, the challenge is that they usually resist adopting Sustainable diet system, what makes it more challenging is that many people lack awareness of home economics, proper dietary and culinary knowledge. Nevertheless, it is promising that many food industry entities started to have commitments towards applying SDGs in their businesses with circular economy multi-sectorial cooperation to reduce food emissions, enhance livelihood of people and raising proper sustainable diet awareness [3].

1.1 Sustainable Diet Double-Burden

An ongoing paradox of Obesity versus. Malnutrition, is considered a double burden in the food system....land fertility, lack of resources, hunger, over consumption and production.

As per recent statistics:

- In 2024: 695 million people, approximately 10% suffer from mild to severe hunger [4].
- 1 billion people suffer from obesity (diabetes 2 and/or CVDs) which costs $2 trillion annually, it is projected to reach 1.53 billion by 2035 [5].
- 2 billion people suffer from hidden hunger (lack of one or more essential micronutrients), with continuing rise in mortality rate related to diet [6].

In developed countries, there are massive supply chain production with advanced food production system. However, health benefit is not only about availability of food, it goes beyond that to the quality of food and availability of essential micronutrients (Vitamins and Minerals), macronutrients (Protein, Fats and Carbohydrates) and other functional food elements for chronic diseases care and prevention, i.e. fruit flow, beta glucan, probiotics and prebiotics...etc.

A research undertaken by Dr. August Dunning, founder of Eco Organics; Illustrated that, to get the amount of iron you used to get in 1950 from one apple by 1998 you need to eat 25 apples, nowadays in order to get the same amount of iron you need to eat 36 apples. This reflects the continuing deterioration in nutritional quality due to chemical usage in industrial farming. This is compounded by bad eating habits that people tend to eat more refined sugars and saturated fats, which is associated with rise in chronic diseases like: Cardiovascular disease and Diabetes 2 [7].

In Developing countries: people suffer from mild to severe food insecurity and hunger; food is neither available nor healthy. World population is projected to reach 9.7–9.8 billion by 2050 [8]. There will be 2 more billion who wouldn't find enough food resources, simply because the number of natural resources is fixed. So, the challenge is to increase food production in the same land size without causing environmental impacts or climates changes; Deforestation, land exploitation, destroying ecosystem and reducing biodiversity [9].

Fortunately, sustainability awareness highly considered, many industries began to adopt sustainability commitment in their industries, such as Unilever which is committed to enhance wellbeing and health for > 1 billion people, reduce environmental impacts by half and enhance livelihood for millions, Tesco is also committed to have a 300% sales increase in meat alternatives by 2025 [10] (Fig. 1).

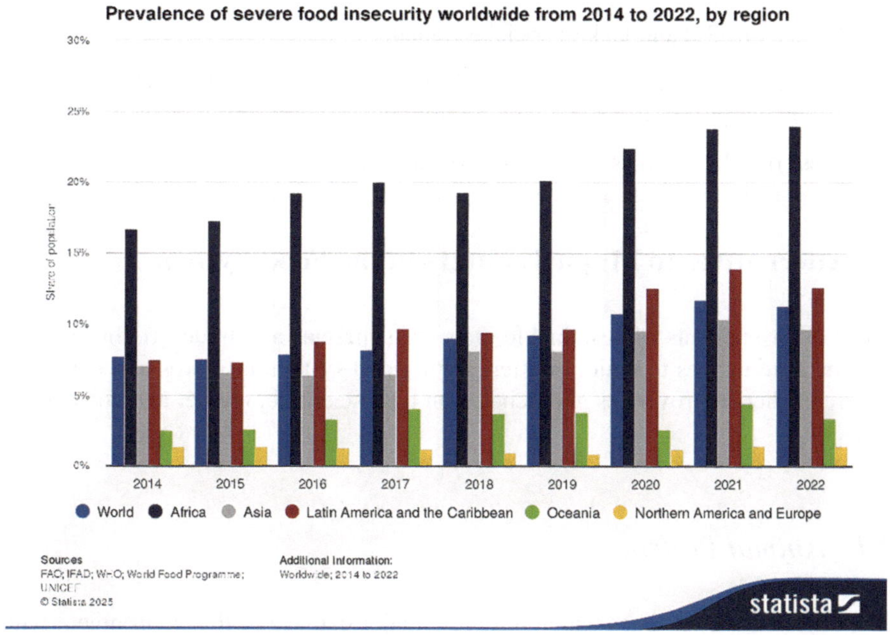

Fig. 1 A histogram illustrating food insecurity Prevalence by region, where Africa is in continuous deterioration, in contrast; North America and Europe are considered to be highly food secured [11]

2 Sustainable Diet and Cultural Taboos

Population in different zones have their own traditional food with it's own identity, ingredients and specific culinary skills. Here is a difficulty that citizens usually oppose the change, gathering the entire universe to certain food items is undeniably a tough mission to accomplish. For most people food is about culture, beliefs, social activity, joy and traditions.

As per Professor Claude Fischer: "Man feeds not only on proteins, Fats, carbohydrates, but also on symbols, myths, fantasies" [12].

There are gender and age preferences and stigmas; for instance: male teenagers relate meet eating to masculinity. Furthermore, religious beliefs profoundly affect food choices; Muslims do not drink alcohol or eat pork as it is Haram in Islam Religion.

Geographical Locations also have influence on how food is cooked or served; in USA and UK food is individualized, in the contrary in France, Italy and Switzerland: food is about joy and gathering, eating takes times as family gather at a table.

Paul Morand (French), in 1937 described New Yorkers lunch:

They eat standing up, fast, with their hat on, in a row "like in a stable."

- Absence of rules and lack of social exchange.

Daniel Lerner (American), in 1956 described French eating habits to be "rigid". And he was surprised to see them eating ritually and at set times: "like in a zoo".

- Excess of rules, seen as a hindrance to choice [13].

3 Alternatives to Major Emitters in the Food System

Some other elements of sustainable diets are synergies and trade offs in the food system. Alternatives to major emitters in the food system have pros and cons, this generates diet controversies and debates on both scientific, market and consumer's level.

3.1 Animal Protein

Livestock production is considered to be a major damage to the environment with extremely high levels of GHG emissions along with deforestation and endangering biodiversity, it is estimated that around 60% of environmental resources are already wasted or used unsustainably. In many western countries, people consume more meat than daily requirements, by 2050 there have to be 70% more food production in the limited food system resources to cover population dietary needs [14].

Meat alternatives could possibly be a highly recommended solution to this conundrum. Plant- based proteins are an option as in Soy, Legumes and Cereals, Edible Insects are also a recent trend as a sustainable option to livestock production. Aquaculture, Seaweeds and Algae are a possible replacement that are part of global diets as in Okinawa Diet in Japan and other Asian countries. Finally, Cultured meat is widely considered in recent scientific research as a sustainable source of animal protein that gives the same organoleptic as natural meat [15].

All of the abovementioned choices are manufactured in environmentally sustainable ways. However, there is a controversy amongst the population that some options are repulsive, others are not tasty or may be culturally unacceptable.

Edible insects:

Entomophagy: eating insects by humans was popular since the prehistoric age. There are over 1900 species of edible insects, insects are very high in protein and taste like chicken. Insects breeding is environmentally friendly and sustainable without using fertilizers and more importantly with much lower feed than livestock, what's more interesting about the fodder is that it comes from organic waste, so it applies 4R Framework of Circular-economy with reducing, reusing, recycling and recovering.

Insects are introduced in the market as crushed, powder, burger, bars, pasta or as a whole. The only barrier nowadays is public acceptance and cultural taboos, that many people find eating insects repulsive and culturally unacceptable [16].

Cultured meat:

In vitro cultured meat burger was produced for the first time in 2013 with minimum environmental impacts and less land and water consumption compared to livestock breeding. Stems cells of animals are extracted then they are growing further into muscle tissues.

Burgers, minced meat and meat balls are produced to mimic slaughtered livestock meat, but without slaughtering and more importantly, cultured meat can be tailored to be low in fat and to meet nutritional needs [17].

The challenge in this business is to convince consumers to replace livestock meat products with cultured meat. There is a debate that although it is promising, it is still coming from animals and production takes long time and money than conventional methods, at the same time the available options in the market are still limited.

4 Food from Waste into Value:

Reducing food waste is urgently fundamental to create a sustainable food system, especially because global food system is responsible for 20-30% of greenhouse gas emissions,

When food is wasted or lost, other critical resources are also wasted such as: land, water, labor, energy and capital. For example: half of bread purchased in Britain is almost wasted, which means that all resources used in wheat planting, flour milling, and bread baking are also wasted, along with manpower, transportation…etc.

As per FAO Annual Food report, it is estimated that a third of food produced annually is wasted or lost during one or more stations of the food supply chains. Food waste is considered a huge economic burden; by numbers: 143 billion euros are wasted as food in EU in 2012 [18].

4.1 Statistics on Food Waste Challenge

WRAP 2020: If people in UK stop wasting food for one day, it could have an impact on GHG emissions as planting half a million trees!

4.4 G tonnes of CO_2 are emitted because of food waste cycles; production, processing, distribution, transportation and disposal [19].

In 2007, 1.4 billion hectares of agricultural land was exploited and wasted to produce food that wasn't actually consumed. Furthermore, Food waste is one main cause of biodiversity loss and deforestation, since 74% of annual deforestation happens to be used in food production.

70% of the earth's fresh water goes to agriculture to produce food, so when one third of food is wasted, water is also wasted, i.e. just one cup of coffee uses 125 liters of embedded water.

WRAP 2020: If people in UK stop wasting food for one day, it could have an impact on GHG emissions as planting half a million trees!

820 million people are undernourished, 1 billion people are overfed and one third of food production is wasted or lost annually, that can feed all hungry people in the globe.

Now the challenge is not to let food waste reach the landfills, since it releases mainly methane which is 23 times stronger than CO_2. Currently food system stakeholders are moving towards regenerative circular-economy approaches by (reducing, reusing, recovering and recycling).

5 Upcycling Technique of Agro-waste and Household Waste: (Patency Project)

Agro waste upcycling into artworks and furniture pieces, afterwards they can be turned into competitive markets: (From Waste to Value).

Each year 140 billion metric tons of agro waste are generated, they can be of a huge economic benefit: industrial residues, agricultural residues and livestock waste [20].

Fig. 2 Steps of waste capsulation and transformation into artwork blocks by two major phases of capsulation to form a crystallized form, then covering the waste capsule to form the final product [19]

Concept: an innovative artistic and engineering technique based on eco-friendly elements from land and natural chemical reactions to convert and capsulate Non-recyclable Agro waste into massive artworks and blocks. Most importantly, artworks will be used as a Bank to save precious agro waste products for future use.

This new type of art will open a new competitive market, artworks will form themed museums, artworks can be sold, exported or to be set as tourist attractions.

General Outcome: low CO_2 emissions, resource management and Cities Decoration with Art.

Environmental benefit: Low CO_2 and Methane emissions. Social benefit: Decorated agro waste-free cities.

Economic benefit: Removing economic burden from agro waste control, as well as launching new emerging products in the market (Figs. 2, 3 and 4).

6 Microorganisms as a Definitive Solution to Food Insecurity

Microorganisms such as Bacteria and Algae can play a profound role in shaping a new emerging sustainable food sources. Bacteria sufficiently generate short chain fatty acids (SCFAs), which directly contribute to healthy gut and overall microbiome

Fig. 3 Longitudinal section of the waste artwork to show the layers deep inside the body of unbreakable covering, crystallized capsule and finally waste at the heart of the block [19]

Fig. 4 Prototype of the final product; contemporary Art of Waste. It can be generated in different geometrical sculptures and styles [19]

actions. Intriguingly, bacteria with genetic engineering are exploited to produce critical food and energy active components.

Algae as in Micro and Macro algae have undeniable benefits, not only for humans, but also environmentally; for instance: Seaweeds are a great source of Macro and micro nutrients, more importantly this food source is affordable, available throughout the years and sequester gigantic amounts of CO_2 [21].

7 Conclusion

Sustainable Diet is hard to achieve but not impossible, it can be a key to achieving food security and lower environmental impacts, the food system is responsible for one third of total environmental issues, the paradox is that with the extreme food insecurity there is extreme obesity and overfeeding amongst population. Furthermore, food waste can feed all hungry people around the world.

Stakeholders of the food system are the key to achieving sustainable diet goals with collaboration and effective communication agenda.

One legislation can save the planet; Legislations usually take long time with many debates amongst policymakers. However, in the time being a decisive legislation is an urgent necessity. The more plants are grown, the better public health with more awareness, citizenship and lower emissions.

A new contemporary Art of Waste can be a competitive new market in the countries to not only decorate, but also upcycle waste with circular economy perspective.

Edible Agro waste is of a great health benefits and nutritional value, brans, peels and pulps are examples of free of charge edible waste that can be upcycled into new affordable food products to provide more fibers, nutraceuticals, antioxidants and beneficial oils.

References

1. National Academies of Sciences, Engineering, and Medicine; Health and Medicine Division; Food and Nutrition Board; Food Forum; Leslie Pray, Rapporteur(2019). Sustainable diets, food, and nutrition: proceedings of a workshop. National Academies Press. https://www.ncbi.nlm.nih.gov/books/NBK540292/#:~:text=According%20to%20Drewnowski%2C%20the%20FAO,%2C%20and%20(4)%20environmental
2. Herens MC, Pittore KH, Oosterveer PJ (2022) Transforming food systems: Multi-stakeholder platforms driven by consumer concerns and public demands. Glob Food Sec 32:100592. https://doi.org/10.1016/j.gfs.2021.100592
3. Private sector engages with FAO on improving food security and agriculture sector sustainability (2024, August 5) REU. https://www.fao.org/europe/news/detail/private-sector-engages-with-fao-on-improving-food-security-and-agriculture-sector-sustainability/en
4. Hunger Map Live (n.d.) https://hungermap.wfp.org/
5. Prevalence of obesity | World Obesity Federation (2024) World Obesity Federation. https://www.worldobesity.org
6. Micronutrient Forum (2023, February 7) New global hidden hunger estimates. Micronutrient Forum. https://micronutrientforum.org/new-global-hidden-hunger-estimates/
7. Mineral Deficiencies—Team Dream Extreme (2019, June 6) Team dream extreme. https://masterthenewway.com/mineral-deficiencies/
8. Worldometer. Worldometer—real time world statistics. https://www.worldometers.info/

9. The Challenge—Global Food Security (2023) https://www.foodsecurity.ac.uk/challenge/
10. Smithers R (2020, September 28) Tesco sets 300% sales target for plant-based alternatives to meat. The Guardian. https://www.theguardian.com
11. World Hunger: Key Facts and statistics (26 Nov 2024). https://www.deliveryrank.com/blog/world-hunger-facts-statistics
12. Fischler C (1988) Food, self and identity. Soc Sci Inf 27(2):275–292. https://doi.org/10.1177/053901888027002005
13. Fiscler C (2012, May 22) France, Europe, the United States: what eating means to us: an interview with Claude Fischler and Estelle Masson, about the new volume published by Odile Jacob. Le mangeur Ocha. Le Mangeur Ocha. https://www.lemangeur-ocha.com/france-europe-the-united-states-what-eating-means-to-us-an-interview-with-claude-fischler-and-estelle-masson-about-the-new-volume-published-by-odile-jacob/
14. Valli C, Maraj M, Prokop-Dorner A, Kaloteraki C, Steiner C, Rabassa M, Solà I, Zajac J, Johnston BC, Guyatt GH, Bala MM, Alonso-Coello P (2022) People's values and preferences about meat consumption in view of the potential environmental impacts of meat: a mixed-methods systematic review. Int J Environ Res Public Health 20(1):286. https://doi.org/10.3390/ijerph20010286
15. Andreani G, Sogari G, Marti A, Froldi F, Dagevos H, Martini D (2023) Plant-based meat alternatives: technological, nutritional, environmental, market, and social challenges and opportunities. Nutrients 15(2):452. https://doi.org/10.3390/nu15020452
16. Thinking About the Future of Food Safety (2022) FAO eBooks. https://doi.org/10.4060/cb8667en
17. Wu J (2022) Emerging sources and applications of alternative proteins: an introduction. Adv Food Nutr Res. https://doi.org/10.1016/bs.afnr.2022.06.001
18. Food Waste Index Report 2024 | FAO (2024) https://www.fao.org/family-farming/detail/en/c/1681058/
19. Abdelrazeq MM (2025, May 5) Household and agro-waste upcycling into sustainable unbreakable artworks and furniture pieces by eco-friendly organic chemical reactions. Multidisciplinary adaptive climate insights. https://apc.aast.edu/ojs/index.php/MACI/article/view/MACI.2025.02.1.1190
20. Business Waste (2024, April 10) Agricultural waste facts and statistics | Farming waste facts. https://www.businesswaste.co.uk/sectors/agricultural-waste-management/agricultural-waste-facts/
21. Qadir M, Hussain A, Iqbal A, Shah F, Wu W, Cai H (2024, August 24) Microbial utilization to nurture robust agroecosystems for food security. MDPI. https://www.mdpi.com/2073-4395/14/9/1891

New Trends, Technological Advancements, and Obstacles in Food Security Research

Zahida Bano, Rachid Benlamri, and Farhat Abbas

Abstract Over one billion people face hunger worldwide, as the growing population exerts immense pressure on the existing food resources. Conventional farming methods often require lengthy production processes. To address these challenges, technological advancements have led to the adoption of smart agriculture practices, including precision farming, vertical farming and genetically modified organisms (GMOs) to crop seeds. Integrating global positioning system (GPS), geographical information system (GIS), remote sensing, robotics, internet of things (IoT), and drones can support smart agricultural operations. These technologies enable the cultivation of plants in controlled environments, regulating factors such as temperature, humidity, irrigation water, supplemental light, and the control of nutrients, fertilizers, and pesticides. This paper reviews the advancement, utilization, and challenges associated with implementing these tools in smart agriculture tools and technologies.

Keywords Smart agriculture · IoT · WSN · GMOs · Food security · AI

1 Introduction

Food security is a historical global concern, encompassing the availability, accessibility, and utilization of sufficient, safe, and nutritious food to meet the dietary needs of individuals worldwide [1, 2]. The complex nature of achieving and maintaining food security is aggravated by factors such as a growing global population, climate

Z. Bano (✉) · R. Benlamri · F. Abbas
University of Doha for Science and Technology, Doha, Qatar
e-mail: zahida.qau@gmail.com

R. Benlamri
e-mail: rachid.benlamri@udst.edu.qa

F. Abbas
e-mail: farhat.abbas@udst.edu.qa

S. Al-Naemi et al. (eds.), *Water and Food Security in the Face of Climate Change: Challenges and Opportunities for Resilience*, World Sustainability Series,
https://doi.org/10.1007/978-3-032-00098-9_30

change, economic uncertainties, geopolitical challenges, and natural catastrophes [3–5]. The latest report of the Food and Agriculture Organization (FAO) in 2024 revealed that around 733 million individuals face chronic hunger daily, with an increase of 152 million people in 2023 compared to 2019 [6]. Additionally, nearly 2 billion people struggle with various degrees of food insecurity, disproportionately affecting vulnerable groups like women, children, and indigenous communities [7]. Despite a 100% increase in food production over the last 30 years, projections from the FAO indicate that a 60% increase will be required by 2050 to meet the escalating demand of the global population [8]. Traditional agricultural systems are deemed insufficient, prompting a transition towards sustainable and precision agriculture, exploiting technological advancements to enhance crop yield while conserving resources.

Smart greenhouses and controlled environment agriculture (CEA), that comprise controlled environments, including growth chambers, rooftop greenhouses, and plant factories could be a significant solution to these challenges [9]. Advanced farming techniques such as precision farming, vertical farming, and biotechnology play a crucial role in increasing crop productivity and promoting resistance to environmental challenges [2, 10–12].

Precision farming (PF), also known as precision agriculture is a technology-based farm management that analyses and manages any variability within the farmlands to optimize various aspects of farming including crop yield, crop health, resource use, and environmental sustainability. It involves the collection and analysis of large set of data related to crop health, presence of pets and weeds, soil nutrients, humidity, air quality, and temperature etcetera. This data enables farmers to tailor the decision-making to spatially and temporally maintain crops healthy. Such as, incorporating GPS into precision farming optimizes precision in applications of pesticides, fertilizers, and water. The GIS stores information such as yield, soil survey map, crop scouting report, remotely sensed data, and soil nutrient report [10, 13].

Vertical farming is an engineering solution to enhance crop yield per unit area by extending crop productivity in the vertical dimension. One of the examples of vertical farming is the construction of tall structures with several levels of growing beds, facilitated with artificial lights [14, 15]. Many countries have already implemented vertical farming on the roof top commercial and residential of new and old buildings, including South Korea, Japan, Singapore, USA (Chicago), and Canada (Vancouver). Many other countries including Germany, China, France, and the UAE also endorsed the concept of vertical agriculture for sustainable food security [14]. It is seen as a sustainable alternative to traditional farming, which involves large-scale outdoor agriculture with heavy irrigation, tillage, and excessive use of chemicals [12].

Agriculture biotechnology, including genetic engineering and crop breeding, plays a crucial role in crop improvement by genetic improvement of plants, which results in enhanced yield production, fruit quality, tolerance to biotic and abiotic stress, and efficient use of nutrients. It is estimated that plant breeding contributed 50% in yield growth between 1980 and 2000 [16]. So far more than 70 countries have accepted genetically modified (GM) crops as a source of food, feed, or for cultivation. These GM crops include maize, soyabean, potato, sugar beet, papaya, zucchini, eggplant, alfalfa, canola, and apple. Recently rice has also been genetically modified for herbicide and insect resistance [17, 18].

2 Current Breakthrough in the Field of Smart Agriculture and Food Security Research

2.1 Technological Advancement in Greenhouses

To overcome global challenges such as population growth, climate changes and urbanization, that has a tremendous impact on food security, the greenhouse industry has undergone various technological advancements. Modern technology has been integrated into greenhouses to enhance resource utilization and to overcome limitations. Some of the key technological advancements integrated into the greenhouses and proposed model for the improvement of greenhouses includes the following.

Automated Climate Control and Artificial Intelligence (AI)

A major issue associated with conventional agriculture is the inadequate application of fertilizer, pesticides, monitoring of light, temperature, humidity and CO_2 level, improper management of irrigation, drainage and weed control etc. The application of artificial intelligence (AI) integrated with the Internet of Things (IoT) and sensors including Wireless Sensors Networks (WSN) in greenhouses significantly enhances agriculture efficiency and productivity by real-time monitoring and automation of various environmental aspects and supports precision farming [19]. Such as IoT-based sensors can allow monitoring of temperature, humidity, and light intensity, and detection of pests and diseases inside the greenhouses in real-time and send the collected data to the central system that allows the farmers to adjust the conditions remotely. IoT connected sensors can also detect and automate climate control (ventilation, heating, and cooling) and irrigation systems and provide optimal conditions for plant growth with precise irrigation based on soil moisture. WSN sensors enable the detection of soil conditions such as soil moisture and nutrient content, which allow automated control of irrigation and fertilizer systems. AI model analyses sensor data to predict crop yield, crop health and recommend action for optimal growth. Researchers have proposed various models for integrating IoT and WSN in greenhouses.

Danita et al. [20] presented a model that integrates IoT with moisture sensors, temperature, and humidity sensor in the greenhouse. After receiving data from the sensors, Raspberry P13 automatically controls moisture, temperature, humidity, and inside the greenhouse by activating irrigating pipes, sliding windows, and cooling fans respectively. The recorded data are stored in a cloud database named ThingSpeak and the results are displayed on a webpage for the viewer. Sushanth and Sujata [21] proposed a model for smart agriculture by integrating Arduino, IoT, and WSN to monitor temperature, humidity, moisture, and animal movement. The IoT-based monitoring system uses a wireless sensor network to collect data, and Arduino sends a short messaging service (SMS) to the mobile or farmer's smartphone using Wi-Fi/3G/4G. The system also allows data inspection and scheduling of irrigation through an Android application. Srinidhi et al. [22] proposed to integrate a hydroponic system in a controlled environment with IoT, Machine Learning Algorithm, and Computer Vision. The author claimed that once the software is developed and integrated, it will enable everyone to grow their vegetables at a lower cost. Herman et al. [23] proposed an innovative framework for integrating hydroponics precision agriculture to IoT and Fuzzy Logic. As IoT facilitates regular monitoring of plant nutrients and water needs, while Fuzzy logic is used to control the supply of the inputs. Virtual presentation of a plant using this technique showed increased width and length of leaf of lettuce and bok choy (Chinese White Cabbage). Bakhtar et al. [24] proposed several types of hydroponic methods including the Nutrient Film Technique, Deep Water Culture, Wick System, EBB and Flow, Drip System, and Aeroponic, just to mention a few. They grow plants without soil by exposing the plant root to a mixture of minerals and water solutions. A hydroponic method is suitable for places where gardening is not feasible. Water and humidity are controlled with the help of a microcontroller kit connected to a WSN and the internet. Integration of IoT technology in the system allows monitoring of plant growth in real-time. Ismail et al. [25] and Gaikwad et al. [26] integrated the WSN with Zigbee technology and implemented it in a greenhouse to monitor and control environmental conditions including sunlight, temperature, and humidity via remote automation. The study also proposed a user-friendly Android application to access the data remotely.

Arif and Abbas [27] suggested an efficient and accurate model for automatic irrigation, which depends on computing several characteristics essential for a greenhouse. The proposed model employed WSN based on servers and client web services for monitoring and controlling irrigation. The proposed framework covers two folds: on one hand it reduces the power and control over long distances and on the other hand it provides accurate tools for auto control on output devices. The proposed design system comprised five sensors, i.e., temperature, humidity, soil moisture, CO_2, and light sensor. Qazi et al. [28] presented an exhaustive survey of IoT and AI advancement in the field of smart agriculture. The review explained the use of deep learning (DL) in various aspects of agriculture including the detection of plant diseases, type of disease, real-time pest detection, weed detection, and plant phenotyping. Yu et al. [29] constructed various deep convolutional neural networks (DCNN) for weed detection growing in perennial ryegrass. Integration of these technologies in greenhouses would enhance yield production with an efficient usage of resources.

Ventilation System

Proper ventilation and air conditioning system allow optimal plant growth with improved yield quality. Numerous studies have focused on assessing the global air exchange rate in natural ventilation, employing tracer gas measurements and energy balance calculations [30, 31]. However, these methods lack comprehensive insights into airflow patterns, humidity levels and temperature profiles, hindering the optimal design of ventilation systems. The advancements in computational fluid dynamics (CFD) allow precise simulation and analysis of airflow patterns, humidity level, temperature, fluid pressure and velocity in the greenhouses. Li et al. [32] provided a comprehensive numerical framework using computational fluid dynamics CFD tools to predict the airflow and thermal behavior patterns on various arched greenhouses. The study formulated the angle and position for natural ventilation and respective roof vent scenarios. Natural ventilation is an effective energy-saving strategy conducive to promoting sustainable agricultural production. A comprehensive numerical simulation was performed to predict the airflow pattern and thermal behavior in different arched greenhouses. The numerical simulations were accurate and efficient as the model was compared with the experimental data. So, the numerical model integrated with CFD tools can be employed to perform simulation on a high-resolution grid using finite element methods. The proposed ventilation scheme has perfect air exchange capacity. Using CFD, Kim et al. [33], conducted a calculation of the wind pressure coefficient of a multi-span greenhouse in South Korea.

Drones, Automation and Robotics

Drones and robotics play a transformative role in agriculture by enhancing productivity, precision and sustainability. Drones equipped with cameras and sensors can provide real-time data on crop health such as disease and pest infestation and soil condition (nutrients and moisture), that allows farmers to make informed decisions on irrigation, application of fertilizer, and pest control etcetera. Robotics and drones connected with the sensors can monitor irrigation systems, planting and seeding, harvesting, spraying, weed and pest control and data collection [34]. Analysis of this data using AI tools allows the farmers to predict crop yield, pest outbreak, and optimal harvesting time. A number of studies have presented robots and drones models that could be used in agriculture practice. For example, Al-Beeshi et al. [35], introduced a robot called iPlant, which is more convenient to use, needs less effort and time efficient. It can control the greenhouse effect individually by interacting with the environment. The iPlant monitors and transfers plant state, humidity, and nutrient concentration. Jayakrisna et al. [36], presented a design of a four-wheel robot prototype that is used for seed sowing.

Gautam and Sarkar [37] proposed a drone model equipped with sensors like hyperspectral, thermal and LiDAR. These sensors could be used to detect various parameters including weed, drought conditions, water stress, and nitrogen content etc. Hijazipoor et al. [38], presented a four-degree freedom spraying robot that calculates the bulk volume of plant and sprays accordingly. Hasuda and Hiryana [39], introduced an autonomous pesticide spraying robot that could move in the greenhouse, identify crops, and accurately apply pesticides on the crop. Bennis et al. [40] proposed a model that integrates moisture sensors in drip Irrigation, sprinkle irrigation, and automated irrigation systems will optimize water usage and enhance crop yield by controlling and regulating soil moisture. These technologies are making agriculture more efficient, data-driven, and sustainable.

Plant Breeding and Genetic Engineering

Plant Breeding and Genetic Engineering are agricultural biotechnology that is involved in the genetic improvement of crops in order to increase crop yield, crop quality, disease diagnosis and resistance and conservation of genetic resources [17, 41–43]. Integrating biotech crops, particularly those resilient to climate stresses is advocated as part of a toolkit for farmers. Currently, 44 countries have approved 385 genetic modification events for use in food products [17]. Traits such as herbicide tolerance and insect resistance are prevalent, aiming to enhance crop resilience. The emergence of new plant breeding procedures (NPBTs), including clustered regularly interspaced short palindromic repeat (CRISPR)/CRISPR associated protein (Cas) (CRISPR/Cas), transcription activator-like effector nuclease (TALEN), and zinc finger are being used to modify genes to speed up the introduction of a desired trait [44–47]. These traits include heat, cold, and herbicide-tolerant and pathogen resistant (fungi, bacteria and virus), improved quality and quantity of commercially valued crops such as rice wheat, tomato, potato, cotton and soyabeans [44]. Besides, genetically modified crops are seen as potential solutions to reduce pesticide use, soil damage, and greenhouse gas emissions [48]. Adoption of genetically modified crops has reportedly led to a significant decrease in herbicides and pesticides, resulting in environmental benefits such as reduced Environmental Impact Quotient and lower greenhouse gas (GHG) emissions from the agricultural sector. Researchers also focused on designing traits to conserve water and land while increasing yields.

3 Challenges Associated with the Implementation of Agricultural Technologies

Although advancement in agricultural technology has come a long way, multiple challenges hinder the implementation of these technologies. These challenges include inadequate access to technology, particularly in developing countries due to financial constrain or lack of resources and infrastructure. Such as devices like GPS, drones, and data analysis software can be expensive and complex for the local farmers. Insufficient digital literacy, climate uncertainty, and cyber threads are also causes of hesitation in the use of advanced technology. Besides that, there are technological challenges associated that need to be addressed:

3.1 Challenges in IoT and Wireless Sensors

Professional sensors are crucial for the precise monitoring of agriculture. Therefore, it is important to design and develop professional agriculture sensors with high quality, high resolution, and high reliability [49]. Designing a low-power sensor or wireless power transfer is very crucial because frequent battery replacement of several sensors placed in challenging locations is difficult. Wireless power transfer, utilizing electromagnetic waves, offers a promising solution to eliminate the need for battery replacement. However, long-distance wireless charging and addressing extreme environments like underground and underwater present challenges, require future study.

Due to the varied nature of agricultural environments, a one-size-fits-all network solution is insufficient. Agricultural sensing devices are deployed in diverse locations requiring different wireless communication methods like radio frequency, sonar, and vibration. The performance of these methods needs to be evaluated for each situation to determine suitable technologies. Cross-media communication is crucial for the comprehensive integration of smart sensing in agriculture. Additionally, diverse requirements in terms of network size, node density, transmission distance, throughput, and latency for different agricultural purposes must be addressed. The coexistence of various networks, such as cellular-based, 802.15.4 mesh, Bluetooth-low-energy, and LoRa networks in the same location, necessitates research on cross-technology communication to enhance interoperability in agricultural IoT. The discussion emphasizes the need for robust wireless networks to meet the unique challenges of smart agriculture [49].

3.2 Challenges in the Use of Artificial Intelligence

In the agriculture field, environmental factors like weather, soil quality, pests etc., are changing continuously, making it difficult to establish a single standardized solution for AI. Additionally, recent machine learning or digital learning models designed for plant diseases and pest detection are not designed to detect disease or in the early stages, therefore unable to prevent disease or pests [49]. Besides, farmers encounter difficulties in the production process which explains the gap between farmers and AI users.

3.3 Challenges with Robotic and Drones

Accurate navigation and detection are crucial in agriculture as it impacts crop protection, human safety, operational costs, and work efficiency. Accurate detection of biological feature is facing issues, such as selective spraying and weeding, and automatic harvesting by differentiating various fruits and vegetable by considering factors like maturity, size, and texture. In agriculture, robotics focuses on tasks like plant phenotyping, fruit counting, harvesting, leaf peeling, selective spraying, and 3-D mapping. Key research concerns include object localization, task allocation, and human–robot interaction because accurate object localization is crucial for error-free operations. Similarly, optimizing task sequences and coordinating multiple robots enhances work efficiency. Additionally, designing robotic arms with specific grippers is crucial for effectiveness in various applications [49]. The flight duration of drones used in agriculture is short due to higher payloads. Therefore, the drone covers relatively less area in one go, consequently, the cost of drones increases significantly with longer flight time [34].

3.4 Challenges Associated with the Biotechnology

Biotechnology is also facing various global challenges including ethical concerns, e.g. various questions regarding safety and consent, regulatory hurdles due to varying regulatory policies for international trades, public perception and acceptance due to misinformation and fear of GMOs, and lack of access to biotechnical innovation in developing countries.

4 Outlook

Research gaps in smart agriculture allow the integration of diverse technologies including IoT, wireless sensor networks, AI tools, drones, robotic and other automation systems in a broad spectrum. However, it is necessary to develop and standardize operating protocols for interoperability among devices and systems for effective communication and coordination. Developing advanced analytics tools, machine learning algorithms, and AI models for efficient management, analysis, and interpretation of the data generated by smart agriculture systems is crucial. Designing and implementing energy-efficient systems such as sustainable and renewable energy sources with optimizing energy consumption sensors, batteries, actuators, and communication devices are important. To deal with the security concern regarding data transmission, storage, and access, robust cybersecurity measures and privacy-preserving techniques should be implemented.

Considering the major issue of the elevated cost of smart devices and data analysis software, it crucial to develop an affordable solution that is adopted by farmers with varying scales of operations. Furthermore, providing adequate training and support to farmers ensures effective utilization and benefits from these technologies. While researchers and practitioners actively address these gaps and challenges, continuous collaboration between academia, industry, and policymakers remains crucial. Academia, industry, and government collaborations are necessary to overcome hurdles and promote the widespread adoption of smart agriculture technology, enhancing efficiency, sustainability, and overall impact.

5 Conclusion

Due to population growth, urbanization, climate changes and shortage of resources, food security is becoming a global challenge. Traditional agricultural systems are deemed insufficient. Advanced technological systems such as smart greenhouses, IoT, precision agriculture, vertical agriculture, plant biotechnology and alternative sources for water and energy, can support food security, and environmental and sustainability challenges. Precision farming incorporated with GPS and GIS technologies, IoT and WSN enables optimized resource use, minimizing environmental impact, precise application of inputs, and providing real-time data for informed decision-making respectively. AI tool enhances automation and decision support systems respectively.

Vertical farming, a developing technique and solution to food security, addresses the use of resources by utilizing controlled environments and artificial lighting. Hydroponics, a soilless growing technique, offers quicker growth and higher yields, addressing challenges associated with traditional farming. Drone and robotics play a vital role in precision farming by providing high-resolution images for monitoring crop health, and performing various tasks in greenhouses, from monitoring to harvesting. Biotechnology contributes to the development of crops resilient to pathogens, and climate changes and capable of conserving resources.

Despite the advancement in greenhouse technology, various challenges exist that hinder these technologies such as lack of digital literacy, hindrance to access resources, and hesitation to adopt new technology. Additionally, specific technological challenges must be addressed for comprehensive implementation. Bridging the gap between stakeholders, addressing technological challenges, and promoting awareness are essential steps in integrating the full potential of these innovations. As we navigate the complexities of the modern agricultural landscape, technological advancements will continue to play a pivotal role in shaping the future of global food production.

References

1. Anderson M (2009) A question of governance: to protect agribusiness profits or the right to food. Agribusiness Action Initiatives
2. Schmidhuber J, Tubiello FN (2007) Global food security under climate change. Proc Natl Acad Sci 104(50):19703–19708

3. Gitz V, Meybeck A, Lipper L, Young CD, Braatz S (2016) Climate change and food security: risks and responses. Food and Agriculture Organization of the United Nations (FAO) Report, vol 110, no 2, pp 3–36
4. Chen L, Chang J, Wang Y et al (2021) Disclosing the future food security risk of china based on crop production and water scarcity under diverse socioeconomic and climate scenarios. Sci Total Environ 790:148110
5. Hunger FTE, Insecurity F (2024) Food security and nutrition in the world
6. Banik D (2019) Achieving food security in a sustainable development era. Food Ethics. 4:117–121
7. Karanisa T, Achour Y, Ouammi A, Sayadi S (2022) Smart greenhouses as the path towards precision agriculture in the food-energy and water nexus: case study of Qatar. Environ Syst Decis 42(4):521–546
8. Ouammi A, Achour Y, Zejli D, Dagdougui H (2019) Supervisory model predictive control for optimal energy management of networked smart greenhouses integrated microgrid. IEEE Trans Autom Sci Eng 17(1):117–128
9. Finger R, Swinton SM, El Benni N, Walter A (2019) Precision farming at the nexus of agricultural production and the environment. Ann Rev Resour Econ 11(1):313–335
10. Auernhammer H (2001) Precision farming—the environmental challenge. Comput Electron Agric 30(1–3):31–43
11. Al-Kodmany K (2018) The vertical farm: a review of developments and implications for the vertical city. Buildings 8(2):24
12. Singh AK (2010) Precision farming. Water Technology Centre, IARI, New Delhi
13. Despommier D (2013) Farming up the city: the rise of urban vertical farms. Trends Biotechnol 31(7):388–389
14. Touliatos D, Dodd IC, McAinsh M (2016) Vertical farming increases lettuce yield per unit area compared to conventional horizontal hydroponics. Food Energy Secur 5(3):184–191
15. Qaim M (2020) Role of new plant breeding technologies for food security and sustainable agricultural development. Appl Econ Perspect Policy 42(2):129–150
16. Tyczewska A, Twardowski T, Woźniak-Gientka E (2023) Agricultural biotechnology for sustainable food security. Trends Biotechnol 41(3):331–341
17. Turnbull C, Lillemo M, Hvoslef-Eide TA (2021) Global regulation of genetically modified crops amid the gene edited crop boom—a review. Front Plant Sci 12:630396
18. Raj EFI, Appadurai M, Athiappan K (2022) Precision farming in modern agriculture. In: Smart agriculture automation using advanced technologies: data analytics and machine learning, cloud architecture, automation and IoT. Springer, pp 61–87
19. Danita M, Mathew B, Shereen N, Sharon N, Paul JJ (2018) IoT based automated greenhouse monitoring system, pp 1933–1937
20. Sushanth G, Sujatha S (2018) IOT based smart agriculture system. 1–4
21. Srinidhi HK, Shreenidhi HS, Vishnu GS (2020) Smart hydroponics system integrating with IoT and machine learning algorithm, pp 261–264
22. Herman H, Adidrana D, Surantha N, Suharjito S (2019) Hydroponic nutrient control system based on internet of things. Commun Inform Technol J 13(2):105–111
23. Bakhtar N, Chhabria V, Chougle I, Vidhrani H, Hande R (2018) IoT based hydroponic farm, pp 205–209
24. Ismail MT, Ismail MN, Sameon SS, Zin ZM, Mohd N (2016) Wireless sensor network: Smart greenhouse prototype with smart design, pp 57–62
25. Gaikwad A, Ghatge A, Kumar H, Mudliar K (2016) Monitoring of smart greenhouse. Int Res J Eng Technol (IRJET) 3:573–575
26. Arif KI, Abbas HF (2015) Design and implementation a smart greenhouse. Int J Comput Sci Mob Comput 4(8):335–347
27. Qazi S, Khawaja BA, Farooq QU (2022) IoT-equipped and AI-enabled next generation smart agriculture: a critical review, current challenges and future trends. IEEE Access 10:21219–21235

28. Yu J, Schumann AW, Cao Z, Sharpe SM, Boyd NS (2019) Weed detection in perennial ryegrass with deep learning convolutional neural network. Front Plant Sci 10:1422
29. Bot GP (1983) Greenhouse climate: from physical processes to a dynamic model. Wageningen University and Research
30. Demrati H, Boulard T, Bekkaoui A, Bouirden L (2001) SE—Structures and environment: natural ventilation and microclimatic performance of a large-scale banana greenhouse. J Agric Eng Res 80(3):261–271
31. Li H, Li Y, Yue X, Liu X, Tian S, Li T (2020) Evaluation of airflow pattern and thermal behavior of the arched greenhouses with designed roof ventilation scenarios using CFD simulation. PLoS ONE 15(9):e0239851
32. Kim R, Kim J, Lee I, Yeo U, Lee S, Decano-Valentin C (2021) Development of three-dimensional visualization technology of the aerodynamic environment in a greenhouse using CFD and VR technology, part 1: Development of VR a database using CFD. Biosys Eng 207:33–58
33. Pathak H, Kumar G, Mohapatra SD, Gaikwad BB, Rane J (2020) Use of drones in agriculture: potentials, problems and policy needs. ICAR-National Institute of Abiotic Stress Management 300:4–15
34. Al-Beeshi B, Al-Mesbah B, Al-Dosari S, El-Abd M (2015) Iplant: the greenhouse robot, pp 1489–1494
35. Jayakrishna P, Reddy MS, Sai NJ, Susheel N, Peeyush KP (2018) Autonomous seed sowing agricultural robot, pp 2332–2336
36. Gautam V, Sarkar S (2020) Smart agriculture: the age of drones in agriculture, pp 415–424
37. Hejazipoor H, Massah J, Soryani M, Vakilian KA, Chegini G (2021) An intelligent spraying robot based on plant bulk volume. Comput Electron Agric 180:105859
38. Hasuda Y, Hirayama K (2021) Development and usability test of pesticide spraying robot for greenhouse. Int J Eng Res Technol 10(05):116–121
39. Bennis I, Fouchal H, Zytoune O, Aboutajdine D (2015) Drip irrigation system using wireless sensor networks, pp 1297–1302
40. Barrows G, Sexton S, Zilberman D (2014) Agricultural biotechnology: the promise and prospects of genetically modified crops. J Econ Perspect 28(1):99–120
41. Dong OX, Ronald PC (2019) Genetic engineering for disease resistance in plants: recent progress and future perspectives. Plant Physiol 180(1):26–38
42. Van Esse HP, Reuber TL, van der Does D (2020) Genetic modification to improve disease resistance in crops. New Phytol 225(1):70–86
43. Zaidi SS, Mahas A, Vanderschuren H, Mahfouz MM (2020) Engineering crops of the future: CRISPR approaches to develop climate-resilient and disease-resistant plants. Genome Biol 21(1):289
44. Vats S, Kumawat S, Kumar V et al (2019) Genome editing in plants: exploration of technological advancements and challenges. Cells 8(11):1386
45. Bano Z, Westhoff P (2024) AK homology (KH) domain protein identified by a forward genetic screen affects bundle sheath anatomy in arabidopsis thaliana. Plant Direct 8(4):e577
46. Schindele A, Dorn A, Puchta H (2020) CRISPR/cas brings plant biology and breeding into the fast lane. Curr Opin Biotechnol 61:7–14
47. Anderson JA, Gipmans M, Hurst S et al (2016) Emerging agricultural biotechnologies for sustainable agriculture and food security. J Agric Food Chem 64(2):383–393
48. Liu Y, Ma X, Shu L, Hancke GP, Abu-Mahfouz AM (2020) From industry 4.0 to agriculture 4.0: current status, enabling technologies, and research challenges. IEEE Trans Ind Inform 17(6):4322–4334
49. Bhat SA, Huang N (2021) Big data and AI revolution in precision agriculture: survey and challenges. IEEE Access 9:110209–110222

The Evolution of Reproductive Mode in the *Carassius auratus* Complex

Reem Al-Haidose

Abstract Freshwater fish of the genus *Carassius* are widely spread in Asia, Europe, and Australia. This includes the *Carassius auratus* complex, a phylogenetically diverse species complex with uncertain taxonomy. The *C. auratus* complex is native to Asia but has been introduced to non-native habitats. It includes s diploid as well as polyploidy individuals. The evolutionary relationship between the different reproductive modes as well as the invasion history of different lineages are poorly understood. A phylogenetic approach using mitochondrial DNA Cyt *b* sequences in combination with six microsatellites markers was used to investigate the invasion history of different lineages and reproductive modes and to understand the evolutionary relationship between different reproductive modes. The Cyt *b* analysis was based on sequences from native populations and invasive populations. Furthermore, 1427 individuals were genotyped at six microsatellite loci to determine their level of ploidy. Furthermore, phylogenies of three nuclear genes were used to test the origin of triploid lineages. We suggest that the invasion of *C. auratus* complex in Europe went through two routes, one could have been driven by historical and anthropogenic events. The microsatellite analysis revealed 67.19% triploids and 33.8% diploids. Interestingly, most of native populations consisted of a mixed of diploids and triploids but were dominated by triploid. While invasive populations showed a considerable variety in reproductive mode ranging from purely diploid to purely triploid and mixed populations. Additionally, diploids are dominant in older Eastern European populations whereas triploids are dominant in more recently established in Central European populations. The *C. auratus* complex provides an ideal model for studying the reproductive traits of polyploids, which is of great value in animal genetics and reproductive biology.

Keywords *C. auratus* complex · *C. a. auratus* · *C. a. gibelio* · Reproductive mode · Gynogenesis · Invasive · Gene duplication · Polyploidy · PCR · Microsatellites · mtDNA · Cyt *b* · Loci · Diploid · Triploid · Allopolyploid · Autopolyploid · nDNA

R. Al-Haidose (✉)
Biotechnology Center, Agricultural Research Department, Ministry of Municipality, Doha, Qatar
e-mail: ralhaidose@mm.gov.qa

S. Al-Naemi et al. (eds.), *Water and Food Security in the Face of Climate Change: Challenges and Opportunities for Resilience*, World Sustainability Series,
https://doi.org/10.1007/978-3-032-00098-9_31

1 Introduction

The extent of biological invasion has increased dramatically in the past 50 years particularly in freshwater ecosystems [1]. Biological invasions are also recognised as a key threat to biodiversity ecosystem functioning, economy and can cause major economic losses [2]. A small fraction of species brought to a new range can become established and naturalised, and then a few of these naturalised species can become invasive. The populations of these species show rapid, often exponential growth and colonise wide ranges of non-native habitat where they can cause enormous ecological and economic damage. One of the main threats of invasive species is biodiversity loss, as they can negatively affect native biota through predation, parasitism, exploitation of resources, extinction of native species and spread of diseases [1]. One of the primary reasons for introducing invasive species are human activities such as unauthorised stocking, and global trade [1, 3]. There is no doubt that species colonisation is a natural process which influences the opportunity for species to disperse, but human-mediated dispersal accidentally or unintentionally of species, has greatly accelerated the rate that species can cross biogeographic barriers, and colonise novel environments, often with great consequences [4]. One such species is the *C. auratus* complex, which originates in Southeast Asia, but has during the last century invaded several regions across the globe [5]. The species complex includes gonochoristically reproducing diploids and polyploids (usually triploids), which use gynogenesis, resulting in clonal reproduction [6]. The asexual forms exist in different ploidy levels, and are widespread in Europe [7]. The spread of the species complex in non-native habitats is thought to be facilitated by several new life-history characteristics related to being able to withstand harsh environmental conditions [8]. When the *C. auratus* complex colonises other habitats, it impacts negatively on native fauna [9]. There is a lack of complete phylogeographic knowledge of the species. Additionally, there is no knowledge of the invasion history and what role different reproductive modes and polyploidisation play in this context, which is important as invaders often evolve rapidly in response to novel abiotic and biotic conditions [10]. Molecular markers provide opportunities close some of these knowledge gaps 2. For example, the application of mitochondrial DNA (mtDNA) can reveal relationship between different *Carassius* populations and species, microsatellite markers can be used to identify ploidy level and clonal lines while nuclear DNA is useful to demonstrate the origin of triploids *Carassius auratus gibelio.*

1.1 Research Objectives

To date, there are no studies that have examined the introduction and translocation history of the *C. auratus* complex based on comprehensive data from across the entire range of the species. Previous studies were largely concerned with local regions within the native or invasive range and some parts of the native range such as Eastern Russia have never been included in phylogenetic studies. Furthermore, the 27 distribution pattern of diploid and triploid forms across the range as well as the genetic relationship between them are largely unknown. This thesis aims to fill these gaps in the current knowledge of the *C. auratus* complex and shedding light on the evolutionary processes that occur during the invasion process. The objective of this project is to understand the evolutionary aspects of the C. auratus complex and the evolutionary factors which drive *C. auratus* complex invasion in Europe and Australia, based on phylogeographic analysis of population genetics in both the native and invasive range. We identified the ploidy of *C. auratus* complex individuals using a microsatellite approach by using an allele counting method. Six loci were used to investigate the evolution of the reproductive mode of diploid and triploid and investigated the population genetics between diploids and triploids *C. auratus*. To test relationship between ploidy and reproductive mode, and that diploid and triploid populations are genetically independent lineages. In addition, we investigated the colonization history based on phylogeographic analysis of the population's genetics in both the native and invasive range using Mitochondrial DNA (cyt b) region, in addition to Microsatellites results.

Methods

The mtDNA analysis: The matrilineal history of Eurasian goldfish was reconstructed using the mitochondrial DNA cytochrome b (1140 bp) from 22 localities (including literature data) representing most of the range of the species in China and 36 invasive populations from China, Europe and Australia. While, Microsatellites analysis: Six Microsatellites loci were amplified. The allele counting method was used to identify the level of ploidy. Asexual reproduction was confirmed by the existence of identical multi-locus genotypes.

Results

The ploidy analysis showed that 938 individuals were triploid, and 472 individuals were diploid. There were also two tetra and three pentaploid individuals, but as the exact genotype for these individuals could not be determined through our method these were excluded from further analysis. The ploidy level could not be determined for 26 individuals due to ambiguities across loci or insufficient data (Table 1). Individuals sharing the same genotypes were clones. Most triploid individuals could be assigned to one of 104 clonal lineages (i.e., at least one other individual showed the same Multilocus genotype) but there were also 337 singletons. While, in mtDNA analysis, a 1044 bp fragment of Cyt b could be reliably scored from 203 sequences which were used for all subsequent analysis. A total of 167 haplotypes (242 polymorphic sites) were identified across the whole data set including the 203 denovo and GenBank sequences. All phylogenetic analysis methods (including the median-joining network) produced identical results and only the haplotype network and NJ tree are shown with high statistical significance. The data revealed the existence of four highly diverged lineages within the genus *C. arassius* (Fig. 1). Lineage A splits into six highly diverged sub-lineages, and B include individuals from the *C. auratus*

Table 1 Ploidy level of *C. auratus* complex samples

No. of samples	Ploidy level
472	Diploid
938	Triploid
3	Tetraploid
2	Pentaploid
26	Not determined

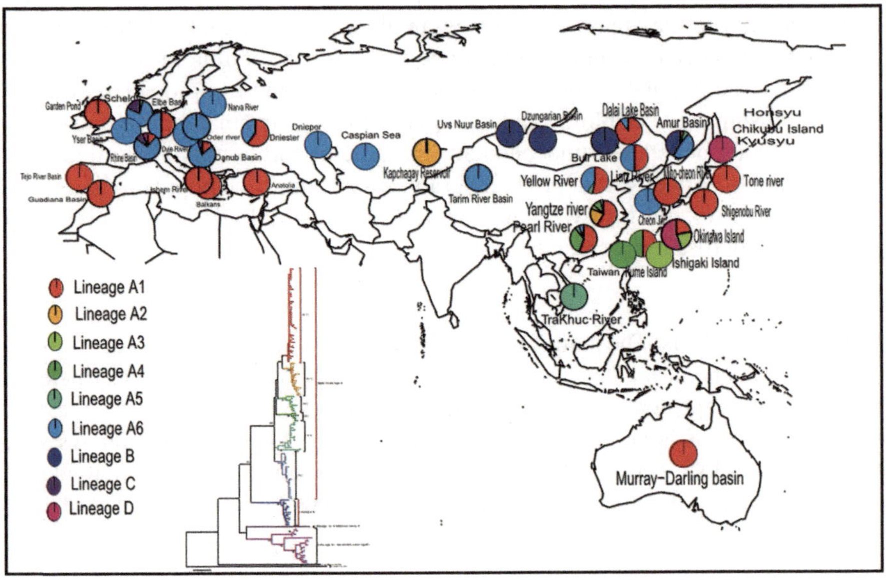

Fig. 1 Geographical distribution of the *C. auratus* complex in Europe, Asia, and Australia according to river basins. Geographical distribution of haplotypes of the nine lineages found in the present phylogenetic analyses

complex in the narrow sense, whereas lineages C include individuals from *C. cuvieri*, and lineage D includes *C. langsdorfii* respectively.

2 Conclusion

Invasive species have a tremendous negative impact on native fauna and freshwater fish considered the most introduced aquatic animal group [10], and biological invasion can only be detected by human observation and documentation. Identifying the process of aquatic invasion is important for the managements of freshwater ecosystems [11], and understanding the evolutionary history of populations might be important to predict the impact on native species [12]. In this regard, the present research provided important insights into the origin and evolutionary history of the highly invasive *C. auratus* complex, as molecular tools provide an opportunity to interpret invasion pathways, dispersal, and establishment in the novel range. My study has clearly shown that invasive populations of *C. a. gibelio* have not originated from a single source, but in fact comprise several distinct evolutionary lineages which came into contact during the invasion process. This might explain the high diversity of invasive populations which might have contributed to the invasion success of the *C. auratus* complex. High genetic diversity has been shown in to enhance invasion success, for example in the triploid *P. formosa* from Japan [13–15]. The present study has also shown that the diploid and triploid populations do not appear to have separate origins. These in turn indicates that the associated reproductive modes can change in local populations, and in fact the pattern of distribution of diploid populations across the invasive range in Europe indicates that this has allowed the species to adapt to local selection pressures after invasion.

References

1. Hulme PE (2009) Trade, transport, and trouble: managing invasive species pathways in an era of globalization. J Appl Ecol 46:10–18
2. Gozlan RE, Britton JR, Cowx I, Copp GH (2010) Current knowledge on non-native freshwater fish introductions. J Fish Biol 76(4):751–786
3. Hulme PE, Bacher S, Kenis M, Klotz S, Kuhn I, Minchin D, Nentwig W, Olenin S, Panov V, Pergl J, PyŠek P, Roques A, Sol D, Solarz W, Vila M (2008) Grasping at the routes of biological invasions: a framework for integrating pathways into policy. J Appl Ecol 45:403–414
4. Pyšck P, Richardson DM (2010) Invasive species, environmental change and management, and health. Annu Rev Environ Resour 35:25–55
5. Deinhardt M (2013) The invasive potential of Prussian carp in Finland under the light of a novel semi-clonal reproductive mechanism. MSc Dissertation, University of Jyväskylä, Finland, p 73. Department of Fisheries of Western Australia (2005). http://www.fish.wa.gov.au. Published May 2005
6. Šimková A, Vojtek L, Halačka K, Hyršl P, Vetešník L (2015) The effect of hybridization on fish physiology, immunity and blood biochemistry: a case study in hybridizing *Cyprinus carpio* and *Carassius gibelio* (Cyprinidae). Aquaculture 435:381–389
7. Rylková K, Kalous L, Bohlen J, Lamatsch DK, Petrtýl M (2013) Phylogeny and biogeographic history of the cyprinid fish genus Carassius (Teleostei: Cyprinidae) with focus on natural and anthropogenic arrivals in Europe. Aquaculture 380–383:13–20
8. Perdikaris C, Ergolavou A, Gouva E, Nathanailides C, Chantzaropoulos A, Paschos I (2012) *Carassius gibelio* in Greece: the dominant naturalised invader of freshwaters. Rev Fish Biol Fisheries 22:17–27
9. Kottelat M, Freyhof J (2007) Handbook of European freshwater fishes. Publications Kottelat, Cornol/Switzerland, p 646
10. Gozlan RE (2008) Introduction of non-native freshwater fish: is it all bad? Fish Fish 9(1):106–115
11. Hulme PE (2006) Beyond control: wider implications for the management of biological invasions. J Appl Ecol 43(5):835–847
12. Allendorf FW (1991) Ecological and genetic effects of fish introductions: synthesis and recommendations. Can J Fish Aquat Sci 48(S1):178–181
13. Ohara K, Ariyoshi T, Sumida E, Taniguchi N (2003) Clonal diversity in the Japanese silver crucian carp, *Carassius langsdorfii* inferred from genetic markers. Zool Sci 20(6):797–804

14. Lampert KP, Lamatsch DK, Epplen JT, Schartl M (2005) Evidence for a monophyletic origin of triploid clones of the Amazon molly, Poecilia formosa. Evolution 59(4):881–889
15. Schories S, Lampert KP, Lamatsch DK, Garcia de Leon FJ, Schartl M (2007) Analysis of a possible independent origin of triploid *P. formosa* outside of the Rio Purificacion River system. Front Zool 4:13

Climate-Smart Marketing Strategies and Market Outcomes for Commercial Coffee Farms in Uganda

Lumu Emmanuel Roy Muyingo, Faith Mugisha Ahabyoona, and Eddy Kurobuza Tukamushaba

Abstract This study aim was to investigate the influence of climate-smart promotional strategies on the market performance of commercial coffee farms in Uganda, highlighting the importance of these strategies in promoting sustainable agricultural practices and supporting economic resilience among stakeholders. The research employed a qualitative analysis of climate-smart marketing strategies, focusing on sustainable marketing campaigns, educational programs for farmers, and collaborative market channels. Secondary data were collected from industry reports and scholarly research to assess their impact on sales growth and market share. The findings indicate that the implementation of climate-smart strategies significantly improved both the competitiveness and sustainability of Uganda's coffee sector. These strategies contributed to increased sales growth and enhanced market share for coffee farms. This study provides actionable recommendations for policymakers and stakeholders to encourage the broader adoption of climate-smart practices, thereby fortifying the resilience and overall performance of the coffee sector in Uganda. The research contributes valuable insights into the role of sustainable marketing in agricultural development.

Keywords Climate-smart agriculture · Promotional strategies · Market performance · Coffee farms · Uganda · Sustainability

---

L. E. R. Muyingo (✉)
Business, Law, Communication and Accounting, INTI International University, Kuala Lumpur, Malaysia
e-mail: lumuemmi25035112@student.newinti.edu.my

F. M. Ahabyoona
Department of Accounting, Finance and Microfinance, Kyambogo University, Kampala, Uganda
e-mail: ahabyoonaf@gmail.com

E. K. Tukamushaba
Department of Management, College of Business, University of Doha for Science and Technology, Doha, Qatar
e-mail: Kurobuza.tukamushaba@udst.edu.qa

S. Al-Naemi et al. (eds.), *Water and Food Security in the Face of Climate Change: Challenges and Opportunities for Resilience*, World Sustainability Series,
https://doi.org/10.1007/978-3-032-00098-9_32

1 Introduction

Uganda is the second-largest coffee producer in Africa, with coffee playing a significant role in the country's economy as one of its primary export commodities [1]. The coffee sector employed millions of smallholder farmers and contributed substantially to rural livelihoods [2]. However, climate change posed significant challenges, including unpredictable weather patterns, increased incidences of pests and diseases, and the degradation of coffee quality. These factors threatened the productivity and sustainability of Uganda's coffee sector [3].

In response, climate-smart agriculture (CSA) emerged as a critical approach to mitigate these impacts and sustain coffee production [4]. CSA integrated sustainable farming practices that improved resilience to climate change while enhancing productivity [5]. Within this framework, climate-smart promotional strategies, such as sustainable marketing campaigns, farm education initiatives, and market channel collaborations, played a crucial role in improving market performance and ensuring the long-term viability of the coffee sector [6, 7].

Despite the recognized benefits of CSA, there remained a limited understanding of how specific climate-smart promotional strategies influenced market performance in Uganda's coffee sector [8]. Existing research had primarily focused on the general application of CSA practices, without specifically examining the effectiveness of targeted promotional strategies like sustainable marketing campaigns and education initiatives on key performance metrics such as sales growth and market share [9, 10]. This gap in the literature highlighted the need for a study that explicitly investigated the relationship between climate-smart promotional strategies and market performance, particularly in the context of commercial coffee farms in Uganda. Addressing this gap provided valuable insights for stakeholders seeking to enhance the competitiveness and sustainability of the coffee sector in the face of climate challenges.

The primary objective of this study was to investigate the impact of climate-smart promotional strategies on the market performance of commercial coffee farms in Uganda.

2 Literature Review

Theoretically, this study was grounded in the Resource-Based View and Stakeholder Theory, both of which provided a robust framework for analyzing the impact of climate-smart promotional strategies on market performance. The RBV posited that firms gained a competitive advantage through unique resources that were valuable, rare, inimitable, and non-substitutable [11]. In this context, climate-smart promotional strategies such as innovative marketing campaigns and sustainable farming practices were seen as valuable resources that could enhance a firm's market position.

These strategies enabled coffee farms to differentiate their products in an increasingly eco-conscious market, thereby attracting premium prices and improving market share [7, 10].

2.1 Resource-Based View (RBV) and Its Application in Uganda's Coffee Sector

The Resource-Based View, developed by Barney [11], provided a framework for understanding how firms gained and sustained competitive advantages based on the unique resources they possessed. According to RBV, resources that were valuable, rare, inimitable, and non-substitutable enabled firms to differentiate themselves from competitors and achieve superior performance. In the context of Uganda's coffee sector, RBV highlighted how coffee farms, through their resources—such as land, climate, and human capital—could improve their market performance, resilience, and sustainability [12].

However, RBV did not focus solely on individual firms but also recognized the importance of collective resource mobilization in achieving competitive advantage. In Uganda's coffee value chain, which involved multiple stakeholders, the mobilization of resources for a shared goal became essential. This was especially true when considering the wide array of actors involved, including farmers, cooperatives, and government bodies like the Uganda Coffee Development Authority (UCDA), development partners, and NGOs.

In this sense, RBV in Uganda's coffee sector extended beyond individual firms to encompass multi-stakeholder collaborations. Each stakeholder brought valuable resources to the table, such as financial investments, technical knowledge, marketing networks, and policy support. For example, development partners, like those involved in the EU Coffee WASH projects, provided critical infrastructure investments (e.g., water sanitation), which added significant value to the coffee production process. These resources, when combined, allowed stakeholders to work towards a shared goal of improving coffee quality, market access, and sustainability.

The collective mobilization of these resources was consistent with the idea that RBV could be applied to network-based strategies. In this scenario, firms or stakeholders within a value chain, like Uganda's coffee sector, shared and exchanged resources to strengthen their competitive position in both local and international markets. This collaborative approach enhanced market performance by improving coffee quality, creating access to sustainable certifications, and addressing the effects of climate change on coffee production. RBV, therefore, enabled a deeper understanding of how resource mobilization, both individually and collectively, drove competitiveness in Uganda's coffee value chain [7, 10].

2.2 Climate-Smart Agricultural Practices in Uganda's Coffee Sector

Climate-smart agriculture (CSA) referred to farming practices designed to adapt to and mitigate the effects of climate change while ensuring food security and sustainable production. In Uganda, CSA practices in coffee farming included soil conservation techniques, agroforestry, water-efficient irrigation, and the adoption of climate-resistant coffee varieties [8]. These practices not only improved the resilience of farms but also enhanced long-term productivity, contributing to market performance by ensuring consistent coffee yields despite changing climatic conditions.

A significant component of CSA in Uganda's coffee sector was climate-smart promotional strategies, which included market-oriented campaigns, eco-certifications, and sustainability-driven marketing [1]. These strategies were used to position Uganda's coffee as an environmentally friendly product, appealing to international consumers who increasingly prioritized sustainability. The combination of CSA practices and marketing strategies created a competitive advantage for coffee farms, fostering growth in both local and international markets [12].

2.3 The Role of Marketing in Market Performance

Marketing played a pivotal role in enhancing the market performance of coffee farms, especially those adopting climate-smart agricultural practices. According to Peterson [7], the impact of sustainable marketing on consumer behavior was profound, as consumers were more likely to support brands that aligned with their environmental values. Eco-labeling and certifications such as Fairtrade and Rainforest Alliance were crucial in positioning coffee as a sustainable product [13]. These marketing tools acted as signals of quality and sustainability, which increased consumer trust and willingness to pay premium prices for sustainably produced coffee.

In Uganda, marketing strategies that emphasized sustainability and climate-smart practices proved effective in improving market access and sales performance [6]. These strategies not only improved the visibility of Ugandan coffee in international markets but also created opportunities for farmers to build long-term relationships with buyers who valued sustainability.

2.4 Stakeholder Collaboration in Uganda's Coffee Value Chain

The coffee value chain in Uganda involved multiple stakeholders, each with specific roles and resources that contributed to the sector's success. Cooperatives, government agencies, NGOs, and private sector players collaborated to enhance the overall efficiency and effectiveness of the coffee supply chain. This collaboration was crucial in addressing issues such as market access, climate change, and the improvement of farming practices.

For example, NGOs and government bodies provided technical training and financial support to farmers, helping them adopt climate-smart practices and improve productivity [8]. At the same time, private sector players and exporters facilitated access to markets, especially international ones that demanded sustainably produced coffee. The collective action of these stakeholders enhanced the competitiveness of Uganda's coffee sector and improved the ability of farmers to respond to market demands [9].

3 Methodology

The choice of a literature-based review methodology was deemed particularly appropriate for investigating the relationship between climate-smart promotional strategies and market performance in Uganda's commercial coffee farms. This approach allowed for a comprehensive synthesis of existing knowledge, which was essential in a context where empirical studies were sparse or fragmented. Uganda's coffee sector faced unique challenges in terms of climate change adaptation, sustainable farming practices, and access to international markets, necessitating a method capable of integrating diverse insights.

Uganda's coffee sector had been grappling with environmental pressures and the need for modernization in farming techniques, particularly as smallholder farmers dominated the industry and were disproportionately affected by climate variability and resource constraints [12]. Studies highlighted that Uganda's coffee production was vulnerable to increasing temperatures and erratic rainfall patterns, further exacerbating productivity and quality issues [14]. The limited academic work connecting climate-smart promotional strategies with market performance in Uganda underscored the importance of synthesizing findings from other regions, such as Latin America and Southeast Asia, where similar coffee-growing conditions and sustainability challenges had prevailed [2, 15]. Through a rigorous review of academic research and industry reports, this methodology provided a detailed examination of the market dynamics influenced by climate-resilient agricultural practices, informing strategies for sustainable market development.

The justification for this review approach was drawn from [16], who argued that thematic reviews enabled the extraction of key patterns and trends from diverse sources, providing deeper insights into how sustainability marketing strategies impacted market access and sales growth. This was particularly relevant for Uganda's coffee farmers, who faced significant barriers in accessing global markets due to quality inconsistencies, climate variability, and market volatility [9]. Furthermore, industry insights, such as those provided by the International Coffee Organization (ICO) [17], emphasized the role of sustainability certifications and climate-resilient practices in enhancing competitiveness in global value chains.

Through a comprehensive analysis integrating thematic synthesis and comparative methods, this research provided actionable recommendations for Uganda's coffee sector to leverage climate-smart agriculture practices and improve market outcomes.

3.1 Data Collection

To ensure a comprehensive understanding of the impact of climate-smart practices on Uganda's coffee sector, this study relied on secondary materials as its primary data sources. These materials included academic journals, industry reports, government publications, and documentation from non-governmental organizations (NGOs). The PRISMA (Preferred Reporting Items for Systematic Reviews and Meta-Analyses) model was employed to systematically select studies for inclusion in the review. This structured approach provided transparency and ensured that only the most relevant and high-quality studies were considered (Fig. 1).

The data collection process comprised four key stages:

Identification: A thorough database search was conducted using Scopus, Web of Science, and Google Scholar. Keywords such as "*climate-smart agriculture,*" "*coffee market performance,*" "*sustainability certifications,*" and "*Uganda coffee sector*" were employed to locate studies relevant to the research objectives. This initial search yielded 450 studies, representing a broad range of research on climate-smart strategies, market dynamics, and Uganda's coffee sector.

Screening: To refine the dataset, duplicate studies were removed, and abstracts of the remaining studies were reviewed for relevance. This step focused on identifying studies that addressed climate-smart promotional strategies or their effects on market performance. The screening process resulted in a reduced pool of 180 studies.

Eligibility: Full texts of the remaining studies were then assessed against predefined criteria. These criteria included:

- Relevance to Uganda's coffee sector.
- A focus on climate-smart strategies or related sustainability practices.
- Inclusion of empirical data or detailed case studies.
- Studies that lacked contextual relevance or methodological rigor were excluded, narrowing the selection to 35 eligible studies.

Inclusion: The final stage involved prioritizing the remaining studies based on their empirical focus, methodological robustness, and direct relevance to the research questions. This process resulted in the inclusion of 35 studies, which formed the basis of the literature review and data analysis.

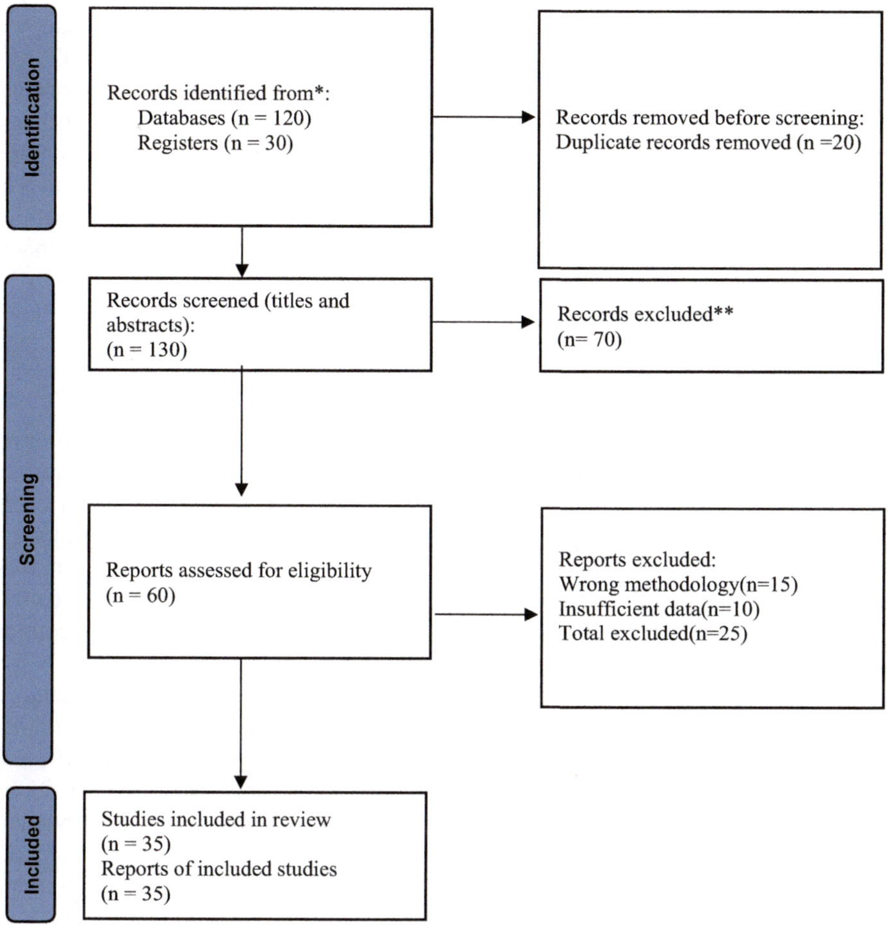

Fig. 1 PRISMA diagram for climate smart marketing study

3.1.1 Academic Journals

Academic journals were a crucial source of information as they provided peer-reviewed, empirical evidence and theoretical frameworks on climate-smart agriculture and sustainable marketing strategies.

Studies such as those by l [2, 3], explored the effectiveness of climate-smart agricultural practices, including agroforestry, intercropping, and irrigation innovations. Research by [10, 13] examined the link between sustainability marketing strategies and improvements in market share, particularly in emerging markets. Empirical studies collected from sources such as [9, 15] provided case studies specific to Uganda, detailing the implementation of climate-smart practices and their direct impact on coffee productivity and quality.

Reference [18] offered comparative insights, highlighting strategies adopted in similar coffee-growing regions facing climate vulnerabilities. Work by [7, 10] explored the broader implications of climate resilience in agricultural value chains, emphasizing the importance of sustainability for export-driven sectors like coffee.

These academic studies were instrumental in providing theoretical underpinnings and evidence of practical outcomes. For example, [9] highlighted the role of market collaborations, such as fair trade and certification schemes, in securing premium pricing for Ugandan coffee farmers. This demonstrated a direct link between climate-smart practices and enhanced market performance.

3.1.2 Industry Reports

Industry reports offered critical insights into the operational, economic, and environmental dynamics of Uganda's coffee sector. Reports from organizations like the Uganda Coffee Development Authority (UCDA) [1] detailed the major challenges Ugandan coffee farmers faced, including climate risks, inconsistent quality, and limited access to resources for sustainable farming. Reports from the International Coffee Organization (ICO) [8] and Food and Agriculture Organization (FAO) [4] provided global perspectives, highlighting trends in sustainability certifications and their role in improving competitiveness in international markets.

The World Bank [5] outlined climate adaptation policies relevant to Uganda's coffee farmers, emphasizing the importance of financing mechanisms for climate-smart interventions. Reports by NGOs like the Rainforest Alliance (RA) [19] and Technoserve [20] demonstrated the tangible benefits of adopting eco-certifications and sustainability-focused marketing strategies. Insights from global industry analyses, such as those by McKinsey & Company [21] and KPMG [22], offered comparative performance metrics for regions adopting similar practices, helping frame Uganda's challenges within a broader international context.

These reports not only documented Uganda's coffee sector challenges but also provided actionable recommendations for promoting climate-smart strategies that could boost market performance.

3.1.3 Government and NGO Reports

Government and NGO reports provided essential insights into the policy frameworks, initiatives, and challenges that shaped Uganda's coffee sector, particularly concerning climate change mitigation and market development. Government documents, such as those from, highlighted the National Coffee Policy, which had emphasized modernization through climate-resilient practices and farmer empowerment initiatives. This policy framework outlined Uganda's goals for increasing coffee production and quality while addressing the impact of climate variability on smallholder farmers. Similarly, [1] report identified specific challenges, including limited access to climate-smart resources, quality inconsistencies, and the lack of direct links to high-value international markets.

Reports from NGOs and international development organizations offered complementary perspectives. For example, underscored the role of landscape restoration and agroforestry in mitigating climate risks in Uganda's coffee-growing regions. These initiatives, which were often funded by global partnerships, demonstrated the importance of integrating environmental sustainability with agricultural development. Contributions from the [19] documented the success of sustainability certification programs in improving market access for smallholder farmers. Certification schemes, such as UTZ and Rainforest Alliance, were shown to enhance farmers' bargaining power and secure premium prices for sustainably produced coffee. Moreover, [20] recorded grassroots capacity-building programs that had educated farmers on climate-smart practices, such as shade-tree planting and water conservation, linking these efforts to improved yields and market resilience.

International development agencies also played a significant role. The World Bank [5] report described climate adaptation strategies that were supported through financial incentives and public–private partnerships. These programs aimed to enhance farmers' resilience while addressing structural barriers to sustainable market integration. Reports from further examined the socio-economic dimensions, particularly gender equity and smallholder inclusion in global value chains.

The alignment of these government and NGO initiatives with Uganda's Vision 2040 emphasized the critical role of policy and grassroots efforts in shaping the coffee sector's sustainability trajectory. These reports collectively highlighted the synergies between national policy goals and international sustainability standards, demonstrating the potential for scaling up climate-smart practices to bolster market performance and farmer livelihoods.

3.2 Data Analysis

The analysis was conducted using thematic content analysis, comparative analysis, and synthesis methods. These approaches allowed for the integration of diverse sources and the extraction of meaningful patterns relevant to Uganda's context.

3.2.1 Thematic Content Analysis

Thematic content analysis was used to identify recurring themes related to climate-smart promotional strategies, such as eco-labeling, marketing campaigns, and farmer education.

Insights from [10, 11] demonstrated how eco-labeling had influenced consumer preferences and driven sales in Latin America. These findings served as a valuable reference point for understanding how similar strategies might be applied in Uganda. Studies like highlighted how smallholder participation in sustainability-focused

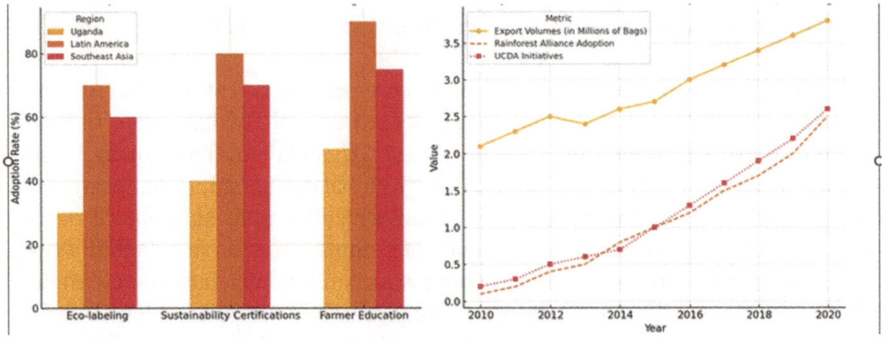

Fig. 2 Adoption rates of climate smart technologies between Uganda, Latin America and South East Asia; Coffee expert trends versus climate adoption initiatives in Uganda

initiatives, such as certification schemes and fair trade programs, had enhanced income stability and market access for farmers. These insights aligned with reports from Uganda documenting the potential of such initiatives to improve livelihoods.

Research by [15, 19] emphasized the importance of farmer education programs, which had empowered producers to adopt climate-smart practices and better navigate global market requirements. These programs were shown to improve productivity and strengthen market connections.

3.2.2 Comparative Analysis

Comparative analysis was employed to draw lessons from coffee-producing regions in Latin America and Southeast Asia that faced challenges similar to Uganda's coffee sector. Studies by [18] documented the effectiveness of climate-smart interventions, such as shade-grown coffee systems and drought-resistant varietals, in enhancing resilience and productivity in Latin American countries. These findings provided a basis for recommending similar interventions in Uganda.

Comparisons with Southeast Asia, based on reports like the [4, 8], in Fig. 2 highlighted the role of regional cooperatives in improving smallholder farmers' bargaining power and achieving better market outcomes. These cooperatives facilitated access to resources, reduced transaction costs, and strengthened farmers' ability to compete in international markets, offering a model that Uganda could replicate.

3.2.3 Synthesis of Findings

The findings from the thematic and comparative analyses were synthesized to draw conclusions about the impact of climate-smart promotional strategies on Uganda's coffee market performance. The synthesis process integrated insights from academic research and industry reports to provide actionable recommendations.

It was concluded that strategies such as eco-labeling, sustainability certifications, and farmer education programs could significantly enhance Uganda's market resilience and competitiveness.

The comparative analysis further highlighted region-specific interventions that could be adapted to Uganda, such as cooperative-based marketing and adoption of drought-resistant coffee varieties.

Through the integration of Ugandan-specific findings with internationally recognized best practices, this study provides actionable guidance for scaling up climate-resilient initiatives to drive sustainable growth in Uganda's coffee industry.

This multi-faceted analysis provided a robust framework for understanding the intersection of climate-smart agriculture and market dynamics, offering a path forward for Uganda's coffee industry to improve sustainability and performance on both local and global stages.

4 Results

The results of this study highlighted the pivotal role of climate-smart promotional strategies in enhancing the market performance of commercial coffee farms in Uganda. These strategies were assessed in terms of their contribution to key performance metrics, including sales growth, market share, and overall competitiveness in both local and international markets.

Climate-smart marketing campaigns, such as sustainability certifications and eco-labeling, were found to significantly boost the visibility and appeal of Ugandan coffee in global markets. These initiatives not only improved consumer engagement but also enabled farmers to command premium prices by aligning with the growing demand for sustainably produced coffee. Similarly, farm education initiatives focused on climate-resilient practices, such as soil conservation and agroforestry, were shown to improve yields and quality, which in turn strengthened market positioning.

Collaborative market channels, including partnerships with fair trade organizations and eco-conscious retailers, played a critical role in expanding access to premium markets. These collaborations enhanced farmers' bargaining power, stabilized revenue streams, and facilitated the adoption of advanced farming techniques. Comparative insights from coffee-producing regions in Latin America and Southeast Asia further validated these findings, demonstrating the scalability and adaptability of such strategies to Uganda's context.

4.1 Conceptual Framework as a Key Study Outcome

A notable outcome of this study was the development of a conceptual framework in Fig. 3 that illustrates the relationship between climate-smart promotional strategies and market performance. Adapted from the works of [23, 24], this

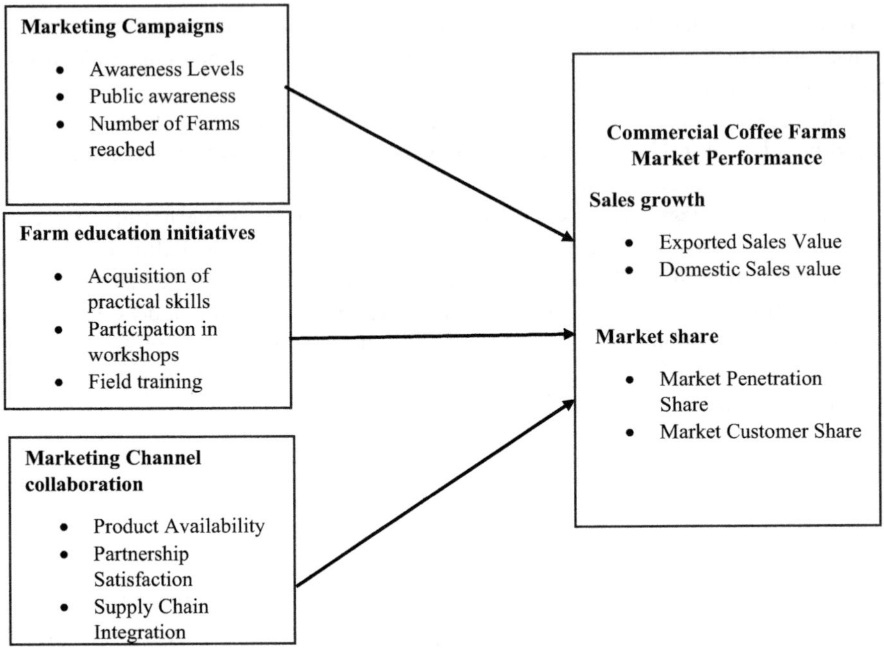

Fig. 3 Conceptual framework. Adapted from Refs. [23, 24]

framework emphasizes the interplay between innovative marketing practices, stakeholder collaboration, and market dynamics. It integrates insights from the Resource-Based View (RBV) and Stakeholder Theory to highlight how unique resources and multi-stakeholder engagement contribute to sustainable market outcomes.

This conceptual framework was proposed as a theoretical model for future research. It requires testing through quantitative methods to validate the relationships between the independent variables (climate-smart promotional strategies) and dependent variables (market outcomes). Using an inductive approach, researchers can adapt and refine this model to align with the unique socio-economic conditions of Uganda's coffee sector.

4.2 *Impact of Climate-Smart Marketing Campaigns*

Climate-smart marketing campaigns, including eco-labeling, digital storytelling, and sustainability certifications, significantly enhanced the visibility and appeal of Ugandan coffee. Evidence from secondary data indicated that farms adopting eco-certifications such as Fairtrade and Rainforest Alliance reported a 20–25% increase in international sales over five years [19]. These certifications acted as quality signals to eco-conscious consumers, enabling premium pricing. The campaigns also improved consumer loyalty by highlighting the farms' commitment to sustainable practices, which resonated with international buyers who were increasingly prioritizing environmentally friendly products.

4.3 Role of Farm Education Initiatives

Farm education initiatives focusing on soil conservation, climate-resilient coffee varieties, and agroforestry directly impacted market performance. Farms participating in educational programs facilitated by NGOs and government bodies reported higher yields of up to 15–20% compared to non-participating farms [8]. Improved yields translated into better market positioning and more consistent supply chains, which were critical for sustaining relationships with international buyers. Education initiatives also enhanced farmer knowledge of international quality standards, thereby improving the competitiveness of Ugandan coffee.

4.4 Benefits of Market Channel Collaborations

Collaborations with market channels emphasizing sustainability, such as fair-trade organizations and eco-friendly retailers, created access to premium markets. Data showed that coffee farmers engaging with these channels saw a 30% increase in average export prices compared to traditional market channels [9]. These partnerships not only offered stable revenue streams but also facilitated the adoption of advanced farming technologies and better market information.

4.5 Comparative Insights from Other Regions

Insights from Latin America and Southeast Asia indicated that similar climate-smart promotional strategies resulted in comparable benefits. For instance, marketing campaigns emphasizing shade-grown coffee in Costa Rica increased export volumes by 40% over a decade [2]. The adoption of such strategies in Uganda was expected to yield similar results, given the comparable agricultural conditions and global market trends. Figure 4 shows the impact of climate- smart strategies on market performance.

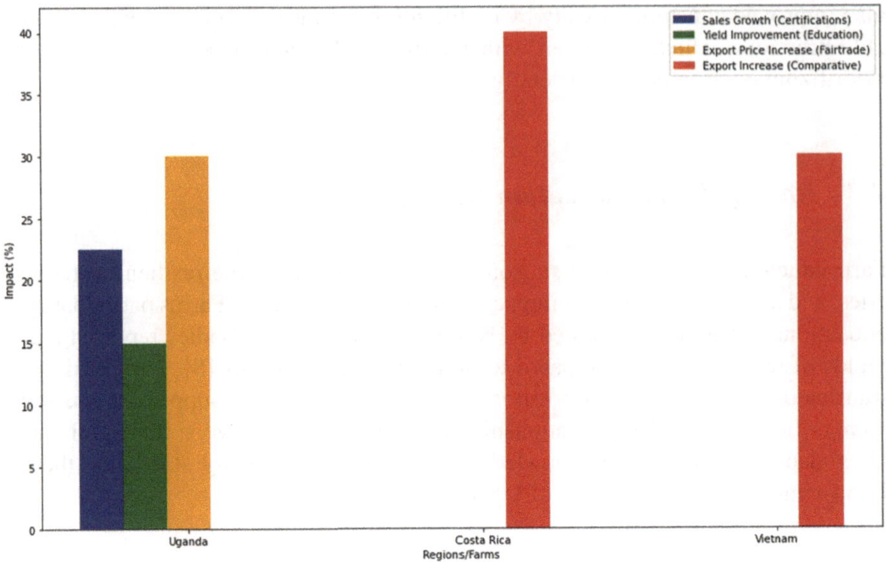

Fig. 4 The impact of climate- smart strategies on market performance

4.6 *Challenges and Limitations*

Despite the notable benefits of climate-smart promotional strategies, several challenges and limitations were encountered. One of the primary challenges was inadequate funding for marketing campaigns, which hindered the full potential of these strategies in expanding market reach. Stakeholders highlighted the need for more targeted investments in climate-smart marketing and the scaling of existing initiatives. Limited access to international markets remained another critical issue, as many smallholder coffee farmers struggled with the cost of certifications and the logistical barriers to entering premium markets.

Moreover, there were concerns about the sustainability of these efforts without stronger government support. Farmers often lacked the resources needed to meet international quality standards, and thus, faced difficulty in positioning themselves as competitive players in the global market. These challenges underscored the importance of a coordinated effort between the government, development partners, and the private sector to create an enabling environment for growth.

5 Conclusion and Recommendations

5.1 Conclusion

This study concluded that climate-smart promotional strategies, such as marketing campaigns, farm education, and market channel collaborations, played a pivotal role in enhancing the market performance of Uganda's commercial coffee farms. These strategies contributed significantly to improving sales growth, market share, and international competitiveness, all while addressing key challenges posed by climate change and market volatility. By adopting sustainable farming practices and strategic market positioning, Uganda's coffee sector improved its resilience to external shocks and attracted premium prices from eco-conscious consumers. The findings highlighted the necessity of an integrated approach that combined climate-smart agricultural practices with strong marketing and international market engagement to ensure the long-term sustainability of Uganda's coffee industry.

5.2 Stakeholder-Based Recommendations

To the Farms: Farmers were encouraged to continue adopting eco-certifications and climate-smart practices to improve the quality and marketability of their coffee. Participation in educational programs was recommended to ensure that farmers remained updated on sustainable farming practices and international market standards, which would increase their competitiveness and open new market avenues.

To the Government and Policymakers: It was crucial for the government to increase funding for climate-smart promotional campaigns and provide subsidies for eco-certification processes, which would reduce the financial burden on farmers seeking to enter premium markets. Additionally, the government was urged to strengthen partnerships with international buyers, facilitating better access to high-value markets and ensuring that Uganda's coffee was well-positioned on the global stage.

To the Development Partners: Development organizations were advised to expand farmer training programs to include a broader range of climate-smart techniques that go beyond just coffee cultivation. Collaborative initiatives that connected farmers to sustainability-focused market channels, such as fair trade cooperatives, were recommended to increase farmers' bargaining power and access to stable revenue streams.

To the Private Sector: Private sector players, especially coffee exporters, were encouraged to invest in marketing campaigns that highlight the sustainability and quality of Ugandan coffee. This would attract environmentally conscious consumers who were increasingly prioritizing ethically sourced products. Long-term partnerships with coffee cooperatives and farmer associations were also recommended to help build stable and reliable supply chains, which were essential for sustaining Uganda's competitiveness in global markets.

To the Academia: Further studies were encouraged to explore the long-term impacts of climate-smart promotional strategies on market performance in Uganda. Understanding consumer behavior trends, particularly in response to sustainability marketing, could provide valuable insights for shaping future marketing campaigns and refining best practices. Academic research could also contribute to developing new climate-smart techniques that further improve coffee productivity and resilience to climate change.

References

1. UCDA (2019) Uganda coffee profile: export trends and market share. Uganda Coffee Development Authority
2. Läderach P, Rahn E, Bunn C (2017) Vulnerability of coffee production to climate change. Glob Environ Chang 23(5):89–104
3. Alwang J, Siegel PB (2009) Does global warming threaten Uganda's coffee sector? Economic perspectives on the sustainability of Uganda's coffee production under climate change. J Clim Policy 12(3):245–260
4. FAO (2013) Climate-smart agriculture sourcebook. Food and Agriculture Organization of the United Nations
5. ADB (2015) Climate-smart agriculture: a policy perspective. Asian Development Bank
6. Kirungi C (2023) Climate-smart marketing: strategies for sustainable growth in the Ugandan coffee sector. J Mark Sustain 15(3):221–238
7. Peterson RA (2021) The impact of sustainable marketing campaigns on consumer behavior. Mark Res J 35(4):408–423
8. Mugisha J, Byamukama E, Tumushabe G (2019) Adoption of climate-smart agriculture in Uganda: an empirical analysis. J Agric Econ 39(2):157–173
9. Tumushabe G (2023) Market channel collaborations and market performance: a case of Ugandan coffee farms. Dev Econ Sustain 15(2):207–224
10. Zhang S, Zhou H (2019) Market share and consumer loyalty in the coffee industry: the role of sustainable marketing strategies. Int J Mark Stud 12(3):115–131
11. Barney J (1991) Firm resources and sustained competitive advantage. J Manag 17(1):99–120
12. Hachum A, Mulinganiza M (2014) Climate-smart agriculture in Uganda: practices, challenges, and future opportunities. Afr J Agric Stud 28(1):44–59
13. Rockström J, Sachs J (2015) Resilient agriculture for a changing climate: sustainable strategies for growing coffee. J Agric Sustain 19(3):201–219
14. UCDA (2018) Annual report 2017/18. Uganda Coffee Development Authority
15. Rahn E, DaMatta FM (2007) Climate-smart strategies for resilient coffee production. Int J Coffee Stud 12(2):45–60
16. Braun V, Clarke V (2006) Using thematic analysis in psychology. Qual Res Psychol 3(2):77–101
17. Sachs J, Rockström J (2016) The future of farming in a warming world: leveraging climate-smart strategies for sustainable coffee production. Springer
18. Schaltegger S, Wagner M (2017) Managing the transition to a sustainable business model. Bus Sustain Rev 12(2):119–134
19. UCDA (2020) Sustainability in Uganda's coffee sector: an analysis of current practices and future opportunities. Uganda Coffee Development Authority

20. World Bank (2021) The economics of climate-smart agriculture in Sub-Saharan Africa. World Dev Rev 33(7):451–467
21. McKinsey & Company (2023) Consumers care about sustainability wallets. McKinsey & Company
22. KPMG Uganda (2025) Uganda's corporate ESG outlook. KPMG East Africa. https://kpmg.com/ke/en/home/insights/2025/06/uganda_corporate_esg_outlook.html
23. Creswell JW, Plano Clark VL (2017) Designing and conducting mixed methods research. SAGE Publications
24. Tornatzky LG, Fleischer M (1990) The processes of technological innovation. Lexington Books
25. Tumushabe G (2014) The effect of fair trade and climate-smart agriculture in Uganda's coffee sector. Afr J Sustain Agric 9(4):321–336
26. Tumushabe G (2023) Market collaborations in sustainable coffee farming: the case of Uganda. J Agric Collab 8(1):43–60

Assessing the Role of Gender in Food Security Amidst Climate Change: A Study of Rural Farmers in Ekiti State, Nigeria

Matthew Olufemi Adio, Olufemi Oyedokun Fakunle, Matthew Durojaiye Ayeni, Oyepeju Abosede Ajetunmobi, Ayodele Olarewaju, and Akindele James Akintola

Abstract The study evaluates the roles of gender on food security in the context of climate change: a study of rural farmers in Nigeria's Ekiti State. A three-stage sampling process was used to choose 120 respondents. Data from the respondents was gathered using a structured questionnaire, and descriptive statistics, the Food Security Index, and a logistic regression model were utilised for analysis. The findings showed that 45.76% of respondents were men and 44.26% were women, with the mean age of male respondents being 39.2 years and that of female respondents being 37.3 years. The majority of respondents (45.76%) have been married for an average of 12 years, regardless of gender. The estimated monthly mean per capita food expenditure was N23376.74, indicating that 62.30% of families headed by women and 77.97% of male respondents experienced food insecurity. The results also showed that among the rural agricultural households in the study area, some of the signs cited indicating climate changes were flooding, irregular rainfall patterns, extended droughts, severe temperatures, insect/pest infestation, storms, and extreme weather events. Education ($p < 0.05$), household size ($p < 0.05$), floods ($p < 0.010$), and pest infestation ($p < 0.05$) were the factors that affected food security among agricultural households in the area. Therefore, the study suggests that in order to lessen the severity of food insecurity, household heads should be encouraged to keep their households small and that gender-sensitive government policies against climate change be implemented.

M. O. Adio (✉) · O. O. Fakunle · A. Olarewaju · A. J. Akintola
Department of Agricultural Economics and Farm Management, Faculty of Agriculture, Federal University, Oye-Ekiti, Ekiti State, Nigeria
e-mail: matthew.adio@fuoye.edu.ng

M. D. Ayeni
Department of Agricultural Economics and Extension, College of Agricultural Sciences, Landmark University, Omuaran, Kwara State, Nigeria

O. A. Ajetunmobi
Department of Agricultural Economics, Faculty of Agricultural Sciences, Ladoke Akintola University of Technology, Ogbomoso, Nigeria

S. Al-Naemi et al. (eds.), *Water and Food Security in the Face of Climate Change: Challenges and Opportunities for Resilience*, World Sustainability Series,
https://doi.org/10.1007/978-3-032-00098-9_33

Keywords Gender · Food security · Climate change · Logistic regression

1 Introduction

Climate change poses a significant threat to global food security, especially in the rural agricultural sector where the impact is more pronounced. Climate change and its impacts are increasing the vulnerability of already at-risk populations, while creating new hurdles in ensuring food production and access. Food security remains a key issue in over most developing countries, with one of the highest percentage of rural population depending directly on agriculture as a means of livelihood [1], Food and Agriculture Organization (FAO) [2, 3]. In Nigeria, irregular rainfall patterns, prolonged droughts and extreme climate events mean that climate conditions play a significant role in food security levels. Climate change has made these conditions more by uncertain. Such changes dramatically heighten the threats to food security, especially for smallholder farmers who largely rely on the ecosystem and traditional agriculture techniques [4].

Rural farmers—largely smallholder farmers, who rely heavily on rain-fed agriculture and natural resources—are at the forefront of this dilemma. Gender inequality under these circumstances worsens the vulnerability of rural populations. Women play a vital role in not only household food security but also agricultural productivity, yet there are often systemic barriers to access on a land right to women or access to critical resources such as credit or extension services that link impacting food security at the household level [3]. Food security in communities and households is further threatened by the fact that women's decision-making and adaptability to changing environmental conditions are often restricted by gender roles and cultural norms. These gender-specific barriers may hinder their ability to adapt to climate change, which could affect global food security. However, there is evidence that empowering women in agriculture boosts output and leads to food security [4, 5]. Research that specifically examines how food security and climate change adaptation intersect in the setting of rural Nigeria is lacking, despite the fact that gender issues are recognized as being significant in these areas.

Food security, according to the FAO [6, 7], is the state in which everyone, everywhere, has physical, social, and financial access to enough wholesome food that meets their dietary requirements and food preferences for an active and healthy life. Due to a variety of factors, including poverty, inadequate infrastructure, and deteriorating environmental conditions, food security in rural Nigeria is typically unpredictable [1]. Due to their limited access to resources and financial resources, rural households—especially those led by women—are more likely to experience food insecurity [8]. Research in Ekiti State and other parts of Nigeria has found that food insecurity is associated with a number of factors, including farm size, crop diversification, household income, and educational attainment. Furthermore, because rural farmers now face greater risk and unpredictability in their farming operations, the effects of climate change have made achieving food security much more challenging. When creating inclusive and practical solutions, it is particularly critical to comprehend the gendered impacts of climate change on food security in Ekiti State, where agriculture continues to be a significant economic driver. In light of the difficulties presented by climate change, this study aims to close this gap by investigating the ways in which gender affects food security outcomes among rural farmers in Ekiti State. Understanding the gendered effects of climate change on food security in Ekiti State, where agriculture remains a major economic driver, is especially important for developing inclusive and workable solutions. By examining the ways in which gender influences food security outcomes among rural farmers in Ekiti State, this study seeks to narrow this gap in light of the challenges posed by climate change.

2 Review of Literatures

Research has demonstrated that socio-cultural norms that give precedence to male ownership and management of land and productive assets limit women's access to resources [9]. Women's productivity and ability to earn an income are impacted by this unequal access, which in turn affects household food security. In addition, women frequently manage the home and farm simultaneously, which adds to the time and energy demands on them [10, 11]. These difficulties highlight how important it is for agricultural development plans and interventions to take gender into account. Global food security is becoming more and more threatened by climate change, with rural communities in developing nations being especially vulnerable, International Panel on Climate Change [12, 13]. Ajetomobi, et al. [14] reported that the Nigeria's agricultural sector is extremely vulnerable to changes in the climate because it depends heavily on rain-fed farming techniques. It has been shown that altered rainfall patterns, a rise in the frequency of droughts, and extreme weather events lower crop yields, interfere with planting seasons, and make food shortages worse. Climate change has a complex effect on food security that affects food supply, accessibility, stability, and consumption. Higher food costs, greater poverty, and a higher risk of malnutrition can result from decreased agricultural productivity, particularly for vulnerable groups like women and children [15]. To lessen these effects, adaptation measures include the use of climate-resilient crops, better water management, and diversifications of livelihoods are crucial [16]. Nevertheless, the effectiveness of these tactics frequently depends on having equal access to information and resources, which is not the case for both genders.

In recent years, there has been a growing corpus of work showing the ways in which men and women feel the impacts of climate change differently. This has brought attention to the intersection of gender and climate change. Women are more susceptible to the effects of climate change because of their reliance on natural resources and limited ability to adapt, especially in rural areas [17]. Researches [1, 18, 19] indicate that due to their roles in family administration, food production, and water collection, women are particularly vulnerable to shocks connected to climate change. In Nigeria, gender-based differences in the availability of climate knowledge, technology, and financial services make women farmers even more vulnerable [19]. For example, studies by Akinola and Adeyemo [20] have found that compared to men, women are less likely to engage in climate adaptation decision-making processes or to obtain services from agricultural extension. Their inability to use practical adaptation measures is hampered by their lack of access, which has an impact on the food security of their households. Given the gendered aspects of food security and climate change, there is a growing consensus on the necessity of gender-sensitive methods to agricultural growth and climate adaptation. According to Akinola and Adeyemo [20], these strategies involve acknowledging the various needs, duties, and responsibilities that men and women have as well as making sure that intervention and policies are inclusive and equitable. Enhancing women's resilience to climate change and improving food security outcomes requires empowering them through access to resources, education, and decision-making processes (Quisumbing et al. 2014). Various initiatives have attempted to incorporate gender considerations into climate adaptation programs, particularly in Nigeria, with varying degrees of success. Although women's access to agricultural inputs and training has increased, there are still obstacles to be overcome in order to overcome ingrained gender biases and make sure that women's opinions are heard when policy is being discussed [4]. Thus, in the context of food security and climate change, there is a need for ongoing research and policy development that gives gender equity priority.

3 Methodology

The study took place in Gbonyin Local Government Area (LGA) Ekiti State, Nigeria. The Local Government as it is made up of eight (8) towns with other small settlements. The 2006 figures from the National Population Census puts the overall population of the LGA at 147,999 (75,459 males and 72,734 females). Most of the people are occupied with farming activities and some engage in trading and civil service. Major arable crops grown in the area include; yam, maize, cassava, millet, guinea corn, cowpea, groundnut and tree crops such as oil palm, citrus, mango and cashew. Cattle, sheep, goat, and poultry [21] are the major livestock kept in the area.

This study employed three stage sampling technique to select the respondents. To begin with in the first stage, Gbonyin LGA was purposively selected on the basis that it falls under the category of one of the rural Local Government Areas of Ekiti state Nigeria. In the second stage, twelve communities were purposively sampled from the LGA, namely Ode Ekiti, Agbado Ekiti, Imesi Ekiti, Aisegba Ekiti, Egbe Ekiti, Ijan Ekiti, Iluomoba Ekiti, Iro Ekiti, Aba-Baale, Oke-Afa, and Ologoji which were picked at random. Finally, the last stage was the selection of ten (10) respondents (5 male and 5 female) proportionately and randomly from each of the communities selected earlier bringing the total number of respondents to 120. The studyuse of primary data collected through the use of a well- structured questionnaire. Non Proportional Random Sampling was used and the samples number arrived were based on the formula by Slovin method [22]

$$n = \frac{N}{1 + N(e)^2}$$

Note: n = Number of Population, N = Number of Samples; and E = Margin of Error 10%.

The stated objectives were analyzed through the use of logistic regression, descriptive statistics such as frequency and percentages, and a particular index called Food Security Index (FSI). In calculating FSI, the level of food security was estimated according to the purpose of the FSI methodology. Any household whose food expenditure exceeds two-thirds of the cut-off was said to be food insecure, while any household that spends below the two-thirds cut-off was said to be food secure. The level of food security poverty line was estimated from the amount of money that a respondent spent on food items in the previous week. The food security index formula is as follows:

$$F_i = \frac{per\ capita\ food\ expenditure\ for\ the\ ith\ households}{2/3\ mean\ per\ capita\ food\ expenditure\ in\ all\ households}$$

where

$$per\ capita\ food\ expenditure(PCE) = \frac{Total\ food\ Expenditure}{Household\ Size}$$

$$mean\ per\ capita\ household\ food\ expenditure(MPCFHE) = \frac{Total\ household\ PCE}{Total\ number\ of\ Households}$$

$$F_i = Food\ security\ index$$

When

$$F_i \geq 1 = food\ secure\ ith\ household$$

$$F_i \leq 1 = food\ insecure\ ith\ household.$$

In order to compute the mean per capita household expenditure index, the following equations were adopted:

$$mean\ per\ capita\ household\ expenditure(MPHE) = \frac{\sum_i^n K_i}{\sum_i^n H_i}$$

where K_i is the monetary value of all food items that were consumed per day by the household, H_i is sum of members in the household. Therefore, the proportion of households that were food secured index (FSI) was computed as:

$$FSI = \frac{m}{N}, m$$

m is the number of food secured households, N is sample population. When this is multiplied by 100, it gives the percentage of households that were food secured.

A food secure household is therefore that whose per capita monthly food expenditure falls above or is equal to two-third of the mean per capita food expenditure. On the other hand, a food insecure household is that whose per capita food expenditure falls below two-third of the mean monthly per capita food expenditure.

4 The Logistic Regression (LR) Model

The LR gives each predictor a coefficient which measures its independent contribution to variation in the dependent variable. The dependent variable Y takes the value 1 if the response is "food secure" and takes a value 0 if the response is otherwise (food insecure).

The Logit equation is written as [23].

$$P_i = \frac{e^{\beta x}}{1 + e^{\beta x}}$$

where $P_i = 1$(For food secure) and 0 for food insecure. Since Logistic regression calculates the probability of success (p) over the probability of failure (q), the results of the analysis are in the form of an odds ratio (p/q).

The logistic regression model employed is expressed as follows:

$$Y = \beta_0 + \beta_1 X_1 + \beta_2 X_2 + \beta_3 X_3 \cdots \beta_n X_n + \varepsilon$$

$$\varepsilon Y = 1(food\ secure), 0 = otherwise$$

$X_1 = age, X_2 = Marital\ status, X_3 = education, X_4 = household\ size,$
$X_5 = age, X_6 = experience, X_7 = membership\ of\ association, X_8 = extension\ visit,$
$X_9 = flooding, X_{10} = erratic\ rainfall\ pattern, X_{11} = drought, X_{12} = pest\ infestation,$

5 Results and Discussions

The socioeconomic distribution by gender of the respondents is presented in Table 1. According to the results of the age distribution in the Table, the average age of the male respondents was 39.2 years, whereas the females had an average of 37.3 years. This shows that the respondents are still very much able to carry out agricultural production that would contribute to household food security. This is consistent with the findings of [24], who noted that household heads at the active working age spare a time to incorporate innovations that will boost their productivity and income. The data from the marital status analysis indicated that the most people who responded were 45.76% male respondents and 44.26% female respondents who were married. This suggested that most of this households heads were matured and responsible enough to take care of their households and were well aware of their welfare. Marriage usually comes with business responsibility, as family needs have to be met as established by Assa et al. [25], so there's an inherent motivation to work harder. Therefore, this would increase the productivity and the level of food security. All of the respondents had some sort of education, according to the Table, which also revealed that the majority of respondents (average of 12 years of school for both male and female respondents) were literate, which can improve their food security position. This outcome confirms the findings of [26], who found that improved farm productivity and management practices are significantly influenced by the family head's educational attainment. It also establishes the ability to earn money and the cost of meals. Even though the dependency ratio will be high, a large household can be a source of family labour. Because large households typically have lower per capita food expenditures, they are more likely to experience food insecurity. The findings showed that both male and female respondents had an average household size of four. The average farming experience of male respondents was 10 years, compared to 7 years for female respondents. This suggests that male respondents have more farming experience than their female counterparts, and that the majority of household heads have sufficient work experience, which could improve their productivity, income, and food security status. The above corroborates the findings by Omonona et al. [27].

6 Gender Disaggregation on Food Security Situation in the Study Area

The food security status of each home was determined by classifying the respondents into food secure and food insecure households using the spending approach, which is the same method used by Refs. [28–30] to estimate the food security index. Therefore, a household is considered food secure if its monthly food spending per capita is at least two-thirds of the average monthly food expenditure per capita. Alternatively, a household is considered food insecure if its monthly per capita food spending is less than two-thirds of the average monthly per capita food expenditure. The food security

Table 1 Socioeconomic characteristics of the respondents

	Male		Female		Pooled	
	F	%	F	%	F	%
Age						
≤ 20	1	1.69	2	3.28	3	2.50
21–40	36	61.02	40	65.57	76	63.33
41–60	19	32.20	14	22.95	33	27.50
61–80	3	5.08	5	8.20	8	6.67
Mean	39.17		37.26		38.20	
Marital status						
Single	26	44.07	27	44.26	53	44.17
Married	27	45.76	27	44.26	54	45.00
Divorced	3	5.08	1	1.64	4	3.33
Widowed	3	5.08	6	9.84	9	7.50
Education						
No formal	0	0	27	44.26	53	44.17
Primary	6	10.17	27	44.26	54	45.00
Secondary	25	42.37	1	1.64	4	3.33
Tertiary	28	47.46	6	9.84	9	7.50
Years spent in school						
< = 6	3	5.08	4	6.56	7	5.83
7–12	23	38.98	27	44.26	50	41.67
> 12	33	55.93	30	49.18	63	52.50
Mean	12.58		12.36		12.46	
Household size						
< 10	59	100	60	98.36	119	99.17
11–20	0	0	1	1.64	1	0.83
Mean	4.14		4.25		4.19	
Source of income						
Farming	13	22.03	14	22.95	27	22.50
Civil servant	5	8.47	10	16.39	15	12.50
Self employed	30	50.85	32	52.46	62	51.67
Entrepreneur	9	15.25	5	8.20	14	11.67
Others	2	3.39	0	0.00	2	1.67
Experience						
< = 10	39	66.10	48	78.69	87	72.50
11–20	13	22.03	12	19.67	25	20.83
21–30	7	11.86	1	1.64	8	6.67
Mean	10.63		7.74		9.16	

(continued)

Table 1 (continued)

	Male		Female		Pooled	
	F	%	F	%	F	%
Association membership						
Yes	12	20.34	16	26.23	92	76.67
No	47	79.66	45	73.77	28	23.33
Extension visit						
Yes	46	77.97	48	78.69	94	78.33
No	13	22.03	13	21.31	26	21.67

Source Filed Survey, 2024

Table 2 Distributions of food security situations of respondents

	Male		Female		Pooled	
	F	%	F	%	F	%
Food insecure(0)	46	77.97	38	62.30	84	70.00
Food secure(1)	13	22.03	23	37.70	36	30.00
Total	59	100	61	100	120	100

Source Field Survey, 2024

index was calculated using the expected monthly mean per capita food expenditure of N23376.74. In other words, a respondent was considered food insecure if their weekly mean food spend was less than N23376.74 (the food security index). According to Table 2, just 22.03% of male respondents and 37.70% of female respondents were food secure, whereas 77.97% of male respondents and 62.30% of families headed by women were food insecure. This suggested that compared to households headed by men, those headed by women had greater food security. By Wudil et al. [31], similar outcomes were achieved.

7 Observable Climatic Effects

According to an expanding body of research, climate-related disasters have impacted human populations in a variety of areas, such as public health, food security, agricultural production, and water management [32]. The extent of impacts and coping strategies used by populations are strongly impacted by socioeconomic position, poverty, socio-cultural norms, gender, and resource availability. The extent of impacts and coping strategies used by populations are strongly impacted by socioeconomic position, poverty, socio-cultural norms, gender, and resource availability. Since women and children are among the groups most at risk, research has also uncovered more evidence that the repercussions are not gender neutral. Important factors that

Table 3 Summary statistics

Variable	Obs	Mean	SD	Min	Max
Climatic effects					
Flooding	120	1.42	0.90	1	5
Erratic rainfall patterns	120	3.03	1.37	1	5
Prolonged droughts	120	2.47	0.95	1	5
Extreme Temperature	120	2.13	0.85	1	5
Insect/Pest infestation	120	2.03	0.56	1	5
Storm	120	2.11	0.64	1	5
Extreme weather events	120	2.32	1.00	1	5

Source Field Survey, 2024

explain the disparities in vulnerability between men and women to climate change risks include gender-based differences in time use, access to assets and credit, treatment by formal institutions, which can limit women's opportunities, limited access to policy discussions and decision making, and a lack of sex-disaggregated data for policy change. The study's findings showed that flooding, irregular rainfall patterns, protracted droughts, extreme temperatures, insect and pest infestation, storms, and extreme weather events were among the indicators of climate change that the rural farming households in the study area reported (Table 3).

8 Factors Influencing Food Security Status of Male and Female Respondents in the Study Area

Logistic regression analysis was used to examine the factors affecting food security among agricultural households in the region; the results are shown in Table 4. Age, marital status, education, household size, experience, access to extension agents, association membership, and climate change factors (drought, flooding, unpredictable rainfall patterns, and pest infestation) were the variables that were fitted for the regression. The findings indicated that the factors that most significantly impacted the food security status of both male and female respondents in the research area were education, household size, flooding, and insect infestation. A change in any of these factors is likely to have an impact on the food security of the respondents in the research area, according to the odd ratio. Food insecurity in households is positively correlated with the education coefficient (0.959). The findings indicate that among male and female household heads in the research area, food insecurity declines with education. This may be because education is linked to the human potential and technical know-how that advance such households' productivity and, in turn, their level of food security. This supports research by Adio et al. [30] on how financial capital affects food security among Nigerian cassava processors in Oyo and Ogun States.

The findings also showed that the food insecurity of catfish producers in the research area is increased by household size ($p < 0.001$). Keeping all other factors equal, the likelihood of being food secure falls by 0.268 for every incremental increase in household size. This suggests that the likelihood of a home becoming food secure decreases as household size increases. Increases in household size inevitably result in higher food expenses, particularly when many other household members are solely dependent on the heads of the household and do not earn any revenue. The study conducted by Haruna et al. [33] on the gender aspects of food security status among rural farming households in Dekina Local Government, Kogi State, Nigeria, is in line with this. Furthermore, there is a negative correlation between the food security status of rural households and the respondents' coefficient climate variables (pest infestation and flooding). Flooding ($p < 0.10$) has a higher likelihood of causing food insecurity in rural households, which is in line with the apriori expectation. Furthermore, pest infestation ($p < 0.05$) is more likely to result in food insecurity for rural households, which is similarly in line with the apriori hypothesis.

9 Coping Strategies Employed by Households in Mitigating the Effects of Food Insecurity on Household

According to the results in Table 5, the main coping mechanisms used by both male and female respondents to lessen the effects of food insecurity include: reducing the quantity and quality of food consumed, skipping meals to extend the food supply, eating less expensive foods, and borrowing food from neighbours or family when necessary. In order to lessen the impact of food insecurity, the households in the study area also used these important techniques. The deployment of these two tactics was motivated by the high market price of staple goods. In order to feed and generate more food, households are turning to consuming less expensive and favoured foods due to rising food prices and decreased income.

10 Conclusion and Recommendations

According to the study's findings, homes headed by men were more food secure than those headed by women in the study area, despite the fact that food security is a major concern for both types of households. Households headed by women experience greater levels of food insecurity than households headed by men. These findings imply that factors impacting food security in the research area included meal frequency and household size. The study comes to the conclusion that although food insecurity affects households headed by both men and women, it has a bigger effect on the former. Nigeria should implement gender-responsive food policies, initiatives, and institutional arrangements.

Table 4 Summary Result of Logistic Regression Analysis

	Male			Female			Pooled		
	Odds ratio	Std. Error	*p*-value	Odd ratio	Std. Error	*p*-value	Odd ratio	Std. Error	*p*-value
Age	1.037	0.660	0.572	1.087	0.083	0.275	1.078	0.047	0.084
Marital status	0.624	0.565	0.603	1.035	1.208	0.977	0.733	0.488	0.640
Education	0.986	0.142	0.020**	0.865	0.135	0.055**	0.959	0.091*	0.060**
Household size	0.314	0.126	0.004**	0.190	0.098	0.001***	0.268	0.079	0.000***
Experience	0.958	0.082	0.618	0.940	0.114	0.604	0.919	0.061	0.202
Association membership	9.379	15.641	0.180	1.168	3.477	0.958	5.949	7.635	0.165
Extension	0.220	0.323	0.303	0.280	0.811	0.660	0.195	0.225	0.156
Flooding	– 4.312	4.557	0.067*	– 12.252	13.139	0.019**	– 6.198	3.859	0.003***
Erratic rainfall pattern	0.748	0.978	0.824	5.617	6.865	0.158	1.980	1.461	0.354
Drought	0.743	0.357	0.537	0.435	0.251	0.149	0.711	0.234	0.301
Pests infestation	– 0.735	0.675	0.037**	– 2.498	2.586	0.076*	– 1.492	0.942	0.026**
Constant term	26.423	107.078	0.419	1.254	3.861	0.941	0.986	2.1076	0.995
LR chi^2	15.73			43.010			56.33		
Prob > chi^2	0.151			0.000			0.000		
Pseudo R^2	0.253			0.532			0.384		
Log likelihood	23.247			18.915			45.140		
Number of obs	59			61			120		

Source Author's Computation. ***; ** and * represents 1%, 5%and 10%, significant levels
Std. Err = Standard Error

Therefore, the study suggests that in order to lessen the severity of food insecurity, household heads should be urged to maintain a small household size. Additionally, in order to improve their food security status, female household heads should be encouraged to develop their employment skills so they can eat enough meals each day and avoid skipping any. Lastly, when designing and implementing policies for climate mitigation initiatives, government policy should take gender equality into consideration.

Table 5 Coping strategies employed by households in mitigating the effects of food insecurity on household

			Female		Pooled	
	Yes	No	Yes	No	Yes	No
Borrow food from neighbors or relatives in time of needs	25 (42.37)	34 (57.63)	31 (50.82)	30 (49.18)	56 (46.67)	64 (53.33)
Skipping meals to stretch food supply	20 (33.90)	39 (66.10)	27 (44.26)	34 (55.74)	47 (39.17)	73 (60.83)
Changing dietary habit to cope with food insecurity	31 (52.54)	28 (47.46)	27 (44.26)	34 (55.74)	58 (48.33)	62 (51.67)
Reduced in portion size to make food last longer	37 (62.71)	22 (37.29)	38 (62.30)	23 (37.70)	75 (62.50)	45 (37.50)
Engaging in asset selling to buy food	21 (35.59)	38 (64.41)	19 (31.15)	42 (68.85)	40 (33.33)	80 (66.67)
Access to community-based initiatives or support networks that household benefits from	30 (50.85)	29 (49.15)	29 (47.54)	32 (52.46)	59 (49.17)	61 (50.83)
Engagement in food preservation techniques to make food last longer	40 (67.80)	19 (32.20)	45 (73.77)	16 (26.23)	85 (70.83)	35 (29.17)
Switching to cheaper food option to manage food expenses	35 (59.32)	24 (40.68)	42 (68.85)	19 (31.15)	77 (64.17)	43 (35.83)
Community garden or shared food resources that you household participant	30 (50.85)	29 (49.15)	26 (42.62)	35 (57.38)	56 (46.67)	64 (53.33)
Prioritizing certain household member for food consumption during shortage	23 (38.98)	36 (61.02)	30 (49.18)	31 (50.83)	53 (44.17)	67 (55.83)
Soughing additional employment opportunities to increase income for food	29 (49.15)	30 (50.85)	40 (65.57)	21 (34.43)	69 (57.50)	51 (42.50)
Availability of traditional or cultural practices related to food sharing or support	18 (30.51)	41 (69.49)	23 (37.70)	38 (62.30)	41 (34.17)	79 (65.83)
Reliance on informal credits or loans to purchase food during lean periods	17 (28.81)	42 (77.19)	26 (42.62)	35 (57.38)	43 (35.83)	77 (64.17)
Participation in food assistance programs or receive food aid	37 (62.71)	22 (37.29)	28 (45.90)	33 (54.10)	65 (54.17)	55 (45.83)

Source Field Survey, 2024. *Note* Figures in Parenthesis are the Percentages

References

1. Oni SA (2016) Food security in Nigeria: the role of agriculture. Agric Econ J 7(1):55–67
2. Food and Agriculture Organization (FAO) The state of food and agriculture 2010–11: women in agriculture—closing the gender gap for development. Food and Agriculture Organization of the United Nations, Rome
3. Peng W, Berry EM (2019) The concept of food security. In: Ferranti P, Berry EM, Anderson JR (eds) Encyclopedia of food security and sustainability, vol 2, pp 1–7. Elsevier. ISBN: 9780128126875
4. Nwokoro CV (2017) Gender and climate change policy in Nigeria: challenges and prospects. Glob J Hum-Soc Sci Res 17(5):15–23
5. Quisumbing AR, Meinzen-Dick R, Raney TL, Croppenstedt A, Behrman JA, Peterman A (eds) (2014) Gender in agriculture: closing the knowledge gap. Springer
6. Food and Agriculture Organization (FAO) (2015) The state of food insecurity in the world. Food and Agricultural Organization for the United Nations, pp 64–71
7. Food and Agriculture Organization (FAO) (1996) Declaration on world food security and world food summit plan of action. Food and Agriculture Organization of the United Nations, Rome
8. Adebayo OO (2012) Effects of family size on household food security in Nigeria. Asian J Agric Rural Dev 2(2):136–141
9. Doss C (2013) Intrahousehold bargaining and resource allocation in developing countries. World Bank Res Observer 28(1):52–78
10. Boserup E (1970) Woman's role in economic development. George Allen & Unwin Ltd., London
11. Udry C, Hoddinott J, Alderman H, Haddad L (1995) Gender differentials in farm productivity: implications for household efficiency and agricultural policy. Food Policy 20(5):407–423
12. International Panel on Climate Change (IPCC) (2014) Climate change 2014: impacts, adaptation, and vulnerability. Cambridge University Press
13. Adebayo AA (2010) Food security status of rural households in Ekiti State, Nigeria. J Agric Food Inf 11(3):245–255
14. Ajetomobi JO, Abiodun A, Hassan R (2010) Economic impact of climate change on food security in sub-Saharan Africa: evidence from Nigeria. J Dev Agric Econ 2(2):41–50
15. Schmidhuber J, Tubiello FN (2007) Global food security under climate change. Proc Natl Acad Sci 104(50):19703–19708
16. Haddabi AS, Ndehfru NJ, Aliyu A (2019) Analysis of food security status among rural farming households in Mubi North local government area of Adamawa State, Nigeria. Int J Res 7(7):226–246
17. Dankelman I (2010) Gender and climate change: an introduction. Earthscan
18. Carr ER, Thompson MC (2014) Gender and climate change adaptation in agrarian settings: current thinking, new directions, and research frontiers. Geogr Compass 8(3):182–197
19. Tambo JA, Abdoulaye T (2013) Smallholder farmers' perceptions of and adaptations to climate change in the Nigerian Savanna. Reg Environ Change 13(2):375–388
20. Akinola BO, Adeyemo R (2013) Effects of climate change on food security among small scale farmers in Ekiti State, Nigeria. J Environ Manag Saf 4(1):59–65
21. National Population Commission (NPC) (2006) Population and housing census of the Federal Republic of Nigeira. Priority Tables Volume I, Distribution of population by state and sex. National Population Commission, Abuja, Nigeria, p 1
22. Sugiyono (2015) Educational research methods, quantitative and qualitative approaches. Bandung Alphabeta
23. Greene HW (2002) Limdep version 8.0: econometric modelling guide, vol 2. Econometric Software Inc., USA
24. Yusuf SA, Balogun OL, Falegbe OE (2015) Effect of urban household farming on food security status in Ibadan Metropolis Oyo State, Nigeria. Agric Sci 60(1):61–75
25. Ayoade AR, Adeola RG (2012) Effects of poverty on rural household welfare in Oyo State. Glob J Sci Front Res Agric Biol 12(4):24–32

26. Assa MM, Maonga BB, Gebremariam GG (2015) Non-price determinants of households' choice of cooking energy in Malawi. Int J Dev Sci 4(1):18–28
27. Ambali OI, Adegbite DA, Ayinde IA, Awotide DO (2012) Analysis of production efficiency of food crop farmers in Ogun State, Nigeria. ARPN J Agric Biol Sci 7(9):680–688
28. Omonona B, Agoi T, Adetokunbo G (2007) An analysis of food security situation among Nigerian urban households: evidence from Lagos State, Nigeria. J Central Eur Agric 8(3):399–406
29. Zubairu EA, Maurice DC (2014) Analysis of household food insecurity among food crop farmers in Ardo-Kola local government area of Taraba State, Nigeria. Taraba J Agric Res 2(2):20–25
30. Adio MO, Olarinde LO, Ogunwale AB, Fakunle OO (2023) Determinant of food insecurity among urban farming households in Oyo State. In: Awoyemi TT, Okoruwa VO, Rahji MAY, Yusuf SA, Omonona BT, Adeoti AI, Oni OA, Adenegan KO, Salman KK (eds) 22nd National conference of The Nigerian association of agricultural economists (NAAE), Held at The University of Ibadan (UI), Oyo State, Nigeria, held from April 3rd to 6th, 2023. Theme: Agricultural Productivity: Gateway to Sustainable Economic Recovery from Recession andCOVID-19 Pandemic in Nigeria
31. Wudil AH, Ali A, Aderinoye-Abdulwahab S, Raza HA, Mehmood HZ, Sannoh AB (2023) Determinants of food security in Nigeria: empirical evidence from beneficiaries and non-beneficiaries rice farmers of the Kano River Irrigation Project. Front Sustain Food Syst 7:999932
32. World Bank Group (2016) Gender equality, poverty reduction, and inclusive growth. http://document1.worldbank.org/curated/en/820851467992505410/pdf/102114.REVISED-WBG-Strategy.pdf
33. Haruna OI, Gbenga O, Haruna Ezekiel O, John D (2019) Gender dimensions of food security status among rural farming households in Dekina local government Kogi State, Nigeria. Agriculture 1(2):109–110

Sustaining Crops, Securing Futures: A Lifecycle and Regional Analysis of Climate Change Impacts on Food Systems

Mohd Basheer and Luma Abuorabi

Abstract Climate change is an escalating threat to global food security, driving up temperatures, altering rainfall patterns, and intensifying extreme weather events. This study offers a comprehensive assessment by integrating Lifecycle Assessment (LCA) with advanced climate modeling to evaluate the impacts of climate variables on each stage of the food production cycle—from agriculture to consumption and waste management. Utilizing high-resolution climate data from global models and agricultural statistics from the Food and Agriculture Organization (FAO), the analysis reveals significant vulnerabilities, including a projected 7% reduction in maize yields per 1 °C temperature increase and a 20% rise in water usage needed for irrigation in arid regions. Additionally, greenhouse gas emissions from agriculture are anticipated to increase by 15–25% by 2050 under high-emission scenarios, underscoring the urgency for sustainable interventions. Key points in the food lifecycle—such as production and processing—are identified where eco-friendly farming practices and renewable energy integration could mitigate adverse impacts. The study's findings offer actionable insights for policymakers, agricultural practitioners, and regional planners, especially in climate-sensitive regions, by emphasizing the necessity of tailored adaptation strategies that enhance food system resilience. Beyond immediate applications in agriculture, this research has broader implications for global environmental policy and food security frameworks. By advocating for structured approaches to sustaining food systems in a warming world, this study establishes a foundation for future LCA research to encompass a wider range of environmental impacts, offering a comprehensive roadmap for resilience in the face of climate change.

M. Basheer (✉)
University of Jordan, Amman, Jordan
e-mail: mhm8220041@ju.edu.jo

L. Abuorabi
Food and Agriculture Organization (FAO), Amman, Jordan
e-mail: luma.Abuorabi@fao.org

S. Al-Naemi et al. (eds.), *Water and Food Security in the Face of Climate Change: Challenges and Opportunities for Resilience*, World Sustainability Series,
https://doi.org/10.1007/978-3-032-00098-9_34

Keywords Climate change · Food security · Lifecycle assessment · Climate modeling · Sustainable agriculture

1 Introduction

1.1 Overview of Current Research

Climate change is increasingly recognized as one of the most critical threats to global food security. It disrupts temperature regulation, alters precipitation patterns, and intensifies the frequency of extreme weather events, directly impacting agricultural productivity, food system stability, and food availability, particularly in vulnerable regions [1]. Despite substantial research into specific impacts—such as fluctuations in crop yields and localized supply chain disruptions—most studies affect a segmented approach, often limiting their focus to isolated stages of the food production cycle, such as agricultural production or distribution processes [2, 3]. Consequently, the lack of a comprehensive view across the entire food lifecycle constrains the development of integrated strategies that address climate change's compounded effects on all stages of food production, distribution, consumption, and waste management [4]. Given the increasing urgency of climate adaptation, there remains a pressing need for research that provides an integrated perspective on how climate variability impacts the entire food lifecycle, enabling more effective, real-world applications [5, 6].

1.2 Research Focus and Motivation

This study aims to address this research gap by integrating Lifecycle Assessment (LCA) with advanced climate modeling, creating a holistic framework to examine how critical climate variables—particularly temperature and precipitation—affect the all food lifecycle, from raw production to waste management. Unlike prior studies that focus on specific stages, this cradle-to-grave approach identifies critical intervention points across the food system, fostering a deeper understanding of vulnerabilities and resilience opportunities. Such an approach is especially valuable for regions currently facing climate-related disruptions, as it enables the development of location-specific adaptation strategies tailored to regional climate challenges.

The motivation for this study arises from the urgent need to design robust adaptation strategies that enhance food systems' resilience to climate variability. By expanding the conventional LCA framework to incorporate climate projections and consider multiple impact categories, this research provides actionable insights for a range of stakeholders—including policymakers, agricultural practitioners, and regional planners—who are in a position to implement policies and practices that mitigate climate risks. These findings are essential for fostering adaptive strategies that can promote long-term food security and system stability in regions highly susceptible to climate impacts.

1.3 Research Questions and Objectives

This study is guided by the following primary research questions:

How does climate change impact each stage of the food cycle, from agricultural production to waste management?
Which critical points in the food system would benefit most from targeted interventions aimed at enhancing resilience to climate impacts?

Research Objectives

Objective 1: Conduct a comprehensive LCA integrated with climate modeling to assess the impacts of climate variables across all food lifecycle stages. While this study initially focuses on CO_2 emissions, it establishes a foundation for future studies to incorporate other environmental impact categories, such as land use, water use, and eutrophication.
Objective 2: Identify actionable strategies to strengthen food system resilience in response to projected climate variability. The study emphasizes the development of region-specific policies and practices that can be implemented to reduce climate-related risks effectively and sustainably.

2 Methodology

2.1 Overview

This study adopts an integrated approach, combining climate modeling with Lifecycle Assessment (LCA) to evaluate climate change impacts across the food system comprehensively. This dual-methodology framework examines how critical climate variables—such as temperature and precipitation—affect food production, distribution, and consumption stages, offering a holistic perspective on climate risks. Each data source's inclusion and processing techniques are detailed to ensure clarity, transparency, and reproducibility.

2.2 Data Collection

Climate data were sourced from the Coupled Model Intercomparison Project Phase 6 (CMIP6), using models like HadGEM3, GFDL-CM4, and CCSM4, chosen for their precision in capturing regional climate patterns [1]. These models provided temperature and precipitation data under two Representative Concentration Pathways (RCPs): RCP 2.6 (low emissions) and RCP 8.5 (high emissions), capturing a spectrum of potential future scenarios essential for assessing diverse climate risks across food systems. Each RCP pathway facilitated comparisons across different emissions scenarios, allowing the study to examine impacts on food security under both optimistic and pessimistic climate futures.

Agricultural indicators, including crop yields, soil moisture, and irrigation needs, were extracted from the FAOSTAT database [2]. Covering a wide geographical range, this dataset allowed for detailed analysis of climate factors' effects on agricultural productivity and crop resilience. Each data point underwent processing to calculate regional averages and assess crop-specific vulnerabilities under varying climate conditions. This provided a baseline for evaluating how different crops respond to climate changes, creating a basis for region-specific adaptation strategies.

Socioeconomic metrics, including GDP, population growth, and food demand projections, were gathered from the World Bank's World Development Indicators [3]. These metrics offer insights into how environmental shifts might impact economic and social factors critical to food security. This data was aligned in terms of timeframe and geographical granularity with the climate and agricultural datasets, ensuring a consistent basis for comparative analyses.

2.3 Data Preprocessing

Normalization: To ensure consistency, all datasets were normalized using max scaling, computed as follows:

$$X' = \frac{X - X_{\min}}{X_{\max} - X_{\min}} \tag{1}$$

where X′ represents the normalized value. This standardization enabled effective comparisons across datasets from varied sources, facilitating integration across different scales and units.

Interpolation: Linear interpolation was applied to fill in data gaps, which helped maintain continuity within time-series data, especially in climate projections and agricultural trends. This preprocessing step minimized distortions and improved the robustness of the results, enhancing the accuracy of climate projections.

2.4 Climate Modeling

Model Selection and Downscaling: This study employed a multi-model ensemble approach, combining outputs from CMIP6 General Circulation Models (GCMs) to improve the reliability of projections. Statistical downscaling and bias correction techniques enhanced spatial resolution, making projections more locally relevant. Bias-corrected local temperatures $Tlocal(x, y, t)$, were calculated as:

$$Tlocal(x, y, y) = TGCM\,(x, y, t) + \Delta Tbias(x, y) \tag{2}$$

where $\Delta Tbias$ is a correction factor derived from historical data, enhancing projection accuracy for diverse regions.

Scenario Analysis: Climate impacts were assessed under both RCP 2.6 and RCP 8.5, reflecting low and high emission scenarios, respectively. This comprehensive approach enabled an accurate risk assessment across a range of climate futures, accommodating uncertainties inherent to climate models.

2.5 Lifecycle Assessment (LCA)

System Boundary and Functional Unit: The LCA followed ISO 14040/44 standards, encompassing all stages of the food lifecycle—from raw material extraction to waste management [4]. The functional unit, defined as "one kilogram of consumable food product," ensured comparability across food types. Environmental impacts were assessed for greenhouse gas emissions, resource use, and potential expansions to other impact categories such as water use and land footprint in future research, addressing the limitations of focusing only on CO_2 emissions.

Global Warming Potential (GWP) Calculation: The study calculated GWP over 100 years GWP_{100} using:

$$GWP_{100} = \sum_{i=1}^{n} EF_i * Q_i \tag{3}$$

where EF_i is the emission factor for each greenhouse gas, and Q_i is the emission quantity. Emission data from IPCC guidelines served as the baseline for this calculation. Regional adjustments were made to Table 2 values, and the dataset was verified against standard emissions factors to ensure accuracy.

2.6 Statistical Analysis

Multiple Regression Analysis: To quantify the relationship between climate factors (temperature, precipitation) and agricultural outcomes (e.g., crop yields), a multiple regression model was applied:

$$Y = \beta_0 + \beta_1 X_1 + \beta_2 X_2 + \cdots + \beta_n X_n + \epsilon \quad (4)$$

where Y represents crop yield, and X_1, X_2,...X_n are the independent climate variables, and ϵ is the error term. This model provided quantitative insights into each variable's influence on crop yield.

Monte Carlo simulations were used to factor in uncertainties, in climate predictions. How they could affect food systems following this simulation formula:

$$X_j = X_{mean} + Z_j * \sigma \quad (5)$$

where X_j is the simulated value, Z_j is a standard normal random variable, and σ is the standard deviation. This probabilistic approach strengthened the study's resilience to variability in data, allowing for more reliable climate impact projections on food security.

3 Results and Discussion

3.1 Overview of Findings

This study integrates Lifecycle Assessment (LCA) with advanced climate modeling to provide a comprehensive analysis of climate change impacts on food security. Findings reveal specific vulnerabilities within global food systems due to shifts in temperature and precipitation patterns. These insights emphasize the urgent need for adaptive, region-specific strategies to strengthen resilience across all stages of the food lifecycle, from agricultural production to waste management.

3.2 Climate Modeling Results

Temperature and Precipitation Projections

Climate projections using models such as HadGEM3 and GFDL-CM4 indicate a global temperature increase of 2.5–4.7 °C by 2100 under the high-emission scenario (RCP 8.5). Regionally, tropical areas could experience up to a 15% increase in rainfall, while semi-arid areas face up to a 20% decrease, escalating drought risks.

These findings align with IPCC projections, underscoring the necessity of climate adaptation strategies for these vulnerable regions [1, 5] (Fig. 1).

Interpretation: The projected temperature and precipitation changes suggest significant impacts on crop yields, particularly in semi-arid zones where water scarcity is already a concern. Adaptive strategies such as water-efficient farming and enhanced irrigation systems are essential in these areas to mitigate potential food security challenges. Table 1 illustrates these regional vulnerabilities, reinforcing the need for regionally tailored adaptation strategies aligned with IPCC recommendations.

Interpretation: The projections of temperature and rainfall shifts indicate substantial impacts on agricultural productivity, particularly in regions vulnerable to climate variability, such as semi-arid areas. These areas, which already face water scarcity, are

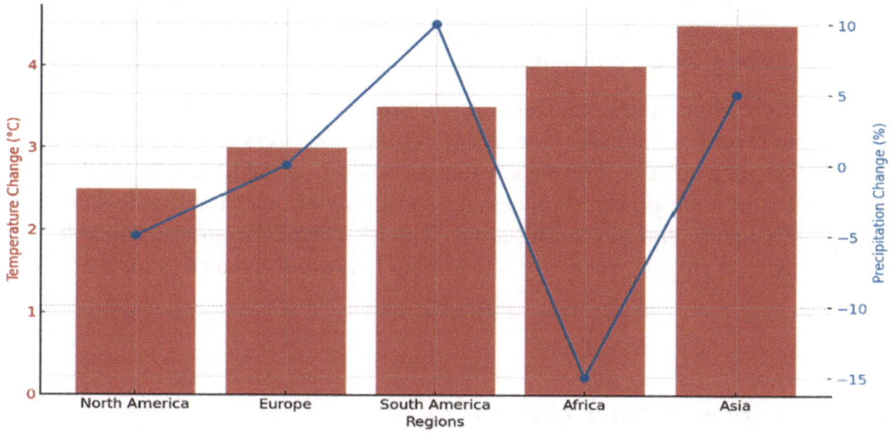

Fig. 1 Global temperature and precipitation projections (RCP 8.5, 2050)

Table 1 Temperature increase and precipitation change across different regions under RCP 8.5

Region	Temperature increase (°C)	Precipitation change (%)
Tropical	+ 2.5 °C to + 4.7 °C	+ 15% rainfall increase
Semi-arid	+ 2.5 °C to + 4.0 °C	− 20% rainfall decrease
Temperate	+ 2.0 °C to + 3.0 °C	Stable or slight increase

Table 2 Breakdown of greenhouse gas emissions across different stages of the food production lifecycle

Lifecycle stage	Greenhouse gas emissions (CO_{2e}, %)
Agricultural production	60
Processing and packaging	25
Distribution and transport	15

expected to experience additional strain on irrigation systems, potentially leading to declines in crop yields. The findings emphasize the urgency of implementing localized adaptation strategies, such as water-efficient irrigation methods and drought-resistant crops, tailored to the specific needs of these climate-sensitive regions. This interpretation aligns with IPCC recommendations for region-specific climate adaptation.

3.3 Lifecycle Assessment (LCA) Results

Greenhouse Gas Emissions: The LCA findings show that agricultural production is responsible for about 60% of greenhouse gas emissions throughout the food lifecycle, largely due to synthetic fertilizer use and energy-demanding farming practices. Processing and packaging contribute an additional 25%, while distribution and transport account for the remaining 15%.

Interpretation: High emissions from agricultural production highlight the need for sustainable practices, including the use of organic fertilizers and renewable energy sources. The data in Table 2 reveal key emission points in the lifecycle, emphasizing the potential of solar-powered processing facilities and other renewable sources to significantly reduce emissions. This aligns with IPCC guidelines on sustainable agriculture, underscoring the necessity of reducing fossil fuel reliance in high-emission areas.

3.4 Energy and Water Usage

Energy Use: Food processing is highly energy-intensive, especially for rice and wheat crops. Transitioning to renewable energy sources could potentially reduce energy use by around 20%, lessening the overall environmental impact.

Water Consumption: Regions reliant on irrigation for crop growth show particularly high water usage. Efficient irrigation systems, like drip irrigation, can reduce water consumption by approximately 30%, while lowering emissions by around 20%—a crucial measure for water-scarce regions [3].

Interpretation: Findings in Table 3 emphasize the importance of resource-efficient technologies, such as solar energy and drip irrigation, in mitigating climate impacts. Adopting these practices can contribute to more resilient food systems, particularly in regions facing severe climate risks. This aligns with international resilience strategies and offers actionable insights for policymakers.

Table 3 Energy and water usage for key staple crops under current farming practices

Crop	Energy consumption (kWh)	Water usage (liters per kg)
Rice	600	2500
Wheat	450	1500
Maize	300	1200

Table 4 Yield decrease per 1 °C temperature increase for key crops

Crop	Yield decrease (%) per 1 °C increase
Maize	−7
Wheat	−5

3.5 *Statistical Analysis of Agricultural Data*

Impact on Crop Yields: The regression analysis indicates that each 1 °C rise in temperature correlates with a 7% reduction in maize yields and a 5% reduction in wheat yields, highlighting climate change's adverse effects on essential crops [3].

Interpretation: Table 4 illustrates the critical need for adaptive crop management, including heat-resistant crop varieties and improved water management, in regions that depend heavily on maize and wheat. These strategies are essential to preserving crop productivity under increasing climate stress, aligning with resilience-building recommendations in global food security research.

3.6 *Regional Variability*

Regional Crop Yield Impacts: The study's analysis shows that climate change will variably affect crop yields across different regions. Temperate zones may experience slight yield increases due to extended growing seasons, while tropical regions may see significant declines due to heightened heat stress and altered precipitation patterns [2, 3].

Interpretation: The variability in regional yield impacts stresses the need for tailored adaptation strategies. While temperate zones may benefit from longer growing seasons, tropical areas should prioritize heat-resistant crops and water-saving technologies. Table 5 underscores the necessity for region-specific interventions, which are vital for addressing unique climatic challenges in each area.

Table 5 Regional crop yield impacts due to climate change

Region	Expected crop yield impact
Tropical	Significant decrease (up to 15%)
Semi-arid	Moderate decrease (up to 10%)
Temperate	Slight increase (+5%)

3.7 Limitations of the Study

This study acknowledges limitations, including uncertainties inherent in climate forecasts due to assumptions specific to certain models. Furthermore, the LCA results rely on data availability and quality, particularly in regions with limited data access. Enhancing data collection and refining models in future research could improve the accuracy and relevance of climate impact projections on food systems [5].

3.8 Practical Implications

The findings provide policymakers, agricultural practitioners, and researchers with actionable strategies to mitigate climate impacts on food systems. Recommendations include implementing sustainable farming practices, optimizing water usage, and incorporating renewable energy into agricultural operations. These regional adaptation strategies should be tailored to address local challenges, ensuring greater effectiveness and strength in diverse areas.

3.9 Recommendations for Future Research

Future studies should focus on developing region-specific adaptation strategies and exploring emerging technologies like precision agriculture and vertical farming. Improved climate model accuracy and expanded data collection, especially in insufficient representation regions, will be crucial for supporting informed policy decisions and enhancing global food system resilience.

4 Conclusion

4.1 Summary of Key Findings

The study provides actionable insights to support climate adaptation policies and enhance agricultural resilience. By pinpointing the most climate-sensitive stages of the food lifecycle, it lays a foundation for targeted policies addressing climate impacts, productivity, and socio-economic stability. While focused on CO_2 emissions, this approach allows future research to include other environmental impact categories, offering a more comprehensive view of sustainability in food systems. Additionally, this framework guides future socio-economic assessments, steering efforts toward climate-resilient and sustainable practices in food systems.

4.2 Implications of the Study

The study provides insights to shape climate adaptation policies and strengthen agricultural resilience. By identifying the most climate-sensitive stages of the food lifecycle, it supports targeted policies that integrate climate impacts, agricultural productivity, and socio-economic stability. Although the study centers on CO_2 emissions, it creates a foundation for future research to expand into other environmental impact areas, offering a more holistic view of food system sustainability. Additionally, the study's framework serves as a reference for future socio-economic evaluations of environmental changes, guiding policies and practices toward climate-resilient and sustainable food systems.

4.3 Future Research Directions

To improve the predictive accuracy of climate models, future research should prioritize data collection in underrepresented regions with sparse data. Advances in artificial intelligence and machine learning offer promising ways to enhance precision agriculture, optimize resource allocation, and boost sustainability in food production. Expanding LCA methodologies to include more environmental impact categories and adaptation strategies, like precision farming and water-saving technologies, is key to building resilience against climate impacts. Additionally, addressing data limitations and refining model assumptions will be vital for reliable climate risk assessments. This research path holds substantial promise for tackling the challenges climate change poses to global food security.

References

1. Eyring V et al (2016) Overview of the coupled model intercomparison project phase 6 (CMIP6) experimental design and organization. Geosci Model Dev 9:1937–1958
2. Lobell DB, Schlenker W, Costa-Roberts J (2011) Climate trends and global crop production since 1980. Science 333(6042):616–620
3. Food and Agriculture Organization of the United Nations (FAO) (2023) FAOSTAT database. Available at: http://www.fao.org/faostat/en/#data
4. International Organization for Standardization (ISO) (2006) ISO 14040: environmental management—life cycle assessment—principles and framework. ISO, Geneva, Switzerland
5. Intergovernmental Panel on Climate Change (IPCC) (2006) Guidelines for National Greenhouse Gas Inventories. Available at: https://www.ipcc-nggip.iges.or.jp/public/2006gl/index.html
6. Wang J, Brown DG, Chen J (2018) The impacts of climate change on crop yields: a study on global agricultural productivity. J Environ Sci 34(7):123–136

Population Growth as a Climate Change Stressor: Strategic Options Towards Mitigating Its Impact on Food Security in Africa

Anthony Kola-Olusanya, Hakeem A. Ajonbadi, and Peter Adewale

Abstract Africa's growing population has been identified as a climate change stressor that will significantly negatively impact the continent due to projected warming and rainfall deficits and the population's vulnerability, especially regarding food insecurity. Based on the exigencies of climate change and its low capacity for adaptation, Africa appears to be the most exposed to the vulnerabilities of extreme weather and climate effects because it relies on rain-fed agriculture. This paper links the fundamental causes of food insecurity to inadequate food production compounded by the fragility of food production systems and climate risk, water for irrigation and information access. Tragically, projections over the next century indicate that an increase in temperature and concomitant warming in Africa will exceed the global annual mean, with an average increase of 3–4 °C. To address this impact of climate risk on Africa and ensure food security, this paper also explored the paradox that African nations are among the lightest polluters worldwide yet suffer the most from climate change in their pursuit of water, food security, and sustainable development. Therefore, in this paper, efforts were made to establish a linkage and relationships between climate change and population, climate change and food security. It also presented strategic options for climate action in Africa and the need for climate action centred around climate change information, adaptation and mitigation.

Keywords Climate change · Food insecurity · Population · Climate action

A. Kola-Olusanya (✉) · P. Adewale
Osun State University, Osogbo, Nigeria
e-mail: anthony.olusanya@uniosun.edu.ng

P. Adewale
e-mail: peter.adewale@uniosun.edu.ng

H. A. Ajonbadi
University of Doha for Science and Technology, Doha, Qatar
e-mail: hakeem.ajonbadi@udst.edu.qa

S. Al-Naemi et al. (eds.), *Water and Food Security in the Face of Climate Change: Challenges and Opportunities for Resilience*, World Sustainability Series,
https://doi.org/10.1007/978-3-032-00098-9_35

1 Introduction

It has been 45 years since the first world meeting on climate change and 32 years since the first Conference of Parties (COP) and the building of the global coalition against climate change. The global alliance is known as the United Nations Framework Convention on Climate Change (UNFCCC). The Framework focused specifically on limiting the average global temperature increase, the resultant climate change and efforts to mitigate the effects of climate phenomenon. Establishing the linkage between human activities and global warming in 1997 marked the focused turnaround in global efforts toward curtailing climate change. Thus, over the last 27 years, we have witnessed a renewed commitment to global efforts to mitigate climate change. The global agreement limits the temperature increase to below 2 degrees Celsius above the pre-industrial level and the responsibility to strengthen countries' capabilities to address the impacts of climate change [1].

Despite these efforts, the consequence of escalating anthropogenic activities on global greenhouse gas emission, warming, and climate change patterns has assumed the most discussed issue between the last decade of the twentieth century and the first decade of the twenty-first century worldwide, from global, regional and national institutions including all levels of society [2]. Consequently, over the years, climate change has assumed prominence among various human race problems. This assertion is inherently implied, especially considering the global nature of its causes, consequences, and potential solutions. This consideration is perhaps responsible for Thiele's [3: 42–43] position that "no one country can solve the climate crisis on its own". The principal cause of climate change is linked to high consumption in developed countries, with the attendant impact the poor people in developing countries face. Buttressing this, Stephenson, Newman, and Mayhew [4] argued that it is already clear that the effects of climate change will disproportionately fall upon developing countries [5] and the most impoverished sectors within all nations. The implications include exacerbating inequities in health status and access to adequate food, clean water and other resources [4, 6]. As a result, it has been clearly understood that humans in different world regions will experience climate change to varying degrees. Furthermore, Samson et al. [7] noted with current rapid demographic growth, the human impacts of climate change are likely to be most severe in the world, thereby amplifying and expanding conditions currently supporting low human population densities [7, 8].

Thus, it will be in the right place to say that the dimensions of the climate change consequences are also inexorably linked to population-driven activities. Therefore, this linkage has made it possible to suggest that the situation will affect the world's poor people and emerging developing economies more than the wealthy and affluent nations of the developed economies. This is because emerging countries are believed to lack the capacity and capabilities to deal with the challenges associated with the climate crisis [9]. Meanwhile, evidence shows climate change will most negatively affect humans in regions where populations strongly depend on climate when favourable climatic conditions decline [7]. Therefore, predicting the strongly negative impacts of climate change for much of Africa won't be out of place. As a result, a few studies have argued that countries of the sub-Saharan Africa (SSA) region with the least to do with growing emissions will likely experience the most severe impacts [10]. In addition, government agencies have inadequate technical capacity to show the benefits of integrating population, climate change and development goals, weak technical ability in government agencies to design integrated programs, and insufficient funding for developing and implementing integrated programs [11]. Thus, this paper will examine the linkages between climate change and population and African food crises to establish a substantial nexus between climate change and population on food Security in sub-Saharan Africa. This paper's contribution is threefold: First, we discuss the linkage between climate change and population. We also discuss the African continent by presenting its need for climate action. Finally, case studies of best practices related to strategic options for climate action for addressing food insecurity and improving agricultural yield across the African continent are presented.

2 Linking Climate Change to Population Growth

Scholars have argued that making a clear and direct linkage between population change and climate change is complex because anthropogenic effects are products of a range of driving forces, including economic growth, technological changes, and population growth [12]. However, the Intergovernmental Panel on Climate Change (IPCC), in its fourth assessment report in 2007, provided unequivocal evidence linking the planet's warming to human-induced activities. IPCC asserts consistent indication of a 9 out of 10 chance of positiveness that the globe is warming at an alarming rate. This warming is due primarily to greenhouse gas emissions related to human activity [13]. As such, it concludes that if the warming of the planet, which is the precursor of climate change, is the product of human-related activities, then it presupposes that climate change, the by-product of this phenomenon, can be linked to human activities [14–17]. Supporting this assertion, Abbas et al. [18] noted that there is general agreement that human industrial activity has released vast quantities of greenhouse gases, about 900 billion tonnes of carbon dioxide, 450 of which have stayed in the atmosphere. Industrialisation accounts for about 80% of carbon dioxide emissions; the remaining 20% is caused by land use, such as deforestation. There is strong evidence that burning fossil fuels since the beginning of the Industrial Revolution has already caused a 0.75 °C rise in global temperatures and a 22 cm rise in sea level during the twentieth century.

A follow-up report (fifth Assessment report) of the IPCC moves further to confirm the assertions contained in the 2007 report (see Fig. 1) [19]. The report concludes the affirmation that it is highly likely (95–100% probability) that most of the rising temperatures since 1950 have been due to human influences [13, 19, 20]. According to Jarraud and Steiner [21], the IPCC report demonstrates that we must significantly reduce global emissions to avoid the worst effects of climate change.

Besides the report of the IPCC linking anthropogenic activities like fossil fuel burning to climate change, as highlighted above, the link between population and climate change is more prominent in that the arguments have been neither here nor there. More often, the discussions have been largely hypothetical [12] because no established linkage connects the population in a particular area or territory compared to the induced climate change consequences. Another dimension of the discussion is that growing population numbers like what exists in SSA do not necessarily translate to worse outcomes than what is experienced in Western Europe, which has comparatively lower population numbers. This implies that climate change resulting from

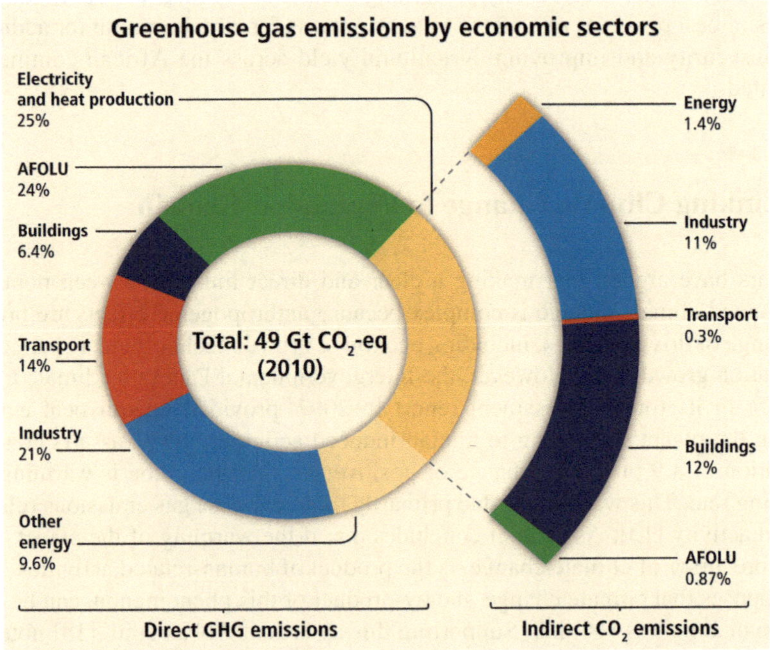

Fig. 1 Total anthropogenic greenhouse gas (GHG) emissions (gigatonne of CO_2-equivalent per year, $GtCO_{2\text{-eq/yr}}$) from economic sectors 2010. The circle shows the shares of direct GHG emissions (in % of total anthropogenic emissions) from five economic sectors in 2010. The pull-out shows how shares of indirect CO_2 emissions (in % of total anthropogenic GHG emissions) from electricity and heat production are attributed to sectors of final energy use. "The energy" refers to all GHG emission sources in the energy sector as defined in WGIII Annex II, other than electricity and heat production {WGIII Annex II.9.1}. The emission data on agriculture, forestry, and other land use (AFOLU) includes land-based CO_2 emissions from forest fires, peat fires and peat decay that approximate the net CO_2 flux from the sub-sectors of forestry and other land use (FOLU) as described in Chap. 11 of the WGIII report. Emissions are converted into CO_2 equivalents based on the 100-year Global Warming Potential (GWP_{100}) from the IPCC Second Assessment Report (SAR). Sector definitions are provided in WGIII Annex II.9. [WGIII Figure SPM.2]. *Source* Observed Changes and their Causes. https://ar5-syr.ipcc.ch/topic_observedchanges.php

global warming is more directly linked to consumption habits or consumer behaviour than huge or lower population numbers. The implication of a higher population is the increase in emissions and the attendant vulnerability of many people to climate-related impacts. For example, in many of the world's most impoverished areas, including SSA, shifting temperatures and changing precipitation patterns damage agricultural production and make scarce water supplies even more difficult to manage [22].

According to Bryant et al. [23], nearly all (37 of 40) reports of the National Adaptation Programs of Action (NAPA) document the most critical vulnerabilities to climate change impacts. The most urgent adaptation priorities refer to population growth as a significant factor that exacerbates the harmful effects of climate change [23]. The reports conclusively linked the impact of population growth to climate change; that is, rapid population growth acts in tandem with climate change to deplete vital natural resources, such as water, fuel, and soil fertility [24]. Furthermore, the reports also point to how population growth heightens human vulnerability to climate change in numerous ways and may force people to migrate to areas more at risk of the negative impacts of climate change [25]. Finally, the reports warn that rapid population growth leads to the mismanagement and decline of natural resources due to environmental variability and climate change [4]. For example, population growth in Ethiopia results in soil degradation, dwindling landholdings and low agricultural productivity, which increases pressure on poor people to move either to environmentally marginal or urban areas [4]. This suggests that more people from SSA will be exposed to the projected impacts of climate change.

3 Conceptualizing Food Insecurity

The conversation around food (in)security focuses on food poverty and lack of populace access to sufficient, safe and nutritious food. The complex concept of food poverty in global-scale narratives is about the levels of development and economic disenfranchisement in the global South's developing countries. It is increasingly recognised as a place-based outcome relevant to food systems everywhere [26]. Food security also speaks to the dichotomy in the understanding and access to food guarantees between the global North and South. Food security describes a situation that exists when all people, at all times, always have physical, social and economic access to sufficient, safe and nutritious food to meet every individual's diet and food preferences for an active and healthy life [27, 28], Accordingly, Rohr et al. [27] summarises the context of food security at different levels as availability—national; accessibility—household; utilisation—individual; stability [29] may be considered a time dimension affecting all levels Generally speaking, it, therefore, suggests that all the highlighted four dimensions must be intact to be regarded as total food security [27]. Naheed [30] further broke down food accessibility at the household level into five indicators: self-assessment measurements, experiential indicators, consumption behaviours, spending on food, and dietary diversity and food frequency.

On the other hand, food insecurity describes the lack or absence of the active ingredients of food security, best described as hunger. Food insecurity is dynamic, and the resultant status depends on the interaction between the stressors of food insecurity and the coping strategies to deal with them. Sustainable Development Goal (SDG) 2 recognises achieving food security and ending hunger as critical drivers for sustainable socioeconomic growth and development, especially in developing economies [31]. According to Glazebrook and Kola-Olusanya [32], 216 million fewer people today than in 1990 suffer from global hunger [32: 26]. Although world leaders at the 1996 World Food Summit in Rome pledged to reduce the population of hungry people in developing countries from a billion in 1990 to 515 million by 2015 [32: 26], however, this target was missed by 260 million while global South continues by far to carry most of this hunger burden with 780 million (over 98%) in the global South. Concerning Africa, 220 million (27%) lacking access to food security are in sub-Saharan Africa, where a massive food security crisis is underway.

Despite the inconclusive evidence of the linkage between Goal 2 of the Sustainable Development Goals, food security in Sub-Saharan Africa continues to be a priority in global development discourse. High climate vulnerability still poses severe challenges to food security in Africa [33]. Where the four indicators are incomplete, the dimension of food security becomes questionable, and it can then be concluded that hunger or food insecurity is present.

4 Nexus Between Climate Change and Food Insecurity

Development in Africa can be impaired by the record hike in global greenhouse gas emissions, with profound implications for the ability of the continent to deal with the consequences of the warming of the continent's climate at unprecedented rates. According to Shepard [34], sub-Saharan Africa has witnessed more frequent and more intense and extreme weather variabilities in the last decade, suggesting the projected increase in temperature is to be higher than the global mean temperature within 15 degrees of the equator and pointing to a rise in night temperature as and longer and more frequent heat waves [34]. There are numerous pathways through which climate change impacts food security, including accelerating land degradation via droughts and water and wind erosion [35]. Inextricably, African food crises have been linked to many diverse and often intertwined causes, suggesting that climate change is not the only driving factor but has played the leading role alongside other factors in the long term. According to Mbaye and Gueye [35], as a result of the exponential warming of the periods between 1961–1990 and 1991–2022, some parts of the African continent, notably the Sahel region, experienced a higher than the global average temperature increase between 0.1 °C and 0.3 °C leading to serious drought and severe flooding causing massive fatalities of more than 5000 deaths and other adverse consequences predicted to deteriorate further [35].

The worsening of food insecurity in Africa has primarily been linked to climate-related shocks such as drought, war and conflicts. African countries' dependence on rain-fed agriculture makes the continent susceptible to the vagaries of accelerating increases in temperature. Agriculture accounts for many African countries' Gross Domestic Product (GDP), given that the agricultural sector offers more than 50% of full-time employment. Such dependence on rain-fed agriculture exposes the countries' frailty of agricultural production capability. This over-dependence thus makes African countries volatile locations of food insecurity, especially considering the impact of climate change and weather variabilities resulting from the projected increase in global warming and intermittent high rain intensity across the continent [36, 37]. According to Adesete et al. [38], food inflation in the region will continue to experience an exponential increase in the face of multiple challenges, including food supply constraints, high levels of poverty, and high food costs. Recent empirical evidence on the impact of climate change on food security shows that an increase in temperature negatively impacts access to food and its availability [38, 39].

There are numerous pathways through which climate change impacts food security, including accelerating land degradation via droughts and water and wind erosion [35]. Inextricably, African food crises have been linked to many diverse and often intertwined causes, suggesting that climate change is not the only driving factor but has played the leading role alongside other factors in the long term. According to Mbaye and Gueye [35], as a result of the exponential warming of the periods between 1961–1990 and 1991–2022, some parts of the African continent, mainly the Sahel region, experienced a higher than the global average temperature increase between 0.1 °C and 0.3 °C leading to serious drought and severe flooding causing massive fatalities of more than 5000 deaths and other adverse consequences predicted to deteriorate further [35].

The African continent's population will continue to grow beyond 2100 [40], thus making Africa the only world region with a projection of strong population growth for the rest of this century. For example, it has been projected that the African population will increase to 4.3 billion by 2100. This projection shows this gain will be mainly in sub-Saharan Africa, where its population is expected to spike by 2100 (Fig. 2).

Rapid population growth can significantly affect human development [4] by magnifying the effects by compromising people's capacity to adapt. Accordingly, the interactions between population growth and climate change risks have grave consequences on human food needs and well-being. This rapid increase in population is stretching the need for agricultural expansion and human settlements across the continent, thus exacerbating the consequential food insecurity. The combination of the growing population resulting from its very high fertility rates and climate change has been noted as a threat to the continent's food security. For example, the need for food and the worsening hunger occasioned by farmland shrinking and persistent drought has put Africa's 1.3 billion population in a food security crisis, with many unable to afford healthy diets. Previous studies have researched the impact of climate change and population as part of Africa's strategic drivers of food insecurity. In more than six decades, Africa (especially countries of the sub-Saharan region) has seen a reduction of 34% in its total factor productivity with low agricultural productivity due to human-induced climate change on the continent [42]. In 2020, about 282 million

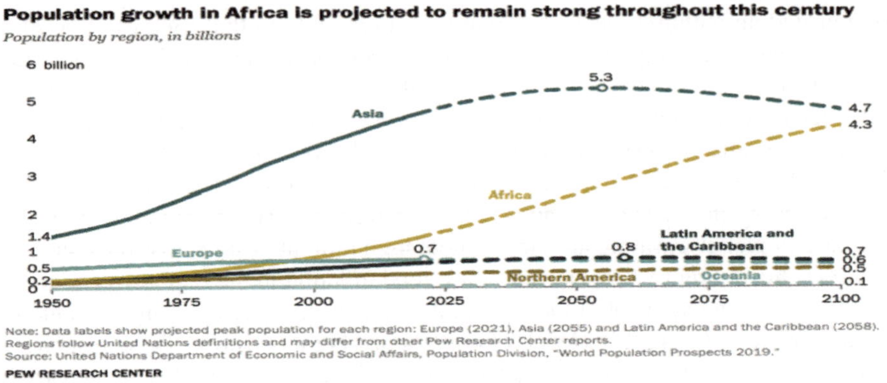

Fig. 2 Population growth in Africa throughout the century. *Source* Cilluffo and Ruiz [41]. The world's population is projected to stop growing by the end of the century nearly

(21%) Africans were affected by malnutrition and food insecurity due to climate-related shocks and low productivity [43]. In addition, Beltran-Peña and D'Odorico [43]. Today, many African countries, especially those in the Sahel region, depend on food imports [44, 45]; to address food insecurity, Beltran-Peña et al. [46] suggested an increase in crop production to address the incidence of population outstripping food supply.

5 African Continent and Need for Climate Action

Although the African continent is not known to have contributed seriously to anthropogenic factors causing climate change, Africa is the worst-hit [47]. The consequences of climate change are already driving changes in freshwater and marine ecosystems in the eastern and southern parts of Africa and the terrestrial ecosystems in south and western Africa [47]. The abundance and interaction among species have also changed [19, 48]. Literature is replete with how climate change is increasing the vulnerability of people and communities to different degrees. For instance, it will cause extreme disruption in manners in which people will be unable to cope with its overwhelming consequences on their lives and communities [49]. Hence the need for climate action. Goal 13 of the Sustainable Development Goals described climate action as the means of stepping up efforts towards reducing the emission of greenhouse gases and strengthening resilience and adaptive capacity to climate-induced impacts, including climate-related hazards in all countries; integrating climate change measures into national policies, strategies and planning; and improving education, awareness-raising and human and institutional capacity concerning climate change mitigation, adaptation, impact reduction and early warning [50].

Climate action implies the urgent need for action because there is more to lose if nothing is done to halt and reverse the change to avert severe and debilitating effects on humanity and the planet. This suggests that any chance of inaction on the part of human beings will translate to self-annihilation. This indescribable consequence can only be likened to the extinction of dinosaurs. Doing nothing will exacerbate the ongoing threats such as food and water scarcity experienced by many SSA countries. It will also lead to conflicts, while the cost of the inaction will be higher and burdensome for the continent. However, the African continent is not a substantial contributor to greenhouse gases. According to Shinn [51], the direct effect of climate change on Africa will continue to increase due to increased climate and weather variability characterised by higher temperatures, drought, and changing rainfall patterns [52].

For instance, in the early decade of the twenty-first century, we have witnessed climate catastrophes such as prolonged and intensified droughts, unprecedented floods, depletion of rain forests, and increased ocean acidity across the African continent [53]. The vastly altered weather patterns and climate extremes threaten agricultural production, food security, health, water, and energy security, undermining Africa's ability to grow and develop [54, 55]. Owing to the above-stated issues, it is also evidently clear more than ever that climate change has already and will continue to affect Africa's ecological sustainability [56]. According to Kola-Olusanya [57] and Smith and Taylor [58], many of Africa's major economic sectors (e.g. agriculture, forestry, fisheries, tourism, construction, etc.) are sensitive in various ways to weather and climate conditions [58]. Also, to varying degrees, large portions of the continent's population rely on subsistence livelihoods, which makes them susceptible to the vagaries of climate and other environmental changes [52, 58, 59]. Besides, the debilitating consequences of climate have begun to affect the African forest, with the forest with carbon dioxide-absorbing capacity of forests being threatened as it is occurring in the Congo Basin. For example, the prediction is that between 2030 and 2035, the African jungle's capacity to absorb carbon dioxide will drop between 14% and zero [60]. This situation occurred due to the increasing heat and drought caused by decreased tree cover. This situation calls for urgent action, especially considering the IPCC's description of the continent as the most vulnerable continent when considering the capacity to cope with the effects of climate change occasioned by widespread poverty, weak capacity and overly dependence on rain-fed agriculture [15, 48].

6 Strategic Options for Climate Action in Africa

The African continent, as predicted, will have the worst experience of climate change based on four critical reasons according to APG [61]; the closely coupled nature of the population and its dependence on rain-fed agricultural practices. Second is the capability of all sorts of weather and climate system surprises, which is largely vastly understudied. The third is the large degree of expected climate change, such as the projected extensive decreases in rainfall by the century's end. The final is the low capacity for climate change adaptation, as well as the general failure to governance, prioritise and act on climate change [62].

Thus, critical and good knowledge of the African climate demands that it be based on extended periods and spatially coherent records to derive accurate and realistic climate change scenarios for Africa. Its critical nature lies in its importance because, without proper and practical data, what strategic options are available cannot be answered. In response, the African Union Summit in 2009 adopted a decision to have a comprehensive African Strategy on Climate Change. According to the African Union [61, 63]. This strategy aims to provide a framework for integrated and coordinated mechanisms and strategic direction to AU member states and other stakeholders in addressing the climate change challenges and opportunities associated with the continent to improve the livelihoods of the African people and the environment in which they live [61].

The key reasons why the continent needed this strategy include the following.

(a) The need to fill the climate change governance gap to achieve optimal responses;
(b) Reduction of vulnerability and impacts of climate change on the continent, hence the need to integrate climate change into development and to climate-proof African economies
(c) Dealing with the challenges and exploitation of the opportunities of climate change via distribution, transmission and the use of knowledge; technology, and climate change finance;
(d) There is a need to promote practical cooperation and collaboration at the regional and international levels to address climate change [61, 63].

The four key reasons above are also described as the four pillars of action, which reflect the continental climate action priorities aimed at mitigating climate change and its impacts. These four pillars are presented as follows;

i. Climate change governance improvement.
ii. Enhancing the means of climate change program implementation.
iii. Integrating climate change considerations into African economies' development for climate-proofing.
iv. Promoting climate change cooperation and actions at various continental levels.

The inability to take action against climate change will result in unanticipated devastating impacts on the African continent, while a delay in action will make attaining lower climate risk levels impossible [63, 64]. The above-listed speaks to the urgency of now and that Africa, as a continent, needs to take bold steps towards climate change rather than see it as a problem in faraway lands/places. This can be done by focusing on significant components of the African Union Strategy on Climate Change, centring around four priorities: adaptation, mitigation, financing, and strengthening the negotiation process [61].

6.1 Capacity Development

Capacity development is one of the significant components of the African Union Strategy on Climate Change and an effective strategy to respond to, build, and strengthen capacities in climate change adaptation and mitigation at various levels across the continent [61]. The goal of this strategy is to promote and enhance the capacity of member states to deal with climate change [63]. Such capacity development will facilitate and develop tools, methods, and technologies to support adaptation and mitigation [61].

6.2 Technology Development and Transfer

Developing and transferring appropriate technology will be integral to Africa's climate change *adaptation* and *mitigation efforts*. This strategy provides for the establishment of the Climate Technology Centre Network (CTCN) to facilitate the development and transfer of technology in the following critical areas:

(a) Enhancing complex technologies for water harvesting, drought-resistant crop varieties, renewable energy technologies, drip irrigation, and soft technologies, such as best practices anchored on knowledge, systems, and procedures.
(b) Breaking down barriers to policies limiting technology transfer, such as trade tariff rules, intellectual property rights and technical trade barriers, like standards and eco-labelling [61].
(c) Boosting the capacity for research and development across the continent as part of Africa's mitigation and adaptation efforts to foster the growth and local cleaner technologies manufacturing [61].
(d) The enhancement of technological cooperation between and among African countries, involving a common African position on Technology development and transfer, includes the need for developed countries to commit to the deployment, diffusion, and transfer of technologies to developing countries based on the principles of accessibility, affordability, appropriateness, and adaptability, as well as to addressing the barriers to technology transfer among countries [61, 63].

6.3 Finance and Resource Mobilisation

According to the African Union [61, 63], Africa needs to take direct and urgent responsibility in mobilising climate finance to implement climate change programs in all sectors of the economy. In this regard, the financing mechanisms should spell out what Africa can do from its resources, even as it has to be given new, additional, and massive complimentary support from external sources [63] and resource mobilisation maximisation [61].

i. Improvement of financial access to rationalise the ever-growing number of funds;
ii. Harmonization of the funds' governance;
iii. Reduction in disbursement conditionalities;
iv. Streamlining bureaucratic procedures and
v. Reducing transaction costs [63].

In this regard, providing adequate financing means a lot for climate change adaptation and mitigation.

6.4 Communication Framework

Building a solid climate communication strategy for Africa requires establishing a robust communication framework consisting of regional and national media bodies across the continent for better awareness of climate change issues at the grassroots level [61]. This speaks to enabling an adequate understanding of critical issues like information sharing and collaboration, attracting support, and evaluating the impacts of the strategy. The climate change strategy must link all communication and dissemination stakeholders. The communication framework must consider the role of institutions such as the Climate Change and Desertification Unit (CCDU) to enhance cooperation, collaboration advocacy, education, and awareness about climate change issues [61, 63]. The AU Climate Change Communication strategy must identify priorities, potential partners and collaborators, and their specific needs by leveraging a high-impact medium for the dissemination of messages about climate change to the targets and stakeholders. Further, ensuring that the design, implementation and evaluation mechanism of impacts/effects align with set goals and actions [61, 63].

6.5 Implementation, Roles, and Responsibilities.

According to APG [61], climate change's multi-disciplinary and cross-cutting nature requires implementing and close coordination of intervention strategies and partnerships with all key stakeholders, such as member states, civil societies, NGOs and CBOs, faith-based organisations and media associations, each with specific roles at continental levels. The implementation of the strategy should also take advantage of all current and future opportunities guided by the AU's common positions on UNFCCC and the associated international and regional conventions and treaties [63]. Each partner should have specific roles and responsibilities within the AUC coordination framework [63].

7 Conclusion

It is essential to recognise that adaptation is an overriding priority when preparing a climate change strategy for Africa. Based on this, implementing mitigation and adaptation measures calls for urgent action. This strategy recognises Africa's capability to contribute to mitigation by developing green economic pathways to mitigate climate change impacts. In essence, Africa's capacity to adapt and mitigate against the growing climate menace would be contingent upon its ability to acquire relevant climate-friendly technologies strategically. It is instructive to note that these adaptation priorities are meant to achieve sustainable development, alleviate poverty associated with hunger among the most vulnerable groups, and integrate climate change adaptation measures into policies and strategies. This paper has attempted to address the strategic options available for Climate Action for Africa's growing population and food insecurity. The question of population as a stressor and food insecurity in Africa has been identified and tied to climate change—one of the major problems facing the African continent, its people and their governments and the continental body, the African Union (AU). Undoubtedly, the rising population has been linked to climate change, recognised as the leading environmental crisis of the twenty-first century, and scientists have pointed to the grave danger Africa is likely to experience due to higher-than-normal temperatures, changing rainfall patterns, increased climate variability and rising sea levels. Essentially threatening the African population is the effect of climate change on agricultural production -viz- higher temperatures, the drying up of soils, increased pest and disease pressure, shifts in suitable areas for growing crops and livestock, increased desertification in the Sahara region, floods, deforestation, and erosion, all of which impact Africa and its people's ability to be food secured. This paper also explored the paradox that African nations are among the lightest polluters worldwide yet suffer the most from climate change in their pursuit of water, food security, and sustainable development. Therefore, in this paper, efforts were made to establish a linkage and relationships between climate change and population, climate change and food security. It also presented a case for the African continent and the need for climate action centred around climate change information, adaptation and mitigation. In addition, the paper presented efforts towards addressing the climate crisis and its retarding effects on the continent's ability to food secured and develop sustainably. Finally, this paper has shown that climate change brings a cascade of risks, from physical impacts to Africa, such as impairing agricultural production, hunger, and food security. Thus, mitigating the impact of climate change in Africa is crucial to reducing the net effects on food insecurity and nutrition. Increasing food security resilience in the face of climate change and population increase requires multiple interventions, such as adaptation and mitigation to climate change for food security and nutrition as part of the effective interventions, integrated strategies and plans. Although human or societal adaptive capacity is low for the continent, it is now better understood, and several case studies of current and future adaptation options support this understanding. Finally, despite Africa's weak adaptive capacity, this paper showed that Africa has policies for coping with the vagaries of weather variability and climate change over the years. However, such adaptations may be insufficient to deal with future consequences of climate change.

References

1. Lara de la Fuente D (2025) The climate movement and the meaning of sustainability in the anthropocene. J Int Polit Theory. https://doi.org/10.1177/17550882251322136
2. Bang G (2024) The U.S. Inflation Reduction Act: climate policy as economic crisis response. Environ Politics. Epub ahead of print
3. Thiele LP (2013) Sustainability. Polity Press
4. Stephenson J, Newman K, Mayhew S (2010) Population dynamics and climate change: what are the links? J Public Health 32(2):150–156
5. Dube K (2021) A review of strategies used by non-governmental organisations (NGOs) to reduce vulnerability to poverty in Zimbabwe. Afr J Soc Work 11(6):371–378. https://www.ajol.info/index.php/ajsw/article/view/220362
6. Berglund O, Bailey DJ (2023) Whose system, what change? A critical political economy approach to the UK climate movement. Environ Politics 32(6):1012–1032
7. Samson J, Berteaux D, McGill BJ, Humphries MM (2011) Geographic disparities and moral hazards in the predicted impacts of climate change on human populations. Glob Ecol Biogeogr 20(4):532–544. https://doi.org/10.1111/j.1466-8238.2010.00632.x
8. Jacobsen S, Hunt O (2022) Climate justice and global politics. In: Pellizzoni L, Leonardi E, Asara V (eds) Handbook of critical environmental politics. Edward Elgar Publishing, Cheltenham, pp 192–205
9. UNCTAD (2019) Commodities and development report 2019. https://unctad.org/webflyer/commodities-and-development-report-2019
10. Mhembwe S (2025) Climate change and the feminization of poverty in Africa: insights from rural Zimbabwe. J Dev Soc 41(1):85–107. https://doi.org/10.1177/0169796X241305062
11. Sithole PK, Mawere M, Mubaya TR (2023) Socio-economic impacts of climate change on indigenous communities in the Save Valley area of Chipinge District, Zimbabwe. Front Environ Econ 2:1–16. https://doi.org/10.3389/frevc.2023.1135831
12. Henderson JV, Jang BY, Storeygard A, Weil DN (2024) Climate change, population growth, and population pressure. NBER Working Paper No. w32145. Available at SSRN: https://ssrn.com/abstract=4731054
13. IPCC (2013) IPCC confirms that human activity will further warm the Earth, with dramatic effects on weather, sea levels and the Arctic. https://www.unenvironment.org/news-and-stories/press-release/ipcc-confirms-human-activity-will-further-warm-earth-dramatic
14. IPCC (2007) Climate change 2007: impacts, adaptation and vulnerability. Summary for policymakers. Retrieved June 12 from http://www.ipcc.ch/pdf/assessment-report/ar4/syr/ar4_syr_spm.pdf
15. IPCC (2007) Fourth assessment report. Intergovernmental Panel on Climate Change Secretariat. IPCC. Retrieved October 23 from http://www.ipcc.ch/
16. IPCC (2007) Intergovernmental Panel on Climate Change, Climate Change Synthesis Report. http://www.ipcc.ch/pdf/assessment-report/ar4/syr/ar4_syr.pdf
17. IPCC (2007) Warming is very 'likely human' made. IPCC. Retrieved February 2 from http://news.bbc.co.uk/2/hi/science/nature/6321351.stm
18. Abbas S, Haider A, Kousar S et al (2025) Climate variability, population growth, and globalisation impacting food security in Pakistan. Sci Rep 15:4225. https://doi.org/10.1038/s41598-025-88916-2
19. IPCC (2014). Fifth assessment report. https://www.ipcc.ch/assessment-report/ar5/
20. IPCC (2013) The fifth assessment report (AR5). http://www.climatechange2013.org/images/uploads/WGIAR5-SPM_Approved27Sep2013.pdf

21. Jarraud M, Steiner A (2013) Joint press release WMO/UNEP—IPCC confirms that human activity will further warm the Earth, with dramatic effects on weather, sea-levels and the Arctic. https://public.wmo.int/en/media/press-release/joint-press-release-wmounep-ipcc-confirms-human-activity-will-further-warm-earth
22. Skeldon R (2024) Migration, climate change and the future of global population redistribution. N Z Econ Pap 1–5. https://doi.org/10.1080/00779954.2024.2315329
23. Bryant L, Carver L, Butler CD, Anage A (2009) Climate change and family planning: least developed countries define the agenda. Bull World Health Organ 87:852–857
24. Adger WN, Safra de Campos R (2021) Climate-change disruptions to migration systems. In: Bastia T, Skeldon R (eds) Routledge handbook of migration and development. Routledge, London, pp 382–395
25. Huang L (2023) Climate change 101: an explainer. Migration Policy Institute, Washington, DC. Retrieved from https://www.migrationpolicy.org/article/climate-migration-101-explainer?eType=EmailBlastContent&eId=eedf3c10-39fd-4b5b-a65d-f0a9bc26135d
26. Loring PA, Whitely C (2019) Food security and food system sustainability in North America
27. Rohr V, Blakley J, Loring P (2021) A framework to assess food security in regional strategic environmental assessment. Environ Impact Assess Rev 91:106674
28. de Pee S (2013) Food security in encyclopaedia of human nutrition, 3rd edn
29. Peng W, Berry EM (2019) Encyclopedia of food security and sustainability, vol 2. Elsevier, pp 1–7
30. Naheed S (2023) An overview of the influence of climate change on food security and human health. Arch Food Nutr Sci 7:001–011
31. Rani P, Reddy RG (2023) Climate change and its impact on food security. Int J Environ Clim Chang 13(3):104–108. https://doi.org/10.9734/ijecc/2023/v13i31687
32. Fanzo J, Davis C, McLaren R, Choufani J (2018) The effect of climate change across food systems: implications for nutrition outcomes. Glob Food Policy 18:12–19
33. Lefe YDH, Asare-Nuamah P, Njong AM, Kondowe J, Musakaruka RR (2024) Does climate variability matter in achieving food security in Sub-Saharan Africa? Environ Chall 15:100870
34. Shepard D (2018) Global warming: severe consequences for Africa: new report projects greater temperature increases. Africa Renewal
35. Mbaye AA, Gueye F (2024) Climate change and food security in the Sahel
36. Thompson HE, Berrang-Ford L, Ford JD (2010) Climate change and food security in sub-Saharan Africa: a systematic literature review. Sustainability 2(8):2719–2733
37. Christensen JH, Christensen OB (2007) A summary of the PRUDENCE model projections of changes in European climate by the end of this century. Clim Change 81(Suppl 1):7–30
38. Adesete AA, Olanubi OE, Dauda RO (2022) Climate change and food security in selected Sub-Saharan African countries. Environ Dev Sustain 1-19. https://doi.org/10.1007/s10668-022-02681-0
39. Affoh R, Zheng H, Dangui K, Dissani BM (2022) The impact of climate variability and change on food security in sub-Saharan Africa: perspective from panel data analysis. Sustainability 14(2):759. https://doi.org/10.3390/su14020759
40. UN (2011) World population prospects, the 2010 revision
41. Cilluffo A, Ruiz NG (2019) World population growth projected to nearly halt by end of century
42. Ortiz-Bobea A, Ault TR, Carrillo CM, Chambers RG, Lobell DB (2021) Anthropogenic climate change has slowed global agricultural productivity growth. Nat Clim Chang 11(4):306–312. https://doi.org/10.1038/s41558-021-01000-1
43. Beltran-Peña A, D'Odorico P (2022) Future food security in Africa under climate change. Earth's Future 10:e2022EF002651. https://doi.org/10.1029/2022EF002651
44. D'Odorico P, Carr JA, Laio F, Ridolfi L, Vandoni S (2014) Feeding humanity through global food trade. Earth's Future 2(9):458–469. https://doi.org/10.1002/2014ef000250
45. Graves A, Rosa L, Nouhou AM, Maina F, Adoum D (2019) Avert catastrophe now in Africa's Sahel. Nature 575(7782):282–286. https://doi.org/10.1038/d41586-019-03445-z
46. Beltran-Peña A, Rosa L, D'Odorico P (2020) Global food self-sufficiency in the 21st century under sustainable intensification of agriculture. Environ Res Lett 15(9):095004. https://doi.org/10.1088/1748-9326/ab9388

47. 350Africa (2014) 8 ways climate change is already affecting Africa. https://350africa.org/8-ways-climate-change-is-already-affecting-africa/
48. Niang I, Ruppel OC, Abdrabo MA, Essel A, Lennard C, Padgham J, Urquhart P (2014) Africa. In Barros VR et al (eds) Climate change 2014: impacts, adaptation, and vulnerability. Part B: Regional Aspects. Cambridge University Press, pp 1199–1265. https://www.ipcc.ch/site/assets/uploads/2018/02/WGIIAR5-Chap22_FINAL.pdf
49. Ani KJ, Anyika VO, Mutambara E (2022) The impact of climate change on food and human security in Nigeria. Int J Clim Change Strateg Manag 14(2):148–167. https://doi.org/10.1108/IJCCSM-11-2020-0119
50. UNDP (2020) Goal 13: climate action. https://www.sdfinance.undp.org/content/sdfinance/en/home/sdg/goal-13--climate-action.html
51. Shinn D (2015) China, Africa and food security. https://intpolicydigest.org/china-africa-and-food-security/
52. Kola-Olusanya A (2016) Climate change and environmental conflict in Nigeria: implications on national security. Afr J Soc Sci 7(6):3–16
53. Adishi E, Oluka NL (2018) Climate change, insecurity and conflict: issues and probable roadmap for achieving sustainable development goals in Nigeria. Int J Soc Sci Manag Res 4(8):12–20
54. Kola-Olusanya A (2015) Climate and African food crisis. Afr J Soc Sci 6(2):51–68
55. Lisk F (2009) Overview: the current climate change situation in Africa. In Besada H, Sewankambo NK (eds) Climate change in Africa: adaptation, mitigation and governance challenges (CIGI Special Report). The Centre for International Governance Innovation (CIGI)
56. Annan K (2007) The human face of climate change. Mail & Guardian
57. Kola-Olusanya A (2016) Climate-induced resource conflict risks in Nigeria and Sudan: recognizing and coping with increasing climate impacts on scarce resources. J Law Dev Hum 1(1):6–17
58. Smith B, Taylor A (2011) Overview of regional climate change in Africa—weADAPT. https://weadapt.org/knowledge-base/national-adaptation-planning/africa-regional-strategy/
59. Kola-Olusanya A (2014) Indigenous knowledge and climate change risks among smallholder farmers: a case-study of Southwest Nigeria. UNESCO J Afr Cult Int Underst 10:37–42
60. Grossman D (2020) The Congo rainforest is losing the ability to absorb carbon dioxide. That's bad for climate change. Washington Post
61. APG (2020) Making Africa's population an asset. Africa Progress Group Report 2020. https://africaprogressgroup.org/APG-2020-Report-Zero-Draft-December-1-2020.pdf
62. Washington R, BBC (2019) How Africa will be affected by climate change. Retrieved August 22 from https://www.bbc.com/news/world-africa-50726701
63. AU (2014) African Union strategy on climate change
64. Lisinge-Fotabong E, Diakhité M, Ababio K, N'dongo CT (2016) Climate diplomacy in Africa NEPAD

Interactive Effects of *Prosopis* Species Biostimulants and Drought Stress on Wheat Germination, Growth, and Resilience

Najma Nur Islam, Attiat Elnaggar, Kareem A. Mosa, and Ali El-Keblawy

Abstract This study evaluated the effects of biostimulants derived from *Prosopis cineraria* and *P. juliflora* on wheat (*Triticum aestivum* L.) germination and growth under drought stress induced by polyethylene glycol (PEG) at varying concentrations (−0.9, −1.1, and −1.3 MPa). Aqueous extracts of *Prosopis* species were applied at 0.5, 1, and 2% concentrations. Results showed that *P. cineraria* biostimulants consistently enhanced seed germination, shoot, and root growth across all drought levels, with the most substantial effects observed at higher concentrations. Conversely, *P. juliflora* biostimulant exhibited variable effects, providing benefits depending on the drought severity. It performed better under moderate drought stress, significantly increasing root length and shoot dry weight. At a 2% concentration, it also showed some improvement under severe drought conditions. Both biostimulants influenced the root-to-shoot length ratio, promoting root growth to mitigate drought stress. These findings underscore the potential of *P. cineraria* and *P. juliflora* extracts as biostimulants to enhance wheat growth and resilience under water-limited conditions. Optimizing biostimulant concentrations for different stress levels is essential to maximize their effectiveness in improving plant productivity.

N. N. Islam · K. A. Mosa · A. El-Keblawy (✉)
Department of Applied Biology, College of Sciences, University of Sharjah, Sharjah, United Arab Emirates
e-mail: akeblawy@sharjah.ac.ae

A. Elnaggar
Research Institute for Science and Engineering, University of Sharjah, Sharjah, United Arab Emirates

Department of Botany and Microbiology, Faculty of Science, Alexandria University, Alexandria, Egypt

K. A. Mosa
Department of Biotechnology, Faculty of Agriculture, Al-Azhar University, Cairo, Egypt

A. El-Keblawy
Faculty of Pharmacy, Al Salam University, Tanta, Egypt

S. Al-Naemi et al. (eds.), *Water and Food Security in the Face of Climate Change: Challenges and Opportunities for Resilience*, World Sustainability Series,
https://doi.org/10.1007/978-3-032-00098-9_36

Keywords Abiotic stress · Biostimulants · Drought tolerance · *Prosopis* species · Wheat germination

1 Introduction

Ensuring food security for the growing global population is one of the most pressing challenges of the twenty-first century. By 2050, rising food demand and climate change will threaten global food security [1]. Wheat (*Triticum aestivum* L.), a key staple crop, is highly vulnerable to drought, heat, and nutrient stress, risking reduced yields. Ensuring stable wheat production is essential to feeding the growing population [2].

Climate change intensifies water scarcity, highlighting the need for strategies to boost plant drought resilience [3]. One promising approach involves the use of biostimulants (BS), which are natural or synthetic substances that, when applied in small amounts, can significantly enhance plant growth, yield, and stress tolerance [4]. Biostimulants comprise many organic compounds, including amino acids, humic substances, seaweed extracts, protein hydrolysates, and beneficial microorganisms. These components stimulate natural plant processes, leading to enhanced nutrient uptake, improved water use efficiency, and increased production of stress-related proteins and enzymes [5].

Prosopis species, notably *P. cineraria*, and *P. juliflora* show promise as biostimulants for enhancing crop resilience and productivity under drought conditions. While low concentrations of *P. juliflora* extracts can stimulate germination and growth, higher concentrations have adverse allelopathic effects. In contrast, *P. cineraria* consistently promotes growth even at higher concentrations, making it more reliable for arid environments [6, 7]. Optimizing their use is key to maximizing benefits and minimizing drawbacks.

This research aims to evaluate the potential of biostimulants derived from *P. juliflora* and *P. cineraria* to enhance wheat growth and resilience under drought conditions. By investigating the optimal concentrations and effects of these biostimulants on wheat germination and growth parameters, this study seeks to provide valuable insights into developing more resilient and sustainable agricultural systems. Such strategies are crucial for ensuring stable wheat production and meeting the demands of the growing global population.

2 Materials and Methods

2.1 *Plant Material and Collection of* Prosopis *Species*

Seeds of *Triticum aestivum* L., cultivar Yecora Rojo, were obtained from the Department of Agriculture and Livestock in Maleha, Sharjah, UAE. Branches of *Prosopis* species (*P. cineraria* and *P. juliflora*) were collected from various sandy areas in Sharjah, UAE. The collected branches were dried under shade at room temperature. Once dried, the leaves were separated from the branches and ground into fine powder for subsequent biostimulant preparation.

2.2 *Preparation of Biostimulant Extracts*

Aqueous extracts of *P. cineraria* and *P. juliflora* were prepared, according to Mehar [8], with some modifications, such as mixing 4 g of the dried leaf powder in 1 L of distilled water. The mixtures were kept at room temperature and placed on a shaker for 48 h to extract bioactive compounds thoroughly. The resulting solutions were filtered through double-layered muslin cloth to remove solid residues. The filtrate was then diluted to concentrations of 0.5, 1, and 2% (w/v) for subsequent experimental treatments.

2.3 *Seed Germination Assay*

The effects of the *Prosopis* extracts on seed germination were evaluated using *Triticum aestivum* seeds. Wheat seeds were exposed to varying concentrations of the aqueous extracts (0.5, 1, and 2%) in combination with drought stress simulated by polyethylene glycol (PEG) solutions at osmotic potentials of −0.9, −1.1, and −1.3 MPa. For each treatment, 10 wheat seeds were placed in Petri dishes lined with filter paper, to which 5 mL of the corresponding *Prosopis* extract and 5 mL of the PEG solution were added. The plates were incubated in a growth chamber with controlled temperatures of 25 °C/30 °C (night/day) to assess the influence of biostimulants under varying drought conditions. Ten replicates were prepared for each treatment.

2.4 *Growth Parameters Assessment*

Germinated seeds were monitored daily for 20 days. The Petri dishes were supplemented with distilled water as needed to maintain a total liquid volume of 10 mL.

After 20 days, the seedlings were harvested, and the following growth parameters were recorded: shoot length, root length, and fresh and dry weights of both shoots and roots. The dry weights were determined by drying the plant material in an oven at 70 °C for 72 h.

2.5 Statistical Analysis

All data were analyzed using three-way ANOVA to evaluate the effects of *Prosopis* species, biostimulant concentration, PEG concentration, and their interactions on seed germination and growth parameters. Post-hoc comparisons of treatment means were conducted using Fisher's Least Significant Difference (LSD) test at a significance level of $p < 0.05$. Germination percentages were arcsine-transformed to meet the assumptions of ANOVA and enhance the normality of the data distribution.

3 Results and Discussion

3.1 Final Germination Percentage (FGP)

The final germination percentage (FGP) of wheat seeds was significantly affected by different levels of biostimulants (BS) from *P. cineraria* and *P. juliflora* under varying PEG concentrations (Fig. 1). The highest FGP was recorded under non-drought conditions (0 MPa PEG), with FGP decreasing as PEG concentration increased, simulating drought stress.

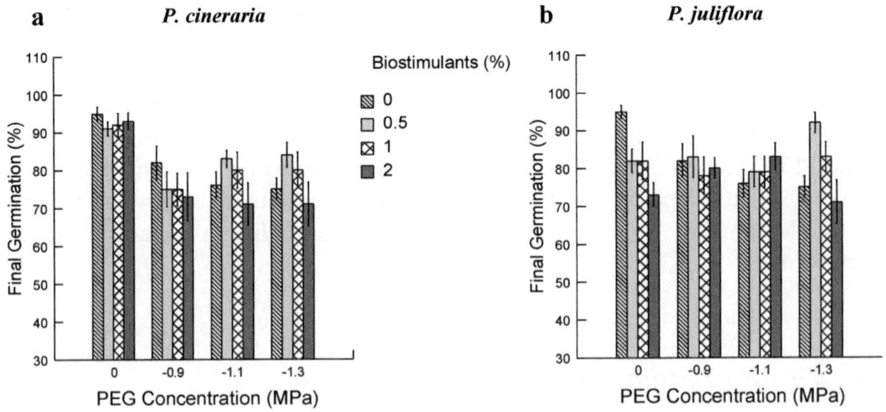

Fig. 1 Effect of **a** *P. cineraria* and **b** *P. juliflora* biostimulants on the final germination percentage of wheat seeds under different levels of drought stress induced by PEG

There were significant effects of BS and PEG ($P < 0.001$) but no significant effect of the *Prosopis* species ($P > 0.05$) on FGP (Table 1). For *P. cineraria*, BS levels had no significant impact under 0 MPa PEG. BS showed no significant effect at moderate drought stress (−0.9 MPa), but at −1.1 MPa, 0.5% BS increased FGP by 9.2% compared to the control. Under severe drought (−1.3 MPa), the same concentration increased FGP by 12%.

BS of *P. juliflora* reduced FGP under non-drought conditions but increased it by 1.2% at −0.9 MPa and 22.7% under severe drought (−1.3 MPa). These results suggest that BS from both species can improve germination under drought, with optimal concentrations varying by stress level. Comparable findings were reported by Wadduwage et al. [9], where biostimulants increased lettuce seed germination under drought, achieving rates of 58–62%. Similarly, Ali et al. [10] demonstrated improved wheat germination under osmotic stress with optimal effects at 20% *Cuscuta reflexa* extract, while higher concentrations inhibited germination. Mehar [8] also observed enhanced rice seed germination (75–100%) with low concentrations (0.1–1%) of *P. juliflora* extract. These results align with the current study, confirming the role of biostimulants in mitigating drought effects on seed germination.

Table 1 Results of three-way ANOVA (F-values) assessing the effect of biostimulant and PEG application and their interaction on wheat germination and growth

Variable	Species (S)	Biostimulant conc (BS)	PEG conc (PEG)	S × BS	S × PEG	BS × PEG	S × BS × PEG
FGP	0.176 ns	3.520**	12.505***	0.622 ns	6.110***	3.068**	1.342 ns
GRI	0.821 ns	5.031**	158.543***	2.599 ns	3.008*	2.562**	2.166*
Shoot Length	0.488 ns	107.932***	11.997***	0.663 ns	1.988 ns	5.075***	1.896 ns
Leaf length	1.256 ns	63.217***	4.046**	0.288 ns	1.314 ns	6.875***	1.084 ns
Shoot dry weight	0.015 ns	3.244*	5.795***	3.737*	0.535 ns	3.155**	0.979 ns
Root length	0.001 ns	83.752***	8.462***	17.109***	4.895**	7.534***	8.162***
Root dry weight	8.119**	14.458***	10.742***	2.466 ns	0.444 ns	3.102**	3.528***
R:S length ratio	1,642.38***	29.115***	4.004**	22.216***	2.965*	9.62***	10.295***

*: < 0.05, **: $P < 0.01$, ***: $P < 0.001$, ns: insignificant difference at $P \geq 0.05$

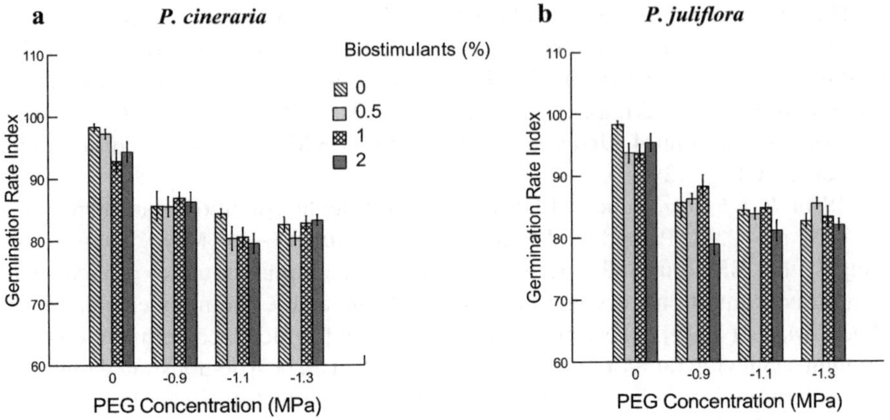

Fig. 2 Effect of biostimulants from **a** *P. cineraria* and **b** *P. juliflora* on the germination rate index of wheat seeds under different levels of drought stress induced by PEG concentrations

3.2 Germination Rate Index (GRI)

GRI showed a similar pattern to FGP, decreasing with higher PEG concentrations, indicating slower germination under stress (Fig. 2). BS of *P. cineraria* reduced GRI slightly under non-drought conditions (0 MPa), with a 5.6% decrease at 1% BS. Under moderate drought (−0.9 MPa), GRI dropped 12.9% without BS, but at −1.1 MPa with 1% BS, GRI stayed at 80% of the control, showing some stress mitigation.

BS of *P. juliflora* reduced GRI by 4.7% at 1% under non-drought conditions. However, under moderate drought (−0.9 MPa), 1% BS increased GRI by 1.9%. Under severe drought (−1.3 MPa), GRI decreased by 8.3% with 2% BS, but 0.5% BS raised GRI by 3.7%. ANOVA confirmed significant BS and PEG interactions (Table 1). These results indicate that moderate BS levels improve germination under drought stress, consistent with Shah et al. [11], who found similar benefits of biostimulants under osmotic stress in dry regions.

3.3 Effects on Shoot Parameters

ANOVA revealed significant effects of BS and PEG ($P < 0.001$) on shoot height, leaf length, and shoot dry weight, with a significant PEG-BS interaction ($P < 0.001$) but no significant difference between *Prosopis* species ($P > 0.05$) (Table 1). BS application increased shoot length under varying drought stress levels. At 0 MPa (no drought stress), BS from both species increased shoot length by 22.1%. Under moderate and severe drought (−1.1 MPa and −1.3 MPa), shoot length increased by 58.2% and 26.2%, respectively, depending on BS concentration (Fig. 3). Leaf length showed a similar trend, with a 20.1% increase for *P. cineraria* and 19.4% for *P. juliflora* at

0 MPa. At −1.1 MPa, 2% BS increased leaf length by 62.8% for *P. cineraria* and 45.0% for *P. juliflora*.

Shoot dry weight was significantly influenced by BS type. For *P. cineraria*, 0.5% BS at −0.9 MPa increased shoot dry weight by 63.8%, while 2% BS at −1.1 MPa increased it by 48.3%. For *P. juliflora*, 2% BS at −1.1 MPa resulted in a 37.7% increase in the shoot dry weight. These results show that BS from both species

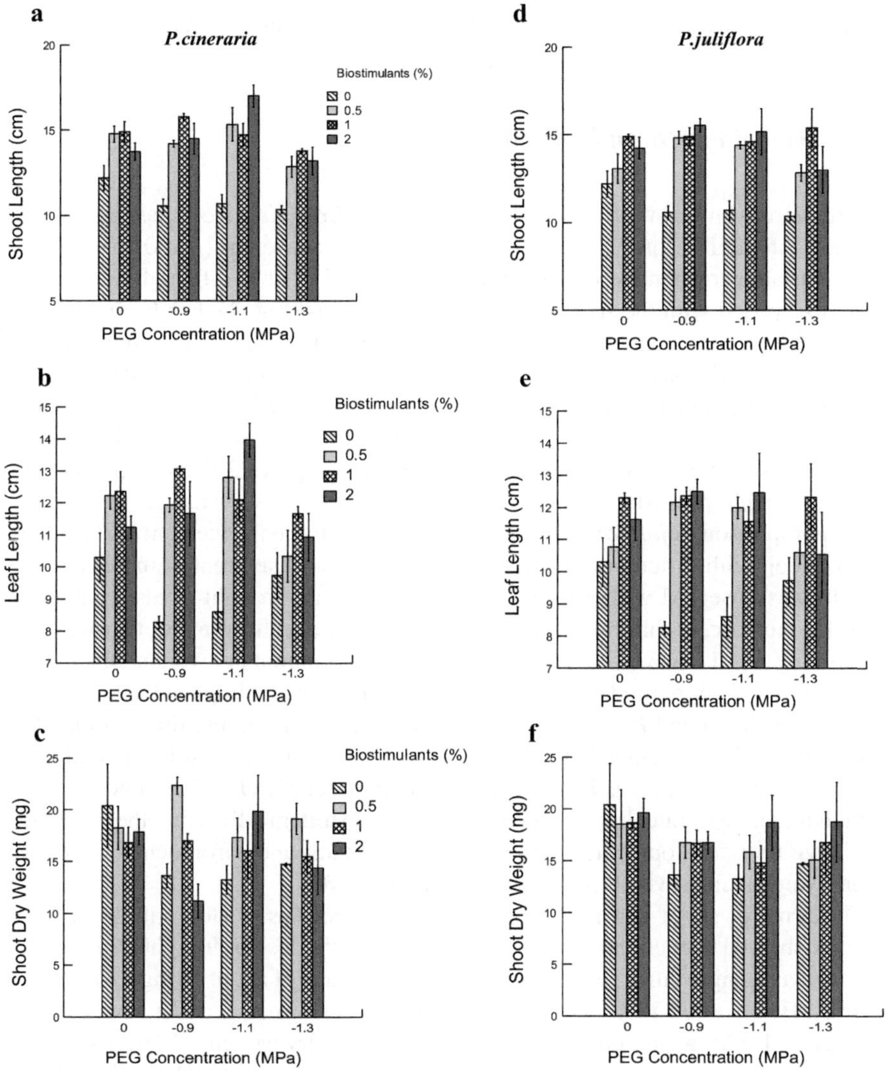

Fig. 3 Effects of biostimulants of *P. cineraria* and *P. juliflora* and drought simulated by PEG on shoot length (**a**, **d**), the longest leaf length (**b**, **e**) and dry weights (**c**, **f**) of the shoots of wheat seedlings

mitigates drought stress, with *P. cineraria* generally having a stronger effect on shoot dry weight.

These findings highlight the role of *Prosopis*-based BS in promoting shoot and leaf growth under drought stress, making them valuable for improving plant resilience. Elnajar et al. [12] similarly reported increased shoot length (4.6–12%) and leaf fresh mass (29–51.2%) in wheat treated with *Streptomyces platensis* extract under drought stress. Mehar [8] also noted growth enhancements in rice seedlings treated with *P. juliflora* extract. These results underscore the potential of plant-extract BS in mitigating drought effects and improving plant performance under water stress.

3.4 Root Length and Dry Weight

There were significant effects for BS, PEG, and *Prosopis* species, as well as most of their interactions on wheat root length and root dry weight ($P < 0.05$) (Table 1). Both variables responded positively to the application of BS, though the extent of response varied with the species and stress conditions (Fig. 4). For *P. cineraria*, the most substantial increase in root length was observed with a 2% biostimulant under no drought conditions, showing a 309.1% increase compared to the control (Fig. 4a). This was complemented by a significant 43.3% increase in root dry weight under the same conditions (Fig. 4b). Under moderate drought stress (−0.9 MPa), the root length increase remained notable with a 1% BS leading to a 227.8% increase, while root dry weight peaked with a 0.5% BS, showing an 81.9% increase.

In comparison, *P. juliflora* also demonstrated effective enhancement of root parameters, albeit with different optimal BS concentrations. The greatest increase in root length was observed with a 0.5% BS under moderate drought stress, leading to a 329.4% increase, highlighting the species' efficiency at enhancing root growth under stress (Fig. 4c). However, the corresponding increase in root dry weight under similar conditions was more modest, at 27.6% ((Fig. 4d). These findings indicate that while both *P. cineraria* and *P. juliflora* BS are beneficial in mitigating drought effects, *P. cineraria* tends to enhance both root length and dry weight more robustly under non-stress conditions, whereas *P. juliflora* appears particularly effective in boosting root length under moderate drought. This suggests that strategically using these BS could optimize root development, depending on the prevailing environmental conditions, thereby enhancing overall plant resilience and growth.

In terms of shoot and root biomass, wheat seedlings treated with *P. cineraria* extract showed higher shoot dry weight under all the tested drought stresses, with values averaging around 20 mg, compared to an average of about 18 mg for *P. juliflora* extract. However, *P. juliflora* demonstrated better performance under severe drought stress at −1.3 MPa, particularly in the shoot and root dry weights (18 mg), compared to *P. cineraria*'s 15 mg. This difference in biomass allocation, where *P. juliflora* is better in root growth under severe drought, aligns with Zhang et al. [13], who found that the seaweed extract-based biostimulant enhanced root length and biomass under

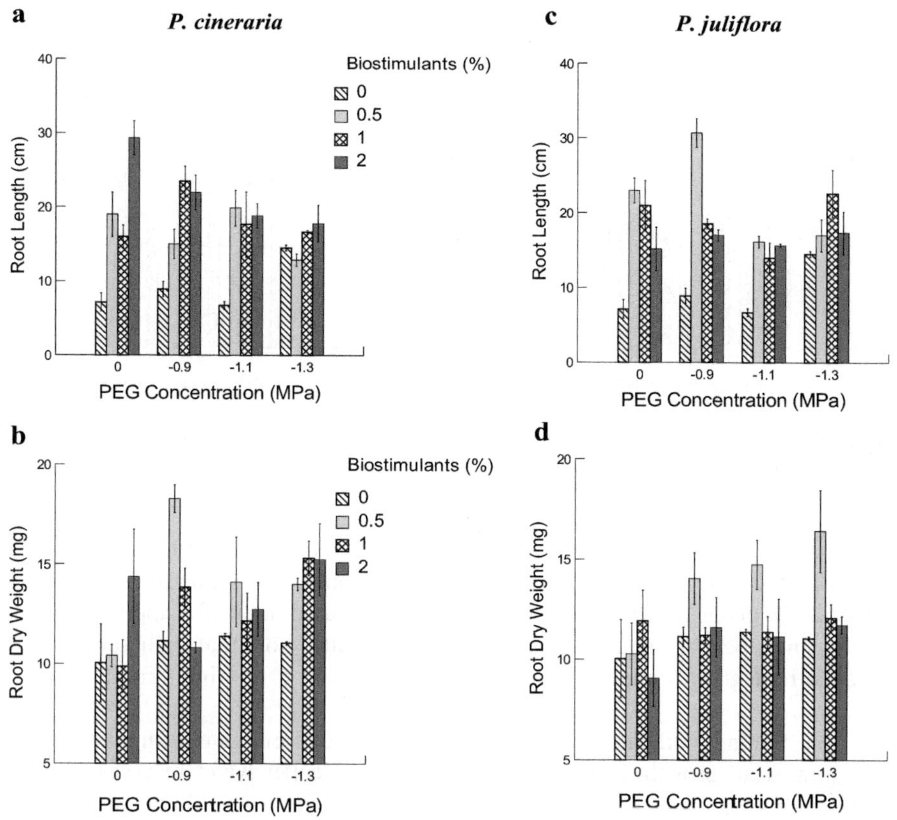

Fig. 4 Effects of biostimulants from *P. cineraria* (**a**, **b**) and *P. juliflora* (**c**, **d**) and drought simulated by PEG on root length (**a**, **c**) and dry weight (**b**, **d**) of the roots of wheat seedlings

heat and drought conditions, reflecting an adaptive strategy to improve water uptake in stressed environments.

3.5 Root-to-Shoot Length Ratio

The root-to-shoot length ratio was significantly influenced by all factors and their interactions ($P < 0.05$) (Table 1). For *P. cineraria*, a 2% biostimulant (BS) under non-drought conditions (0 MPa) led to a 255% increase in the ratio compared to the control (Fig. 5a). Under moderate drought (−0.9 MPa), a 1% BS increased the ratio by 148.3%, highlighting its effectiveness in promoting root growth. At severe drought (−1.1 MPa), a 0.5% BS resulted in a 105% increase, suggesting that lower concentrations may be optimal under higher stress.

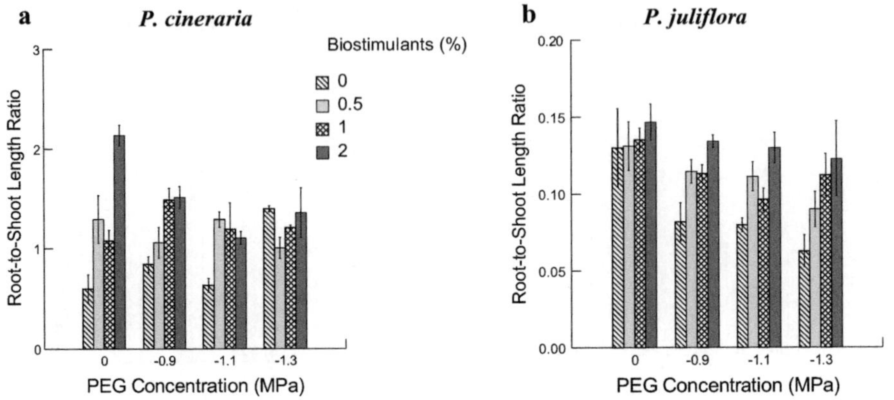

Fig. 5 Root-to-shoot length ratio of wheat seedlings treated with biostimulants from **a** *P. cineraria* and **b** *P. juliflora* at different PEG concentrations

For *P. juliflora*, the response was less pronounced but still significant. A 2% BS increased the ratio by 15.4% under non-drought conditions, while 0.5% BS raised it by 62.5% under moderate drought. At severe drought, a 2% BS application also resulted in a 62.5% increase (Fig. 5b). These results show that both species' BS enhance root growth under drought, with *P. cineraria* exhibiting stronger overall effects.

These findings indicate that *Prosopis*-based BS can mitigate drought stress by promoting root growth, enabling plants to access deeper water reserves. Similar observations by Bacher et al. [14] showed increased root-to-shoot ratios under drought, where prioritizing root growth minimized evaporation and improved water uptake.

The results emphasize the importance of optimizing BS concentrations based on stress levels and suggest that differences between *P. cineraria* and *P. juliflora* may stem from their unique phytochemical profiles. *P. cineraria*, rich in phenolic compounds, offers antioxidant protection, while *P. juliflora*, high in alkaloids and secondary metabolites, likely enhances stress tolerance through hormonal regulation [15–17]. This aligns with research showing plant-extract BS as effective tools for improving drought resilience and soil fertility, particularly in water-scarce regions.

4 Conclusion

The results demonstrate that biostimulants derived from *Prosopis* species have the potential to enhance wheat germination and growth under drought stress, although their effectiveness varies depending on concentration and species. *P. cineraria* biostimulants consistently improved plant growth across all drought levels, making it a more reliable option for mitigating drought stress. In contrast, *P. juliflora* biostimulants showed variable effects, offering varying benefits depending on the drought level. The difference in the effects observed between *P. cineraria* and *P. juliflora* emphasizes the importance of conducting further studies to optimize the selection of appropriate biostimulants based on the specific stress conditions and desired growth outcomes. These findings also highlight the need to carefully optimize biostimulant concentration and application conditions to maximize plant resilience and improve agricultural sustainability in water-scarce environments. Future research should be conducted in order to explore the underlying molecular mechanisms of these biostimulants, exploring the potential for these biostimulants to be used in combination to maximize their benefits across different stages of plant development and their implementation in larger farming plans, considering the ongoing challenges posed by climate change, global rise in population and demand for food.

References

1. Bahar NH, Lo M, Sanjaya M, Van Vianen J, Alexander P, Ickowitz A, Sunderland T (2020) Meeting the food security challenge for nine billion people in 2050: what impact on forests? Glob Environ Chang 62:102056
2. Hossain A, Skalicky M, Brestic M, Maitra S, Ashraful Alam M, Syed MA, Hossain J, Sarkar S, Saha S, Bhadra P, Shankar T, Bhatt R, Chaki AK, EL Sabagh A, Islam T (2021) Consequences and mitigation strategies of abiotic stresses in wheat (*Triticum aestivum* L.) under the changing climate. Agronomy 11(2):241
3. Rey D, Holman IP, Knox JW (2017) Developing drought resilience in irrigated agriculture in the face of increasing water scarcity. Reg Environ Change 17:1527–1540
4. De Vasconcelos ACF, Chaves LHG (2019) Biostimulants and their role in improving plant growth under abiotic stresses. In: Biostimulants in plant science, pp 3–16
5. Bulgari R, Cocetta G, Trivellini A, Vernieri PAOLO, Ferrante A (2015) Biostimulants and crop responses: a review. Biol Agric Hortic 31(1):1–17
6. El-Keblawy A, Abdelfatah MA (2014) Impacts of native and invasive exotic *Prosopis congeners* on soil properties and associated flora in the arid United Arab Emirates. J Arid Environ 100:1–8
7. Tsombou FM, El-Keblawy A, Elsheikh EA, AbuQamar SF, El-Tarabily KA (2022) Allelopathic effects of native and exotic *Prosopis congeners* in Petri dishes and potting soils: assessment of the congeneric approach. Botany 100(3):329–339
8. Mehar SK (2011) Assessment of effect of *Prosopis juliflora* litter extract on seed germination and growth of rice. Assessment 2
9. Wadduwage J, Egidi E, Singh BK, Macdonald CA (2024) Impacts of biostimulants on crop yield and biological activity under drought conditions. J Sustain Agric Environ 3(1):e12093
10. Ali Q, Perveen R, El-Esawi MA, Ali S, Hussain SM, Amber M, Iqbal N, Rizwan M, Alyemeni MN, El-Serehy HA, Al-Misned FA, Ahmad P (2020) Low doses of Cuscuta reflexa extract act as natural biostimulants to improve the germination vigor, growth, and grain yield of wheat grown under water stress: photosynthetic pigments, antioxidative defense mechanisms, and nutrient acquisition. Biomolecules 10(9):1212
11. Shah NA, Ullah S, Khan MN, Ali S, Adnan M, Ali A, Ali M, Khan A, Hassan S, Khan WM (2021) Sugar beet extract acts as a natural bio-stimulant for physio-biochemical analysis of Hordeum Vulgare L. under induced salinity stress. bioRxiv 2021-08
12. Elnajar M, Aldesuquy H, Abdelmoteleb M, Eltanahy E (2024) Mitigating drought stress in wheat plants (*Triticum Aestivum* L.) through grain priming in aqueous extract of spirulina platensis. BMC Plant Biol 24(1):233.

13. Zhang X, Taylor Z, Goatley M, Booth J, Brown I, Kosiarski K (2022) Seaweed extract-based biostimulant impacts on nitrate reductase activity and root viability of ultradwarf bermudagrass subjected to heat and drought stress. HortScience 57(10):1328–1333
14. Bacher H, Sharaby Y, Walia H, Peleg Z (2022) Modifying root-to-shoot ratio improves root water influxes in wheat under drought stress. J Exp Bot 73(5):1643–1654
15. Asati V, Deepa PR, Sharma PK (2022) Desert legume Prosopis cineraria as a novel source of antioxidant flavonoids/isoflavonoids: biochemical characterization of edible pods for potential functional food development. Biochem Biophys Rep 29:101210
16. Sabarisenthil B, Abdul Kader Muaatdh MA, Ahamed Asfak KK, Khan AR, Ahamed Syedh K, Jena D (2023) An assessment of the pharmacognostic properties of Prosopis juliflora. Int J Pharmacognosy Life Sci 4(1):85-90
17. Lei L, Zhao Y, Shi K, Liu Y, Hu Y, Shao H (2021) Phytotoxic activity of alkaloids in the desert plant Sophora alopecuroides. Toxins 13(10):706

A Novel Predictive Model for Integrating Pest Detection and Environmental Factors for Food Security Assessment

Muhammad Qasim, Danish Mehmood, Asifa Bibi, Ghufran Ahmed, and Adnan Akhunzada

Abstract In response to the growing challenges of climate change and pest infestations, we have successfully developed and implemented a novel system that integrates advanced pest detection techniques with real-time climate data to enhance food security, specifically within rice crops. Our approach employs the ResNet50 model, a deep learning architecture known for its robust image classification capabilities, fine-tuned to accurately detect pests within complex crop images. This model's high precision in identifying and classifying various pests is complemented by the incorporation of environmental data, specifically temperature and precipitation records from the AgMERRA dataset. By integrating these climate variables, our system not only detects pests but also contextualizes their potential impact based on prevailing and forecasted weather conditions. This allows for more informed and climate-aware decision-making regarding pest management practices. For instance, the system can determine the optimal timing for pesticide application, avoiding periods where high temperatures might reduce efficiency or when impending rainfall could wash away treatments. This strategic approach minimizes unnecessary pesticide use, reduces environmental impact, and significantly enhances crop preservation. Our methodology has been rigorously tested and demonstrated a 95% accuracy rate in pest detection, leading to the preservation of 50–55% of crop yield that would otherwise be lost to pest damage. This performance not only surpasses traditional pest management techniques but also outperforms other prominent pre-trained models, such as VGG-16, VGG-19, and ResNet101, in the specific context of agricultural disease detection. The success of our system underscores its potential as a critical tool in the

M. Qasim · D. Mehmood
SZABIST University, Islamabad, Pakistan

A. Bibi
Bahria University, Islamabad, Pakistan

G. Ahmed
National University of Computer and Emerging Sciences, Karachi, Pakistan

A. Akhunzada (✉)
University of Doha for Science and Technology, Doha, Qatar
e-mail: adnan.adnan@udst.edu.qa

S. Al-Naemi et al. (eds.), *Water and Food Security in the Face of Climate Change: Challenges and Opportunities for Resilience*, World Sustainability Series,
https://doi.org/10.1007/978-3-032-00098-9_37

global effort to secure food supplies against the dual threats of pest infestations and climate variability.

Keywords Image processing · Pest detection · Deep learning · Food security · Crop management

1 Introduction

Rice is a staple food for over half of the world's population, making its cultivation critical for global food security. Rice crops are highly susceptible to pest infestations, which can cause significant losses and threaten the livelihoods of millions of farmers. Traditional methods of pest management, while effective to some extent, often rely on blanket pesticide applications that can lead to environmental harm, pesticide resistance, and unnecessary costs. Moreover, these methods fail to consider the complex interplay between pest activity and climate conditions, which can significantly influence the effectiveness of pest control measures. As climate change continues to alter weather patterns, there is an urgent need for innovative solutions that integrate pest detection with climate responsive decision-making to enhance the resilience of rice production systems.

The need for a more precise and sustainable approach to pest management in rice farming is increasingly evident. Traditional pest management strategies often rely on static methods that fail to account for the dynamic nature of pest behavior and the significant influence of climate conditions. Our approach addresses this gap by integrating state-of-the-art deep learning models with climate data to provide a more accurate and responsive system for pest detection and management. This integration allows for a more informed decision-making process that considers both the immediate pest threat and the broader environmental context.

We have implemented a comprehensive system that combines the enhanced ResNet50 model with fully connected, softmax, and classification layers for high accuracy pest detection with temperature and precipitation data from the dataset. This system not only detects pests but also predicts the potential impact of climate conditions on pest activity and crop health. By doing so, it informs the optimal timing and dosage of pesticide application, ultimately leading to better pest control and significant crop yield preservation. The system assesses whether pesticide use is necessary, taking into account climate factors that may affect its efficacy, such as high temperatures or impending rainfall.

We show through rigorous experimentation and comparative analysis that our methodology outperforms existing approaches in terms of precision, recall, and accuracy. Our research highlights how our proposed methodology transforms agricultural disease management strategies and enables more productive and sustainable pear crop farming. In summary, this study represents a significant step forward in leveraging cutting-edge AI technologies to address critical challenges in agricultural disease detection and management, with implications for enhancing productivity and resilience in pear cultivation worldwide. The major contributions of the proposed framework are presented as follows,

- The authors have proposed a predictive model that integrates environmental factor and enhanced deep learning.
- The authors have proposed a decision logic for applying pesticides to crops.
- The authors have computed features using state-of-the-art deep learning features such as VGG-16, VGG-19, ResNet101, Inception V3 and DenseNet and compared with the proposed framework.

This paper unfolds as follows: Sect. 2 delves into related work, highlighting the existing challenges and gaps in current approaches. Section 3 provides proposed methodology. Section 4 presents experimental setup and Sect. 5 contains experimental results. Finally, Sect. 6 concludes the paper, summarizing key findings and outlining avenues for future research.

2 Related Work

Study shows that climate change presented significant challenges to agriculture by altering insect pest dynamics. Rising temperatures, increased CO_2 levels, and changing precipitation patterns lead to expanded pest distributions, increased survival rates, and altered interactions between plants and pests. These changes posed a threat to crop productivity and food security. To address these challenges, adaptive management strategies and enhanced monitoring systems are necessary. Future research should focus on developing modified integrated pest management tactics, implementing climate and pest population monitoring, and utilizing modeling tools for accurate predictions [1]. Adler et al. [2] highlighted the impact of global warming on the distribution of stored product pests in Europe. He emphasized the need for enhanced surveillance and monitoring of these pests to ensure food security. His study proposed a joint European initiative to track the movement and spread of stored product pests and advocates for increased support for the World Food Program to mitigate the effects of climate change on food availability. Additionally, the study recommended a review of quarantine regulations for serious stored product pests to strengthen food safety measures.

Another study highlighted the significant impacts of climate change on terrestrial and oceanic ecosystems. It emphasized the need for modeling approaches to assess long-term changes in climatic variables and their effects on natural ecosystems. The study warns of potential consequences such as extreme heat events, altered rainfall patterns, increased flooding and drought, food and water scarcity, and health risks. It also emphasizes the importance of considering the effects of increased greenhouse gas levels on future climates [3]. This study investigated the impact of climate change on insect pest distribution and damage in alpine ecosystems. Using species distribution modeling, the researchers simulated the current and future distribution probabilities of 58 insect pest species in Qinghai Province, China. By analyzing the relationship between distribution probability and damage occurrence, the study identified potential risks posed by insect pest expansion under various climate change scenarios. The findings highlight the need for considering both distribution and damage assessment in risk assessment efforts to effectively manage insect pests in changing environments [4].

Climate changes have posed a significant threat to the agricultural system and food security. While pre-harvest pest management has received considerable attention, the impact of climate change on post-harvest pest management has been relatively neglected. This study highlighted the need for in- creased research and adaptation of integrated pest management strategies to address the challenges posed by climate change in the post-harvest supply chain [5]. This study investigated the significant impacts of climate change on wheat and rice, two staple crops. It reviewed the direct effects of climate change on crop growth, yield, and distribution, as well as the indirect effects through pests and pathogens. The paper offered insights into potential management strategies to mitigate these impacts, providing valuable information for agriculture practitioners and researchers [6]. Insect pests pose a significant threat to food security, causing substantial economic losses. This study highlighted strategies for mitigating the negative impacts of in- sect pests on agricultural systems. Recommended approaches include integrated pest management, agroecological farming, improved monitoring, resistant crop varieties, farmer support networks, financial safety nets, post-harvest infrastructure, value-added processing, and consumer awareness initiatives. A combination of these measures is crucial for enhancing the resilience of food systems and protecting vulnerable populations from the adverse effects of pest outbreaks [7].

Yang et al. [8] investigated the impact of GM Bt maize hybrids on maize production in China. The results demonstrated a significant reduction in pest feeding damage, yield loss, and mycotoxin contamination compared to non-Bt hybrids. The use of Bt maize can contribute to sustainable agriculture by reducing the reliance on chemical pesticides and enhancing food security. Chu et al. [9] presented a novel approach for accurate detection and identification of granary pests. A multi- scale granary pest recognition model based on YOLOv5 with BiFPN, DIOU NMS, and ECA modules was developed. The model achieved an average accuracy of 98% on a dataset of 5231 images, outperforming other state-of-the-art models. Ablation studies confirmed the effectiveness of the proposed enhancements, demonstrating the model's robustness and practical significance for granary pest management. Another study presented an improved YOLOv5m model for plant pest detection. By incorporating SWin Transformer, C3TR, ResSPP, and WConcat, the model enhances feature extraction and detection capabilities. The proposed model achieved superior performance compared to previous versions, demonstrating improved accuracy, recall, F1-score, and mAP in detecting various plant pests [10].

Crop pests and diseases pose a significant threat to agricultural productivity and economic stability. Traditional recognition methods have limitations in accuracy and efficiency. This study proposed an improved Vision Transformer (ViT) method for crop pest image recognition. By effectively selecting regions of interest and utilizing self-attention mechanisms, the proposed method achieves high accuracy in classifying crop diseases and pests. This research provides a valuable reference for agricultural workers in developing effective disease and pest control strategies [11]. The study proposed a novel fruit quality assessment system utilizing computer vision and machine learning. This system aims to assist visually impaired individuals in identifying fruit quality and prevent food waste due to unsanitary storage and transportation practices. The system incorporates a camera unit for capturing fruit images and machine learning models for predicting fruit quality and freshness [12].

The study presented a novel framework for paddy insect detection and classification using deep learning techniques. The framework incorporates image preprocessing, feature ex- traction using pre-trained CNN models, feature selection, and machine learning algorithms. Experimental results demonstrate the effectiveness of the proposed approach, achieving high accuracy in paddy insect identification [13]. Another study reviews examines the significant impact of climate change on agricultural systems and insect pest-crop interactions. It highlights the direct and indirect effects of rising temperatures and CO_2 levels on insect physiology, population dynamics, and crop damage. The study emphasized the need for integrated pest management strategies that consider the two-way interactions between insect pests and crops, enabling more accurate yield loss forecasting and effective interventions [14].

Climate change pests poses significant challenges to agricultural systems and food security. To address these challenges, it is crucial to implement effective pest management strategies, develop climate-resilient crop varieties, and invest in sustain- able agricultural practices. By taking proactive measures, we can enhance the resilience of our food systems and ensure a sustainable future for agriculture.

3 Methodology

Our methodology involves a multi-step process where raw images of rice crops with complex backgrounds are captured and processed using a fine-tuned ResNet50 model for pest detection. The model extracts high-level features, enabling precise identification of pests and the extent of infestation. This information is then combined with temperature and precipitation data from the AgMERRA dataset to contextualize the pest activity within the current and forecasted climate conditions. The integrated data informs climate-aware decision-making, optimizing pesticide application to ensure effectiveness while minimizing environmental impact. This system is supported by a user-friendly decision support interface, providing real-time alerts and recommendations to farmers. Continuous feedback and adaptive learning mechanisms ensure the model's accuracy and reliability improve with each season, and scalability options allow for application to other crops and agricultural contexts.

Figure 1 illustrates the architecture of our methodology. It consists of three parts: the data collection and preprocessing layer, segmentation and feature extraction and finally classification layer.

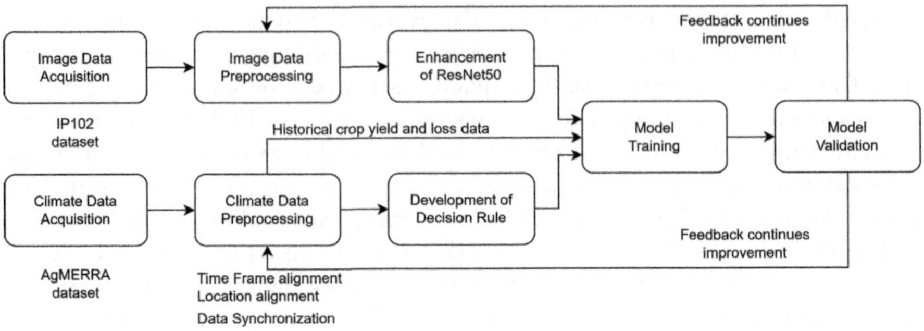

Fig. 1 Proposed methodology

3.1 Data Acquisition

The initial module involves capturing raw images of rice crops under diverse field conditions to ensure a representative dataset. These images are collected from different rice paddies with varying backgrounds, lighting conditions, and pest infestations. The complexity of the backgrounds, including variations in soil, plant density, and lighting, reflects the real-world challenges of pest detection. This stage aims to gather a comprehensive and varied set of images that accurately represent the conditions under which rice crops are grown. The dataset thus obtained forms the foundation for the subsequent analysis, providing the raw material needed for effective feature extraction and model training.

In order to obtain the climate data the AgMERRA dataset was used. This includes daily and historical records of temperature, precipitation, and potentially other variables such as relative humidity, solar radiation, and soil moisture. These variables temperature and precipitation are selected based on their known impact on pest behavior and plant health. The climate data must be synchronized with the pest detection images. Each image in the dataset is associated with the corresponding climate data for the same time period or growing season. This ensures that the pest detection results are contextualized with accurate and relevant environmental conditions.

3.2 Data Preprocessing

Three distinct steps are taken during the pre-processing phase, including contrast correction, picture enhancement, and scaling. It cannot be utilized directly in image processing jobs because of the uneven and incorrect size of the pictures. The photos that are gathered come from various backgrounds, to detect plant diseases photos must be scaled-up. There is no degradation of image quality when the image is resized. The scaled picture that is produced thereafter has more or less pixels. Subsequently, the color and contrast of the dark background photos are adjusted to facilitate processing.

Deep learning models excel in classification and prediction tasks, especially with high-dimensional image inputs, due to their high learning capacity. However, complex models can suffer from overfitting when trained on small datasets [15]. By using random transformations including rotation, cropping, and color variation, data augmentation lessens this problem by producing more training data. Standard augmentation methods are the main focus of this work to improve the model's functionality and capacity for generalization.

Raw climate data is often noisy and may require cleaning and normalization. This involves handling missing values, removing outliers, and scaling the data to a consistent range to ensure that it integrates seamlessly with the feature extraction and model inputs. Depending on the granularity of the climate data and the frequency of pest detection images, temporal aggregation may be necessary. For example, daily temperature and precipitation data might be aggregated into weekly or monthly summaries to match the temporal resolution of the pest detection data.

3.3 Feature Extraction

We have utilized the ResNet50 model as a core component of our pest detection system. After fine-tuning ResNet50 on a specialized dataset of rice crop images, it has been instrumental in extracting high-level features from the images, enabling us to detect pests with a high degree of accuracy. The model's ability to identify subtle signs of pest infestation has allowed us to pinpoint affected areas in the crops early, ensuring timely intervention. Our implementation of ResNet50 has effectively classified various types of pests and assessed the extent of infestations. This capability has been crucial in detecting pest outbreaks early, providing an opportunity to act before the infestation becomes widespread, thereby protecting most of the crop yield.

From climate data, several key features are extracted that are relevant for pest management. These might include.

- Average Temperature and Precipitation: Aggregated over specific periods (e.g., weekly or monthly) to assess all climate conditions.
- Temperature Extremes: Indicators of high or low temperature events that might influence pest activity.
- Precipitation Patterns: Analysis of recent rainfall and its impact on soil moisture and pest behavior.
- Climate Anomalies: Identification of unusual climate patterns (e.g., sudden temperature spikes or extended drought periods) that might affect pest dynamics.

3.4 Integration of Temperature and Precipitation Data

The core of the methodology lies in the ResNet model, which integrates the features extracted in the previous module using a fusion technique. ResNet employs sophisticated architecture that includes fully connected layers to combine features from both handcrafted and deep learning sources, a SoftMax layer to convert combined features into probability scores for different pest classes, and a classification layer to output the final pest detection results. This model is designed to leverage the strengths of both feature extraction techniques, allowing it to achieve high accuracy in pest detection by capturing both fine-grained details and high-level patterns in the images.

Climate Contextualization was done by integrated temperature and precipitation data from the AgMERRA dataset directly into our pest detection pipeline. By synchronizing this climate data with the pest detection outputs from ResNet50, we have been able to contextualize pest activity within the environmental conditions. For instance, our system has successfully correlated certain pest activities with specific temperature and humidity levels, enabling more accurate predictions of pest behavior under varying climate scenarios.

The incorporation of temperature and precipitation data has allowed us to model pest behavior dynamically. For example, we observed that warmer temperatures accelerated the life cycle of some pests, leading to more rapid infestations, while heavy rainfall affected both pest survival and pesticide efficacy. This modeling has significantly improved the accuracy of our pest outbreak predictions and has informed our pest management strategies.

3.5 Climate-Aware Decision-Making

Our system leverages the integrated climate data to make informed decisions about pesticide application. For instance, we have been able to adjust pesticide recommendations based on temperature and precipitation forecasts. When high temperatures were expected to reduce pesticide efficacy, or when rain was forecasted, our system advised delaying application, thereby reducing waste and minimizing environmental impact. This approach has ensured that pesticides are used only when they are most effective, conserving resources and protecting the environment.

By combining pest detection with climate data, our system has made more informed predictions about potential crop losses and the most effective strategies to mitigate damage. For instance, in scenarios where pests were detected and climate conditions were conducive to their proliferation, we recommended immediate intervention, which resulted in significant preservation of the crop yield. Conversely, when climate conditions were not favorable for pest survival, our system suggested a more conservative approach, further optimizing resource use. Decision logic is given below in Fig. 2.

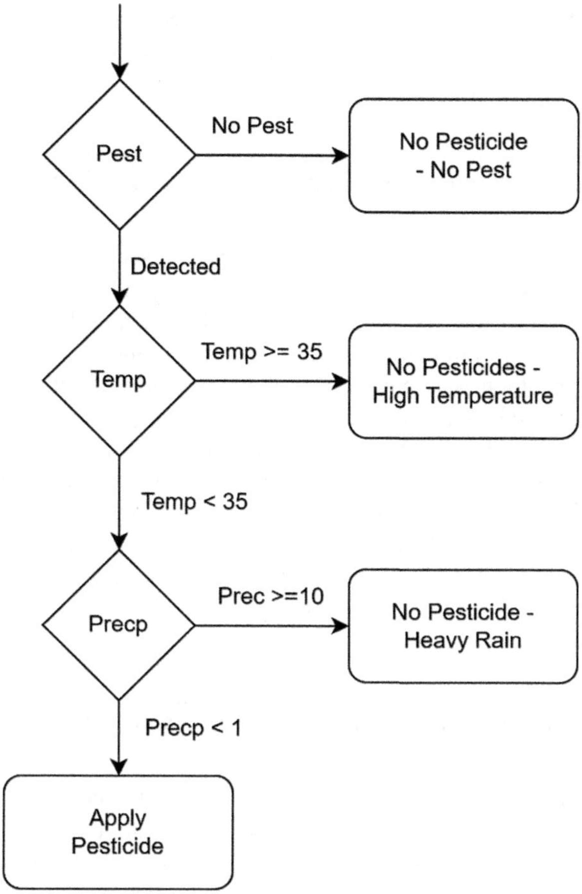

Fig. 2 Proposed decision system

Developing algorithms that use climate data to influence pest management decisions. These algorithms analyze the integrated data to determine:

- Pesticide Application: Whether or not to apply pesticides based on climate conditions (e.g., avoiding application before expected rainfall).
- Application Rates: Adjusting the amount of pesticide to apply based on environmental factors like temperature (e.g., reducing application rates during hot periods when pesticide efficacy may be reduced).
- Crop Loss Estimation: Estimating potential crop loss and deciding on preventive measures based on the combination of pest detection results and climate data.

Table 1 Rice yield loss due to pest and climate change

Factor	Yield in tons	Loss in tons (due to pest)	Loss in tons (due to climate)	Loss in %
Rice	763 million [16]	1152.47 [17]	998.63 [17]	50–55 [18]

3.6 Predictive Modeling for Crop Management

Our integration of ResNet50's pest detection capabilities with climate data has enabled the development of long-term predictive models. These models have been used to forecast potential pest outbreaks and their impact on crop yield, based on seasonal climate trends and historical data. As a result, farmers have been able to plan their pest management strategies more effectively, adjusting their actions based on predicted weather patterns and pest behaviors.

We have implemented a user-friendly decision support system that provides real-time alerts and recommendations. This system has been deployed on mobile devices and computers, offering farmers actionable insights such as the optimal timing for pesticide application, the appropriate amount to use, and whether alternative pest management strategies should be considered based on upcoming weather conditions. Table 1 shows rice yield and loss.

3.7 Feedback and Continuous Improvement

Our system has been designed with adaptive learning capabilities, continuously refining its predictions based on new data and the outcomes of implemented pest management strategies. This feedback loop has ensured that our model becomes more accurate and reliable with each growing season.

Although initially focused on rice, our methodology has been structured to allow for scalability. By retraining the ResNet50 model on datasets specific to other crops and integrating relevant climate data, our system can be adapted to a variety of agricultural contexts, broadening its impact on food security.

4 Experiment Setup

This section outlines the dataset description, preprocessing steps, and evaluation metrics used in the experiments.

4.1 Dataset Collection

The IP102 dataset is a valuable resource for researchers and practitioners working on insect pest recognition and classification. This extensive dataset comprises over 75,000 images belonging to 102 distinct categories of insect pests, providing a rich and diverse collection for model training and evaluation. The dataset's diverse categories encompass a wide range of insect pests commonly found in agricultural and natural environments, ensuring that models trained on IP102 can effectively handle real-world scenarios. Additionally, the dataset's hierarchical taxonomy organizes insect pests into meaningful categories, facilitating efficient classification and analysis. We have used 5 classes of rice pest containing Pest 1 as rice roller, Pest 2 as rice caterpillar, Pest 3 as rice borer, Pest 4 as rice weevil and Pest 5 as rice hopper.

The AgMERRA dataset is a valuable resource for agricultural modeling and climate research, providing high-resolution, daily meteorological data from 1980 to 2010. It includes variables such as temperature, precipitation, solar radiation, wind speed, and relative humidity, which are essential for understanding the impact of climate on agricultural systems. AgMERRA is widely used in various agricultural applications, including crop growth modeling, yield estimation, and climate change impact assessment.

4.2 Experimental Settings

Python tool and deep learning framework are used to simulate the proposed system and accelerated using the GPU. Hardware contains Intel Quad-Core i7 2820QM CPU at 2.30 GHz and NVIDIA Quadro with Samsung 850 Pro 256 GB SSD. The model was trained and tested using these specifications. The details of the training parameters used using the suggested approach is given below. Batch size 32, Epoch 100, Momentum 0.9, Learning rate 0.01–0.00001, weight decay 0.005 and Adam optimizer was used.

4.3 Metrics Used

In order to evaluate our model we have used most commonly used four metrics precision, recall and F1-Score. The number of diseased leaf images that are classified as such in this case is denoted by TP (True Positive), the number of authentic healthy images that are classified as healthy is denoted by TN (True Negative), the number of authentic healthy images that are classified as diseased is denoted by FP (False Positive), and the number of diseased images that are classified as healthy is represented by FN (False Negative).

$$\text{Accuracy} = \frac{\text{TP} + \text{TN}}{\text{TP} + \text{TN} + \text{FP} + \text{FN}} \times 100 \quad (1)$$

It has been proved that only accuracy is a poor performance evaluation metric for real-world application domains. Accuracy may be good enough to use when you're dealing with balanced datasets. That's why we would like to consider other evaluation metrics given below;

$$\text{Precision} = \frac{\text{TP}}{(\text{TP} + \text{FP})} \times 100 \quad (2)$$

$$\text{Recall} = \frac{\text{TP}}{(\text{TP} + \text{FN})} \times 100 \quad (3)$$

$$\text{F1-Score} = 2 \cdot \frac{\text{Precision} \cdot \text{Recall}}{\text{Precision} + \text{Recall}} \quad (4)$$

Our methodology has been rigorously tested and demonstrated a 95% accuracy rate in pest detection, leading to the preservation of 60–70% of crop yield that would otherwise be lost to pest damage. This performance not only surpasses traditional pest management techniques but also outperforms other prominent pre-trained models, such as VGG-16, VGG-19, and ResNet101, in the specific context of agricultural disease detection.

5 Experiment Results

In our comprehensive evaluation, we conducted experiments to assess the performance of our deep learning model using precision, recall, accuracy and F1-score. Rice pest images were used to conduct the experiments, annotated with diverse disease conditions and severity levels. Leveraging a fine-tuned ResNet50 backbone for feature extraction and climate data such as temperature and precipitation integrated to, achieved an outstanding accuracy of 95% in accurately identifying and categorizing pest symptoms on rice leaves. This performance surpasses traditional methods and other pre-trained models, demonstrating its effectiveness in preserving crop yield. By contextualizing pest detections with climate data, the system enables informed decision-making for targeted pest management, reducing pesticide use and environmental impact.

The training and validation accuracy of the proposed approach is displayed in Fig. 3. For 100 epochs, the greatest accuracy for training and validation was 98% and 95%, respectively. The training and validation loss of the proposed approach is displayed in Fig. 4.

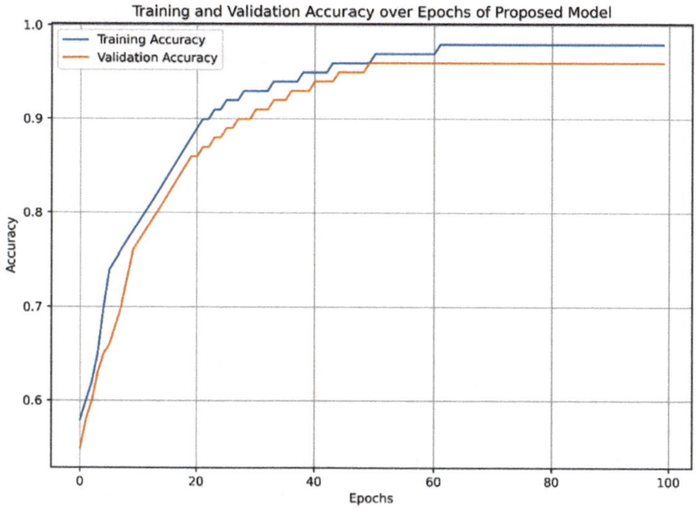

Fig. 3 Training and validation accuracy over epochs

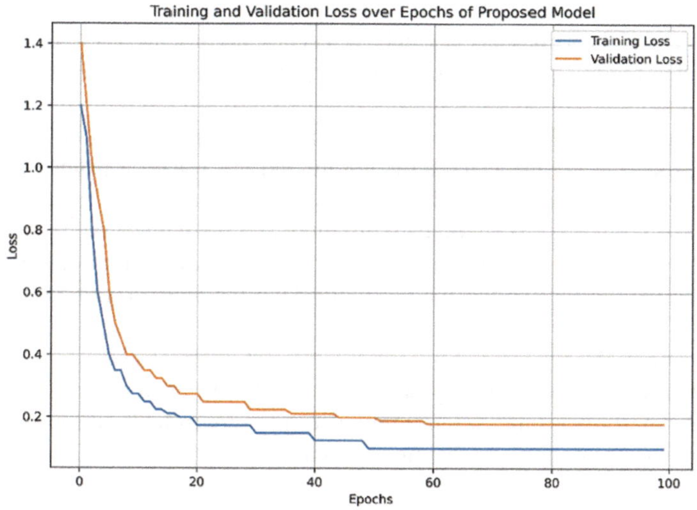

Fig. 4 Training and validation loss over epochs

The superiority of our proposed methodology was tested by comparing with several well-established pre-trained models: VGG-16, VGG-19, ResNet101, InceptionV3, and DenseNet. This superiority can be attributed to the synergistic combination of advanced pest detection techniques and real-time climate data integration. The system's ability to provide contextually relevant information for decision-making sets it apart from existing approaches. as shown in Table 2.

Table. 2 Comparison of Proposed Methodology with State-of-the-art techniques

Technique	Precision (%)	Recall (%)	F1-score (%)
VGG-16	76	73	76
VGG-19	73	74	75
ResNet101	69	70	69
InceptionV3	78	80	80
DenseNet	76	78	79
Proposed methodology	94	95	96

Beyond accuracy, we analyzed precision, recall and F-1 Score metrics to evaluate the model's ability to precisely detect rice pests, respectively. Our model exhibited high precision, indicating a low rate of false positives in disease classification. Moreover, the recall metrics demonstrated the model's capability to effectively identify and capture rice pests, highlighting its comprehensive disease detection capabilities.

The proposed system demonstrated exceptional performance in accurately detecting and classifying plant diseases within rice crops. By integrating advanced pest detection techniques with real-time climate data, the system achieved a remarkable 95% accuracy rate, significantly outperforming traditional methods and other pre-trained models. This success can be attributed to the synergistic combination of the ResNet50 model and climate data, which provides valuable insights for informed decision-making.

The integration of climate data into the pest detection process is a key differentiator of our approach. By considering temperature and precipitation patterns, the system can assess the potential impact of pests on crop health and recommend appropriate management strategies. This contextual understanding enables farmers to optimize pesticide applications, reducing unnecessary chemical use and minimizing environmental impact. Figure 5 shows the normalized confusion matrix of 5 classes of rice pests.

Furthermore, the system's ability to predict pest outbreaks based on climate data allows for proactive measures to be taken, potentially preventing significant crop losses. This early warning system empowers farmers to make informed decisions regarding crop protection and resource allocation.

6 Conclusion and Future Work

Our implementation of an integrated pest detection and management system represents a significant advancement in agricultural technology, particularly in the context of rice crop management. By harnessing the power of the ResNet50 model, we have achieved a high degree of precision in detecting pests within complex agricultural environments. The incorporation of real-time climate data from the AgMERRA dataset further enhances the system's utility by enabling informed, climate-aware

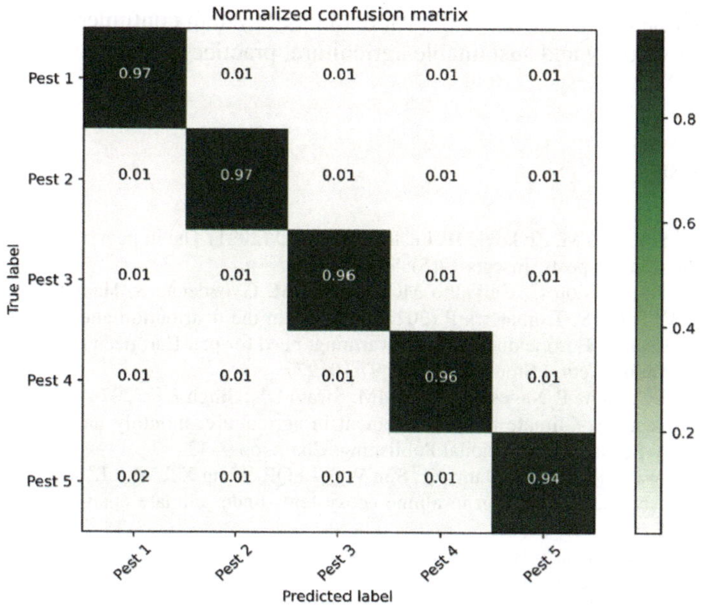

Fig. 5 Confusion matrix of proposed methodology

decision-making. This dual approach combining state-of-the-art machine learning with environmental data has proven to be highly effective, as evidenced by a substantial reduction in potential crop losses ranging from 50 to 55%. These results not only highlight the system's ability to improve pest detection accuracy but also demonstrate its critical role in optimizing pesticide application timing, thereby minimizing environmental impact and preserving crop health.

In comparative analyses, our system consistently outperformed other prominent pre-trained models, such as VGG-16, VGG-19, and ResNet101, in the specific domain of agricultural disease detection. This superior performance underscores the robustness and efficacy of our approach, which leverages both advanced deep learning techniques and contextual climate information. The success of our system has significant implications for agricultural practices, particularly in regions where climate variability and pest infestations pose a continuous threat to food security. By offering a more precise and adaptable solution, our system contributes meaningfully to the ongoing efforts to secure global food supplies in the face of growing environmental challenges.

Future efforts will focus on expanding the system's capabilities, including the integration of additional climate variables and real-time data processing, to further enhance its accuracy and applicability across different crops and regions. We also plan to improve the user interface, enabling more detailed visualizations and predictive analytics, and to scale the system for broader deployment in various agricultural settings. These enhancements will ensure that our system continues to contribute to global food security and sustainable agricultural practices.

References

1. Skendžić S, Zovko M, Živković IP, Lešić V, Lemić D (2021) The impact of climate change on agricultural insect pests. Insects 12(5):440
2. Adler C, Athanassiou C, Carvalho MO, Emekci M, Gvozdenac S, Hamel D, Riudavets J, Stejskal V, Trdan S, Trematerra P (2022) Changes in the distribution and pest risk of stored product insects in Europe due to global warming: need for pan-European pest monitoring and improved food-safety. J Stored Prod Res 97:101977
3. Mohanty UC, Sinha P, Nageswara Rao MM, Swain DK, Singh KK (2024) Climate variability and food security. Climate risk management in agriculture: monthly and seasonal forecast application. Springer International Publishing, Cham, pp 9–32
4. Wang CJ, Wang R, Yu CM, Dang XP, Sun WG, Li QF, Wang XT, Wan JZ (2021) Risk assessment of insect pest expansion in alpine ecosystems under climate change. Pest Manag Sci 77(7):3165–3178
5. Gerken AR, Morrison WR III (2022) Pest management in the postharvest agricultural supply chain under climate change. Front Agron 4:918845
6. Wang BX, Hof AR, Ma CS (2022) Impacts of climate change on crop production, pests and pathogens of wheat and rice. Front Agric Sci Eng 9(1)
7. Ali MA, Abdellah IM, Eletmany MR (2023) Towards sustainable management of insect pests: protecting food security through ecological intensification. Int J Chem Biochem Sci 24(4):386–394
8. Yang X, Zhao S, Liu B, Gao Y, Hu C, Li W, Yang Y, Li G, Wang L, Yang X, Yuan H, Liu J, Liu D, Shen X, Wyckhuys KAG, Lu Y, Wu K (2023) Bt maize can provide non-chemical pest control and enhance food safety in China. Plant Biotechnol J 21(2):391–404
9. Chu J, Li Y, Feng H, Weng X, Ruan Y (2023) Research on multi-scale pest detection and identification method in granary based on improved YOLOv5. Agriculture 13(2):364
10. Dai M, Dorjoy MMH, Miao H, Zhang S (2023) A new pest detection method based on improved YOLOv5m. Insects 14(1):54
11. Fu X, Ma Q, Yang F, Zhang C, Zhao X, Chang F, Han L (2024) Crop pest image recognition based on the improved ViT method. Inf Process Agric 11(2):249–259
12. Suguna M, Prakash D, Kumar CO, Revathi M, Dhanaraj RK (2024) AI-based pest detection and recovery model for cyber-physical agricultural systems. In: Agri 4.0 and the future of cyber-physical agricultural systems. Academic Press, pp 237–256
13. Elmagzoub MA, Rahman W, Roksana K, Tarequl Islam Md, Saifullah Sadi AHM, Rahman MM, Rajab A, Rajab K, Shaikh A (2024) Paddy insect identification using deep features with lion optimization algorithm. Heliyon
14. Tonnang HE, Sokame BM, Abdel-Rahman EM, Dubois T (2022) Measuring and modelling crop yield losses due to invasive insect pests under climate change. Curr Opin Insect Sci 50:100873
15. Ewuzie U, Bolade OP, Egbedina AO (2022) Application of deep learning and machine learning methods in water quality modeling and prediction: a review. In: Current trends and advances in computer-aided intelligent environmental data engineering, pp 185–218
16. Khush GS (2013) Strategies for increasing the yield potential of cereals: case of rice as an example. Plant Breeding 132(5):433–436
17. Shanmugam TR, Sendhil R, Thirumalvalavan V (2006) Quantification and prioritization of constraints causing yield loss in rice (Oryza sativa) in India. Agric Trop Subtrop 39(3):194–201
18. Kaushik B, Singh K, Tiwari DK, Singh UK (2023) Impact of climate change on crop yield due to pests and crop diseases: future projections

Food Waste Valorization: Transforming Corn Distillers Soluble into High-Value Carbon-Rich Products

Aneela Hayder and Animesh Dutta

Abstract Fuels derived from food waste offer a promising alternative to conventional fossil fuels. Converting food waste into valuable energy resources not only reduces waste disposal challenges but also mitigates carbon emissions and decreases industrial reliance on finite energy resources, thereby advancing sustainable development and supporting circular economy principles. The generation of high-value bioproducts from corn condensed distiller soluble (CDS), a byproduct of corn ethanol industry and an agri-food waste, provides a new perspective on utilizing by-products to produce high value carbon products within the framework of a circular economy. Exploring the alternative technological routes for energy and nutrient recovery from CDS is essential to support decision-making based on economic and environmental criteria. This chapter aims to assess both traditional and advanced technological approaches to convert this low-value product (CDS) into high-value, carbon-rich product. The focus of this study is to explore the potential of hydrochar derived from CDS as a high-quality hydrochar, hybrid carbon system, adsorbent, or as a precursor to high-surface-area activated carbon for advanced applications.

Keywords Food waste · CDS valorization · Hydrothermal carbonization · High value products

1 Introduction

Food loss and waste are major concerns for the environment, society, economy, and health; therefore, they are recognized as one of the ten indicators of a circular economy. Circular economy refers to an economy involving industrial processes and economic activities focused on restoring or regenerating products and reducing waste while maintaining their highest value [1]. In the United Kingdom (UK), 13.1 million tonnes of food waste emit 27.0 million tonnes of GHG [2]. However, in the

A. Hayder · A. Dutta (✉)
University of Guelph, Guelph, ON, Canada
e-mail: adutta@uoguelph.ca

S. Al-Naemi et al. (eds.), *Water and Food Security in the Face of Climate Change: Challenges and Opportunities for Resilience*, World Sustainability Series,
https://doi.org/10.1007/978-3-032-00098-9_38

United States (US), food waste prevention and recovery initiatives contribute to an annual net economic value of over $10 billion [3]. Valorization of food waste can lead to economic benefits and reduce greenhouse gas (GHG) emissions. Food waste has recently been recognized as a valuable, underutilized resource with significant potential for energy generation. Hence, energy recovery from food waste presents an appealing option, especially from an energy security perspective [4, 5]. This motivates fundamental research into advanced technologies for extracting valuable fuels from food waste, with the aim of minimizing the environmental impact of waste disposal, preserve natural resources, minimize health risks, and maintain balance in the ecosystem.

Food waste costs about $100 billion annually, with $49 billion worth of waste ending up in landfills or being composted, resulting in significant environmental, societal, and economic impacts [6]. Among various types of food waste, valorization of Corn Condensed Distillers Solubles (CDS) presents a unique opportunity to produce high value bioproducts, bioenergy, and renewable resources. CDS produces as a by-product in corn-based ethanol production industry with its potential of valorization remained largely unexplored in the existing literature. Although CDS was utilized as a cattle feed supplement, but once mycotoxins contaminate crops, CDS ends up in landfills cause additional pollution [7]. CDS is a nutrient-rich liquid residue composed of proteins, lipids, fibers, and minerals, making it a valuable feedstock for numerous applications, such as bioproducts, renewable energy, and food applications [8]. Due to high nitrogen and carbon content, CDS can be considered as a reasonable and sustainable resource to produce high value bioproducts.

This chapter provides innovative valorization pathways to produce sustainable and environmentally friendly bioproducts and renewable energy resources from this underused and low-value waste. The traditional valorization approaches of CDS, along with recent advancements in those approaches have been explored. The primary objective of this chapter includes the valorization of CDS using hydrothermal carbonization technology to convert this byproduct into a carbon-rich product with advanced functional and structural properties. Hydrothermal carbonization (HTC) is an effective technique in treating food waste [9]. Food waste contains high moisture content that acts as a reaction substrate, offering the advantages of short reaction time and mild reaction conditions while retaining key functional groups in the carbon matrix. CDS contains high moisture content, making HTC a feasible option to process CDS material. Additionally, CDS can provide an excellent carbon support for the development of advanced hybrid carbon systems [10].

2 Overview of Corn Condensed Distillers Solubles (CDS)

CDS is a viscous, syrup-like product generated during the fermentation and distillation process in corn ethanol production. It is derived from thin stillage that remains after ethanol is separated from the mash and is concentrated by evaporating water to produce CDS [8]. The quality of feedstock is a key factor in successfully valorizing food waste byproducts. The primary components of CDS produced from different techniques, such as dry milling and wet milling used in bioethanol industries, are illustrated in Table 1.

CDS contains a higher protein content, making it a valuable addition for food applications. The food industry incorporates CDS to enhance the quality of products in terms of protein and energy content [20]. However, this traditional application limits the broader potential of CDS. Therefore, novel valorization pathways need to be investigated to fully exploit the value of CDS in alignment with sustainable and circular economy principles.

3 Traditional Valorization Approaches of CDS

The diverse composition of CDS unveils numerous valorization pathways. This section describes various approaches for transforming CDS into valuable bioproducts and recent advances. CDS can be used as a substrate in anaerobic digestion (AD) processes to generate a renewable energy source such as biogas. Due to the high organic content of CDS, methane yield can be improved, making CDS an efficient feedstock for biogas production. The anaerobic digestion of thin stillage containing CDS has been studied for its potential to produce biogas with high methane content [21]. In another study by Ziero et al., an integrated approach involving the AD of thin stillage (processed to produce corn solubles) and combined heat and power (CHP) system. AD-CHP scenario demonstrated that 1 ton of thin stillage could produce 10.37 m^3 of biogas, which in turn, could generate, 119.9 MJ of heat and 24.7 kWh of electricity in a CHP system, promoting green energy production. Additionally, the authors concluded that the energy recovered from biogas contributes significantly to industrial operations, supplying 7.15% of the heat needed for steam generation and 52.74% of the electricity required for industrial processes [22].

The lipids present in CDS can be extracted and trans esterified to produce biodiesel. This not only provides a potential source of renewable fuel but also adds value to the underutilized lipid fraction of CDS [14, 23]. CDS can serve as a promising feedstock for producing bioproducts, including bioplastics and bio-based chemicals. CDS can serve as a potential resource for producing bio-based chemicals such as succinic acid, lactic acid, and other organic acids through microbial fermentation [24, 25]. These chemicals have various industrial applications, which includes pharmaceuticals, food additives, and bioplastics [26, 27]. Moreover, CDS has been explored as a high-value addition to nutraceuticals and functional food ingredients. The high protein content in CDS can be extracted and concentrated to produce protein-rich isolates used in functional foods, dietary supplements, and sports nutrition products [28]. Additionally, CDS contains bioactive compounds with antioxidant properties that can be harnessed for nutraceutical applications. Extracting these valuable compounds can add value to CDS as a source of health-promoting ingredients [29]. Beyond its traditional use, CDS can be formulated into high-value feed additives that

Table 1 Chemical composition of CDS derived from various milling techniques* [11, 12]

Source	Carbohydrates (%)	Moisture (%)	Protein (%)	Oil (%)	Glycerol (%)	Ash (%)	References
CDS	29.78	64.79 ± 3.13	17.53 ± 0.23	12.34 ± 0.84	27.88 ± 0.38	10.89 ± 0.39	[13]
CDS	19.5 ± 1.0	67.5 ± 1.3	23.0 ± 0.9	18.8 ± 2.2	26.5 ± 1.3	12.2 ± 0.5	[14]
CDS	24.42 ± 2.21	69.60 ± 0.42	24.29 ± 0.72	16.38 ± 0.34	22.98 ± 1.06	11.92 ± 0.48	[15]
CDS	–	65.9 ± 0.12	14.1 ± 0.1	17.9	–	11.83	[16]
CDS	19.5	71.17	22.87	18.23	21.89	7.79	[17]
CDS	24.2	–	32.5	7.09	94.8	4.01 ± 0.01	[18]
CDS**	6.53 ± 0.26	–	7.48 ± 0.03	6.62 ± 0.20	7.41 ± 0.12		[19]

* Calculated on dry basis (%), ** g/100 g

support animal performance and health, including functional feeds or prebiotics that improve nutrient absorption and gut health [30].

However, to maximize the valorization potential of CDS, its efficient processing is crucial. Processing technologies such as spray drying, membrane filtration, fermentation, and extraction can be used to process CDS effectively. Fibers, proteins, and other valuable components can be separated from CDS using membrane filtration, enhancing the quality of the extracted products [31]. Spray drying converts CDS into a stable, dry powder, facilitating its use in food and industrial applications [32]. Fermentation and extraction can be employed for bio-based chemical production and lipid extraction, followed by transesterification for biodiesel production, respectively [12, 33].

4 Advanced Valorization of CDS Using Hydrothermal Carbonization (HTC)

Hydrothermal carbonization (HTC) is a thermochemical technique used to transform biomass or waste into valuable products by submerging the material in water, sealing it, and involves heating at temperatures ranging from 180 °C to 300 °C under self-generated (autogenous) pressures. The HTC process typically yields a partially carbonized solid product known as hydrochar, along with a liquid byproduct called process water. A wide range of biomass sources, such as food waste [32], miscanthus [33], spent grains [34], algae [35], sewage sludge [36], and black liquor from industrial waste [37], have been extensively tested in the literature under various conditions. HTC is particularly effective to process the biomass with high moisture content [38], as it does not require pre-drying step. Additionally, the relatively low operating temperatures help to retain heteroatoms such as nitrogen and sulfur [39, 40], which can be beneficial for applications such as soil amendments [41] or for producing activated carbons for adsorption or supercapacitor applications [42]. HTC is currently limited to small batch processing; however, numerous recent efforts have been made to develop continuous hydrothermal reactors [43, 44], aiming to scale up the process for broader industrial applications.

Many studies have explored the valorization of corn waste products via HTC, such as the utilization of corn husks [45], corn stalks [46, 47], and corn stovers [48] for solid fuel and adsorption applications. Furthermore, biochar derived from distillers dried grains with soluble (DDGS), a byproduct of the corn-based ethanol production industry, has been effectively utilized in the production of activated carbons for supercapacitor applications. However, CDS have received limited research attention. The focus of this section is to describe the value and versatility of CDS and to showcase its potential as a rich carbon source for high end applications.

In a study by Kevin et al., CDS was processed at temperatures of 180 °C, 200 °C, and 220 °C with residence time of 30 min [49]. Raw CDS and hydrochar derived from CDS were analyzed to determine their physical, chemical, functional and structural characteristics. The ultimate analysis and mass yield (MY) are presented in Table 2. The moisture content of CDS was measured as 64.80%. Despite the relatively low reaction temperatures, the mass yield of CDS was modest between 20 and 30%. This suggests that a substantial amount of the carbon was retained in the liquid phase resulting in high carbon content and yield in the aqueous phase.

The ultimate analysis reveals that CDS contains high carbon content after HTC process. Higher HTC temperatures led to a highly carbonized material. Besides, the hydrogen and nitrogen content of the resulting hydrochar was higher at lower HTC temperatures. However, the nitrogen content of the hydrochar varies greatly, depending on the biomass source used. For example, Loblolly Pine has nearly zero nitrogen content [50], while microalgae can contain over 6% nitrogen [35], whereas CDS has a moderate nitrogen content. Likewise, the oxygen and hydrogen content of hydrochar can vary considerably, even when produced under the same operating conditions. Hoekman et al. [50] conducted HTC experiments using corn stover at 215 °C and obtained a hydrochar with 5.16% hydrogen and 35.98% oxygen content. In another study by Zhang et al., HTC was performed on rotten apples at 225 °C, producing the hydrochar with 5.33% of hydrogen content and 29.41% of oxygen content [51]. These examples illustrate that CDS can also produce hydrochar with a high hydrogen content.

The atomic ratios of O/C and H/C were plotted on a Van Krevelen diagram, as shown in Fig. 1, to compare the products with common carbon-based materials. The hydrochar derived from CDS deviates from the typical range for biomass, lignite, or peat, which are the common sources for biomass-derived hydrochar. This deviation may be related to a prevalence of decarboxylation reactions due to dehydration or may be influenced by the 8.49% fat content present in the feedstock [44], as fats are dense in hydrogen. A low O/C and high H/C ratio show a higher heating value (HHV), with the increased hydrogen content potentially useful for hydrogen production through gasification. Notably, the ash content of CDS was not calculated in this study, assuming no ash. As a result, the actual O/C ratio is likely lower than the values measured here.

Table 2 The ultimate analysis and mass yield of CDS and hydrochars

Sources	C (%)	H (%)	N (%)	S (%)	O (%)	MY (%)
R-CDS	44.00 ± 0.80	7.26 ± 0.30	3.50 ± 0.02	1.59 ± 0.08	43.65 ± 1.20	–
CDS-180	64.10 ± 1.70	9.10 ± 0.24	3.99 ± 0.52	0.59 ± 0.03	2.22 ± 2.49	25.65
CDS-200	69.44 ± 0.92	9.40 ± 0.40	3.62 ± 0.74	0.49 ± 0.04	17.05 ± 2.10	22.51
CDS-220	70.50 ± 1.60	9.28 ± 0.33	3.01 ± 0.48	0.35 ± 0.03	16.86 ± 2.44	27.56

Reproduced from [49]

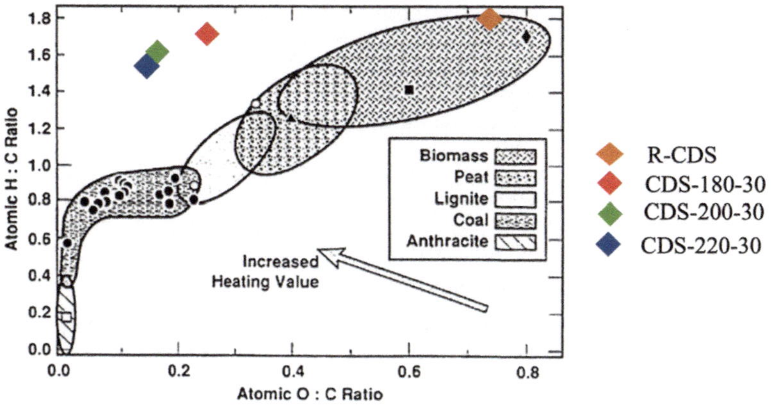

Fig. 1 Van Krevelen diagram with CDS and its hydrochar. Modified from [49]

4.1 Valorization of CDS as a Source of Novel Carbon Hybrid System

Advanced carbonaceous materials with novel characteristics have attracted significant interest from researchers across various fields, particularly in wastewater treatment [52], catalysis [53], energy storage [54], adsorption [55], and environmental and industrial remediation applications [56]. The combination of carbon materials with other valuable materials, such as inorganic nanoparticles and polymers, can enhance their mechanical, thermal, sorptive, and catalytic properties [57]. In this context, hybrid materials are promising candidates for enhancing the physico-chemical, and structural characteristics at atomic as well as macroscopic levels. Hybrid materials typically possess a distinct structure compared to their individual components, while retaining their original properties. The physicochemical properties of the hybrid material can be tailored by controlling its internal structure. Recently, modifying carbon materials with various nanomaterials such as inorganic salts, metals, and metal oxides, has been explored to meet industrial demands. Hybrid systems comprising of carbon-based materials and ZnO offer the potential to further enhance the overall properties of the resulting product. Hence, carbon materials supported by metals or other substrates exhibit enhanced adsorption capacity, mechanical durability, thermal stability, and significantly higher specific surface area compared to their non-supported carbon materials [58]. However, the synthesis of ZnO typically involves expensive equipment, complex processes, and rigorous reaction conditions. The use of additives and/or catalysts in biomass hydrothermal carbonization has recently garnered considerable interest.

In a study by Hayder et al., catalytic HTC in the presence of $ZnCl_2$ was performed to modify the carbon in terms of the structural and morphological enhancements, as well as introducing ZnO particles onto the CDS to form ZnO/C carbon hybrid systems [10]. During HTC, $ZnCl_2$ played a dual role as both a catalyst and a source of zinc

Table 3 Ultimate, proximate and HHV analysis of the raw CDS

Ultimate (%)	
C	39.00 ± 0.90
N	3.62 ± 0.02
H	6.03 ± 0.30
S	1.03 ± 0.04
O	41.65 ± 1.20
Proximate (%)	
Moisture	67.79 ± 2.17
Volatile matter *	59.47 ± 0.35
Fixed Carbon *	15.02 ± 0.15
Ash *	13.63 ± 0.21
Heat of combustion (MJ/kg)	
HHV	14.31 ± 0.51

* Dry basis

oxide during HTC process. Furthermore, CDS was explored as a novel precursor, providing a carbon source for the synthesis of carbon spheres in this study. Ultimate and proximate analysis of CDS were performed prior to the experiments, as shown in Table 3.

This study showed promising results, demonstrating that CDS can be an excellent carbon precursor for synthesizing carbon hybrid systems for catalytic and adsorption applications. Additionally, CDS can be further activated chemically or physically at higher temperatures to achieve a higher surface area carbon product.

To design and optimize the HTC experiments, authors employed the Design Expert Software v13. This software provided the design of experiments at 180, 210 and 240 °C and 60, 240 and 480 min, while optimized the experiments using Response Surface Methodology (RSM) [10]. Additionally, to assess the statistical significance, analysis of variance (ANOVA) was applied. The ultimate analysis of CDS derived hydrochar revealed that the carbon content increased with higher temperature and longer residence time. Notably, the carbon content in raw CDS increased from 39 ± 0.9 to 64.97 ± 0.4% in hydrochar produced at 240 °C and 240 min. However, both nitrogen and hydrogen content decreased under more severe hydrothermal conditions. The hydrogen content reduced from 7.74 to 5.11%, whereas nitrogen content dropped from 7.28 to 5.37% under hydrothermal conditions of 240 °C and 480 min [10]. Most nitrogen containing functional groups begin to decompose at 150 °C, accompanied by the release of volatile matter [59]. HHV results of this study showed significant improvement in energy content of hydrochar upto 27.892 MJ/kg as compared to raw CDS of 14.311 ± 0.51 MJ/kg. After getting the optimized results, catalytic HTC was performed at optimized conditions (225 °C and 60 min) and various ratios of CDS to $ZnCl_2$ (1:10, 1:5, 1:7, 1:3, 1:2 and 1:1). The detailed methodology is presented in the study by [10]. The promising results of raw CDS, hydrochar without $ZnCl_2$ at optimized conditions of RSM (Opt-HC) and with $ZnCl_2$ at various ratios (designated as ZnO/C hybrid carbon materials) are represented in this section.

Figure 2 shows the Scanning Electron Microscopy (SEM) images, which validated the effective inclusion of $ZnCl_2$ resulting in the development of 3D, stone like ZnO particles on the spherical carbon matrix. Conversely, increasing the $ZnCl_2$ to biomass ratio led to the accumulation of ZnO particles. Catalytic HTC in the presence of $ZnCl_2$ (acting both as a catalyst and a ZnO precursor) significantly improved the overall properties of ZnO/C carbon hybrid material, enhancing aromaticity and surface characteristics as well as the presence of nitrogen-containing functional groups. However, as $ZnCl_2$ to CDS ratio increased, it reduces the carbonization and slower down the surface modification process, as the excessive ZnO particle agglomeration disrupted the desired properties of the carbon matrix.

Additionally, Fourier Transmission Infrared Radiations (FT-IR) spectra in Fig. 3a. evidenced significant changes in the functional groups of R-CDS, Opt-HC and the

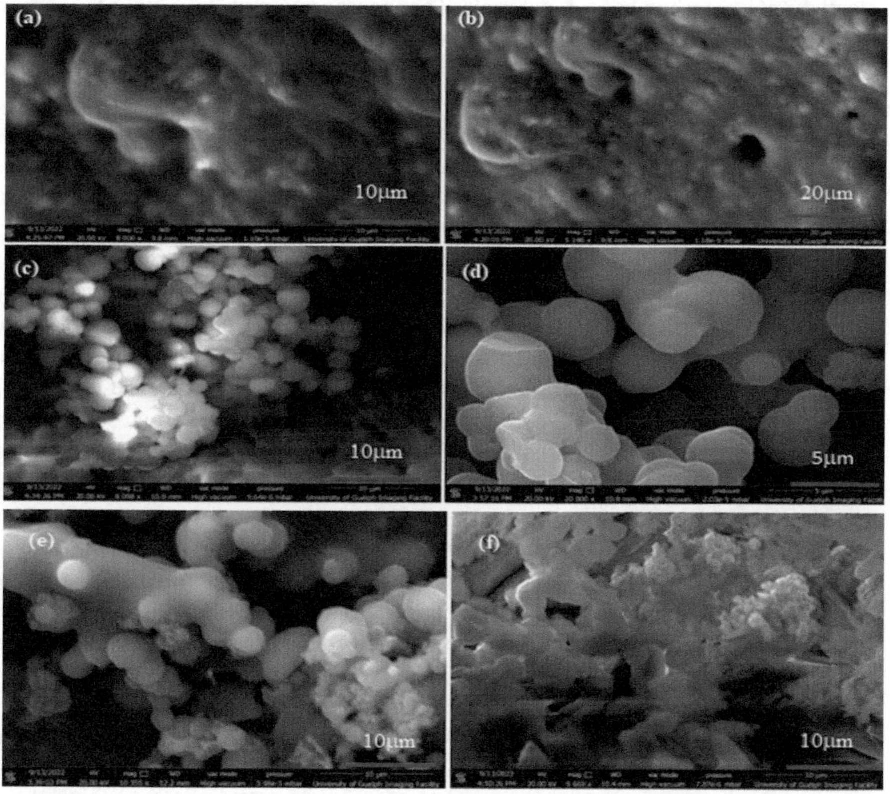

Fig. 2 SEM images of **a**, **b** R-CDS, **c** Opt-HC, **d**, **e** ZnO/C-1:7, **f** ZnO/C-1:3. Reprinted with permission from [10]. Copyright 2024 Elsevier (License Number: 5862170257656)

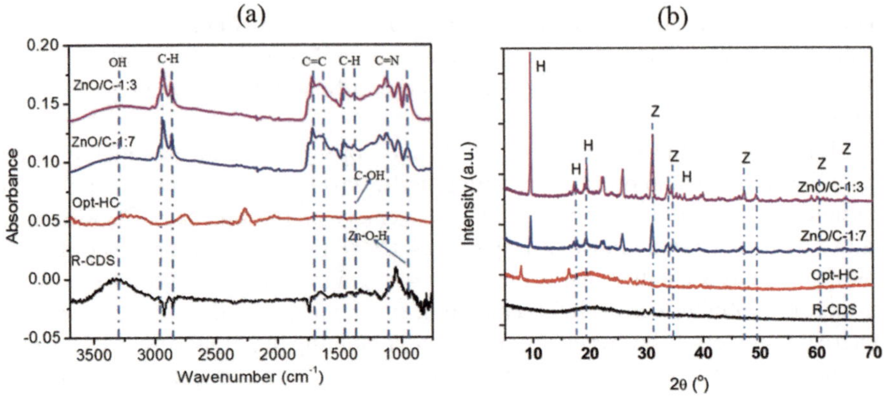

Fig. 3 **a** FTIR spectra of R-CDS, Opt-HC and selected ZnO/C hybrid systems, **b** XRD diffractogram of R-CDS, Opt-HC and selected ZnO/C hybrid systems Reprinted with permission from [10]. Copyright 2024 Elsevier (License Number: 5862170257656)

selected ZnO/C carbon hybrid materials. The incorporation of $ZnCl_2$ into the carbon hybrid materials introduced new functional peaks, indicating a shift in the interactions of the functional moieties. Notably, -OH peak in R-CDS and Opt-HC shifted to a broader peak, playing an important role to form ZnO particles which is also supported by SEM images. The doublet at 2925 and 2853 cm^{-1} along with the peak at 1478 cm^{-1}, corresponded to the stretching and bending vibrations of C-H, respectively, which appeared in the presence of $ZnCl_2$. These results were in consistent to another study for similar carbon system [60].

The ZnO/C carbon hybrid systems demonstrate higher aromaticity and enhanced interaction among diverse functional moieties, contributing to increased crystallinity and stability, as further confirmed by XRD analysis (see Fig. 3b). These functional groups, particularly those associated with C–OH and C–C bonds, showed significant improvement in the spectrum after incorporating $ZnCl_2$. The presence of various absorption bands in the ZnO/C carbon material facilitates the adsorbates binding to the adsorbent [61]. Consequently, these CDS based carbon materials exhibit strong potential as effective adsorbents, owing to the existence of hydrophilic functional groups.

X-ray Diffraction (XRD) pattern showed the significant structural changes resulting from catalytic hydrothermal carbonization, providing a valuable insight into the surface modification after $ZnCl_2$ incorporation. R-CDS illustrated completely amorphous structure due to the presence of amorphous carbon implying the low degree of graphitization [62]. Additionally, the overall structure of Opt-HC remains predominantly amorphous, as indicated by the broad peak at $2\theta = 23.5°$. Interestingly, new crystalline phases and minerals were formed after the addition of $ZnCl_2$ on the surface. The sharp peaks were primarily attributed to Hopeite ($Zn_3(PO_4)_2.4H_2O$) and Zincite crystal phase of hexagonal [10]. Hopeite ($Zn_3(PO_4)_2.4H_2O$) at 2θ about 9.75°, 17.89°, 19.70°, 34.23°, 38.5°, 49.85° and the peaks at $2\theta = 31.78°$, 34.47°,

36.26°, 47.70°, 56.60°, 62.81°, and 65.93° were indexed to the (100), (002), (101), (102), (110), (103) and (112) planes of the zincite crystal phase with hexagonal symmetry. These findings are consistent with the standard ZnO card (JCPSD card number 36-1451), confirming zincite as the primary crystalline phase in the ZnO/C carbon hybrid materials. Additionally, new crystalline phases emerged due to $ZnCl_2$ incorporation particularly at $2\theta = 22.54°$ and 26.03°. The enhanced interaction among functional groups, as evidenced by FTIR spectra (Fig. 3a.), contributing higher aromaticity, promoting the formation of crystalline structures in the ZnO/C hybrid system through $ZnCl_2$ assisted catalytic hydrothermal carbonization. These results align with FTIR findings, confirming the correlation between the functional group composition (intensity and IR peaks) and the structural properties of the ZnO/C carbon hybrid materials. The phase of ZnO particles identified in this research was aligned with the findings from previous studies [63–65].

5 Challenges and Future Directions/Perspectives

Despite the potential of CDS valorization, this material faces challenges such as variability in its composition, high moisture content, and the need for cost-effective processing technologies. Hydrothermal carbonization (HTC) can be a viable option to produce a carbon product from CDS with high carbon content, high energy content, and moderate nitrogen and hydrogen content. However, HTC of CDS produced lower solid yields and higher liquid yields, underscoring its potential for applications such as hydrothermal liquefaction and anaerobic digestion. Given the limited research on CDS, further studies are needed to optimize its valuable processing pathways as well as advancing the development of industrial-scale HTC reactors. To address the challenges related to composition, there is a need to advance separation technologies to isolate useful fractions of CDS, improve process integration, and develop scalable methods for product recovery.

Future research should be mainly focused on the following:

- Developing novel catalysts and enzymes for enhanced conversion and extraction of CDS components.
- Exploring the synergistic use of CDS with other food waste streams or biomass to increase solid yield and optimize valorization.
- Advancing CDS-derived hydrochar by synthesizing high surface area biocarbon for high-end applications.
- Conducting life cycle assessments is essential for assessing the environmental impacts and economic viability of CDS-derived products.

6 Conclusion

This chapter discussed the valorization of CDS as a high-value product for high-end applications. It presented various technologies for CDS processing to fully exploit the value of this underused biomass waste, as well as its associated challenges. The unique chemical composition of this raw material offers a promising opportunity to transform this low-value byproduct into a diverse range of high-value products. It was noted that HTC is a viable and an emerging technology for converting this nutrient-rich product into valuable carbon material. The HTC process efficiently handles the high moisture content of CDS, eliminating the need for pre-drying, and retains key functional groups and heteroatoms, such as nitrogen and sulfur. The resultant hydrochar derived from CDS exhibits tunable functional groups and improved porosity, making it suitable for high-end applications such as catalysts, soil amendments, adsorbents, and precursors for high surface area activated carbon. The process water produced during the HTC of CDS contains soluble organic compounds that can be valorized, treated, or recycled, contributing to a circular and sustainable approach to managing CDS waste. Additionally, CDS can play a crucial role in reducing food waste by enhancing resource recovery and supporting sustainable development goals. These innovative pathways can be further explored with continued research and technological advancements, particularly in the fields of bioenergy production, nutraceuticals, bioplastics, and activated biocarbon.

References

1. Sarker A, Raju Ahmmed SM, Ahsan JR, Ghosh MK, Nandi R (2023) A comprehensive review of food waste valorization for the sustainable management of global food waste. Sustain Food Technol 2(1):48–69. https://doi.org/10.1039/d3fb00156c
2. Jeswani HK, Figueroa-Torres G, Azapagic A (2021) The extent of food waste generation in the UK and its environmental impacts. Sustain Prod Consum 26:532–547
3. National Zero Waste Council. The high cost of food waste. National Zero Waste Council. http://www.nzwc.ca/focus-areas/food/issue/Pages/default.aspx. Accessed 22 Jul 2021
4. Saqib NU, Sharma HB, Baroutian S, Dubey B, Sarmah AK (2019) Valorisation of food waste via hydrothermal carbonisation and techno-economic feasibility assessment. Sci Total Environ 690:261–276
5. Antonopoulou G, Alexandropoulou M, Ntaikou I, Lyberatos G (2020) From waste to fuel: energy recovery from household food waste via its bioconversion to energy carriers based on microbiological processes. Sci Total Environ 732:139230
6. Parizeau K, von Massow M, Martin R (2015) Household-level dynamics of food waste production and related beliefs, attitudes, and behaviours in Guelph, Ontario. Waste Manag 35:207–217
7. Macdermid-Watts K, Adewakun E, Abhi TD, Pradhan R, Dutta A (2021) Hydrothermal carbonization valorization as an alternative application for corn bio-ethanol by-products. J Environ Chem Eng 9(4). https://doi.org/10.1016/j.jece.2021.105431
8. Sharma S, Pradhan R, Manickavasagan A, Thimmanagari M, Dutta A (2021) Valorization and potential of condensed corn distillers solubles fractions from selective milling technology. Biomass Convers Bioref 0123456789. https://doi.org/10.1007/s13399-021-01614-7

9. Akarsu K, Duman G, Yilmazer A, Keskin T, Azbar N, Yanik J (2019) Sustainable valorization of food wastes into solid fuel by hydrothermal carbonization. Biores Technol 292:121959. https://doi.org/10.1016/j.biortech.2019.121959
10. Hayder A, Norouzi O, Sharma S, Santos R, Dutta A (2024) A novel approach for the facile synthesis of zinc oxide/carbon hybrid systems from corn distillers soluble: surface modification and characterization for sustainable remediation. Chemosphere 357(141864)
11. Liu K (2011) Chemical composition of distillers grains, a review. J Agric Food Chem 59(5):1508–1526. https://doi.org/10.1021/jf103512z
12. Sharma S, Hayder A, Pradhan R, Manickavasagan A, Thimmanagari M, Singam SS, Adya AD, Dutta A (2023) Corn distillers solubles as a plant-based bioresource for proteins and bioactive peptides: current status and bioprospects—a critical review. Food Rev Int 40:574–619
13. Reis CER, Rajendran A, Hu B (2017) New technologies in value addition to the thin stillage from corn-to-ethanol process. Rev Environ Sci Biotechnol 16(1):175–206. https://doi.org/10.1007/s11157-017-9421-6
14. Liu K (2017) Fractionation of condensed distillers solubles and compositional characterization of its coproducts. J Am Oil Chem Soc 94(1):161–168. https://doi.org/10.1007/s11746-016-2923-5
15. Milczarek RR, Liu K (2015) Drum drying performance of condensed distillers solubles and comparison to that of physically modified condensed distillers solubles. Food Bioprod Process 94:208–217. https://doi.org/10.1016/j.fbp.2014.01.004
16. Majoni S, Wang T, Johnson LA (2011) Physical and chemical processes to enhance oil recovery from condensed corn distillers solubles. J Am Oil Chem Soc 88(3):425–434. https://doi.org/10.1007/s11746-010-1681-z
17. Liu K, Barrows FT (2013) Methods to recover value-added coproducts from dry grind processing of grains into fuel ethanol. J Agric Food Chem 61(30):7325–7332. https://doi.org/10.1021/jf402226r
18. Tanghe S, de Boever J, Ampe B, de Brabander D, de Campeneere S, Millet S (2015) Nutrient composition, digestibility and energy value of distillers dried grains with solubles and condensed distillers solubles fed to growing pigs and evaluation of prediction methods. Anim Feed Sci Technol 210:263–275. https://doi.org/10.1016/j.anifeedsci.2015.10.015
19. Kingsly ARP, Ileleji KE, Clementson CL, Garcia A, Maier DE, Stroshine RL, Radcliff S (2010) The effect of process variables during drying on the physical and chemical characteristics of corn dried distillers grains with solubles (DDGS)—plant scale experiments. Bioresour Technol 101(1):193–199. https://doi.org/10.1016/j.biortech.2009.07.070
20. Mosqueda MRP, Tabil LG, Christensen C (2013) Effect of drying conditions and level of condensed distillers solubles on protein quality of wheat distillers dried grain with solubles. Dry Technol 31(7):811–824. https://doi.org/10.1080/07373937.2013.765446
21. Anaerobic digestion of thin stillage for energy recovery and water reuse in corn-ethanol plants. Bioresour Technol 102(21):9891–9896 (2011). https://doi.org/10.1016/j.biortech.2011.08.028
22. Di Domenico Ziero H, Ampese LC, Buller LS, Costa JM, Berni M, Forster-Carneiro T (2023) Corn ethanol production with thin stillage anaerobic digestion for bioenergy recovery: a technical and economic evaluation. Ind Crops Prod 206:117618
23. Xu Y et al (2010) Production of biodiesel from various bio-oils and evaluation of the reaction efficiency of lipase. Biores Technol 101(5):1583–1588
24. Soares JA, Stein HH, Singh V, Shurson GS, Pettigrew JE (2012) Amino acid digestibility of corn distillers dried grains with solubles, liquid condensed solubles, pulse dried thin stillage, and syrup balls fed to growing pigs. J Anim Sci 90(4):1255–1261. https://doi.org/10.2527/jas.2010-3691
25. Yang X, Kim J, Pangeni D, Tekeste A, Ren P, Manu H, Baidoo SK (2019) Ileal digestibility of amino acids in corn condensed distiller's solubles and whole stillage for pigs. Anim Feed Sci Technol 255. https://doi.org/10.1016/j.anifeedsci.2019.114202
26. Sasikala-Appukuttan AK, Schingoethe DJ, Hippen AR, Kalscheur KF, Karges K, Gibson ML (2008) The feeding value of corn distillers solubles for lactating dairy cows. J Dairy Sci 91(1):279–287. https://doi.org/10.3168/jds.2007-0250

27. Krishnan P (2016) Production of food grade distillers dried grains (Patent No. US 2016/0135485 A)
28. Sharma S, Pradhan R, Manickavasagan A, Thimmanagari M, Dutta A (2020) Application of analytical pyrolysis to gain insights into proteins of condensed corn distillers solubles from selective milling technology. Food Bioprod Process 124:354–368. https://doi.org/10.1016/j.fbp.2020.09.011
29. Lu L, Liu G, Wang XH, Lou SJ (2016) Effects of supplement of corn peptides combined with aerobic exercise on lipolysis key enzymes: adipose triglyceride lipase and lipoprotein lipase of obese rats. Chin J Appl Physiol. 32(4):326–331. https://doi.org/10.13459/j.cnki.cjap.2016.04.010
30. Lou S, Shi R, Wang R, Wang X, Xu L (2013) Corn peptides ingestion with exercise attenuate body fat mass and enhance lean body mass in obese rats. FASEB J Nutr 27:861–912
31. Wang Z et al (2015) Separation and purification of proteins from corn distillers grains by ultrafiltration. J Chem Technol Biotechnol 90(5):856–862
32. Rupnow JH, Shurson GC (2000) Processing of corn co-products for use in food and feed: methods and applications. Cereal Chem 77(3):232–239
33. Li C et al (2014) Biochemical conversion of corn ethanol distillers grains and solubles (DGS) into bio-based chemicals and biofuels. J Chem Technol Biotechnol 89(6):837–843
34. Fu MM, Mo CH, Li H, Zhang YN, Huang WX, Wong MH (2019) Comparison of physicochemical properties of biochars and hydrochars produced from food wastes. J Clean Prod 236:117637. https://doi.org/10.1016/j.jclepro.2019.117637
35. Kambo HS, Dutta A (2014) Strength, storage, and combustion characteristics of densified lignocellulosic biomass produced via torrefaction and hydrothermal carbonization. Appl Energy 135:182–191. https://doi.org/10.1016/j.apenergy.2014.08.094
36. Olszewski MP, Nicolae SA, Arauzo PJ, Titirici MM, Kruse A (2020) Wet and dry? Influence of hydrothermal carbonization on the pyrolysis of spent grains. J Clean Prod 260:121101. https://doi.org/10.1016/j.jclepro.2020.121101
37. Falco C, Sevilla M, White RJ, Rothe R, Titirici MM (2012) Renewable nitrogen-doped hydrothermal carbons derived from microalgae. Chemsuschem 5:1834–1840. https://doi.org/10.1002/cssc.201200022
38. Wang L, Chang Y, Zhang X, Yang F, Li Y, Yang X, Dong S (2020) Hydrothermal co-carbonization of sewage sludge and high concentration phenolic wastewater for production of solid biofuel with increased calorific value. J Clean Prod 255:120317. https://doi.org/10.1016/j.jclepro.2020.120317
39. Al-Kaabi Z, Pradhan R, Thevathasan N, Gordon A, Chiang YW, Dutta A (2019) Bio-carbon production by oxidation and hydrothermal carbonization of paper recycling black liquor. J Clean Prod 213:332–341. https://doi.org/10.1016/j.jclepro.2018.12.175
40. Heidari M, Salaudeen S, Norouzi O, Acharya B, Dutta A (2020) Numerical comparison of a combined hydrothermal carbonization and anaerobic digestion system with direct combustion of biomass for power production. Processes 8:43. https://doi.org/10.3390/pr8010043
41. Liu Y, Huang B, Lin X, Xie Z (2017) Biomass-derived hierarchical porous carbons: boosting the energy density of supercapacitors: via an ionothermal approach. J. Mater. Chem. A 5:13009–13018. https://doi.org/10.1039/c7ta03639f
42. Liu S, Cai Y, Zhao X, Liang Y, Zheng M, Hu H, Dong H, Jiang S, Liu Y, Xiao Y (2017) Sulfur-doped nanoporous carbon spheres with ultrahigh specific surface area and high electrochemical activity for supercapacitor. J Power Sour 360:373–382. https://doi.org/10.1016/j.jpowsour.2017.06.029
43. Bargmann I, Martens R, Rillig MC, Kruse A, Kücke M (2014) Hydrochar amendment promotes microbial immobilization of mineral nitrogen. J Plant Nutr Soil Sci 177:59–67. https://doi.org/10.1002/jpln.201300154
44. Hu Y, Liu H, Ke Q, Wang J (2014) Effects of nitrogen doping on supercapacitor performance of a mesoporous carbon electrode produced by a hydrothermal soft-templating process. J Mater Chem A 2:11753–11758. https://doi.org/10.1039/c4ta01269k

45. Kumabe K, Itoh N, Matsumoto K, Hasegawa T (2017) Hydrothermal gasification of glucose and starch in a batch and continuous reactor. Energy Rep 3:70–75. https://doi.org/10.1016/j.egyr.2017.04.001
46. Heidari M, Norouzi O, MacDermid-Watts K, Acharya B, Zhang Y, Dutta A (2020) Product evaluation of hydrothermal carbonization of biomass: semi-continuous vs. batch feeding. Biomass Convers Bioref. https://doi.org/10.1007/s13399-020-00932-6
47. Minaret J, Dutta A (2016) Comparison of liquid and vapor hydrothermal carbonization of corn husk for the use as a solid fuel. Bioresour Technol 200:804–811. https://doi.org/10.1016/j.biortech.2015.11.010
48. Lei Y, Su H, Tian R (2016) Morphology evolution, formation mechanism and adsorption properties of hydrochars prepared by hydrothermal carbonization of corn stalk. RSC Adv 6:107829–107835. https://doi.org/10.1039/c6ra21607b
49. Kang K, Nanda S, Sun G, Qiu L, Gu Y, Zhang T, Zhu M, Sun R (2019) Microwave-assisted hydrothermal carbonization of corn stalk for solid biofuel production: optimization of process parameters and characterization of hydrochar. Energy 186:115795. https://doi.org/10.1016/j.energy.2019.07.125
50. Tu R, Sun Y, Wu Y, Fan X, Wang J, Cheng S, Jia Z, Jiang E, Xu X (2019) Improvement of corn stover fuel properties via hydrothermal carbonization combined with surfactant. Biotechnol Biofuels 12:249. https://doi.org/10.1186/s13068-019-1581-x
51. Macdermid-Watts K, Eniola A, Trishan DA, Ranjan P, Dutta A (2021) Hydrothermal carbonization valorization as an alternative application for corn bio-ethanol by-products J Environ Chem Eng 9(4):105431
52. Hoekman SK, Broch A, Robbins C, Zielinska B, Felix L (2013) Hydrothermal carbonization (HTC) of selected woody and herbaceous biomass feedstocks. Biomass Convers Bioref 3:113–126. https://doi.org/10.1007/s13399-012-0066-y
53. Zhang B, Heidari M, Regmi B, Salaudeen S, Arku P, Thimmannagari M, Dutta A (2018) Hydrothermal carbonization of fruit wastes: a promising technique for generating hydrochar. Energies 11:1–14. https://doi.org/10.3390/en11082022
54. Kamaraj M, Srinivasan NR, Assefa G, Adugna AT, Kebede M (2020) Facile development of sunlit ZnO nanoparticles-activated carbon hybrid from pernicious weed as an operative nano-adsorbent for removal of methylene blue and chromium from aqueous solution: extended application in tannery industrial wastewater Environ Technol Innov 17:100540
55. Abd Hamid SB, The Swe J, Lim S (2015) You Catalytic hydrothermal upgrading of α-cellulose using iron salts as a Lewis acid. Bioresources 10(3):5974–5986
56. Fahimi Z, Omran M (2020) Fabrication of ZnO@C foam: a flexible free-standing electrode for energy storage devices. Mater Des 189:108525
57. Giraldo L, Diana PV, Moreno-Piraján Juan C (2020) Study of CO_2 adsorption on chemically modified activated carbon with nitric acid and ammonium aqueous. Front Chem 8:1–11
58. Maniscalco MP, Maurizio V, Antonio M (2020) Hydrothermal carbonization as a valuable tool for energy and environmental applications: a review. Energies 13(16):4098
59. Chen D, Ciao X, Cheng G (2000) Hydrothermal synthesis of zinc oxide powders with different morphologies. Solid State Commun 113:363–366. https://doi.org/10.1016/S0038-1098(99)00472-X
60. Fahimi Z, Moradlou O (2020) Fabrication of ZnO@C foam: a flexible free-standing electrode for energy storage devices. Mater Des 189. https://doi.org/10.1016/j.matdes.2020.108525
61. Mustapha S, Tijani JO, Ndamitso MM, Abdulkareem SA, Shuaib DT, Mohammed AK, Sumaila A (2020) The Role of Kaolin and Kaolin/ZnO nanoadsorbents in adsorption studies for tannery wastewater treatment. Sci Rep 10(1):1–22. https://doi.org/10.1038/s41598-020-69808-z
62. Nakason K, Panyapinyopol B, Kanokkantapong V, Viriya-empikul N, Kraithong W, Pavasant P (2018) Hydrothermal carbonization of unwanted biomass materials: effect of process temperature and retention time on hydrochar and liquid fraction. J Energy Inst 91(5):786–796. https://doi.org/10.1016/j.joei.2017.05.002
63. Iaiche S, Djelloul A (2015) $ZnO/ZnAl_2O_4$ nanocomposite films studied by X-ray diffraction, FTIR, and X-ray photoelectron spectroscopy. J Spectrosc 2015:836859

64. Pełech I, Piotr S, Daniel S, Ewelina KN, Antoni WM, Joanna KK, Urszula N (2022) The effect of the modification of carbon spheres with $ZnCl_2$ on the adsorption properties towards CO_2. Molecules 27(4):1387
65. Cao D, Gong S, Shu X, Zhu D, Liang S (2019) Preparation of ZnO nanoparticles with high dispersibility based on oriented attachment (OA) process. Nanoscale Res Lett 14(1):210

Towards Sustainable Aquaculture in a Changing Climate: Impact of Wave and Current Interactions on Offshore Fish Pen Mooring Systems

Woon Hong, Nagi Abdussamie, Shuhong Chai, Eric Gubesch, Hassan Mahasneh, and Yasen Harrye

Abstract As climate change drives aquaculture into more energetic and exposed offshore environments, the sustainability of fish farming increasingly relies on resilient mooring systems. In this study, a numerical model of an offshore fish pen was developed and analysed using OrcaFlex and the Morison force model, accounting for flow effects such as blockage and wake to better predict hydrodynamic forces due to wave-current interactions. The model was validated against experimental data, though some deviations were noted, likely due to limitations in the net model mesh and the Morison model's inability to fully capture floating collar deformations. Despite these limitations, important insights were gained. Results show a strong relationship between mooring line tension and floating collar motion, with heave increasing alongside wave height and period, highlighting the need for mooring designs that withstand high-amplitude waves. Surge motion also significantly impacted mooring tension. A catenary mooring system outperformed a taut mooring system, displaying lower tension under high wave and current conditions but larger collar motions, particularly in surge. The most significant wave-current interactions occurred when waves and currents were in opposite directions, while perpendicular interactions had minimal impact.

Keywords Fish farming · Offshore aquaculture · Climate change · Mooring integrity · Wave-current interaction

---

W. Hong · S. Chai · E. Gubesch
Centre for Maritime Engineering and Hydrodynamics, Australian Maritime College, University of Tasmania, Launceston, TAS, Australia

N. Abdussamie (✉) · H. Mahasneh · Y. Harrye
College of Engineering and Technology, University of Doha for Science and Technology, Doha, Qatar
e-mail: nagi.abdussamie@udst.edu.qa

S. Al-Naemi et al. (eds.), *Water and Food Security in the Face of Climate Change: Challenges and Opportunities for Resilience*, World Sustainability Series,
https://doi.org/10.1007/978-3-032-00098-9_39

1 Introduction

The FAO (2024) reported that global fisheries and aquaculture production in 2022 reached 223.2 million tonnes, including 185.4 million tonnes of aquatic animals and 37.8 million tonnes of algae [1]. In addition to wild fisheries, nearshore aquaculture farms play a key role in fish production. Specifically, inland aquaculture contributed 62.6% of farmed aquatic animal output, while marine and coastal aquaculture accounted for the remaining 37.4%, as noted in the FAO (2024) report. For example, in Australia, inshore areas are frequently utilized for farming species such as bluefin tuna and salmon, benefiting from their protection against harsh sea conditions. These aquaculture farms typically employ fish pens, which include essential components such as a floating collar, containment net, bottom net, mooring system, and a lower collar that serves as a sinker.

The development of aquaculture farming is constrained by the shortage of suitable marine water spaces and poor water circulation in inshore areas. This has led to the relocation of farms to offshore locations in countries like Norway and Chile, resulting in larger-scale installations. Offshore aquaculture farms, unlike oil and gas platforms, are highly flexible and complex, with many interacting components. These farms face challenges from greater loads due to waves, currents, and winds, making operation and maintenance difficult. Therefore, numerical and experimental investigations are necessary to understand these structures before installation.

Experimental [2–5] and field studies [6–8] have been conducted to simulate realistic environmental conditions for aquaculture fish pens, overcoming the limitations of laboratory experiments. Numerical studies, though sometimes met with skepticism, are valuable for understanding interactions between hydrodynamic loadings and fish pen structures. These studies are more time and cost-effective, gaining popularity with advancements in computational resources. Common numerical methods include the lumped-mass approach and finite-element analysis, both of which have shown close predictions to experimental results [9].

In real-life situations, fish pens experience combined loads from both waves and currents. The floating collar, being near the surface, bears most of the wave loads and is flexible. Larger deformations and stresses occur at the upper part of the net compared to the lower part [5]. Waves have minimal impact on the volumetric reduction of a fish pen [10] but significantly affect mooring line forces due to the movement of the floating collar. Studies have shown that mooring line tension correlates with the heave and surge motions of the fish pen in irregular waves, with the floating collar's heave motion synchronized with wave elevation [10, 11].

Early works [12] have shown that net panel deformation and the influence of Reynolds number (Re) on drag coefficient (C_d) are crucial for accurate drag force predictions. Experiments comparing simple drag formulas with actual results revealed significant discrepancies due to over-prediction of drag forces [13]. Similarly, the over-estimation of forces was observed in [14] where a screen-type force model was used for numerical modelling of current loads on aquaculture nets.

This implies that effects from shielding in the net are important considerations where a noticeable reduction in velocity was observed downstream from the net model [2, 15, 16]. Field tests performed at four farms in Norway also demonstrated large current reduction as the flow passed through the cages [17]. Net solidity was identified as the primary factor influencing flow velocity downstream from the net plane [2]. With a range of numerical, field and experimental investigations showing the flow reduction characteristics, it is therefore important to include the shielding effect in numerical models. Even when accounting for Re (using the drag coefficient formulated by DeCew et al. [18]) and shielding effects using wake models, drag force is still over-predicted when exposed to high current velocity, high S_n, and high net deformation [19].

Due to the blockage effects, most numerical models struggle to accurately replicate the hydrodynamic loads and behaviours observed in experimental modelling for high S_n nets at high current velocities [20]. Cifuentes [21] successfully formulated a semi-empirical C_d model to accommodate the blockage effect where C_d was defined as a function of Re and S_n. Evaluation against physical testing showed significant improvement across a wide range of current velocities and S_n values, even when compared to the results from the screen force model proposed by Kristiansen and Faltinsen [14]. Analysis indicates that characterising C_d for different types of nets is crucial when using a Morison force model [21].

While many studies have progressed the accuracy of modelling aquaculture nets, few have focused on modelling aquaculture moorings. Hence, this article will evaluate the effect of aquaculture moorings when exposed to the above-mentioned flow conditions. Orcina Orcaflex was used to model and analyse the aquaculture net and moorings in this study. The Morison force model was implemented to solve the dynamic responses, whereas the C_d formulation developed by Cifuentes [21] and wake model established by Huse [22] were included in this study to account for the flow behaviours. The study examined the drag force on the net, volume reduction, mooring line tensions, and the motions of the net pen collar. It also investigated the effects of taut and catenary moorings on the tension of a typical aquaculture net. Additionally, the effects of combined wave and current directions on mooring line tensions were analysed.

2 Numerical Model and Simulation

2.1 Equivalent Net Model

Since it is not feasible to model every twine of the physical net, an equivalent numerical net was developed to represent the properties of the physical net. The large number of twines in the physical net can be simplified by representing large groups of twines using smaller number of equivalent twines in the numerical model. As the detailed motions at each knot of the twines are not required, single segments of line were used to model the net connected by 3-degree (3D) of freedom buoys acting as the knots in the numerical model. Bending stiffness and hence moment transfer is considered to be negligible where linear displacements are being transferred by the 3D buoys. The 3D buoys do not add any drag or inertia forces to the system [23]. The floating collar was modelled using line segments connected to 6-degrees of freedom buoys, which can transfer both rotational and translational motion, representing the bending stiffness of the collar. This is important as the collar withstands the majority of the wave forces. The ballast system at the bottom of the net was modelled by increasing the mass of the respective 3D buoys to the specific weight used in the validation data. The motions and deformation of the numerical net (modelled with line elements) were represented by matching the projected area, wet mass and axial stiffness (refer to [24]) of the physical net [25], whereas the twine diameter was based on the solidity ratio (S_n). The mass of the 3D buoys in the body of the net were set to the lowest allowable value in the software, and their volume was adjusted to best represent the buoyancy of the net.

The hydrodynamic forces acting on the cage are calculated with the modified Morison equation (Eq. (1) [26]), whereas the dynamics of the system is calculated with the equation of motion (Eq. (2) [23]). The equation for the global response of the system has the same form as Eq. (2), with global loads and vectors.

$$F_w(t) = \frac{1}{2}\rho C_d dl[v(t) - u(t)]|v(t) - u(t)| + \rho C_M \frac{\pi}{4} d^2 l \dot{v}(t) - \rho(C_M - 1)\frac{\pi}{4} d^2 l \dot{u}(t) \quad (1)$$

Here, $F_w(t)$ is the fluid force, ρ is the fluid density, d is the effective diameter of the element, l is the length, C_d and C_M are the drag and inertia coefficients, $u(t)$ and $v(t)$ are the element and fluid velocity, $\dot{u}(t)$ and $\dot{v}(t)$ are the element and fluid accelerations.

$$M(p, a) + C(p, v) + K(p) = F(p, v, t) \quad (2)$$

Here, $M(p, a)$ represents the system inertia load, $C(p, v)$ the system damping load, $K(p)$ the system stiffness load, $F(p, v, t)$ the external load, t the simulation time and p, v and a are the position, velocity and acceleration vectors respectively.

2.2 Numerical Fish Pen Model and Mooring Lines

The fish pen modelled in this study was based on data presented in [14]. Refer to Fig. 1 and Table 1 for details. The fish pen consists of a floating collar, four mooring lines (pretension of 14 N), netting and ballast masses attached to 16 points at the bottom of the fish pen. The net was modelled with 5 rings made up of 16 segments,

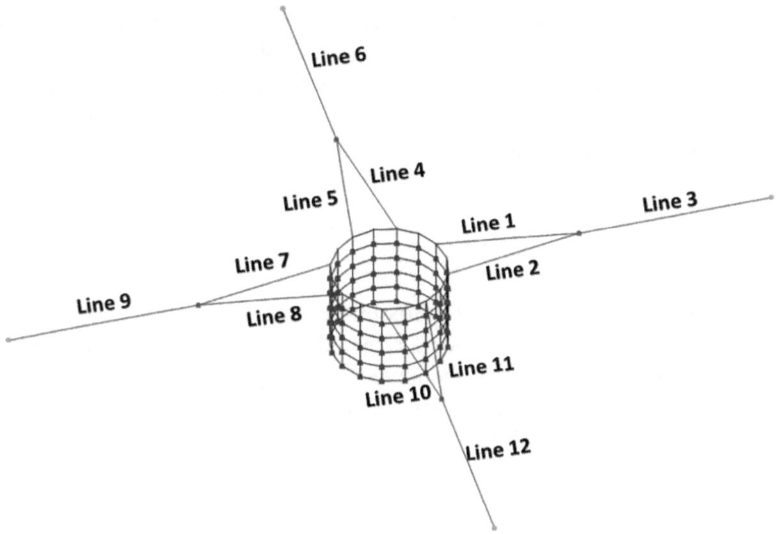

Fig. 1 Numerical fish pen model with mooring lines

Table 1 Net pen properties for numerical and physical models

Properties	Numerical model	Physical model [14]
Fish pen's diameter (m)	1.5	1.5
Fish pen's draft (m)	1.3	1.3
Solidity ratio	0.265	0.265
Twine diameter (m)	0.0277	0.0006
Bar length (m)	0.275	0.006
Collar pipe diameter (m)	0.03	0.03
Collar bending stiffness (Nm^2)	0.136	0.136
Net density (kg/m^3)	1100	1100
Ballast (kg)	16 × 0.075	16 × 0.075
Water depth (m)	1.5	1.5

resulting in 80 × 3D buoys, 16 × 6-D buoys and 160-line elements. The top and bottom of the fish pen are opened to simplify the analysis.

2.3 *Wake Effect and Drag/Inertia Coefficient*

Inertia loading on the net is negligible since the diameter of the line elements are small. Hence, the drag coefficient is critical in establishing hydrodynamic forces.

The drag coefficient formulation developed by Cifuentes [21] was applied in this study. This formulation varies with the product of Reynolds number (Re) and the squared of solidity ratio ($S_n{}^2$) which considers the blockage effect of the flow at high solidity ratios where flow deviates below and around the fish pen. This drag coefficient expression is given in Eq. (3) where Re is calculated using the cross-flow method, considering the angle of attack [23].

$$C_d = -3.2892 \times 10^{-5}\left(\mathrm{Re} \times S_n^2\right)^2 + 0.00068\left(\mathrm{Re} \times S_n^2\right) + 1.4253 \tag{3}$$

The shielding effect occurs as flow crosses the upstream portion of the net resulting in a loss of kinetic energy and hence a reduced flow velocity over the downstream portion of the net. This effect is accounted for by using a riser interaction model developed by Huse [22], which was applied using the same method as [27]. Velocity reduction is calculated but wake lift forces are not included in this wake model [23] as it has been shown to be negligible [28]. This wake model is mathematically defined in Eqs. (4–6).

$$v_d(x, y) = v_{do}(x, y) - k_2 v_{uo}\left[\frac{C_{du} d_u}{x_s}\right]^{\frac{1}{2}} \exp\left[-k_3\left(\frac{y}{b}\right)^2\right] \tag{4}$$

$$x_s = x + 4d_u / C_{du} \tag{5}$$

$$b = k_1 [C_{du} d_u x_s]^{\frac{1}{2}} \tag{6}$$

Here, (x, y) represents the position of the downstream cylinder with respect to the upstream cylinder, v_d the disturbed fluid velocity at downstream position (x, y), v_{do} the undisturbed fluid velocity at (x, y), v_{uo} the undisturbed fluid velocity at the upstream cylinder center, C_{du} the drag coefficient for the upstream cylinder, d_u the upstream cylinder diameter and the non-dimensional model parameters k_1, k_2 and k_3 having values of 0.25, 1.0 and 0.693 respectively [22]. Unlike the net elements, the inertia term in the Morison Eq. (1) is important for the response of the floating collar under wave and current conditions. Re and KC numbers were obtained for each wave and current condition using Eqs. (7) and (8) respectively. With the known Re and KC numbers, the C_M values were determined by interpolating the values provided [29]. A drag coefficient of $C_d = 1.0$ was taken as a nominal guess for the drag coefficient on the collar.

$$\mathrm{Re}_{collar} = \frac{(U_M + U_C)d}{\nu} \tag{7}$$

$$KC_{collar} = \frac{(U_M + U_C)T}{\nu} \tag{8}$$

Here, U_M is the maximum amplitude of the undisturbed water particle velocity at $z = 0$ given by linear potential flow theory, U_C is the uniform current flow velocity, d is the floater diameter, v is the fluid's kinematic viscosity and T is the wave period.

2.4 Hydrodynamic Load and Mooring Analysis

To study the deformation and hydrodynamic drag force on the net, the fish pen was exposed to current velocities ranging from 0 to 2.5 m/s in the direction parallel to Line 3. Simultaneously, combined regular wave and current conditions traveling in the same direction were implemented to investigate the hydrodynamic response of the fish pen, including the floating collar and mooring line tensions. The combined conditions featured a current velocity of 0.1 m/s and a wave steepness of 1/15. Both the current and wave travelled in the same direction, as illustrated in Fig. 2a. Regular waves were generated using Airy wave theory, with wave heights (H) ranging from 0.026 to 0.26 m and periods (T) ranging from 0.5 to 1.6 s. The mooring analysis examined the effects of taut and catenary mooring configurations on mooring line tensions and the motions of the floating collar. For the taut mooring configuration, a mooring line scope of 1.67:1 was implemented, while a scope of 5:1 was used for the catenary mooring configuration. Various wave and current conditions were tested. In this case, irregular waves were generated using the JONSWAP spectrum with a peak enhancement factor of 3.3. Refer to Fig. 2 and Table 2.

3 Results and Discussion

3.1 Hydrodynamic Loadings on a Fish Pen

The Orcaflex simulation was compared with experimental and numerical screen force model results obtained in [14] (see Fig. 3, left). The graph shows that the mean forces in the front mooring were over-predicted by the Orcaflex numerical model. Differences between the experimental and numerical results increase as the waves get longer, with significant variation observed around 1.25–1.75 λ/D ratios. For shorter waves, the simulation results are relatively close to the experimental results. There is close agreement between both numerical results outside the 1.25–1.75 λ/D ratio range. Additionally, larger forces in the front mooring can be observed for longer waves. Potential reasons for the variation between the present numerical results and the published experimental results include inaccuracies from the net modeling, and the inability of the Morison force model to accurately model the deformation of the floating collar, as suggested by Kristiansen and Faltinsen [14]. Overall, significant improvements are still needed for closer prediction accuracy.

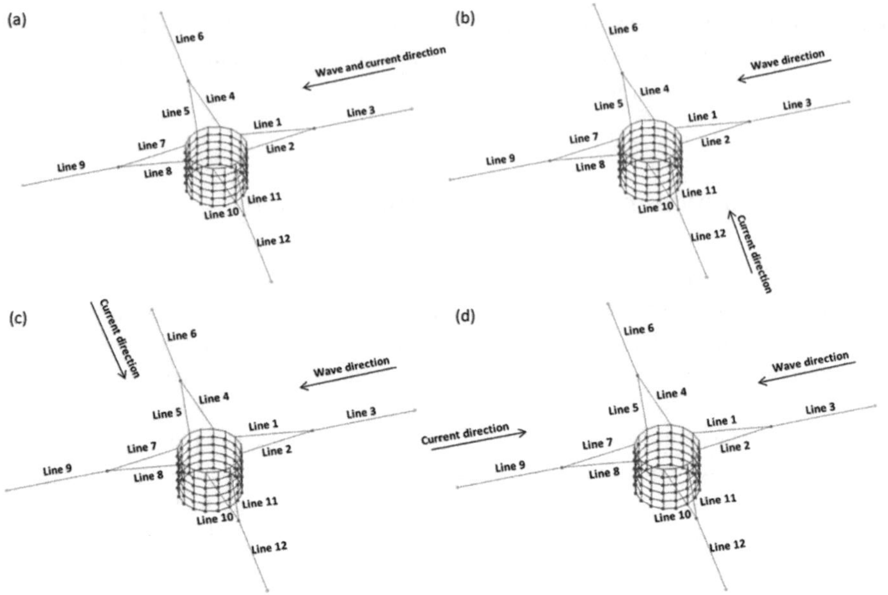

Fig. 2 Test conditions **a** 180° wave and current direction (R1C-R3C and IR1C). **b** 180° wave direction and 90° current direction (IR2C). **c** 180° wave and 270° current direction (IR3C). **d** 180° wave direction and 0° current direction (IR4C and IR5C)

Table 2 Mooring analysis test conditions for regular and irregular wave and current

Condition	Current velocity (m/s)	Current direction (°)	H or H_s (m)	T or T_p (s)	Wave type
R1C	0.20	0	0.50	2.00	Regular
R2C	0.10	0	0.30	1.50	Regular
R3C	0.05	0	0.10	0.85	Regular
IR1C	0.10	180	0.14	1.60	JONSWAP
IR2C	0.10	90	0.14	1.60	JONSWAP
IR3C	0.10	270	0.14	1.60	JONSWAP
IR4C	0.10	0	0.14	1.60	JONSWAP
IR5C	0.05	0	0.14	1.60	JONSWAP
IR6	0.00	–	0.14	1.60	JONSWAP

Even so, little conclusion can be drawn regarding the trend in mean mooring force with the wavelengths.

Figure 3 (right) shows the maximum heave and surge motions of the floating collar. Both heave and surge motions increase with the λ/D ratios, although there are some variabilities, especially for the surge motions. A significant increase in heave motions can be noticed as wave heights and periods increase, indicating that the effects of

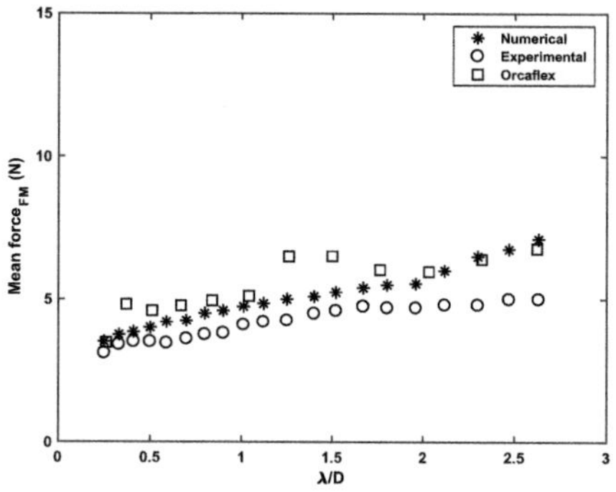

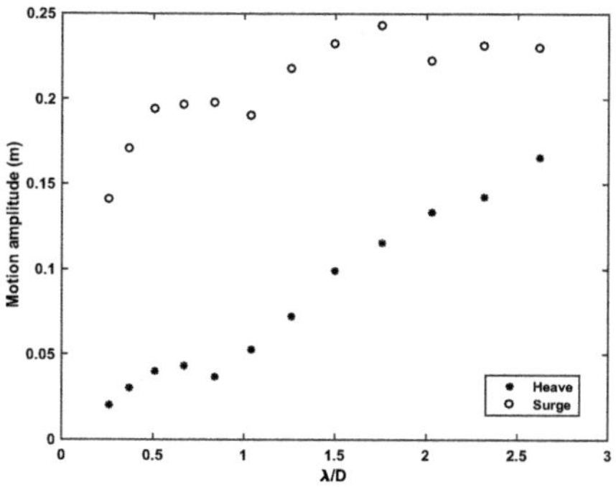

Fig. 3 Left: Mean tension in front mooring line (Line3). Right: Maximum heave and surge motions of floating collar

high amplitude and period waves need to be considered in the mooring design of the fish pen. The variation of surge motions with λ/D ratios is smaller compared to heave motions. However, surge motion has a large impact on the mooring forces, with the highest mooring forces and surge motions occurring in the 1.25–1.75 λ/D ratio range. Furthermore, the dependency relationship between the force in the mooring line and the motion of the floating collar is evident from the results, where greater motions of the floating collar develop greater forces in the mooring line.

3.2 Mooring Analysis of a Typical Fish Pen Arrangement

Regular Wave Conditions

The maximum tension in mooring Line 3, as well as the maximum heave and surge motions of the floating collar, were obtained for both catenary and taut mooring configurations (see Table 3). The results indicate that the catenary mooring configuration performs better than the taut mooring configuration under extreme conditions (R1C), where the maximum tension is lower. However, under calmer conditions (R3C), the taut mooring configuration performs better, with a lower maximum tension of around 20 N compared to approximately 50 N for the catenary mooring configuration. Regarding the motion responses of the floating collar, both heave and surge motions are greater with the catenary mooring. The heave motions do not differ significantly between the two mooring types, but the surge motions are notably larger with the catenary mooring. This can be a major concern, as large motions of the fish pen can affect the well-being of the fish. However, the surge motion is expected to be much lower in a fish farm as the flow across numerous fish pens is expected to reduce. Overall, both catenary and taut mooring configurations have their benefits. The catenary mooring is better in terms of mooring line performance under higher

Table 3 Maximum tension (T_{max}) in Line 3, maximum heave and surge motion for catenary and taut mooring configuration under combined regular wave and current at 180°

Condition	Catenary mooring			Taut mooring		
	T_{max} (N)	Heave (m)	Surge (m)	T_{max} (N)	Heave (m)	Surge (m)
R1C	32.38	0.45	6.91	37.59	0.30	1.02
R2C	24.94	0.34	6.30	24.63	0.17	0.79
R3C	49.71	0.18	6.16	19.83	0.13	0.56

wave height and current conditions, while the taut mooring is better in terms of motion response and reducing the footprint of the entire structure.

Irregular Wave Conditions with Varying Current Direction

Irregular wave simulations were run for 100 s and a summary of the results are shown in Table 4 and Fig. 4. When a current of 0.1 m/s was induced in the same direction as the wave (IR1C), a noticeable increase in the tension of mooring Line 3 was observed, with a mean tension of 9.8 N. Mean mooring line tensions for the other lines did not differ much from the wave-only condition, where the tensions remained around 0 N. As anticipated, the line parallel to the current direction experienced the greatest mean tension, as shown in Fig. 4 (left). The mean mooring line tension in Line 3, parallel to the wave direction, dropped slightly from 3.6 N to around 3.2 N in both IR2C and IR3C conditions, indicating that the cancelling effect is not significant when the current direction does not oppose the wave direction. When the current opposed the wave, as represented by IR4C and IR5C test conditions, cancelling effects were observed. In the IR4C condition, Line 9 experienced the greatest mean tension, while Line 3's mean tension dropped from 3.6 to 0.2 N, exhibiting the cancelling effect. When a lower current of 0.05 m/s was applied in the direction opposite to the wave in the IR5C test condition, cancelling effects of the wave and current were also noticed, with the mean mooring line tension in Line 3 dropping from 3.6 to 0.3 N. At the same time, tensions in other mooring lines were close to 0 N, representing a nearly equilibrium system. Although the line parallel to the current direction experienced the greatest mean tension, this was not the case for the maximum mooring line tension under IR2C and IR3C conditions, where the current came from a direction perpendicular to the wave direction (see Fig. 4, right). This further shows that current flows from directions other than the opposite direction parallel to the wave are less capable of restraining the wave effects on mooring line tension.

4 Conclusions

In conclusion, a numerical model of a fish pen with mooring lines was developed to study the hydrodynamic forces acting on the structure. The simulation used an empirical drag coefficient formulation to account for flow behaviours, including

Table 4 Mean and maximum mooring line tension under irregular wave and current conditions

Condition	Mean tension				Maximum tension			
	Line3 (N)	Line6 (N)	Line9 (N)	Line12 (N)	Line3 (N)	Line6 (N)	Line9 (N)	Line12 (N)
IR1C	9.83	0.11	0.58	0.05	17.87	6.29	9.25	6.98
IR2C	3.33	0.26	0.31	5.69	12.95	6.59	10.57	11.95
IR3C	3.19	5.71	0.28	0.30	12.45	11.19	9.72	6.77
IR4C	0.17	0.08	3.20	0.05	8.53	7.33	11.56	8.26
IR5C	0.33	0.03	0.30	0.03	10.69	8.87	10.22	7.46
IR6	3.55	0.01	0.36	0.01	12.07	7.14	9.19	7.28

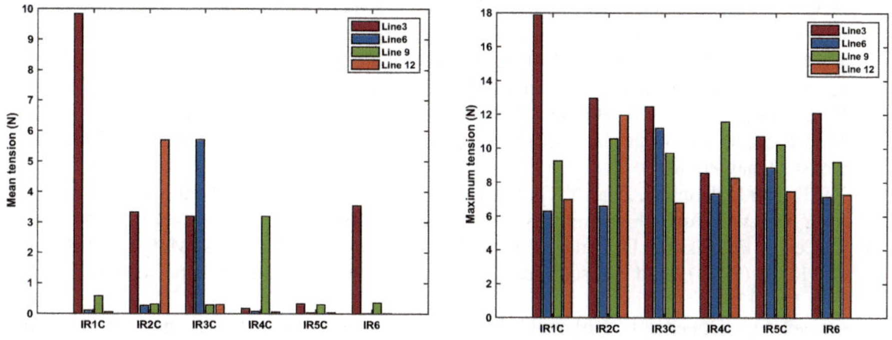

Fig. 4 Left: Mean tensions for each mooring line and condition. Right: Maximum tensions for each mooring line and condition

blockage and wake effects. The simulation results were validated against published experimental and numerical findings, revealing some variations. These discrepancies may be due to smaller mesh sizes, errors in the equivalent net model, and limitations of the Morison force model in accurately modelling the floating collar's deformation. Further improvements to the numerical model and the use of more mesh elements are needed for better alignment with experimental results. Key conclusions from the study include:

- Heave motions of the floating collar increase significantly with higher wave heights and periods, affecting mooring design. Surge motions greatly impact mooring line forces, confirming their dependency on the floating collar's movements.
- Catenary mooring performs better under high wave and current conditions, with lower maximum mooring line tension compared to taut mooring. However, in calmer conditions, taut mooring shows lower tension. Catenary mooring induces greater surge motions in the floating collar.
- Cancelling effects are most significant when wave and current directions are opposite, and less so when the current is perpendicular to the wave direction.

- Mooring line tensions are heavily influenced by wave propagation and current direction.

Acknowledgements The authors are grateful to the University of Doha for Science and Technology (UDST) and the Centre for Maritime Engineering and Hydrodynamics of the Australian Maritime College (AMC), University of Tasmania (UTAS) for facilitating this study.

References

1. FAO (2024) The state of world fisheries and aquaculture. FAO, Rome
2. Bi CW, Zhao YP, Dong GH, Xu TJ, Gui FK (2013) Experimental investigation of the reduction in flow velocity downstream from a fishing net. Aquacult Eng
3. Kristiansen T, Faltinsen OM (2014) Experimental and numerical study of an aquaculture net cage with floater in waves and current. J Fluids Struct
4. Sulaiman OO, Magee A, Bahrain Z, Kader ASA, Mainmun A, Pauzi AG, Othman K (2013) Mooring analysis for very large offshore aquaculture ocean plantation floating structure. Ocean Coast Manag 80–88
5. Xu T, Hou H, Dong G, Zhao Y, Guo W (2016) Structural analysis of float collar for metal fish cage in waves. Turk J Fish Aquat Sci. https://doi.org/10.4194/1303-2712-v17_2_04
6. DeCew J, Fredriksson D, Lader Pl, Chambers M, Howell WH, Osienki M, Celikkol B, Frank K, Hoy E (2013) Field measurements of cage deformation using acoustic sensors. Aquacult Eng 114–125
7. Gansel LC, Oppedal F, Birkevold J, Tuene SA (2018) Drag forces and deformation of aquaculture cages—full-scale towing tests in the field. Aquacult Eng 46–56
8. Lader P, Dempster T, Fredheim A, Jensen J (2008) Current induced net deformations in full-scale sea-cages for Atlantic salmon (Salmo salar). Aquacult Eng 52–65
9. Zhao Y, Wang X, DeCew J, Tsukrov I, Bai X, Bi C (2014) Comparative study of two approaches to model the offshore fish cages. Chin Ocean Eng Soc. https://doi.org/10.1007/s13344-015-0032-0
10. Li L, Fu S, Xu Y (2013) Nonlinear hydroelastic analysis of an aquaculture fish cage in irregular waves. Mar Struct
11. Zhao Y, Xu T, Bi C, Dong G, Liu S (2012) The numerical simulation of hydrodynamics of fishing net cage
12. Tsukrov I, Eroshkin O, Fredriksson D, Swift MR, Celikkol B (2000) Finite element modeling of net panels using a consistent net element
13. Lader P, Enerhaug B (2005) Experimental investigation of forces and geometry of a net cage in uniform flow. IEEE J Ocean Eng 30(1)
14. Kristiansen T, Faltinsen OM (2012) Modelling of current loads on aquaculture net cages. J Fluids Struct
15. Loland G (1993) Current forces on, and water flow through and around, floating fish farms. Aquacult Int 1:72–89
16. Lader P, Enerhaug B, Fredheim A, Krokstad J (2003) Modelling of 3D net structures exposed to wave and current. Paper presented at the 3rd international conference on hydroelasticity in marine technology
17. Johansson D, Juell JE, Oppedal F, Stiansen JE, Rouhonen K (2007) The influence of the pycnocline and cage resistance on current flow, oxygen flux and swimming behaviour of Atlantic salmon (Salmo salar L) in production cages. Aquacult Eng 271–287
18. DeCew J, Tsukrov I, Risso A, Swift MR, Celikkol B (2010) Modeling of dynamic behavior of a single-point moored submersible fish cage under currents. Aquacult Eng

19. Cifuentes C, Kim MH (2015) Numerical simulation of wake effect in nets under steady current. Paper presented at the ASME 2015 34th international conference on ocean, offshore and arctic engineering, OMAE2015
20. Moe-Føre H, Endresen PC, Aarsæther KG, Jensen J, Føre M, Kristiansen D, Fredheim A, Lader P, Reite K-J (2014) Structural analysis of aquaculture nets: comparison and validation of different numerical modeling approaches. Offshore Mech Arctic Eng. https://doi.org/10.1115/1.4030255
21. Cifuentes C (2016) Dynamic analysis of cage systems under waves and current for offshore aquaculture. Texas A&M University
22. Huse E (1993) Interaction in deep-sea riser arrays. Paper presented at the offshore technology conference
23. Orcina (2020) Orcaflex manual
24. Fredheim A (2005) Current forces on net structures. Norwegian University of Science and Technology
25. Tsukrov I, Eroshkin O, Fredriksson D, Swift MR, Celikkol B (2003) Finite element modeling of net panels using a consistent net element. Ocean Eng
26. Haritos N, He DT (1992) Modelling the response of cable elements in an ocean environment. Finite Elem Anal Des
27. Cifuentes C, Kim MH (2017) Hydrodynamic response of a cage system under waves and currents using a Morison-force model. Ocean Eng
28. Cifuentes C, Kim MH (2017b) Numerical simulation of fish nets in currents using a Morison force model. Ocean Syst Eng
29. Sarpkaya T, Isaacson M (1981) Mechanics of wave forces on offshore structures. Van Nostrand Reinhold Co., New York

Invisible Networks: How Supply Chain Innovation is Securing Qatar's Food Future

Mohd Basheer and Luma Abuorabi

Abstract This research investigates the impact of supply chain advancements, such as blockchain and cold chain systems, on enhancing Qatar's food security. These technologies are critical for a country that relies on imports for approximately 90% of its food supply and is vulnerable to disruptions, such as the 2017 blockade. The study employs a mixed-methods approach, combining quantitative regression analysis and qualitative interviews with industry experts to provide a comprehensive understanding. Regression analysis demonstrates a positive correlation between Supply Chain Efficiency (SCE) and Food Security (FS), showing that a 1-point increase in SCE can lead to a 0.42-point increase in FS. Cross-validation between quantitative data and interview themes was conducted to ensure reliability. The qualitative analysis highlights barriers, such as high costs and limited technical expertise, and demonstrates the benefits of blockchain in reducing food waste by 10% through improved transparency and traceability. Cold chain systems have ensured a steady supply of high-quality perishable items. Despite these advancements, Qatar's reliance on imported food remains a significant risk. To address this challenge, the study advocates for a balanced approach combining technological innovations with strategies for boosting local food production, ensuring long-term resilience. These findings provide actionable insights for policymakers to design strategies integrating supply chain advancements with localized agricultural development. The study's approach can be compared with blockchain initiatives in regions like sub-Saharan Africa, which have enhanced food supply transparency. Future research should explore the socioeconomic impacts of blockchain technology on suppliers and conduct comparative studies with nations facing similar food security challenges.

Keywords Supply chain efficiency · Blockchain · Food security · Cold chain management · Qatar

---

M. Basheer (✉)
University of Jordan, Amman, Jordan
e-mail: mhm8220041@ju.edu.jo

L. Abuorabi
Food and Agriculture Organization (FAO), Amman, Jordan
e-mail: Luma.abuorabi@fao.org

S. Al-Naemi et al. (eds.), *Water and Food Security in the Face of Climate Change: Challenges and Opportunities for Resilience*, World Sustainability Series,
https://doi.org/10.1007/978-3-032-00098-9_40

1 Introduction

1.1 Overview of Supply Chain Innovations and Food Security

In a world where ensuring food security is becoming increasingly tricky globally. In areas that heavily depend on imports. Supply chain management technology advancements like cold chain systems are vital for improving effectiveness and reducing waste [1]. These new technologies are crucial for managing and lessening the impact of supply chain disruptions caused by tensions and natural disasters [2]. The FAO reports that 14% of the food produced globally is wasted before it gets to consumers, highlighting the significance of supply chain solutions in addressing this issue. As climate change continues to impact production and threaten supply chains with risks, robust supply chain technologies become even more crucial in maintaining consistent food availability and ensuring global food security.

1.2 The Role of Supply Chain Innovation in Qatar

The blockade, in 2017, highlighted how dependent Qatar is on imports for around 90% of its food needs [3]. This incident prompted Qatar to rethink and improve its food security plans by investing in technologies like blockchain for transparency and cold chain systems to prevent food wastage. These changes in policies and technology were made to ensure food supplies and better preparation for any disturbances. There has been a study on the advantages of supply chain technologies overall; however, knowledge is lacking on how these particular technologies enhance resilience in countries reliant on imports like Qatar amidst challenges posed by climate change and geopolitical tensions. Alternative approaches have been used to establish food hubs and broaden supplier bases. Encounter obstacles when put into practice [4].

This research seeks to close these disparities by exploring the impact of cold chain systems on enhancing supply chain effectiveness and investigating their effects on policymaking and strategic efforts in Qatar. Moreover, it underscores how insights gained from Qatar's journey can shape approaches in areas confronting interdependencies and vulnerabilities.

1.3 Research Questions and Motivation

This research look into how supply chain advancements can secure Qatar's food future, focusing on the following key questions:

1. How do blockchain and cold chain technologies improve Qatar's food security?
2. What weaknesses exist in the current supply chain systems in Qatar?

3. How might these technological advancements reduce Qatar's dependence on imported food supplies?

Motivated by the need to ensure resilience in Qatar's food supply chain infrastructure, this study aims to provide insights that not only support the development of a sustainable food system within the region but also contribute to the broader discourse on global food security. It underscores the importance of integrating technological solutions with policy measures and local production strategies to create a comprehensive response to food security challenges. By examining the policies and technological adoptions that emerged after the 2017 blockade, this research connects the findings to larger regional and global efforts to build resilient food systems capable of withstanding climate change and geopolitical disturbances.

2 Literature Review

2.1 *Overview of Supply Chain Technologies*

The literature extensively documents the transformative potential of supply chain innovations such as blockchain and cold chain systems in enhancing food security [1]. Blockchain technology, known for its transparency, traceability, and ability to combat fraud, has been highlighted as a vital tool for building trust and efficiency in global food supply chains [2]. Cold chain management ensures the preservation of perishable goods through temperature-controlled logistics, reducing spoilage and food loss [3]. However, while these technologies have demonstrated significant promise, the body of literature often overlooks the nuanced challenges and counterarguments associated with their implementation, particularly in developing countries or import-reliant nations.

2.2 *Practical Applications and Examples*

Practical applications of these technologies reveal both opportunities and limitations. For instance, blockchain pilot programs in sub-Saharan Africa have shown that blockchain can reduce food loss by up to 20% by enhancing transparency and traceability [4]. Yet, these programs often face significant challenges related to infrastructure, technological literacy, and scalability. Similarly, cold chain systems have proven essential in countries like India and Brazil, where investments in refrigerated logistics have reduced food spoilage and enhanced food availability. However, these implementations come with obstacles such as high energy costs and the need for reliable electricity [5]. These examples underscore that while technology can drive food security improvements, real-world applications can be complex, especially in regions with limited resources.

2.3 Challenges and Limitations in the Context of Climate Change

Despite the promising benefits of blockchain and cold chain technologies, their limitations must be considered, particularly in the context of climate change. The integration of blockchain technology, for instance, relies on robust digital infrastructure and significant energy consumption, posing a challenge for developing nations [6]. Cold chain systems, while effective in reducing waste, also face sustainability challenges due to their high energy requirements and environmental impact. For example, studies have shown that maintaining cold chain systems contributes to greenhouse gas emissions, which can counteract climate resilience efforts [7]. These aspects are crucial to consider as climate change intensifies, placing additional pressure on already strained food systems.

2.4 Resilience and Building Sustainable Food Systems

The literature supports the view that blockchain and cold chain systems contribute to resilience in food supply chains. Blockchain technology enhances traceability, allowing stakeholders to respond swiftly to disruptions, which is invaluable in scenarios affected by climate change or geopolitical instability [8]. Cold chain systems maintain the quality of perishable items, a critical factor for food security in dry and arid regions where heat can quickly degrade food quality [9]. By integrating these technologies with IoT sensors and AI-driven predictive analytics, nations can build adaptive and resilient food systems capable of withstanding external shocks. However, to maximize long-term benefits, it is essential to address sustainability and affordability. Research highlights the importance of government incentives and public–private partnerships in overcoming cost and infrastructure barriers [10].

2.5 Addressing Research Gaps

While existing research highlights the advantages of supply chain innovations, there is a void in understanding the full range of challenges associated with their carrying out and how these technologies can be optimized to support resilience against climate change. This study aims to address these gaps by examining practical applications and limitations of blockchain and cold chain systems, especially in import-dependent and resource-limited countries. Additionally, it investigates how these innovations contribute to building sustainable, climate-resilient food systems in dry regions, providing new visions that can inform policies and strategic investments.

3 Methodology

3.1 *Study Design*

This research employs a mixed-methods approach, integrating both qualitative interviews and quantitative data analysis to assess the influence of supply chain advancements on food security in Qatar. This comprehensive approach merges quantitative data insights with experiential knowledge from industry professionals for a holistic understanding.

3.2 *Quantitative Analysis*

Data Collection and Variables

Secondary data were collected from reputable sources, including the FAO, World Bank, and Qatar's Ministry of Municipality [1, 2]. The key variables analyzed include:

- Supply Chain Efficiency (SCE): Measured using the Logistics Performance Index (LPI) and cold chain infrastructure ratings to reflect the operational effectiveness of the supply chain.
- Food Security (FS): Based on the FAO's Food Security Index, assessing food availability, accessibility, and utilization.
- Blockchain Adoption (BA): Defined as the percentage of blockchain-enabled transactions in the food supply chain, representing technological integration.
- Import Rate: The ratio of food imports to total food consumption, highlighting reliance on external supply.
- Local Production (Local Prod): The percentage of local food production in relation to total food demand.

Rationale for Factor Selection: Variables were chosen for their critical roles in reflecting the performance and resilience of food supply chains. Blockchain adoption was included due to its potential to enhance transparency, traceability, and reduce waste—essential for Qatar's import-dependent food system. The Import Rate and Local Production factors were selected to understand the balance between dependency on imports and domestic food production.

Regression Models

A multiple regression model was applied to evaluate the impact of SCE and blockchain adoption on food security:

$$FS = \alpha + \beta 1\ SCE + \beta 2\ Import\ Rate + \beta 3\ Local\ Prod + \beta 4\ Blockchain + \epsilon \quad (1)$$

where

- FS: Food Security Index
- SCE: Supply Chain Efficiency (measured via LPI)
- Import Rate: Ratio of food imports to total food consumption
- Local Prod: Local food production as a percentage of demand
- Blockchain: Blockchain adoption rate
- $\alpha, \beta 1, \beta 2, \beta 3, \beta 4$: Regression coefficients
- ϵ: Error term.

Efficiency Forecasting

A time-series forecasting model was used to predict future food security under potential supply chain disruptions:

$$FSt + 1 = FSt + \gamma 1 \Delta SCE\ t + \gamma 2 \Delta Import\ Rate\ t + \gamma 3 \Delta Local\ Prod\ t + \epsilon t \quad (2)$$

This model assesses how projected changes in supply chain efficiency and local production might influence food security over time.

Correlation and Sensitivity Analysis

To further analyze the effect of specific innovations on supply chain efficiency, correlation coefficients between Blockchain Adoption (BA) and SCE were calculated:

$$r = \frac{\Sigma\left(BA - BA^{-}\right)\left(SCE - SCE^{-}\right)}{\sqrt{\sum (BA - BA^{-})^2 \sqrt{(SCE - SCE^{-})^2}}} \quad (3)$$

where r is the Pearson correlation coefficient between blockchain adoption and supply chain efficiency. This equation helps determine how closely these variables are linked in practice.

Sensitivity analysis was performed to assess how small changes in supply chain variables (e.g., SCE and LocalProd) influence food security:

$$\frac{\partial \mathrm{FS}}{\partial \mathrm{SCE}} = \frac{\Delta FS}{\Delta SCE}$$

This derivative quantifies the sensitivity of food security to changes in supply chain efficiency.

3.3 *Qualitative Analysis*

Data Collection

Semi-structured interviews were conducted with 20 supply chain professionals, including logistics managers, policymakers, and technology experts. These participants were chosen for their direct involvement in managing Qatar's food supply chains and implementing technology solutions. Efforts were made to reduce bias by ensuring diverse perspectives and backgrounds among participants.

Thematic Analysis

Interview responses were analyzed using thematic coding, performed manually and cross-validated using NVivo software to ensure accuracy and consistency. The thematic analysis helped identify patterns related to technology adoption, logistical challenges, and improvements in local production. Key themes included technological barriers, cold chain logistics challenges, and the role of blockchain in enhancing supply chain transparency.

3.4 Handling Data Limitations

Data limitations were acknowledged, particularly the availability of specific local data and the depth of qualitative insights due to the sample size. To mitigate these limitations, data were cross-referenced with external industry reports and third-party studies to verify consistency. Triangulation was also employed, comparing quantitative data with qualitative interview findings to strengthen reliability and validity.

3.5 Definitions and Replicability

- Supply Chain Efficiency (SCE): The operational effectiveness of food supply chain processes, assessed using LPI.
- Blockchain Adoption (BA): The extent of blockchain technology use to enhance transparency and combat fraud in food supply chains.
- Food Security (FS): Evaluated using the FAO's Food Security Index, which measures food availability, accessibility, and utilization.

4 Results and Discussion

4.1 Key Findings

Our study confirms a significant relationship between Supply Chain Efficiency (SCE) and Food Security (FS) in Qatar. The regression analysis demonstrated that enhancing SCE by 1 point results in a 0.42-point increase in the Food Security Index ($p < 0.05$). This result emphasizes the importance of adopting logistics improvements and integrating innovative technologies such as blockchain and cold chain systems to bolster supply chain performance.

Technological integration within Qatar's supply chain has led to notable improvements, including a 10% reduction in food waste. This is attributed to enhanced monitoring and transparency measures facilitated by blockchain implementation.

Figure 1 demonstrates the direct impact of blockchain technology on reducing food waste. The graph shows a downward trend in food loss, correlating with increased blockchain adoption. This visual representation emphasizes how real-time tracking and transparency in the supply chain mitigate inefficiencies and contribute to food security.

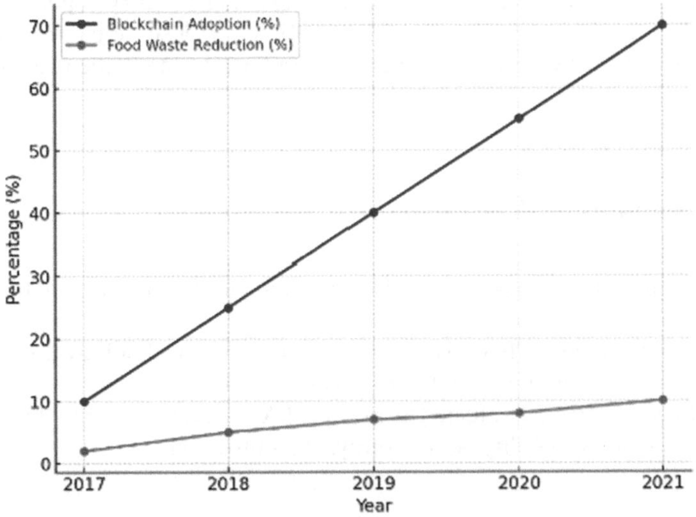

Fig. 1 Impact of Blockchain Adoption on Reducing Food Waste

4.2 Analysis of Qualitative Interview Findings

The qualitative component of this study involved insights from 20 supply chain professionals, which added depth to the quantitative data. Interviewees consistently highlighted the advantages of blockchain for increasing transparency and cold chain systems for preserving perishable goods. However, they also pointed out significant challenges, such as the high cost of implementation and the need for specialized training to operate and maintain these systems effectively.

Themes that emerged from the interviews included:

- Technological Barriers: Limited technical expertise and high costs associated with blockchain and cold chain systems were mentioned as obstacles.
- Infrastructure Concerns: Participants noted that robust infrastructure is essential to maximize the potential of these technologies.
- Policy Recommendations: Interviewees suggested that government support, through subsidies and training programs, could accelerate technology adoption and enhance resilience.

These qualitative insights support the quantitative findings by confirming that while blockchain and cold chain systems contribute to reduced food waste and improved supply chain efficiency, strategic investments and policy interventions are necessary for long-term sustainability.

Figure 2 pick out the role of local food production during periods of supply chain disruption, such as the 2017 blockade. Although local production has bolstered food security during crises, the models tell that its contribution remains secondary compared to the reliance on food imports. This insight underscores the need for a balanced approach that combines technological advancements with strategic local production to build a more resilient food system.

4.3 Broader Implications

The study's findings have significant implications for both policy and practice. The demonstrated link between SCE and FS in Qatar serves as an example for other import-dependent nations facing similar challenges. Embracing technologies like blockchain and cold chain systems can enhance resilience against disruptions. However, it is crucial for policymakers to consider complementary measures, such as supporting local production and infrastructure development, to build comprehensive food security strategies.

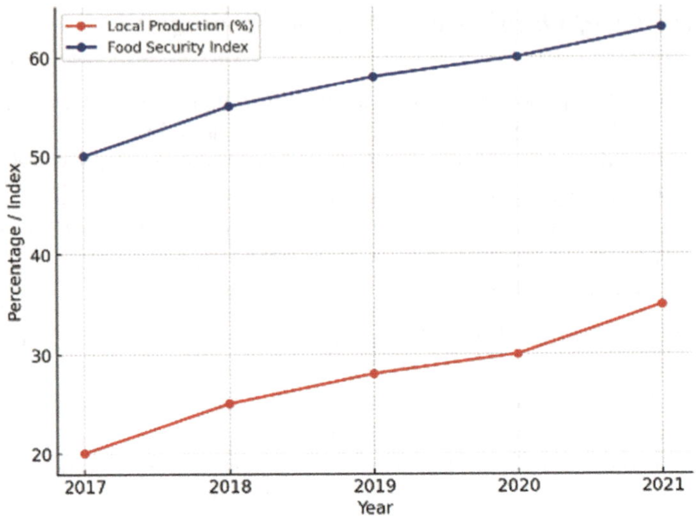

Fig. 2 Local production's contribution to food security during supply disruptions

5 Conclusion

5.1 Summary of Key Findings

This research confirms the significant role of supply chain advancements, such as blockchain technology and cold chain systems, in enhancing Qatar's food security by improving supply chain efficiency and reducing waste. These technologies contribute to better food availability in the region. However, the findings also underscore that Qatar's heavy reliance on imports poses a considerable risk to its food security, emphasizing the need for a multifaceted approach that integrates both technological solutions and local food production strategies.

5.2 Broader Implications

The implications of this study extend beyond Qatar, offering insights for other import-dependent nations, particularly those in dry regions facing similar vulnerabilities. The demonstrated benefits of blockchain for transparency and cold chain systems for preserving food quality highlight how strategic investments in supply chain technology can enhance food security resilience. However, solely depending on technology is insufficient. Supportive policies that promote sustainable domestic food production are crucial for long-term stability. Policymakers should prioritize initiatives that blend technological advancements with infrastructure development and localized agricultural strategies to mitigate the risks posed by global supply chain disruptions and climate change. For example, India's solar-powered cold chain initiatives have proven effective in sustaining perishable goods while reducing carbon footprints, serving as a model for renewable energy solutions in food security strategies.

5.3 Benefits and Limitations

This research provides valuable perspectives on the impact of technology on food security. The study's benefits lie in its combined analysis of qualitative and quantitative data, presenting a holistic view of the challenges and opportunities in Qatar's food supply chain. However, limitations include a lack of detailed local data and a limited number of interview participants, which may have affected the scope of insights. Additionally, the data used in this research did not fully encompass the potential long-term impacts of climate change on global food systems.

5.4 Recommendations for Future Research and Policy

Future research should go deeper into enhancing local food production capabilities and assess the long-term socioeconomic impact of blockchain and cold chain technologies on small-scale suppliers and farmers. Comparative studies with other nations facing similar food security challenges and climate-related risks could offer valuable lessons on managing food supply effectively. For instance, Kenya's blockchain projects aimed at increasing agricultural transparency provide an example of how innovative technologies can be harnessed to strengthen food security. Future studies could use longitudinal methodologies to evaluate the sustained impacts of these technologies and explore pilot programs in collaboration with government bodies and private stakeholders.

Policymakers should focus on fostering public–private partnerships and providing subsidies that facilitate the adoption of sustainable technologies and local food production initiatives. Focusing attention to renewable energy sources, such as solar power, to support cold chain systems can align with United Nations' Sustainable Development Goals (SDGs), particularly Goal 2: Zero Hunger, Goal 12: Responsible Consumption and Production, and Goal 13: Climate Action. Such approaches ensure that food security solutions are not only resilient but also environmentally sustainable.

References

1. Taniguchi E, Thompson RG, Yamada T (2002) Recent trends and innovations in city logistics. Transp Rev 22(1):65–76
2. Petersen C (2021) Blockchain technology and supply chain transparency. Int Supply Chain J 15(3):145–157
3. Christopher M (2016) Logistics & supply chain management, 5th edn. Pearson, UK
4. El-Sayed E, El-Sayed H (2019) Food security in the Arabian Gulf: threats and opportunities. J Food Secur 7(4):162–169
5. FAO (2020) Food supply chain analysis: enhancing food security. FAO J 6(1):20–33
6. World Bank (2021) Food import reliance and supply chain strategies. World Bank Report, pp 45–67
7. Smith J, Kumar P (2022) The environmental impact of cold chain systems in emerging markets. Sustain Logist J 10(2):120–135
8. Global Food Security Initiative (2021) Resilience building through blockchain technology. Policy Brief, pp 22–30
9. Qatar National Vision 2030. https://www.gco.gov.qa/en/about-qatar/national-vision2030/. Last accessed 16 Sep 2024
10. Jones L, Patel R (2023) Public-private partnerships in supply chain resilience: case studies and policy insights. J Glob Logist Policy 12(4):234–250

Analysis of Growth Parameters of Three Genotypes of Turkeys Using Applicable Non-linear Models

Akano Kayode

Abstract Growth rates vary throughout an animal's life, making linear models inadequate for capturing major growth phases. Instead, robust non-linear models should be used. This study analyzed growth data from 400 progenies of four turkey genotypes: Nicholas White Turkey (NWT) × Locally Adapted Turkey (LAT), LAT × NWT, NWT × NWT, and LAT × LAT over 20 weeks. Data analysis was conducted using SAS (REML with fixed effects) and WinBUGS for the Markov Chain Monte Carlo (MCMC) process. Model selection was based on Deviance Information Criteria (DIC), Bayesian Information Criteria (BIC), and Akaike Information Criteria (AIC). Results showed that NWT exhibited superior growth (7465 g at 20 weeks). Across genotypes and sexes, an inverse relationship was observed between asymptotic weight (A) and maturation rate (K), with the strongest negative correlation in female exotic turkeys (− 0.866, − 0.985, − 0.967) for Logistic, Gompertz, and von Bertalanffy models, respectively. The von Bertalanffy model, along with REML and Bayesian approaches (normal and student t-distributions), best predicted turkey growth parameters A, B, and K. The Gompertz model produced estimates close to von Bertalanffy, making both suitable for turkey growth modeling. Additionally, Gompertz was the best fit for female exotic and female crossbred turkeys. The study concludes that non-linear growth model estimates should be used for growth comparison and selection in Nigerian Locally Adapted turkeys, with von Bertalanffy providing the best fit.

Keywords Deviance Information Criteria (DIC) · Bayesian Information Criteria (BIC) · Akaike Information Criteria (AIC) · Markov Chain Monte Carlo (MCMC) · Asymptotic weight (A) · Maturation rate (K) · Nicholas White Turkey (NWT) · Locally Adapted Turkey (LAT) · Gompertz · Von bertalanffy models

---

A. Kayode (✉)
Lagos State University of Education (LASUED), Oto-Ijanikin (Epe Campus), Lagos State, Nigeria
e-mail: ak.samkay@gmail.com

Department of Agricultural Education, College of Vocational and Entrepreneurship Education (COVED), Lagos State, Nigeria

S. Al-Naemi et al. (eds.), *Water and Food Security in the Face of Climate Change: Challenges and Opportunities for Resilience*, World Sustainability Series,
https://doi.org/10.1007/978-3-032-00098-9_41

1 Introduction

Poultry birds are the most numerous of the farm animals held by livestock producers in Nigeria, and they might thus contribute significantly to reducing protein shortfall in most Nigerian diets. This is due to their high level of biological and economic efficiency [3]. Turkeys flourish in arid environments, handle heat better, wander farther, and produce higher quality meat than chickens [10]. Despite its aforementioned greater potential than the chicken, its production has not been vastly exploited in the developing countries [3, 5].

Growth is a fundamental feature of biological systems that involves a rise in body size per time unit [7] Whereas growth curves are defined as the variations in growth over time. However, an animal's growth curve could be influenced by several factors which may include its breed, sex, management factors amongst others [6, 9]. It is critical to identify a model that best depicts the growing state of birds in a given environment. Observable qualities such as point of inflection and growth rate, the effects of different management systems, food requirements, and breeding application results can all be analyzed and used to improve growth characteristics [2]. Growth parameters are important not only as selection criteria, but also in terms of feed management techniques used during the production period to optimize animal production management and efficiency in order to determine nutrient utilization efficiency or to predict daily energy, protein, and mineral dietary requirements [1].

2 Materials and Method

2.1 Experimental Site

This study was conducted at the Turkey Breeding Unit, Directorate of University Farms (DUFARMS), Federal University of Agriculture, Alabata Road, Abeokuta, Ogun State, Nigeria. The university is located in the southwestern portion of Nigeria, at latitude 7o 10! N and longitude 3o 26! E, in a tropical climate with an annual rainfall of about 1037 mm. Between October and March, average monthly ambient temperatures ranged from 28 °C to 36 °C, with a yearly average relative humidity of roughly 82%. The vegetation at the University is intermediate between tropical rainforest and derived savannah [4].

2.2 Experimental Birds

These consists of 30 locals (5 males, 25 females), 30 exotics (5 males, 25 females of Nicolas White breeds of turkey). These were mated as follows to generate the following progenies i.e. crosses from the two genotypes:

Table 1 Three non-linear models used for the prediction of growth curve functions

Model	$BW_{ij} = f(\theta_i, t_{ij}) + e_{ij}$	t_{inf}	W_{inf}
Logistic	$BW_{ij} = A_i\,(1 + B_i \exp\{-K_i\, t_{ij}\})^{-1} + e_{ij}$	$\frac{-\log(1/B_i)}{K_i}$	$\frac{A_i}{2}$
Gompertz	$BW_{ij} = A_i \exp\{-B_i \exp\{-K_i\, t_{ij}\}\} + e_{ij}$	$\frac{-\log(B_i)}{K_i}$	$\frac{A_i}{2.7182}$
Von bertalanffy	$BW_{ij} = A_i\,(1 - B_i \exp\{-K_i\, t_{ij}\})^3 + e_{ij}$	$\frac{\log(B_i) + \log(3)}{K_i}$	$\frac{8A_i}{27}$

i. 100 (male Nicolas White turkey × female local)
ii. 100 (male local × female Nicolas White turkey)
iii. 100 (male local turkey × female local turkey)
iv. 100 (male Nicolas White × female Nicolas White turkey).

2.3 Statistical Model for REML Approach

The Logistic, Gompertz (Gompertz, 1825), and von Bertalanffy (von Bertalanffy, 1957) growth models with fixed parameters A, B, and K were fitted to Turkey growth data (Table 1). In these models, BW_{ij} is the body weight of turkey at age (week), t_{ij}; A is the asymptotic weight or an estimation of mature weight as age approaches infinity ($t_i = \infty$); B is the integration constant defining the degree of maturity at $t_i = 0$; K is the rate of maturing and refers to growth rate relative to mature weight. E_i is the residuals with the assumption of $e_i \sim N\,(\mathbf{0}, \sigma^2\mathbf{I}_i)$, where $\sigma^2\mathbf{I}_i$ is the residual variance structure for all subjects, assuming that no covariance structure exists between the residual variance structure for all subjects, assuming that no covariance structure exists between the residuals of the model t_{inf} and W_{inf} are the age and weight at the inflection point of the growth model, respectively.

The non-linear model for growth data for animal i is as follows:

$$BW_{ij} = f\left(\theta_i,\, t_{ij}\right) + e_{ij} \quad i = 1,\, N \text{ and } j = 1$$

where f is the non-linear function that connects the response variable (BW_{ij}) to time (t_{ij}), i is a vector containing the non-linear function's parameters, N is the number of animals, and n_i is the number of measurements obtained from animal i.

3 Results

Results of the research are graphically shown in Tables 2, 3, 4, 5, 6, 7 and 8.

Table 2 Parameter estimates of three growth models fitted for three turkey genotypes based on Bayesian approach with normal error model

Model	Sex	Genotype	A	B	K	t_{inf}	W_{inf}
Logistic	Male	Local	3244 ± 33.27	18.53 ± 0.5812	0.2668 ± 0.004557	10.94 ± 0.1122	1622 ± 16.63
		Exotic	4755 ± 73.74	19.32 ± 1.056	0.2816 ± 0.007838	10.51 ± 0.1701	2378 ± 36.87
		Cross	3752 ± 85.64	18.91 ± 0.8927	0.2419 ± 0.007119	12.15 ± 0.2515	1876 ± 42.82
Gompertz		Local	3877 ± 64.42	3.938 ± 0.06133	0.1394 ± 0.003224	9.834 ± 0.1495	1426 ± 23.7
		Exotic	5525 ± 131.4	4.066 ± 0.1112	0.1519 ± 0.0054	9.235 ± 0.2072	2033 ± 48.35
		Cross	4822 ± 181.5	3.889 ± 0.07878	0.1172 ± 0.004829	11.6 ± 0.3786	1774 ± 66.78
Von bertalanffy		Local	4519 ± 107.9	0.8097 ± 0.009031	0.09524 ± 0.002896	9.324 ± 0.1958	1339 ± 31.96
		Exotic	6268 ± 202.4	0.8336 ± 0.01715	0.107 ± 0.00482	8.575 ± 0.2459	1857 ± 59.98
		Cross	6081 ± 365.2	0.7946 ± 0.011	0.07427 ± 0.004507	11.73 ± 0.5963	1802 ± 108.2
Logistic		Local	2493 ± 38.43	15.23 ± 0.6411	0.254 ± 0.006566	10.72 ± 0.1743	1247 ± 19.21
		Exotic	4622 ± 102.9	21.23 ± 1.252	0.2609 ± 0.008286	11.71 ± 0.2388	2311 ± 51.47
		Cross	3638 ± 99.55	22.68 ± 1.297	0.2442 ± 0.008116	12.78 ± 0.2901	1819 ± 49.78
Gompertz	Female	Local	2931 ± 72.9	3.64 ± 0.07966	0.1371 ± 0.004907	9.432 ± 0.2267	1078 ± 26.82
		Exotic	5738 ± 211.1	4.123 ± 0.1131	0.1302 ± 0.005734	10.89 ± 0.3453	2111 ± 77.66
		Cross	4954 ± 247.3	4.154 ± 0.0967	0.1115 ± 0.005517	12.79 ± 0.5067	1822 ± 90.97
Von bertalanffy		Local	3392 ± 115.4	0.7694 ± 0.01114	0.09456 ± 0.004165	8.857 ± 0.2821	1005 ± 34.19
		Exotic	6930 ± 402	0.8308 ± 0.01697	0.08602 ± 0.005524	10.65 ± 0.5149	2053 ± 119.1
		Cross	6552 ± 489.3	0.8252 ± 0.01209	0.06785 ± 0.00469	13.41 ± 0.7878	1941 ± 145

A = asymptotic weight (g), B = integration constant, K = maturing rate, t_{inf} = age at inflection point (wks), W_{inf} = weight at inflection point (g)

Table 3 Bayesian model choice criteria (Deviance Information Criteria, DIC) for the growth models

Growth curve model	Normal	Error model		
	Dbar	Dhat	pD	DIC
Bertalanffy	52,348.2	52,332.3	15.901	52,364.1
Gompertz	52,474.1	52,456.4	17.696	52,491.8
Logistic	53,024.9	53,006.5	18.367	53,043.3

Model choice criteria (smallest value indicates better fitted model)

Estimates of Degree of Freedom by Growth Models

Von Bertalanffy growth model had the least standard deviation (0.09504) and MC error (0.001340) followed by Gompertz growth model with the standard deviation of 0.12390 as shown in Table 5.

Table 4 Parameter estimates of three growth models fitted for three turkey genotypes based on Bayesian approach, student's t-error model

Model	Sex	Genotype	*A*	*B*	*K*	t_{inf}	W_{inf}
Logistic	Male	Local	3144 ± 33.05	20.88 ± 0.6262	0.2879 ± 0.004901	10.55 ± 0.1117	1572 ± 16.53
		Exotic	4735 ± 67.27	20.27 ± 1.017	0.2866 ± 0.007174	10.5 ± 0.1534	2367 ± 33.63
		Cross	3705 ± 74.84	21.46 ± 0.9178	0.2549 ± 0.006557	12.03 ± 0.2155	1852 ± 37.42
Gompertz		Local	3698 ± 54.42	4.145 ± 0.04954	0.1524 ± 0.002984	9.334 ± 0.1288	1360 ± 20.02
		Exotic	5506 ± 95.92	4.179 ± 0.07572	0.155 ± 0.003844	9.231 ± 0.1548	2026 ± 35.29
		Cross	4847 ± 143.9	4.081 ± 0.05747	0.1206 ± 0.0037	11.67 ± 0.2925	1783 ± 52.94
Von bertalanffy		Local	4432 ± 98.95	0.8226 ± 0.006654	0.09955 ± 0.002634	9.078 ± 0.1802	1313 ± 29.32
		Exotic	6356 ± 142.2	0.8357 ± 0.01034	0.1059 ± 0.003173	8.679 ± 0.1774	1883 ± 42.13
		Cross	6484 ± 293.1	0.8073 ± 0.006243	0.07162 ± 0.002992	12.37 ± 0.4613	1921 ± 86.85
Logistic		Local	2243 ± 29.02	16.38 ± 0.6234	0.2871 ± 0.006461	9.739 ± 0.1343	1121 ± 14.51
		Exotic	4588 ± 96.21	23.72 ± 1.352	0.271 ± 0.008091	11.68 ± 0.2213	2294 ± 48.11
		Cross	3552 ± 81.58	24.66 ± 1.128	0.2568 ± 0.006832	12.48 ± 0.2335	1776 ± 40.79
Gompertz	Female	Local	2471 ± 38.98	3.813 ± 0.06184	0.1647 ± 0.003967	8.128 ± 0.1278	909 ± 14.34
		Exotic	5862 ± 178.7	4.312 ± 0.08873	0.1311 ± 0.004513	11.16 ± 0.2884	2157 ± 65.75
		Cross	4858 ± 186.7	4.279 ± 0.06454	0.1163 ± 0.004097	12.52 ± 0.3737	1787 ± 68.69
Von bertalanffy		Local	2790 ± 54.93	0.7903 ± 0.008237	0.1152 ± 0.003156	7.496 ± 0.1425	826.6 ± 16.27
		Exotic	7465 ± 340.9	0.8376 ± 0.01041	0.0812 ± 0.003786	11.37 ± 0.4305	2212 ± 101
		Cross	6937 ± 476.8	0.8281 ± 0.007067	0.06561 ± 0.003783	13.91 ± 0.7322	2056 ± 141.3

A = asymptotic weight (g), *B* = integration constant, *K* = maturing rate, t_{inf} = age at inflection point (wks), W_{inf} = weight at inflection point (g)

Table 5 Estimates of degree of freedom by growth models

Growth curve model	Df-mean	Sd	MCerror
Von bertalanffy	2.240	0.09504	0.001340
Gompertz	2.338	0.12390	0.002195
Logistic	3.979	0.30080	0.005061

4 Conclusion

This study compared two analytical approaches, Bayesian using normal error model and student's t-error and fixed effects model in fitting weight-age data of three genotypes of turkey in Nigeria. Growth models are essential for implementing a decision support tool for turkey producers, providing extra information when vital feeding, slaughter, and breeding decisions are made. The study found that the fixed effects model using von Bertalanffy growth model described well the turkey data set for asymptotic final weight, while for model parameters, both the fixed effects model and Bayesian approach with student's t-error model were better for turkey data set in terms of integration constant a. However, at the point of inflection, any of the methodologies and models can be used to forecast age and weight.

Using any approach, the Logistic model reached 50% of the asymptotic mature weight at the point of inflection, Gompertz achieved 37%, and the von Bertalanffy

Table 6 Parameter estimates of three growth models fitted for three turkey genotypes based on fixed effects model

Model	Sex	Genotype	A	B	K	t_{inf}	W_{inf}
Logistic	Male	Local	3365.7 ± 72.6	22.0715 ± 1.6120	0.275096 ± 0.007119	11.3 ± 0.2	1682.8 ± 36.3
		Exotic	4744.5 ± 63.3	19.9329 ± 1.1258	0.284670 ± 0.005311	10.5 ± 0.2	2372.2 ± 31.7
		Cross	4029.5 ± 152.7	20.9130 ± 1.0510	0.2377521 ± 0.007996	13.0 ± 0.5	2014.7 ± 76.3
Gompertz		Local	4178.9 ± 138.7	4.2807 ± 0.1473	0.144879 ± 0.005291	10.5 ± 0.3	1537.4 ± 51.0
		Exotic	5499.3 ± 105.5	4.1433 ± 0.1118	0.154660 ± 0.003827	9.2 ± 0.2	2023.1 ± 38.8
		Cross	5378.1 ± 363.3	4.0141 ± 0.0848	0.115228 ± 0.005578	12.7 ± 0.8	1978.5 ± 133.7
Von bertalanffy		Local	5222.2 ± 271.2	0.8515 ± 0.0141	0.099325 ± 0.004422	10.5 ± 0.4	1066 ± 43.0
		Exotic	6271.2 ± 160.3	0.8467 ± 0.0163	0.109814 ± 0.003582	8.5 ± 0.2	2173.2 ± 193.9
		Cross	6763.1 ± 426.4	0.8063 ± 0.0097	0.074026 ± 0.004764	12.9 ± 0.9	2251.7 ± 190.4
Logistic		Local	2541.6 ± 57.5	16.4338 ± 0.6774	0.258756 ± 0.005219	10.8 ± 0.2	1270.8 ± 28.8
		Exotic	4636.9 ± 129.2	23.2705 ± 1.9117	0.268468 ± 0.008408	11.7 ± 0.4	2318.4 ± 64.6
		Cross	3806.7 ± 133.0	25.3976 ± 1.2959	0.245381 ± 0.008673	13.3 ± 0.5	1903.3 ± 66.5
Gompertz	Female	Local	3046.4 ± 92.1	3.7213 ± 0.0688	0.139727 ± 0.003644	9.7 ± 0.3	1120.7 ± 33.9
		Exotic	5844.2 ± 311.7	4.2831 ± 0.1499	0.134778 ± 0.007458	11.0 ± 0.7	2150 ± 114.7
		Cross	5183.7 ± 252.1	4.3163 ± 0.0929	0.114477 ± 0.005893	13.2 ± 0.6	1907 ± 92.7
Von bertalanffy		Local	3598.2 ± 145.3	0.7849 ± 0.0099	0.098305 ± 0.003309	9.2 ± 0.3	1066.1 ± 43.0
		Exotic	7334.7 ± 654.3	0.8487 ± 0.0187	0.088949 ± 0.007366	11.1 ± 1.0	2173.2 ± 193.9
		Cross	7599.6 ± 642.6	0.8417 ± 0.0107	0.068102 ± 0.005308	14.9 ± 1.1	2251.7 ± 190.4

A = asymptotic weight (g), B = integration constant, K = maturing rate, t_{inf} = age at inflection point (wks), W_{inf} = weight at inflection point (g)

model achieved 30% at the time of inflection. Von Bertalanffy model had superior fit for the turkey data based on the selection criteria for the best fit model based on Bayesian method with normal error and student's t-error model utilizing DIC. Using AIC and BIC in a fixed effects model, it was determined that the von Bertalanffy growth model had the smallest projected value in both criteria and was the best match for all male turkeys and local female turkeys, but the Gompertz growth model was the best fit for female exotic and crossbred turkeys. However, the difference between von Bertalanffy and Gompertz growth models are very minimal thus making them fit for modeling growth curve of domestic turkey.

Negative correlation between A and k showed that early maturing turkey will attain smaller weight or ligher asymptotic weight and associated with high inflection point. The Bayesian approach using von Bertalanffy model is better in describing the growth pattern of locally adapted turkeys in Nigeria. Variations in the genetic background of the different genotypes accounted for the observed differences in the growth curve parameters estimated for the turkey used in this study. Exotic turkey had the heaviest mature weight and weight at the point of inflection among the examined genotypes because it has the genetic ability to grow quickly and attain a higher mature weight than the other turkey genotypes.

Table 7 Model choice criteria for the three growth models

Growth curve model	Sex	Genotype	AIC	BIC
Von Bertalanffy	Male	Local	19,804	19,825
		Exotic	2600	2613
		Cross	5838	5854
	Female	Local	16,717	16,737
		Exotic	2556	2569
		Cross	4903	4919
Gompertz	Male	Local	19,835	19,856
		Exotic	2638	2651
		Cross	5974	5991
	Female	Local	16,763	16,783
		Exotic	2507	2520
		Cross	4896	4912
Logistic	Male	Local	19,992	20,014
		Exotic	2773	2787
		Cross	6437	6454
	Female	Local	16,925	16,946
		Exotic	2557	2570
		Cross	5314	5330

Model choice criteria (smallest value indicates better fitted model)

Table 8 Pearson correlation between A and K parameters within each model

Growth curve model	Sex	Genotype	R	p-value
Von Bertalanffy	Male	Local	– 0.792	0.0001
		Exotic	– 0.844	0.0001
		Cross	– 0.866	0.0001
	Female	Local	– 0.800	0.0001
		Exotic	– 0.967	0.0001
		Cross	– 0.907	0.0001
Gompertz	Male	Local	– 0.775	0.0001
		Exotic	– 0.715	0.02
		Cross	– 0.761	0.0001
	Female	Local	– 0.717	0.0001
		Exotic	– 0.985	0.0001
		Cross	– 0.895	0.0001
Logistic	Male	Local	– 0.686	0.0001
		Exotic	– 0.360	0.3073
		Cross	– 0.761	0.0001
	Female	Local	– 0.457	0.0002
		Exotic	– 0.866	0.0025
		Cross	– 0.727	0.0004

References

1. Darmani-Kuhi H, Porter T, Lopez S, Kebreab E, Strathe AB, Dumas A, Dijkstra J, France J (2010) A review of mathematical functions for the analysis of growth in poultry. World's Poultry Sci J 66:227–240
2. Narine D, Aksoy T, Karaman E, Curek DI (2010) Analysis of fitting growth models in medium growing chicken raised indoor system. Trends Animal Veterinary Sci J 1:12–18
3. Ngu GT, Etchu KA, Butswat ISR, Woogeng IN (2014) Semen and microbial characteristics of two breeds of turkeys in an arid tropical environment of Bauchi State. Nigeria Afri J Microbiol Res 8(21):2174–2182
4. Odjugo PAO (2005) An analysis of rainfall pattern in Nigeria. Global J Environ Sci 4(2):139–145
5. Pérez-Lara E, Camacho-Escobar MA, Ávila-Serrano NY, Arroyo-Ledezma J, Sánchez-Bernal EI, Torre MR, Reyes-Borques V (2013) Productive evaluation of slow growing Mexican turkeys with different diets in confinement. Open J Animal Sci 3:46–53
6. Porter T, Kebreab E, Darmani-Kuhi H, Lopez S, Strathe AB, France J (2010) Flexible alternatives to the Gompertz equation for describing growth with age in turkey hens. Poult Sci 89:371–378
7. Rizzi C, Contiero B, Cassandro M (2013) Growth patterns of Italqin local chicken populations. Poult Sci 92:2226–2235
8. SAS Institute Inc. (2000) The SAS system for windows. Released 8.2. SAS Inst., Cary, NC
9. Segura-Correa JC, Santos-Ricalde RH, Palma-Ávila (2016) Non-linear model to describe growth curves of commercial Turkey in the tropics of Mexico. Brazilian J Poultry Sci 19(1):27–32

10. Yakubu A, Salako AE, Ige AO (2007) Effects of genotype and housing system on the laying performance of chicken in different seasons in the semi-humid tropics. Int J Poult Sci 6:434–439

Efficiency Enhancement of PV Energy System for Aquaponics in Qatar and GCC Countries

Ahmed Badawi, Muhammad Zubair, I. M. Elzein, Walid Kh Alqaisi, and Alhareth Zyoud

Abstract In recent years, there has been a significant increase in the installation of solar photovoltaic (PV) systems worldwide. This paper presents a comprehensive analysis of the cost-effectiveness of this renewable energy source based on tilt and azimuth angles. An economic study is conducted to analyze the net present value (NPV) in relation to the tilt angle. In a smart grid power system, reliability performance is crucial and requires additional attention. Furthermore, the integration of Battery Energy Storage (BES) schemes, solar photovoltaic (SPV), and wind systems in the smart grid system offers significant efficiency and reliability to utilities. However, grid coordination with PV and other resources can lead to issues such as power interruptions or outages. The objective is to demonstrate how the tilt angle, along with other PV factors can optimize the overall net present value of the system. The findings from this study will contribute to a better understanding, management, planning, forecasting, and stabilization of solar power generation.

Keywords Solar PV array · Wind system · Net present value · Smart grid · Optimization · Control method · Reliability

---

A. Badawi (✉) · M. Zubair · I. M. Elzein · W. K. Alqaisi
University of Doha for Science and Technology, Doha, Qatar
e-mail: ahmed.badawi@udst.edu.qa

M. Zubair
e-mail: muhammad.zubair@udst.edu.qa

I. M. Elzein
e-mail: 60101973@udst.edu.qa

W. K. Alqaisi
e-mail: walid.alqaisi@udst.edu.qa

A. Zyoud
Department of Electrical and Computer Engineering, Faculty of Engineering, Birzeit University, Birzeit, Ramallah, Palestine
e-mail: alhmtz@gmail.com

S. Al-Naemi et al. (eds.), *Water and Food Security in the Face of Climate Change: Challenges and Opportunities for Resilience*, World Sustainability Series,
https://doi.org/10.1007/978-3-032-00098-9_42

1 Introduction

Solar power can be harvested in mainly two ways: photovoltaic (PV) systems, and concentrated solar power (CSP) systems. PV systems offer benefits such as lower capital costs, shorter installation periods, easier maintenance, standardized equipment, large-scale manufacturing facilities, and scalability [1–3]. However, the disadvantages of PV systems include their intermittent nature and expensive batteries. On the other hand, CSP systems have the advantage of cheaper storage energy in the form of heat [4].

The levelized cost of energy (LCOE) of PV systems has already achieved grid parity [5–7]. In Qatar, the cost of energy for residential customers is 0.033 USD/kWh, while the power purchase agreement (PPA) of an 800 MW PV system in Qatar was announced at 0.01567 USD/kWh [8], which is 52% lower than the residential customers' sale price. PV panels can be installed on rooftops using fixed-axis, single-axis tracking, or dual tracking regimes. Dual tracking systems produce the highest annual energy output but require a more complex design structure [9–11]. The wind speed increases with height, so rooftop systems require stronger structures. A single-axis tracking system provides better annual output than a fixed-axis tracking system, but it comes with higher costs and maintenance requirements. Fixed systems can be positioned at specific tilt and azimuth angles to optimize PV energy generation based on the load profile [12].

The government of Qatar has not yet implemented a renewable energy policy for on-grid systems. Residential customers in Qatar have the option to install stand-alone PV systems, as the cost of electricity is low compared to European countries [13, 14], there is no load-shedding, and all residents are connected to the national grid [15]. However, there are no national objectives in Qatar for residential customers to install PV systems. PV systems for smaller loads without storage or grid connection experience significant losses during peak generation times, while PV energy systems for larger loads with load profile matching result in higher net present values.

Aquaponics is an energy-intensive agricultural system that requires uninterrupted electrical energy for nutrients, artificial lighting, water pumps, aerators, heating, cooling, ventilation (HVAC), and dehumidifiers [16]. In water-scarce Qatar, dehumidification in aquaponics cooling systems is essential to conserve water. The load profile of aquaponics systems necessitates a combination of PV systems and grid connection to provide energy after sunset, as pumps and cooling systems must operate throughout the day. Aquaponics systems are economically feasible in countries with higher energy costs [17, 18], but Qatar's lower energy costs and higher prices for fish and vegetables provide an advantage. Implementing a PV energy system will further enhance the net present value of aquaponics projects.

2 Methodology

2.1 Economic Analysis

Economic analysis of the aquaponics PV system installed on a rooftop will be conducted using metrics such as net present value, levelized cost of energy, and payback period. The combined economics of aquaponics and the PV system will be studied to assess the project's feasibility. The net present value of a project is calculated based on the project's cash flow using the following equation [19]:

$$\mathrm{NPV} = \sum_{\mathrm{n}=0}^{\mathrm{N}} \frac{\mathrm{C_n}}{(1 + \mathrm{d_{nominal}})^{\mathrm{n}}} \tag{1}$$

The levelized cost of energy (LCOE) is another important parameter to analyze the feasibility of the project. This cost represents the total cost of energy over the project's lifespan and is calculated using the equation [20–22]

$$real\ LCOE = \frac{NPV}{\sum_{n=1}^{N} \frac{Q_n}{(1+d_{real})^n}} \tag{2}$$

$$nominal\ LCOE = \frac{NPV}{\sum_{n=1}^{N} \frac{Q_n}{(1+d_{nominal})^n}} \tag{3}$$

The payback period provides the information of recovering the capital in number of years out of total number of years of the project presented by the equation [22].

$$\sum_{n} \frac{\Delta I_n}{(1 + d_{nominal})^n} \le \sum_{n} \frac{\Delta S_n}{(1 + d_{nominal})^n} \tag{4}$$

where

Cn	Cash flow in year n after paying tax [USD]
Q_n	Electricity generated by the system in a year n [kWh].
Co	Equity investment amount of the project [USD]
$d_{nominal}$	Nominal discount rate [%].
d_{real}	Real discount rate [%].
ΔI	The incremental investment cost [USD]
ΔS	The annual saving net of future annual costs [USD]
n	year
N	Analysis period.

2.2 PV System Energy Analysis

The energy generation by PV panels is directly proportional to incoming irradiance. The empirical model that will be used here is the one proposed by Perez et al. in 1990 [23].

$$I_b + I_{d,sky} + I_{d,ground} \tag{5}$$

where I_{poa} is plane of array of irradiance, I_b is direct irradiance, $I_{d,sky}$ is diffused irradiance from the sky and $I_{d,ground}$. is diffused irradiance from ground.

The cell temperature degrades the PV output by a factor γ_{PV}. The energy equation of PV is shown as [24].

$$P_{PV} = \frac{I_{tr} P_{dc0}}{1000}\left[1 + \gamma_{PV}\left(T_{cell} - T_{ref}\right)\right]. \tag{6}$$

Here P_{PV} represents the output power of the PV panel, I_{tr} is the incoming irradiance, $T_{cell} and T_{ref}$ are the cell temperature and test temperature of the PV panel, and P_{dc0} is the rated power of the PV panel.

The total energy generated by a PV panel in a year can be calculated as [24]:

$$E_p = \sum_{x=1}^{8769} P_{pv}(x) \tag{7}$$

2.3 PV Tilt Angle and Azimuth Angles

Determining the azimuth and tilt angles for photovoltaic (PV) panels is crucial for effectively generating electricity from solar energy. Research in this area often focuses on maximizing radiation values on stationary panels. However, improvements can be made by concentrating on optimal positioning that maximizes the cumulative radiation incident on PV panels with adjustable azimuth and tilt angles.

Various systems are used to harness solar energy for electricity generation, including fixed tilt angle systems and solar tracking systems. In solar power plant deployment, PV panels are typically aligned with the specific tilt angle of the installation site. Accurately assessing both tilt and azimuth angles is essential for the effective arrangement of PV panels. While much existing research emphasizes determining optimal tilt angles to maximize radiation on stationary panels, it is equally important to investigate how PV panels perform under different azimuth and tilt angle configurations.

Studying the tilt angle and azimuth angles is necessary to position PV panels according to the load profile of the aquaponics greenhouse. Matching the load profile with tilt angles and azimuth angle will enhance the net present value of the project. The solar declination angle δ, solar altitude angle 'α' [25]. and solar azimuth angles γ are shown in Eqs. 6 and 7 respectively [12]. Figure 1 illustrates the solar angles of solar altitude,azimuth, and zenith.

$$\delta = 23.45 \sin\left(\frac{2\pi(284 + n)}{365}\right), \quad (8)$$

$$= \begin{cases} \pi/2 & if\ \alpha_{od} + r > 90 \\ \frac{\pi}{180}(\alpha_{od} + r) & if\ \alpha_{od} + r \leq 90 \end{cases} \quad (9)$$

$$\gamma = \begin{cases} b & if\ HA < -\pi \\ \pi - b & if\ -\pi \leq HA \leq 0,\ of\ if\ HA \geq \pi \\ \pi + b & if\ a > 1 \end{cases} \quad (10)$$

In a stationary photovoltaic (PV) panel installation, both the azimuth and tilt angles remain constant. Conversely, in a single-axis tracking system, the tilt angle is fixed while the azimuth angle is adjusted through motorized tracking. A dual-axis tracking system allows for both the azimuth and tilt angles to be dynamically adjusted via a motorized mechanism, optimizing the annual energy output of the PV panels. Figure 2 illustrates the tilt and azimuth angles relevant to the positioning of the PV panel.

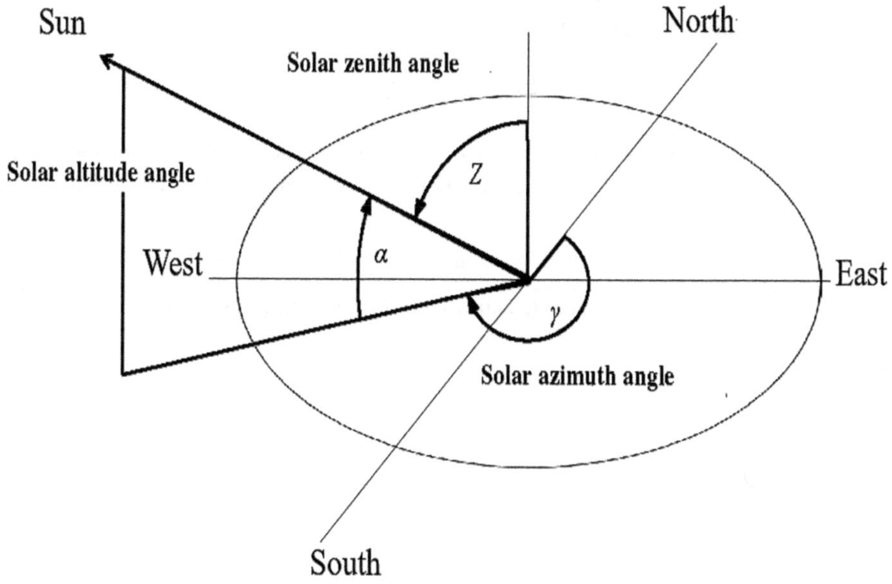

Fig. 1 Illustration of solar angles of solar altitude, solar azimuth, and solar zenith

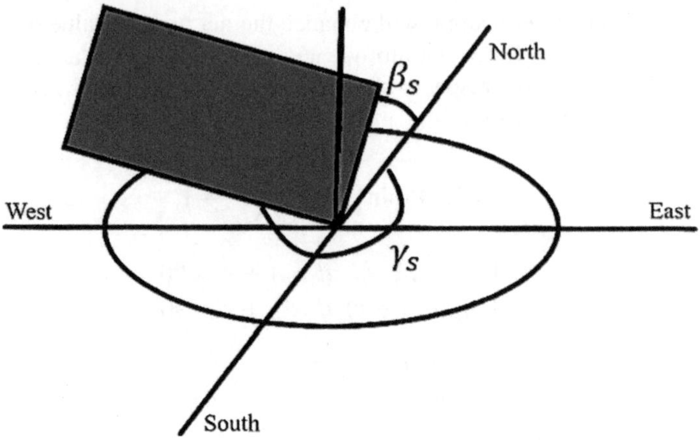

Fig. 2 PV placement with respect to tilt angle and azimuth

The assessment of solar radiation on an inclined surface depends on the values of direct radiation (HB), diffuse radiation (HD), and reflected radiation (HR). The total radiation (HT) can be calculated as follows:

$$H_T = H_B + H_{TD} + H_R \tag{11}$$

In this context, HB represents the monthly average daily radiation received on a horizontal surface, defined as follows:

$$H_B = (H - H_D)R_b \tag{12}$$

In this scenario, H represents the average monthly total radiation per day at the horizontal plane, while Rb serves as a functional cloudiness index defined as:

$$R_b = \frac{\cos(\varphi - \beta)\cos(\delta)\sin\left(\omega'_s\right) + \left(\frac{\pi}{180}\right)\omega'_s\ \sin(\varphi - \beta)\sin(\delta)}{\cos(\varphi)\cos(\delta)\sin(\omega_s) + \left(\frac{\pi}{180}\right)\omega_s\sin(\varphi)\sin(\delta)} \tag{13}$$

where;

φ Latitude
δ Angle of declination
ωs Angle of sunrise
ω'_s Tilted surface sunset angle

The angle of declination is set as;

δ 23.45sin[(360D + 284)/365]
ω_s cos 1 (tan(δ)tan(φ))

$$\omega' \mathrm{s} = \min[\omega \mathrm{s} = \cos 1(\tan(\varphi)\tan(\delta) \cos 1(\tan(\varphi\beta)\tan(\delta) \quad (14)$$

Surface reflection radiations along with the falls on the inclined surface is defined by;

$$\mathrm{HR} = \mathrm{H}\rho(1 - \cos\beta)/2 \quad (15)$$

Here,

β is the angle of tilt of the panels, and ρ represents the ground reflectance, which is commonly selected as 0.2.

The diffuse radiation on horizontal planes can be calculated using the following expression:

$$\mathrm{HD} = \mathrm{H}(1 + 1.13\mathrm{H}/\mathrm{H0}) \quad (16)$$

2.4 Net Present Value with Respect to Tilt Angle

The concept of net present value (NPV) encompasses the time value of money related to an investment. For instance, when analyzing a commercial solar investment, NPV is calculated to compare the projected cash flows over a 30-year period with their present-day value.

This assessment is significant as it takes into account future considerations such as inflation, associated risks, and the potential costs of foregoing other investment opportunities. The intrinsic value of solar energy systems is highlighted by their capacity for long-term financial returns. Solar panels can endure for many years, consistently generating savings.

However, it is crucial to understand that a dollar saved in the future, such as ten years hence, does not possess the same value as a dollar today, due to inflationary effects and the potential returns from investing that dollar elsewhere. This is where Net Present Value (NPV) becomes relevant.

NPV assesses various factors, including discount rates, inflation, and projected energy savings, to convert the future value of these savings into a single, present-day figure.

The net present value (NPV) is an established financial metric that articulates the worth of future income or savings derived from a solar installation.

This value, expressed in dollars, is computed by taking the total discounted future cash flows over the investment's duration and subtracting the initial investment cost, thereby yielding the present value of the expected future cash flows, adjusted for the discount rate.

Table 1 Illustrates the title of the angle the net present value

Tilt angle	Net present value
0	297
5	327
10	320
15	281
20	210
25	109

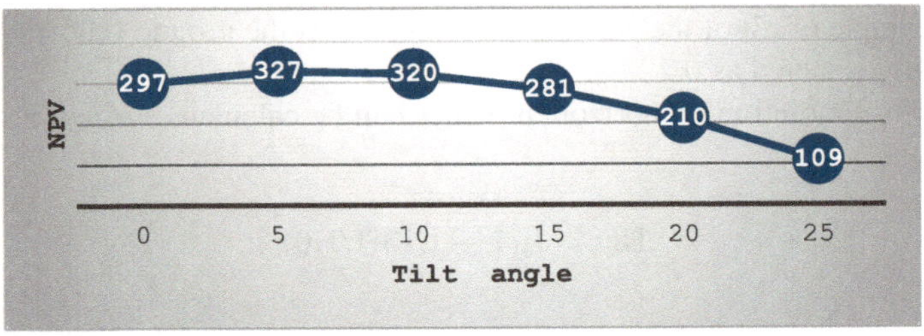

Fig. 3 Net present value with respect to tilt angle

In our analysis we selected a range of tilt angles and evaluated through a mathematical model the net present value. This is shown in both Table 1 and Fig. 3.

When the tilt angle was set at 0, the net present value showed savings of 297, whereas when the tilt angle was set at 25, we noticed a drop in the net present value of 109, indicating a crucial deduction of the NPV by 188. Therefore, from the obtained results it is crucial to carefully select the tilt angle to optimize the optimal NPV.

The analysis of Table 1 results was based on the accumulated data from Table 2.

Table 2 Parameters of PV panels

Location	Doha
DC PV size	21.4774 kW
AC size	18.138
Modules	39
Strings	3
Inverters	3
DC to AC ratio	1.18
Modules 39	5368.55
Inverters 3	4294.84
Balance of equipment	3221.13
Total capital cost	16,105.64
Cost per capacity	0.75
Operating cost	10 $/kW/y
Debt	75%
Loan terms	25 years
Loan rate	7%
Grid purchase price	
1–2000 kWh	0.0301372
2000 4000	0.0356164
4000 6000	0.0493151
6000 8000	0.0712329

3 Conclusion

The orientation of photovoltaic (PV) panels, specifically their tilt and azimuth angles, is a significant consideration during the installation of solar energy systems. To maximize the efficiency of PV panels, it is crucial to position them at angles that optimize their exposure to solar radiation. This paper conducts a comprehensive experimental and theoretical study to identify the most effective azimuth and tilt angles for PV panels. Current and voltage measurements were taken from the experimental setup, allowing for real-time calculations of the energy generated by the panels. Furthermore, a theoretical calculation was executed to determine the optimal tilt angle, based on both the collected data and the analytical findings of this research.

The objective of this research is to explore how consumer evaluations of the financial benefits associated with investments in solar PV and EES may differ when the future trajectory of the energy system is considered. Utilizing an electricity dispatch model to overcome the shortcomings in accuracy and perspective identified in earlier studies, we analyze the importance of factoring in possible future changes within the broader energy system. The results demonstrate that it is crucial to account for both short-term and long-term developments in the electricity system to effectively ascertain the value of distributed energy technologies for consumers making investment decisions.

References

1. Bilal Awan A, Khan MN, Zubair M, Bellos E (2020) Commercial parabolic trough CSP plants: research trends and technological advancements. Solar Energy 211:1422–1458
2. Taiea EAMA (2021) Simulation of the thermal power station at Alkuraymat using solar parabolic trough concentrators with thermal storage. Helwan University
3. Badawi A, Ali H, Elzein IM, Zyoud A, Abu-Hudrouss A (2023) Highly efficient pure sine wave inverter using microcontroller for photovoltaic applications. In: 2023 International symposium on networks, computers and communications (ISNCC). IEEE
4. Zubair M, Awan AB, Baseer MA, Khan MN, Abbas G (2021) Optimization of parabolic trough based concentrated solar power plant for energy export from Saudi Arabia. Energy Rep 7:4540–4554
5. Papaefthimiou S, Souliotis M, Andriosopoulos K (2016) Grid parity of solar energy: imminent fact or future's fiction?. Energy J Int Assoc Energy Econ 263–276. https://doi.org/10.5547/01956574.37.SI2.spap
6. Sarasa-Maestro CJ, Dufo-López R, Bernal-Agustín JL (2019) Evaluating the effect of financing costs on PV grid parity by applying a probabilistic methodology. Appl Sci 9(3):425
7. Yan J, Yang Y, Elia Campana P, He J (2019) City-level analysis of subsidy-free solar photovoltaic electricity price, profits and grid parity in China. Nat Energy 4(8):709–717
8. Bellini E (2020) Qatar's 800 MW tender draws world record solar power price of $0.01567/kWh
9. Başaran Filik Ü, Filik T, Gerek ÖN (2018) A hysteresis model for fixed and sun tracking solar PV power generation systems. Energies 11(3):603
10. Eldin SAS, Abd-elhady MS, Kandil HA (2016) Feasibility of solar tracking systems for PV panels in hot and cold regions. Renew Energy 85:228–233
11. Huang BJ, Huang YC, Chen GY, Hsu PC, Li K (2013) Improving solar PV system efficiency using one-axis 3-position sun tracking. Energy Proc 33: 280–287
12. Zubair M, Bilal Awan A (2021) Optimization of photovoltaic energy systems for residential customers in hot climate areas based on seasonal and average daily load profile. Energy Technol 9(7):2100036
13. Boßmann T, Staffell I (2015) The shape of future electricity demand: exploring load curves in 2050s Germany and Britain. Energy 90:1317–1333
14. Siddiqui JM (2015) Grid parity analysis of stand-alone hybrid microgrids: a comparative study of Germany, Pakistan, South Africa and the United States. Rochester Institute of Technology
15. Al-Aali I, Modi V (2022) Decarbonizing the electricity sector in Qatar using PV combined with ice thermal and battery storage. Energy Strategy Rev 44:101014
16. Khandakar A, Elzein IM, Nahiduzzaman M, Ayari MA, Ashraf AI, Korah L, Zyoud A, Ali H, Badawi A (2024) Smart aquaponics: an innovative machine learning framework for fish farming optimization. Comput Electr Eng 119:109590
17. Love DC et al (2015) Commercial aquaponics production and profitability: findings from an international survey. Aquaculture 435:67–74
18. Souza SV, Gimenes RMT, Binotto E (2019) Economic viability for deploying hydroponic system in emerging countries: a differentiated risk adjustment proposal. Land Use Policy 83:357–369
19. Mihoub S, Chermiti A, Beltagy H (2017) Methodology of determining the optimum performances of future concentrating solar thermal power plants in Algeria. Energy 122:801–810

20. Liu G, Rasul MG, Amanullah MTO, Khan MMK (2011) Feasibility study of stand-alone PV-wind-biomass hybrid energy system in Australia. In: Asia-Pacific power and energy engineering conference, APPEEC, no. March, pp 3–9. https://doi.org/10.1109/APPEEC.2011.5749125
21. Darling SB, You F, Veselka T, Velosa A (2011) Assumptions and the levelized cost of energy for photovoltaics. Energy Environ Sci 4(9):3133–3139
22. Zubair M, Awan AB, Praveen RP, Abdulbaseer M, Praveen RP, Abdulbaseer M (2019) Solar energy export prospects of the Kingdom of Saudi Arabia. J Renew Sustain Energy 11(4)
23. Perez R, Seals R, Zelenka A, Ineichen P (1990) Climatic evaluation of models that predict hourly direct irradiance from hourly global irradiance: prospects for performance improvements. Sol Energy 44(2):99–108
24. Hasan K, Yousuf SB, Tushar MS, Das BK, Das P, Islam MS (2022) Effects of different environmental and operational factors on the PV performance: a comprehensive review. Energy Sci Eng 10(2):656–675
25. Manzoor HU, Aaqib SM, Manzoor T, Azeem F, Ashraf MW, Manzoor S 2025) Effect of optimized tilt angle of PV modules on solar irradiance for residential and commercial buildings in different cities of Pakistan: simulation-based study. Energy Sci Eng

Leveraging Emerging Technologies for Enhanced Food Security: Precision Agriculture, Vertical Farming, and Blockchain-Enabled Supply Chains

Ahmed A. Abdalla, Yasen Harrye, Hassan Ali, and Hassan Mahasneh

Abstract This paper examines developing technologies to address universal food security challenges, concentrating on precision agriculture, vertical farming, and blockchain-enabled supply chains. Through a comprehensive literature analysis and review of case studies, we examine how these technologies can increase agricultural productivity, optimize resource use, and enhance supply chain transparency. We discuss the synergies created by integrating these technologies and identify potential barriers to their widespread adoption. Our conclusion recommends that while these technologies offer significant promise, their implementation must be accompanied by careful consideration of economic, social, and ethical implications, particularly in developing world contexts.

Keywords Precision agriculture · Vertical farming · Blockchain · Food security · Sustainable agriculture

1 Introduction

According to the definition of the United Nations, food security is a state in which all people always have physical, social, and economic access to sufficient safe and nutritious food continues to be a major challenge for humanity. As the global population rises to 9.7 bn in 2050, up by more than a third from an estimated 7.3bn today, the food supply will need to grow by about 70% worldwide of what was produced in 2010 [1]. These evolving technologies offer promising solutions to enhance productivity, sustainability, and resilience in our food systems [2]. This paper explores three emerging technologies that show significant promise in addressing challenges in food security: Precision Agriculture (PA), Vertical Farming (VF), and Blockchain-Enabled

A. A. Abdalla (✉) · Y. Harrye · H. Ali · H. Mahasneh
College of Engineering and Technology, University of Doha for Science and Technology, Doha, Qatar
e-mail: ahmed.abdalla@udst.edu.qa

S. Al-Naemi et al. (eds.), *Water and Food Security in the Face of Climate Change: Challenges and Opportunities for Resilience*, World Sustainability Series,
https://doi.org/10.1007/978-3-032-00098-9_43

Supply Chains. Precision Agriculture utilizes data-driven techniques and technologies to optimize crop yields and resource use. Vertical Farming implies producing food in vertically stacked layers, frequently in controlled urban environments [2]. Blockchain al-lowed Supply Chains to interact with distributed ledger technology to improve transparency and traceability in food distribution networks [3].

The paper begins with a comprehensive literature review covering the current state of food security and the three focal technologies. As explained by the United Nations, food security is a state in which all people always have physical, social, and economic access to sufficient, safe, and nutritious food. Nevertheless, reaching this state globally remains a significant challenge, with climate change posing a severe threat to food systems through rising temperatures and unpredictable rainfall patterns that stress traditional agricultural practices. Food security utilizes technologies such as IoT sensors, drones, and artificial intelligence to optimize resource use and maximize crop yields.

Author [4] the responsibility of big data in smart farming is highlighted, emphasizing its potential to enhance decision-making processes in agriculture. Vertical farming is presented as a potential solution to land scarcity and urban food production challenges. The thought of farming up the city is outlined, along with a review of developments and implications for vertical cities. Key technologies enabling vertical farming include LED lighting systems, soilless growing techniques, and advanced climate control systems (Fig. 1).

2 Precision Agriculture

Precision Agriculture (PA), also known as precision farming or smart farming, represents a paradigm shift in agricultural practices. It is an approach to farm management that uses information technology and a wide array of items such as control systems, sensors, robotics, drones, GPS, autonomous vehicles, variable rate technology, GPS-based soil sampling, automated hardware, telematics, and software [2–4, 6, 7]. The main goal of PA is to guarantee profitability, efficiency, and sustainability while protecting the environment [8]. This is achieved through the precise application of agricultural inputs with respect to soil and crop requirements [9]. Open ocean aquaculture, also known as offshore aquaculture, involves cultivating fish and other aquatic species in the open sea, typically several kilometers from the coast and in deeper waters compared to traditional coastal aquaculture. Open ocean aquaculture offers the potential to overcome many of the constraints associated with coastal and land-based systems, while aligning with Qatar's goals of sustainable development and food security. The move towards offshore aquaculture is also in line with global trends in the sector, where many countries are looking to deeper waters to sustainably increase their aqua-culture production [10].

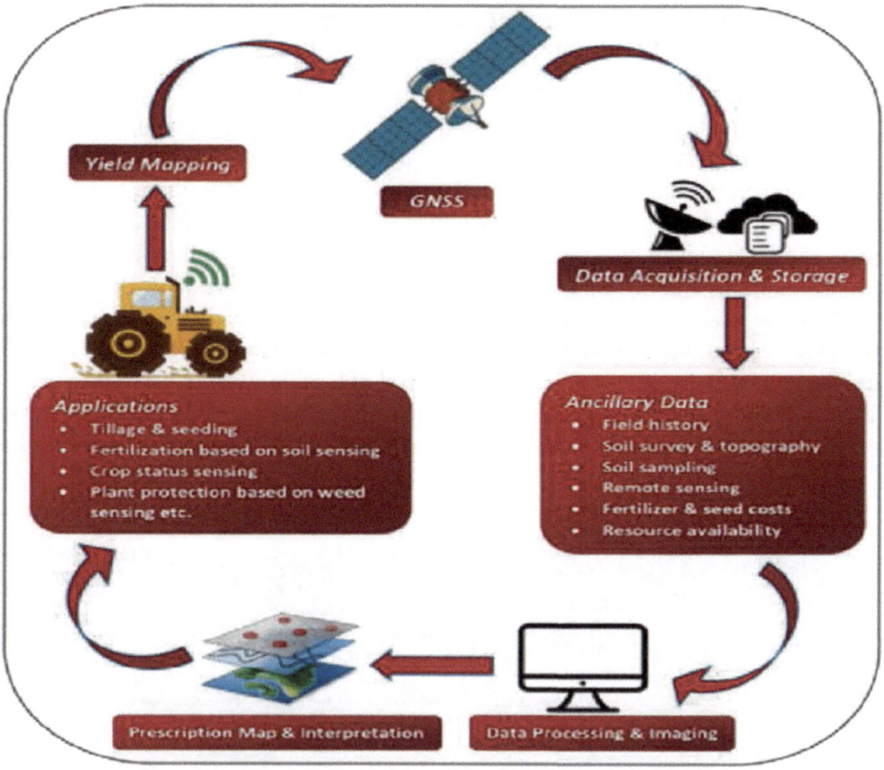

Fig. 1 Tool for precision farming [5]

2.1 Key Technologies and Applications

Precision Agriculture (PA) is therefore a shift in the way farming is managed through an application of information technology to crop production and resource use. Some of the components defining them include Internet of Things (IOT) sensors, drones and satellite imagery, artificial intelligence (AI) and machine learning (ML), Global positioning System (GPS) and Geographic Information System (GIS) technologies. Together these technologies enable real time control of the soil conditions, crop health, yield prediction and location specific application of inputs. Now, advanced generation PA technologies as mentioned earlier have sparked towards the enhancement of PA efficiency. For instance, one research showed that the use of AI in handling crops in PA systems was discovered capable of delivering 20% yield increment and only 15% resource input [2, 5].

2.2 Case Studies and Impact on Food Security

In this section, several examples have been given to establish how PA impacts food security. Researchers in the United Stated Corn belt region variable rate application (VRA) of nitrogenous fertilizer depending on the data gathered from the analysis of the soil and crop to enhance yields cut down overall nitrogen fertilizer usage across the trials [6]. In order to save up to 20% of water of Californian almond orchards, new types of precision irrigation technologies were used, including soil moisture sensors, weather stations, and raw images gathered by drones to increase the yield, or at least not decrease it [7]. An Intrusion Detection Systems (IDS) based on an IoT device for early warning system on blister blight disease in Indian Tea Plantation reduced disease incidence by 60% and improved overall of tea quality by 15% [8–11]. These cases show how PA can improve productivity, minimize resource consumption and promote food production systems that are more robust, thereby and thus supporting food security, directly (Fig. 2).

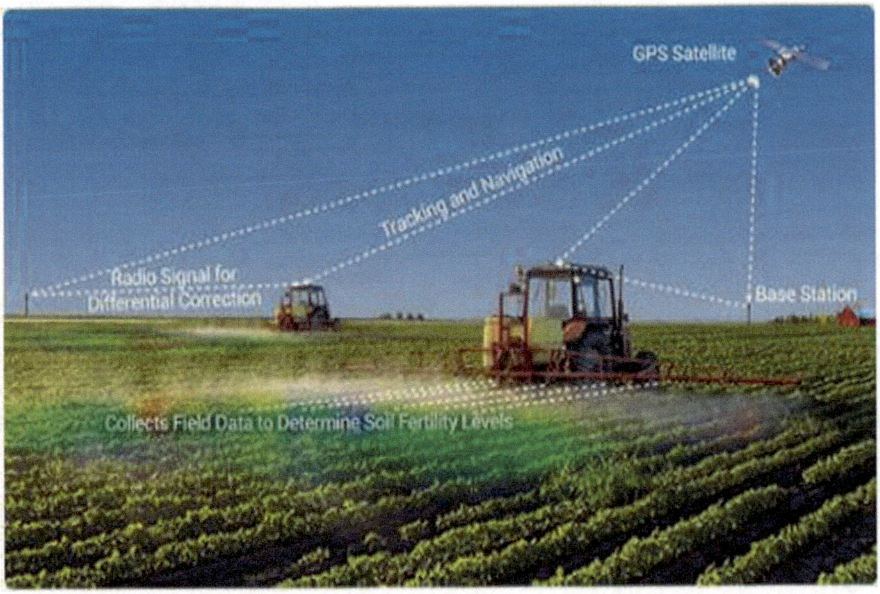

Fig. 2 Precision agriculture [6]

3 Vertical Farming

Vertical farming represents a revolutionary approach to agriculture that aims to produce food in vertically stacked layers, often integrated into other structures such as skyscrapers, repurposed warehouses, or shipping containers. The core principle of vertical farming is the maximization of crop output per square foot of land by growing crops in vertically stacked layers, often without soil, using various techniques such as hydroponics, aeroponics, or aquaponics [3, 12, 13].

3.1 Principles and Enabling Technologies

Vertical farming refers to a superior kind of farming that suits the climatic conditions of urban areas. It holds them as a way of growing crops from one above the other layers and can be in controlled systems. Main technologies being used in vertical farming includes light emitting diode light, hydroponics and aeroponics, climate control, and automation and Robotics [14].

Recent research has revealed a lot of progress in these technologies. A study showed that customized LED spectra could enhance the yield of lettuce by 30% to nor-mal lighting [15]. Scientists also discovered that hydroponic systems in vertical farms required 90–95% less water as compared to traditional soil-based agriculture for lettuce [16, 17]. Intelligent system solutions for climate control based on IoT have been demonstrated to have versatility and save up to 30% on energy while producing the optimum growing temperatures [18]. In the automation domain, Researchers designed a robotic harvesting system for the leafy greens in vertical farms with a better yield rate of success and shrinking the time required for harvesting as compared to manual operations.

3.2 Benefits for Urban Food Production

By applying this technologically advanced system of growing crops that is compatible with dense population structures, there are several benefits that vertical farming has. Conventional agriculture ceases to obtain the productivity per unit area, while integrated organic agriculture does so for some crops [19, 20]. People say that controlled environment systems enable production of crops even when the weather is unfavorable. Closeness to markets eliminates the need for transportation thus reducing carbon footprint tremendously, and a comparative analysis shows that vertical farms could potentially cut transport costs on locally consumed [20]. Further, the use of recirculation water systems in place of open hydroponic systems employed in vertical farming, Water use may be cut in traditional soil based agricultural production [21,

22]. These advantages make Vertical farming as an important key in developing food security in cities where lands are limited but population density is high.

4 Blockchain-Enabled Supply Chains

Blockchain is already being identified as a valuable means to transform supply chains in the food sector. Such a distributed ledger offers an effective means of tracking food products from producers to consumers through a sequence of transactions in the network and cannot be easily altered. With regards to food supply chain management, blockchain presents the promise to solve problems like traceability, food fraud, ineffective logistics, and inventory tracking. In essence, the decentralized ledger, which records every interaction and process surrounding a food item, can offer transparency and credibility among all the parties involved, and provide consumers with necessary details regarding the product and how it has been processed [23].

4.1 Applications in Food Traceability and Safety

The experience shows that Blockchain technology is one of the effective tools to introduce novelties in such essential and critical for consumers' health products as food supply chains targeting more effective traceability and transparency along with Food safety. Some significant applications include identifying and locating through the chain of supply, in smart contacts, and for customers. However, a study at the beginning of this year identified that using blockchain technology in food supply chain networks would significantly support the fight against food frauds and enhance customers' confidence [24].

4.2 Case Studies and Impact on Efficiency

This segment provides several examples of how food supply chain effectiveness is affected by blockchain. Walmart's implementation of the IBM Food Trust blockchain for tracing mangoes reduced the time required to trace the fruit's origin from 7 days to 2.2 s [24]. The BeefChain initiative in Wyoming enables consumers to access detailed information about an animal's life cycle by scanning a QR code, with research indicating an increase in consumer willingness to pay for products with blockchain-verified sustainability claims [25]. Nestlé's blockchain system for palm oil traceability could trace palm oil to its source with 99.9% accuracy, while reducing the time required for traceability checks by 70% [24]. These case studies highlight blockchain's potential to enhance transparency, reduce food fraud, and improve overall supply chain efficiency, contributing to more secure and resilient food systems.

5 Integration and Synergies

5.1 Potential for Combined Impact on Food Security

The integration of precision agriculture technology combined with vertical farms and blockchain supply chains exhibits potential approaches towards tackling food insecurity. An insightful analysis predicted that the expansion of both precision agriculture integrated with vertical farming techniques together with block chain supported supply chains would uplift the condition of food self-sufficiency in most of the world's mega cities [26]. Integrated systems were also less vulnerable to supply chain disruption than traditional agricultural systems: reasons included localized production and improved food tracking [27, 28].

5.2 Future Research Directions

The following major areas seem to emerge as important for future endeavor. These include artificial intelligence and machine learning, IoT and 5G, improved sensor system, and autonomous systems. Farther development of complex AI models to address gain inherent integrated agricultural systems issues, from diseases forecasts to supply chain issues must also be a goal. Another key finding has to do with exploring how High-Speed Link (HSL) networks can be optimistically used to develop more improved agricultural systems. However, to realize the full potential of the vertical farm and develop profitability, it is necessary to create new sensors based on advanced technologies that are cheap, but also resistant to the conditions of specific facilities, and allow obtaining current and accurate data [9, 29]. Finally, further developments of integrating the autonomous systems with blockchain and AI technologies for fully automated and self-managing agricultural environment is possibly the most appealing idea for the future works.

6 Economic Considerations and Feasibility

The adoption of precision agriculture (PA), vertical farming (VF), and blockchain-enabled supply chains presents significant economic considerations, particularly in developing countries. A comprehensive cost–benefit analysis of implementing these technologies reveals both challenges and opportunities. For precision agriculture, the initial investment can be substantial for a small farm. However, studies show that PA can reduce input costs and increase crop yields potentially resulting in a return on investment within few years [9]. This suggests that despite high upfront costs, PA can offer significant economic benefits in the medium to long term.

Vertical farming, while requiring a high initial capital, can produce up to 350 times more crop per square meter than traditional farming [30]. This remarkable increase in productivity could potentially justify the high startup costs, especially in urban areas with limited agricultural land. However, energy costs remain a significant challenge for VF operations. This highlights the need for continued research and development in energy-efficient growing systems and the potential integration of renewable energy sources to make VF more economically viable.

Blockchain-enabled supply chains present a different set of economic considerations. Implementation costs can vary widely pending and location of the company [31]. While this represents a significant investment, the potential benefits are substantial. Block-chain technology can help reduce food fraud, estimated at $30–40 billion annually glob-ally, and improve traceability, potentially saving billions in recall costs [32–34]. These economic benefits extend beyond individual companies to the entire food system, enhancing overall efficiency and trust.

Despite the potential benefits, scalability remains a key challenge, particularly in developing countries. Factors affecting adoption include limited access to capital and financing options, lack of technical expertise and training, inadequate infrastructure (such as reliable electricity and internet connectivity), and regulatory barriers and lack of standardization. To address these challenges, public–private partnerships and targeted government initiatives are crucial. For example, the African Development Bank's "Technologies for African Agricultural Transformation" (TAAT) program has shown promise in scaling up agricultural technologies across the continent [35]. Such initiatives can help bridge the gap between the potential of these technologies and their practical implementation in diverse economic contexts.

7 Social and Ethical Implications

Labor market disruption is identified as a major concern, with automation potentially affecting agricultural employment significantly by 2030, though this can be mitigated through worker training programs, entrepreneurship opportunities in agritech, and equitable management strategies. Data rights and protection present another crucial challenge, particularly regarding farmer ownership, fair competition, and the risk of corporate monopolies in the agricultural sector, with potential regulation through frameworks like GDPR. Environmental concerns form the third major issue, including soil disruption from PA sensors, high energy consumption in VF operations, and blockchain's carbon footprint, necessitating life cycle assessments (LCA) and circular economy strategies to ensure these technologies truly contribute to sustainability rather than harm it.

8 Policy Recommendations

The successful implementation of digital agricultural technologies requires a comprehensive policy framework that encompasses several key elements: country specific national strategies with adequate funding for research and development, robust data protection rules that balance innovation with farmers' rights, investments in rural digital infrastructure including communication and electricity access, enhanced educational and vocational training programs, international knowledge sharing on agritech adoption, measures to ensure fair transition for affected workers through retraining and wage subsidies, support for standardized technological interfaces to prevent vendor lock-in, and policies to encourage smallholder adoption through subsidies and loans. If these economic, social, and ethical considerations are effectively addressed through proper policies regarding Precision Agriculture, Vertical Farming, and Blockchain-enabled Supply chains, it can lead to improved global food security while ensuring fairness and value creation throughout the food supply chain.

9 Conclusion

Precision agriculture, vertical farming, and blockchain supply chains offer innovative solutions to global food insecurity by optimizing resource use, addressing land scarcity, and ensuring food trust, but face significant implementation challenges. These include high initial costs, potential job displacement, data privacy concerns, and environmental implications, particularly affecting developing countries. While these technologies could create efficient, traceable, and sustainable food networks, their successful implementation requires a comprehensive policy system that includes national digital agriculture strategies, rural infrastructure investment, educational initiatives, and data protection frameworks, with special attention to small-scale farmers. The success of these emerging technologies ultimately depends on balancing technological possibilities with economic viability, social impact, and environmental sustainability, requiring international collaboration, scientific research, and coordinated policy approaches to create equitable and sustainable food systems worldwide.

References

1. Galanakis CM (2024) The future of food. Foods 13.4:506
2. Erekath S et al. (2024) Food for future: exploring cutting-edge technology and practices in vertical farm. Sustain Cities Soc 105357

3. Shaikh TA, Rasool T, Lone FR (2022) Towards leveraging the role of machine learning and artificial intelligence in precision agriculture and smart farming. Comput Electron Agricul 198:107119
4. Fuentes-Peñailillo F et al. (2024) Transformative technologies in digital agriculture: leveraging Internet of things, remote sensing, and artificial intelligence for smart crop management. J Sensor Actuator Netw 13.4:39
5. Mani PK (2020) Remote sensing and geographic information system: a tool for precision farming. https://doi.org/10.1007/978-981-15-6864-0_2
6. vervetronics.com (n.d) precision-agriculture http://www.vervetronics.com/solutions/iot-smart-agriculture/precision-agriculture/
7. Ahmad L, Mahdi SS (2018) In: Satellite farming. Basel, Switzerland,: Springer International Publishing
8. Soussi A et al. (2024) Smart sensors and smart data for precision agriculture: a review. Sensors 24.8:2647
9. Karunathilake EMBM et al. (2023) The path to smart farming: Innovations and opportunities in precision agriculture. Agriculture 13.8:1593
10. Abbas A et al. (2023) Drones in plant disease assessment, efficient monitoring, and detection: a way forward to smart agriculture. Agronomy 13.6:1524
11. Liu Z et al. (2022) Internet of Things (IoT) and machine learning model of plant disease prediction–blister blight for tea plant. IEEE Access 10:44934–44944
12. Konar M, Ding H (2023) A sustainable ocean economy for 2050: approximating its benefits and costs. In: The blue compendium: from knowledge to action for a sustainable Ocean economy. Cham,: Springer International Publishing, pp 681–714
13. Deshmukh S, Thejo Kalyani N, Raghorte S (2024) Novel approach of using LEDs to grow indoor lettuce hydroponically. In: Sustainable food systems (Volume I) SFS: framework, sustainable diets, traditional food culture and food production. Cham,: Springer Nature, Switzerland, pp 443–452
14. Ahamed MS et al. (2023) A critical review on efficient thermal environment controls in indoor vertical farming. J Cleaner Prod 138923
15. Gumisiriza MS et al. (2022) Building sustainable societies through vertical soilless farming: a cost-effectiveness analysis on a small-scale non-greenhouse hydroponic system. Sustain Cities Soc 83:103923
16. Gao Y, Wang J, Yiu TW (2024) Multi-information integration-based life cycle analysis of greenhouse gas emissions for prefabricated construction: a case study of Shenzhen. Environ Impact Assessm Rev 104:107330
17. Nižetić S et al. (2020) Internet of things (IoT): opportunities, issues and challenges to- wards a smart and sustainable future. J Cleaner Prod 274:122877
18. Steensland A, Zeigler M (2021) Productivity in agriculture for a sustainable future
19. Fabiani S et al. (2020) Assessment of the economic and environmental sustainability of variable rate technology (VRT) application in different wheat intensive European agricultural areas. A Water energy food nexus approach. Environ Sci Policy 114:366–376
20. Jia X et al. (2024) Chitin and chitosan: pioneering sustainable substrates for next-generation soilless vertical farming. Trends in Food Sci Technol 104599
21. Kisekka I (2023) Orchard water management. In: Advanced automation for tree fruit or-chards and vineyards. Cham,: Springer International Publishing, pp 59–74
22. Sharma R et al. (2024) Rapid and sensitive approaches for detecting food fraud: a review on prospects and challenges. Food Chem 139817
23. Matloob R et al. (2024) Food traceability 4.0. In: Food industry 4.0. Academic Press, pp 83–97
24. Zhao Q, Fan Z-P, Sun M (2024) Sales mode selection and blockchain technology adoption decisions in a platform supply chain. Int J Prod Econ 272:109255
25. Smit L et al. (2021) Human rights due diligence in global supply chains: evidence of corporate practices to inform a legal standard. The Int J Human Rights 25.6:945–973
26. Kabir MSN et al. (2023) Technological trends and engineering issues on vertical farms: a review. Horticulturae 9.11:1229

27. Duan Y (2024) Blockchain and enhancing online retail consumer value. IEEE Eng Managem Rev
28. Merfield CN (2023) Integrated weed management in organic farming. In: Advances in resting-state Functional MRI. Woodhead publishing, pp 31–109
29. Anirudh KC, Dixit M, Karat S (2024) Exploring the applications of wireless sensor networks and the internet of things in agricultural and livestock production and management. In: Emerging technologies and the application of WSN and IoT. CRC Press, pp 209–227
30. Parameswari P et al. (2024) Vertical farming: revolutionizing sustainable agriculture in the 21st century. J Scient Res Reports 30.5:917–930
31. Srikanth M, Jagan Mohan RNV (2020) Stop spread corona based on voice, face and emo-tional recognition using machine learning, query optimization and block chain technology. Solid State Technol 63.6:3512–3520
32. Birkby J (2016) Vertical farming. ATTRA Sustain Agricul 2(1):1–12
33. Zhang Y, Lee S, van de Ligt JLG (2019) Blockchain technology and the potential applicability in the feed industry
34. Akram MW et al. (2024) Blockchain technology in a crisis: advantages, challenges, and lessons learned for enhancing food supply chains during the COVID-19 pandemic. J Cleaner Prod 434:140034
35. Mulei WM, Maiyo JK (2024) Transitioning technology deployment to adoption: implementation challenges of the TAAT program in Western Kenya. Gates Open Res 8.66:66

Open Ocean Aquaculture Vessels: An Emerging Frontier for Sustainable Fish Production that Can Meet Sea Food Demand in the State of Qatar

Yasen Harrye, Ahmed A. Abdalla, and Hassan Mahasneh

Abstract This paper explores the potential of open ocean aquaculture as a sustainable solution to meet the growing demand for seafood in Qatar. The study analyses the current state of aquaculture in Qatar, the challenges, opportunities presented by offshore farming, and the technological and regulatory frameworks necessary for its successful implementation. Findings suggest that open ocean aquaculture could significantly contribute to Qatar's food security and economic diversification goals while minimizing environmental impacts associated with traditional coastal aquaculture.

Keywords Offshore aquaculture · Qatar · Sustainable seafood production · Food security · Marine farming · Aquaculture technology · Environmental impact · United nations sustainable development goals (SDGs)

1 Introduction

Qatar, like many countries in the Gulf Cooperation Council (GCC) countries, faces significant challenges in ensuring food security due to its arid climate and limited arable land. The country's "Qatar National Vision 2030" emphasizes the need for sustainable development and economic diversification [1]. Qatar's fishing industry plays a crucial role in the country's economy and food security strategy. As a peninsula nation with a 563 km coastline, Qatar has a long-standing tradition of fishing [2]. In recent years, the industry has undergone significant transformations driven by technological advancements, changing consumer preferences, and increasing environmental concerns. The fishing industry in Qatar includes both capture fisheries and aquaculture, with the latter gaining increasing importance in recent years. The total fish production in Qatar reached 17,171 tons in 2020, with capture fisheries

Y. Harrye (✉) · A. A. Abdalla · H. Mahasneh
College of Engineering and Technology, University of Doha for Science and Technology, Doha, Qatar
e-mail: yasen.harrye@udst.edu.qa

S. Al-Naemi et al. (eds.), *Water and Food Security in the Face of Climate Change: Challenges and Opportunities for Resilience*, World Sustainability Series,
https://doi.org/10.1007/978-3-032-00098-9_44

accounting for most of this production [3]. According to recent market analysis, the Qatar aquaculture market is projected to grow at a Compound Annual Growth Rate (CAGR) of 6.2% during the forecast period of 2023–2028 [4]. This growth is driven by several factors, including government initiatives to enhance food security, increasing demand for seafood, and the need to reduce reliance on imports.

Fish and seafood occupy a central position in the culinary landscape and nutritional profile of Qatar's population, reflecting a deep-rooted cultural affinity with marine resources that has been shaped by the nation's geographical context and historical development. The Arabian Gulf's rich marine biodiversity has long been a cornerstone of Qatari cuisine, with traditional dishes such as machboos, featuring locally sourced fish as primary ingredients. This culinary heritage not only preserves cultural identity but also contributes significantly to the nutritional well-being of the populace. Recent data from the Food and Agriculture Organization (FAO) corroborate the substantial role of fish in the Qatari diet, indicating an average per capita fish consumption of approximately 24 kg per year, which notably exceeds the global average of 20.5 kg per year [5]. This elevated consumption pattern underscores the nutritional and cultural significance of fish in Qatar, while simultaneously highlighting the potential pressures on local and regional fish stocks.

The high fish consumption rate in Qatar can be attributed to a confluence of factors, including cultural preferences, nutritional awareness, and economic prosperity. The country's rapid economic development in recent decades has led to increased purchasing power and access to a diverse range of seafood products, both locally sourced and imported. Moreover, the growing emphasis on healthy eating habits and the well-documented nutritional benefits of fish consumption, particularly in terms of omega-3 fatty acids and high-quality proteins, have further reinforced the importance of seafood in Qatari diets. However, this substantial reliance on fish resources also presents challenges in terms of sustainability and food security. The high demand places considerable pressure on local fish stocks, potentially leading to overfishing and ecosystem disturbances in the Gulf waters. This situation is further complicated by the impacts of climate change, including rising sea temperatures and ocean acidification, which pose threats to the long-term viability of marine ecosystems in the region.

In response to these challenges and to ensure a sustainable supply of fish to meet domestic demand, Qatar has been actively pursuing strategies to enhance its aquaculture sector. The development of sustainable aquaculture practices, including the exploration of open ocean aquaculture technologies, represents a proactive approach to reconciling the nation's high fish consumption rates with environmental sustainability goals. Furthermore, the integration of traditional fishing practices with modern aquaculture techniques offers potential pathways for preserving cultural heritage while addressing contemporary food security challenges. As Qatar continues to navigate the complexities of maintaining its fish-centric dietary patterns in an era of environmental change and resource constraints, the interplay between cultural traditions, nutritional needs, and sustainable resource management will remain a critical area for ongoing research and policy development.

As of 2020, the domestic fish production only accounted for about 20% of its total fish consumption, with the remaining 80% being met through imports [6]. This heavy reliance on imports not only impacts food security but also exposes the country to potential supply chain disruptions and price volatilities in the global seafood market. Besides, the fish stocks in the Arabian Gulf, Qatar's traditional fishing grounds, are under increasing pressure. Overfishing, habitat degradation, and climate change induced alterations in marine ecosystems have led to declining catches in recent years [7]. The FAO has reported that several key commercial fish species in the Gulf are either fully exploited or overexploited, raising concerns about the long-term sustainability of traditional capture fisheries in the region [8]. These challenges highlight the urgent need for Qatar to develop sustainable alternatives to meet its growing seafood demand. The country's limited coastline and the environmental sensitivities of its near shore areas further complicate the expansion of traditional aquaculture methods. In this context, open ocean aquaculture emerges as a promising avenue to enhance domestic food production while aligning with country sustainability goals.

The establishment of Qatar's first commercial fish farm in 2015 marked a significant milestone in the country's aquaculture journey [9]. This initial venture, as shown in Fig. 1a is a collaboration between the Ministry of Municipality and Environment and a private sector partner, focused on the production of sea bream and sea bass in offshore cages near Ras Matbakh. Since then, the sector has expanded considerably, with offshore aquaculture project, Samkna, using floating-cage technology to produce seabass for the local market [10]. The fish farm is located 50 km off the Qatari coast, northeast of Ruwais, covering an area of 900,000 square meters with 16 floating cages as shown in Fig. 1b. To support this operation, the company has built a 10,000 square meter factory in Ruwais for final production stages, including packaging and wrapping. This growth is primarily driven by land-based and near-shore operations, which currently dominate the country's aquaculture landscape.

The selection of species for aquaculture in Qatar represents a strategic confluence of environmental adaptability, market demand, and technological capability. This careful curation of cultivated species reflects the nation's commitment to developing a robust and sustainable aquaculture sector that aligns with both ecological constraints and consumers' preferences. The primary species currently under cultivation in Qatar's aquaculture facilities have been chosen based on a comprehensive assessment of their suitability to local environmental conditions, market acceptability, and economic viability [12, 13].

Sea Bream (Sparus Aurata) has emerged as a cornerstone species in Qatar's aquaculture portfolio. Its popularity in regional cuisines, coupled with its adaptability to both land-based and near-shore cage systems, positions it as an adaptable option for various cultivation methodologies. The species' tolerance to the high salinity levels characteristic of the Arabian Gulf makes it particularly well-suited to Qatar's marine environment. Additionally, the established market demand for Sea Bream in the region ensures a stable economic foundation for its cultivation, contributing to the overall sustainability of the aquaculture sector [12]. Barramundi (Lates Calcarifer), also known as Asian sea bass, represents another significant species in Qatar's aquaculture landscape. The selection of Barramundi is notable for its euryhaline

(a) (b)

Fig. 1 **a** Ras Matbakh, aquatic research center [11] **b** Samkna, floating-cage technology and offshore housing units for the staff [10]

nature, allowing it to thrive across a spectrum of salinity levels. This physiological adaptability renders Barramundi suitable for cultivation in both brackish and marine environments, offering flexibility in site selection and production methods. The species' rapid growth rate and high market value further enhance its appeal for commercial aquaculture operations. Thus, the successful cultivation of Barramundi in Qatar demonstrates the potential for introducing non-native species that are well adapted to local environmental conditions, thereby diversifying the nation's aquaculture output [13].

The inclusion of Tilapia in Qatar's aquaculture portfolio showcases the country's commitment to leveraging advanced technologies in freshwater aquaculture. Cultivated primarily in land-based recirculating aquaculture systems (RAS), Tilapia production exemplifies the integration of innovative aquaculture technologies with species selection. The use of RAS for Tilapia cultivation is particularly significant in the context of Qatar's arid climate and limited freshwater resources, demonstrating how technological solutions can overcome environmental constraints. Tilapia's rapid growth rate, efficient feed conversion, and broad market acceptance make it an economically viable option for intensive aquaculture production [13]. In addition to these primary species, Qatar's aquaculture sector has demonstrated a commitment to diversification through experimental cultivations of other species such as grouper, sobaity bream, and hammour. These initiatives reflect a forward-thinking approach to aquaculture development, aimed at expanding the sector's resilience and market offerings. The cultivation of these species, which hold significant cultural and culinary importance in the region, aligns with efforts to cater to local preferences while potentially reducing pressure on wild fish stocks.

The pivotal role of the Aquatic Research Centre in supporting the growth and diversification of Qatar's aquaculture sector cannot be overstated. This institution serves as a hub for research, development, and knowledge dissemination, facilitating the translation of scientific insights into practical aquaculture applications. The Centre's work incorporates a wide range of activities, including broodstock management, larvae rearing techniques, feed formulation, and disease management strategies tailored to the specific needs of Qatar's aquaculture species and environmental conditions. The cultivation of these species will play a crucial role in achieving the nation's goals of food security, economic diversification, and environmental sustainability.

2 Challenges

Despite the promising start, the expansion of coastal aquaculture in Qatar faces several significant limitations. The coastline, while extensive, is not uniformly suitable for aquaculture development. Factors such as water depth, currents, and proximity to sensitive ecosystems (e.g. mangroves and coral reefs) restrict the areas available for near-shore aquaculture [14]. The coastal zones are increasingly busy, with competing interests from various sectors. Urban development, particularly in areas around Doha, continues to expand towards the coast. In addition, near-shore aquaculture operations can have significant environmental impacts if not properly managed, due to effluent discharge, habitat iteration, genetic interactions, water quality issues and limited fresh water. The release of nutrient rich water from fish farms can lead to eutrophication in coastal areas, potentially harming local ecosystems [15], while construction of aquaculture facilities and associated infrastructure can disrupt coastal habitats, including sensitive areas like seagrass beds. The potential impact of farmed fish and issues of it interacting with wild populations was discussed in [16]. In addition, the coastal water is characterized by high temperatures and varied salinity levels. These conditions can be challenging for aquaculture, particularly during summer months when temperatures can exceed optimal levels for many species [17], and due to scarce freshwater resources, posing challenges for land-based aquaculture systems requiring a significant water input, particularly for freshwater species like tilapia [18].

Open ocean aquaculture, also known as offshore aquaculture, involves cultivating fish and other aquatic species in the open sea, typically several kilometers from the coast and in deeper waters compared to traditional coastal aquaculture. Open ocean aquaculture offers the potential to overcome many of the constraints associated with coastal and land-based systems, while aligning with Qatar's goals of sustainable development and food security. The move towards offshore aquaculture is also in line with global trends in the sector, where many countries are looking to deeper waters to sustainably increase their aquaculture production [19, 20].

Figure 2, shows Ocean Farm 1 deployed about 5 km off the central Norwegian coast, showcasing engineering principle adapted from submersible offshore units. With an impressive diameter of 110 m and a volume of 250,000 m^3, Ocean Farm 1 can accommodate up to 1.5 million salmon. The structure is designed for installation in water depths ranging from 100 to 300 m and boasts a 25-year lifespan.

Fig. 2 Semi-submersible ocean farm 1 fish farm rig [21]

Its advanced features include over 20,000 sensors and more than 100 monitors and control units, enabling precise management of the farming environment. The pen can be submerged by filling its ballast tanks and is securely moored using eight lines attached to the lower part of its side vertical columns. Ocean Farm 1 utilizes Econet, an innovative netting material using polyethylene terephthalate. This material offers advantages such as resistance to marine fouling, ease of cleaning, improved durability, and enhanced prevention of fish escapes. Ocean Farm 1 stands as a testament to the potential of large-scale, technologically advanced offshore aquaculture solutions [21].

The transition of aquaculture operations from coastal to offshore environments presents a complex array of challenges that span environmental, operational, and design considerations. These challenges necessitate innovative solutions and interdisciplinary approaches to ensure the viability and sustainability of offshore aquaculture. The environmental challenges inherent in offshore fish farming are particularly pronounced, given the dynamic and often unpredictable nature of open ocean conditions. Offshore fish pens must contend with greater water depths, stronger currents, and more severe wave action than their coastal counterparts. These factors not only influence the structural integrity of the aquaculture installations but also have significant implications for fish welfare and overall production efficiency [22].

The design of offshore aquaculture systems must account for varied seabed conditions, which can affect the anchoring and stability of fish pens. This variability necessitates adaptive engineering solutions that can withstand the diverse and often extreme conditions encountered in offshore environments. Moreover, the design challenges extend beyond basic structural considerations to include the creation of a conducive environment for fish growth and welfare within these high energy marine settings. This includes maintaining optimal water quality parameters, ensuring adequate feed distribution, and mitigating the stress factors associated with exposure to open ocean conditions [23]. Operational challenges in offshore aquaculture are numerous and complex. The remote nature of offshore sites complicates routine operations, including feeding, monitoring, and maintenance activities. This remoteness not only increases operational costs but also requires the development of robust and reliable automated systems to manage day-to-day operations with minimal human intervention. Furthermore, the potential for vessel collisions and marine animal invasions presents additional operational risks that must be mitigated through careful planning and advanced monitoring technologies [24].

The case of Ocean Farm 1, illustrated in Fig. 2, exemplifies both the potential and the limitations of current offshore aquaculture technology. While representing a significant advancement in offshore fish farming, Ocean Farm 1 faces several challenges that highlight the complexities of scaling up offshore aquaculture operations. The structure's limited mobility complicates site management and rotation practices, which are crucial for maintaining environmental sustainability and optimizing production cycles. The large scale of the installation, while offering potential economies of scale, also presents challenges in terms of replication and regulatory approval across different jurisdictions [25]. The entity responsible for the operation of the innovative Ocean Farm 1 salmon aquaculture installation "Ocean Farming AS" reported a significant financial deficit exceeding NOK 100 million in the previous fiscal year. This substantial operating loss coincided with a period during which the offshore structure was non-operational, undergoing essential maintenance procedures and technological enhancements. The temporary cessation of production activities for these necessary upgrades highlights the complex economic challenges inherent in pioneering offshore aquaculture technologies [25].

Despite the incorporation of advanced technologies, Ocean Farm 1 and similar offshore installations remain vulnerable to extreme weather events and technical failures. These vulnerabilities could have significant impacts on fish welfare and production outcomes, highlighting the need for robust contingency plans and fail-safe mechanisms. Environmental concerns persist, particularly regarding the potential effects of large-scale offshore aquaculture on wild fish populations and marine ecosystems. These concerns necessitate ongoing research and monitoring to assess and mitigate the ecological impacts of offshore fish farming [26]. The design specificity of Ocean Farm 1 to salmon farming highlights another challenge in offshore aquaculture development: the need for versatile systems that can be adapted to different species and environmental conditions. This lack of versatility could limit the global applicability of such systems, particularly in regions where salmon farming is not economically or ecologically viable. Additionally, the extensive electrical power requirements for operation, embedded sensors and control systems in offshore installations present logistical challenges in ensuring consistent and reliable power supply in remote marine environments [27].

Regulatory challenges further complicate the widespread adoption of offshore aquaculture technologies. The development of appropriate regulatory frameworks that balance economic development with environmental protection is crucial for the sustainable growth of the offshore aquaculture sector. These regulatory considerations must address issues such as site selection, environmental impact assessments, and the integration of offshore aquaculture with other marine activities [28].

In conclusion, while installations like Ocean Farm 1 represent significant technological advancements in offshore aquaculture, they also highlight the challenges that must be addressed for this sector to realize its full potential. Future research and development efforts should focus on enhancing the resilience, versatility, and environmental sustainability of offshore aquaculture systems. This will require interdisciplinary collaboration among engineers, marine biologists, environmental scientists, and policymakers to develop holistic solutions that address the complex interplay of technological, ecological, and socio-economic factors in offshore aquaculture. As the global demand for seafood continues to grow, overcoming these challenges will be crucial in establishing offshore aquaculture as a viable and sustainable component of future global food production systems.

This paper investigates the viability and potential impact of intelligent aquaculture vessels, an innovative approach to marine farming, within the specific context of Qatar's unique geographical, economic, and environmental landscape. Despite its relative nascency, Qatar's aquaculture industry has exhibited remarkable growth trajectories and untapped potential in recent years. This rapid expansion aligns with Qatar's strategic objectives, particularly its commitment to enhancing food security and diversifying its economy, as delineated in the Qatar National Vision 2030. The emergence of smart aquaculture vessels represents a potentially transformative development in this sector, offering new avenues for sustainable marine resource utilization and technological innovation. By examining this advanced approach within Qatar's specific national context, this paper aims to contribute valuable insights to the ongoing dialogue surrounding the country's aquaculture development strategies and their role in achieving broader national goals. Future research directions may also include exploring the potential for cultivating additional native species, optimizing production techniques for existing species, and investigating the long-term ecological impacts of aquaculture operations in Qatar's marine and coastal environments. The ongoing collaboration between research institutions, government agencies, and private sector entities will be essential in navigating the complex challenges and opportunities presented by the expansion of aquaculture in Qatar.

3 Smart Aquaculture Vessels: The Case Study of Havfarm 1 and Guoxin 1 Fish Farm Vessels

Offshore aquaculture demands robust equipment capable of withstanding harsh open ocean conditions [29], necessitate advanced monitoring system, specialized vessels for maintenance and harvesting [30]. High upfront costs for infrastructure and technology and the need for clear policies governing offshore aquaculture operations in Qatar's exclusive economic zone [31, 32]. While advancement in cage-based systems, offshore aquaculture presents numerous opportunities for Qatar. It is crucial to consider the associated potential risks, particularly oil spills and geopolitical conflicts. These factors add layers of complexity to the implementation and management of offshore aquaculture in Qatar.

The Arabian Gulf is one of the world's busiest oil transportation routes, with numerous offshore oil and gas installations. This concentration of petrochemical activity poses a significant risk to offshore aquaculture operations [33]. An oil spill in proximity to offshore cage systems could lead to direct contamination of farmed fish, potentially rendering entire stocks unmarketable and causing significant economic losses [34]. Even if cages are not directly impacted, oil pollution can degrade water quality over a wide area, affecting fish health and growth rates [35]. This can have long-lasting effects on marine ecosystems, potentially disrupting the environmental conditions necessary for successful aquaculture operations [36]. Other challenges involve logistic supports infrastructure that involves specialized vessels for maintenance, feeding, and harvesting.

In order to better understand the potential for offshore aquaculture in Qatar, it is valuable to examine advanced fish vessels developed in other countries. Two notable examples are Norway's Havfarm 1 and China's mobile deep-sea aquaculture ship Guoxin 1 shown in Fig. 3. These vessels represent a flexible and mobile solution for offshore fish farming that could provide valuable insights for Qatar's offshore aquaculture development. Havfarm 1, also known as "Jostein Albert," is a pioneering offshore fish farm developed by Norwegian company Nordlaks [37]. This massive structure (shown in Fig. 3a) represents a significant advancement in offshore aquaculture technology. The vessel is 385 m long and 59.5 m wide, with a height of 37 m. It can accommodate up to 10,000 tonnes of salmon, equivalent to the production of five conventional fish farms [37]. It is designed to withstand waves up to 10 m high, making it operate in exposed waters, thereby allowing it to access deeper, cleaner waters with stronger currents [38]. The feeding and monitoring fish health is automated, which reduces the need for constant on-site personnel and thus, improving operational efficiency.

Mobile deep-sea aquaculture ship, "Guoxin 1" [39] (shown in Fig. 3b), has a unique and advanced approach to offshore aquaculture. The vessel can handle the entire process of breeding, feeding, harvesting and processing of the fish, whilst producing up to 3200 tonnes of fish annually. Guoxin 1 can move to optimal locations based on water quality, temperature, and weather conditions unlike Have farm 1. This mobility allows for year-round production by following ideal environmental conditions and reduces the need for frequent trips to shore, which enhances operational efficiency [39]. It uses AI-driven systems to monitor fish behaviour, automate feeding, and maintain optimal water quality [40]. To demonstrate adaptability to market demand, the ship is designed to cultivate multiple high-value species including yellow croaker, grouper, and Atlantic salmon. In extreme weather scenarios, such as typhoons or red tides, the ship can safely move out of the path into safer waters.

Adopting technologies from Havfarm 1 and Guoxin 1 could address these challenges and contribute to Qatar's food security goals. Havfarm 1 large-scale, vessel-shaped design and ability to withstand significant wave heights could significantly boost Qatar's offshore fish production capacity and resilience to harsh marine

(a) (b)

Fig. 3 Smart aquaculture vessels **a** Havfarm 1 [37] **b** Guoxin 1 [39]

conditions. Meanwhile, Guoxin 1's innovations could provide valuable insights for developing locally adapted solutions. Implementing these technologies could offer multiple benefits for Qatar including increased fish production, enhanced environmental resilience, technology transfer opportunities, and economic diversification through job creation in the high-tech aquaculture sector [41]. Guoxin 1 represents a paradigm shift in aquaculture practices, embodying the potential for technology-driven solutions to address global food security challenges. Its successful operation in the East China Sea serves as a proof of concept for smart, mobile aquaculture platforms, potentially heralding a new era in sustainable marine resource utilization. A particularly noteworthy aspect of Guoxin-1 operation is its ability to deliver packaged fish products to shore within a six-hour timeframe post-harvest [41]. This rapid turnaround from production to distribution has profound implications for supply chain efficiency and product quality. The minimization of time between harvest and market availability could potentially enhance the freshness and nutritional value of the aquaculture products, while simultaneously reducing the need for extensive preservation methods.

A mobile deep-sea aquaculture ship will represent a significant advancement in sustainable fish farming technology for the state of Qatar and neighboring countries. This innovative vessel will operate in open ocean environments, allowing for the cultivation of fish stocks far from coastal areas. Its mobility is a key advantage, enabling it to navigate to optimal locations based on water conditions, temperature, and nutrient availability. The ship will likely incorporate state-of-the-art aquaculture systems, including large, submersible fish pens that can be lowered to various depths to mimic natural habitats and protect fish from surface-level disturbances such as storms or oil spills. Advanced filtration and waste management systems onboard would help maintain water quality within the pens, while automated feeding mechanisms could ensure efficient and precise nutrition delivery. Similar or more advanced than Guoxin 1 design should include laboratories for monitoring fish health and conducting research, as well as processing facilities to handle harvests at sea. This mobile approach to aquaculture could help reduce the environmental impact on coastal areas, distribute the economic benefits of fish farming more widely, and potentially aid in the replenishment of wild fish stocks in depleted areas. However, operating such a vessel would come with challenges, including navigating within international waters, managing the logistics of deep-sea operations, and ensuring the economic viability of mobile aquaculture on this scale.

Regulating smart fish vessels presents unique challenges due to their mobile nature and advanced technological capabilities. These vessels would likely need to comply with both traditional maritime regulations and specialized aquaculture standards. Key areas of regulation would include vessel classification and registration, navigation and positioning rules, environmental impact assessments, and fish welfare standards. Strict biosecurity protocols would be necessary to prevent disease spread between different water bodies as the vessel moves. Additionally, regulations would need to address the advanced technology onboard, including automation systems, cybersecurity requirements, and data collection and sharing protocols.

The mobile nature of these vessels also necessitates careful consideration of international operations. Regulations would need to cover protocols for operating in different national waters, compliance with regional fisheries management organizations, and harmonization of standards across jurisdictions. Product traceability systems would be crucial to track fish from vessels to market. Safety standards for crew members operating these specialized vessels would also be paramount, including specific training requirements and emergency response plans. As this is an emerging field, it is important to note that regulations would need to be adaptive and regularly updated to keep pace with technological advancements and new scientific insights into mobile aquaculture practices. The overall goal of such regulations would be to ensure the sustainable and responsible operation of smart fish vessels while maximizing their potential to contribute to food security and economic development.

4 Conclusion

The potential of open ocean aquaculture as a sustainable and innovative solution to address Qatar's increasing demand for seafood, while simultaneously advancing the nation's strategic objectives in food security and economic diversification was addressed in this paper. The findings of this study underscore the significant potential of open ocean aquaculture to revolutionize Qatar's fisheries sector. By leveraging advanced technologies and implementing robust regulatory frameworks, Qatar can overcome the limitations of traditional coastal aquaculture, which often faces constraints related to land use, coastal ecosystem impacts, and production scalability. The offshore model presents a promising avenue for expanding aquaculture production while mitigating environmental pressures on sensitive coastal habitats.

Crucially, this paper delineates the technological prerequisites for successful open ocean aquaculture implementation in Qatar. These include the development and deployment of resilient offshore structures capable of withstanding the Gulf's challenging marine conditions, advanced remote monitoring systems, and innovative fish health management protocols. The study also emphasizes the need for cutting-edge feeding technologies and waste management systems to ensure the environmental sustainability of offshore operations. From a regulatory perspective, the paper outlines a comprehensive framework necessary to govern this emerging sector effectively. This encompasses clear licensing procedures, stringent environmental impact assessments, and the establishment of aquaculture zones that harmonize with existing maritime activities. The proposed regulatory structure aims to foster industry growth while ensuring adherence to international best practices in sustainable aquaculture. Moreover, this research contextualizes open ocean aquaculture within Qatar's broader national development agenda, particularly the Qatar National Vision 2030. The findings suggest that investment in this sector could significantly contribute to the diversification of Qatar's economy, reducing dependence on hydrocarbon resources and fostering the growth of a knowledge-based economy through aquaculture-related research and innovation.

The study also shows the potential socio-economic impacts of developing a robust open ocean aquaculture sector in Qatar. Opportunities for job creation, skill development in high-tech aquaculture operations, and the potential for Qatar to emerge as a regional leader in sustainable marine farming practices is highlighted in this study. While acknowledging the promising potential of open ocean aquaculture, this research also identifies areas requiring further investigation. These include the need for long-term environmental monitoring to assess the cumulative impacts of offshore farming, economic feasibility studies to determine the optimal scale of operations, and research into species selection that aligns with both market demand and the environmental conditions of Qatar's offshore waters.

In conclusion, this study provides a foundational analysis for policymakers, industry stakeholders, and researchers to advance the implementation of open ocean aquaculture systems tailored to Qatar's unique maritime conditions and national priorities. By embracing this innovative approach to aquaculture, Qatar has the opportunity to not only enhance its food security and economic diversification but also to position itself at the forefront of sustainable marine resource utilization in the GCC region. Future research directions should focus on pilot projects to validate the findings of this study and on developing international collaborations to share knowledge and best practices in this rapidly evolving field.

References

1. General Secretariat for Development Planning (2008) Qatar National Vision 2030
2. Ministry of Municipality and Environment, Qatar (2021) Fisheries Statistics Annual Report.
3. Food and Agriculture Organization of the United Nations (2022) Fishery and Aquaculture Country Profiles: The State of Qatar
4. Mordor Intelligence (2023) Aquaculture in Qatar—Growth, Trends, Covid-19 Impact, and Forecasts (2023–2028). [Online]. Available: https://www.mordorintelligence.com/industry-reports/aquaculture-in-qatar-industry
5. Food and Agriculture Organization of the United Nations (2022) The State of World Fisheries and Aquaculture 2022, Rome
6. Ministry of Municipality and Environment, Qatar (2018) Qatar National Food Security Strategy 2018–2023, Doha
7. Elzaki RM (2024) Does fish production influence the GDP and food security in Gulf Cooperation Council countries? Evidence from the dynamic panel data analysis. Aquaculture 578:740058, ISSN 0044-8486. https://doi.org/10.1016/j.aquaculture.2023.740058
8. Food and Agriculture Organization of the United Nations (2021) Regional review on status and trends in aquaculture development in the Near East and North Africa—2020. FAO Fisheries and Aquaculture Circular No. 1217/4, Rome
9. https://thepeninsulaqatar.com/article/26/01/2020/Aquatic-Research-Centre-opened-in-Ras-Matbakh
10. Samakna International: https://www.samknainternational.com/
11. Mordor Intelligence: https://www.mordorintelligence.com/industry-reports/aquaculture-in-qatar-industry
12. https://agrico.qa/qatar-aquatic-research-centre/
13. Food and Agriculture Organization of the United Nations (2022) National Aquaculture Sector Overview, Qatar
14. Al-Jamali F, Bishop JM, Osment J, Jones DA, LeVay L (2005) A review of the impacts of aquaculture and artificial waterways upon coastal ecosystems in the Gulf (Arabian/Persian) including a case study demonstrating how future management may resolve these impacts. Aquat Ecosyst Health Manage 8(1):81–94
15. Read PA, Fernandes TF (2003) Management of environmental impacts of marine aquaculture in Europe. Aquaculture 226(1–4):139–163
16. Blohm D, Francois B, Gary C, Donatella C, Tom C, Geir D (2007) Genimpact—Evaluation of genetic impact of aquaculture activities on native populations. Final scientific report. In: 6th Framework plan of the European Commission. A European network (EU contract n. RICA-CT-2005–022802)
17. Hereher ME (2020) Assessment of climate change impacts on sea surface temperatures and sea level rise—The Arabian Gulf. Climate 8(4):50
18. Ibrahim LA, Abu-Hashim M, Shaghaleh H, Elsadek E, Hamad AAA, Alhaj Hamoud Y (2023) A comprehensive review of the multiple uses of water in aquaculture-integrated agriculture based on international and national experiences. Water 15(2):367
19. Ministry of Development Planning and Statistics (2018) Qatar Second National Development Strategy 2018–2022
20. Buck BH et al. (2017) Offshore and multi-use aquaculture with extractive species: seaweeds and bivalves. In: Aquaculture perspective of multi-use sites in the open Ocean, Springer, Cham, pp 23–69
21. https://www.fishfarmingexpert.com/cages-china-fish-farm/worlds-first-offshore-fish-farm-arrives-in-norway/1338984
22. Buck BH, Troell MF, Krause G, Angel DL, Grote B, Chopin T (2018). State of the art and challenges for offshore integrated multi-trophic aquaculture (IMTA). Front Marine Sci 5:165
23. Fredheim A, Langan R (2009) Advances in technology for offshore and open ocean finfish aquaculture, In: Woodhead publishing series in food science, technology and nutrition, new technologies in aquaculture, Woodhead Publishing, pp 914–944, ISBN 9781845693848. https://doi.org/10.1533/9781845696474.6.914

24. Holmer M (2010) Environmental issues of fish farming in offshore waters: perspectives, concerns and research needs. Aquacul Environ Interact 1(1):57–70
25. https://www.fishfarmingexpert.com/ocean-farm-1-salmar-aker-ocean-as/ocean-farm-1-cost-owner-nok-100m-last-year/1537677.
26. Froehlich HE, Smith A, Gentry RR, Halpern (2017) Offshore aquaculture: i know it when I see it. Front Marine Sci 4:154
27. Gentry RR, Froehlich HE, Grimm D et al (2017) Mapping the global potential for marine aquaculture. Nat Ecol Evol 1:1317–1324. https://doi.org/10.1038/s41559-017-0257-9
28. Lester SE, Stevens JM, Gentry RR et al (2018) Marine spatial planning makes room for offshore aquaculture in crowded coastal waters. Nat Commun 9:945. https://doi.org/10.1038/s41467-018-03249-1
29. Bridger CJ, Costa-Pierce BA (2003) Open Ocean aquaculture: from research to commercial reality. World Aquaculture Society
30. Telfer T, Cardia D (2009) Offshore aquaculture development in the Mediterranean. In: Aquaculture planning: policy formulation and implementation for sustainable development, FAO Fisheries and Aquaculture Technical Paper No. 542, pp 141–162
31. IntraFish (2021) Offshore aquaculture: the next big thing or pipe dream?
32. Rust MB et al (2015) US department of commerce aquaculture policy. Mar Policy 54:71–77
33. Sheppard C, Al-Husiani M, Al-Jamali F, Al-Yamani F, Baldwin R, Bishop J, Benzoni F, Dutrieux E, Dulvy NK, Durvasula SRV, Jones DA, Loughland R, Medio D, Nithyanandan M, Pilling GM, Polikarpov I, Price ARG, Purkis S, Riegl B, Saburova M, Namin KS, Taylor O, Wilson S, Zainal K (2010) The gulf: a young sea in decline. Marine Pollut Bullet 60(1):13–38, ISSN 0025-326X, https://doi.org/10.1016/j.marpolbul.2009.10.017
34. Danish EY (2010) Ecological impact from chemicals in the Arabian Gulf due to Gulf oil spill. Water Environ J 24:65–73. https://doi.org/10.1111/j.1747-6593.2008.00145.x
35. Mustafa SA, Al-Rudainy AJ, Salman NM (2024) Effect of environmental pollutants on fish health: an overview. Egypt J Aquatic Res 50(2):225–233, ISSN 1687-4285. https://doi.org/10.1016/j.ejar.2024.02.006
36. Peterson CH, Rice SD, Short JW, Esler D, Bodkin JL, Ballachey BE, Irons DB (2003) Long-term ecosystem response to the Exxon Valdez oil spill. Science 302(5653):2082–2086. https://doi.org/10.1126/science.1084282. PMID: 14684812
37. https://www.nskshipdesign.com/project/havfarm-1/
38. https://www.fishfarmingexpert.com/the-first-salmon-arrive-in-havfarm/1382310
39. https://www.undercurrentnews.com/2024/07/04/chinas-giant-aquaculture-ship-hits-3200t-annual-harvest-milestone/
40. https://news.cgtn.com/news/2024-07-01/World-s-1st-floating-fish-farming-ship-yields-3-200t-of-fish-yearly-1uSDfoyAraE/p.html
41. https://www.bastillepost.com/global/article/3956951-worlds-first-tanker-sized-fish-farming-ship-guoxin-1-yields-3200-tons-of-yellow-croaker-yearly

AI and ML in Food Industry: A Review Study

Walid Kh Alqaisi, Ahmed Badawi, Suzan Ali, Mostafa Soliman, and I. M. Elzein

Abstract Advancements in technology in various fields have significantly expanded the potential applications of robotics. This paper focuses on a recent area of interest: the use of robotics in the food industry. Robots have the capacity to revolutionize processes such as food processing, handling, palletizing, packing, and serving. As a result, there has been a notable increase in the deployment of robots within this sector in recent years. Consequently, key topics such as robot kinematics, dynamics, hygiene, economic efficiency, human–robot interaction, safety, and maintenance are critically examined in this review. Additionally, a comparison of the robots currently utilized in the industry is included. The findings indicate that the food serving sector represents a promising area for research, where integrating advancements from various technological domains could yield significant opportunities. It is expected that the broader dissemination of research in "Robo-food" will encourage greater collaboration within the research community and drive further advancements in the field.

Keywords Artificial intelligence · Machine learning · Robotics application in food industry · Agricultural image processing · Plant disease prediction · Food serving robots

W. K. Alqaisi (✉) · A. Badawi · M. Soliman · I. M. Elzein
College of Engineering and Technology, University of Doha for Science and Technology, Doha, Qatar
e-mail: wkaq@hotmail.com

S. Ali
College of Computing and Information Technology, University of Doha for Science and Technology, Doha, Qatar

S. Al-Naemi et al. (eds.), *Water and Food Security in the Face of Climate Change: Challenges and Opportunities for Resilience*, World Sustainability Series,
https://doi.org/10.1007/978-3-032-00098-9_45

1 Introduction

Artificial Intelligence (AI) systems play a crucial role in enhancing food security by improving agricultural efficiency and productivity. They enable precision farming, optimizing resource use like water and fertilizers, which reduces waste and increases crop yields. AI-driven predictive analytics help farmers anticipate market demand and weather patterns, leading to better planning. Robotics streamline processes such as planting, harvesting, and packaging, minimizing labor shortages and enhancing food distribution. Overall, these technologies foster a more resilient food supply chain, ensuring access to nutritious food for growing populations. AI systems include the use of Machine Learning (ML), Natural Language Processing, computer vision, and robotics (see Fig. 1). Machine Learning (ML) essentially teaches a machine how to learn automatically from the given training data and improves its performance based on learned experiences. The machine then learns to predict things without being explicitly programmed. ML can be defined as a scientific study of algorithms and statistical models that enable computers to perform specific tasks without explicit programming. Solutions based on machine learning can help food companies estimate the demand for their products.

As a result, companies can manage various tasks such as optimizing pricing, transportation, and inventory strategy. Machine learning solutions can efficiently automate rigorous and time-consuming processes, such as food sorting and packaging, saving companies significant costs on human labor and expediting production. AI system components in the food industry are paving the way for promising advancements in food security, both now and in the future, as shown in Fig. 2.

AI and ML in food security have been used to develop new methods in various related aspects. Crop monitoring and yield prediction, including the use of satellite imagery and drones, have been implemented. A vision-based algorithm is utilized in [1] and [2] to precisely approach a palm tree crown for pollination. Deep learning is employed to assess palm tree health [3], while image recognition assigns labels to images of orange trees [4]. Predictive analytics are being conducted in interconnected supply chains to address complex disruptions in global value chains caused by factors such as natural disasters, climate events, geopolitical conflicts, and economic

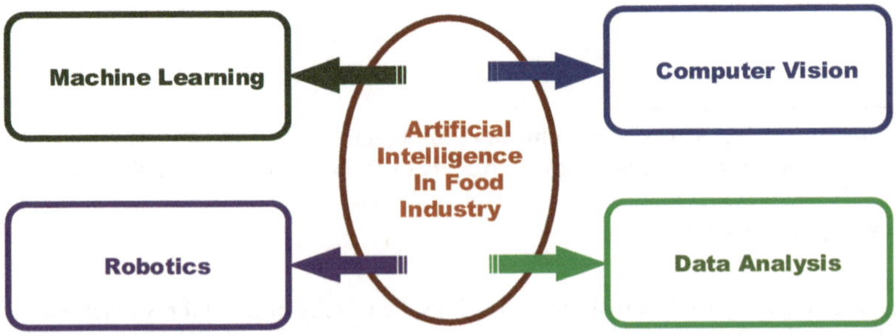

Fig. 1 AI system components in food industry

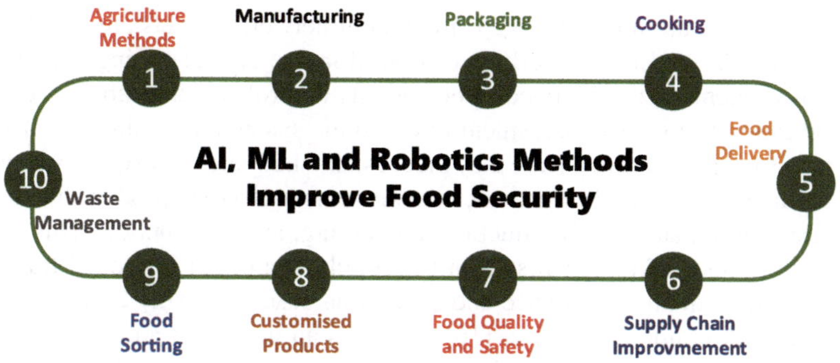

Fig. 2 AI is developing food security industries in many aspects

crises. Artificial intelligence, machine learning, blockchain, and big data analytics offer new possibilities for forecasting and managing these disruptions effectively [5]. Predictive analytics also involve applying machine learning procedures to forecast future crop profits and optimize resource allocation. Precision agriculture integrates cutting-edge sensors, GPS mapping, and data analytics to optimize crop management practices, revolutionizing agricultural practices by providing farmers with timely information for decision-making on seeding, fertilization, and harvesting [6]. Precision agriculture, including smart irrigation using the Internet of Things and sensory systems, leads to improvements in irrigation systems crucial for water-use efficiency and sustainable development. Automated irrigation systems play a vital role in conserving water and minimizing water usage [7]. Smart irrigation systems based on IoT and machine learning are utilized [8]. Predictive models for precise nutrient management have been developed, with remote sensing and imaging technologies playing a crucial role in real-time monitoring for informed decision-making [9]. Precision agriculture based on convolutional neural networks in rice production nutrient management using machine learning genetic algorithms is explored [10]. Machine learning in supply chain optimization for demand forecasting and logistics management is extensively studied. The integration of deep learning and machine learning approaches in supply chain management presents transformative potential for more efficient supply chain management, including supplier selection, production, inventory control, transportation, demand estimation, and sales forecasting. An article [11] analyzes the 'Food Demand Forecasting' dataset released by Genpact, compares the effects of various factors on demand, extracts characteristic features with potential influence, and proposes a comparative study of seven regressors for forecasting orders.

Neural networks are developed to incorporate both time series and explanatory factors for demand forecasting in supply chain management. The method is applied and evaluated in the context of functional products and a steel manufacturer, showing statistically significant supply chain performance improvements with ML-based demand forecasting methods [12]. Pest and disease management is crucial for early detection and integrated pest management. Deep learning models for plant disease detection are utilized [13], while neural network models are developed for plant disease detection [14]. Early detection of maize diseases using deep learning models is implemented [15]. Food safety, quality control, and traceability are vital in the food industry. Quality assessment refers to the systematic evaluation of various aspects of food products to ensure they meet established safety and quality standards. Traceability refers to the ability to track the movement of a food product and its ingredients through all stages of production, processing, and distribution. This involves documenting and linking each step in the supply chain. An improved traceability system for food quality assurance and evaluation based on fuzzy classification and neural network is implemented in [16]. AI plays an important and effective role in waste management and waste prediction to reduce waste. In the food industry, AI can help forecast demand more accurately, reducing overproduction and spoilage, while also optimizing inventory management in restaurants and grocery stores. Machine learning is used for predictive analytics, sorting and recycling, product design, supply chain optimization, consumer behavior insights, food waste reduction, and energy efficiency. The role of artificial intelligence in waste management [17]. Intelligent approaches for sustainable management of food waste are implemented in [18]. Improving customer experience and business profits are more possible with AI and ML. Autonomous agricultural robots are an example of production enhancement using technology. AI and ML algorithms are used in greenhouse yield production [19], autonomous harvesting open farm productivity [20], AI robot form monitoring [21], smart farming with drones [22], and an autonomous robot weeder [23].

2 Crop Monitoring and Yield Prediction

Satellite Imagery and Drones: Satellite imagery and drones provide real-time data on crop health, soil conditions, and moisture levels. For instance, a farmer can use drone technology to capture high-resolution images of their fields, allowing them to identify stressed areas that may need additional water or nutrients. Satellite images can also monitor large fields for variations in growth patterns over time, enabling farmers to make data-driven decisions. Vision-based algorithm is used in [1] to provide an exact approach to a palm tree crown for pollination. Crop Monitoring and Yield Prediction including satellite imagery and drones utilize, a vision-based algorithm in [2], as shown in Fig. 3a, to precisely approach a palm tree crown for pollination. Deep learning is used to assess palm tree health [3], as depicted in Fig. 3c. Image recognition assigns a label to an image of orange trees [4], as shown in Fig. 3d. Predictive analytics involves using historical data, weather patterns, and machine learning algorithms to forecast future yields. For example, a farm may analyze past crop yields alongside weather forecasts to predict how much corn they can harvest this season. By employing predictive analytics, farmers can optimize planting schedules and make better financial decisions regarding resource allocation.

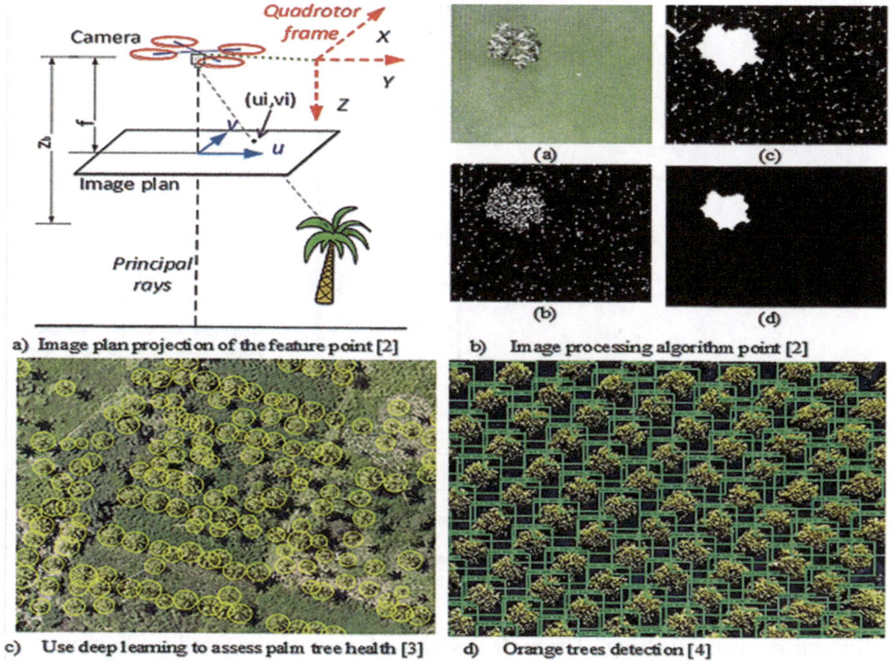

Fig. 3 Tree detection application

3 Precision Agriculture

Smart irrigation and nutrient management are key components of precision agriculture practices. Smart irrigation systems utilize sensors and weather data to determine the optimal amount of water needed for crops. For example, a vineyard can use soil moisture sensors that trigger irrigation only when moisture levels drop below a certain threshold, reducing water waste and ensuring the vines receive the right amount of hydration for optimal growth. The components of a smart irrigation system used in AI systems include data acquisition (sensors), irrigation control, wireless communication, data processing, and fault detection. Each of these components can be used in IoT devices (see Fig. 4). The benefits of using IoT in irrigation systems are summarized in Fig. 5. After ensuring the storage of the data using Node-Red and MongoDB, we grouped months of data to train the system to predict the required data [8]. Nutrient Management: Precision agriculture enables customized nutrient application based on soil health and crop requirements. By analyzing soil samples, farmers can pinpoint specific nutrient deficiencies in different field sections and apply fertilizers selectively. This targeted approach reduces the overuse of chemicals, resulting in healthier crops and a lower environmental impact. Significant progress has been made in developing predictive models for precise nutrient management. Remote sensing and imaging technologies are essential for real-time monitoring, enabling

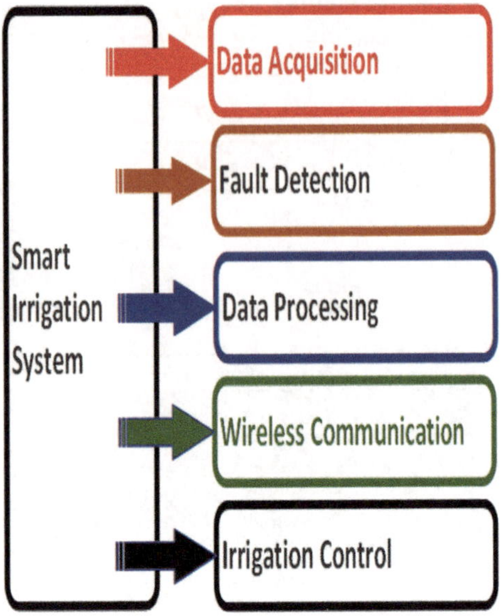

Fig. 4 Smart irrigation systems [7]

informed decision-making. Through a series of case studies and evaluations, we showcase the effectiveness and potential benefits of AI-enabled crop health monitoring and nutrient management in modern agriculture. AI algorithms have been developed for soil nutrient management, such as identifying Nitrogen, Phosphorus, or Potassium deficiencies in soil through leaf analysis. (Reference: [10], Fig. 6).

4 Supply Chain Optimization

Demand forecasting utilizes data analysis to predict consumer demand for agricultural products. For example, a fruit distributor may analyze historical sales data and seasonal trends to predict how many apples will be sold in the upcoming months. This information helps farmers and suppliers adjust production and stock levels accordingly, minimizing waste and maximizing profits. Logistics Management: Effective logistics management ensures efficient delivery of products from farms to consumers. For instance, a dairy company may implement route optimization software to streamline milk delivery to stores, reducing fuel costs and ensuring fresher products. This system can track inventory levels and adjust routes in real-time based on demand fluctuations.

Accurate demand forecasting has become crucial, especially in the food industry, as many products have a short shelf life, and improper inventory management can lead to significant waste and losses for the company. Several machine learning and deep learning techniques have shown significant improvements in handling time-dependent data [11].

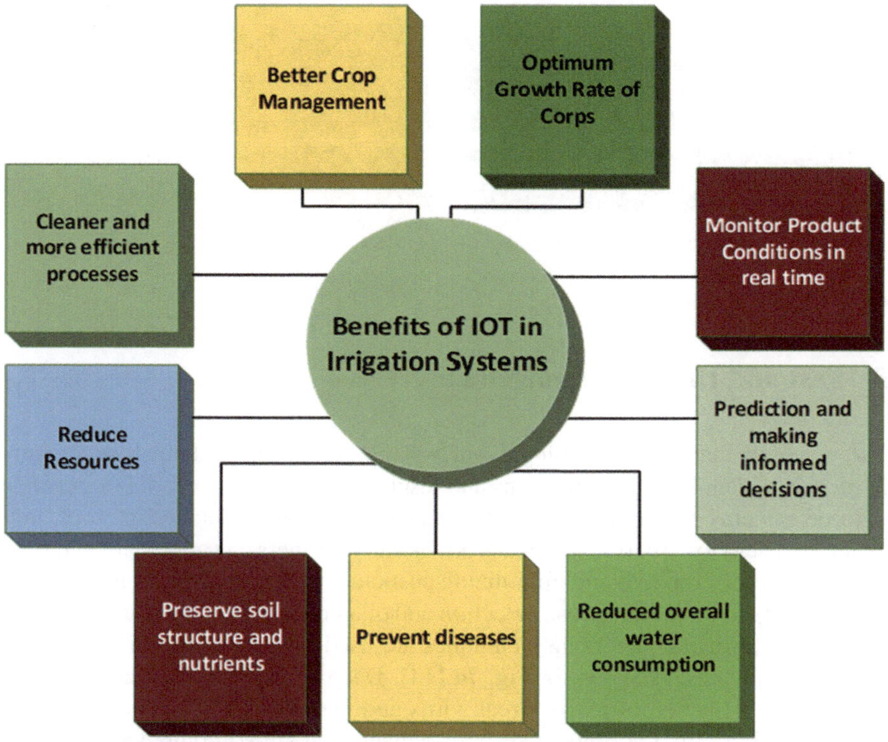

Fig. 5 Benefits of smart irrigation systems [7]

Fig. 6 AI for soil nutrient management [10]

Different types of pathogens:
Bacteria (A), Viruses (B), Fungi (C) [14].

Maize dataset: (A) Healthy (B) Maize Lethal Necrosis (C) Maize Streak Virus [15]

Fig. 7 AI disease detection

5 Pest and Disease Management

Through early detection technologies such as remote sensing and machine learning, farmers can identify pest infestations and diseases before they spread. For example, infrared cameras can be used to detect unusual temperature changes in crops, indicating pest activity. By detecting these issues early, farmers can apply targeted treatments, reducing crop loss and minimizing pesticide use. Deep learning models have been developed for plant disease detection and diagnosis, achieving a success rate of 99.53% [13]. Image processing techniques are used for identifying different types of plant pathogens, as shown in Fig. 7a [14]. Deep learning models are also used for the early detection of maize streak virus and maize lethal necrosis diseases, as depicted in Fig. 7b [15]. Integrated pest management (IPM) combines biological, cultural, and chemical practices to manage pests sustainably. For example, farmers may introduce natural predators like ladybugs to control aphid populations and use crop rotation to disrupt pest life cycles. This holistic approach reduces reliance on chemical pesticides and promotes a healthier ecosystem.

6 Food Production: Safety, Production and Quality Control

In Quality Assessment, advanced technologies such as sensors and machine vision systems are used for real-time quality assessment of food products. For example, an apple packing facility might use cameras to analyze the color, size, and surface defects of apples, ensuring only high-quality fruits are shipped to consumers, thus maintaining brand reputation and reducing waste. Traceability systems track food products from farm to table, providing transparency in the supply chain. For instance, a beef producer might use RFID tags to track each animal's health and feeding history throughout its life. This information can be accessed quickly in case of a food safety concern, allowing for swift action and improving consumer trust in the product.

An artificial neural network is utilized to determine the final grade of food quality based on the results of the stage quality evaluation. Additionally, a fuzzy classifier algorithm is employed in pattern recognition to assess the stage food quality with five levels: very bad, bad, medium, good, and very good. This method utilizes five symmetrical fuzzy sets with triangular membership functions to partition the output space and determine the quality of meat at each step of the supply chain [16]. The evaluation of food quality at each stage of the supply chain is crucial for ensuring the overall quality of the product.

7 Waste Management and Prediction

In personalized nutrition, advances in technology allow for tailored dietary recommendations based on individual health data and preferences. For example, a food company might use consumer health data to suggest specific products rich in nutrients beneficial for specific dietary needs, such as high-fiber snacks for individuals looking to improve gut health. AI-powered marketing tools are enabling food companies to deliver personalized experiences to consumers. By analyzing consumer data, AI can suggest tailored nutrition plans and recipes, enhancing customer satisfaction and loyalty. This personalized approach not only helps in meeting individual dietary needs but also boosts sales and engagement for food companies. The core focus of any business establishment is to improve its customer experience through innovative services. In the food industry, it matters a lot when poor footfalls cost dear to any business in the food industry. AI offers significant assistance to food companies in improving customer experience by helping them make their customer service better and effectively manage employee schedules. In addition, the technology underscores its relevance for food companies by helping them free up human resources and employ them to work on more critical tasks. It consequently results in better customer service and a more delightful experience overall. Regarding waste reduction, AI is important for reducing waste and waste management. For example, AI sensors in waste bins can monitor fill levels and optimize collection routes, reducing unnecessary pickups and associated emissions. Furthermore, machine learning can develop dynamic scheduling by analyzing traffic patterns and historical data to schedule waste collection during off-peak hours, improving efficiency. Enhanced recycling processes help minimize food waste. For example, a grocery store may promote imperfect produce at a discount, encouraging consumers to buy them rather than letting them go unsold. This not only reduces waste but also makes healthy food more accessible. AI technologies enhance efficiency and reduce emissions in waste management. Figure 8 shows the improvements in recycling processes, organic waste processing, and emission reduction due to AI implementation [17, 18].

Smart Waste Management is an AI-powered sensor in bins that can monitor fill levels, optimizing collection routes and schedules, reducing fuel consumption and operational costs. Other aspects of waste reduction, as shown in Fig. 9, include:

- **Predictive Analytics**: By analyzing data on waste generation patterns, AI can help businesses predict future waste production, enabling better resource management and waste reduction strategies.

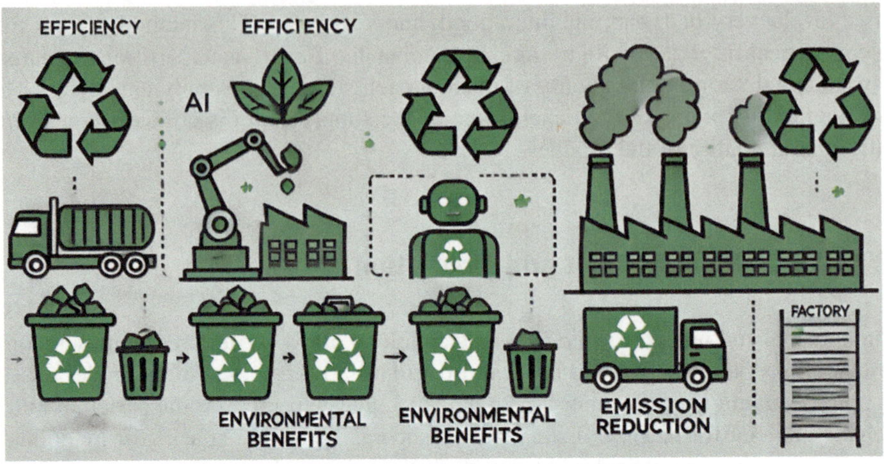

Fig. 8 AI in waste management: efficiency and emission reduction [17]

Fig. 9 AI in waste reduction

- **Sorting and Recycling**: AI-driven robots and computer vision systems can enhance sorting efficiency in recycling facilities by accurately identifying and separating various materials. This advancement can lead to increased recycling rates and reduced contamination.
- **Product Design**: AI can assist companies in designing products with sustainability in mind by suggesting materials and processes that minimize waste throughout the product lifecycle.
- **Supply Chain Optimization**: AI can analyze supply chain data to reduce excess inventory and improve production processes, minimizing waste generated from overproduction or spoilage.
- **Consumer Behavior Insights**: AI tools can analyze consumer data to promote more sustainable purchasing habits, encouraging reduced waste by providing better product recommendations and personalized suggestions.

- **Food Waste Reduction**: In the food industry, AI can help forecast demand more accurately, reducing overproduction and spoilage. It can also optimize inventory management in restaurants and grocery stores.
- **Energy Efficiency**: AI can optimize energy usage in manufacturing processes, which not only reduces energy waste but also minimizes waste generated in the production phase.

8 Improves Customer Experience and Business Profits

Autonomous agricultural robots offer a promising solution to the challenges faced by the agriculture industry, such as increasing food demand and labor shortages in green house yield production [19], as depicted in Fig. 10a. These intelligent machines can automate repetitive tasks, enhancing security of food supply. By integrating autonomous technology, the industry can streamline operations and boost productivity as shown in Fig. 10b [20]. An AI-powered agricultural robot tending to crops is depicted in Fig. 10c [21]. Smart Farming with Drones for Crop Monitoring and Management is illustrated in Fig. 10d [22]. A fully autonomous robot weeder, (Fig. 10e), has been developed to detect and spray over 90% of weeds with 20 times fewer chemicals, providing solar-powered unlimited autonomy [23]. The Swarm Farm spraying robot is shown in Fig. 10f.

9 Conclusion

In conclusion, the integration of artificial intelligence (AI) and machine learning (ML) in the food industry represents a transformative shift that enhances efficiency, sustainability, and consumer satisfaction. This review highlights the diverse applications of AI and ML, from optimizing supply chain management and enhancing food safety to personalizing consumer experiences and improving product development. As these technologies continue to evolve, they hold the potential to address critical challenges such as food waste and resource scarcity. However, it is essential for industry stakeholders to remain mindful of ethical considerations, data privacy, and the need for regulatory frameworks. By fostering collaboration between technology developers and food industry professionals, we can harness the full potential of AI and ML, paving the way for a more innovative and sustainable food system.

a) Green house yield production [19]
b) Autonomous Agricultural Robots [20]
c) AI powered agricultural robot [21]
d) Smart Farming with Drones [22]
e) Robot weeder [23]
f) Swarm farm spraying [23]

Fig. 10 Robotics autonomous systems

References

1. Alqaisi W, Brahmi B, Ghommam J, Saad M, Nerguizian V (2018) Multivariable super-twisting control in a vision-based quadrotor utilized in agricultural application. In: 2018 IEEE International conference on computational intelligence and virtual environments for measurement systems and applications (CIVEMSA), IEEE, pp 1–6. Accessed 21 Sep 2024. [Online]. Available: https://ieeexplore.ieee.org/abstract/document/8439964/
2. Alqaisi WK, Brahmi B, Ghommam J, Saad M, Nerguizian V (2019) Adaptive sliding mode control based on RBF neural network approximation for quadrotor. In: IEEE International

symposium on robotic and sensors environments, ROSE 2019—Proceedings. https://doi.org/10.1109/ROSE.2019.8790423
3. Use deep learning to assess palm tree health | Documentation (2024). Accessed 21 Sep 2024. [Online]. Available: https://learn.arcgis.com/en/projects/use-deep-learning-to-assess-palm-tree-health/
4. Firigato JON (2024) Orange trees detection with YOLO v5 in UAV imagery. Medium. Accessed 21 Sep 2024. [Online]. Available: https://joaootavionf007.medium.com/orange-trees-detection-with-yolo-v5-in-uav-imagery-22ec29db922e
5. Tamasiga P, Houssin Ouassou E, Onyeaka H, Bakwena M, Happonen A, Molala M (2023) Forecasting disruptions in global food value chains to tackle food insecurity: the role of AI and big data analytics—A bibliometric and scientometric analysis. J Agric Food Res 14:100819. https://doi.org/10.1016/j.jafr.2023.100819
6. Farooqui NA, Haleem M, Khan W, Ishrat M (2024) Precision agriculture and predictive analytics. In: Intelligent techniques for predictive data analytics, Wiley & Sons, Ltd, pp 171–188. https://doi.org/10.1002/9781394227990.ch9
7. Obaideen K et al (2022) An overview of smart irrigation systems using IoT. Energy Nexus 7:100124. https://doi.org/10.1016/j.nexus.2022.100124
8. Tace Y, Tabaa M, Elfilali S, Leghris C, Bensag H, Renault E (2022) Smart irrigation system based on IoT and machine learning. Energy Rep 8:1025–1036. https://doi.org/10.1016/j.egyr.2022.07.088
9. Swarnkar SK, Dewangan L, Dewangan O, Prajapati TM, Rabbi F (2023) AI-enabled crop health monitoring and nutrient management in smart agriculture. In: 2023 6th International conference on contemporary computing and informatics (IC3I), September 2023, pp 2679–2683. https://doi.org/10.1109/IC3I59117.2023.10398035
10. Sakthipriya S, Naresh R (2024) Precision agriculture based on convolutional neural network in rice production nutrient management using machine learning genetic algorithm. Eng Appl Artif Intell 130:107682
11. Panda SK, Mohanty SN (2023) Time series forecasting and modeling of food demand supply chain based on regressors analysis. IEEE Access 11:42679–42700. https://doi.org/10.1109/ACCESS.2023.3266275
12. Feizabadi J (2022) Machine learning demand forecasting and supply chain performance. Int J Logist Res Appl 25(2):119–142. https://doi.org/10.1080/13675567.2020.1803246
13. Ferentinos KP (2018) Deep learning models for plant disease detection and diagnosis. Comput Electron Agric 145:311–318. https://doi.org/10.1016/j.compag.2018.01.009
14. Devaraj A, Rathan K, Jaahnavi S, Indira K (2019) Identification of plant disease using image processing technique. In: 2019 International conference on communication and signal processing (ICCSP), April 2019, pp 0749–0753. https://doi.org/10.1109/ICCSP.2019.8698056
15. Mayo FS, Mduma N, Mgala M, Maina C (2024) Deep learning models for the early detection of maize streak virus and maize lethal necrosis diseases in Tanzania. Front Artif Intell 7. https://doi.org/10.3389/frai.2024.1384709
16. Wang J, Yue H, Zhou Z (2017) An improved traceability system for food quality assurance and evaluation based on fuzzy classification and neural network. Food Control 79:363–370. https://doi.org/10.1016/j.foodcont.2017.04.013
17. Hernandez D, Pasha L, Yusuf DA, Nurfaizi R, Julianingsih D (2024) The role of artificial intelligence in sustainable agriculture and waste management: towards a green future. Int Trans Artif Intell 2(2). https://doi.org/10.33050/italic.v2i2.552
18. Said Z et al (2023) Intelligent approaches for sustainable management and valorisation of food waste. Bioresour Technol 377:128952. https://doi.org/10.1016/j.biortech.2023.128952
19. How Artificial Intelligence Is Used in the Food Industry (2024) Food Quality and Safety. Accessed 21 Sep 2024. [Online]. Available: https://www.foodqualityandsafety.com/article/how-artificial-intelligence-is-used-in-the-food-industry/
20. Autonomous Agricultural Robots: Ensuring Food Security and Efficiency (2024) IoT [mesh]. Accessed 24 Sep 2024. [Online]. Available: https://www.iot-mesh.io/autonome-agrarroboter-landwirtschaft/

21. An AIpowered agricultural robot tending to crops | Premium AI-generated image (2024). Freepik. Accessed 24 Sep 2024. [Online]. Available: https://www.freepik.com/premium-ai-image/aipowered-agricultural-robot-tending-crops_57063326.htm
22. Smart Farming with Drones: Revolutionizing Crop Monitoring and Management (2024). Accessed 24 Sep 2024. [Online]. Available: https://www.linkedin.com/pulse/smart-farming-drones-revolutionizing-crop-monitoring-management
23. Tanner S (2016) Can robots contribute to food security?. Ecorobotics. Available online: https://unctad.org/system/files/non-officialdocument/cstd2016_p22_SteveTanner_en.pdf

Biotechnology and Genomics for Developing Climate-Resilient Crops in the Gulf Region

Fahim Ullah Khan, Farhat Abbas, Hassan Ali, Israr Ud Din, Attiq ur Rehman, and Michael Phillips

Abstract The agriculture sector in the Gulf region faces severe challenges of extreme temperatures, water scarcity, and high soil salinity, exacerbated by climate change. These harsh environmental conditions have led to a significant reliance on food imports across Gulf Cooperation Council (GCC) countries. However, recent advancements in biotechnology and genomics offer transformative opportunities to enhance the region's agricultural resilience. Innovative technologies, including genomic selection, marker-assisted breeding, and CRISPR-Cas9 genome editing, enable precise crop trait improvements, specifically targeting drought, salinity, and heat tolerance. Omics technologies, such as genomics, transcriptomics, and proteomics, provide crucial insights into plant stress responses, facilitating the development of crops that can thrive under the region's extreme conditions. In addition to stress tolerance, biotechnological advancements such as gene stacking, genetic manipulation, and metabolic engineering are enhancing water-use efficiency and salt tolerance, addressing critical agricultural challenges in the Gulf. Field trials and case studies demonstrate the potential of biotech solutions, with salt-tolerant, drought-resistant, and heat-tolerant varieties showing promising results in regional contexts. Successful integration of biotech crops in the Gulf requires structured field trials, adoption pathways, and scaling processes. Continued research, investment in infrastructure, and supportive policy frameworks will be essential for scaling these innovations, ultimately contributing to food security and sustainable agricultural practices in the region.

F. U. Khan (✉) · F. Abbas · H. Ali · M. Phillips
University of Doha for Science and Technology, Doha, Qatar
e-mail: fahimbiotech@hu.edu.pk

F. U. Khan
Department of Agriculture, Hazara University Mansehra, Mansehra, Pakistan

I. U. Din
Institute of Biotechnology and Genetic Engineering, The University of Agriculture Peshawar, Peshawar, Pakistan

A. Rehman
Department of Agricultural Sciences, Faculty of Agriculture and Forestry, The University of Helsinki, Helsinki, Finland

S. Al-Naemi et al. (eds.), *Water and Food Security in the Face of Climate Change: Challenges and Opportunities for Resilience*, World Sustainability Series,
https://doi.org/10.1007/978-3-032-00098-9_46

Keywords Agricultural sustainability · CRISPR-Cas9 · Marker-assisted breeding · Salt tolerance · Water-use efficiency

1 Introduction to Climate Challenges in Gulf Agriculture

Harsh environmental conditions, including extreme temperatures, high soil salinity, and limited water resources significantly impact the Gulf region's agriculture. These challenges are exacerbated by climate change, which further strains the already limited agricultural capacity in the Gulf Cooperation Council (GCC) countries—Saudi Arabia, the United Arab Emirates (UAE), Qatar, Oman, Kuwait, and Bahrain. As a result, there is a high dependency on food imports [1, 2]. However, advancements in biotechnology offer promising solutions to these challenges [3, 4]. This dependency underscores the need for region-specific solutions that ensure long-term sustainability.

This chapter aims to explore how advancements in biotechnology can address these challenges by enhancing the region's agricultural resilience and sustainability. By focusing on technologies like genomic selection, marker-assisted breeding, CRISPR-Cas9, and controlled-environment agriculture, this review contextualizes these innovations in the Gulf region's unique agricultural landscape. The primary goal is to outline the role of these technologies in developing crops and farming systems that are not only climate-resilient but also tailored to the Gulf's reliance on drought-tolerant species such as dates and sorghum. Major challenges posed by the climate change in the agriculture sector of Gulf region are summarized in Fig. 1.

Agriculture in the Gulf region faces extreme challenges, including summer temperatures exceeding 50 °C and severe water scarcity, with annual rainfall often below 100 mm [5, 6]. Groundwater depletion, particularly in Saudi Arabia, further strains irrigation systems [7, 8]. Soil salinity, affecting over 60% of arable land, is exacerbated by poor-quality irrigation water and high evaporation rates, particularly in Bahrain, Kuwait, and coastal areas [9, 10]. To address these issues, biotechnological advances like CRISPR-Cas9 are being used to develop drought- and water-efficient crops, such as genome-edited date palms and salt-tolerant sorghum, which show promise in coping with water scarcity and saline conditions [3, 11].

Climate change compounds these challenges, altering rainfall patterns and increasing extreme weather events that threaten agricultural productivity [12, 13]. Modern tools, including CRISPR-Cas9 and genomic selection, are enhancing crop resilience, particularly in drought-tolerant species like sorghum, which is becoming increasingly important for the region [3, 14, 15]. With only 1.63% of land suitable for farming, the Gulf is investing in controlled-environment agriculture, such as vertical farming and hydroponics, to optimize food production on limited land [16–18]. These innovations, alongside biotechnological advancements, are vital for strengthening food security and reducing dependence on food imports in an increasingly unpredictable global market [1, 19, 20].

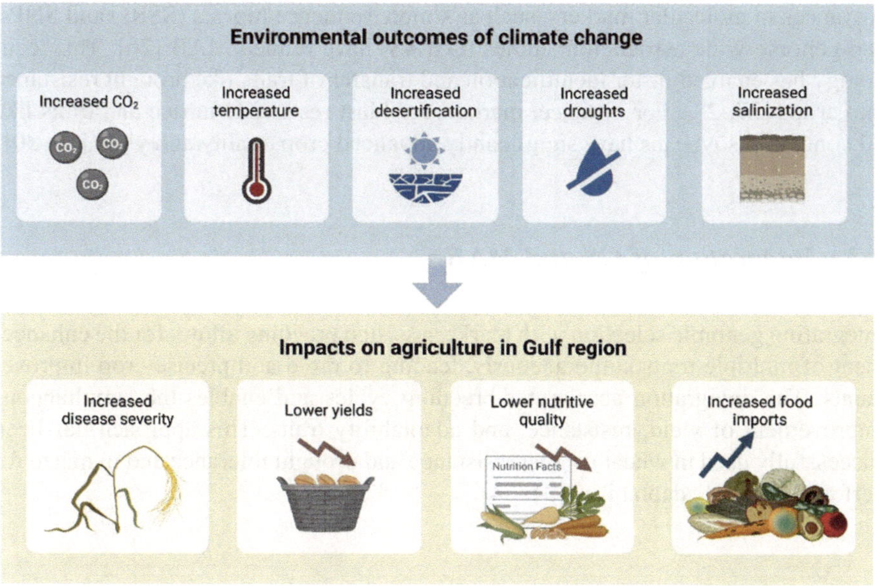

Fig. 1 Most important environmental outcomes of climate change and their respective impact on Gulf region's agriculture

2 Genomic Selection and Marker-Assisted Breeding

2.1 Genomic Selection

Genomic selection (GS) uses genome-wide markers to predict the breeding values of plants, enhancing the efficiency of selecting desired traits. This approach is more effective than traditional phenotypic selection, which requires multiple generations to assess trait performance. Simplified techniques like high-density genotyping and statistical models, such as genomic best linear unbiased prediction (GBLUP), enable more precise trait identification [21]. Single nucleotide polymorphisms (SNPs) provide high-resolution genetic data for predicting complex traits, including yield and stress tolerance [22]. For example, GS has enhanced maize hybrids for yield and stress resilience and accelerated wheat breeding for disease resistance and climate adaptation [23–25].

2.2 Marker-Assisted Breeding

Marker-assisted breeding (MAB) employs genetic markers to track alleles associated with desirable traits, enabling efficient trait incorporation into new varieties.

Advances in molecular markers, such as simple sequence repeats (SSRs) and SNPs, and genome-wide association studies (GWAS) have refined MAB [26]. This technology has enabled faster identification and transfer of traits like drought resistance into crops [27, 28]. For instance, markers for blast resistance in rice and traits like oil content in soybeans have significantly advanced crop quality and yield [29, 30].

2.3 Integration of GS and MAB

Integrating genomic selection with marker-assisted breeding allows for the enhancement of multiple traits simultaneously, leading to rapid and precise crop improvements. This integration accelerates breeding cycles and enables the simultaneous improvement of yield, resistance, and adaptability traits. This approach has been successfully used in wheat for rust resistance and drought tolerance and in maize for performance and adaptability [31–34].

3 CRISPR-Cas9 and Genome Editing for Crop Improvement

CRISPR-Cas9 offers precise genome editing, enabling targeted modifications to enhance stress resilience. This technology addresses challenges such as, drought, salinity, and heat by modifying specific genes. For instance, the DREB gene family plays a critical role in enhancing drought and salt tolerance by regulating water retention and osmotic balance, while HKT1 genes help maintain ion homeostasis under saline conditions [35, 36]. Additionally, heat tolerance has been enhanced by targeting heat-shock proteins like TaHSP17 in wheat and OsHSP70 in rice [15]. CRISPR-Cas9 also improves water-use efficiency by targeting ABA biosynthesis genes [37–39]. However, its implementation faces hurdles such as regulatory frameworks and public skepticism, particularly in the Gulf, where cultural perceptions about genetic modification present challenges. Increasing acceptance of genome editing suggests broader adoption opportunities in the future [40, 41].

4 Genetic Engineering for Stress Tolerance

Genetic engineering has significantly advanced stress tolerance by introducing targeted genetic modifications. Key advances include engineering heat-shock proteins, such as HSP70, to bolster plants' ability to withstand extreme heat, and genes like HKT1 for ion balance in saline soils. For instance, the transfer of salt-tolerant genes like HKT1 has enhanced productivity in saline soils and drought tolerance has been improved through genes affecting osmotic balance [42–46]. Gene stacking has emerged as an effective strategy, combining traits like drought and salinity resistance to improve resilience under complex stress conditions. This technique demonstrates potential for further applications in the Gulf region [47–51].

5 Use of Biotechnology to Enhancing Crop Stress Resilience

5.1 *Role of Omics Technologies in Understanding Stress Responses*

Omics technologies—genomics, epigenomics, transcriptomics, proteomics, and metabolomics—are revolutionizing the understanding of plant stress responses (Fig. 2). Genomics identifies genetic variants linked to stress tolerance, while transcriptomics reveals gene expression changes under stress. Epigenomics explores heritable gene expression changes without altering DNA sequences, offering insights into resilience mechanisms. Proteomics investigates stress-responsive proteins like heat-shock proteins, and metabolomics identifies metabolites crucial for adaptation. Integrating these datasets through systems biology provides a holistic view, aiding the development of stress-resilient crops [15, 21, 22, 45, 46, 52].

5.2 *Biotechnology for Water Use Efficiency in Arid Climates*

Optimizing water use efficiency (WUE) is essential for agriculture in arid regions like the Gulf. Genetic engineering of aquaporins, proteins regulating water transport, enhances water uptake and retention, improving drought tolerance and minimizing water loss [48]. Developing drought-resistant crops by modifying pathways that produce Osmo protectants, such as proline and glycine betaine, has further strengthened stress responses. Enhancements in DREB (dehydration-responsive element-binding) genes have proven effective in improving drought and salinity tolerance in crops like wheat [44, 45].

Gene stacking combines multiple stress-resistance traits, such as drought and water conservation, into a single genotype, with applications in cotton and maize showing promise in arid environments [50]. Metabolic engineering has improved Osmo protectant synthesis and stress-response hormones, while modifications in stomatal conductance reduce water loss without compromising photosynthesis, critical for productivity in water-scarce conditions [48, 49]. Despite advancements, regulatory barriers and public skepticism remain challenges, requiring targeted outreach and policy reforms for broader adoption.

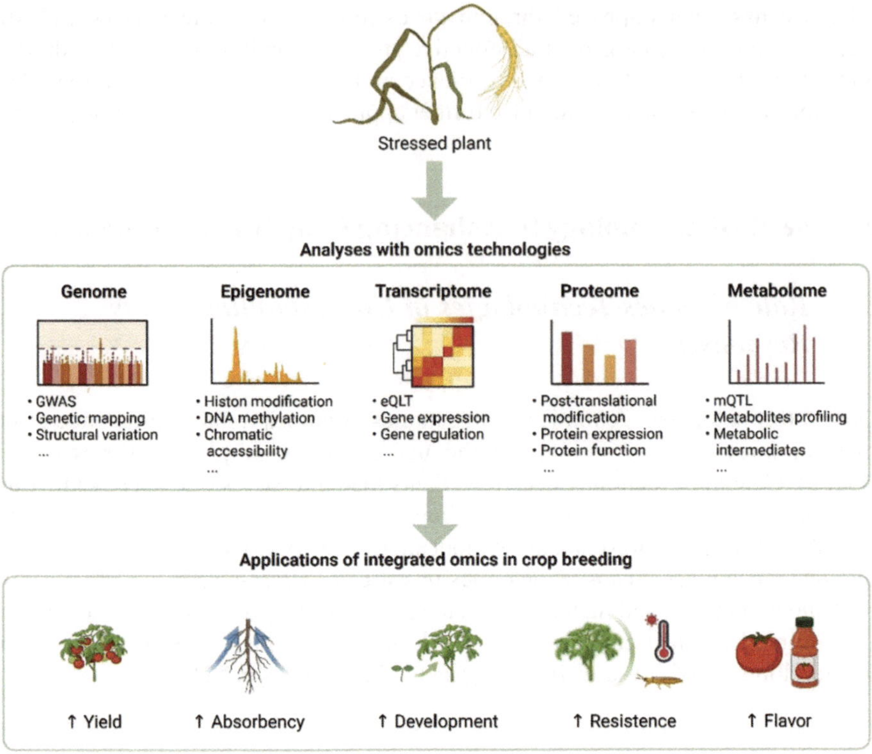

Fig. 2 Multi-omics approaches to study stress responses of plants and strategies crop breeding for better climatic resilience and productivity

5.3 Biotechnology for Developing Salt-Tolerant Crops

Soil salinity significantly affects crop yields, particularly in the Gulf. Genetic studies have identified key genes, such as HKT1, which maintains ion homeostasis, and DREB, which improves tolerance to salinity and drought [44, 46]. Marker-assisted selection (MAS) accelerates the development of salt-tolerant crops by targeting markers linked to salinity tolerance traits, with successes in rice and wheat [26, 27].

CRISPR/Cas9 genome editing allows precise modifications of stress-response genes, such as HKT1, to enhance salinity tolerance. Omics technologies, including transcriptomics and proteomics, provide comprehensive insights into salinity responses, identifying novel targets for genetic improvement. Combining multi-omics data has further advanced strategies for enhancing crop resilience [35, 43, 46, 48, 49].

Gene stacking and transgenic approaches, alongside speed breeding, enable rapid development of crops capable of thriving in high-salinity environments. These innovations represent a significant step toward addressing salinity challenges.

Integrating biotechnological innovations with conventional breeding practices and agronomic strategies offers the most promising path forward. Collaboration among researchers, policymakers, and agricultural stakeholders will be essential to scaling these solutions for improving crop resilience in the Gulf region and beyond.

6 Challenges and Future Directions

Despite these advancements, challenges remain, including ensuring the stability of genetic modifications, navigating regulatory approvals, and addressing public perceptions. The regulatory landscape in the Gulf region regarding genetically modified organisms (GMOs) and genome-edited crops is complex. While some countries like the UAE and Saudi Arabia have started to establish guidelines for the approval and commercialization of biotech crops, there are still significant hurdles, including inconsistent regulations across the region. These frameworks often require extensive testing and approval processes, which can delay adoption. In addition, the regulatory bodies must consider environmental impacts, safety, and ethical concerns surrounding GMOs. The existing policies in some Gulf countries, such as Saudi Arabia's regulations on biotech crops, offer a supportive framework for research but can be slow in translating into practical, large-scale adoption.

Public perception also remains a critical challenge in the region. While there is increasing awareness about biotechnology, there is still a level of skepticism regarding GMOs. Cultural concerns, such as fears about the long-term environmental impact of genetic modifications and potential health risks, play a significant role in shaping public opinion. In response, governments in the region, particularly in the UAE and Saudi Arabia, have initiated public awareness programs to address these concerns. These programs aim to educate the public on the safety, benefits, and environmental sustainability of biotech crops, but overcoming entrenched skepticism requires continuous dialogue and transparent communication.

Integrating biotechnological solutions with conventional breeding practices and effective agronomic strategies is essential for sustainable water management in agriculture. Continued research and collaborative efforts among researchers, policymakers, and agricultural stakeholders are crucial for translating these innovations into practical solutions for improving agricultural sustainability in the Gulf region.

7 Developing Salt-Tolerant Crops

7.1 *Significance of Salt Tolerance*

Soil salinity is a significant barrier to crop production, particularly in the Gulf region where high salinity levels can severely impact crop yields [1, 2]. With climate change and water scarcity exacerbating these challenges, developing salt-tolerant crops is vital for ensuring food security and sustainable agriculture [3, 4].

7.2 *Genetic Insights into Salt Tolerance*

Recent advances in genomics have elucidated the genetic mechanisms underlying salt tolerance. Key genes such as HKT1 (High-Affinity K + Transporter 1) are crucial for maintaining ion homeostasis, which is essential for salt tolerance [46]. The DREB (Dehydration Responsive Element Binding) gene family enhances tolerance to both drought and salinity [44].

7.3 *Marker-Assisted Selection for Salt Tolerance*

Marker-assisted selection (MAS) is a valuable tool in developing salt-tolerant crops. By utilizing molecular markers, linked to salt tolerance traits, the breeders can accelerate the development of improved varieties. MAS has been successfully applied to create salt-tolerant rice and wheat varieties through selection of markers associated with key tolerance genes [26, 27].

7.4 *Innovations in Genomic Selection and Speed Breeding*

Genomic selection (GS) and speed breeding are transformative techniques that enhance crop improvement efficiency. GS predicts breeding values based on genomic data, facilitating the selection of plants with desirable traits such as salt tolerance [21, 25]. Speed breeding complements this by enabling the rapid development of multiple generations per year, thus accelerating the breeding process [22].

7.5 CRISPR/Cas9 and Genome Editing for Salt Tolerance

CRISPR/Cas9 technology has revolutionized genome editing, allowing precise modifications to improve salt tolerance. Targeting specific genes involved in stress responses, such as HKT1, has led to enhanced salinity tolerance in crops like wheat [46]. Additionally, CRISPR/Cas9 was used to engineer stress-responsive regulatory pathways, further enhancing salt tolerance [35, 43].

7.6 Transgenic Approaches and Gene Stacking

Transgenic approaches, including the introduction of genes from salt-tolerant species, have successfully improved salt tolerance in various crops. Gene stacking, which combines multiple stress-tolerance genes into a single genotype, is a promising strategy for achieving higher resistance levels [50]. This approach shows potential for developing crops that can endure extreme salinity conditions.

7.7 Application of Omics Technologies

Omics technologies—genomics, transcriptomics, and proteomics—offer comprehensive insights into the molecular mechanisms of salt tolerance. These technologies can help in dissecting complex stress responses and identify novel targets for genetic improvement [48, 49]. Integrating multi-omics data can enhance our understanding of plant responses to salinity and support the development of more effective crop improvement strategies.

8 Future Directions and Opportunities for Research

As the Gulf region continues facing the dual challenges of climate change and food security, biotechnology and genomics present substantial opportunities for transforming agriculture. This section outlines future research directions and potential opportunities, highlighting areas where advancements could significantly impact the development of climate-resilient crops.

8.1 Advancing Genomic Tools and Techniques

Future research should emphasize the integration and optimization of advanced genomic tools, such as CRISPR/Cas9, for precise genome editing. The recent advancements in CRISPR technology offer opportunities for targeted modifications that can enhance crop resilience to stress. Focusing on adapting CRISPR/Cas9 systems for regional crop varieties in the Gulf could yield significant improvements in drought and salinity tolerance [3, 43, 44].

8.2 Development of Multi-Stress Tolerant Crop Varieties

Developing crop varieties capable of withstanding multiple environmental stress is critical for the Gulf region. Research should focus on combining various stress tolerance traits through approaches like gene stacking and marker-assisted selection. Creating multi-stress tolerant varieties can provide robust solutions to the combined challenges of heat, salinity, and drought prevalent in the region [30, 50].

8.3 Enhancing Genomic Selection Methods

While genomic selection has shown promise, further refinement of these methods for crops in the Gulf region is needed. Integrating genomic selection with speed breeding techniques could accelerate the development of improved crop varieties. Tailoring these methods to address the specific needs and environmental conditions of the Gulf region can enhance breeding efficiency and effectiveness [4, 22, 25].

8.4 Leveraging Advanced Phenotyping Technologies

Advancements in phenotyping technologies, including high throughput and remote sensing methods, offer significant potential for better assessing crop performance under stress conditions. Combining these technologies with genomic data can improve predictions of crop resilience and refine breeding strategies [11].

9 Conclusions

This review underscores the transformative potential of biotechnology and genomics in developing climate-resilient crops tailored for the Gulf region. Addressing the unique agricultural challenges posed by the region's extreme conditions—such as high temperatures, high soil salinity, and limited water availability—requires innovative and targeted solutions. Biotechnology, particularly through genetic engineering and advanced genome editing technologies like CRISPR-Cas9, offers precise and efficient methods to enhance crop resilience. These technologies enable the development of crops with improved drought tolerance, salt resistance, and heat tolerance, which are essential for maintaining productivity in the face of environmental stresses. Advances in genomic selection and marker-assisted breeding can facilitate the identification and enhancement of traits crucial for thriving in the Gulf's harsh conditions.

The application of omics technologies—such as genomics, transcriptomics, and proteomics—has provided valuable insights into plant stress responses, enabling a more comprehensive understanding of how crops can be engineered to withstand abiotic stresses. Furthermore, efforts to improve water use efficiency through genetic modification have shown promise, with successful case studies demonstrating the potential for developing crops that utilize water more effectively. However, scaling up biotech crops in the Gulf region faces several challenges. These include policy-related hurdles, cultural acceptance, and economic considerations. Addressing these challenges will be crucial for the successful adoption and widespread use of climate-resilient crops. Furthermore, there are substantial opportunities for further research and innovation. Emerging biotechnological tools, such as advanced genome editing and phenotyping technologies, combined with collaborative research initiatives, hold promises for advancing crop improvement. Through fostering strong research partnerships and developing supportive policy frameworks, the Gulf region can harness the full potential of biotechnology and genomics. This will be pivotal for achieving sustainable agricultural practices and enhancing food security amid the challenges of climate change.

References

1. Abou-Hadid A (2014) Impact of climate change on food security. In: Arab environment: food security. Annual report of the Arab Forum for Environment and Development, pp 130–154
2. FAO (2020) Regional overview of food security and nutrition in the near East and North Africa. Food and Agriculture Organization of the United Nations
3. Rasheed A, Gill RA, Hassan MU, Mahmood A, Qari S, Zaman QU, Wu Z (2021) A critical review: recent advancements in the use of CRISPR/Cas9 technology to enhance crops and alleviate global food crises. Current Issues in Molecular Biol 43(3):1950–1976
4. Eggen A (2012) The development and application of genomic selection as a new breeding paradigm. Anim Front 2(1):10–15

5. Tarawneh QY (2022) Monitoring climate change in Jordan and its impact on agriculture. The Arab World Geographer 25(4):230–247
6. World Bank (2019) World Development Indicators. World Bank
7. Alharthi S, Alharthi A, Alharthi M (2019) Sustainable development goals in the Kingdom of Saudi Arabia's 2030 vision. WIT Trans Ecol Environ 238:455–467
8. Moossa B, Trivedi P, Saleem H, Zaidi SJ (2022) Desalination in the GCC countries—a review. J Clean Prod 357:131717
9. Abdelfattah MA, Shahid SA, Othman YR (2009) Soil salinity mapping model developed using RS and GIS—a case study from Abu Dhabi, United Arab Emirates. Eur J Sci Res 26(3):342–351
10. Corwin DL (2021) Climate change impacts on soil salinity in agricultural areas. Eur J Soil Sci 72(2):842–862
11. Singh RL, Mondal S (2017) Biotechnology for sustainable agriculture: emerging approaches and strategies. Academic Press, London
12. Field CB, Barros VR (Eds) Climate change 2014–impacts, adaptation and vulnerability: regional aspects. Cambridge University Press
13. Pal JS, Eltahir EA (2016) Future temperature in southwest Asia projected to exceed a threshold for human adaptability. Nat Clim Chang 6(2):197–200
14. Verner D (2012) Adaptation to climate change in the Arab World. World Bank Policy Research Working Paper
15. Hirayama T, Shinozaki K (2010) Research on plant abiotic stress responses in the post-genome era: past, present and future. Plant J 61(6):1041–1052
16. Abahussain A et al (2002) Desertification in the Arab region: analysis of current status. J Arid Environ 51(4):521–545
17. Despommier D (2011) The Vertical Farm: Feeding the World in the 21st Century. Macmillan
18. UAE Ministry of Climate Change and Environment. Vertical Farming Initiatives (2021)
19. Khouri N, Shideed K, Kherallah M (2011) Food security: perspectives from the Arab World. Food Security 3:1–6
20. FAO (2011) Biotechnologies for Agricultural Development
21. Calus MP (2010) Genomic breeding value prediction: methods and procedures. Animal 4(2):157–164
22. Heslot N, Yang HP, Sorrells ME, Jannink JL (2012) Genomic selection in plant breeding: a comparison of models. Crop Sci 52(1):146–160
23. Crossa J, Pérez-Rodríguez P, Cuevas J, Montesinos-López O, Jarquín D, De Los Campos G, Varshney RK (2017) Genomic selection in plant breeding: methods, models, and perspectives. Trends in Plant Sci 22(11):961–975
24. Voss-Fels KP, Cooper M, Hayes BJ (2019) Accelerating crop genetic gains with genomic selection. Theor Appl Genet 132:669–686
25. Watson A, Hickey LT, Christopher J, Rutkoski J, Poland J, Hayes BJ (2019) Multivariate genomic selection and potential of rapid indirect selection with speed breeding in spring wheat. Crop Sci 59(5):1945–1959
26. Collard BC, Mackill DJ (2008) Marker-assisted selection: an approach for precision plant breeding in the twenty-first century. Philosophical Trans Royal Soc B: Biolog Sci 363(1491):557–572
27. Kumar J, Choudhary AK, Solanki RK, Pratap A (2011) Towards marker-assisted selection in pulses: a review. Plant Breeding 130(3):297–313
28. Ma J, Cao Y (2021) Genetic dissection of grain yield of maize and yield-related traits through association mapping and genomic prediction. Front Plant Sci 12:690059
29. Nadim MKA, Islam MM, Hoque MI, Hasan MJ, Uddin MI (2024) Development of blast-resistant rice varieties through marker-assisted selection. Bangladesh J Agricul 49(1):41–51
30. Pantalone VR, Walker DR, Dewey RE, Rajcan I (2004) DNA marker-assisted selection for improvement of soybean oil concentration and quality. Legume Crop Genomics, AOCS Press, Champaign, IL, pp 283–311
31. Anilkumar C, Sunitha NC, Harikrishna, Devate NB, Ramesh S (2022) Advances in integrated genomic selection for rapid genetic gain in crop improvement: a review. Planta 256(5):87

32. Kumar A, Mir RR, Sehgal D, Agarwal P, Carter A (2021) Genetics and genomics to enhance crop production, towards food security. Front Genet 12:798308
33. Crossa J, Perez P, Hickey J, Burgueno J, Ornella L, Cerón-Rojas J, Mathews K (2014) Genomic prediction in CIMMYT maize and wheat breeding programs. Heredity 112(1):48–60
34. Cerrudo D, Cao S, Yuan Y, Martinez C, Suarez EA, Babu R, Trachsel S (2018) Genomic selection outperforms marker-assisted selection for grain yield and physiological traits in a maize doubled haploid population across water treatments. Front Plant Sci 9:366
35. Ndudzo A, Makuvise AS, Moyo S, Bobo ED (2024) CRISPR-Cas9 genome editing in crop breeding for climate change resilience: implications for smallholder farmers in Africa. J Agricul Food Res 101132
36. Bortesi L, Fischer R (2015) The CRISPR/Cas9 system for plant genome editing and beyond. Biotechnol Adv 33(1):41–52
37. Rashid B, Tariq M, Khalid A, Shams F, Ali Q, Ashraf F, Husnain T (2017) Crop improvement: new approaches and modern techniques. Plant Gene Trait 8(3):18–30
38. Gao J, Wang G, Ma S, Xie X, Wu X, Zhang X, Xia Q (2015) CRISPR/Cas9-mediated targeted mutagenesis in Nicotiana tabacum. Plant Molecul Biol 87:99–110
39. Food and Agriculture Organization of the United Nations (FAO) (2011) Biotechnologies for Agricultural Development: Proceedings of the FAO International Technical Conference on Biotechnology in Agriculture
40. James C (2016) ISAAA briefs brief 39 global status of commercialized biotech/GM crops: 2016. ISAAA Briefs
41. Emadi MH, Rahmanian M (2020) Commentary on challenges to taking a food systems approach within the food and agriculture organization (FAO). In: Food security and land use change under conditions of climatic variability: a multidimensional perspective, pp 19–31
42. Herdt RW (2006) Biotechnology in agriculture. Annu Rev Environ Resour 31(1):265–295
43. Kumar M, Prusty MR, Pandey MK, Singh PK, Bohra A, Guo B, Varshney RK (2023) Application of CRISPR/Cas9-mediated gene editing for abiotic stress management in crop plants. Front Plant Sci 14:1157678
44. Hassan S, Berk K, Aronsson H (2022) Evolution and identification of DREB transcription factors in the wheat genome: modeling, docking, and simulation of DREB proteins associated with salt stress. J Biomol Struct Dyn 40(16):7191–7204
45. Khan Z, Shahwar D (2020) Role of heat shock proteins (HSPs) and heat stress tolerance in crop plants. In: Sustainable agriculture in the era of climate change, pp 211–234
46. Shohan MUS, Sinha S, Nabila FH, Dastidar SG, Seraj ZI (2019) HKT1; 5 transporter gene expression and association of amino acid substitutions with salt tolerance across rice genotypes. Front Plant Sci 10:1420
47. Amin I, Rasool S, Mir MA, Wani W, Masoodi KZ, Ahmad P (2021) Ion homeostasis for salinity tolerance in plants: a molecular approach. Physiol Plant 171(4):578–594
48. Blum A (2009) Effective use of water (EUW) and not water-use efficiency (WUE) is the target of crop yield improvement under drought stress. Field Crop Res 112(2–3):119–123
49. Yang JF, Hao GF, Yang GF (2021) Genetic engineering and chemical control related to abscisic acid for improving plant drought tolerance 3563–3565
50. Shailani A, Joshi R, Singla-Pareek SL, Pareek A (2021) Stacking for future: pyramiding genes to improve drought and salinity tolerance in rice. Physiol Plant 172(2):1352–1362
51. Suleiman MK, Shahid SA (2024) Prospective of agricultural farming in Kuwait and energy-food-water-climate nexus. In: Terrestrial environment and ecosystems of Kuwait: assessment and restoration, Cham, Springer Nature Switzerland, pp 363–391
52. Brown JJ, Das P, Al-Saidi M (2018) Sustainable agriculture in the Arabian/Persian Gulf region utilizing marginal water resources: making the best of a bad situation. Sustainability 10(5):1364

The Role of Agricultural Insurance as an Input to Contribute to Achieving Food Security—A Reference to the Algerian Experience

Hocine Hassani, Mohammed Himrane, and Fatima Mekhlouf

Abstract Recently, the world has witnessed numerous crises, including the COVID-19 pandemic and various local and international conflicts, in addition to natural disasters. These events have had a clear impact on global food security and extended security dimensions, as they are considered essential components for the stability and security of nations. These multiple and complex risks pose several challenges regarding their management by the direct actors responsible for food provision. However, these challenges do not affect insurance companies with the same severity, whether in terms of risk assessment and identification or in evaluating and addressing damages and compensations. An insurance system can play a crucial role in further contribution to the food security in the state. In this study, we attempt to demonstrate the importance of insurance in achieving food security by outlining the main risks facing agricultural activities and the guarantees that insurance companies can utilize to support farmers and protect them from the consequences of these risks.

Keywords Agricultural insurance · Food security · Risks

1 Introduction

Most international reports indicate that the world will witness an increase in the number of people facing food insecurity, posing a risk to the stability and security of the countries in which they reside. This is due to multiple factors, including the conflicts, the COVID-19 pandemic, and the climate crisis.

H. Hassani (✉)
University of Chlef, Chlef, Algeria
e-mail: h.hassani@univ-chlef.dz

M. Himrane
University of Jijel, Jijelv, Algeria

F. Mekhlouf
University of Relizane, Relizane, Algeria

S. Al-Naemi et al. (eds.), *Water and Food Security in the Face of Climate Change: Challenges and Opportunities for Resilience*, World Sustainability Series,
https://doi.org/10.1007/978-3-032-00098-9_47

Algeria, like many other nations, is not immune to the repercussions and consequences of these crises, despite having the potential to elevate itself to the ranks of leading countries in achieving food security, both regionally and internationally.

Achieving food security and addressing its implications in Algeria will require the involvement of all stakeholders, particularly financial institutions. The importance and role of the insurance sector, despite its critical function in providing protection against risks associated with agricultural activities. It is imperative for stakeholders to empower insurance companies to activate the mechanisms of this industry at various levels, thereby contributing to the advancement and development of agriculture in Algeria and achieving food security.

Algeria possesses numerous assets that position it among the world's most significant agricultural regions, including vast expanses of land suitable for various types of farming, a broad range of plateaus and major basins, and mountainous chains that can be utilized, particularly for tree planting and animal husbandry. Additionally, the country boasts a long and diverse coastline rich in fishery resources, suitable for aquaculture.

Thus, the importance of insurance lies in its capacity to support and develop the agricultural sector, ultimately contributing to food security.

In this context, Algeria has adopted a series of reforms aimed at aligning with the needs of agricultural stakeholders and providing sufficient insurance products to protect them from various risks.

The latest reform is the Executive Decree No. 21–81, dated on February 28, 2021, which outlined the practice of cooperative insurance in Algeria. This decree serves as an additional measure to relieve farmers from the burden of insuring against various risks within a legal framework, allowing Algerian insurance companies to enter the cooperative insurance industry through established windows or by forming new companies. This adds to the regulatory framework governing the sector, notably Law 06/04 issued on February 20, 2006, which amends and supplements the Order 95/07 issued on January 25, 1995, concerning insurance. This reflects the Algerian legislator's desire to liberalize the insurance sector across all domains to contribute more effectively to comprehensive development.

Through this study, we will address the following main question: **What is the importance and the role that the insurance sector can play in supporting the agricultural sector to achieve food security over the ongoing crises and international changes?**

To tackle this issue, the researcher will discuss the following points:

- The impact of current international crises on food security.
- Agricultural insurance as a means to support agricultural development and enhance food security.

2 Impact of Current International Crises on Food Security

Food security is understood as a community's ability to provide adequate food for its citizens in both the short and long term, in terms of quality and at reasonable prices. It encompasses various economic, social, and political dimensions and is influenced by several factors, including population size and growth rates, natural resources (land, water, climate, etc.), citizens' incomes, food prices, and global market conditions.

The issue of food security has become a global phenomenon subject to evaluation and analysis at all levels, locally and internationally, using specific indicators, notably the Integrated Food Security Phase Classification (IPC). This classification aims to align methodologies with other approaches addressing food security. The following table summarizes this classification (Table 1).

2.1 *The Impact of the COVID-19 Pandemic on Food Security*

Reports from international organizations have revealed a significant increase in food insecurity for millions around the world during 2020, following the onset of the COVID-19 pandemic in late 2019. The number of people suffering from hunger rose to approximately 828 million, while those facing acute food insecurity surged from 135 to 345 million since 2019. Around 50 million people in 45 countries are on the brink of famine (Global, Annual Review 2021, 2022).

Additionally, many sectors have been negatively affected by the pandemic, most notably tourism, aviation, restaurants, and even the energy sector and oil exports due to decreased global demand, which could create additional crises for countries that rely on these industries (Al-Hussein 2020).

Table 1 Integrated food security phase classification (IPC)

Phase number	Phase	Description
1	Generally food secure	More than 80% of households can meet basic food needs
2	Insecure food security	At least 20% of households reduce food consumption
3	Acute food and livelihood crisis	At least 20% of households face significant gaps in food consumption
4	Humanitarian assistance	At least 20% of households experience severe food consumption gaps
5	Famine/ humanitarian disaster	At least 20% of households face complete food shortages and severe needs, with malnutrition exceeding 30%

The pandemic also significantly impacted food prices, which reached all-time highs. The monthly operational costs for the World Food Programme exceeded $73.6 million above the average for 2019, reflecting a remarkable 44% increase (Global 2021).

2.2 The Impact of Climate Change on Food Security

The increasing global food insecurity is largely attributed to climatic phenomena that affect weather patterns, leading to heatwaves, heavy rainfall, and droughts. The ongoing droughts have resulted in water scarcity, reduced harvest yields, and livestock deaths, forcing hundreds of thousands to abandon their land in search of livelihoods, while increasing the risk of resource-based conflicts among local communities (Mondial 2022).

Studies indicate a continuous rise in annual atmospheric temperatures worldwide, particularly in the Mediterranean region, where annual increases have reached over 1.5 degrees Celsius (MedEC 2020). This trend is illustrated in the following statement (Fig. 1).

Estimates indicate that Algeria will experience intense heat waves similar to those in the Mediterranean region, with temperature increases projected between 3 and 5 degrees Celsius during the summer from 2080 to 2100.

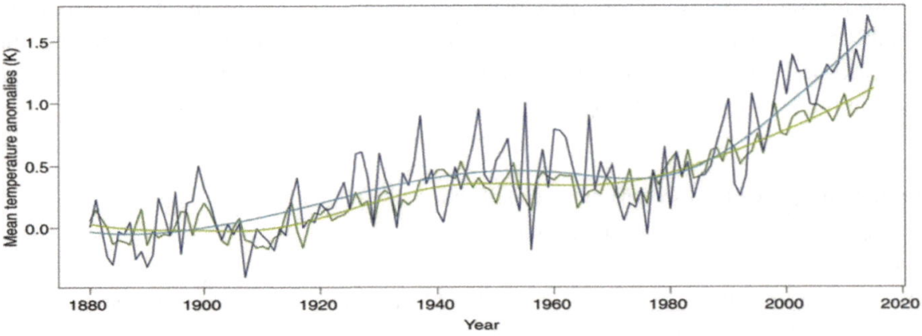

Fig. 1 Rise in annual atmospheric temperature. *Source* Mediterranean Climate Experts Network (MedEC), Climate and Environmental Change Risks in the Mediterranean Region, France, 2020

3 Agricultural Insurance to Support Agricultural Development and Food Security

Although Algeria is relatively distant from the food and malnutrition issues faced by some countries, as shown in the accompanying map of global malnutrition, it is crucial to take appropriate measures in all directions to support its food security and fully utilize its resources and capabilities to enhance public security.

Algeria's agricultural potential is so rich. Algeria possesses numerous assets, including an area exceeding 2 million square kilometers, making it the largest country in Africa. It features two major mountain ranges: the Tell Atlas in the north and the Saharan Atlas in the south, dividing the country into three distinct climatic regions: the northern coastal area, the high plateaus, and the vast Sahara. This geographical diversity can be leveraged to diversify agricultural resources. If Algeria provides the necessary support and protection against various risks faced by agricultural stakeholders, it could become a regional leader in the agricultural sector [1].

3.1 Importance of Agricultural Development for Food Security in Algeria and the Role of Insurance

The significance of agriculture in Algeria cannot be overstated, as it plays a crucial role in achieving food security and public safety. Agriculture contributes to public spending on consumer goods, expanding industries and services related to this sector, and securing additional financial resources that improve the balance of payments. Developing agricultural areas, especially those with favorable natural and climatic conditions, can enhance job opportunities and raise living standards by ensuring food availability.

Algeria has made efforts to prioritize the agricultural sector due to its economic and social importance. Key reforms include:

Organizing land use through Law 03–10 (August 15, 2010) that defines agricultural land use conditions [5].
Encouraging agricultural investments.
Combating desertification and drought while modernizing production methods [4].
Establishing the Natural Disaster Compensation Fund (FGCA) to assist and compensate farmers affected by disasters [2].
Creating the National Agricultural Investment Development Fund (FNDIA) [4].

Despite these efforts, Algeria still faces numerous challenges in converting its agricultural potential to meet global market demands, including:

- Difficult climate changes.
- Frequent agricultural disasters in various regions [6].

Table 2 Revenue of the Algerian insurance market for 2022

		In millions of DZD
Branch	Amount	Share (%)
Automobile	65,243	47
Damage insurance	64,367	46
Transport risk	7407	5
Agricultural risk	**2445**	**2**
Total	**139,507**	**100**

Source Annual Insurance Report in Algeria, Ministry of Finance, Algeria, 2022, p. 12

- Insufficient water resources for irrigation.
- Unclear mechanisms for addressing agricultural risks.
- Technological advancements may introduce new risks that are difficult to predict.

Thus, the insurance industry plays a vital role in supporting farmers and agricultural activities, especially in managing various risks related to workplace accidents, illnesses, and crop or livestock issues, despite challenges in risk assessment and pricing of insurance products.

3.2 *Reading Agricultural Insurance Indicators in Algeria*

The insurance sector has recently been impacted by the COVID-19 pandemic, with a 5% decline in revenue in 2020 compared to 2019, dropping from 144 billion DZD to 137 billion DZD. Agricultural insurance in Algeria is experiencing weak demand from farmers, despite numerous available products, due to high premium costs, misalignment with agricultural activity characteristics, and inadequate compensation.

The agricultural sector accounted for only 2% of total insurance revenue by the end of 2020, reaching just 2.4 billion DZD, as indicated in the following Table 2.

The weakness of agricultural insurance's share in the Algerian market has necessarily led to a low value of compensations compared to other insurance branches, as it represents only about 1% of total compensations, as illustrated in the following Table 3.

3.3 *Agricultural Insurance and Natural Disaster Risk*

The agricultural activities are often exposed to natural disasters (droughts, hurricanes, floods, hail, snow, landslides), which significantly affect agricultural products. All these risks are insurable. Following a series of natural disasters in various regions,

Table 3 Structure of insurance compensation in Algeria for 2022

		In millions of DZD
Branch	Amount	Share
Automobile	41,411	59%
Damage insurance	17,612	25%
Transport risk	893	3%
Agricultural risk	**7488**	**1%**
Total	**69,685**	**100%**

Source Annual Insurance Report in Algeria, Ministry of Finance, Algeria, 2022, p. 23

particularly after the Boumerdes earthquake on May 21, 2003, the Algerian government enacted laws requiring different stakeholders to subscribe to insurance contracts against natural disaster risks.

While experts in natural disaster risk management have yet to develop a specific model for estimating these disasters, some indicators can be gleaned through:

- Direct observation of damage from previous natural disasters related to building types, disaster intensity, and population density.
- Identification of areas prone to various natural phenomena such as earthquakes, volcanoes, and hurricanes.
- Understanding the ground and source of natural risks, which vary by location.
- Utilizing models to assess the degree and extent of energy concentration and speed, enabling the evaluation of earthquake impacts.

The relationships among different elements can provide new data for estimating natural risks. Insurance specialists have collaborated with natural phenomenon researchers to create a mechanism for evaluating and estimating these risks in Algeria, considering the location of agricultural areas and the level of risk from earthquakes, floods, and hurricanes. All 1541 municipalities in the country have been classified by the relevant authorities for pricing purposes, as shown in the following Table 4.

Z_SIS: This refers to the seismic zone type, which takes values of 0, 1, 2a, 2b, and 3, according to the degree of exposure of the area to earthquakes. **R_INON** indicates flood risk, with O for yes and N for no. **R_TEMP** indicates hurricane risk in the area, with O for yes and N for no.

Table 4 Classification of natural disaster risk in certain areas of Algiers province

Tourism zone	WILAYA	Z_SIS	R_INON	R_TEMP
ZERALDA	ALGER	3	O	N
BIRTOUTA	ALGER	3	N	N
STAOUELI	ALGER	3	O	N

Source Adapted from the Table of Natural Disaster Risks and Seismic Zones, National Insurance Council (CNA), Algeria

3.4 *Fire Risks and Civil Liability for Actors in Agricultural Activities*

Among the most significant risks that can affect agricultural activities of all kinds is the risk of fire and the consequences of civil liability for those involved in agricultural activities.

Fire insurance covers fire, explosions, and lightning. The assessment of fire risk for agricultural facilities is related to the type of activities conducted within their buildings, as well as the area of crops. The risk is also connected to the preventive measures and tools in place, the type of heating used, the proximity or distance from civil protection units, and the materials used in construction (wood or stone, etc.), among other factors that can affect this type of risk.

This type of insurance can cover buildings based on a valuation equivalent to the cost of reconstruction, in addition to other expenses such as architect fees, cleaning, and other costs. It can also include all movable property within the agricultural buildings at the cost of restoring them to their original state on the day of the incident.

Furthermore, within the context of agricultural activities, any person acting within their functions is considered a civil liability for the consequences of their mistakes toward others. Insurance companies can provide coverage for their civil liability to compensate for damages they may cause to others, as will be explained in some examples later. The purpose of civil liability insurance is to cover material losses resulting from the insured's liability due to damages caused to others, for which they are legally responsible [3]. Civil liability insurance aims to provide protection through the guarantee offered by the insurance company, allowing the farmer to compensate the affected party for all losses incurred without bearing the losses resulting from their liability.

The civil liability insurance contract for farmers considers the insured amount or maximum coverage, which can be related to various factors such as the nature and type of their agricultural activity, the area of land used, the number of livestock or poultry raised, the number of beehives, and other factors depending on the case.

When assessing civil liability risk, insurance companies follow certain elements according to the civil code articles 124, 134, to 140 from the Algerian Civil Law of 1975, including the maximum insured amount, the total wage mass of the agricultural enterprise's workers, and the expected income of the farmer.

Various actors in the agricultural sector can benefit from numerous guarantees related to their activity, such as insurance against overseas travel risks, vehicle insurance, and regular health monitoring provided by insurance companies in cooperation with veterinary services at no cost to the insured. Farmers can also benefit from some advantages related to the preventive care of their products, receiving advice on fertilizers, as well as insurance for loans aimed at agricultural investment.

4 Conclusion

The rise of food insecurity in most countries worldwide has been exacerbated by the impacts of the COVID-19 pandemic, the ongoing conflicts and climate change, all of which have significantly affected food and fertilizer supplies. This situation is likely to impact public security in many nations and complicate matters further in the near future.

These numerous risks, along with additional threats that can affect agricultural activities, such as damages from natural disasters, crop fires, and diseases or epidemics affecting livestock and/or crops, as well as civil liability for various actors in the sector towards individuals or their properties, present several challenges related to how to manage them to contribute to achieving food security.

However, these difficulties in managing emerging risks do not pose the same severity for insurance companies, whether in terms of assessing and identifying risks or evaluating and handling damages and compensations. An insurance system can play a crucial role in better supporting various actors in the agricultural sector, thus contributing to achieving food security and public safety in a country like Algeria, which heavily relies on the hydrocarbons sector. This is despite the noted shortcomings that need to be addressed through the following recommendations:

- In addition to the necessity of utilizing all necessary funding tools and mechanisms to encourage agricultural investment, it is essential to also use all available tools and mechanisms for protecting against various risks that agricultural actors may face.
- Providing the necessary support to farmers to manage the risks associated with their activities by exempting them from fees and taxes related to insurance for the risks tied to their operations.
- Offering greater support to small farmers to increase production by accompanying them and providing technical support for managing all risks related to their activities.
- Alongside the need to diversify food sources and provide the necessary funding to implement adaptation measures for climate change, financial mechanisms must be established to provide necessary measures to adapt to climate changes and mitigate their impacts through an insurance system against these changes' damages. This may include utilizing market mechanisms to enhance risk management in food production sectors.
- Greater emphasis should be placed on training for both agricultural actors and specialists working in the insurance industry to more accurately identify various risks and seek suitable insurance products to cover them.
- Raising awareness among farmers about the importance of insurance in managing risks related to their professional lives.
- Simplifying insurance products to facilitate understanding for farmers.
- Working towards creating a national risk map related to agricultural activities, involving insurance companies to assist in finding necessary strategies for managing them.

References

1. Derouiche G (2007) Climate risks and Algerian agriculture. B.N.E.D.E. Algéria
2. JORADP (1990) Article 44 and 45, Executive Decree No. 90–58 of May 26, 1990, relating to the organization and operation of the FGCA (agricultural disaster guarantee fund)
3. JORADP (1995) Art 56, Ordinance 95–07 of January 25, 1995 relating to insurance, official journal of the Algerian Republic. Algeria
4. JORADP (2005) Executive Decree No. 05–413 of October 25, 2005, establishing the operating conditions of the special allocation account No. 302–067 entitled "national fund for the development of agricultural investment, JORADP No. (72) of November 2, Algeria
5. JORADP (2010) Law 03–10 of August 15, 2010 stipulates the conditions and methods for exploiting agricultural lands. Algeria
6. Saida Z (2007) Insurable agricultural potential and development prospects. Algeria
7. World Bank (2022) What to know about food security and climate change. Retrieved 10 Jan 2023, from https://www.albankaldawli.org/: https://www.albankaldawli.org
8. World Food Programme (2022) Annual review 2021. Rome, Italy, World Food Programme
9. Belasal bint Nabi Y, Al-Hussein A (2020) The repercussions of the Corona pandemic on food security. Al-Mishkat J Econ Developm Law 05(12):168
10. Fatima TBA (2012) Economic integration and strategies for achieving Arab food security in light of international regional challenges. Chlef
11. Kristialina G, Sosa S, Rutter B (2022) The global food crisis requires support for citizens, trade openness, and increasing local harvests. Washington, USA, International Monetary Fund

Irrigation Water Management Challenges and Solutions for Arid, Saline, and Unfertile Lands

Farhat Abbas, Rachid Benlamri, and Aitazaz A. Farooque

Abstract This article explores the complexities of irrigation water management in difficult environments and examines the innovative solutions being implemented worldwide and the potential for adapting such techniques in GCC countries. Effective irrigation water management in arid, saline, and infertile lands has been focused in the recent in locally produced literature, reflecting the growing importance of sustainable agricultural practices in challenging environments. This review synthesizes findings from recent publications mostly addressing issues of GCC countries, providing insights into the current state of research and emerging solutions. By understanding and addressing the unique challenges of arid, saline, and unfertile lands, more resilient agricultural systems, capable of withstanding the pressures of climate change and growing global food demands, must be developed. Irrigation water management challenges and solutions for arid, saline, and unfertile lands need a clear understanding for developing sustainable solutions. Integrating traditional knowledge with modern scientific research can offer valuable insights into sustainable irrigation practices tailored to specific local conditions as being explored by a project funded by the Qatar National Research Fund to the University of Doha for Science and Technology. This project is about the development of smart agricultural technologies to optimize resource allocation to ensure food security as a pathway towards sustainable vegetables and date palm production in Qatar.

Keywords Precision agriculture · Irrigation systems · Water conservation

F. Abbas (✉) · R. Benlamri
University of Doha for Science and Technology, Doha, Qatar
e-mail: farhat.abbas@udst.edu.qa

A. A. Farooque
Canadian Centre for Climate Change and Adaptation, University of Prince Eeward Island, Charlottetown, PE, Canada

S. Al-Naemi et al. (eds.), *Water and Food Security in the Face of Climate Change: Challenges and Opportunities for Resilience*, World Sustainability Series,
https://doi.org/10.1007/978-3-032-00098-9_48

1 Context

Irrigation water management is a critical component of agricultural sustainability, particularly in regions characterized by arid climates, saline soils, and infertile lands. These challenging environments demand innovative and modern water management techniques to ensure crop yield leading to food security. Arid regions suffer from low and erratic rainfall, which necessitates the efficient use of limited water resources. Saline soils, often the result of improper irrigation and drainage practices, further complicate crop growth due to high salt concentrations that hinder plant uptake of water and nutrients. Similarly, infertile lands, which lack essential nutrients and organic matter, require strategic irrigation methods to enhance soil fertility and crop yield. Najm et al. [1] investigated saline agriculture techniques, including cultivation of salt-tolerant crops like quinoa and barley. These crops are more suited to the saline soils of the GCC region and can help enhance food security.

1.1 Challenges and Opportunities

Major challenges involve (i) integrated water and water resource management for water conservation, (ii) soil salinity and soil management for soil health, and (iii) climate change impacts on agriculture.

Water and Water Resource Management. Al-Khalaf et al. [2] examine the impact of rapid urbanization on water resources in the GCC region, stressing the need for integrated urban and agricultural water management approaches to ensure water conservation. Al-Nasr et al. [3] emphasized the importance of challenges such as IWRM in the GCC region. The IWRM approaches consider the entire water cycle and promote the coordinated development and management of water, land, and related resources. Community-based water management is one of the approaches often discussed in professional meetings and scientific literature. For example, Al-Kuwari et al. [4] highlighted the significance of community involvement in water management practices in Qatar. Engaging local communities in water conservation efforts and decision-making processes can enhance the effectiveness of irrigation practices.

Adaptation of innovative technologies and techniques is challenging for new farmers. However, precision agriculture technologies are key to smart agriculture. Al-Rashidi et al. [5] explored innovative technologies like soil moisture sensors and automated irrigation systems being implemented in Kuwait and Bahrain. These technologies optimize water use and reduce labor requirements, crucial for efficient water management. Soil management needs sustainable soil conservation techniques such as using both organic and inorganic sources of soil amendments for improved and healthier soils.

Soil Salinity and Soil Management. From an agricultural point of view, saline soils are characterized by sufficient neutral soluble salts that are detrimental to the growth of most crops. Electrical conductivity of saturated extract of saline soils at 25 °C is greater than 4 dS/m [6]. Benefits of organic and inorganic soil amendments to improve soil fertility and water retention in the infertile lands of the UAE and Saudi Arabia were demonstrated by Al-Hamadi et al. [7].

Climate Change and Food Security. The major challenge facing the world is climate change leading to food insecurity. Despite this, the solution lies in the development and use of sustainable food production systems. About 30% of the global emissions come from agricultural activities and food production systems across the globe. For about 10 billion global population of the world, it looks a challenge to produce food on a sustainable basis. Increased production of food items will need expansion in the use of natural resources leading to increased emissions causing further global warming and climate change.

The work of Brown et al. [8] examined adverse impacts of changing climates on the availability of water to schedule irrigation in arid and saline regions. Understanding these impacts is crucial for developing adaptive management strategies. Al-Maamari et al. [9] examine the impacts of climate change on water resources in the GCC region, emphasizing the need for adaptive irrigation strategies. Their research underscores the importance of resilient agricultural practices to cope with increasing temperatures and changing precipitation patterns.

1.2 Potential Solutions

Potential solutions for the above challenges have been discussed in the literature. For example, a multidisciplinary approach that combines advancements in technology, agronomy, and sustainable practices. Solutions such as precision irrigation, the use of drought-resistant crop varieties, and soil amendments play a pivotal role in optimizing water use efficiency, improving soil health, and combating adverse climate change impacts. Modern irrigation water management techniques and their adaptation require economic analysis. Al-Hajri et al. [10] provided an economic analysis of various irrigation technologies in Qatar, assessing their cost-effectiveness and potential return on investment. This helps policymakers and farmers make informed decisions about technology adoption.

System Automation and Integrated Water Resource Management (IWRM). Recent publications such as those by Wang et al. [11] explored innovative technologies like soil moisture sensors and automated irrigation systems that optimize water use and reduce labor requirements. These publications collectively highlight the multifaceted nature of irrigation water management challenges and the diverse array of solutions being explored. By synthesizing the latest research, this review underscores the importance of continued innovation and collaboration in addressing the pressing issues of water scarcity, soil salinity, and infertility in agriculture. For example, Green et al. [12] discussed the principles of Integrated Water Resource Management (IWRM) and its application in managing water resources in arid and saline regions. IWRM promotes a holistic approach to water management, considering all aspects of the water cycle.

Martinez et al. [13] emphasized the importance of community-based approaches to water management, highlighting the benefits of local participation and collective action in managing water resources sustainably. Jones et al. (2023) [14] discussed the role of policy and institutional frameworks in promoting sustainable irrigation practices. Effective policies and supportive institutions are essential for the successful implementation of water management strategies.

Johnson et al. [15] provided an economic analysis of different irrigation systems, evaluating their cost-effectiveness and potential return on investment in arid and saline regions. This helps farmers make informed decisions about adopting new technologies and methods of water resource generation. Lopez et al. [16] reviewed sustainable water use practices, including deficit irrigation and the use of treated wastewater, as viable options for addressing water scarcity in agriculture. Hassan et al. [17] investigated the feasibility of using desalinated water for irrigation in saline regions. Advances in desalination technologies offer a potential solution to water scarcity and salinity problems.

In Qatar and the UAE, treated wastewater is increasingly being used to irrigate landscapes and agricultural fields, reducing the dependency on desalinated water. Al-Mashaqbeh et al. [18] highlighted the critical role of desalination in addressing water scarcity in Qatar and other GCC countries. Desalination plants provide a significant portion of the freshwater used for irrigation, despite the high energy costs and environmental impacts associated with this technology. Al-Katheeri et al. [19] explored the use of treated wastewater for irrigation as a sustainable alternative to freshwater.

Desalination of Irrigation Water. Water scarcity is the largest threat to agricultural production in dry regions around the world. The potential solution for water availability lies in desalination to provide enough water for agriculture. Desalination removes salts to purify saline water. Facing extreme droughts as well as population growth, the affected countries are planning to desalinize sea and brackish water for domestic and agriculture use. The latter, uses about 70% of the Planet Earth's freshwater, despite water scarcity is a major challenge for the production of food items of more than 40% of the population of world. Estimated needs of water for the agricultural sector will reach to about 19% more than the current requirements by 2050.

Water Harvesting Techniques. Rainwater harvesting is the practice of collecting and storing rainwater in a surface or underground aquifer. An aquifer is an underground rock formation that contains groundwater. Rainwater must be collected before it is lost as surface runoff. This man-made water resource can be harvested at any time. Rainwater usually serves as a natural recharge for groundwater. Al-Ali et al. [20] review traditional and modern water harvesting techniques employed in Oman and the UAE. Methods like terracing and the construction of water reservoirs are essential for capturing and storing rainwater in these arid regions. Recent publications like that of Ahmed et al.) [21] reviewed traditional and modern water harvesting techniques, such as rainwater harvesting and the construction of check dams, which are crucial for capturing and storing water in arid regions.

Soil Salinity and Soil Management. Moderate salinity, especially when uniform across the field, often goes undetected because it does not cause visible damage other than restricted growth. Leaves of plants growing in saline areas may be smaller than normal leaves and dark blue-green in color. Increased juiciness is often due to salinity, especially when there is a high concentration of chloride ions in the soil solution. Plants in saline soils often look the same as plants growing under water stress (drought), but wilting is much rarer because the osmotic pressure of the soil solution usually changes gradually and the plants adjust the internal salinity sufficiently to maintain turgor pressure and avoid wilting. Li et al. [22] explored methods for mitigating soil salinity, such as gypsum application and improved drainage systems. These interventions help to reduce salt accumulation in the soil, promoting healthier crop growth.

Organic Amendments. Soil organic amendments are sourced from natural resources including chicken manure, cow dung, and other animal waste [23]. Composting of such natural wastes is necessary for bringing the organic fertilizer to par with acceptability as a soil amendment. The use of raw animal waste is harmful to the crops and causes methane emissions to the atmosphere.

Inorganic or mineral additives. These additives typically contain minerals associated with soil fertility. A combination of composted sewage sludge and steel mill slag has also been used as a potential amendment resulting in improved soil fertility and plant growth [24]. The work of Patel et al. [25] demonstrated the benefits of organic amendments, including compost and biochar, in improving soil fertility and water retention in infertile lands. These amendments enhance soil structure and nutrient availability, leading to better crop performance.

Climate-Smart Agriculture to Combat Climate Change Impacts. A study by Smith et al. [26] underscores the importance of integrating climate-smart agricultural practices, such as conservation tillage and agroforestry, to enhance resilience to climate variability and improve water management in challenging environments. Climate-smart agriculture takes help from remote sensing and geographical information system (GIS) applications. Al-Shammari et al. [27] highlighted the application of remote sensing and GIS technologies in monitoring and managing irrigation practices in Kuwait and Bahrain. These tools provide accurate data on soil moisture and crop health, aiding in precise irrigation planning. Chen et al. [28] highlighted the use of remote sensing and GIS for monitoring soil moisture levels and planning irrigation schedules. These tools provide real-time data, facilitating more efficient water management.

2 Food Security Research at the University of Doha for Science and Technology, Qatar

The University of Doha for Science for Science and Technology (UDST) in Qatar is pioneering a food security project funded by the Qatar National Research Fund to promote irrigation system automation for IWRM, modern technology and artificial intelligence for agricultural activities, soil health improvement through circular bioeconomy, and capacity building of agriculture workforce in Qatar under the concept of climate-smart agriculture to combat food insecurity climate change impacts [29].

2.1 System Automation

The ESP32 microcontroller (Espressif Systems Co. Ltd., Brno, Czech Republic), which is frequently used for Internet of Things applications like robotics, industrial automation, and high-performance sensor networks, is connected to a capacitive soil moisture sensor, specifically the SM100 WaterScout soil moisture sensor (Source: Spectrum Technologies; Aurora, Illinois, USA). According to the methodology followed [29], the soil moisture sensor must function intelligently for an irrigation system to operate in real-time in an agricultural area. The components of a smart irrigation system include drip irrigation lines with emitters or popup sprinklers placed at predetermined lengths, wirelessly integrated with soil moisture sensors placed in vegetable or fruit tree orchards, and water supply lines from a water tank or freshly pumped source, connected via a solenoid valve.

A wireless SM-100 soil moisture sensor (A), an irrigation solenoid valve controller (B), a single channel relay (C), a solenoid valve (D), a step-down transformer 240/24VAC (E), and soil state representations (Fig. 1) comprised the laboratory experimental setup for this work.

The solenoid valve controller was set up to read the volumetric water content from the database every 15 s, and the wireless soil was set up to transmit data to the Firebase database every 15 s. It determines whether to turn the solenoid valve on or off after reading the data. The wilting point and field capacity of the sandy loam soil in the UDST landscape were considered to be 30% and 70%, respectively, of the soil's volumetric water content. Program codes were created to activate the solenoid valve at the wilting point's volumetric water content and to deactivate it at the field capacity's volumetric water content.

The two microcontrollers are shown in Fig. 1 as components A and B. Microcontroller B is used for wireless irrigation control, whereas microcontroller A is used for wireless soil moisture sensing. Micro USB cables (orange for microcontroller B and black for microcontroller A) are used to power the two microcontrollers. Microcontrollers A and B communicate wirelessly rather than through physical wiring. The two microcontrollers can communicate regardless of their distance from one another

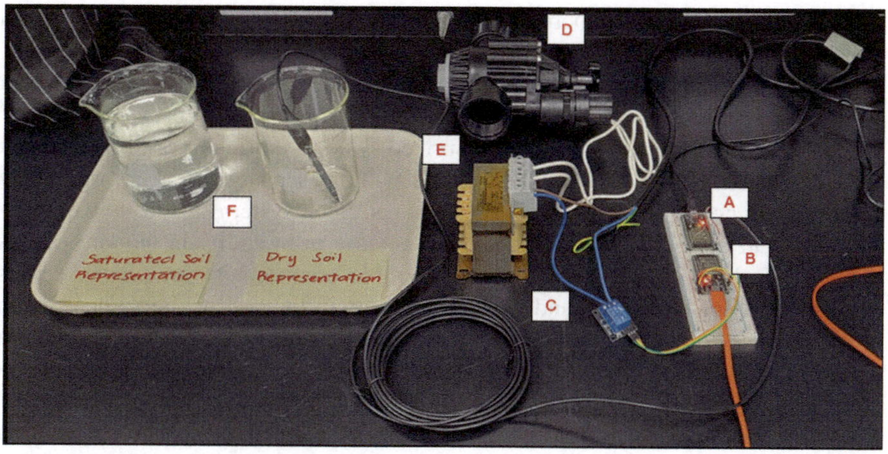

Fig. 1 Laboratory experimental setup showing its contents labeled as **a** a wireless SM-100 soil moisture sensor **b** a wireless irrigation solenoid valve controller, **c** a single channel relay, **d** a solenoid valve, **e** a step-down transformer 240/24VAC, and **f** soil state representations [29]

thanks to the availability of Wi-Fi systems. As a result, this prototype design cannot function without a Wi-Fi connection.

2.2 *Smart Irrigation as a Precision Agriculture Technique*

Precision agriculture is a farm management concept based on observing, measuring, and responding to inter and intra-field variability in crops. Precision agriculture technologies permit the use of technological innovations such as GPS-guided drones and agriculture machinery, system automation based on sensors network, and optimization of natural resources via variable-rate applications of crop inputs.

Climate-smart agriculture practices have gained popularity and are at an exploratory stage in Qatar. Smart agriculture precisely manages crop inputs (fertilizer, pesticides, and irrigation water) and operations (mechanical harvesting and post-harvest handling) both in spatial and temporal dimensions. The novel, innovative, and cost-effective technologies need to be evaluated for their adoption in Qatar's indoor and outdoor agriculture.

Digital agriculture is the use of digital technology to integrate agricultural production from the paddock to the consumers. Digital agriculture, also termed as smart farming or e-agriculture comprises tools that digitally collect, store, analyze, and share electronic data and/or information in a farming system. The Food and Agriculture Organization (FAO) of the United Nations described the digitalization process of agriculture as the digital agricultural revolution. FAO harnesses the power of digital technologies to pilot, accelerate and scale innovative ideas with high potential, for impact on agricultural production, transforming digital solutions and services into global public goods.

The current automation of irrigation water application systems in indoor and outdoor agriculture of Qatar uses time-controlled operations of the systems. Particularly, the irrigation valves for the pressurized systems (sprinkler or drip) open and close based on pre-set fixed time intervals without sensing the need, right time, and correct amount of plant water requirements. Such systems result in wastage of precious water in outdoor systems and an increase in humidity in indoor systems, if not properly ventilated, causing a susceptible environment for plant diseases. Accumulation of hazardous and unused nutrients in the soil or the soilless media of greenhouses (e.g., coco coir, perlite, and rock wool), sourced from fertigation with drip lines, create environmental degradation and crop yield loss. However, the crops grown hydroponically have less risk of plants being exposed to soil diseases and allow the saving of water and agrochemicals. For Qatar, precision agriculture initiatives can include evaluating Qatar's existing outdoor and indoor water management systems to suggest innovations for their automation to deliver irrigation water and plant nutrients at right time and in the correct amount based on crop water and nutrient requirements. A comprehensive and interactive approach is shown in Fig. 2.

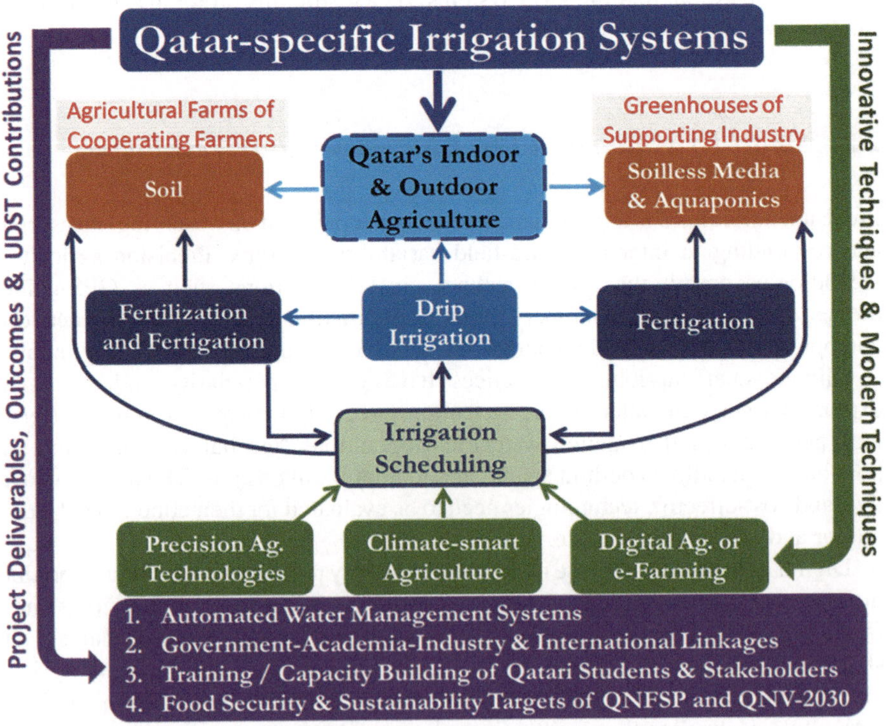

Fig. 2 Phases of the proposed project, anticipated activities, and subsequent outcomes

As described in Fig. 2, the use of innovative technologies and modern techniques including precision agriculture technologies, climate-smart agriculture concepts, and digital agriculture approach for both outdoor (field) and indoor (greenhouse) cultivation scenarios of Qatar can lead to solutions for Qatar's irrigation water management challenges. Interactive irrigation scheduling will automate drip irrigation systems operating in fertilized or fertigated agricultural fields and fertigated soilless media or aquaponics in greenhouses.

Evaluation of the existing irrigation water management system in place in Qatar will help determine their suitability and/or poor performance in indoor and outdoor agriculture of Qatar. Improvements and innovations will be suggested aiming at fine-tuning water and the related nutrient management practices for vegetable cultivation in Qatar. Development of innovative and automated irrigation water management systems based on scientific investigations, engineering skills, and lab/field evaluations will train early career researchers and students in the fields of climate-smart agriculture production technologies. The Qatari students trained in this project will be able to secure employment in industry, government, and academia. Research and development in emerging areas of climate-smart agriculture practices will make Qatar a regional leader in technology development and its use to sustain agriculture.

3 Conclusion

Irrigation water management in Qatar and the other GCC countries presents unique challenges due to the arid climate, scarce water resources, and saline soils prevalent in the region. This review synthesizes findings from recent studies, providing insights into the current state of research and emerging solutions specifically tailored to these countries. These publications collectively underscore the critical challenges and innovative solutions for irrigation water management in GCC countries. The region's unique environmental conditions necessitate tailored strategies that integrate technological advancements, sustainable practices, and strong policy support to ensure water security and agricultural sustainability. For policy and institutional support, Al-Saidi and Elagib [30] discussed the role of policy frameworks and institutional support in promoting sustainable irrigation practices across the GCC countries. Al-Mohannadi et al. [31] reviewed sustainable water use practices, such as deficit irrigation and the use of brackish water, as viable options for addressing water scarcity in Qatar and other GCC countries. Effective water management for innovative irrigation solutions, soil health improvements, and smart agricultural technology are pivotal for addressing food security issues in GCC countries. The UDST project MME03-1121-210025 explores natural resource optimization and allocation using smart agricultural technologies as part of food security research in Qatar [29].

Acknowledgements QNRF project MME03-1121-210025 titled, "Development of Smart Agricultural Technologies to Optimize Resource Allocation to Ensure Food Security: A Pathway Towards Sustainable Vegetables and Date Palm Production in Qatar".

References

1. Abou Najm M, Al-Ghamdi SG, Al-Kuwari S (2022) Saline agriculture in the GCC: cultivating salt-tolerant crops. Agronomy 12(3):681
2. Al-Khalaf N, Al-Obaid A, Al-Rashed A (2023) Impact of urbanization on water resources in the GCC region. Urban Water J 20(3):293–305
3. Al-Nasr N, Al-Saadi R, Al-Wahaibi T (2022) Integrated water resource management in the GCC region. Water Policy 24(4):591–606
4. Al-Kuwari H, Al-Sulaiti M, Al-Naimi A (2023) Community-based water management in Qatar. Water International 48(2):172–186
5. Al-Rashidi M, Al-Shehhi S, Al-Najjar M (2022) Technological innovations in irrigation in Kuwait and Bahrain. J Agric Food Chem 70(14):4268–4279
6. USDA (1954) Diagnoses and improvement of saline and alkali soils. In: Richards LA (ed) Agricultural handbook No 60, Riverside CA, USSL
7. Al-Hamadi M, Al-Khulaifi M, Al-Dosari A (2023) Organic and inorganic soil amendments to improve soil fertility in the UAE and Saudi Arabia. Soil Sci 188(2):102056
8. Brown A, Green R, Liu Z (2023) Impact of climate change on water availability and irrigation needs. Clim Risk Manag 39:100437
9. Al-Maamari S, Al-Hamidi S, Al-Rashidi Y (2023) Climate change adaptation strategies for water resources in the GCC. Climate Policy 23(1):98–112
10. Al-Hajri S, Al-Mohannadi M, Al-Suwaidi K (2023) Economic feasibility of irrigation technologies in Qatar. Water Resour Manage 37(4):1345–1361
11. Hassan A, Rahman M, Ismail S (2023) Feasibility of using desalinated water for irrigation in saline regions. Desalination 543:116103
12. Johnson D, Jones E, Smith K (2022) Economic analysis of different irrigation systems in arid and saline regions. Irrigat Drainage Syst Eng 11(3):194–206
13. Jones P, Brown C, Taylor R (2023) Policy frameworks for promoting sustainable irrigation practices. J Environ Manage 318:115584
14. Martinez L, Garcia F, Fernandez R (2023) Community-based approaches to water management in agriculture. Agric Syst 202:103422
15. Wang Q, Chen Y, Li J (2022) Innovative technologies in irrigation: soil moisture sensors and automated systems. Sensors 22(21):8087
16. Lopez R, Smith J, Nguyen T (2023) Sustainable water use practices in agriculture: deficit irrigation and treated wastewater. Agric Water Manag 249:107148
17. Green T, White S, Black M (2023) Integrated water resource management: principles and application in arid regions. Water Resour Manage 37(5):1289–1305
18. Al-Mashaqbeh O, Al-Omari A, Al-Zubi A (2022) The role of desalination in addressing water scarcity in GCC countries. Desalination 537:115587
19. Al-Katheeri H, Al-Mansoori M, Al-Sayed S (2023) The use of treated wastewater for irrigation in Qatar and UAE. Environ Sci Pollut Res 30(8):10927–10936
20. Al-Ali A, Al-Tamimi N, Al-Kindi A (2022) Traditional and modern water harvesting techniques in Oman and UAE. J Arid Environ 189:104463
21. Ahmed S, Al-Ghamdi S, Al-Najjar H (2023) Review of traditional and modern water harvesting techniques in arid regions. Water 15(1):72
22. Li W, Zhang S, Zhao J (2023) Methods for mitigating soil salinity: gypsum application and improved drainage systems. Soil Sci Soc Am J 87(1):58–68
23. Fares A, Abbas F, Ahmad AA, Deenik JL, Safeeq M (2008) Response of selected soil physical and hydrological properties to manure amendment rates, level and types. Soil Sci 173:522–533
24. Farooque A, Zaman Q, Madani A, Abbas F, Percival DC, Esau TJ (2011) Ecological impacts of the N-Viro biosolids land-application for wild blueberry (Vaccinium angustifolium. Ait) production in Nova Scotia. J Environ Sci Health Part B 46(4):366–379
25. Patel M, Thakkar R, Desai P (2022) Organic amendments for improving soil fertility and water retention in infertile lands. Compost Sci Utilizat 30(4):281–292

26. Smith L, Johnson P, Thompson D (2023) Integrating climate-smart agricultural practices for enhanced resilience. Climate Change Agricul 11(2):204–218
27. Al-Shammari Z, Al-Bahrani H, Al-Mutairi S (2022) Remote sensing and GIS applications in irrigation management in Kuwait and Bahrain. Remote Sens 14(10):2420
28. Chen H, Zhao X, Liu Y (2022) Remote sensing and GIS for monitoring soil moisture and irrigation planning. Remote Sens Appl: Soc Environ 25:100670
29. Benlarmi R et al. (2022) Development of smart agricultural technologies to optimize resource allocation to ensure food security: a pathway towards sustainable vegetables and date palm production in Qatar, MME03–1121–210025 funded by Qatar National Research Fund
30. Al-Saidi M, Elagib NA (2023) Policy frameworks and institutional support for sustainable irrigation in GCC countries. Water Resour Res 59(2):e202200418
31. Al-Mohannadi M, Al-Obaid S, Al-Ruwaili H (2022) Sustainable water use practices in Qatar and GCC countries. Sustain Water Resourc Managem 8(1):12

Water Scarcity Threats to Qatar's Agriculture: A Hindrance to Food Security Initiatives

Hafsa Chohan, Farhat Abbas, Rachid Benlamri, and Aitazaz A. Farooque

Abstract This article explores the water scarcity and thus food security challenges faced by the Gulf Cooperation Council (GCC) region countries, particularly Qatar, providing some efficient solutions to effectively mitigate this issue. The arid climate and limited freshwater resources have necessitated innovative strategies for water management and agricultural sustainability. Reviewing the literature on recent scientific research and applying it to the ongoing food security project of the University of Doha for Science and Technology (UDST) has helped explore the current state of water resources, the implications for food production, and the effectiveness of current food security initiatives in the context of Qatar. By analyzing policy frameworks, technological advancements, and regional cooperation efforts, this paper aims to identify key strategies for mitigating water scarcity and enhancing food security. The findings intend to inform future research and policy development, contributing to more resilient and sustainable solutions for the GCC's water and food challenges.

Keywords Qatar · Water scarcity · Irrigation system · Water management

1 Introduction

Water scarcity is a profound and escalating challenge for Qatar and the broader Gulf Cooperation Council (GCC) region. The region is characterized by its extreme arid conditions and minimal natural freshwater resources. The GCC countries, including Saudi Arabia, the United Arab Emirates, Kuwait, Oman, Bahrain, and Qatar, face unique and pressing water issues due to their desert climates, rapid population growth,

H. Chohan (✉) · F. Abbas · R. Benlamri
University of Doha for Science and Technology, Doha, Qatar
e-mail: farhat.abbas@udst.edu.qa

A. A. Farooque
Canadian Centre for Climate Change and Adaptation, University of Prince Eeward Island, Charlottetown, PE, Canada

S. Al-Naemi et al. (eds.), *Water and Food Security in the Face of Climate Change: Challenges and Opportunities for Resilience*, World Sustainability Series,
https://doi.org/10.1007/978-3-032-00098-9_49

and increasing industrial and agricultural demands. In Qatar, the situation is particularly acute. The country receives minimal annual rainfall (< 100 mm/year), and its groundwater resources are limited and largely non-renewable (Abbas et al. 2023). Desalination, which provides a substantial portion of Qatar's water supply, comes with its own set of environmental and economic concerns. The reliance on desalinated water, while technologically advanced, is costly and energy-intensive, presenting additional sustainability challenges. This water scarcity poses significant obstacles to food security. It is a critical concern for the GCC nations as they strive to reduce their reliance on food imports and enhance domestic agricultural production. Water is a fundamental input for agriculture, and the limited availability impacts crop yields and farming practices. The GCC countries are investing in various strategies to address these challenges, including advancements in irrigation technology, the development of drought-resistant crops, and the implementation of water-saving practices. Ajjur et al. [1] presented that the big challenge for water resources is to stop the enduring depletion and manage aquifer storage while at the same time meeting the increased food demand. Managed aquifer recharge and utilizing the treated sewage effluent might be feasible solutions to augment and sustainably manage Qatar aquifers.

1.1 Major Threats and Potential Solutions

1.1.1 Limited Water Resources

Baalousha et al. [2] stated that several large projects such as the underground train network and football stadiums for the 2022 FIFA World Cup have created a large influx of expats into the country. This puts high stress on the limited water resources of the country. Darwish et al. [3] said that the increasing demand for food cannot be met by the scarce and dwindling conventional groundwater resources. These conventional resources have already been over-exploited by irrigation practices. The water problem should be solved by sustainable water management technologies to cope with ever-increasing water demands. Qatar depends on desalinated water to keep up with the increasing demand for water supply. As a result of limited freshwater resources, Dawoud et al. [4] indicated that to meet the present and future water demands of the region, the available options are limited to either long-distance water imports or transfers from neighboring countries or investment in large-scale seawater desalination technology.

The desalination process is highly energy intensive. Qureshi et al. [5] estimated that energy use for water production in the UAE and Qatar has reached 30% of their total energy consumption, whereas 15% of daily oil production in Saudi Arabia is consumed by desalination plants. Hence, precision water conservation technologies are key to smart water management strategies.

1.1.2 Agricultural Challenges and Food Insecurity

The reuse of wastewater is afflicted with agricultural challenges. In most of the GCC countries, wastewater is treated at the tertiary level. This situation may lead to elevated risks for human health by propagating pathogenic germs, especially for those who are in direct contact with the wastewater during the irrigation activity. Therefore, before proposing irrigation with wastewater, a comprehensive evaluation of its impacts on soil, plant, and human health needs to be accomplished. Darwish [6] stated that more advanced treatment technology would be required for irrigating vegetables and other crops for human consumption. Despite the majority of GCC groundwater being used in agriculture to increase food security, GCC countries remain net importers of agricultural goods [7].

1.1.3 Environmental Impact

Food insecurity caused due to water scarcity indirectly leads to environmental degradation. Over-extraction of groundwater leads to aquifer depletion, land subsidence, and increased salinity, which renders soil unfit for agriculture. Desalination plants discharge brine, which contains high salt concentrations and chemicals, back into the ocean, harming marine ecosystems. Once the soil is saline or degraded, it becomes challenging and costly to restore it for future agricultural use. The transport of food from distant regions increases the carbon footprint, contributing to global climate change. The effects of climate change on rainfall and temperature trends have already begun.

The work of Abulibdeh et al. [8] examined that wastewater requires a certain amount of energy for proper treatment, which in turn makes the region's water supply increasingly dependent on available energy. At the national level, Qatar will need to continue to improve technological solutions, reduce food waste, and expand trade relations and agricultural investments [9].

1.2 Potential Solutions

Literature has addressed potential remedies for the above problems. One example might be a combined strategy that integrates advances in agronomy, technology, and environmental practices. Improving soil health, reducing the negative effects of climate change, and maximizing water use efficiency all depend heavily on solutions like precision irrigation, the adoption of drought-resistant crop varieties, and soil additives. Integrating automated systems into agricultural practices can help manage water scarcity in an efficient way in countries that experience a hot and arid climate. Karanisa et al. [10] provided advanced technological solutions such as the Internet of Things (IoT), metering and communication infrastructure, artificial intelligence, advanced monitoring and control techniques, and smart greenhouses targeting ways to optimize resource use, maximize agricultural productivity, adaptation and mitigation against climate change.

1.2.1 Smart Irrigation System to Combat Limited Water Resources

Recent literature such as those by Zaina et al. [11] explored innovative technologies like smart irrigation systems that optimize water use and reduce labor requirements. Smart irrigation systems provide frameworks that improve plant management by scheduling controllers and sensors that efficiently use water. Due to the lack of cost-effective smart irrigation systems, developing countries use more water than developed countries that achieve the exact yield. Thus, there is a need to create an advanced technology-based and smart strategic system to utilize water effectively. The development of smart irrigation systems in hot-arid climates is essential in addressing water security. For example, Touati et al. [12] developed a ZigBee wireless system to monitor system performance in real time. Key environmental and climatic real-time factors along with the time and duration of irrigation as well as the amount of water being sourced were wirelessly recorded.

Al-Naemi et al. [13] emphasized that the controlled irrigation system delivered the right amount of irrigation levels based on water sensing in the soils and, therefore, improved the quality of the plant by delivering the just needed amount of water. Et-Taibi et al. [14] discovered that by seamlessly integrating multiple smart farms and linking them through a centralized control unit, these systems offer immense potential for optimizing resource usage and enhancing agricultural efficiency.

Chowdhury et al. [15] suggested using a cost-effective automated vertical hydroponic system using the Internet of Things (IoT) platform, which maintains and controls the environmental parameters and the efficient supply of nutrition and water for healthy growth of the plants. Aljughaiman et al. [16] suggested that the main step toward reducing irrigation water needs is to choose crop varieties that have less water demand but still provide sufficient added value. Greeshma et al. [17] suggested using IoT-based smart farming and irrigation systems that can automate the watering of plants, which can lead to increased irrigation efficiency, decreased water waste, and increased agricultural yields.

Baalousha et al. [18] highlighted some advancements in lysimeter hardware, data storage, and processing that have facilitated the accurate determination of water mass balance, allowing the continuous and efficient monitoring of soil evaporation. Smart agriculture addresses many issues related to crop production as it allows monitoring of the changes in climate factors, soil characteristics, soil moisture, etc. The Internet of Things (IoT) technology can link various remote sensors such as robots, ground sensors, and drones, as this technology allows devices to be linked together to operate wirelessly. Mohamed et al. [19] suggested the use of smart technologies that should be supported by governments in third-world countries at the level of small farms, as they aim to increase production and improve the efficient use of land and water resources.

Drip Irrigation System. Drip irrigation helps manage water by delivering it directly to the plant roots in a slow, controlled manner. This minimizes water loss through evaporation, runoff, or deep percolation. The system uses pipes, tubing, and emitters to target specific plants, reducing the amount of water needed compared to traditional methods like flood irrigation. Additionally, it ensures that plants receive water efficiently, even in hot, arid climates, promoting better water conservation and more sustainable agricultural practices.

Water Harvesting Techniques. Water harvesting techniques help manage water for food security by capturing and storing water during rainfall, ensuring a reliable water supply during dry periods. This stored water can be used for irrigation, enhancing crop growth even in drought-prone areas. By reducing dependence on unpredictable rain and improving water availability, these techniques help stabilize agricultural productivity. Water harvesting also prevents soil erosion and improves groundwater recharge, maintaining soil moisture levels necessary for sustainable farming. This increases agricultural resilience, helping farmers produce food consistently, thus promoting food security in regions with scarce water resources.

1.2.2 Introduction of Drought-Resistant Crops to Combat Agricultural Challenges

Introducing or developing drought-resistant and salt-tolerant crop varieties that can thrive in arid conditions is vital for the GCC. These crops require less water and can withstand harsh environmental conditions, making agriculture more feasible in water-scarce regions. Promoting native plants and crops naturally adapted to the arid climate could increase agricultural sustainability without excessive water use. Encouraging local farmers to grow crops that require less water or switching from traditional crops to more sustainable alternatives is essential. Expanding food imports from regions with greater water resources can alleviate the need for water-intensive domestic food production. Muhammed et al. [20] found that date palms and barley are some of the best-adapted crops for the desert climate, which can contribute to ensuring food security. The date palm, as in other Arabian countries, is a popular and traditional fruit in Qatar.

Hydroponic Farming. In arid environments, assembling hydroponic systems inside protected environments such as buildings or greenhouses is common. An automated nutrient supply system supplies plants with nutrients. Thus, the process facilitates the supply of plants with optimal concentrations of nutrients. Qatar will particularly benefit from aquaponics given; its arid climate, lack of an extensive amount of arable land, the need to diversify the economy while meeting food self-sufficiency, and the urgent need to reduce its environmental footprint [21].

Desalination Technology Advancements. Sources of renewable energy for desalinating sea and brackish water in urban cities in the GCC must be utilized to meet the increasing demand of water for food production and reduce gas emissions. Economic constraints can be greatly reduced with solar-operated reverse-osmosis units and intensive agriculture of high-value horticultural crops [22]. The work of Moossa et al. [23] demonstrated that an environmental impact assessment should be carried out before building a new desalination plant or prolonging the capacity of the prevailing one to limit the adverse impact of the facility as GCC countries hold 43% of the total global share of desalination capacity.

1.2.3 Climate-Resilient Agriculture to Combat Climate Impacts

Climate-controlled greenhouses conserve water by reducing evaporation and can help in growing a variety of crops with minimal irrigation water. The agriculture sector in the region needs immense support, guidance, and capacity development to adopt practices, approaches, and technologies that are indicative of sustainability and resilient to climate change. A study by Jonathan et al. [24] concludes that the climate-smart initiatives practiced in the GCC countries can be expanded to other Middle Eastern countries to address the local food demand which can subsequently reduce GHG emissions. Climate-controlled greenhouses allow for the growth of crops in a protected environment where humidity, temperature, and water usage are regulated, conserving water and increasing crop yield. They can be designed to use less water by capturing and recycling evaporated moisture. Scalet et al. [25] highlighted that at the level of national planning, policy should focus on food security, water security, climate-resilient health systems, and resilience to climate-driven migration based on a robust and scientific understanding of the drivers behind these threats. Iqbal et al. [26] highlighted that the goal of climate-resilient agriculture is to increase long-term agricultural yields and productivity by sustainably exploiting natural resources. These methods eliminate GHG emissions while also responding to climate change and fostering resilience in the agricultural sector.

2 Automated Irrigation Systems: A Technical Solution to Water Scarcity in Agriculture

Automated irrigation systems are a key technological achievement in solving the difficulties of water scarcity, particularly in arid regions such as Qatar. These systems are intended to maximize water usage by ensuring that crops receive the exact amount of water they require at the proper time, with no operator intervention. The use of such technology is crucial in Qatar's agriculture sector, where water resources are already scarce and traditional irrigation methods frequently result in water waste.

An automated irrigation system uses sensors, controls, and actuators to keep track of soil moisture levels, weather conditions, and crop water requirements. By doing so, it reduces human mistakes and assures effective water distribution. Soil moisture sensors, for example, can detect when water levels fall below a specific threshold and initiate irrigation, whereas weather sensors prevent watering during periods of heavy rainfall or high humidity, reducing water consumption.

Automated irrigation systems support Qatar's food security objectives by encouraging sustainable agriculture and minimizing reliance on water-intensive practices. They assist in alleviating the effects of droughts and erratic rainfall, both of which are becoming more common as a result of climate change. Furthermore, by reducing water use, these systems reduce farmers' operational costs, making them a cost-effective long-term solution.

Southeast Asian countries have been experimenting with sophisticated irrigation technologies, which show potential benefits for water management in similar climes. Smart irrigation, drip irrigation, and sensor-driven systems are all effective strategies.

Saif et al. [32] mention that in many regions of Asia, automated irrigation systems maximize water use by using data from soil moisture sensors, meteorological stations, and satellite imaging. For example, soil moisture monitors allow farmers to irrigate just when necessary, reducing water waste. In Vietnam, these technologies have cut water usage in rice cultivation by up to 30% while preserving yields [32].

Shao et al. [33] highlight that solar-powered pumps, particularly in off-grid rural regions, offer a sustainable alternative to fuel. Cooperatives in India, for example, use solar-powered irrigation to reduce energy costs and promote renewable energy. Indoor farms and vertical farming setups in Singapore and Thailand use automated hydroponics and LED lighting to save water usage by up to 95%. These systems recirculate water, minimizing runoff and evaporation, which can be crucial for water conservation in countries with few supplies [32].

Southeast Asian countries such as Vietnam and Myanmar have adopted telemetry-based gravity irrigation systems to modernize and improve water management. These devices use solar-powered sensors and communication modules to send real-time data every ten minutes. Farmers and managers may monitor water flow and soil conditions using smartphones, which improves irrigation efficiency [34]. Saif et al. [35] conclude that modern drip systems, such as those produced by Netafim, have sensors that monitor soil moisture and meteorological data, allowing for automatic water delivery modifications. These systems have shown significant water savings, with usage down by up to 70% compared to regular flood irrigation.

Precision agriculture instruments are becoming more prevalent in Southeast Asia, notably in major rice-based irrigation operations. For example, the Food and Agriculture Organization (FAO) funds studies that examine water management needs in multiple nations, allowing for specific modifications. Farmers in Thailand, Cambodia, and Laos use sensor networks for automated water distribution, which includes remote control features that optimize irrigation to meet real-time crop needs [35]. Idrica [36] state that another technological achievement is the implementation of smart metering and data integration systems. These systems provide precise control over water distribution, increasing efficiency by monitoring usage and finding leaks in real-time. In Singapore and Malaysia, smart meters linked with data analysis tools assist farmers and authorities in making timely decisions that match sustainability goals, seeking to reduce water wastage and assure a more consistent supply for agriculture.

Amhamed et al. [30] mention the Strategic Reserves of the Qatar National Food Security Program (QNFSP) that include the following initiatives: (a) to cope with any water crisis, it is essential to increase the reserve of potable water that could be used during any crisis. (b) Focus on decreasing the water salinity and net depletion in aquifers by improving the water management system for agriculture. Al-Addous et al. [31] highlight that solar-powered reverse osmosis (RO) desalination plants offer a promising pathway toward addressing the pressing water challenges faced by the region.

2.1 Working of Automated Drip Irrigation System

Soil Moisture Sensors: These sensors are inserted into the soil near the plant roots. They constantly monitor soil moisture levels and send data to a central control unit. When the moisture level goes below a specified threshold, the system initiates watering [29].

Weather Sensors: Many automated systems have weather sensors that monitor exterior variables such as rainfall, temperature, and humidity. This minimizes unneeded watering during rainy or humid weather, thereby conserving water [29].

Control Unit: The control unit processes data from the sensors and determines when and how much water is required. If the moisture levels are low and no rain is forecast, the machine will indicate that the valves should be opened [29].

Water Distribution: When the control unit gives the signal, the water valves open and allow water to flow through the irrigation pipes. Drip irrigation provides water directly to plant roots via small, precise emitters, reducing water evaporation and runoff [29].

Monitoring and Adjustment: As water is given, soil moisture sensors continue to monitor the soil. When the target moisture level is attained, the control unit turns off the water supply by closing the valves, which stops the irrigation [29].

Automation and Scheduling: Some systems can be designed to irrigate crops at specific periods, allowing for greater flexibility and water efficiency [29]. One such system is shown as a flowchart in Fig. 1.

3 Conclusion

In conclusion, water scarcity in Qatar and other GCC countries presents a critical challenge, particularly in its profound impact on agriculture and food security. The region's limited freshwater resources, coupled with agricultural challenges and climate change, have exacerbated the situation, threatening both economic stability

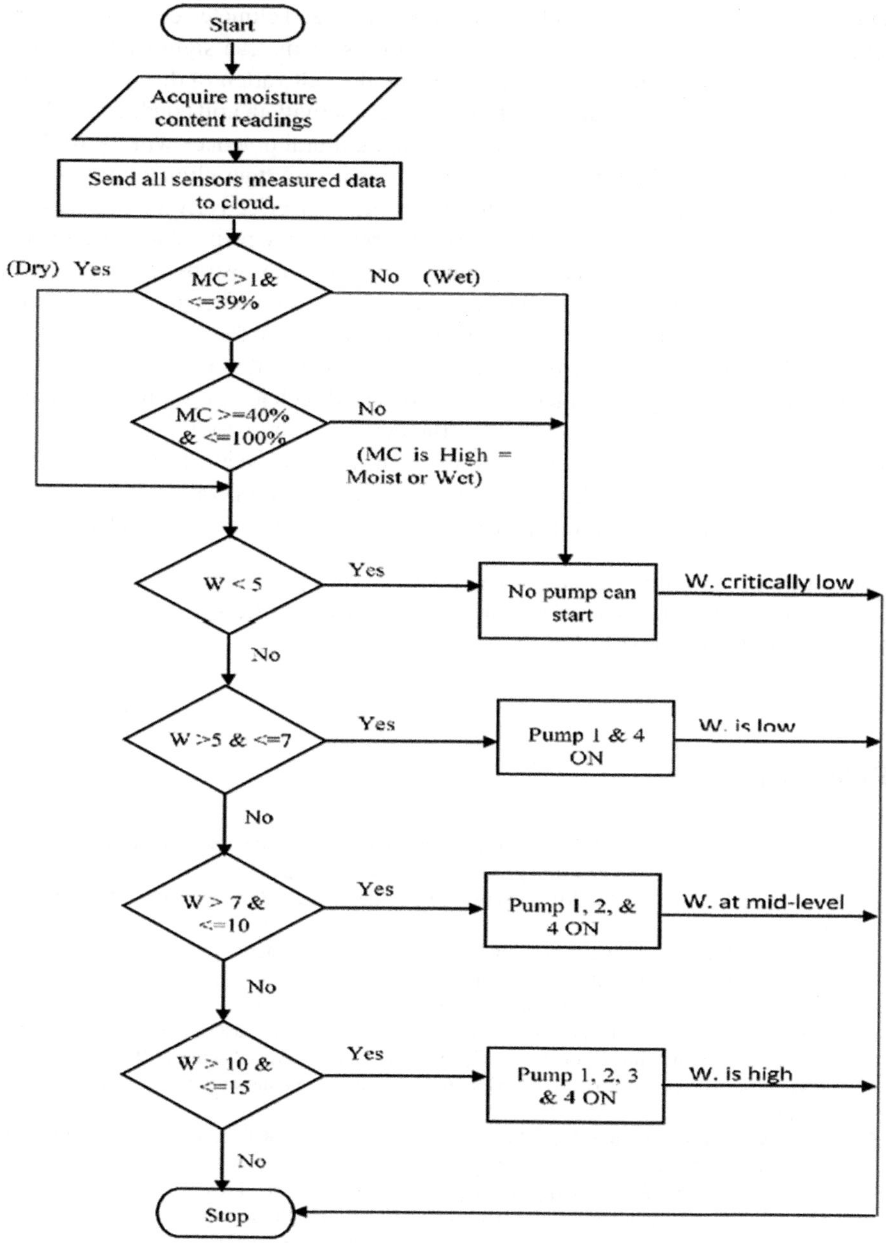

Fig. 1 Flowchart demonstrating automated irrigation system [28]

and food supply. However, the adoption of innovative and climate-resilient solutions offers a promising path forward. Smart irrigation systems can significantly reduce water wastage and optimize resource use, while the cultivation of drought-resistant crops will help ensure sustainable agricultural productivity in arid conditions. Moreover, the implementation of climate-resilient agricultural practices will fortify food security and safeguard the region's ecosystems against further climate impacts. By embracing these strategies, Qatar and the wider GCC can effectively address the dual crises of water scarcity and food insecurity, fostering a more resilient and sustainable future. Saif et al. [7] discussed that the high rates of water use by the agricultural sectors of GCC countries should lead to increased irrigation efficiency coupled with better crop choices. Ultimately, addressing water scarcity and food insecurity in Qatar and the GCC is not just an environmental issue but also a matter of economic stability and human well-being. The integration of advanced technologies with sustainable practices and a forward-thinking policy framework can transform these challenges into opportunities for a more sustainable, resilient, and food-secure future. The UDST project MME03-1121-210025 explores natural resource optimization and allocation using smart agricultural technologies as part of food security research in Qatar [27].

Acknowledgements QNRF project MME03-1121-210025 titled, "Development of Smart Agricultural Technologies to Optimize Resource Allocation to Ensure Food Security: A Pathway Towards Sustainable Vegetables and Date Palm Production in Qatar".

References

1. Ajjur SB, Al-Ghamdi SG (2022) Towards sustainable energy, water and food security in Qatar under climate change and anthropogenic stresses. Energy Rep 8:514–518
2. Baalousha HM, Ouda OKM (2017) Domestic water demand challenges in Qatar. Arab J Geosci 10:537
3. Darwish M, Hassan A, Sayeed M, Rabi M (2015) The role of energy to solve water scarcity in Qatar. Desalination and Water Treatment 57:1–29
4. Dawoud MA (2007) Challenges and opportunities. ResearchGate
5. Qureshi AS (2020) Challenges and prospects of using treated wastewater to manage water scarcity crises in the Gulf Cooperation Council (GCC) countries. Water 12:1971
6. Darwish M, Hassan A, Yasir M, Mohamed D (2014) Qatar and GCC water security. Desalination 1–24
7. Saif O, Mezher T, Arafat HA (2014) Water security in the GCC countries: challenges and opportunities. J Environ Stud Sci 4:329–346
8. Abulibdeh A (2019) Water-energy nexus challenges and opportunities in Qatar
9. Ajibade I, Egge M, Pallathadka (2020) A climate change and the sustainable development goal on food security: barriers and opportunities for progressive realization in Qatar and Nigeria. J Sustain Developm Law Policy (The) 10:158
10. Karanisa T, Achour Y, Ouammi A, Sayadi S (2022) Smart greenhouses as the path towards precision agriculture in the food-energy and water nexus: case study of Qatar. Environ Syst Decisions 42:521–546
11. Zaina SM, Fadli F, Hosseini SM (2022) Evaluation of smart irrigation systems in hot-arid climates for green roofs and walls: case of Doha, Qatar. Smart Sustain Built Environ 11:346–367

12. Farid T, Mohammed A-H, Kamel B (2012) A fuzzy logic based irrigation management system in arid regions applied to the State of Qatar. WIT Trans Ecol Environ 168
13. Al-Naemi S, Al-Otoom A (2023) Smart sustainable greenhouses utilizing microcontroller and IOT in the GCC countries; energy requirements and economical analyses study for a concept model in the state of Qatar. Results in Eng 17:100889
14. Et-Taibi B, Abid MR, Boufounas E-M, Morchid A, Bourhnane S, Hamed TA, Benhaddou D (2024) Enhancing water management in smart agriculture: a cloud and IoT-Based smart irrigation system. Results in Eng 22:102283
15. Chowdhury MEH, Khandakar A, Ahmed S, Al-Khuzaei F, Hamdalla J, Haque F, Reaz MBI, Shafei AA, Al-Emadi N (2020) Design, construction and testing of IoT based automated indoor vertical hydroponics farming test-BED in Qatar. Sensors 20:5637
16. Aljughaiman AS (2019) Water and irrigation systems management at Kingdom of Saudia Arabia, a review. Plant Archives 63–77
17. Revolutionizing Farming with IoT: Smart Irrigation System for Sustainable Agriculture (2023) In: IEEE Conference Publication, IEEE Xplore
18. Baalousha HM, Ramasomanana F, Fahs M, Seers TD (2022) Measuring and validating the actual evaporation and soil moisture dynamic in arid regions under unirrigated land using smart field lysimeters and numerical modeling. Water 14:2787
19. Mohamed ES, Belal Aa, Abd-Elmabod SK, El-Shirbeny MA, Gad A, Zahran MB (2021) Smart farming for improving agricultural management. The Egyptian J Remote Sens Space Sci 24:971–981
20. Muhammed NH, Ahmed OE, Ahmed TA, Al-Yafai MS (2015) Date palm status and perspective in Qatar. In: Springer eBooks. pp 323–351
21. Abusin SAA, Mandikiana BW (2020) Towards sustainable food production systems in Qatar: assessment of the viability of aquaponics. Glob Food Sec 25:100349
22. Al Jabri S, Mushtaque A (2018) Use of renewable energy for desalination in urban agriculture in the GCC countries: possibilities and challenges. J Agricul Marine Sci [JAMS] 22:48
23. Moossa B, Trivedi P, Saleem H, Zaidi SJ (2022) Desalination in the GCC countries—a review. J Clean Prod 357:131717
24. Jonathan H, Magd H, Al Salhi A (2022) Climate smart agriculture and mitigation techniques for sustainable resilient farming in middle East Region. Global Business Managem Res: An Int J 14
25. Sustainable Security of the Middle East: The emerging climate-security nexus (2022)
26. Climate-Resilient Agriculture, Vol 1. In: Google Books
27. Benlarmi R et al. (2022) Development of smart agricultural technologies to optimize resource allocation to ensure food security: a pathway towards sustainable vegetables and date palm production in Qatar; MME03–1121–210025 funded by Qatar National Research Fund
28. Ogidan O, Abiodun O, Oluwabukola A (2019) Smart irrigation system: a water management procedure. Agricul Sci
29. Sivagami A, Kandavalli MA, Yakkala B (2021) Design and evaluation of an automated monitoring and control system for greenhouse crop production
30. Amhamed A, Genidi N, Abotaleb A et al (2023) Food security strategy to enhance food self-sufficiency and overcome international food supply chain crisis: the state of Qatar as a case study. GRN Tech Res Sustain 3:3
31. Al-Addous M, Bdour M, Rabaiah S, Boubakri A, Schweimanns N, Barbana N, Wellmann J (2024) Innovations in solar-powered desalination: a comprehensive review of sustainable solutions for water scarcity in the Middle East and North Africa (MENA) Region. Water 16:1877
32. Saif. Latest Agricultural Equipment Innovations In Southeast Asia | Jebsen & Jessen: Turf & Irrigation. In: Jebsen & Jessen Technology Turf & Irrigation (2024)
33. Husfarm Innovative Irrigation: smart water management solutions for Asia's farms—HusFarm. In: Husfarm agriculture platform (2024)
34. Efficiency Enhancement of Regional Irrigation Water Management through Gravity Irrigation System | Japan-ASEAN Integration Fund. In: Japan-ASEAN Integration Fund (2023)

35. Saif (2024) Boost your yields with drip irrigation: the future of agriculture in Southeast Asia. In: Jebsen & Jessen Technology Turf & Irrigation
36. Idrica (2023) The ten challenges of water management in Southeast Asia. Smart Water Magazine

GPSR Compliance

The European Union's (EU) General Product Safety Regulation (GPSR) is a set of rules that requires consumer products to be safe and our obligations to ensure this.

If you have any concerns about our products, you can contact us on ProductSafety@springernature.com

In case Publisher is established outside the EU, the EU authorized representative is:

Springer Nature Customer Service Center GmbH
Europaplatz 3
69115 Heidelberg, Germany

Batch number: 10755324

Printed by Printforce, the Netherlands

World Sustainability Series

Series Editor

Walter Leal Filho, European School of Sustainability Science and Research, Research and Transfer Centre "Sustainable Development and Climate Change Management", Hamburg University of Applied Sciences, Hamburg, Germany

Due to its scope and nature, sustainable development is a matter which is very interdisciplinary, and draws from knowledge and inputs from the social sciences and environmental sciences on the one hand, but also from physical sciences and arts on the other. As such, there is a perceived need to foster integrative approaches, whereby the combination of inputs from various fields may contribute to a better understanding of what sustainability is, and means to people. But despite the need for and the relevance of integrative approaches towards sustainable development, there is a paucity of literature which address matters related to sustainability in an integrated way.

Notes on the quality assurance and peer review of this publication

Prior to publication, the works published in this book are initially assessed and reviewed by an in-house editor. If suitable for publication, manuscripts are sent for further review, which includes a combined effort by the editorial board and appointed subject experts, who provide independent peer-review. The feedback obtained in this way was communicated to authors, and with manuscripts checked upon return before finally accepted. The peer-reviewed nature of the books in the "World Sustainability Series" means that contributions to them have, over many years, been officially accepted for tenure and promotion purposes.

Salem Al-Naemi · Rachid Benlamri ·
Walter Leal Filho · Jeffrey McDonnell ·
Rehan Sadiq · Haruna Musa Moda
Editors

Water and Food Security in the Face of Climate Change: Challenges and Opportunities for Resilience

Proceeding of the International Conference Water and Food Security in the Face of Climate Change: Challenges and Opportunities for Resilience (WFCC2025)

Volume 2

Springer

Editors
Salem Al-Naemi
UNEVOC Centre
University of Doha for Science
and Technology
Doha, Qatar

Rachid Benlamri
Academic Food Security Research Cluster
University of Doha for Science
and Technology
Doha, Qatar

Walter Leal Filho
Research and Transfer Centre
"Sustainability and Climate Change
Management"
Hamburg University of Applied Sciences
Hamburg, Germany

Jeffrey McDonnell
National Hydrology Research Centre
University of Saskatchewan
Saskatoon, SK, Canada

Rehan Sadiq
Academic University of British Columbia
Vancouver, BC, Canada

Haruna Musa Moda
Environmental Health and Safety
University of Doha for Science
and Technology
Doha, Qatar

ISSN 2199-7373 ISSN 2199-7381 (electronic)
World Sustainability Series
ISBN 978-3-032-00097-2 ISBN 978-3-032-00098-9 (eBook)
https://doi.org/10.1007/978-3-032-00098-9

This Springer imprint is published by the registered company Springer Nature Switzerland AG
The registered company address is: Gewerbestrasse 11, 6330 Cham, Switzerland

Foreword

What we eat, drink, and breathe should be safe, clean, and good for our health. My interest in the topics explored in this excellent collection of scholarly articles is reflected in the concept of elemental metabolomics that my colleagues and I described a decade ago. We defined elemental metabolomics as the measurement and analysis of the overall levels of chemical elements in biological samples, along with the observation of how these levels vary under different biological or environmental conditions. This method uses advanced mass spectrometry and other techniques for high-throughput elemental analysis, making it an invaluable tool in medical diagnostics, nutrition, food science, agriculture, and environmental monitoring of pollutants.

However, we should not focus only on food, nutrition, and human health. Our concerns must expand to the entire planet—we have only one Earth. We must take care of it and leave it to future generations in better shape than it was when we received it. That is why I care deeply about environmental issues, and I am pleased to see that the environment plays a big part in this book.

The main topics—food, water, health, and the environment—are all connected. The One Health approach recognizes that the health of people, animals, and environment are closely interconnected. We cannot protect human health without considering the health of animals and the ecosystems we all share. Diseases, pollutants, and food systems cross these boundaries, so the solutions must be systemic. One Health encourages collaboration across fields of medicine, veterinary science, agriculture, and environmental science to create healthier communities, ecosystems, and a more sustainable planet.

Science has a big role to play. It gives us facts, not guesses or wishful thinking. Scientifically verified facts help leaders make informed decisions—including laws, policies, and strategies for managing our shared resources.

This book also reflects the values of the University of Doha for Science and Technology. We proudly support the United Nations Sustainable Development Goals (SDGs) and Qatar's Vision 2030. The research and discussions in this book are well-aligned with all 17 SDGs neatly grouped across the four main tracks of the book. Through research, education, and responsible action, we are working together to build a future that is fair, healthy, and sustainable for everyone.

May 2025

Dr. Vladimir Brusic
University of Doha for Science
and Technology
Doha, Qatar

Preface

In an era defined by unprecedented climatic shifts, the interplay between water resources and food security has become increasingly critical. As we face the many interrelated challenges posed by climate change, we are compelled to explore how alterations in weather patterns, water availability, and agricultural productivity threaten the delicate balance that sustains our global food systems. This book, *Water and Food Security in the Face of Climate Change: Challenges and Opportunities*, part of scientific papers presented at the international conference on water and food security, University of Doha for Science and Technology, Qatar, seeks to illuminate these pressing issues while offering insights into viable pathways for resilience and adaptation.

Water is the lifeblood of agriculture; it is essential for crop growth, livestock health, and food production. However, as climate change accelerates, we witness rising temperatures, altered precipitation patterns, and increased frequency of extreme weather events, all of which strain our water resources. In this context, understanding the nexus between water and food security is paramount, as communities worldwide strive to ensure access to nutritious food amidst growing uncertainties.

This volume assembles contributions from a diverse array of experts across disciplines ranging from environmental science to agricultural economics and public policy. Together, they tackle a spectrum of themes, including the implications of dwindling freshwater supplies on agricultural practices, the role of innovative technologies in improving water use efficiency, and strategies for fostering sustainable practices that bolster both water and food security.

Moreover, this book highlights the opportunities that arise from collaborative efforts and innovative approaches to adapt to climate challenges. By fostering sustainable water management, promoting resilient agricultural practices, and implementing policies that prioritize food security, we can create a more secure future for all.

As you delve into the chapters that follow, we invite you to reflect on the profound interconnectedness of water and food systems. Our collective ability to respond to the challenges of climate change will depend on our understanding of these systems and our commitment to building adaptive capacity across communities and regions.

In the face of uncertainty, we have the power to transform challenges into opportunities. It is our hope that this book serves as a valuable resource for policymakers, researchers, practitioners, and students who envision a world where water and food security are secured for future generations, even amidst the changing climate.

Doha, Qatar Salem Al-Naemi
Doha, Qatar Rachid Benlamri
Hamburg, Germany Walter Leal Filho
Saskatoon, Canada Jeffrey McDonnell
Vancouver, Canada Rehan Sadiq
Doha, Qatar Haruna Musa Moda

Reviewers

Ravi Rangarajan, University of Doha for Science and Technology, Qatar, e-mail: ravi.rangarajan@udst.edu.qa
Amin Esmaeili Khalil Saraei, University of Doha for Science and Technology, Qatar, e-mail: amin.esmaeili@udst.edu.qa
Farhat Abbas, University of Doha for Science and Technology, Qatar, e-mail: farhat.abbas@udst.edu.qa
Bahman Boostani, University of Doha for Science and Technology, Qatar, e-mail: bahman.boostani@udst.edu.qa
Aparna Krishna, University of Doha for Science and Technology, Doha, Qatar, e-mail: aparna.krishna@udst.edu.qa
Raj Kumar Dahal, University of Doha for Science and Technology, Qatar, e-mail: raj.dahal@udst.edu.qa
Norr Hassan, University of Doha for Science and Technology, Qatar, e-mail: norr.hassan@udst.edu.qa
Ashraf Chamseddine, University of Doha for Science and Technology, Doha, Qatar, e-mail: ashraf.chamseddine@udst.edu.qa
Haruna Musa Moda, University of Doha for Science and Technology, Qatar, e-mail: haruna.moda@udst.edu.qa
Azad Ashraf, University of Doha for Science and Technology, Qatar, e-mail: azad.ashraf@udst.edu.qa
Hassan Hijazi, University of Doha for Science and Technology, Qatar, e-mail: hassan.hijazi@udst.edu.qa
Zahida Bano, University of Doha for Science and Technology, Qatar, e-mail: Zahida.qau@gmail.com
Christopher Fellows, The University of New England, Australia, e-mail: cfellows@une.edu.au
Nourhan Ghoneim, International Maritime College Oman, IMCO, Oman, e-mail: norhan@imco.edu.om
Venkatesh Meda, University of Saskatchewan, Saskatoon, Canada, e-mail: vem281@campus.usask.ca

Polices, Legislations and Society

Contributors

Farhat Abbas University of Doha for Science and Technology, Doha, Qatar

Amina Abbaz Laboratory of Environmental Engineering Process, University Constantine 3, Salah Boubnider, Ali Mendjeli, Constantine, Algeria

Ahmed A. Abdalla College of Engineering and Technology, University of Doha for Science and Technology, Doha, Qatar

Noor Abdelaziz University of Doha for Science and Technology, Doha, Qatar

Merdjaoui Abdelkadir Laboratory of Biological Oceanography and the Marine Environment (LOBEM), University of Science and Technology of Houari Boumedien (USTHB), El Alia, Bab Ezzouar, Algiers, Algeria

Mai Magdy Abdelrazeq National Research Centre, Dokki, Cairo, Egypt; The Rowett Institute, University of Aberdeen, Aberdeen, UK

Djamal Abdessemed Laboratory of the Industrial Science Process Engineering, University of the Sciences and Technology Houari Boumediene (USTHB), Algiers, Algeria

Zahrah Abdulghafour Department of Civil Engineering, College of Engineering, Australian University, West Mishref, Safat, Kuwait

Nagi Abdussamie College of Engineering and Technology, University of Doha for Science and Technology (UDST), Doha, Qatar

Luma Abuorabi Food and Agriculture Organization (FAO), Amman, Jordan

Bishnu Acharya Saskatchewan Ministry of Agriculture and University of Saskatchewan, Saskatoon, SK, Canada

Peter Adewale Osun State University, Osogbo, Nigeria

Peter Sanjo Adewale Osun State University, Osogbo, Nigeria

Matthew Olufemi Adio Department of Agricultural Economics and Farm Management, Faculty of Agriculture, Federal University, Oye-Ekiti, Ekiti State, Nigeria

Faith Mugisha Ahabyoona Department of Accounting, Finance and Microfinance, Kyambogo University, Kampala, Uganda

Ghufran Ahmed National University of Computer and Emerging Sciences, Karachi, Pakistan

Mourad Taleb Ahmed Laboratory Valorization and Recycling of Matters and Sustainable Development, University of Science and Technology Houari Boumediene, Bab Ezzouar, Algeria

Halima Aissaoui Laboratory of Environmental Engineering Process, University Constantine 3 Salah Boubnider, Ali Mendjeli, Constantine, Algeria

Oyepeju Abosede Ajetunmobi Department of Agricultural Economics, Faculty of Agricultural Sciences, Ladoke Akintola University of Technology, Ogbomoso, Nigeria

Hakeem A. Ajonbadi University of Science and Technology, Doha, Qatar

Adnan Akhunzada University of Doha for Science and Technology, Doha, Qatar

Akindele James Akintola Department of Agricultural Economics and Farm Management, Faculty of Agriculture, Federal University, Oye-Ekiti, Ekiti State, Nigeria

Ali Akyol Business Informatics University of Oldenburg, Oldenburg, Germany

Marwa Al-Ani Department of Chemical Engineering, College of Engineering, Qatar University, Doha, Qatar

Mustafa Khaled Al-Ani Ministry of Water Resources—National Water Resources Center, New Delhi, India

Ala Al-Ardah Department of Materials Science and Physics, College of Arts and Sciences, Qatar University, Doha, Qatar

Meera Alazem University of Doha for Science and Technology, Doha, Qatar

Sara Al-Baker Department of Chemical Engineering, College of Engineering, Qatar University, Doha, Qatar

Ali Hussein Ibraheem Al-Bayati Department of Biology, College of Education for Woman, University of Anbar, Ramadi, Iraq

Alzulfa Albulushi International Maritime College Oman, National University of Science and Technology, Suhar, Oman

Reem Al-Haidose Biotechnology Center, Agricultural Research Department, Ministry of Municipality, Doha, Qatar

Mohamed Saleh Al-Haj Faculty of Humanities, Sana'a University, Sanaa, Republic of Yemen

Naser Alhakmani International Maritime College Oman, National University of Science and Technology, Suhar, Oman

Ali Hamad Alhamadi Qatar Science and Technology Secondary School for Boys, Doha, Qatar

Naval Al-Hamdan Department of Chemistry and Earth Sciences, College of Arts and Sciences, Qatar University, Doha, Qatar

Ali A. Al-Hamzah Water Technologies Innovation Institute and Research Advancement, Saudi Water Authority, Jubail, Kingdom of Saudi Arabia;
School of Science and Technology, The University of New England, Armidale, Australia

Maathir Alhasani International Maritime College Oman, National University of Science and Technology, Suhar, Oman

Salim Alhashmi International Maritime College Oman, National University of Science and Technology, Suhar, Oman

Yousef Alhorr Gulf Organisation for Research and Development, Ar-Rayyan, Qatar

Hassan Ali College of Engineering and Technology, University of Doha for Science and Technology, Doha, Qatar;
Centre of Excellence for Sustainability and Food Security, University of Doha for Science and Technology, Doha, Qatar;
Department of Electrical Engineering, University of Doha for Science and Technology, Doha, Qatar

Lamia Ali Laboratory of Biological Oceanography and the Marine Environment (LOBEM), University of Science and Technology of Houari Boumedien (USTHB), Algiers, Algeria

Suzan Ali College of Computing and Information Technology, University of Doha for Science and Technology, Doha, Qatar

Saad Alkentar Homs University, Homs, Syria

Shahed AlNawaiseh School of Computing and Informatics, Al Hussein Technical University, Amman, Jordan

Ghoson Abdulaziz AL-Obaidly University of Doha, Doha, Qatar

Huthaifa Al-Omari School of Computing and Informatics, Al Hussein Technical University, Amman, Jordan

Noora Al-Qahtani Center for Advanced Materials (CAM), Office of the Vice President for Research and Graduate Studies (VPRGS), Qatar University, Doha, Qatar

Mansour Al Qubeissi College of Engineering and Technology, University of Doha for Science and Technology, Doha, Qatar

Walid Kh Alqaisi College of Engineering and Technology, University of Doha for Science and Technology, Doha, Qatar

Ahmad M. Alqudah Department of Biological and Environmental Sciences, College of Art and Science, Qatar University, Doha, Qatar

Bassem Alsahwa HIAST, Damascus, Syria

Rashid Alsawai International Maritime College Oman, National University of Science and Technology, Suhar, Oman

Ali Alseari Qatar Science and Technology Secondary School for Boys, Doha, Qatar

Abdulaziz Al Shandoudi International Maritime College Oman, National University of Science and Technology, Suhar, Oman

Mohammad Alsheyab National Planning Council, Doha, Qatar

Fahad Jassim Altamimi Qatar Science and Technology Secondary School for Boys, Doha, Qatar

Alshayma N. Al-Thani Department of Biological and Environmental Sciences, College of Art and Science, Qatar University, Doha, Qatar

Faisal Alzyoud Department of Computer Science, Faculty of IT, Isra University, Amman, Jordan

Abdulla Ansari Environmental Health and Safety, University of Doha for Science and Technology, Doha, Qatar

Sihem Arris Laboratory of Environmental Process Engineering, Faculty of Process Engineering, University of Constantine 3 Salah Boubnider, Constantine, Algeria

Summayya Asghar Department of Biochemistry, Shaheed Benazir Bhutto Women University (SBBWU), Peshawar, Pakistan

Hasna Ashraf University of Doha for Science and Technology, Doha, Qatar

Selin Aslantaş University of Health Sciences, Istanbul, Turkey

Abdulkareem Assalem Homs University, Homs, Syria

Asma Ayat Environmental Process Engineering Laboratory, University of Constantine3 Salah Boubnider, Constantine, Algeria;
University Setif 1; Ferhat ABBAS, Faculty of Technology, Setif, Algeria

Matthew Durojaiye Ayeni Department of Agricultural Economics and Extension, College of Agricultural Sciences, Landmark University, Omuaran, Kwara State, Nigeria

Nour El Islam Bachari Laboratory of Biological Oceanography and the Marine Environment (LOBEM), University of Science and Technology of Houari Boumedien (USTHB), Algiers, Algeria

Ahmed Badawi College of Engineering and Technology, University of Doha for Science and Technology, Doha, Qatar;
Department of Electrical Engineering, University of Doha for Science and Technology, Doha, Qatar

Evelyn H. Bandonill Rice Chemistry and Food Science Division, Science City of Munoz, Philippines;
Phillipine Rice Research Institute, Science City of Munoz, Nueva Ecija, Philippines

Zahida Bano University of Doha for Science and Technology, Doha, Qatar

Mohd Basheer University of Jordan, Amman, Jordan

Mahmoud Bataineh College of General Education, University of Doha for Science and Technology, Doha, Qatar

Scott Daniel Beard AtkinsRéalis, Epsom, Surrey, UK

Riadh Bechir Laboratoire LESOR Institut des Régions Arides de Médenine, Université de Gabès, Medenine, Tunisia

Jesusa C. Beltran Socio Economics Division, Science City of Munoz, Philippines;
Phillipine Rice Research Institute, Science City of Munoz, Nueva Ecija, Philippines

Amel Benhadji Laboratory of Reaction Engineering, University of Science and Technology Houari Boumediene, Algiers, Algeria

Farouk Benhamouda University Constantine 3, Ali Mendjeli, Constantine, Algeria

Rachid Benlamri Centre of Excellence for Sustainability and Food Security, University of Doha for Science and Technology (UDST), Doha, Qatar

Aqeela Bibi Department of Biochemistry, Shaheed Benazir Bhutto Women University (SBBWU), Peshawar, Pakistan

Asifa Bibi Bahria University, Islamabad, Pakistan

Seyedmasnour Bidoki Department of Textile Engineering, Yazd University, Yazd, Iran

Abdullah bin Bilal Al-Subhi Ministry of Environment Water and Agriculture, Riyadh, Kingdom of Saudi Arabia

Bahman Boostani College of Engineering Technology, University of Doha for Science and Technology, Doha, Qatar

Nabil Bougherira Laboratoire Ressource en Eau and Développement Durable, Université Badji Mokhtar Annaba, Annaba, Algeria

Hamza Bouguerra Laboratoire Ressource en Eau and Développement Durable, Université Badji Mokhtar Annaba, Annaba, Algeria

Yasser Boumezough Laboratory of Environmental Process Engineering, Faculty of Process Engineering, University of Constantine 3 Salah Boubnider, Constantine, Algeria

Ahmet Çabuk Karabük University, Karabük, Turkey

Julianne C. Caday University of Doha for Science and Technology, Doha, Qatar

Mohamed Chabi College of Engineering, Qatar University, Doha, Qatar

Hicham Chaffai Laboratoire Ressource en Eau and Développement Durable, Université Badji Mokhtar Annaba, Annaba, Algeria

Rami Chahin Business Informatics University of Oldenburg, Oldenburg, Germany

Shuhong Chai Centre for Maritime Engineering and Hydrodynamics, Australian Maritime College, University of Tasmania, Launceston, TAS, Australia

Supriyo Chakraborty Ministry of Earth Sciences, Indian Institute of Tropical Meteorology, Pune, India

Roneta Chaliulina Gulf Organisation for Research and Development, Ar-Rayyan, Qatar

Madhurima Chatterjee Department of Biological and Environmental Sciences, College of Art and Science, Qatar University, Doha, Qatar

Anis Chekirbane National Institute of Agronomy (INAT), Tunis, Tunisia

Chris Chin Australian Maritime College, Launceston, TAS, Australia

Hafsa Chohan University of Doha for Science and Technology, Doha, Qatar

Raj Kumar Dahal University of Doha for Science and Technology, Doha, Qatar

Larbi Dajbri Laboratoire Ressource en Eau and Développement Durable, Université Badji Mokhtar Annaba, Annaba, Algeria

Shaimaa R. Dalal Mansoura University, Mansoura, Egypt

Ali Demir Department of Textile Technologies and Design Engineering, Istanbul Technology University, Istanbul, Turkey

Nenita V. Desamero Plant Breeding and Biotechnology Division, Science City of Munoz, Philippines;
Phillipine Rice Research Institute, Science City of Munoz, Nueva Ecija, Philippines

Maha A. Dewidar University of Doha for Science and Technology, Doha, Qatar

Israr Ud Din Institute of Biotechnology and Genetic Engineering, The University of Agriculture Peshawar, Peshawar, Pakistan

Mouad Djelloudi Laboratory of Environmental Engineering Process, University Constantine 3, Ali Mendjeli, Constantine, Algeria

Animesh Dutta University of Guelph, Guelph, ON, Canada

Taghreed El-Kasaby Mansoura University, Mansoura, Egypt

Ali El-Keblawy Department of Applied Biology, College of Sciences, University of Sharjah, Sharjah, United Arab Emirates;
Faculty of Pharmacy, Al Salam University, Tanta, Egypt

Sohayla El Mokadem College of Engineering and Technology, University of Doha for Science and Technology, Doha, Qatar

Attiat Elnaggar Research Institute for Science and Engineering, University of Sharjah, Sharjah, United Arab Emirates;
Department of Botany and Microbiology, Faculty of Science, Alexandria University, Alexandria, Egypt

I. M. Elzein College of Engineering and Technology, University of Doha for Science and Technology, Doha, Qatar

Amin Esmaeili College of Engineering Technology, University of Doha for Science and Technology (UDST), Doha, Qatar

Anika Fajutagana College of Health Sciences, University of Doha for Science and Technology, Doha, Qatar

Olufemi Oyedokun Fakunle Department of Agricultural Economics and Farm Management, Faculty of Agriculture, Federal University, Oye-Ekiti, Ekiti State, Nigeria

Wajid Farhad University College Dera Murad Jamali, Dera Murad Jamali, Pakistan;
Lasbela University of Agriculture, Water and Marine Sciences, Uthal, Pakistan

Ahmed Faris Qatar Science and Technology Secondary School for Boys, Doha, Qatar

Aitazaz A. Farooque Canadian Center for Climate Change and Adaptation University of Prince Edward Island, St Peter's Bay, PEI, Canada;
Faculty of Sustainable Design Engineering, University of Prince Edward Island, Charlottetown, PEI, Canada

Christopher M. Fellows Water Technologies Innovation Institute and Research Advancement, Saudi Water Authority, Jubail, Kingdom of Saudi Arabia;
School of Science and Technology, The University of New England, Armidale, Australia

Valeria Freundt University of Doha for Science and Technology (UDST), Doha, Qatar

Milad Sabouri Gasgarei Carleton University, Ottawa, ON, Canada

Nourhan I. Ghoneim International Maritime College Oman, National University of Science and Technology, Suhar, Oman

Jorge Marx Gómez Business Informatics University of Oldenburg, Oldenburg, Germany

Abooali Golzary School of Environment, College of Engineering, Faculty of Environment, University of Tehran, Tehran, Iran

Giuliana Gorrasi Department of Industrial Engineering, University of Salerno, Fisciano, Italy

Ala Gouissem University of Doha for Science and Technology (UDST), Doha, Qatar

Eric Gubesch Centre for Maritime Engineering and Hydrodynamics, Australian Maritime College, University of Tasmania, Launceston, TAS, Australia

Yasmin A. Hageer Department of International Affairs, College of Arts and Sciences, Qatar University, Doha, Qatar

Fahima Halimi Laboratoire de Recherhe en Géologie, Université Badji Mokhtar Annaba, Annaba, Algeria

Hafiz Mohkum Hammad Department of Agronomy, MNS University of Agriculture Multan, Multan, Pakistan;
Department of Agricultural and Biological Engineering, University of Florida, Gainesville, FL, USA

Azzedine Hani Laboratoire Ressource en Eau and Développement Durable, Université Badji Mokhtar Annaba, Annaba, Algeria

Andrew Harris BMT Australia, Adeliade, SA, Australia

Yasen Harrye College of Engineering and Technology, University of Doha for Science and Technology, Doha, Qatar

Khaled El-Sayed Hassan Former Demographic Consultant at UNDP-Kuwait, Kuwait City, Kuwait;
Vice-president of the Egyptian Society for Migration Studies (EGYMIG), Cairo, Egypt

Norr Hassan Department of Environmental Health and Safety, University of Doha for Science and Technology, Doha, Qatar

Hocine Hassani University of Chlef, Chlef, Algeria

Aneela Hayder University of Guelph, Guelph, ON, Canada

Massoomeh Hedayati Marzbali Department of Urban Planning, School of Housing, Building and Planning, Universiti Sains Malaysia, Penang, Malaysia

Hassan Y. Hijazi University of Doha for Science and Technology, Doha, Qatar

Mohammed Himrane University of Jijel, Jijelv, Algeria

Woon Hong Centre for Maritime Engineering and Hydrodynamics, Australian Maritime College, University of Tasmania, Launceston, TAS, Australia

Fouzia Houma Marine and Coastal Ecosystems Laboratory (ECOSYSMarL), National Higher School of Marine Sciences and Coastal Management (ENSSMAL), Algiers, Algeria

Majid Ibrahim College of General Education, Doha University for Science and Technology, Doha, Qatar

Masoud Irannezhad Department of Civil Engineering, College of Engineering, Australian University, West Mishref, Safat, Kuwait;
Water, Energy and Environmental Engineering Research Unit, Faculty of Technology, University of Oulu, Oulu, Finland

Bachari Nour El Islam Laboratory of Biological Oceanography and the Marine Environment (LOBEM), University of Science and Technology of Houari Boumedien (USTHB), El Alia, Bab Ezzouar, Algiers, Algeria

Najma Nur Islam Department of Applied Biology, College of Sciences, University of Sharjah, Sharjah, United Arab Emirates

Seyed Behbood Issa-Zadeh School of Maritime Science, University of Gibraltar, Campus Europa Point, Gibralta GX11 1AA, UK

Asma Ghafoor Jan Department of Biological and Environmental Sciences, College of Art and Science, Qatar University, Doha, Qatar

Mohamed Jaoued Laboratoire LESOR Institut des Régions Arides de Médenine, Université de Gabès, Medenine, Tunisia

Abitha John University of Doha for Science and Technology (UDST), Doha, Qatar

Matilda Kaliba University of Zambia, Lusaka, Zambia

Muhammad Nawaz Kandhro Department of Agronomy, Sindh Agriculture University Tandojam, Tandojam, Pakistan

Akano Kayode Lagos State University of Education (LASUED), Oto-Ijanikin (Epe Campus), Lagos State, Nigeria;
Department of Agricultural Education, College of Vocational and Entrepreneurship Education (COVED), Lagos State, Nigeria

Carol Khadra QAFCO (Qatar Fertiliser Company), Mesaieed, Qatar

Fahim Ullah Khan University of Doha for Science and Technology, Doha, Qatar; Department of Agriculture, Hazara University Dodhial Mansehra, Khyber Pakhtunkhwa, Pakistan

Khalaf Khatatneh Prince Abdullah bin Ghazi for Communication and IT College, AlBalqa Applied University, Amman, Jordan

Chaima Khayri Environmental Process Engineering Laboratory, University of Constantine3 Salah Boubnider, Constantine, Algeria

Djihane Khechine Environmental Process Engineering Laboratory, University of Constantine3 Salah Boubnider, Constantine, Algeria

Khaoula Khellaf Environmental Process Engineering Laboratory, University of Constantine3 Salah Boubnider, Constantine, Algeria

Anthony Kola-Olusanya Osun State University, Osogbo, Nigeria

Aparna Krishna Applied Research, Innovation and Economic Development Directorate, University of Doha for Science and Technology, Doha, Qatar

Anurag Kumar Birbal Sahni Institute of Palaeosciences, Lucknow, India

Jamal Kurdi Chemical Engineering, University of Doha for Science and Technology, Doha, Qatar

Ahmet Kurklu Department of Agricultural Mechanization and Technologies Engineering, Akdeniz University, Antalya, Turkey

Ayaz Ahmed Lashari Department of Agronomy, Sindh Agriculture University Tandojam, Tandojam, Pakistan

Latif-ur-Rahman Institute of Chemical Sciences, University of Peshawar, Peshawar, Pakistan

Mohamed Said Lebad Laboratory of the Industrial Science Process Engineering, University of the Sciences and Technology Houari Boumediene (USTHB), Algiers, Algeria

Sherzod Madrahimov Department of Sciences, Texas A&M University, Doha, Qatar

Mohammad Javad Maghsoodi Tilaki Department of Geography, School of Humanities, Universiti Sains Malaysia, Penang, Malaysia

Hassan Mahasneh College of Engineering and Technology, University of Doha for Science and Technology, Doha, Qatar

Tahira Jehan Makda University of Doha for Science and Technology (UDST), Doha, Qatar

K. M. Malabe Department of Agricultural Extension Services, University of Maiduguri, Maiduguri, Nigeria

Hind Mamlouk Department of Sciences, Texas A&M University, Doha, Qatar

Dana Khider Mawlood Centre for Environmental Studies, University of Kurdistan Hewlêr, Erbil City, Kurdistan Region, Iraq

Mohamed Meddi Ecole Nationale Supérieure d'Hydraulique (GEE), Guerrouaou, Algeria

Danish Mehmood SZABIST University, Islamabad, Pakistan

Fatima Mekhlouf University of Relizane, Relizane, Algeria

A.-H. Meniai Laboratory of Environmental Engineering Process, University Constantine 3, Ali Mendjeli, Constantine, Algeria

Saeed Mohajernasab Delmar Systems, Perth, WA, Australia

Asma Mohamed Department of Biological and Environmental Sciences, College of Arts and Sciences, Qatar University, Doha, Qatar

Raneem Mohamed Department of Biological and Environmental Sciences, College of Arts and Sciences, Qatar University, Doha, Qatar

Bryant Montenegro Gulf Organisation for Research and Development, Ar-Rayyan, Qatar

Kareem A. Mosa Department of Applied Biology, College of Sciences, University of Sharjah, Sharjah, United Arab Emirates;
Department of Biotechnology, Faculty of Agriculture, Al-Azhar University, Cairo, Egypt

Haruna Musa Moda Department of Environmental Health and Safety, University of Doha for Science and Technology, Doha, Qatar

S. B. Mustapha Department of Agricultural Extension Services, University of Maiduguri, Maiduguri, Nigeria

Lumu Emmanuel Roy Muyingo Business, Law, Communication and Accounting, INTI International University, Kuala Lumpur, Malaysia

Solomon Mwampikita Centre for Trade Policy and Development (CTPD), Olympia Park, Lusaka, Zambia

Lamri Nacef Laboratory of Biological Oceanography and the Marine Environment (LOBEM), University of Science and Technology of Houari Boumedien (USTHB), El Alia, Bab Ezzouar, Algiers, Algeria

Muhammad Asif Nawaz Department of Biotechnology, Shaheed Benazir Bhutto University (SBBU), Sheringal, Upper Dir, Pakistan

Jonathan M. Niones Genetic Resource Division, Science City of Munoz, Philippines;
Phillipine Rice Research Institute, Science City of Munoz, Nueva Ecija, Philippines

Oluwaseun Oyebisi Odetola Osun State University, Osogbo, Nigeria

Ayodele Olarewaju Department of Agricultural Economics and Farm Management, Faculty of Agriculture, Federal University, Oye-Ekiti, Ekiti State, Nigeria

Mohammed Ouda Electrical Engineering, University of Doha for Science and Technology, Doha, Qatar

Omar K. M. Ouda Ministry of Environment Water and Agriculture, Riyadh, Kingdom of Saudi Arabia

Imeldalyn G. Pacada Plant Breeding and Biotechnology Division, Science City of Munoz, Philippines;
Phillipine Rice Research Institute, Science City of Munoz, Nueva Ecija, Philippines

Loida M. Perez Plant Breeding and Biotechnology Division, Science City of Munoz, Philippines;
Phillipine Rice Research Institute, Science City of Munoz, Nueva Ecija, Philippines

Sidra Pervez Department of Biochemistry, Shaheed Benazir Bhutto Women University (SBBWU), Peshawar, Pakistan

Michael Phillips University of Doha for Science and Technology, Doha, Qatar

K. Prasanna Birbal Sahni Institute of Palaeosciences, Lucknow, India

Muhammad Qasim SZABIST University, Islamabad, Pakistan

Hadia Qureshi Department of Biochemistry, Shaheed Benazir Bhutto Women University (SBBWU), Peshawar, Pakistan

Ravi Rangarajan Department of Environmental Health and Safety, College of Health Sciences, University of Doha for Science and Technology (UDST), Doha, Qatar

Attiq ur Rehman Department of Agricultural Sciences, Faculty of Agriculture and Forestry, The University of Helsinki, Helsinki, Finland

Juliet P. Rillon Crop Protection Division, Science City of Munoz, Philippines;
Phillipine Rice Research Institute, Science City of Munoz, Nueva Ecija, Philippines

Harid Roumaissa Marine and Coastal Ecosystems Laboratory (ECOSYSMarL), National Higher School of Marine Sciences and Coastal Management (ENSSMAL), Algiers, Algeria

Nassila Sabba Laboratory Valorization and Recycling of Matters and Sustainable Development, University of Science and Technology Houari Boumediene, Bab Ezzouar, Algeria;
Laboratory of Recovery and Recycling of Matter for Sustainable Development, USTHB, University of Science and Technology Houari Boumediene, Algiers, Algeria

Milad Sabouri Department of Civil and Environmental Engineering, Carleton University, Ottawa, Canada

Lamine Salim Laboratory of Biological Oceanography and the Marine Environment (LOBEM), University of Science and Technology of Houari Boumedien (USTHB), El Alia, Bab Ezzouar, Algiers, Algeria;
Higher School of Saharan Agriculture Adrar, Adrar, Algeria

Haitem Sekhri Laboratory of Environmental Engineering Process, University Constantine 3, Ali Mendjeli, Constantine, Algeria

Farah Seray Laboratory Valorization and Recycling of Matters and Sustainable Development, University of Science and Technology Houari Boumediene, Bab Ezzouar, Algeria

Farzin Shabani Department of Biological and Environmental Sciences, College of Art and Science, Qatar University, Doha, Qatar

Anupam Sharma Birbal Sahni Institute of Palaeosciences, Lucknow, India

Chris Shearer BMT Australia, Melbourne, VIC, Australia

Yusuf Shogbesan Osun State University, Osogbo, Nigeria

Mohamed R. Shouman Faculty of Maritime Transport and Technology, Arab Academy for Science, Technology and Maritime Transport, Alexandria, Egypt

Mostafa H. Sliem Center for Advanced Materials (CAM), Office of the Vice President for Research and Graduate Studies (VPRGS), Qatar University, Doha, Qatar

Mostafa Soliman College of Engineering and Technology, University of Doha for Science and Technology, Doha, Qatar

Sara Soussy College of Engineering Technology, University of Doha for Science and Technology, Doha, Qatar

Roel R. Suralta Crop Biotech Center, Santo Domingo, Philippines;
Phillipine Rice Research Institute, Science City of Munoz, Nueva Ecija, Philippines

Salah Eddine Tachi Laboratoire Ressource en Eau and Développement Durable, Université Badji Mokhtar Annaba, Annaba, Algeria

Mourad Taleb Ahmed Laboratory of Recovery and Recycling of Matter for Sustainable Development, USTHB, University of Science and Technology Houari Boumediene, Algiers, Algeria

Monther Tarawneh Department of IT, Faculty of IT and Communications, Tafila Technical University, Tafila, Jordan

Omar Tarawneh Software Engineering Department, Amman Arab University, Amman, Jordan

Michel Tchotsoua Department of Geography and Geomatic, University of Ngaoundéré, Ngaoundéré, Cameroon

Norbert F. Tchouaffe-Tchiadje Forestry, Water and Environment, Higher Institute for Agriculture, University of Ebolowa, Ebolowa, Cameroon;
Kesmonds International University, Ngaoundéré, Cameroon

Serap Tepe University of Health Sciences, Istanbul, Turkey

Nick Tighe BMT Australia, Melbourne, VIC, Australia

Eddy Kurobuza Tukamushaba Department of Management, College of Business, University of Doha for Science and Technology, Doha, Qatar

R. Uthra SRM Institute of Science and Technology, Kattankulathur, Tamil Nadu, India

Gianluca Viscusi Department of Industrial Engineering, University of Salerno, Fisciano, Italy

Murad A. Yaghi School of Computing and Informatics, Al Hussein Technical University, Amman, Jordan

Raziq Yaqub Department of Electrical Engineering and Computer Science, A&M University, Huntsville, AL, USA

Noor Kh. Yashooa Centre for Environmental Studies, University of Kurdistan Hewlêr, Erbil City, Kurdistan Region, Iraq

Anteneh Mesfin Yeneneh International Maritime College Oman, National University of Science and Technology, Suhar, Oman

Rokia Youcef Laboratory Valorization and Recycling of Matters and Sustainable Development, University of Science and Technology Houari Boumediene, Bab Ezzouar, Algeria;
Laboratory of Recovery and Recycling of Matter for Sustainable Development, USTHB, University of Science and Technology Houari Boumediene, Algiers, Algeria

Florence Adeoti Yusuf Osun State University, Osogbo, Nigeria

Muhammad Zubair University of Doha for Science and Technology, Doha, Qatar

Aiste Zukaityte Gulf Organisation for Research and Development, Ar-Rayyan, Qatar

Alhareth Zyoud Department of Electrical and Computer Engineering, Faculty of Engineering, Birzeit University, Birzeit, Ramallah, Palestine

Water Resources Management

Smart Sahara Farming

Smart Agriculture Application Based on 5G Smart Country Project

Rami Chahin, Ali Akyol, and Jorge Marx Gómez

Abstract This research paper explores innovative methods for autonomous Sahara farming using advanced technologies to address pressing global challenges such as food security, climate change, and pollution. Focusing on the integration of smart agriculture applications based on the methods and technology which are used in the 5G Smart Country project in lower saxony in Germany. The study presents a comprehensive approach to transforming desert landscapes into productive agricultural hubs. Key solutions include autonomous farming systems, solar-powered desalination, and IoT-enabled monitoring to optimize water and soil management. By integrating these technologies within the Smart Country Project framework, this paper presents a scalable model that mitigates the impacts of global warming, enhances sustainable food production, reduces pollution, promotes greater climate stability, and strengthens environmental resilience in arid regions.

Keywords Smart farming · Sahara farming · Automated planting · Desalination · AI

1 Introduction to Smart Sahara Farming

The global population is growing at an unprecedented rate. In the early eighteenth century, the world's population was approximately 1 billion [1]. Today, it has surged to over 8 billion and is projected to reach nearly 9.8 billion by 2050. This rapid population growth places immense pressure on global food production systems. Compounding this issue are environmental challenges that threaten the world's ability to maintain sufficient food supplies. Among these challenges, desertification stands out as one of the most critical [2]. The Sahara desert, covering an expansive area of about 9.2 million square kilometres, is the largest hot desert in the world, and its reach continues to expand. The harsh conditions of the Sahara present numerous challenges

R. Chahin (✉) · A. Akyol · J. M. Gómez
Business Informatics University of Oldenburg, Oldenburg, Germany
e-mail: rami.chahin@uni-oldenburg.de

S. Al-Naemi et al. (eds.), *Water and Food Security in the Face of Climate Change: Challenges and Opportunities for Resilience*, World Sustainability Series,
https://doi.org/10.1007/978-3-032-00098-9_50

Nile River offers a critical dual-source water supply for desert agriculture. Egypt's use of desalination is particularly notable for its large-scale agricultural projects in regions like Sinai and the Nile Delta. The country has pursued advanced desalination technology to mitigate the growing gap between water availability and agricultural needs [24].

Sudan, with its vast desert areas and limited water resources, stands to benefit from both Nile waters and desalination technology from the Red Sea. However, large-scale infrastructure development is needed to make desalination a practical solution in agricultural sectors [25].

3.4 Challenges and Environmental Considerations

While desalination holds promise for agricultural irrigation, it also faces several challenges:

- **High Energy Costs**: Desalination is an energy-intensive process. In a region where energy infrastructure can be limited, particularly in rural and desert regions, desalination may not be cost-effective without significant investment in renewable energy sources, such as solar power. Solar desalination plants are increasingly being considered in regions like Morocco and Tunisia [26].
- **Brine Disposal**: Desalination produces brine, a concentrated saltwater byproduct. Safe disposal of this brine is crucial to avoid environmental damage, particularly to coastal ecosystems. Brine management solutions, such as deep-sea discharge and brine mining for valuable minerals, are being explored to mitigate these environmental risks [27].
- **Infrastructure and Maintenance**: Large desalination plants require substantial investment in infrastructure, including pipelines to transport freshwater to inland agricultural areas. Maintenance of these systems in harsh desert environments is another challenge, as sandstorms, high temperatures, and corrosion can affect the longevity and efficiency of desalination plants [28].

3.5 The Nile River's Role in Sahara Agriculture

Beyond desalination, the Nile river plays an indispensable role in sustaining agriculture in Egypt and Sudan. The Nile irrigates vast tracts of farmland in Egypt, enabling the country to maintain agricultural productivity in desert regions. However, rising populations, climate change, and geopolitical issues surrounding water rights among Nile Basin countries have placed increasing pressure on the river's resources [29]. Egypt's agricultural sector relies heavily on the Nile for its desert farming projects, such as the Toshka Project, which aimed at reclaiming desert land for farming. The success of these initiatives hinges on sustainable water management, combining Nile water with desalinated water and efficient irrigation practices [30].

Desalination offers a viable solution for addressing water scarcity in Sahara-bordering nations, particularly those with access to the Atlantic Ocean, Mediterranean Sea, and Red Sea.

As water demands continue to rise, desalination technologies, when combined with energy-efficient solutions such as solar power, hold the potential to transform desert landscapes into productive agricultural regions. Countries like Morocco, Egypt, and Tunisia are leading the way in leveraging their coastlines for agricultural development. Meanwhile, the Nile river remains a critical lifeline for farming in Egypt and Sudan, underscoring the need for integrated water management strategies that combine both river and desalinated water resources for long-term agricultural sustainability [31].

However, large-scale deployment of desalination for agriculture in the Sahara faces significant challenges, including energy costs, environmental impact, and infrastructure development. Addressing these challenges will require coordinated efforts, international investment, and the continued development of new technologies to ensure that desert agriculture thrives in one of the harshest environments on earth.

4 Suitable Trees and Bushes for the Sahara Desert: Water Requirements, Soil, and Growing Conditions

The Sahara desert presents one of the most challenging environments for plant growth due to its extreme temperatures, scarcity of water, and sandy, nutrient-poor soils. However, various species of trees and bushes have adapted to these harsh conditions, some of which are suitable for large-scale planting to combat desertification, improve biodiversity, and support sustainable agricultural projects. This section explores the most resilient trees and bushes that can thrive in the Sahara desert, their water needs, irrigation frequency, soil requirements, temperature tolerance, and spacing recommendations. Additionally, comparisons are drawn to similar afforestation efforts in China and the United States [32].

4.1 Key Trees and Bushes Suitable for the Sahara Desert

Several drought-resistant trees and bushes have been identified as viable options for desert afforestation. These plants have evolved mechanisms to cope with water scarcity and extreme temperatures. Below are some of the most important species:

1. **Acacia (*Acacia* species)**
    - **Water Requirements**: Acacia trees are known for their exceptional drought resistance. They require approximately 200 to 600 L of water per year, depending on the specific species and the local environment [33].

- **Irrigation Frequency**: Acacias can survive with 1–2 irrigation sessions per year, particularly if planted during the rainy season or given early establishment support.
- **Soil**: Acacias prefer well-drained sandy or loamy soils, making the Sahara's sandy substrate suitable for many of its species.
- **Temperature Tolerance**: *Acacia* trees can withstand temperatures up to 50 °C and occasional cold snaps.

2. **Argan (*Argania spinosa*)**

- **Water Requirements**: Argan trees, which is native to Morocco, need approximately 400 L of water per year, and can tolerate drought conditions.
- **Irrigation Frequency**: They should be irrigated once or twice per year to ensure survival during prolonged drought periods.
- **Soil**: Argan trees thrive in sandy and rocky soils with minimal nutrients, making them a suitable choice for the Sahara Desert.
- **Temperature Tolerance**: Argan can tolerate temperatures between 0 and 45 °C, making them resilient to both heat and occasional cold weather [34].

1. **Jujube (*Ziziphus jujuba*)**

- **Water Requirements**: The jujube tree requires about 500–800 L of water per year to thrive, though some varieties are more drought-tolerant.
- **Irrigation Frequency**: These trees can survive with 3–4 irrigation sessions per year during the dry season.
- **Soil**: Jujube trees prefer sandy, loamy, or slightly saline soils, making them a potential option for the Sahara.
- **Temperature Tolerance**: They can endure temperatures ranging from – 15 °C to 45 °C, making them one of the more temperature-resilient options [34].

2. **Date Palm (*Phoenix dactylifera*)**

- **Water Requirements**: Date palms need approximately 800 to 1500 L of water per year, making them one of the more water-demanding desert plants.
- **Irrigation Frequency**: They require regular irrigation around 4–5 times per year especially during the growing season.
- **Soil**: Date palms are highly adaptable but prefer deep, sandy loam soils with good drainage, which the Sahara can provide in many areas.
- **Temperature Tolerance**: They can tolerate extreme heat, up to 50 °C, but they need some protection against prolonged freezing temperatures [35].

4.2 Soil Requirements and Suitability of Sahara Sand

The Sahara's sandy soils have both advantages and challenges for plant growth. Sandy soils allow for good drainage, which is beneficial for desert plants adapted to low water conditions, as it reduces the risk of root rot. However, these soils often lack organic matter and essential nutrients, making it necessary to enrich the soil with organic compost or fertilizers, particularly for less hardy species [36].

While the Sahara's sand is well-suited for species like Acacia and Date palms, which thrive in nutrient-poor, well-drained soils, others, like Argan, benefit from slightly more fertile conditions. In regions where the sand is too loose, measures such as mulching or the addition of clay can help retain moisture and improve soil structure [34].

4.3 Temperature Resilience

Most desert-adapted trees and bushes can withstand the Sahara's extreme temperatures. Species like Acacia and Date palms thrive in environments where temperatures can exceed 45–50 °C. Additionally, these species are well-equipped for temperature fluctuations that include cold nights and occasional frosts, particularly in regions at higher elevations or during winter [35].

4.4 Spacing and Plantation Density

The appropriate spacing between trees is crucial for maximizing growth and resource use efficiency in desert environments. The root systems of desert plants tend to spread widely to capture as much water as possible [36]. For instance, acacias and argan trees should be planted at least 5 m apart, while Date palms and Jujube trees require longer spacing, typically around 8–10 m, to avoid competition for water and nutrients [37].

4.5 Comparisons with Chinese and American Afforestation Efforts

Both China and the United States have extensive experience with afforestation in desert regions, which provides valuable insights for Sahara afforestation efforts.

- **China's "Great Green Wall"**: China has implemented a large-scale afforestation project to combat desertification, primarily in the Gobi Desert. The project focuses on drought-resistant trees like *Haloxylon, Ammodendron* and *Populus Euphratica,* which require 300–600 L of water per year. China's use of drip irrigation and water-efficient practices could be a model for the Sahara, especially in minimizing water use for large-scale desert planting [38].
- **American Southwest**: In the US, the American Southwest has implemented afforestation and desert reclamation using native species such as Mesquite

(*Prosopis spp.*) and Creosote bush (*Larrea tridentata*), which thrive in saline soils and require minimal water [39]. The American experience highlights the importance of using native plants that are adapted to the local soil and climate, a lesson that can be applied to the Sahara by focusing on indigenous species like acacias and date palms [40].

The Sahara Desert, despite its harsh conditions, can support a variety of resilient trees and bushes that play a role in combating desertification and promoting sustainable agriculture. Species such as Acacias, Argan trees, and Date palms are well-adapted to the desert's sandy soils, high temperatures, and low water availability [41]. By adopting best practices from global afforestation efforts, including China's large-scale desert planting and the US's experience with desert-native species, afforestation in the Sahara can become a reality. Careful attention to water management, soil enhancement, and planting techniques will be crucial in transforming these arid lands into productive ecosystems [42].

5 5G Smart Country Project

The 5G Smart Country Project is a three-year initiative funded by the Federal Ministry for Digital Affairs and Transport (BMDV), under its 5G Innovation Competition. This collaborative effort, involving universities, research institutions, and companies in Lower Saxony, Germany, focuses on advancing smart agriculture, particularly the cultivation of sugar beet in Schickelsheim Domain. The project aims to leverage 5G technologies to streamline agricultural practices through the use of a comprehensive dashboard that provides essential real-time information to farmers. This dashboard enables users to monitor machine conditions, manage equipment, and access live data streams. Key features include crop analysis (e.g., moisture and protein content), real-time weed and animal detection, and the automation of irrigation, planting, and harvesting. A significant aspect of the project is the application of artificial intelligence (AI) for real-time detection and analysis. AI technologies enhance the precision of outcomes, which are also utilized by government and private institutions for research on improving agricultural and environmental conditions.

There are two primary AI techniques used for weed detection. The first method involves identifying areas that require treatment across the entire field. Unmanned Aerial Vehicles (UAVs) perform this task by capturing images and transmitting them to the software's backend for analysis. Subsequently, Unmanned Ground Vehicles (UGVs) are responsible for autonomously removing the weeds in these identified areas, as well as applying appropriate fertilizers and/or biopesticides. See Fig. 1.

The second detection method is executed by Unmanned Ground Vehicles (UGVs). These vehicles use AI technology to differentiate between weeds and crops, allowing them to autonomously remove the weeds from the field. This automated process enhances efficiency in weed management and minimizes the need for manual labor. See Fig. 2.

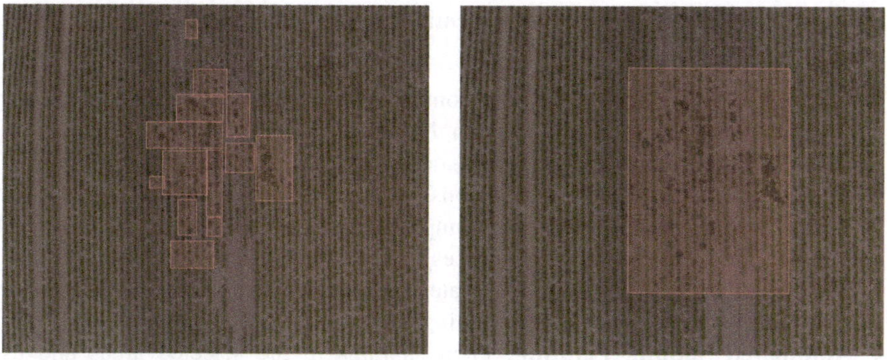

Fig. 1 Detected weed with UAV (own source)

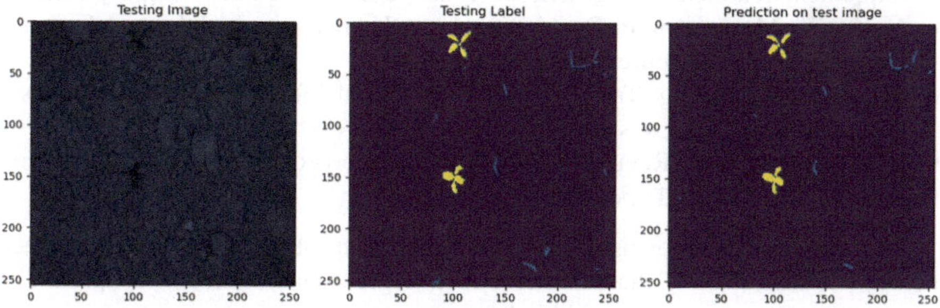

Fig. 2 Plant and weed detection with AI (own source)

The 5G Smart Country Project addresses several challenges, including weed detection, monitoring individual plant health, and managing wildlife in agricultural fields. The project also focuses on developing data governance frameworks, ensuring IT security and data protection, and designing a 5G-enabled IT/IoT architecture to support these use cases. The project's data, securely stored and managed using advanced systems, are crucial for further research and environmental sustainability efforts [5].

6 Pre-Project Idea Description

6.1 Pre-Project Idea Objectives Overview

The Smart Sahara Farming Project aims to utilize advanced technology, both software and hardware, in collaboration with governments of nations interested in afforestation initiatives across the Sahara desert. The objective is to achieve these goals with full autonomous operation throughout the entire process. The core steps of the project include:

- **Geological Surveying:** Conduct a comprehensive geological survey to identify suitable areas for afforestation within the desert.
- **Solar-Powered Stations:** Establish stations powered by solar energy in designated work zones to support operations.
- **Water Supply Infrastructure:** Pump water from desalination plants (both marine-based and groundwater sources) to the solar-powered stations.
- **Tree Seedling Cultivation:** Cultivate/raise seedlings and bushes near water storage tanks in preparation for planting.
- **Automated Sapling Planting:** Plant saplings in the selected areas and use automated systems to inject water directly into their roots.
- **Plant Health Monitoring:** Monitor plants for potential diseases or pests, and implement treatments using environmentally friendly methods without pesticides.
- **Irrigation Management:** Develop automated irrigation schedules and determine the specific water requirements for each plant.
- **Wildlife Monitoring**: Track and monitor animal life in the afforested areas to ensure a balanced ecosystem.
- **Reporting System:** Regularly send detailed reports on the status of the stations, plant health, and field operations.

This project focuses on using renewable energy and sustainable agricultural technology to transform the Sahara into a productive, green environment while promoting environmental sustainability and biodiversity.

6.2 Robots Operations in Smart Sahara Farming

Robots are essential in replacing humans for various tasks in Sahara farming due to the following factors:

- **Desert Climate:** The Sahara experiences extreme temperatures, ranging from −11 °C to over 50 °C, making human survival and labor challenging in such harsh conditions [43].
- **Saharan Geography:** The sparse population and lack of urban infrastructure in the Sahara make it difficult for workers to live far from their communities, adding logistical challenges to long-term human deployment [44].
- **Labor Intensity:** Working in the desert for extended periods is physically demanding and nearly impossible for humans, thus making automation more viable [45].
- **Network and Precision Requirements:** The project requires constant data collection and real-time synchronization of operations, which demands punctuality and consistency, best achieved through robotic systems.

Using Unmanned Aerial Vehicles (UAVs), Unmanned Ground Vehicles (UGVs), and sensors will ease the operation. The robots and sensor can be connected to a dashboard a backend and AI through on utilizing 5G mobile communication technology in real-world agricultural scenarios to assist both research and practical applications for farmers [46].

6.3 Communicate with Robots

The two main ways to communicate with robots wirelessly in the field are:

1. **Terrestrial Communication (using towers)**: This method involves the use of cellular networks (e.g., 5G, 4G) and Wi-Fi towers to establish communication between robots and control centres. This method is typically called cellular communication or radio-based communication, as it relies on ground-based infrastructure like towers or antennas.
2. **Satellite Communication**: This method uses satellites to transmit data to and from robots in remote areas where terrestrial networks might not be available. It is referred to as satellite-based communication or simply satellite communication.

Both methods are used in remote sensing, field robotics, and autonomous systems to gather information from and control robots in the field [47].

6.4 Area Setup and Data Collection

The first mission of the UAVs can be manual or automated. The UAVs survey the geographical and biological data in the target land that needs to be cultivated and send simultaneous data about the possibility of cultivation and the area that can be planned or not. The data which have been surveyed will be send to AI modules to be analysed and stored in a High Performance Object Storage like MinIO which can be placed in a governmental institute far from the working areas [48]. In addition to UAVs, one of the UGVs mission also to collect data from the soil and plants.

7 Integration of Hardware and Software

In the framework of smart farming, particularly in challenging environments such as the Sahara desert, integrating various technological elements is essential for improving efficiency, accuracy, and scalability. The use of UAVs (Unmanned Aerial Vehicles), UGVs (Unmanned Ground Vehicles), and advanced software systems, including AI modules, data storage, and real-time analysis platforms, is crucial to support agricultural decision-making. This section explores how these components

8.1 Employee Costs

A diverse range of expertise is required for this project, including specialists in energy, mechatronics, agricultural engineering, and software development. Salaries for these employees will make up a significant portion of the budget. This includes hiring:

- **Energy experts** to design and optimize renewable energy solutions for water desalination and power needs.
- **Mechatronics engineers** to oversee the development and integration of robotics systems, including UAVs, UGVs and sensors.
- **Agricultural engineers** to manage the design and operation of sustainable farming practices in desert conditions.
- **Programmers** and data scientists to manage software systems, AI, and data analytics that will drive the automation and monitoring processes.

This expense is expected to remain consistent throughout the project's lifecycle due to the need for ongoing research, system maintenance, and improvements in automation and AI systems, in addition to other experts and workers.

8.2 Energy Requirements

Since the project aims to use renewable energy sources, solar power will be the primary energy input. Solar energy will be harnessed for several critical operations:

- **Water desalination** through solar-powered reverse osmosis (RO) systems. The cost of desalinated water is estimated to range from $0.50 to $2.00 per cubic meter, depending on the efficiency of the system and the geographic location [55, 56].
- **Pumping water** from desalination plants to different parts of the desert.
- **Charging robots** and powering other essential services such as data centres and monitoring equipment.

Solar infrastructure, including Photovoltaic (PV) panels and energy storage systems, will represent a significant upfront cost but will reduce operating expenses in the long term due to low ongoing energy costs [57].

8.3 Tree Seedling and Fertilizer Costs

The project's agricultural component will focus on growing desert-adapted plants, such as Acacia, Argania Spinosa, Jujube, and Date Palm, which have been proven to thrive in arid conditions. These trees require careful planning and investment in:

- **Seedlings**, which are crucial for reforestation efforts.

- **Fertilizers**, though the use of organic or locally-sourced fertilizers could reduce costs and environmental impact.

The initial investment in seedlings and fertilizers is a critical step in establishing a self-sustaining ecosystem in the desert [58].

The exact number of trees that can be planted per square kilometre depends on the type of tree, soil quality, and the availability of water for irrigation. When planning afforestation in the Sahara, a combination of these species with appropriate irrigation and soil management strategies can yield favorable results. the total number of trees per square kilometre varies significantly [59]:

- **Acacia**: 15,000–30,000 trees
- **Argania spinosa**: 5000–10,000 trees
- **Jujube**: 40,000–50,000 trees
- **Date Palm**: 10,000–15,000 trees

8.4 Robotics Infrastructure

Robots play an essential role in the automation of agricultural tasks in harsh desert conditions. Expenses in this category include the fabrication and maintenance of UAVs, UGVs, and a network of sensors:

- UAVs will conduct aerial surveys of the land and provide data on soil conditions, water needs, and the health of plants.
- UGVs will handle tasks such as soil sampling, planting, and monitoring.
- Sensors will continuously track environmental conditions such as temperature, humidity, fire alarm and soil moisture.

To minimize costs, the project should focus on local fabrication of these robotic systems, which could significantly reduce the price compared to importing commercial equipment. By using locally available resources and expertise, the long-term operational costs of the robots and their maintenance can be reduced.

8.5 Water Desalination

Water supply is crucial for the project, and it will rely on solar-powered desalination plants to provide freshwater for irrigation and other needs. The cost of desalinating water ranges from $0.50 to $3 per cubic meter, depending on the specific technology and operational conditions. This variability is influenced by factors such as:

- **Scale of the desalination plant:** Larger plants benefit from economies of scale, reducing the cost per cubic meter.
- **Water salinity:** Higher salinity levels require more energy to desalinate, raising costs.

- **Local energy availability:** The efficiency and scale of the solar energy system used will also affect the overall cost This is one of the most critical and cost-sensitive aspects of the project since desert environments typically have very limited freshwater resources. As a result, the project's success is heavily dependent on cost-effective and scalable desalination solutions powered by renewable energy [55].

9 Conclusion

The Smart Sahara Farming project is a forward-looking initiative designed to combat desertification and stabilize the climate while offering economic benefits. The project involves the gradual implementation of advanced technologies, including AI, UAVs, UGVs, and solar-powered systems, to transform desert areas into fertile agricultural lands. By leveraging on automation, the need for human labor in harsh desert conditions is minimized. Collaborative efforts between governments, researchers, and the private sector will be essential to ensure the project's success. It serves as a model for sustainable desert farming and environmental restoration.

The financial framework of the project incorporates a blend of high upfront investment in technology and infrastructure, particularly in renewable energy and robotics, with long-term operational efficiency aimed at minimizing ongoing costs. Strategic planning in sourcing local labor and materials for robotics, combined with the cost advantages of solar-powered desalination, positions the project has a potentially sustainable model for desert farming.

Acknowledgements The authors thank the BMDV for funding this research project (grant number: 45FGU117).

References

1. World Population Clock, 8.1 Billion People (LIVE, 2023). [Online]. Available: https://www.worldometers.info/world-population/#google_vignette. (Accessed 18 Nov 2024)
2. Ehrlich P, Harte J (2015) Opinion, to feed the world in 2050 will require a global revolution. Proc Natl Acad Sci USA 112(48):14743–14744
3. "Sahara Desert" WorldAtlas (2023). [Online]. Available: https://www.worldatlas.com/deserts/sahara-desert.html. (Accessed 18 Nov 2024)
4. Shehadi E (2021) Growing food in the desert: is this the solution to the world's food crisis?. Euronews (2021). [Online]. Available: https://ourworld.unu.edu/en/growing-food-in-the-desert-is-this-the-solution-to-the-worlds-food-crisis. (Accessed 18 Nov 2024)
5. Akyol A, Chahin R, Dillschneider E-M, Gerloff L, Grabenhorst I, Marx Gómez J, Patil A et al. (2023) Development of a smart farming dashboard based on 5G mobile data
6. Arif MM, Rehman A (2021) Autonomous farming: application of 5G and IoT for sustainable agriculture. 9:83609–83623. https://doi.org/10.1109/ACCESS.2021.3087392
7. Salem HS, Wright EP (2002) The Nubian sandstone aquifer system: developments in utilization. J Afr Earth Sc 34(2):103–112. https://doi.org/10.1016/S0899-5362(02)00009-2

8. El-Baz AA, Kamal MAL (2016) The Nubian sandstone aquifer system: hydrogeological study of the largest fossil aquifer in the world. Desalin Water Treat 57(30):14168–14182. https://doi.org/10.1080/19443994.2015.1083759
9. Bakhit TAT, El-Kenawy MMA (2015) Assessment of the groundwater resources of the Nubian sandstone aquifer system. Water Resour Manage 29(8):2873–2890. https://doi.org/10.1007/s11269-015-0912-7
10. Flowers RA, Tuteja MJT, McCarty JG (2013) Salinity tolerance in plants: a review. Plant Sci 203:85–100. https://doi.org/10.1016/j.plantsci.2012.07.017
11. Barhoum DMH (2016) The role of halophytes in sustainable agriculture. Agricul Food Secur 5(3). https://doi.org/10.1186/s40066-016-0075-6
12. Everbright, "Power of Gypsum in Clay Soils," Everbright (2024). [Online]. Available: https://everbrighttr.co/en/power-of-gypsum-in-clay-soils/#:~:text=Benefits%20of%20Gypsum%20for%20Clay%20Soil&text=Reduces%20Sodium%20Levels%3A%20In%20sodic,reducing%20soil%20stickiness%20when%20wet. (Accessed 18 Nov 2024)
13. Arnao D, González-Bulnes MA, de Almeida LRBS (2024) Fostering halophytes in saline agriculture: a sustainable option for crop production under climate change. Front Plant Sci 14(Article 1092885). [Online]. Available: https://www.frontiersin.org/articles/10.3389/fpls.2023.1092885/full. (Accessed 18 Nov 2024)
14. Tarolli P, Luo J, Park E, Barcaccia G, Masin R (2024) Soil salinization in agriculture: mitigation and adaptation strategies combining nature-based solutions and bioengineering. iScience 27(2)
15. Shrivastava P, Kumar R (2015) Soil salinity: a serious environmental issue and plant growth promoting bacteria as one of the tools for its alleviation. Saudi J Biol Sci 22(2):123–131
16. Al-Addous M, Bdour M, Rabaiah S, Boubakri A, Schweimanns N, Barbana N, Wellmann J (2024) Innovations in solar-powered desalination: a comprehensive review of sustainable solutions for water scarcity in the Middle East and North Africa (MENA) Region. Water 16(13):1877
17. Knapp KC, Weinberg M, Howitt R, Posnikoff JF (2003) Water transfers, agriculture, and groundwater management: a dynamic economic analysis. J Environ Manage 67(4):291–301
18. Choukr-Allah R, Ragab R (2023) In: Biosaline agriculture as a climate change adaptation for food security, Springer Nature, Switzerland
19. United Nations Development Programme (UNDP) (2013) Water Governance in the arab region: managing scarcity and securing the future, UNDP, New York
20. "Water Governance in the Arab Region: Managing Scarcity and Securing the Future" (2013) United Nations Development Programme (UNDP), New York
21. Youssef AK (2023) Water resources in Egypt: future challenges and opportunities. J Environ Stud 12(1):34–45
22. Boumediene HR (2022) Desalination initiatives and the future of agriculture in North Africa. J Sustain Water Resour 14(2):56–62
23. Mahmoud A, El-Khattabi B (2023) Water scarcity challenges and desalination solutions in North African countries. J Water Resour Manag 25(3):145–159
24. El-Sadek MA, Abd El-Samie HM (2021) Desalination for agricultural purposes in Egypt: challenges and opportunities. Water Resources Managem 35(5):1425–1440
25. El-Hadad RA, Mohamed AH (2020) The role of desalination in addressing water scarcity in Sudan. Int J Water Resour Dev 36(1):45–60
26. Al-Mansoori MA (2016) Desalination using solar energy in the Middle East. J Renew Sustain Energy 8(2):025901
27. Khan M, Al-Ghouti MA (2021) DPSIR framework and sustainable approaches of brine management from seawater desalination plants in Qatar. J Clean Prod 319:128485
28. Al-Khatib M, Khatib NK, Khatib SM (2018) A review of desalination technologies and their environmental impacts. Water Sci Technol 78(9):2020–2031
29. El-Ramady HR, El-Marsafawy SM, Lewis LN (2013) Sustainable agriculture and climate changes in Egypt. Sustain Agricul Rev 12:41–95
30. Wilo (2024) "Toshka: Reclaiming Land for Sustainable Use" Wilo, [Online]. Available: https://wilo.com/en/Pioneering/Stories/Toshka-reclaiming-land-for-sustainable-use-_28417.html. (Accessed 18 Nov 2024)

Implications of Critical Minerals Mining on Food Security and Environmental Sustainability in Zambia

Solomon Mwampikita and Matilda Kaliba

Abstract The devastating impacts of climate change on Zambia's socio-economic development cannot be overemphasised. However, the abundance of critical minerals offers a significant opportunity for economic development and participation in the global green energy transition. On the other hand, the participation in the global green transition reinforces and/or escalates existing socio-ecological challenges that include food and water insecurities among the communities. It is clear the critical minerals are essential for modern technologies and industries, and Zambia possesses significant reserves of these minerals, and mining is a vital sector for Zambia's economy, but current practices often have negative environmental and social consequences. To that end, responsible mining practices and robust regulations are essential, and stakeholder engagement is vital for finding sustainable solutions. Promoting responsible sourcing, environmental mitigation measures, and investment in innovative technologies can minimize the negative impacts of mining. The government should strengthen the regulatory framework for mining operations, water resources and agriculture through an inclusive process, including enforcement mechanisms and addressing gap as they are critical. Transparency and accountability in the mining sector should be promoted as it plays an essential part for building trust with stakeholders. The Zambian government should enhance research and development of sustainable mining practices are necessary to reduce the environmental footprint of the industry. By addressing these challenges and promoting responsible and sustainable mining practices, Zambia can leverage its critical minerals for economic growth while minimizing environmental and social impacts as it will ensure long-term benefits for the country and its people.

Keywords Climate change · Critical minerals · Food security · Environmental sustainability

S. Mwampikita (✉)
Centre for Trade Policy and Development (CTPD), Olympia Park, Lusaka, Zambia
e-mail: researcher.cce@ctpd.org.zm

M. Kaliba
University of Zambia, Lusaka, Zambia

S. Al-Naemi et al. (eds.), *Water and Food Security in the Face of Climate Change: Challenges and Opportunities for Resilience*, World Sustainability Series,
https://doi.org/10.1007/978-3-032-00098-9_51

1 Introduction

The growing global demand for critical minerals, driven by the transition to renewable energy, electric vehicles, and advanced technologies, has ignited a surge in mining activities in rich resource countries like Zambia. As countries strive to meet their climate goals under the Paris agreement, the demand for these minerals has risen, leading to what is termed a "mineral rush" [10]. While this boom presents significant economic opportunities, it also raises serious concerns about the potential negative consequences for food security and environmental sustainability. The expansion of mining operations often involves the conversion of agricultural land into mining sites, reducing the area available for food production. Additionally, the mining process can lead to soil contamination and water pollution, jeopardizing the quality and availability of resources essential for agriculture [13]. Furthermore, the increased demand for energy associated with mining operations can contribute to deforestation and greenhouse gas (GHG) emissions, exacerbating climate change and its impacts on food production. It is clear, these interconnected challenges highlight the urgent need for sustainable mining practices and comprehensive policies that balance economic growth with environmental protection and food security in Zambia.

2 Overview

The reality of climate change and its link to the use of fossil fuels is now well established [9]. Thus, the need to reduce carbon emissions and associated gases has heightened following the devastating impacts of both global warming and climate change on socio-economic activities and biodiversity. On the other hand, the demand for critical minerals is projected to rise dramatically, more than doubling by 2030 and quadrupling by 2050, with an expected $400 billion in income each year. In response to the global demand for critical minerals, investments have surged by 20% in 2021 and 30% in 2022. In 2022, spending on critical minerals exploration also increased by 20%, primarily due to the world's record-breaking boom in lithium exploration. Lithium investments saw the largest increase at 50%, followed by significant growth in nickel and copper exploration [2].

In Zambia, mining and agriculture sectors have continued to play a significant role in natural resources utilisation, with significant impacts on the livelihoods of local communities. In 2023, real Gross Domestic Product (GDP) averaged at 5.8% was driven by, among other sectors, mining and agriculture on the supply side and by household and corporate consumption on the demand side [1]. The mining sector has offered a lifeline for Zambia's economy for over 100 years. Copper and cobalt, the country's traditional exports, account for well over 70% of export earnings. In the period between 2011 and 2021, growth in the mining sector averaged 2.5% and its contribution to GDP averaged 14.8%. Agriculture sector on the other hand, which provides livelihood to over 70% of the Zambian people, growth averaged 0.4%

while its share of GDP is 5.8% in the same period. In 2021, agriculture sector was one of the major contributors to the real GDP growth of 3.6%, though the mining sector output declined during the period [4] owing to numerous challenges such as Covid19, electricity shortages, and volatile international commodity prices coupled with climate change.

The recent exploration and licensing in the mining sector have expanded into the agriculturally rich Southern, Central, Eastern and North-Western provinces, where new critical minerals deposits have been identified. This expansion highlights Zambia's growing strategic importance in the global mineral supply chain but also raises concerns about the potential conflicts between mining and agricultural land use. To tap from the surge for critical minerals rush, the government has developed the Critical Minerals Strategy as well as the 3 million metric tonnes strategize for copper production annually by 2031. However, the increase of Large-Scale Land based Investments (LSLBI) in Zambia especially in mining and agriculture sectors does not mirror socio-economic wellbeing of communities hosting these investments; there is little to no benefit from the investments [4]. Livelihoods of local communities that account about 70% depend on rainfed agriculture, have negatively been affected by climate change and associated risks such as floods, droughts and diseases.

While critical minerals present a significant opportunity to transition away from carbon-intensive energy systems, their extraction and exploitation may reinforce existing socio-ecological challenges, particularly in resource-dependent countries like Zambia. However, the extent and dynamics of these challenges remain underexplored. This paper aims to contribute to the growing discourse on critical minerals, focusing on their implications for Zambia. The paper examines the dual role of critical minerals as drivers of economic development and contributors to the socio-ecological challenges in the mining communities. First, the paper defines critical minerals and their significance to Zambia's economic development and global supply. This is followed by the discussion on the scale and nature of critical minerals mining operations in Zambia. Subsequently, the paper explores the implications of critical minerals extraction on food security and environmental sustainability. The findings show that critical minerals extraction exacerbate land degradation, water resource depletion and ecological imbalances which in turn heighten food insecurity risks in mining communities. The paper concludes that while critical minerals offer significant opportunities, their extraction can result in far-reaching socio-ecological consequences, including worsening food insecurity in rural agricultural communities whose productivity is already affected by climate change effects. It calls for integrated strategies that balance green economic benefits of critical minerals mining with the need to safeguard environmental and food system resilience.

3 Critical Minerals and Their Significance

Zambia, a mineral-rich nation, is home to several critical minerals that are essential for modern technologies and industries. These critical minerals, which are often scarce or difficult to extract, play a vital role in various sectors such as electronics, energy, and transportation. Important to note that critical minerals form part of the "green minerals or metals". Green minerals further form part of the minerals or metals necessary for the manufacturing of green or clean technologies that include solar panels, wind turbines, electric vehicles, battery storage, hydrogen electrolysers and fuel cells. "Green minerals" are crucial for transitioning to a cleaner, low-carbon economy especially in the transport and industrial economy sectors. Some key examples of green minerals include cobalt, copper, bauxite, chromium, high-purity iron ore, platinum group metals, lithium, and rare earth metals. Critical minerals are a specific type of "green mineral" with two important features: they are essential for the functioning of modern technologies, economies, or national security, and there is a risk that their supply chains could be disrupted [2].

In the Zambian context, critical minerals have been defined as natural resources that are essential for modern technology, economic growth, and national development. These strategic minerals include lithium, graphite, nickel, cobalt, tin, copper, manganese, sugilite, uranium, and Rare Earth Elements (REEs), as declared by the Zambian government [5, 6]. Worth noting, Zambia is among the major producers of copper, a metal widely used in electrical wiring, plumbing, and construction. Additionally, the country possesses significant reserves of cobalt, a key component in lithium-ion batteries that power electric vehicles and consumer electronics. Furthermore, the country Zambia is mining nickel at First Quantum Minerals (FQM) in Kalumbila district of Northwestern province, another valuable mineral used in batteries and stainless steel. The abundance of these critical minerals position Zambia as a strategic player in the global supply chain as well as the manufacturing of advanced technologies that utilise these processed minerals, while also presenting both opportunities and challenges for sustainable development and economic growth. Zambia has declared that all minerals are critical but key ones include copper, cobalt, lithium and nickel. The national critical minerals strategy emphasises on the deliberate policy or legislative intervention by government to guide the management and development of such minerals, as they are susceptible to unsustainable exploitation. Boafo et al. [2] states that the "criticality" of critical minerals is context specific, as each country or region has reasons why these minerals are classified as "critical".

3.1 Economic Importance of Critical Minerals

Zambia's economy is heavily reliant on the extraction and export of critical minerals, particularly copper and cobalt. These minerals constitute a significant portion of the country's foreign exchange earnings, contributing substantially to its overall economic growth. The mining sector has offered a lifeline for Zambia's economy for over 100 years and sets the tempo of other activities. In 2023, real GDP averaged at 5.8% was driven by, among others, mining and quarrying on the supply side and by household and corporate consumption on the demand side [1]. Copper and cobalt, the country's traditional exports, account for well over 70% of export earnings. In the period between 2011 and 2021, growth in the mining sector averaged 2.5% and its contribution to GDP averaged 14.8% [4]. Arising from the global demand, climate agreements, and to cement the position of critical minerals in the country's economic development, the government developed the Critical Minerals Strategy with the aim of harnessing and utilizing the critical minerals to support Zambia's social economic development. It is envisaged to guide investments for explorations, mining, development, processing and supply of goods and services in Zambia's sub-sector [5].

Additionally, Zambia's strategic location in Southern Africa position it as a key player in global supply chains for critical minerals. With the 3 million metric tons copper production strategy developed, the country's abundant resources and competitive mining environment attract foreign investment and multinational corporations [6], further solidifying its role in the international market. Two key mining companies (Konkola Copper Mines and Mopani Copper Mines) have been reopened following legal disputes, and new mining companies are coming up such as Kitumba mine in Mumbwa district and Mingomba mine in Chililabombwe. Aside being a key player in the supply chain, it will realise several socio-economic benefits from the effective management and exploitation such as job creation; enhanced industrialization, increased revenue for national treasury, increased investment opportunities, and development of local services and infrastructure [5]. However, Zambia's dependence on critical minerals also makes it vulnerable to fluctuations in global demand and prices, highlighting the importance of diversifying its economy and developing sustainable mining practices. The lack of adequate and high-resolution geological and geophysical information makes Zambia less attractive to the much-needed private sector investment [4–6]. Other challenges arising from the opportunity include lack of sustainable mining practices which encourage the adoption of environmentally responsible and socially inclusive mining practices for critical minerals, such as cobalt, copper and manganese, to minimize ecological impact. Also, inadequate policies, laws and regulatory frameworks to implement and enforce robust environmental and social regulations governing the critical minerals sector to ensure responsible and ethical extraction practices add to the challenges.

3.2 Mining Activities in Zambia

Even though the mining sector dominates the economy, unsustainable mining practices can be attested to the scale and nature of mining operations in Zambia. Large-scale processing and earthmoving machines are frequently involved in these operations, which can have a large negative influence on the environment. Large-scale,

To mitigate the challenges, there is need to tighten regulations on mine rehabilitation where the Zambia Environmental Management Agency (ZEMA) ensures that comprehensive mine closure plans address terrain restoration and the creation of wetlands to manage pit lakes. Ecotourism co-benefits could also be explored.

4.3 Processing

Minerals processing especially hard rock mining utilises substantial amounts of water that causes significant impacts on both surface and ground water resources that affects ecosystems that depend on it and available nearby communities [2]. Mining operations often use a lot of water, which can make it harder for people in nearby areas to get clean drinking water. This problem is especially bad in places where water is already scarce. Water is also important for everyday life, like farming and recreation. Using too much water for mining can create serious problems for these activities. In Mkushi district, manganese mining firms have started diverting streams of water to their mining operations, affecting local communities access and availability of water for domestic use, and their animals and agriculture activities. Communities adjacent to Barrick Lumwana Mine in Kalumbila district, undertaking the expansion of the mine, have observed drying up of water sources. Women indicated that they fear that they will be undertaking long journeys to fetch water. Thus, using limited freshwater available for mining will make it harder to get water for domestic, farming, and other things in Zambia. This will worsen the problem of not accessing enough clean water. Given the high demand for water in mining, particularly in water-stressed areas, there are significant concerns to water supply security. As a mitigative measures, mining firms can be tasked to implement technologies that minimise water usage in mineral processing. Additionally, water recycling techniques such as introducing closed-loop water systems can reduce consumption.

Another challenge the processing of critical minerals is posing and/or likely to pose is the intense use of energy. The African countries where these mining operations are proposed or currently taking place are already struggling to create a sustainable electricity source [2]. Zambia is currently experiencing energy deficit in the electricity sub-sector that has been attributed to the drought situation. Zambia's installed power generation capacity is at 3356.6 megawatts against about 900 megawatts during the drought situation, and the mining sector uses around 51%. The surge in mining investments coupled with limited infrastructure and over reliance on hydroelectricity, may hamper efforts in improving access to electricity in rural areas that may lead to increased deforestation. Mining firms could be required to invest in alternative energy sources such as solar or wind to power mining operations.

4.4 Mine Rock Stockpiles and Tailings Storage Facilities

Mine rock stockpiles (waste rock dumps) hold a significant volume of waste rocks, as more than 90% of excavated rocks end up in a waste dump facility. In addition, processed rocks (tailings) are released into tailings storage areas. These storage facilities have a substantial impact on surface water and groundwater-dependent ecosystems, reducing their quality [2]. This challenge is not farfetched in the context of Zambia coming from the Copperbelt experiences. In 2005, the tailings facilities at KCM busted and resulted in the pollution of Kafue River and affected many communities that depend on the river. For example, communities involved in vegetable gardens and other crops had their fields were affected by the polluted water that resulted in the loss of their livelihood. Further, stockpiles of residue are likely to be increased like "black mountains" in the Copperbelt province. Potential saline drainage could affect sensitive ecosystems, with impacts remaining in the environments for thousands of years like currently being experienced in Kabwe district due to lead pollution. Agencies such as ZEMA should mandate regular inspections and maintenance to prevent such hazards. Additionally, authorities could encourage innovative uses of rock stockpiles such as construction material to reduce waste.

5 Food Security Implications

From the environmental implications, this section examines the food security implications arising from critical minerals extraction in Zambia. Key among the implications include land rights issues, land use changes, displacement and loss of livelihoods for local communities, and health and safety problems. However, proactive governance and community engagement can mitigate these impacts.

5.1 Land Rights Disputes

Despite significant efforts to address associated land access issues through robust resource governance, access to mineralized lands in Africa remains enmeshed in conflicts. Along with tensions between large-scale and small-scale miners or participants within the agricultural sector, these conflicts involve a variety of stakeholders, including state and non-state actors. These disputes have their origins in the coexistence of a dualized system of land access governance that is regulated by both state laws and customs. The state's ownership of mineral resources on behalf of the public complicates the application and interpretation of both state and customary laws, which is one of the main causes of the persistent resource-related conflicts that persist in many African nations [3]. The hunt for vital minerals and the ensuing geopolitics exacerbates these disputes even further. In 2013, FQM obtained 1000

Agricultural Soil Remediation and Its Tangible Relation with Food Security in Qatar: A Short Review

Milad Sabouri Gasgarei

Abstract Soil health is one of the most vital environmental indices that could affect the whole health chain. Due to various human activities that are direct results of industrialization, the soil is prone to pollution by various pollutants like heavy metals, pesticides, petroleum products, etc. These pollutants are absorbed by plants and finally consumed by humans. Agriculture is one of the most strategic industries in Qatar, with a growing share in GDP after 2017, when the country faced a sea, air, and land blockade by the neighboring countries. Hence, monitoring the soil quality and maintaining its health constantly is significantly important. Various physical, chemical, and biological remediation methods can be employed for soil remediation, and each of them brings its own advantages and disadvantages in terms of energy consumption, possibility of secondary pollution, altering soil community, and efficiency. Compared to conventional remediation methods, bioremediation is an attractive alternative where pollutant degradation occurs through microbial activities, offering cost effectiveness, safety, and sustainability. Some biological agents like yeast, fungi, bacteria, algae, and amendments are employed to consume pollutants as feed and alter them into harmless products like ethane and methane. In this work, we will introduce some major soil pollutants, assess various soil remediation methods, and focus on bioremediation as the sustainable and green alternative. Finally, the promising features of biochar as a sustainable and green alternative for soil remediation in Qatar will briefly be discussed.

Keywords Agriculture soil · Bioremediation · Soil remediation

M. S. Gasgarei (✉)
Carleton University, Ottawa, ON, Canada
e-mail: miladsabourigasgarei@cmail.carleton.ca

S. Al-Naemi et al. (eds.), *Water and Food Security in the Face of Climate Change: Challenges and Opportunities for Resilience*, World Sustainability Series,
https://doi.org/10.1007/978-3-032-00098-9_52

high remediation costs, which makes it unfeasible. Further, vitrification and soil flushing are effective in metal removal in some types of soil, while in others, they have medium removal. Similar to excavation and disposal, they suffer from high costs of remediation [13].

On the other hand, biological remediation has average removal of organic and inorganic pollutants, but they benefit from shorter remediation time and far lower remediation costs. In addition, they produce no secondary pollution and benefit from less equipment. Finally, the technique is still young and many improvements on its efficiency can be figured [13, 20].

4.1 Biochar

Among the various types of biological soil remediation, biochar is a promising alternative due to its vast advantageous features. Biochar is a carbon-rich porous substance which is finely grinded and is obtained from thermal decomposition of biomass under oxygen-limited conditions and usually at temperatures below 700 °C. As far as soil remediation is concerned, biochar can act as a carbon sequester in addition to its ability to improve soil fertility under nutrient retention and moisture elevating. Biochar has promising performance in stabilizing heavy metals such as Cd, Cu, Ni, Pb, and Zn and reduce their bioavailability through either enhanced sorption or chemical precipitation. The first one is based on electrostatic attraction, ion exchange, and surface complexation and the latter occurs under soil pH elevation and ash addition of carbonates and phosphates [21].

The potential of producing biochar for treatment purposes from food waste in Qatar was evaluated in the work of Parthasarathy et al. [22]. With an annual 1.8MT food waste, Qatar has a great potential to recover nutrients and useful compounds from waste in the form of biochar. It was reported that wastepaper has the greatest potential for biochar production with 534,969 T of biochar at 250 °C and a 107 million USD annual revenue potential [22]. Use of biochar for soil remediation in Qatar was further studied by Bansode et al. [23] where they used various blends of wastes for soil remediation purposes. It was reported that biochar yields up to 62% is achievable which is a milestone for a more sustainable remediation purpose.

As previously mentioned, biochar can be obtained through pyrolysis from various feedstock or a blend of them. In order to prepare a blend of biochar with appropriate carbon sequestration capacity in soil, Al-Rumaihi et al. [24] found that a blend with composition of 20.4% Camel manure, 27% date pits, 26.3% low-density polyethylene LDPE, and 26.4% high-density polyethylene (HDPE) can maximize carbon sequestration in soil. This is a strong support of the idea of extensive potential of biochar for remediation and carbon sequestration purposes where infinite sources of biomass can be blended to form an effective biochar [24].

Although the role of biochar has not been studied extensively in Qatar, but its potential for being employed as a promising alternative for soil remediation should be a priority for the further studies.

5 Conclusion and Recommendations for Future Work

Agricultural soil is one of the most strategic mediums that has a growing share of Qatar's GDP. Any imbalance in soil health and quality is a potential threat for the food supplement chain and food security. Agricultural soil in Qatar is prone to heavy metals that can easily enter the food chain through plants and crops. Hence, constant monitoring and remediation of soil is significantly important to meet the safety standards and limits. Although various well-established physical and chemical techniques are practiced for soil remediation, they suffer from high remediation costs, the possibility of secondary pollution, and potential negative effects on the soil community. Hence, bioremediation with biochar as a viable alternative towards a more sustainable future was introduced and efforts to employ it in Qatar was briefly reviewed. It is true that the technique is still young and may have lower performance compared to traditional techniques, there are wide horizons for its development due to its environmentally friendly nature and far lower costs of remediation. Hence, the future work must be guided towards experimental and theoretical studies which aim to improve the efficiency of biochar. This will not only contribute to a circular economy as waste is not disposed of without treatment but also plays a significant role in a green and sustainable remediation of the soil. In addition to governmental financial helps, these studies can help Qatar to reach its objectives for Qatar National Vision 2030.

References

1. Arora NK (2018) Agricultural sustainability and food security. Environ Sustain 1:217–219
2. Viana CM, Freire D, Abrantes P et al (2022) Agricultural land systems importance for supporting food security and sustainable development goals: a systematic review. Sci Total Environ 806. Epub ahead of print 2022. https://doi.org/10.1016/j.scitotenv.2021.150718
3. Kopittke PM, Menzies NW, Wang P et al (2019) Soil and the intensification of agriculture for global food security. Environ Int 132:105078
4. Natalia Rodríguez Eugenio and Marilena Ronzan. Sources of soil pollution. https://openknowledge.fao.org/server/api/core/bitstreams/fe5df8d6-6b19-4def-bdc6-62886d824574/content/src/html/list-of-contributors.html. Accessed 1 Oct 2024
5. worldometers.info. The total land area is (2%2C965%2C657 people in 2024). https://www.worldometers.info/world-population/qatar-population/#:~:text. Accessed 1 Oct 2024
6. Karanisa T, Amato A, Richer R et al (2021) Agricultural production in Qatar's hot arid climate. Sustain 13. Epub ahead of print 2021. https://doi.org/10.3390/su13074059
7. Qatar GDP from agriculture. https://tradingeconomics.com/qatar/gdp-from-agriculture. Accessed 1 Oct 2024
8. Johnson J. Agriculture sector projected to amount over QR622m in 2024. The Peninsula Qatar. https://thepeninsulaqatar.com/article/05/01/2024/agriculture-sector-projected-to-amount-over-qr622m-in-2024#:~:text. Accessed 1 Oct 2024
9. Qatar food import by 2022. https://wits.worldbank.org/CountryProfile/en/Country/QAT/Year/LTST/TradeFlow/Import/Partner/by-country/Product/16-24_FoodProd. Accessed 3 Jan 2025
10. Alsafran M, Usman K, Al Jabri H et al (2021) Ecological and health risks assessment of potentially toxic metals and metalloids contaminants: a case study of agricultural soils in Qatar. Toxics 9:1–15

11. Li S, He L, Zhang B et al (2022) A comprehensive evaluation method for soil remediation technology selection: case study of ex situ thermal desorption. Int J Environ Res Public Health 19. Epub ahead of print 2022. https://doi.org/10.3390/ijerph19063304
12. Almutairi M (2020) The assessment strategy for selecting and evaluating geoenvironmental remediation. Geopolymers Other Geosynth 1–19
13. Remediation Technologies Handbook (2018), pp 27–28
14. Sarker A, Al Masud MA, Deepo DM, Das K, Nandi R, Ansary MWR, Towfiqul Islam ARM, Islam T. Biological and green remediation of heavy metal contaminated water and soils: a state-of-the-art review. Chemosphere 332
15. Liu L, Li W, Song W et al (2018) Remediation techniques for heavy metal-contaminated soils: principles and applicability. Sci Total Environ 633:206–219
16. Sizmur T, Quilliam R, Puga AP et al (2015) Application of biochar for soil remediation. Agric Environ Appl Biochar Adv Barriers 98104:295–324
17. Singh A, Kuhad RC, Ward OP (2009) Biological remediation of soil: an overview of global market and available technologies, pp 1–19
18. Kapahi M, Sachdeva S (2019) Bioremediation options for heavy metal pollution. J Heal Pollut 9. Epub ahead of print 2019. https://doi.org/10.5696/2156-9614-9.24.191203
19. Delangiz N, Varjovi MB, Lajayer BA et al (2020) Beneficial microorganisms in the remediation of heavy metals. INC. Epub ahead of print 2020. https://doi.org/10.1016/b978-0-12-818469-1.00034-1
20. Kondakindi VR, Pabbati R, Erukulla P et al (2024) Bioremediation of heavy metals-contaminated sites by microbial extracellular polymeric substances—a critical view. Environ Chem Ecotoxicol 6:408–421
21. Guo M, Song W, Tian J (2020) Biochar-facilitated soil remediation: mechanisms and efficacy variations. Front Environ Sci 8. Epub ahead of print 21 October 2020. https://doi.org/10.3389/FENVS.2020.521512/FULL
22. Parthasarathy P, Alherbawi M, Shahbaz M et al (2023) Developing biochar from potential wastes in Qatar and its revenue potential. Energy Convers Manag X 20:100467
23. Biochars from Solid Organic Municipal Wastes for Soil Quality Enhancement (2014) 5339
24. Al-Rumaihi A, Alherbawi M, Mckay G et al (2023) Assessing plastic and biomass-based biochar's potential for carbon sequestration: an energy-water-environment approach. Front Sustain 4. Epub ahead of print 2023. https://doi.org/10.3389/frsus.2023.1200094

Impact of Climate Change on Soil Pathogens, Crop Yields, and Food Security

Abdulla Ansari and Ravi Rangarajan

Abstract A fundamental ecological impact of climate change is the alteration of agroecosystems. In particular, soil-pathogen levels and crop health are important as they eventually influence food security. The aim of this short scoping review synthesis focusing on recent updated research (last 5 years) is to better understand the link between climatic factors such as temperature, precipitation, and soil moisture, and their direct and indirect impacts on the distribution and proliferation of pathogens affecting crop yield and food security. The observed outcome of the synthesis was that increase in soil pathogen levels and the resulting decrease in crop yield, coupled with heightened risk levels of mycotoxin contamination requires urgent need for adaptive strategies to safeguard global food systems in the face of anthropogenic climate change.

Keywords Climate change · Food security · Pathogens · Soil moisture

1 Introduction and Study Approach

Global agricultural sector is under tremendous threat due to the impacts of climate change. Rising temperature, increasing frequencies of erratic extreme weather events, are expected to affect agricultural productivity as well as food and food security globally. Among the many consequences of climate change, one key driver is the proliferation of soil-borne pathogens. As global temperature rises, species such as *Fusarium and Alternaria*, have been shown to thrive, leading to an increased incidence of plant diseases and lowered crop yields [1, 2]. The presence of soil microbiome is essential in the distribution of nutrients to plants and increasing their vitality. However, the presence of pathogenic species disrupts nutrient uptake and facilitate secondary infections, leading to loss of plant vigor [3]. This interplay between soil-microbiome and

A. Ansari · R. Rangarajan (✉)
Department of Environmental Health and Safety, University of Doha for Science and Technology, Doha, Qatar
e-mail: ravi.rangarajan@udst.edu.qa

S. Al-Naemi et al. (eds.), *Water and Food Security in the Face of Climate Change: Challenges and Opportunities for Resilience*, World Sustainability Series,
https://doi.org/10.1007/978-3-032-00098-9_53

serious implications for public health, particularly in regions that rely heavily on staple crops for nutrition. Thus, exerting a heavy burden on maintaining global food security [5].

4 Implications for Food Security

The burden of maintaining global food security in the face of climate-driven pathogen proliferation is immense. With a growing global population and increasing demand for food, the agricultural sector is under pressure to produce more while simultaneously grappling with the challenges posed by climate change. The reduction in crop yields due to pathogen pressure exacerbates food insecurity, particularly in vulnerable regions where agricultural productivity is already compromised by environmental stressors. Addressing this challenge requires a multifaceted approach that includes the development of climate-resilient crop varieties, improved pathogen management strategies, and efforts to mitigate the broader impacts of climate change on agricultural ecosystems. The interplay between warmer temperatures, fluctuating soil moisture levels, and pathogen dynamics highlights the complexity of the challenges facing global agriculture. Addressing these challenges will require concerted efforts to develop adaptive strategies that enhance the resilience of agricultural systems to climate change while safeguarding food security for future generations.

Vulnerable population, such as those in low-income regions face a tremendous burden as pathogen-driven crop losses increase. Furthermore, reduction in agricultural productivity, coupled with loss of farmlands due to extreme hydrological events such as floods and droughts further exacerbates food insecurity and malnutrition [8]. It is important to note that developing countries with limited resources face a disproportionate amount of these climate induced challenges. As global temperatures increase and frequency of erratic weather increases, all four pillars of food security, which include availability, utilization, access, and stability, will further deepen global economic inequalities [3].

To overcome the challenges induced by climate change, adaptation strategies should focus on strengthening plant resilience through enhanced soil and water management, beneficial plant–microbe interactions [3, 14], growing period adaptation [15], and use of emerging technologies like artificial intelligence [16]. Integration of these methods provides a pathway for increasing resilience of plants to pathogenic infections under challenging environmental conditions. Furthermore, increasing prevalence of mycotoxins in food systems also requires robust control measures since it has a significant impact on both crop yields and public health. Implementing strategies such as a robust crop management system as well as development of resistant crops will help mitigate the risk of mycotoxin contamination and safeguard food security [9].

5 Conclusion

The recently published literature clearly indicates that climate change will have a significant impact on the proliferation and spread of soil-borne pathogens, which would eventually lead to global food insecurity. Vulnerability of crops to pathogenic infections will increase due to fluctuations in temperature and precipitation, thus decreasing the overall crop yield. Furthermore, the increase in mycotoxin contamination can further complicate global food safety. Prompt action and innovative soil management strategies is needed to maintain higher crop yield as well as strengthen crop resilience to climate-driven pathogenic infections.

Acknowledgements This research was supported by Qatar Research Development and Innovation Council Food Security grant [MME03-1121-210025] awarded to University of Doha for Science and Technology.

References

1. Delgado-Baquerizo M, Guerra CA, Cano-Díaz C et al (2020) The proportion of soil-borne pathogens increases with warming at the global scale. Nat Clim Chang 10:550–554. https://doi.org/10.1038/s41558-020-0759-3
2. Peng Z, Liu Y, Qi J et al (2023) The climate-driven distribution and response to global change of soil-borne pathogens in agroecosystems. Glob Ecol Biogeogr 32:766–779. https://doi.org/10.1111/geb.13662
3. Wang X, Chi Y, Song S (2024) Important soil microbiota's effects on plants and soils: a comprehensive 30-year systematic literature review. Front Microbiol 25(15):1347745. https://doi.org/10.3389/fmicb.2024.1347745
4. Sharath Chandran US, Tarafdar A, Mahesha HS et al (2021). Temperature and soil moisture stress modulate the host defense response in chickpea during dry root rot incidence. Front Plant Sci 12:653265. https://doi.org/10.3389/fpls.2021.653265
5. Nnadi NE, Carter DA (2021) Climate change and the emergence of fungal pathogens. PLoS Pathogens 17(4):e1009503. https://doi.org/10.1371/journal.ppat.1009503
6. Casadevall A (2023) Global warming could drive the emergence of new fungal pathogens. Nat Microbiol 8:2217–2219. https://doi.org/10.1038/s41564-023-01512-w
7. Talley SM, Coley PD, Kursar TA (2002) The effects of weather on fungal abundance and richness among 25 communities in the intermountain West. BMC Ecol 2:7. https://doi.org/10.1186/1472-6785-2-7
8. Owino V, Kumwenda C, Ekesa B et al (2022) The impact of climate change on food systems, diet quality, nutrition, and health outcomes: a narrative review. Front Clim 4:16. https://doi.org/10.3389/fclim.2022.941842
9. Vågsholm I, Arzoomand N, Boqvist S (2020) Food security, safety, and sustainability—getting the trade-offs right. Front Sustain Food Syst 4:21. https://doi.org/10.3389/fsufs.2020.00016
10. Rohr JR, Cohen JM (2020) Understanding how temperature shifts could impact infectious disease. PLoS Biol 18(11):e3000938. https://doi.org/10.1371/journal.pbio.3000938
11. Ribeiro PL, Singh A, Sagervanshi A et al (2024) High soil moisture rather than drying-rewetting cycles reduces the effectiveness of nitrification inhibitors in mitigating N_2O emissions. Biol Fertil Soils 60:627–638. https://doi.org/10.1007/s00374-024-01811-2

of the public services in to order to ensurethe desired durability. The multidimensional, multi-space and intertemporal nature of this evaluation encounters several methodological difficulties which it is necessary to surmount [2].

- the need to take into account the externalities between users and actors at the various compartments of the watershed from the upstream to the downstream area.
- the problems related to the distribution of the water resources between sectoral users.
- the water use and access modes of the local actors in rural zones are not sufficiently known neither in the operation plan nor in the rules plan which govern them. The interactions and the negotiations between economic and social agents within micro watersheds deserve to be apprehended within an integrated and interactive management system.
- the economic efficiency of the agricultural production units in relation to the water resources also deserves to be evaluated.

3 Research Approach and Developed Methodological Tools

Oum Zessar watershed is characterized by the aridity of the climate, and the fragility of the plant communities. The agrarian system itself is characterised by the coexistence of irrigated and dry agriculture (cereal, fruit, etc.) with rangeland. Conversely, land uses have changed with the privatization of the traditional collective lands and the evolution towards a more intensive agro-pastoral system. Due to the increased human needs and the development of agriculture, the pressure on natural resources, and particularly on land resources, is becoming higher and higher. This pressure induces land degradation and a significant decrease in crop yield. The sedentarisation of pastoralism and land privatization, as well as governmental subsidies for some new agricultural activities like irrigated agriculture, have caused land fragmentation and increasing pressure on the land. Environmental threats and the progressive degradation of natural resources pose critical impacts on sustainable development, particularly in the arid regions of Tunisia. Oum Zessar watershed aces severe land degradation problems due to limited water resources, growing population, increasing land competition and agricultural intensification [10, 13]. In response to ongoing degradation problems, the Tunisian government has invested 562 million Dinars (= US $389 mil) since 1990 on soil and water conservation (SWC) measures at the national level [14]. The major goals of the SWC policy are to tackle land degradation in a vulnerable environment and to enhance the capacity of the land for agricultural production through the promotion of both traditional and modern water-harvesting techniques [10]. Om Zessar has witnessed several interventions by the state, where several projects have been completed, including a sustainable water management policy that is being evaluated and considering the possible expansion of the sustainable water management policy in the region. In order to support the role of local officials, a new method was used in this study, we illustrate the use of the Participatory

Impact Assessment (FoPIA) framework by conducting a case study-based sustainability impact assessment of sustainable water management scenarios in the Umm Zsar basin. For this purpose, FoPIA, which was originally developed for the European context [9], was adapted to the regional conditions in Oum Zessar watershed. The Framework for Participatory Impact Assessment (FoPIA) provided an overall structure and guideline for the impact assessment based on DPSIR approach (Drivers, Pressure, State, Impact, Responses). The FoPIA approach requires some available information (scenarios, land-use changes, sustainability issues, case study of informations) and condensed complex and cross-disciplinary information into "handable" workshop (stakeholder friendly) information; allowed for direct interactions and transparent discussions of potential impacts.

4 Main Characteristics of "FoPIA" Approach and Unfolding of the Workshop

FoPIA has been developed to facilitate a stakeholder inclusive sustainability impact assessment [8]. It facilitates the exchange of individual perceptions (knowledge) about regional sustainability issues and structures and stimulate discussions among. The FoPIA is a participation-based, stand-alone framework of sequenced methods for involving national, regional and local stakeholders in assessments of land use policy impacts. The main purpose of the FoPIA is to estimate impacts and changes in sustainable development (SD) (social, economic, environmental) that take place when a certain land-use policy is implemented [6].

4.1 Preparation Phase

During the preparation phase, scenarios and *land use functions* (LUFs) are drafted, stakeholders selected, and workshop facilities (posters, maps) prepared. The preparation phase starts before the FoPIA workshop and includes gathering relevant and quantitative information and it should address the most relevant viewpoints to the problem (decision context and policy actions, land-use changes and driving factors, sustainability issues).

4.2 Stakeholders Selection

The FoPIA uses expert knowledge to perform the impact assessment. A group size of 10–15 participants is recommended that should allow for active discussions and knowledge exchange among workshop participants. The final stakeholder group should consist of experts from social, economic, and ecological disciplines (balanced) in order address all three sustainability dimensions within the impact assessment.—In this context stakeholders (experts) are individuals, groups and organisations that are 'more or less' directly affected by decisions and actions or that have the power to influence the outcomes of these decisions. For the FoPIA, however, it is recommended to consider only experts in the same workshop since this method requires some academic exercises [12].

4.3 Policy and Scenario Development

Scenarios are the basis of the ex-ante impact assessment in the FoPIA and should be designed to facilitation the impact assessment procedure on alternative land use management options. Principle scenario elements include a description of step-wise changes in land use patterns over time, the consideration of driving forces, a base year, the time horizon and target year, and a formulated storyline that describes the causal links between scenario assumptions and likely consequences. The projection year should meet the decision maker's requirements with regard to policy implementation and the duration of the policy program. Driving forces need to be identified that affect the land use activities in the study area for the considered time horizon and that can be estimated with simple statistics or expert knowledge. For the assessment of the impacts of policies, it is required to construct a baseline or so-called counterfactual scenario in order show what would happen in the absence of a policy change.

Following the two national forums organized with the local actors of rural development, two LUP were retained for the analysis of impact by the FoPIA method:

- The Water and Soil Conservation policy (WSC),
- The Water management policy, with their two components. Intersectoral Policy and Pricing Water Policy.

The analysis of the two policies suggested and the discussions that it generated, led participants to evaluate in priority the impacts of the WSC policy by the FoPIA method during this workshop. Being constrained by time (workshop for only one day) and to avoid dispersion, the participants agreed to limit themselves to the WSC policy watershed of Oum Zessar. The scenario of reference is the continuation of these policy actions at the current rhythm of installations (autonomous development) to reach 85% of potentially suitable surfaces until 2015. Two other scenarios are proposed to the participants (Fig. 2).

The last two scenarios are described as follows:

- Reduction of the State intervention (on the level of 70% of arranged surfaces), but with an appropriation by the local populations of 100% of these surfaces,
- "Integral" installation of the watershed area, i.e. 100% of potentially arranged surfaces.

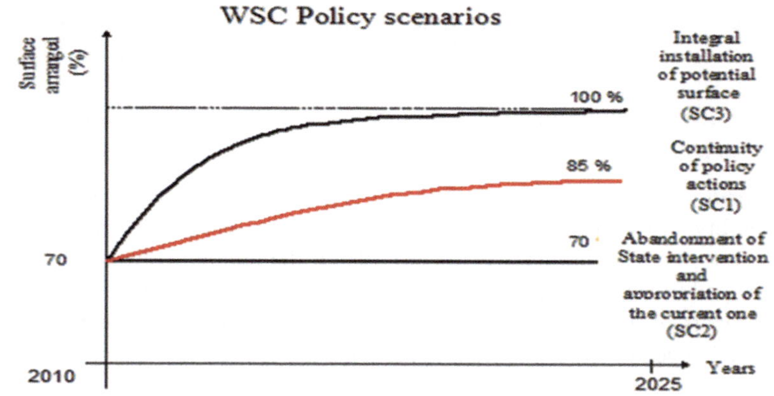

Fig. 2 Scenarios retained for water and soil conservation policy

Compared to the scenario of reference, these two scenarios are extreme scenarios which should make it possible to obtain contrasted results and as much as possible to avoid confusions in the evaluation process.

4.4 Land Use Functions and Indicators of Development

Synthetic indicators representative of each LUF were proposed to the participants for evaluation and adjustment (Table 1).

For the FoPIA approach, a set of nine regional *land use functions* (LUFs) needed to be developed that should reflect key sustainability issues at regional level. As a starting point, a brainstorming session can be used to identify the most relevant sustainability issues related to land use in the study area. The LUFs framework should then be applied to classify and group those sustainability issues into social, economic,

Table 1 Indicators selected for each land use function (LUF)

Dimensions	Land use functions (LUF)	Indicators
SOC 1	Provision of work	Employment rate
SOC 2	Human health	Life expectancy
SOC 3	Cultural heritage	Traditional techniques
ECO 1	Industries and services	Regional investments
ECO 2	Primary sector production	Farmers income
ECO 3	Infrastructure	Road network
ENV 1	Abiotic resources	Water availability
ENV 2	Biotic resources	Biodiversity
ENV 3	Ecosystem conservation	Natural land

environmental categories, three per dimension, and complement if necessary [7]. For the impact assessment each LUF will be represented through one corresponding indicator in order have a more precise criterion for handling the LUFs. These indicators should be defined during the preparation phase to be presented to the stakeholder group during the workshop.

First scoring: a first step, it is to find out stakeholders' perceptions on regional sustainability. In Doing so, individual perceptions were reported by the participants by giving scores of importance on each LUF for the region of the watershed. The scores ranged from 0 (no importance) to 10 (very important). An analysis and discussion of results of the first scoring functions of land use have been conducted was used to exchange ideas and preferences among participants—providing the basis for a second round.

Second scoring: Following an interpretation of the first scoring results, adapting elements of this discussion such that land use functions have been made. A second scoring with the same group of participants was conducted and showed that, through discussions, average scoring results become more homogenous while stakeholders manifested the preferences and importance on regional LUFs.

Identification and assignment of LUF indicators: for each land use function, award a development index deemed most relevant. For each LUF one corresponding indicator was selected to make the LUF more precise for the IA.

5 Workshop Proceedings: Steps in Applying FoPIA Method (IA)

During the impact assessment, stakeholders were asked to give scenario impact-scores on each LUF indicator. This step consists of a scoring of land-use functions indicators. In this step, the scores range from (− 3 to + 3) (− 3: highly significant negative impact and + 3: highly significant positive impact).

The Weighting of land use functions: This step expresses the trade-off between several land-use functions. The assignment of weights for each land use function reflects the degree of importance of each function in the development in the watershed. These weights range from 1 to 9.

6 Results of the Application of the FoPIA Approach

6.1 *Scores Given by Participants*

The results analysis of the LUFs evaluation by participants showed the priority given to employment in the social dimension, or more precisely to employment maintains, and life quality. The importance given to cultural heritage, considered here by the participants in terms of transmission assets. Participants showed little interest in industries and services in the watershed because this area is relatively remote from urban centers and commercial channels. Similarly, the industry has not a great importance because of the improvement of roads network. The majority of participants preferred improving agricultural productivity and the availability of abiotic resources, relative to the preservation of biotic resources and ecosystem. The stakeholder's preference was explained by the priority accorded to the socio-economic situation enhancement of local population related in fact to the Abiotic resources availability and by consequence agricultural productivity. But this result has given rise to debates in which they give importance to preserving and enhancing natural resources for economic activities.

A reassessment of the scores of those LUFs led participants to report the lack of logic in the importance given to work while agriculture in this area contributes only 9% of employment.

6.2 *Analysis of the Impact of Different Scenarios*

Indicators Level

- Life expectancy: This indicator was considered irrelevant and not relate to the water and soil conservation (WSC) policy that to be assessed. The stakeholders suggest using the Human Development Index (HDI) indicator widely used internationally. However, Participants felt that the workshop could keep the proposed indicator.
- The road network in km: this indicator considered irrelevant to evaluate the WSC policy. It has been proposed to consider the length in km of roads covered by the WSC techniques or the number of road protected from erosion.
- Employment. The continuity of the current policy (SC2) scenario and the integral management of the area scenario (SC3) have a positive impact on employment while the abandonment of the state intervention scenario (SC1) has very little effect. It was mentioned here that the WSC works are employments creator to local people and they are indirectly generating new agricultural activities (planting olive trees in particular). But the contradiction was brought by one participant who noted that agriculture currently represents only 9% of the workforce so this indicator has a little weight.

4. Gangbazo G (2004) Management integrated of water by area catchment into Quebec: concepts and applications. Quebec environment, 46 pp
5. Graaff J (1996) The price of soil erosion; year economic evaluation of soil conservation and watershed development. Mansholt studies no. 3. Wageningen University. Backhuys Publishers, Leiden
6. König HJ, Schuler J, Suarma U, McNeill D, Imbernon J, Damayanti F, Helming K (2010) Assessing the impact of land use policy on urban-rural sustainability using the FoPIA approach in Yogyakarta, Indonesia. Sustainability 2(7):1991–2009
7. König HJ, Uthes S, Schuler J, Zhen L, Purushothaman S, Suarma U, Shgaier M, Makokha S, Helming K, Sieber S, Chen L, Brouer F, Morris J, Wiggering H (2013) Regional impact assessment of land use scenarios in developing countries using the FoPIA approach: Findings from five case studies. Integrated land-use and regional resource management—a cross-disciplinary dialogue on future perspectives for a sustainable development of regional resources 127(Supplement):S56–S64
8. Morris J, Camilleri M, Moncada S (2008) Key sustainability issues in European sensitive areas—a participatory approach. In: Helming K, Pérez-Soba M, Tabbush P (eds) Sustainability impact assessment of land use changes. Springer, Berlin, Heidelberg, pp 451–470
9. Morris JB, Tassone V, de Groot R, Camilleri M, Moncada S (2011) A framework for participatory impact assessment: involving stakeholders in European policy making, a case study of land use change in Malta. Ecol Soc
10. Ouessar M, Tâamallah H, Begacem AO (2006) An environment with strong climatic constraints. In: Between desertification and development: Tunisian Jeffara, Cérès Editions, Tunisia, pp 23–32
11. Ouessar M, Sghaier M, Mahdhi N, Abdelli F, De Graaff J, Gabriels D (2003) Integrated approach for impact assessment of toilets harvesting technical in dry areas: the put of wadi Oum Zessar watershed (Tunisia). In: Mouat (ed) Proceedings of the 2002 conference (in press) (J Env Monit Ass)
12. Paas W, Coopmans I, Severini S, Van Ittersum MK, Meuwissen MPM, Reidsma P (2021) Participatory assessment of sustainability and resilience of three specialized farming systems. Ecol Soc 26(2):2. https://doi.org/10.5751/ES-12200-260202 (Research participatory assessment of sustainability and resilience of three specialized)
13. Sghaier M, Genin D (2003) Concepts and methods: diversity of approaches, multiplicity of scales and interaction, report/ratio scientist of synthesis, turning into a desert in Tunisian south-eastern Jeffera. IRA-IRD, pp 6–20
14. Sghaier M, Ouessar M, De Laitre E, Leibovici D, Loireau M, et al (2006) Integrated environmental and socio-economic modeling using LEIS for desertification monitoring and assessment in the observatory of Menzel Habib (South Tunisia). In: The future of drylands. International Science Conference on Desertification and Drylands Research Tunis, Tunisia, pp 19–21

Coagulation-Ultrafiltration of Seawater Enriched with Algal Organic Matter (AOM)

Mohamed Said Lebad and Djamal Abdessemed

Abstract In order to simulate the fouling of ultrafiltration (UF) membranes during an algal bloom, particularly by organic algal matter (AOMs), we undertook filtration tests of real seawater enriched with sodium alginate (SA) via a flat polyacrylonitrile (PAN) membrane integrated in an unstirred AMICON cell operating in frontal mode. The experiments showed that the permeate flux decrease was 60, 80, 85 and 90% for SA concentrations of 5, 10, 15 and 30 ppm, respectively, and this was most likely due to the accumulation of a thicker cake layer on the membrane surface. On the other hand, the addition of iron chloride ($FeCl_3$ $6H_2O$) as a coagulant allowed a reduction in the intensity of fouling when the coagulant concentration was increased. As for the impact of pH on the UF-coagulation process, the results showed that working at an acidic pH of 5 was more favorable in terms of reducing flux drop when the feed solution was dosed with 2.5 and 5 mg Fe^{3+}/L.

Keywords Algal bloom · Coagulation · Ultrafiltration

1 Introduction

Algal blooms are a natural phenomenon that occurs when microscopic algae multiply rapidly in water, creating colored tides on the water surface. However, these blooms can be dangerous for the environment and human health. In Algeria, this natural phenomenon can adversely affect the operation of seawater desalination plants which are a major source of drinking water in the country. Algal blooms can clog filters at the inlet to the treatment process and at the biofouling of reverse osmosis membranes, reducing their efficiency and lifespan. Moreover, algal blooms can release toxins that may contaminate the drinking water produced by desalination plants, posing potentially serious risks to human health. The application of advanced pre-treatment

M. S. Lebad · D. Abdessemed (✉)
Laboratory of the Industrial Science Process Engineering, University of the Sciences and Technology Houari Boumediene (USTHB), Algiers, Algeria
e-mail: adjamal@yahoo.com

S. Al-Naemi et al. (eds.), *Water and Food Security in the Face of Climate Change: Challenges and Opportunities for Resilience*, World Sustainability Series,
https://doi.org/10.1007/978-3-032-00098-9_55

can therefore be considered a sustainable solution to prevent or significantly delay biological clogging in reverse osmosis systems during an algal bloom.

The use of low-pressure membrane filtration (MF/UF) as a pre-treatment system in reverse osmosis desalination plants has grown considerably in recent years. For example, the new Magtaa seawater desalination plant in western Algeria, with a production capacity of 500,000 cubic meters per day, is one of the largest plants in the world to use ultrafiltration as a pretreatment system. MF/UF membranes provide several benefits over traditional pre-treatment systems, including reduced space requirements and consistently high permeate quality in terms of SDI and turbidity (regardless of changes in seawater characteristics) and low overall chemical consumption, etc. However, the performance of MF/UF membranes becomes challenging during algal bloom periods due to a high propensity for fouling. This is caused by the accumulation of microalgae and, particularly, algal organic matter (AOMs), which form a cake layer on the membrane surface. This layer can significantly increase the transmembrane pressure needed to maintain permeate flux [1]. Coagulation is the treatment most commonly for mitigating the negative effects of AOMs on the operation of ultrafiltration membranes. It works by reducing the adherence of AOMs to the membrane surface and promoting the formation of a less compressible cake layer [2]. In order to simulate fouling of ultrafiltration (UF) membranes during an algal bloom, particularly by algal organic matter (AOM), we undertook filtration tests of real seawater enriched with sodium alginate (SA) through a flat-sheet polyacrylonitrile (PAN) membrane mounted in an unstirred AMICON cell operating in frontal mode.

2 Materials and Methods

2.1 Filtration System

Ultrafiltration tests were carried out in frontal mode in an Amicon-type cell (Fig. 1). The cell was filled with the solution to be ultrafiltered, and the initial feed volume was 200 cm^3. The operating pressure was effectively controlled by a pressure-reducing valve on the nitrogen bottle, and readings were taken using a manometer. The magnetic stirrer on which the cell is placed was removed during the coagulation tests to prevent the formation of flocs. The permeate exits the cell through a fine tube, while the solution concentrates in the cell. All experiments were conducted at under a pressure of 1 bar and at a constant temperature $T = 25$ °C.

The AMICON cell is fitted with an IRIS 3038 polyacrylonitrile (PAN) membrane for experiments with seawater. This is a flat, asymmetric, moderately hydrophilic organic membrane with a cut-off of 40 kDa. A new flat sheet of PAN was used for each test. Polyacrylonitrile (PAN) is not a material commonly used on an industrial scale, but [3] showed that the IRIS 3038 type membrane gave satisfactory results in eliminating the microalgae suspension with a limited propensity for fouling.

Fig. 1 Filtration cell, AMICON type

2.2 *The Test Water*

As mentioned previously, the ultrafiltration tests were carried out in order to investigate their fouling in a marine environment contaminated with organic matter from micro-algae. To do this, real seawater from Algiers Bay was spiked with sodium alginate, a substance commonly used in several previous studies as a polysaccharide substitute, considered as the major cause of fouling of low pressure membranes during algal blooms.

Sodium alginate (Fluka 71240) was used in our case. The sodium alginate stock solution was prepared by adding 1 g of alginate in stages to 1 L of deionized water under high agitation for approximately 1 h. The solution was then left under gentle stirring overnight to ensure good dissolution of the alginate.

Finally, the solution was filtered through a 0.45 μm Whatman filter to remove the particulate fraction. Before filtration, the pre-filtered sodium alginate stock solution was added by syringe to seawater under strong agitation for 1 h.

Table 1 lists some of the physico-chemical parameters analysed in the feed water used.

The addition of 10 mg of sodium alginate (SA) increased TOC value from 0.4 to 3 mg/L. This TOC concentration corresponds to that obtained at Oosterschelde in the south-west of the Netherlands during severe efflorescence events that coincided with a very high fouling potential in UF systems [4]. Although this value may seem

Table 1 Physical and chemical properties of the seawater utilized

Parameter	Value
pH	8.0
Turbidity (NTU)	2.5
Conductivity (mS/cm)	54
Total organic carbon (mg/L)	0.4 3.0 (Sea water + 10 mg/L of SA)

insignificant compared to the TOC values found in natural wastewater, it has been reported that a TOC value as low as 2 mg/L can promote biological fouling of reverse osmosis membranes in the case of seawater desalination [5].

3 Results and Interpretation

3.1 Study of Ultrafiltration Membrane Fouling

3.1.1 Effect of Sodium Alginate Concentration

Figure 2 illustrates the experimental results demonstrating the effect of coagulant addition on membrane fouling intensity, as indicated by the normalized flux decline during ultrafiltration. The experiments were conducted using two types of feedwater: unmodified seawater and seawater spiked with sodium alginate at concentrations ranging from 5.0 mg/L to 30.0 mg/L.

Filtration of the Sodium Alginate solution without the application of coagulant resulted in a decrease in flux at four different Sodium Alginate concentrations tested. The rate of fouling increased with the increament of Sodium Alginate concentration,

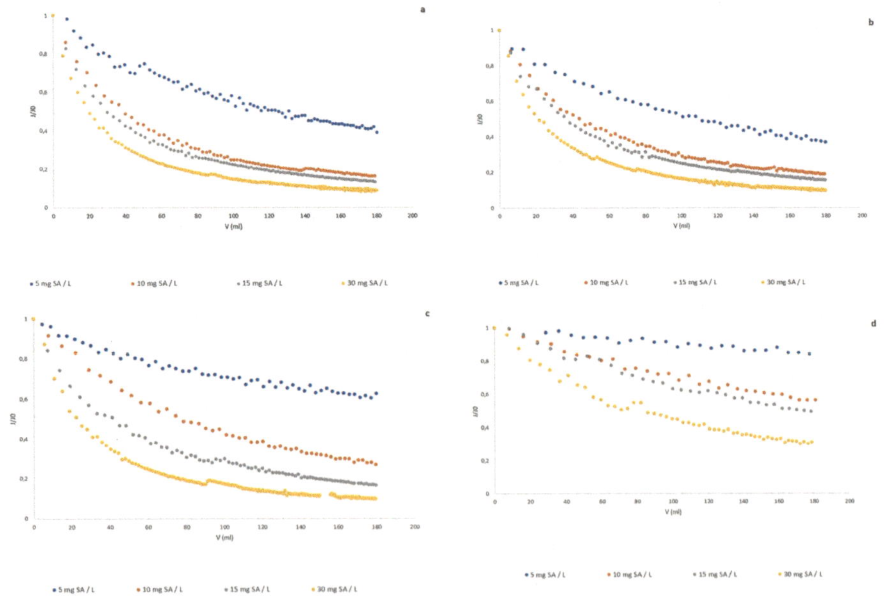

Fig. 2 Filtration of different concentrations of sodium alginate (SA) without coagulation (**a**) and with $FeCl_3$ coagulant at concentrations of 0 mg Fe^{3+}/L (**b**), 1 mg Fe^{3+}/L (**c**), 2.5 mg Fe^{3+}/L and 5 mg Fe^{3+}/L (**d**)

showing a decrease in flux of 60, 80, 85 and 90% for Sodium Alginate concentrations of 5, 10, 15 and 30 ppm, respectively, in the 180 mL filtration of the solution.

The application of a higher concentration of contaminants favours a faster transition from the concentration polarisation layer to a stagnant and much dense cake layer.

Subsequently, pretreatment by coagulation was opted for in order to reduce the intensity of fouling. Firstly, it's important to mention that flocculation might not be required for UF applications. Since enlarging particle size isn't the goal, small flocs are adequate to enhance the performance of UF membranes [2].

As a result, we worked in in-line coagulation mode without any flocculation or sedimentation steps. Coagulation was carried out using a calibrated jar-test apparatus and consisted solely of adding the coagulant to 1 L of feedwater solutions at the start of the rapid mixing stage at 200 rpm for 1 min. The solution in the beaker was then gently transferred to the AMICON cell.

The coagulant chosen in this study was ferric chloride ($FeCl_3$-$6H_2O$).Ferric salts, especially ferric chloride, are ideal for coagulating seawater because of their low solubility across a wide pH range. This results in less residual dissolved iron in the water after pretreatment, thereby preventing scaling issues. The concentration applied in this study was quite low and varied between 1 and 5 mg Fe^{3+}/L.

When coagulation is applied, the drop in flux decreases than in direct filtration (Fig. 2). Konieczny et al. [6] demonstrated that the cake formed during in-line coagulation before the MF/UF system was more porous and adhered better to the membrane surface. This reduced membrane fouling, especially within its pores, leading to fewer hydraulic and chemical cleaning cycles and extending the membranes' operational life.

Nevertheless, the addition of coagulant at a concentration of 1 mg Fe^{3+}/L had no impact on the fouling propensity expressed by the normalised flux reduction (Fig. 2).

At a low coagulant dose of 1 mg Fe^{3+}/L, coagulation likely led to only partial complexation of thealginates, which probably caused a slight improvement in fouling potential and compressibility.

On the other hand, a dose of 5 mg Fe^{3+}/L was able to attenuate fouling resulting in a flux reduction of only 20% and 10% respectively in the case of a sodium alginate concentration of 10 and 5 mg/L, respectively.

It seems that increasing the dosage of coagulant has a positive effect on the filterability of sodium alginate by providing more positive charges. At a coagulant dose of 2.5 mg Fe^{3+}/L and above, the removal of alginate suspensions by the means of the iron hydroxide precipitates prevailed, indicating that the predominant coagulation mechanism is the sweep flocculation, allowing the adsorption of negatively charged functional groups onto the sodium alginate and enmeshment in the iron hydroxide precipitates forming Fe-sodium alginate aggregates.

However, the optimum concentrations observed in this study (2.5 and 5 mg Fe^{3+}/L) remain fairly high compared with those obtained in previous studies coupling the $FeCl_3$ coagulant with the UF membrane for filtration of AOMs.

In fact, Tabatabai [2] and Li et al. [7]showed that a small concentration as low as 0.5 mg Fe^{3+}/L was sufficient to maintain a lower propensity to clog when using an aromatic C–H stretching peaks at 3058.82 and 3025.58 cm^{-1}. These findings demonstrate that MPs are released under typical beverage consumption conditions and highlight the urgent need for safer material alternatives and improved waste management strategies to mitigate the associated health and environmental risks.

Keywords Microplastics · Polyethylene · Polystyrene · Disposable paper cups · Leaching · Contamination

1 Introduction

Microplastics (MPs), defined as synthetic plastic particles less than 5 mm in size, are highly resistant to environmental and biological degradation and are insoluble in water [1–4]. Their toxicity depends on several factors, including size, surface chemistry, polymer type, and hydrophobicity [5, 6]. Once introduced into biological systems, MPs can translocate across cellular barriers, penetrate tissues, and induce inflammation, immune dysregulation, or physical blockages [7, 8]. MPs have been detected in human urine, feces, breast milk, and the bloodstream [9], raising concern due to their association with adverse outcomes such as impaired neurodevelopment, metabolic disturbances, and systemic inflammation [10, 11]. Organs, including the brain, placenta, heart, and gastrointestinal tract, are particularly vulnerable [12], with potential links to neurodegenerative diseases (e.g., Alzheimer's, Parkinson's), respiratory illnesses, diabetes, and obesity [13, 14]. Co-contaminants such as fluoride often exacerbate these risks [15, 16].

A significant yet often overlooked source of MPs is Disposable Paper cups (DP-cups), which are prevalent in everyday beverage consumption. In the United Kingdom alone, approximately 7 million paper cups are discarded daily, contributing to an estimated 30,000 tons of waste annually [17]. Although perceived as eco-friendly, these cups contain inner linings typically made of Poly-Ethylene (PE) or related copolymers [5, 17]. When exposed to heat from hot beverages, these coatings can degrade, releasing MPs into the drink. Studies have reported the release of up to 5,984 MPs may be released into a single hot beverage, translating to a potential daily ingestion of 0.03 ± 0.025 mg per person [5, 7, 16]. This exposure has been associated with oxidative stress, neurotoxicity, endocrine disruption, and immunological alterations [7, 16].

The release of MPs is influenced by multiple factors, including temperature, exposure time, and beverage agitation [18]. Even biodegradable coatings such as Poly-Lactic Acid (PLA) can release harmful plastic particles under realistic usage conditions [19]. Although rinsing with ultrapure water can reduce MPs release by approximately 65%, repeated use continues to degrade the polymer barrier [20]. Moreover, recycling these cups is challenging due to the difficulty of separating the plastic lining from the paperboard, often resulting in incineration or landfill disposal

[21–23]. Innovations in waste utilization—such as converting used cups into cellulose/PE composite films or graphene-based materials—offer promising alternatives [22–24].

The release of MPs from DP-cups varies depending on the polymer type. PE, Poly-Propylene (PP), and Poly-Styrene (PS) coatings are known to leach significant quantities of MPs and Nano-Plastics (NPs) under thermal or acidic conditions, with PE-lined cups exhibiting up to 26 times higher NPs release compared to PLA-coated variants [25, 26]. Sub-nanometer particles have been found in the brain and placenta, implicating them in miscarriages, immune disorders, and neurological decline [27]. While many studies have investigated MPs release under laboratory-controlled environments, this study emphasizes real-world usage scenarios specific to Qatar's climate. Local brands of DP-cups were examined using SEM and FTIR to assess their degradation behavior and MPs release under common consumer practices.

2 Materials and Methods

2.1 General Description and Methodological Area

This study employed an experimental approach to investigate the potential release of MPs from DP-cups when exposed to hot liquids. The primary aim was to assess the associated environmental and human health risks. The methodology comprised two main components: a laboratory-based leaching experiment and a controlled analysis designed to simulate real-world conditions.

2.2 Materials

The materials examined in this study included DP-cups with PE-lined interiors and plastic lids, each with a capacity of 250 mL, sourced from a local supermarket in Qatar. Ultra-purified distilled water was utilized as the test medium to minimize background contamination. The experimental setup comprised the following equipment:

- A kettle for heating water.
- A calibrated laboratory thermometer for temperature verification.
- A vacuum filtration unit equipped with cellulose nitrate filter paper having a pore size of 0.45 μm.
- Sterile Petri dishes and PP tubes for sample storage.

SEM was employed for morphological analysis and analytical evaluation, while FTIR was used for the chemical characterization of the particles and cup materials.

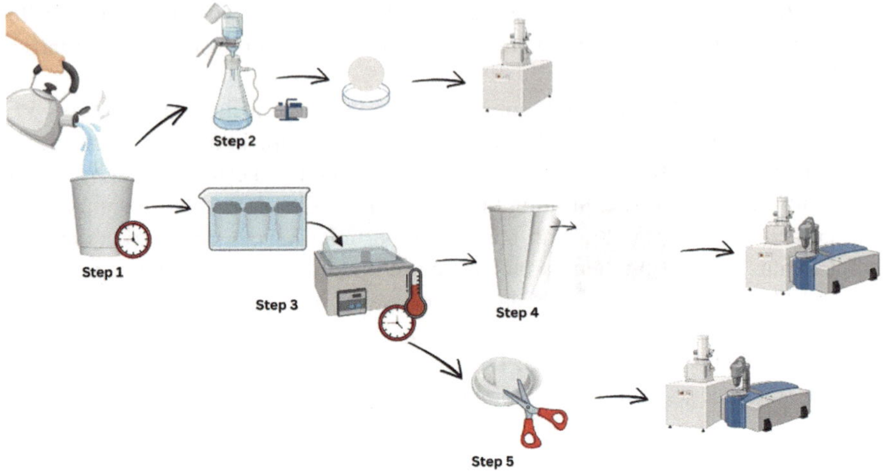

Fig. 1 Schematic representation of the experimental procedure for assessing MPs release from DP-cups, including water exposure, filtration, and analytical characterization

2.3 Experimental Procedure

To simulate real-world usage, boiled distilled water (approximately 100 °C) was poured into the DP-cups and allowed to sit for a predetermined duration. The liquid was then filtered through a 0.45 μm filter using vacuum filtration to collect any leached microplastic particles. The collected filter membranes and fragments of the plastic lids were dried and prepared for SEM and FTIR analyses. These analyses aimed to assess particle morphology, identify polymer types, and detect signs of degradation or leaching. In a supplementary experiment, the DP-cups were immersed in a water bath maintained at 40 °C for at least one hour to facilitate the delamination of the internal plastic film from the paper layer. Separate plastic films were collected and analyzed under the same conditions (Fig. 1).

3 Results and Discussion

3.1 Scanning Electron Microscopy (SEM)

SEM was employed to examine the morphological changes in key components of the DP-cups—namely, the filter paper, plastic lid, and internal PE film—following exposure to hot water. As illustrated in Fig. 2, MPs-like particles were visible on the filter paper (Panel A), suggesting the migration of MPs from the disintegrated polymeric components into the test medium. The accumulation of these particles on the filter paper confirms the leaching pathway and supports concerns regarding potential consumer exposure to MPs and related chemical additives from such disposable containers. The surface of the plastic lid (Panel B) displayed noticeable signs of thermal degradation, including pore formation, increased surface roughness, and micro-fragmentation. These morphological changes indicate polymer breakdown and MPs release. Moreover, the internal PE film layer (Panels C and D) exhibited pronounced structural damage, such as surface cracking, delamination, and fragmentation. These features reflect the thermal sensitivity of the PE coating, which is typically used as a moisture barrier. The observed degradation provides strong evidence of microplastic generation under heat exposure. These findings collectively highlight the thermal instability of polymer-coated DP-cup materials and underscore their role as a potential source of MPs contamination.

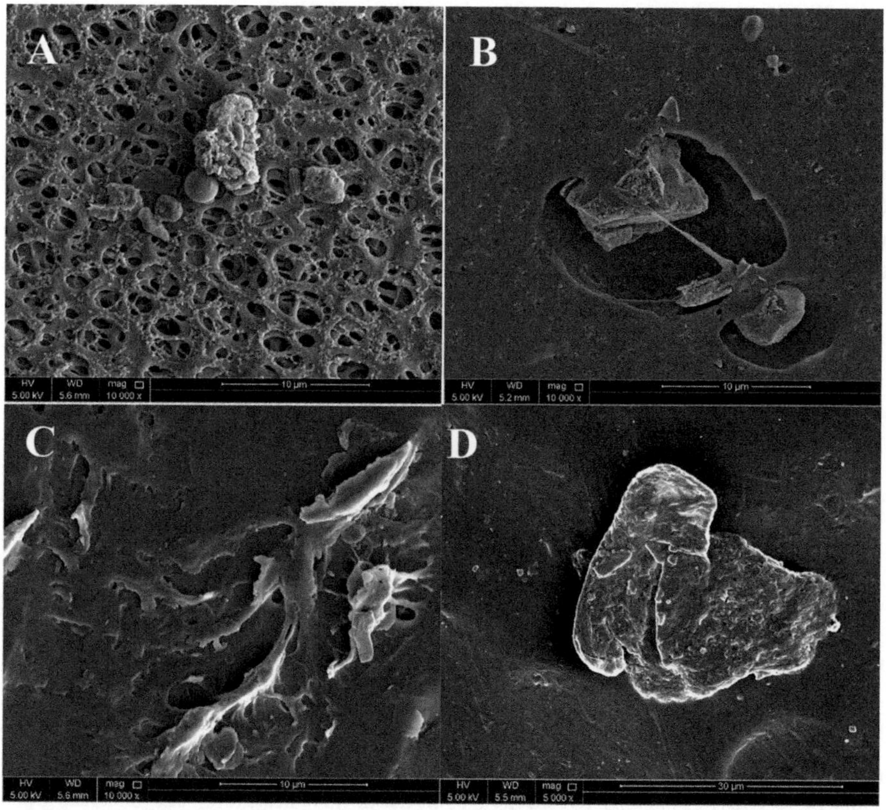

Fig. 2 SEM micrographs showing surface morphology of DP-cup components post hot water exposure: **a** filter paper with retained MPs; **b** plastic lid showing surface degradation; **c** internal PE film with visible cracking and delamination; and **d** PE film at a different magnification/spot highlighting further fragmentation

3.2 Fourier Transform Infrared Spectroscopy (FTIR)

FTIR was utilized to analyze the chemical composition of the' internal film and plastic lid of disposable paper cups. As illustrated in Fig. 3, the FTIR spectrum of the internal film (legend A) reveals distinct absorption bands at 2916.02 and 2848.80 cm^{-1}, which correspond to the asymmetric and symmetric stretching vibrations of methylene (–CH_2–) groups, typically associated with saturated aliphatic hydrocarbons. Additional bands observed at 1463.57 cm^{-1} and 1377.60 cm^{-1} signify CH_2 bending and wagging vibrations, respectively. Moreover, sharp absorptions at 729.95 and 718.66 cm^{-1} indicate CH_2 rocking modes, characteristic features of crystalline PE. These findings confirm that the internal coating of the DP-cups is primarily composed of PE. In contrast, the FTIR spectrum of the plastic lid in Fig. 3 (legend B) displays dominant absorption bands at 3058.82 and 3025.58 cm^{-1}, indicating aromatic C–H stretching vibrations, which suggest the presence of benzene rings within the polymer structure. Peaks at 2921.33 and 2849.74 cm^{-1} correspond to aliphatic C–H stretching, indicating that the polymer backbone contains hydrocarbon chains. Additionally, strong absorptions at 1601.14 and 1492.57 cm^{-1} are attributed to aromatic C=C stretching, while bands at 1451.47 and 1027.86 cm^{-1} are associated with CH bending and deformations of aromatic rings. The prominent out-of-plane C–H bending vibrations at 748.31 and 695.38 cm^{-1} further signify the presence of mono-substituted benzene rings. Collectively, these features are characteristic of PS, thereby confirming its application in the plastic lids.

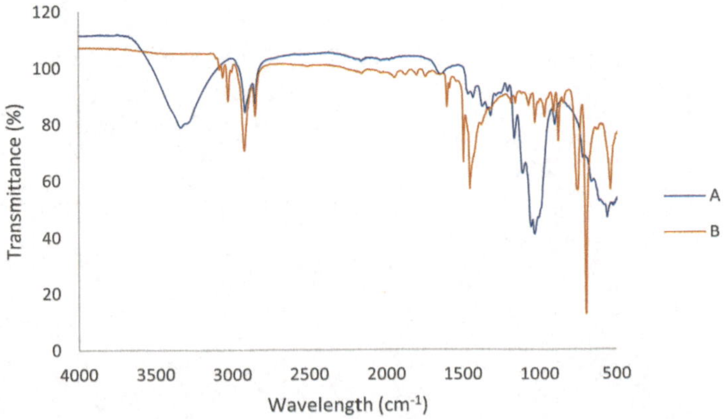

Fig. 3 FTIR spectrum of the PE film shows characteristic methylene group vibrations and crystalline PE features (legend A). FTIR spectrum of the plastic lid indicates aromatic functional groups and confirms PS (legend B)

4 Conclusion

This study reveals that DP-cups emit significant MPs when exposed to hot, boiled water. These MPs originate from both the internal PE coatings and the PS lids of the cups. Structural and chemical analyses, conducted through SEM and FTIR, confirmed the thermal degradation of these polymer materials, resulting in the formation of MPs. The findings highlight the thermal instability of both PE and PS under conditions that mimic typical beverage consumption. Both polymers are implicated in MPs contamination, which has well-established links to adverse health effects, including oxidative stress, respiratory disorders, neurotoxicity, and potential carcinogenicity. This evidence emphasizes the urgent necessity to develop and implement safer, more sustainable alternatives to conventional single-use food packaging. While the study's limitations include a small sample size and possible contamination during sample handling, it nonetheless provides valuable insights into the real-world implications of using disposable cups. To effectively mitigate the risks associated with MP exposure, further large-scale investigations are essential, alongside the establishment of stringent regulatory policies. Such measures are crucial for reducing environmental contamination and safeguarding public health, supporting broader national sustainability and health protection objectives.

Acknowledgements This publication was made possible by a grant from the Qatar National Research Fund (QNRF), awarded under its Undergraduate Research Experience Program (UREP), Grant No. UREP31-074-1-016. The content of this publication is solely the responsibility of the authors and does not necessarily represent the official views of QNRF. The authors would like to thank Qatar University for its continuous support. Special thanks are extended to the Center for Advanced Materials (CAM) and the Central Laboratories Unit (CLU) for providing the essential facilities and technical assistance that were instrumental in completing this research.

Author Contributions **Conceptualization**: M.A., A.A., S.A., R.M., and A.M.; **Methodology**: M.A., A.A., S.A., A.M., and R.M.; **Formal Analysis**: M.A., A.A., S.A., A.M., R.M., and N.A.; **Supervision**: N.A.-Q.; **Validation**: M.H.S.; **Visualization**: N.A.-Q.; **Writing—Original Draft**: A.A., S.A., R.M., and A.M.; **Writing—Review and Editing:** M.A., A.A., and N.A.-Q. All authors have read and agreed to the published version of the manuscript.

Funding This research was supported by a grant from the Qatar National Research Fund (QNRF) under the Undergraduate Research Experience Program (UREP), Grant No. UREP31-074-1-016.

Data Availability Statement All primary data generated or analyzed during this study are included in the published article. For further inquiries, please get in touch with the corresponding author.

Conflicts of Interest The authors declare no conflicts of interest.

Institutional Review Board Statement The research methodology adhered to all applicable institutional guidelines. No formal ethical review was required for this study.

Informed Consent Statement Informed consent was not applicable, as the study did not involve human participants.

References

1. Chen S, Li Y, Mawhorter C, Legoski S (2020) Quantification of microplastics by count, size, and morphology in beverage containers using Nile Red and ImageJ. J Water Health 19(1):79–88
2. Liu X, Shi H, Xie B, Dionysiou DD, Zhao Y (2019) Microplastics as both a sink and a source of bisphenol A in the marine environment. Environ Sci Technol 53(17):10188–10196
3. Zhou G, Wu Q, Tang P, Chen C, Cheng X, Wei XF, Ma J, Liu B (2023) How many microplastics do we ingest when using disposable drink cups? J Hazard Mater 441:129982
4. Jin Y, Xia J, Pan Z, Yang J, Wang W, Fu Z (2018) Polystyrene microplastics induce microbiota dysbiosis and inflammation in the gut of adult zebrafish. Environ Pollut 235:322–329
5. Chen H, Xu L, Yu K, Wei F, Zhang M (2023) Release of microplastics from disposable cups in daily use. Sci Total Environ 854:158606
6. Senathirajah K, Attwood S, Bhagwat G, Carbery M, Wilson S, Palanisami T (2021) Estimation of the mass of microplastics ingested—a pivotal first step towards human health risk assessment. J Hazard Mater 404:124004
7. Wang J, Tan Z, Peng J, Qiu Q, Li M (2016) The behaviors of microplastics in the marine environment. Mar Environ Res 113:7–17
8. Wright SL, Kelly FJ (2017) Plastic and human health: a micro issue. Environ Sci Technol 51(12):6634–6647
9. Zurub RE, Cariaco Y, Wade MG, Bainbridge SA (2024) Microplastics expo-sure: implications for human fertility, pregnancy and child health. Front Endocrinol 14:1330396
10. Darbre PD (2020) Chemical components of plastics as endocrine disruptors: overview and commentary. Birth Defects Res 112(17):1300–1307
11. Maradonna F, Vandenberg LN, Meccariello R (2022) Editorial: endocrine-disrupting compounds in plastics and their effects on reproduction, fertility, and de-velopment. Front Toxicol 4:886628
12. Carvalho JGRD, Augusto HC, Ferraz R, Delerue-Matos C, Fernandes VC (2024) Micro(nano)plastic and related chemicals: emerging contaminants in environment, food and health impacts. Toxics 12(10):762
13. Sofield CE, Anderton RS, Gorecki AM (2024) Mind over microplastics: exploring microplastic-induced gut disruption and gut-brain-axis consequences. Curr Issues Mol Biol 46(5):4186–4202
14. Ghosh S, Ghosh S, Sinha JK, Han S, Bhaskar R, Vashisth K (2023) Microplastics as an emerging threat to the global environment and human health. Sustainability 15(14):10821
15. Cho G, Kim K, Chen W, Son S, Son S, Son S, Jeon T-J, Jeon T-J, Jeon T-J, Kim SM, Kim SM, Kim SM (2024) Nanopore detection of sub-nanosized plastics in PE-coated paper cups and analysis of their inflammatory responses. Chem Eng J 495:153407
16. Joseph A, Parveen N, Ranjan VP, Goel S (2023) Drinking hot beverages from paper cups: lifetime intake of microplastics. Chemosphere 317:137844
17. Foteinis S (2020) How small daily choices play a huge role in climate change: the disposable paper cup environmental bane. J Clean Prod 255:120294
18. Bull MK, Steele RJ, Kelly M, Olivier SA, Chapman B (2010) Packaging under pressure: effects of high pressure, high temperature processing on the barrier properties of commonly available packaging materials. Innov Food Sci Emerg Technol 11(4):533–537
19. Carney Almroth B, Carle A, Blanchard M, Molinari F, Bour A (2023) Single-use take-away cups of paper are as toxic to aquatic midge larvae as plastic cups. Environ Pollut 330:121836
20. Akbulut S, Akman PK, Tornuk F, Yetim H (2024) Microplastic release from single-use plastic beverage cups. Foods 13(10):1564
21. Jia P, Zheng B, Zhang J, Hao W, Ji X, Han W, Xia G, Wang C, Ji X, Zhang J (2022) Eco-friendly and complete recycling of waste bamboo-based disposable paper cups for value-added transparent cellulose-based films and paper plastic composites. Polymers 14(8):1589
22. Wang L, Ji X, Zhou Q, Ji X, Zhang J, Peng J, Zhang J, Nawaz H, Xia G (2022) Fabrication and characterization of transparent and uniform cellulose/polyethylene composite films from used disposable paper cups by the "one-pot method." Polymers 14(6):1070

23. Poortinga W, Whitaker L (2018) Promoting the use of reusable coffee cups through environmental messaging, the provision of alternatives and financial incentives. Sustainability 10(3):873
24. Zhao H, Zhao TS (2013) Graphene sheets fabricated from disposable paper cups as a catalyst support material for fuel cells. J Mater Chem A 1(2):183–187
25. Chen H, Xu L, Yu K, Wei F, Zhang M (2022) Release of microplastics from disposable cups in daily use. Sci Total Environ 854:158606
26. Son J-W, Nam Y, Kim C (2023) Nanoplastics from disposable paper cups and microwaveable food containers. J Hazard Mater 464:133014
27. Chen Q, Ye C, Xie R, Xu H, Jiang C, Wei X, Yang S, Peng C, Su Z, Yili-han G, Shen Y (2024) Placental and fetal enrichment of microplastics from disposable paper cups: implications for metabolic and reproductive health during pregnancy. J Hazard Mater 478:135527

Modeling Groundwater Flow and Chloride Transport from the Dumpsite in Erbil City, Kurdistan Region, Iraq, by Using GMS

Noor Kh. Yashooa and Dana Khider Mawlood

Abstract Groundwater is an essential water resource in many areas of Erbil city. A significant development is noticed in Erbil city, leading to groundwater quality and quantity depletion. This study aims to predict the effect of increasing the pumping rate in the study area and simulate the chloride transport from the dump site to the surrounding area after a different period. Groundwater Modeling System (GMS) version (10.6.2) is used to model groundwater flow and contamination transport; the model is run for steady and unsteady conditions. The dump site is considered a contamination source in the study area. The result indicates that GMS is a powerful software for modelling groundwater flow and contamination transport, and it can help predict aquifer properties and future behaviour. The correlation between the observed and calculated head was $R^2 = 0.9917$, the calibrated value of hydraulic conductivity was 1.4 m/d for the first layer, 0.0025 m/d for the second layer, and 1.35 m/d for the third layer, the porosity of calibrate value of the model was 0.23 for the first layer, 0.095 for the second layer and 0.37 for third layer. The increase of pumping wells by 25% in the study area leads to an increase in the drawdown of about 9.24 m as the maximum value after the year and 14.2 m after increasing pumping by 50% after ten years from 2022. The chloride concentration spread to about 1.6 km with about 113.9 m/year after 50 years from 2022.

Keywords Groundwater · MT3MS · Contamination · Chloride · MODFLOW

N. Kh. Yashooa (✉) · D. K. Mawlood
Centre for Environmental Studies, University of Kurdistan Hewlêr, Erbil City, Kurdistan Region, Iraq
e-mail: noor.khalid@ukh.edu.krd

D. K. Mawlood
e-mail: dana.mawlood@ukh.edu.krd

S. Al-Naemi et al. (eds.), *Water and Food Security in the Face of Climate Change: Challenges and Opportunities for Resilience*, World Sustainability Series,
https://doi.org/10.1007/978-3-032-00098-9_57

1 Introduction

Groundwater is a vital freshwater supply in numerous places globally. Recently, it has become a valuable resource for dry and mostly dry regions where surface water is scarce. Preserving this resource through efficient exploitation is crucial, particularly given the increasing population rate. The need for water in Erbil City is increasing, making groundwater a vital factor for development [1–3].

The drinking water supply in Erbil mainly relies on groundwater wells, which account for over 40% of the total supply. This is a result of the city's rapid population growth, which has led to an increasing demand for water. Erbil City's groundwater has been contaminated due to the release of home waste, industrial pollution, agricultural chemicals, and septic tank leaks [4, 5].

Numerical simulation models depict the different aspects of hydrological processes, including water movement beneath the surface and the transportation of contaminants in groundwater hydrology. Simulations are utilized to examine or predict the long-term consequences of water withdrawals and the movement of pollutants and to explore various management of groundwater alternatives [6, 7].

GMS is a sophisticated software program specifically developed to generate computer models of groundwater and transport contaminants within it. However, GMS can also be obtained commercially from other sources. GMS is utilized to characterize the site environment, construct models, conduct post-processing, calibrate the model, visualize data, and verify the results [8–11].

The MODFLOW module is a computational model that uses finite-difference methods to simulate groundwater flow in three dimensions. It is specifically designed to describe and estimate the behavior of groundwater in aquifer systems. The software offers various packages for simulating many aspects of the hydrogeological system, such as rivers, wells, and hydraulic obstacles. Translating data into different shapefiles and simulating the aquifer through them is a straightforward process [12–15].

The MT3DMS module simulates the dispersion of soluble pollutants with varying concentrations within a groundwater system. Several packages incorporate dispersion, advection, diffusion, and chemical reactions, as well as external sources and sinks that can impact the transport solution [16–18].

There are many studies, some of which can be summarized as follows: Reference [19]—Utilized MODFLOW and MT3D Models to construct a mathematical model for simulating the spatial dispersion of nitrate in the Eocene aquifer in Palestine.. The model's findings indicate that the speed at which nitrate breaks down and the amount of nitrate that seeps into the aquifer substantially influence the levels of nitrate present in the aquifer. Reference [20]—Utilized Processing MODFLOW PMWIN version 5.3 to simulate groundwater flow and pollutant transport at the Astaneh-Kouchesfahan aquifer in Iran. The model horizontal hydraulic conductivity values range from 1 to 23 m/d under transient and steady-state conditions. The specific yield values ranged from 0.03 to 0.25. The salinity levels of groundwater are projected to rise over time, resulting in a more unfavorable groundwater status than the existing one.

Reference [21]—The quality of groundwater for both the Tigress River's left and right sides was assessed to simulate groundwater flow and the movement of solutes. The simulation utilized the Groundwater Modeling System (GMS) with the MODFLOW and MT3DMS models. The findings indicate a consistent rise in total dissolved solids (TDS) concentration across all scenarios over the following 24 years. Specifically, when the total dissolved solids (TDS) concentration in the point source doubles, there is an approximately 13% variation in TDS concentration between 1138 and 1280 ppm over the same period.

The purpose of this study is to model groundwater flow, study the changing pumping rate's effect on groundwater head depletion, and predict the movement of chloride from the dump site to the study area.

2 Method and Theory

2.1 Description of Study Area

The study area was selected in Erbil city, which covers about 579.72 km^2, as presented in Fig. 1. The dump site is located within the Erbil Basin, and many pollutants are emitted from it and transferred to soil layers and groundwater. The dump site is about 10 km west of Erbil, established in 2001. It receives about 1900–2000 tons of waste daily.

2.2 Mathematical Background

The groundwater flow equation can be solved numerically using a Groundwater Modeling System (GMS), including MODFLOW. It depends on Darcy's law and the Mass conservation equation [22]. The groundwater flow equation for unsteady state condition is presented in Eq. 1:

$$\frac{\partial}{\partial x}\left(Kx\frac{\partial h}{\partial x}\right)+\frac{\partial}{\partial y}\left(Ky\frac{\partial h}{\partial y}\right)+\frac{\partial}{\partial z}\left(Kz\frac{\partial h}{\partial z}\right)=S_s\frac{\partial h}{\partial t}\text{ - W} \tag{1}$$

The steady-state groundwater flow equation can be presented in Eq. 2

$$\frac{\partial}{\partial x}\left(Kx\frac{\partial h}{\partial x}\right)+\frac{\partial}{\partial y}\left(Ky\frac{\partial h}{\partial y}\right)+\frac{\partial}{\partial z}\left(Kz\frac{\partial h}{\partial z}\right)=0 \tag{2}$$

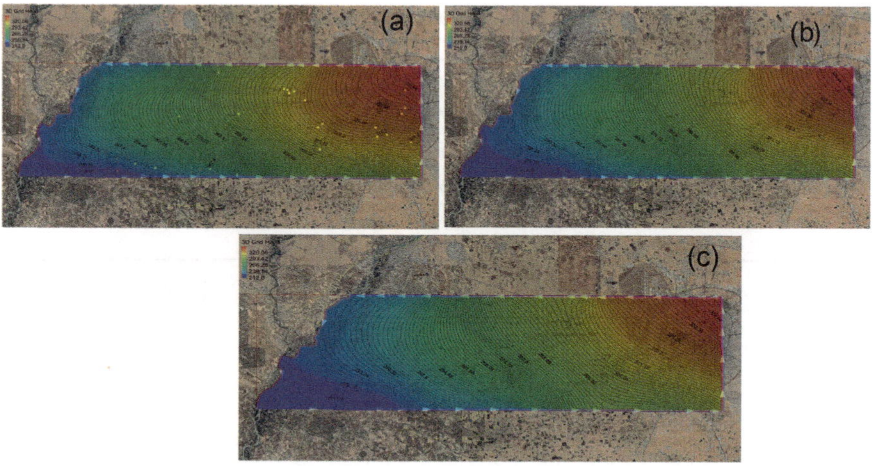

Fig. 4 Couture map for groundwater flow, for a steady state condition, **a** layer 1, **b** layer 2, **c** layer 3

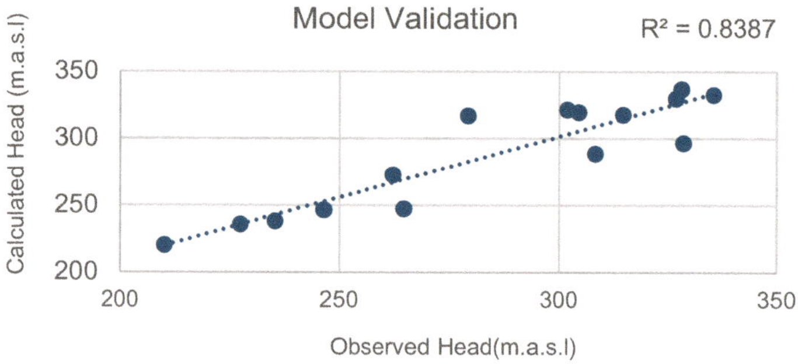

Fig. 5 Best fit line for model validation

The outputs of steady-state simulation are used as initial conditions for the transient state. The model was run for one stress period and ten steps, lasting 365 days. The model runs for the next ten years from 2022 without changing the production rate of wells under the study. Figure 5 shows the estimated groundwater flow direction for the next ten years. The results indicated that, as presented in Fig. 6, some areas in the study area would suffer from a drawdown in groundwater head varying between (1–5) m in the next ten years.

Two scenarios were studied for the effect of production wells on groundwater head. The pumping rate of wells in the study area was increased by 25% and 50%.The model was run with the same properties of steady-state conditions of the calibrated model; the result of increasing pumping wells by 25% scenario indicates that the Maximum increase in drawdown value is about 1 m annually and about 9.24 m after ten years, and there are no drying cells appeared as presented in Fig. 7. Increasing pumping

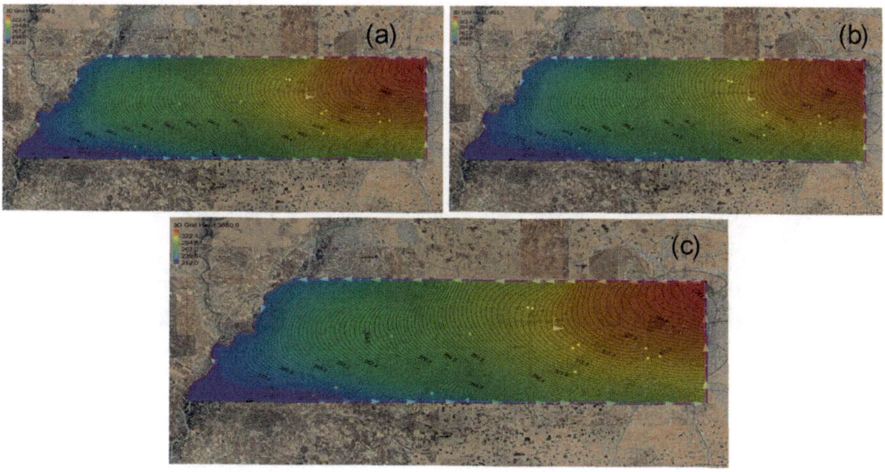

Fig. 6 The estimated direction of groundwater in the study area in **a** 2024, **b** 2027, **c** 2032

wells by 50% for the second scenario indicates that pumping wells will significantly impact groundwater heads. In this case, the highest drawdown will reach about 14.2 m after ten years. Figure 8 shows the variation of groundwater head after increasing the pumping test by 50%.To protect the groundwater in the aquifer, it is essential not to drill new wells and to reduce the pumping rate as much as possible to protect groundwater from depletion.

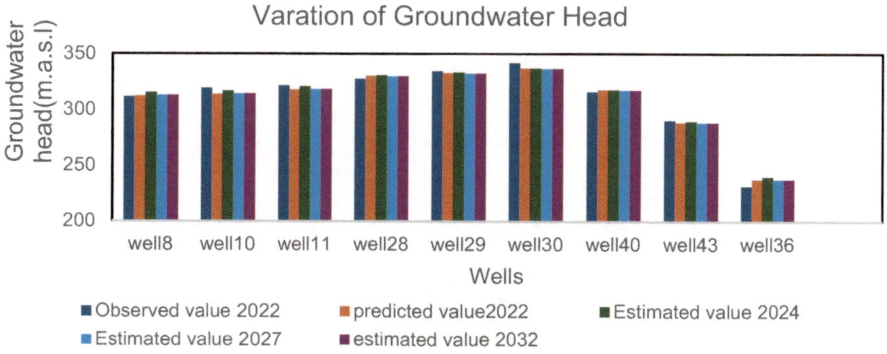

Fig. 7 The estimated groundwater head for subsequent years

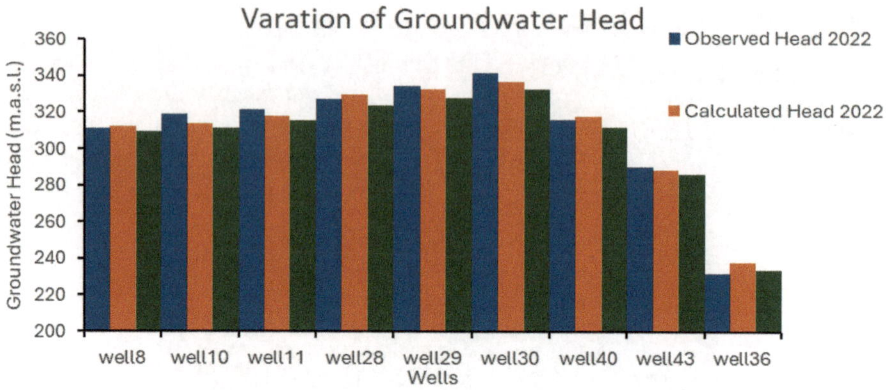

Fig. 8 Variation of groundwater head after increasing the pumping rate by 25%

3.3 *Solute Transport Model (MT3DMS)*

The chloride contamination plume was estimated for the next 50 years from 2022, as presented in Fig. 9. MT3DMS was run with dispersion and advection packages; the result indicated that the contaminant was spread due to dispersion rather than the advection process. The high permeability of the first layer of the study area causes the pollutants to spread vertically and horizontally from the dump site. The chloride concentration was estimated to spread about 1.6 km from the dumpsite after 50 years, with 113.9 m/year as the average pollutants transport rate; the chloride was estimated to reach the third layer after 50 years (Fig. 10).

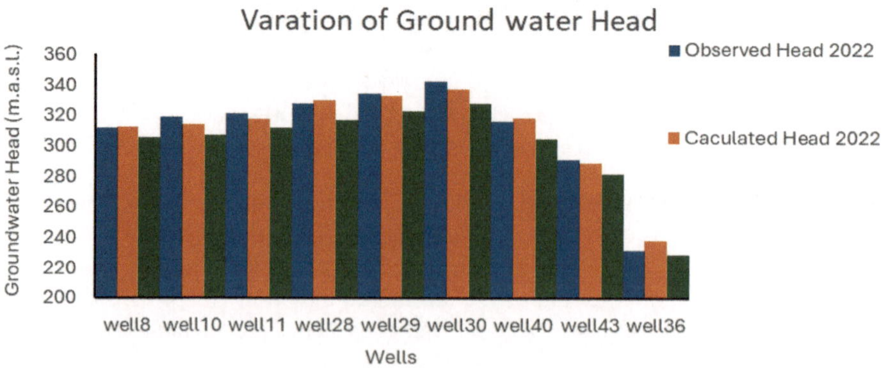

Fig. 9 Variation of Groundwater head after increasing pumping rate by 50%

Fig. 10 Chloride plume after different years from 2022 **a** 10 years, **b** 20 years **c** 30 years, **d** 40 years, **e** 50 years

4 Conclusion

Groundwater is an essential water source in Erbil City, but the quality and quantity of groundwater have noticeably decreased in the last few years. This paper concentrated on studying the effect of increasing pumping wells and the movement of chloride from the Erbil dump site to predict the impact on the surrounding area.

The Groundwater Modeling System (GMS) is robust software designed for simulating the movement of groundwater, predicting the properties of the aquifer, and transferring contaminants in the aquifer. The hydraulic conductivity greatly influences the groundwater flow model throughout the calibration process. A strong correlation between the calculated and observed head is achieved, with an R^2 value of 0.9917.

Two scenarios were used to study the impact of increasing the pumping rate to 25 and 50%. The results indicate that when the pumping rate increases to 25%, the drawdown increases by about 9.24 m as the maximum value after that year and 14.2 m after increasing pumping to 50% after ten years from 2022. To protect groundwater, reducing the production well in the study area is essential to protect the aquifer from depletion.

The process that affects the spread of the chloride aquifer is dispersion, and the impact of the advection process is less than that of the dispersion process. After 50 years from 2022, the chloride concentration spread to about 1.6 km with about 113.9 m/year.

References

1. Nanekely M, Al-Faraj F, Scholz M (2019) Estimating groundwater balance in the presence of climate change impact: a case study of semi-arid area. J Biosci Appl Res 5(4):437–455. https://doi.org/10.21608/jbaar.2019.114536
2. Mawlood DK (2019) Sustainability of aquifer and groundwater condition in Erbil Basin/Iraq. ZANCO J Pure Appl Sci 31(6). https://doi.org/10.21271/ZJPAS.31.6.6
3. Khayyat HAKA, Sharif AJM, Crespi M (2020) Assessing the impacts of climate change on natural resources in Erbil area, the Iraqi Kurdistan using geo-information and landsat data. In: Environmental remote sensing and GIS in Iraq, pp 463–498. https://doi.org/10.1007/978-3-030-21344-2_19
4. Yashooa NKY, Mawlood DK (2023) Numerical modelling of groundwater flow and contamination transport (TDS) in Kaniqurzala area-Central Basin–Erbil City-Kurdistan-Iraq. ZANCO J Pure Appl Sci 35(3):17–29. https://doi.org/10.21271/ZJPAS.35.3.2
5. Yashooa NK, Mawlood DK, Sissakian VK (2024) Assessing water quality index in Kani-Qirzhala area, Erbil City, Kurdistan Region of Iraq. Iraqi Geol J 219–232. https://doi.org/10.46717/igj.57.1D.16ms-2024-4-26
6. Ranjan SP, Kazama S, Sawamoto M (2006) Effects of climate and land use changes on groundwater resources in coastal aquifers. J Environ Manage 80(1):25–35. https://doi.org/10.1016/j.jenvman.2005.08.008
7. Karatzas GP (2017) Developments on modeling of groundwater flow and contaminant transport. Water Resour Manage 31:3235–3244. https://doi.org/10.1007/s11269-017-1729-z
8. Yashooa NK, Mawlood D (2024) A review of the software resources in modeling groundwater contamination transport: case studies in Iraq. https://doi.org/10.5772/intechopen.114933
9. Kumar C (2019) An overview of commonly used groundwater modelling software. Int J Adv Res Sci Eng Technol 6(1):7854–7865
10. Chen C-S, Tu C-H, Chen S-J, Chen C-C (2016) Simulation of groundwater contaminant transport at a decommissioned landfill site—a case study, Tainan City, Taiwan. Int J Environ Res Public Health 13(5):467. https://doi.org/10.3390/ijerph13050467
11. Sathe SS, Mahanta C (2019) Groundwater flow and arsenic contamination transport modeling for a multi-aquifer terrain: assessment and mitigation strategies. J Environ Manage 231:166–181. https://doi.org/10.1016/j.jenvman.2018.08.057
12. Omar PJ, Gaur S, Dwivedi S, Dikshit P (2020) A modular three-dimensional scenario-based numerical modelling of groundwater flow. Water Resour Manage 34(6):1913–1932. https://doi.org/10.1007/s11269-020-02538-z
13. Ehtiat M, Jamshid Mousavi S, Srinivasan R (2018) Groundwater modeling under variable operating conditions using SWAT, MODFLOW and MT3DMS: a catchment scale approach to water resources management. Water Resour Manage 32:1631–1649. https://doi.org/10.1007/s11269-017-1895-z
14. Kumar C (2002) Groundwater flow models. Scientist 'E1' National Institute of Hydrology, Roorkee–247667 (Uttaranchal) Publication
15. Sarma R, Singh SK (2021) Simulating contaminant transport in unsaturated and saturated groundwater zones. Water Environ Res 93(9):1496–1509. https://doi.org/10.1002/wer.1555

16. Zheng C, Wang P (1999) MT3DMS: a modular 3-D multispecies model for simulation of advection, dispersion and chemical reactions of contaminants in groundwater systems. Documentation and user's guide. Contract Report SERDP-99-1, Vicksburg, Mississippi: Army Engineer Research and Development Center
17. Gomaa SM, Hassan TM, Helal E (2021) Assessment of seawater intrusion under different pumping scenarios in Moghra aquifer, Egypt. Sci Total Environ 781:146710. https://doi.org/10.1016/j.scitotenv.2021.146710
18. Yang L, Zheng C, Andrews CB, Wang C (2021) Applying a regional transport modeling framework to manage nitrate contamination of groundwater. Groundwater 59(2):292–307. https://doi.org/10.1111/gwat.13047
19. Najem AAI (2008) Modeling nitrate contamination of the Eocene Aquifer. An-Najah National University, Palestine (Master Thesis)
20. Saatsaz M, Sulaiman WNA, Eslamian S, Javadi S (2013) Development of a coupled flow and solute transport modelling for Astaneh-Kouchesfahan groundwater resources, North of Iran. Int J Water 7(1–2):80–103. https://doi.org/10.1504/IJW.2013.051980
21. Khayyun TS, Sharif MT (eds) (2021) Modelling of groundwater quality of Tigris River Reach-in Baghdad-Iraq using groundwater modeling system software. In: IOP conference series: earth and environmental science. IOP Publishing. https://doi.org/10.1088/1755-1315/779/1/012087
22. Bear J, Cheng AH-D (2010) Modeling groundwater flow and contaminant transport. Springer. https://doi.org/10.1007/978-1-4020-6682-5
23. Javandel I, Doughty C, Tsang C-F (1984) Groundwater transport: handbook of mathematical models. Lawrence Berkeley Lab, CA
24. Al-Kubaisi QY, Hussain TA, Ismail MM, Abd-Ulkareem FA (2019) Estimation of water balance for the central basin of Erbil plain (north of Iraq). Eng Technol J 37(1C):22–28
25. Mustafa JS, Mawlood DK (2024) Assessment of the groundwater in Erbil Basin with support of visual MODFLOW. J Ecol Eng 25(4). https://doi.org/10.12911/22998993/184184
26. Yashooa NK, Mawlood DK (2023) Modeling contamination transport (nitrate) in Central Basin Erbil, Kurdistan region, Iraq with support of MODFLOW software. Iraqi Geol J 234–246. https://doi.org/10.46717/igj.56.1E.18ms-2023-5-28
27. Seeyan S, Merkel B (2015) Groundwater modeling of Harrir plain and Mirawa valley in Shaqlawa-Harrir basins, Kurdistan Region, Iraq. FOG-Freiberg Online Geosci 38
28. Dysli M (2000) Swiss Standard SN 670 010b, characteristic coefficients of soils. Strasse und Verkehr 86:93–94
29. Das BM (2019) Advanced soil mechanics, 1st edn. CRC Press, London. https://doi.org/10.1201/9781351215183
30. Woessner WW, Poeter EP (2020) Hydrogeologic properties of earth materials and principles of groundwater flow. Groundwater Project
31. Kareem HH, Pan S (2016) Modelling of groundwater resources for Al-Najaf City, Iraq. Int J Environ Ecol Eng 10(2):191–201
32. Klaas DK, Imteaz MA, Arulrajah A (2017) Evaluating the impact of grid cell properties in spatial discretization of groundwater model for a tropical karst catchment in Rote Island, Indonesia. Hydrol Res 48(6):1757–1772. https://doi.org/10.2166/nh.2016.250
33. Yahooa NKY, Mawlood DK (2023) Analyzing the groundwater contamination-Central Basin-Erbil City Kurdistan Region-Iraq. ZANCO J Pure Appl Sci 35(2):1–8. https://doi.org/10.21271/ZJPAS.35.2.1
34. Gelhar LW, Welty C, Rehfeldt KR (1992) A critical review of data on field-scale dispersion in aquifers. Water Resour Res 28(7):1955–1974. https://doi.org/10.1029/92WR00607
35. Middlemis H (2001) Murray-Darling Basin Commission groundwater flow modelling guideline. Aquaterra Consulting Pty Ltd., Perth
36. Anderson MP, Woessner WW, Hunt RJ (2015) Applied groundwater modeling: simulation of flow and advective transport, 2nd edn. Academic Press

of the world's population experiences severe water scarcity for part of the year while twenty five percent face 'extremely high' levels of water stress, using over 80% of their annual renewable freshwater supply [1, 2]. Furthermore, about forty five percent of world population live in agricultural areas with high to very high water scarcity, of whom about one-sixth of the world's population live in severely water-scarcity agricultural areas [3].

Agriculture is a key socioeconomic driver for sustainable development, food security, poverty alleviation, livelihoods and labor [4]. Food security is defined as "the state in which people at all times have physical, social and economic access to sufficient and nutritious food that meets their dietary needs for a healthy and active life [5]." Sustainable food production is pre-request to achieve food security. However, Food production faces similar challenges like water globally. Agricultural production is very vulnerable and sensitive to water related risks. In many arid and semi-arid countries, dependence on rain-fed agriculture and lack of access to irrigation water reduces food production potential significantly. Water availability for irrigation stabilizes production, generating direct benefits including increased profitability and reduced risk of crop failure, as well as indirect benefits such as employment, and balanced conditions of food and supply markets [4].

This paper aims to present the status of water resources management in the kingdom of Saudi Arabia and the Kingdom's key achievements since the adaption of the National Water Strategy in 2017 to date. The paper will summarize the experience and key performance indicators for the use of desalinated water in agriculture globally. Finally, a holistic assessment approach to support the decision making process for desalinated water use in agriculture by conducting a comprehensive systematic economic, technical, environmental, and legislative feasibility studies will be developed.

1.1 Water Resources Status in the Kingdom of Saudi Arabia

The Kingdom of Saudi Arabia (KSA) is situated in one of the world's most arid regions, characterized by limited rainfall and high temperatures, inherently restricting natural freshwater availability. The arid climate, coupled with rapid population and economic growth, has placed significant strain on the country's limited water resources where total water demand exceeds the sustainable yield of both conventional and non-conventional water resources in the country [6–11]. The KSA has a reserve of non-renewable groundwater across a group of aquifers with groundwater collected in more than 20 primary and secondary aquifers across the country serving several regions (see Fig. 1). All primary aquifers are located in the Arabian Shelf. In contrast, the Arabian Shield does not include any primary or secondary aquifers due to its non-porous rocky nature. In addition, renewable groundwater is found in shallow and deep aquifers as well as surface water of valleys. The renewable groundwater is currently estimated at 2.8 billion m^3 per year in the Arabian Shield region [8, 12–14]. The total reserve of exploitable water from dams amounts to about 2.5 billion m^3 per year, with the majority of the total exploitable water from dams being in the regions of Aseer et al. [6, 10, 11]. There are more than 30 seawater desalination plants on the eastern and western coasts of the Kingdom with a total production capacity for seawater desalination reaching more than 11 million m^3 per day in 2022 [15]. This positions Saudi Arabia as the world's largest producer of desalinated water. Treated wastewater is produced upon the treatment of sewage to a degree that makes it safe to use for various purposes including industrial processes, cooling, and agriculture. The total quantity of treated wastewater reused across KSA reached more than 1 million m^3 per day in 2022 [8, 12, 14]. In 2016, the total water consumption reached its peak with a total value of about 25.8 billion m^3 where the agriculture sector is the largest consumer (more than 80%) mainly supplied from the non-renewable groundwater aquifers. This is a key challenge for water resources sustainability in the Kingdom.

To manage water challenges, the KSA launched the Saudi National Water Strategy 2030 (NWS2030) in December 2017 [8]. The NWS2030 aims to address all key challenges, leveraging previous and on-going studies, and reform the water and wastewater sector to ensure sustainable development of the kingdom's water resources while providing affordable high-quality services. The NWS2030 vision statement cites "A sustainable water sector, safeguarding the natural resources and environment of the Kingdom and providing cost-effective supply and high quality services. Efficiency contributes to economic and social development" as the central goals. This vision statement can be further detailed into five strategic objectives, as follows [8]:

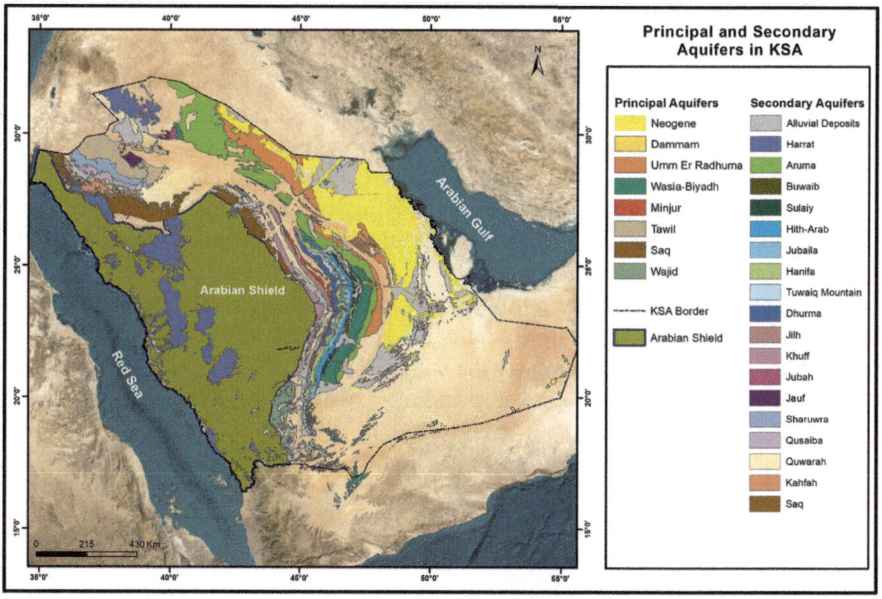

Fig. 1 Principle and secondary aquifers in Kingdom of Saudi Arabia

- Ensure continuous access to adequate quantities of safe water, under normal operations and during emergency situations.
- Enhance water demand management across all uses.
- Deliver cost-effective and high quality water and wastewater services, accounting for affordability.
- Safeguard and optimize the use of water resources, while preserving the local environment for the highest benefit of the Saudi society in this generation and the future.
- Ensure water sector competitiveness and positive contribution to the national economy through promoting effective governance, private sector participation, localization of capabilities and innovation."

The implementation of the NWS2030 programs, initiatives, and projects came out with significant improvement of the water management performance in the Kingdom to date. The NWS2030 implementation boasts doubling production, transport lines capacity, strategic storage and water reuse. Current total water production capacity for domestic use is more than 15 million cubic meters per day of which 11.3 million cubic meters per day from water desalination, and more than 3.7 million cubic meters per day from groundwater and surface water [15]. In addition, the NWS2030 implementation managed to issue and enforce policies and governance system that resulted in reducing agriculture water demand from around 22 billion cubic meters in 2016 to about 12 billion cubic meters per year in 2023. This major achievement is a direct result of phasing out the fodder cultivation and enhancing irrigation management and efficiency [8, 10, 11].

Despite all previously mentioned achievements water resources in the kingdom is far from achieving sustainability. The water demand remains exceeding sustainable water supply capacity of both conventional and non-conventional resources and no doubt, there are needs to additional measures to be adapted in order to balance water demand and supply in the country. The agriculture demand remains the key challenge to the sustainability of water resources in the Kingdom. Desalinated water can be an alternative source to partially meet the agriculture water demand and alleviate the stress on conventional water resources.

This paper aims to propose a holistic assessment approach for the potential use of desalinated water for agriculture in the Kingdom of Saudi Arabia based on international best practice of planning for the use of desalinated water for agriculture and the water resources management setup and framework in the Kingdom.

2 Overview of Global Desalinated Water Industry

The desalination industry has witnessed significant technological achievements over the last decade including improvement in membrane productivity and efficiency, better integration of renewable energy, brine mining, and energy consumption significant reduction [14–18]. For example, a few years ago, the average energy consumption for the production of one cubic meter was about 7 kWh in comparison to less than 3 KWh currently. Furthermore, the cost of production was about one USD per cubic meter while current rate is below 0.4 USD (see Fig. 2) [16–20]. These achievements has shifted the desalination industry from the energy intense thermal desalination to more advanced Saline Water Reverse Osmosis (SWRO) technologies. Accordingly, SWOR became the dominant technology in the last decade. Table 1 presents the key performance indicators for the most recent desalination plants in GCC countries [10, 11].

2.1 Desalinated Water Use in Agriculture

The desalinated seawater and brackish water is currently used for agriculture purposes in five regions in the World (see Fig. 2). The use of desalinated water for agriculture purposes has been started more than forty years ago in Canary Islands and south-eastern Spain to cultivate high value crops, mainly for export. These two regions are currently the largest users of desalinated water for agriculture, 40 and 52% of their total desalinated water production respectively, followed by Abu Dhabi at 21%. It

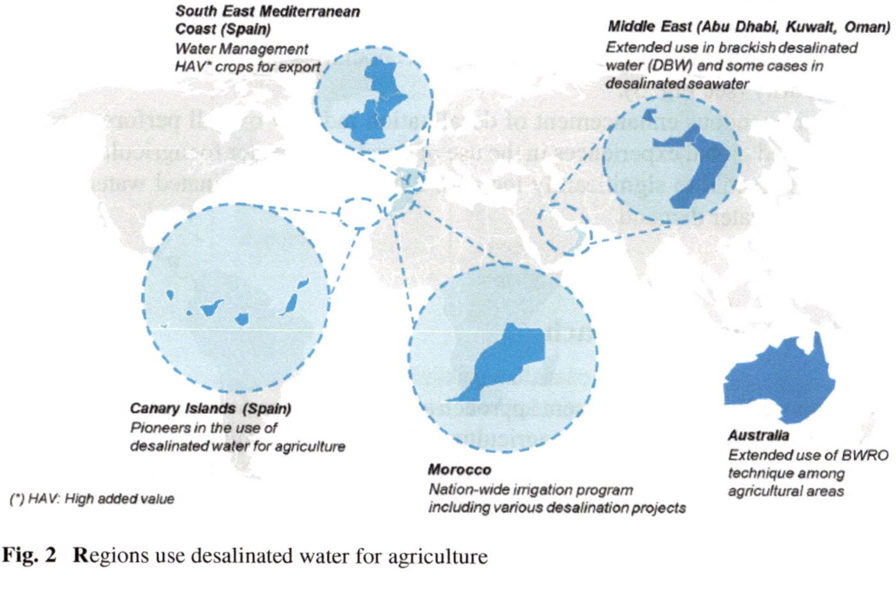

Fig. 2 Regions use desalinated water for agriculture

The benchmarking of global experience phase, all international case studies for the use of desalinated water in agriculture shall be identified. For each case, all technical, environmental, economic, social and business performance indicators shall be assessed and documented as lesson learned. The approach's second phase shall evaluate and assessed the current status of desalinated technologies for the use in agriculture and review international experiences and expertise for overall performance. The technical and economic feasibility of using renewable energy sources or mixed energy sources shall be evaluated too. Furthermore, the possibility of mixing between desalinated water and other water resources such as treated wastewater and renewable ground and surface water shall be assessed.

The approach's third phase aims to determine the environmental impacts and challenges that may arise from the use of desalinated water in agriculture and limit their repercussions. Third phase shall identify the environmental positive and negative impacts of using desalinated water in agriculture, propose appropriate agricultural means and techniques for using desalinated water in agriculture, as well as for the process of producing and transporting desalinated water to reduce its impacts on the environment and maximize its environmental benefits, and evaluate the effects of water quality (desalinated water mixed with proportions from traditional and non-traditional sources) on the production quantity of the crop. Phase four aims to identify feasible crop composition that can be cultivated in KSA based on metrological, soil and environmental conditions. In phase four, the feasible crop composition that can be cultivated in the Kingdom regions will be determined and evaluated. The fifth phase shall analyze the economic and financial viability of the selected crops for cultivation in the different. Phase six will review the existing legal and legislative framework in the kingdom and propose a regulatory and legislative amendments to facilitate the successful implementation of the scheme. Finally, a business model and a roadmap will be developed based on the inputs of all six phases. Stakeholders shall be engaged intensively throughout the assessment process to ensure integrity and support of all stakeholders to the assessment outcomes and the proposed scheme.

4 Conclusion

The agriculture sector is the largest water user globally and accounts roughly for 70% of freshwater withdrawals. This is a key challenge to sustainable water resources management globally and specifically in arid and semi-arid region such as the Kingdom of Saudi Arabia. The advancements of desalination industry overall performance in the last decade and the accumulated global experience of the desalinated water use for agriculture over the last four decades increase the feasibility of the use desalinated water in crop cultivation. The proposed holistic assessment approach in this paper will support information base decision-making process to utilize desalinated water for crop production through conducting a comprehensive systematic economic, technical, legal and environmental feasibility studies.

References

1. United Nations (2024) The United Nations World Water Development Report 2024: water for prosperity and peace, UNESCO, Paris. ISBN 978-92-3-100657-9
2. UN-Water (2021) Summary progress update 2021—SDG 6—water and sanitation for all. Version: July 2021. Geneva, Switzerland
3. FAO (2021) The State of Food and Agriculture 2020, overcoming water challenges in agriculture, Rome, Italy. SBN978-92-5-133441-6
4. FAO, IFAD, UNICEF, WFP and WHO (2023) The State of Food Security and Nutrition in the World 2023. Urbanization, agrifood systems transformation and healthy diets across the rural–urban continuum. Rome, FAO. https://doi.org/10.4060/cc3017en
5. UN. The Global food security index homepage. https://www.unccd.int/resources/knowledge-sharing-system/global-food-security-index. Accessed 6 Aug 2024
6. Ouda OKM (2014) Water demand versus supply in Saudi Arabia: current and future challenges. Int J Water Resour Dev 30(2):335–344
7. Ouda OKM (2015) Domestic water demand in Saudi Arabia: assessment of desalinated water as strategic supply source. Desalin Water Treatment 56(11):2824–2834
8. Ministry of Environment, Water and Agriculture. Saudi National water strategy 2030. Homepage, https://www.mewa.gov.sa/ar/Ministry/Agencies/TheWaterAgency/Topics/Pages/Strategy.aspx. Accessed 4 Aug 2024
9. McIlwaine SJ, Ouda OKM (2020) Drivers and challenges to water tariff reform in Saudi Arabia. Int J Water Resour Dev 26. https://doi.org/10.1080/07900627.2020.1720621
10. Suhail M, Faraj TK, Ahmad W, Ravshanov AX, Khan MN (2024) Issues of water resources in Saudi Arabia: past, present, and future. Sustainability 16. https://doi.org/10.3390/su16104189
11. Alotaibi BA, Baig MB, Najim MMM, Shah AA, Alamri YA (2023) Water scarcity management to ensure food scarcity through sustainable water resources management in Saudi Arabia. Sustainability 15. https://doi.org/10.3390/su151310648
12. Chowdhury S, Al-Zahrani M (2015) Characterizing water resources and trends of sector wise water consumptions in Saudi Arabia. J King Saudi Univ Eng Sci 27:68–82
13. Odnoletkova N, Patzek TW (2023) Water resources in Saudi Arabia: trends in rainfall, water consumption, and analysis of agricultural water footprint. Sustain Agric. https://doi.org/10.1038/s44264-023-00006-w
14. Elzaki RM, Al-Mahish M, Alzahran F (2024) Water–energy–milk nexus: empirical evidence from Saudi Arabia. Water 16. https://doi.org/10.3390/w16111538
15. Ministry of Environment Water and Agriculture (2022) Annual Statistic report. https://www.mewa.gov.sa/ar/search/pages/default.aspx?k=%D8%A7%D9%84%D9%83%D8%AA%D8%A7%D8%A8%20%D8%A7%D9%84%D8%A7%D8%AD%D8%B5%D8%A7%D8%A6%D9%8A. Accessed 3 Aug 2024
16. Barron O, Ali R, Hodgson G, Smith D, Qureshi E, McFarlane D, Campos E, Zarzo D (2015) Feasibility assessment of desalination application in Australian traditional agriculture. Desalination 364:33–45
17. Ahmed FE, Khalil A, Hilal N (2021) Emerging desalination technologies: current status, challenges and future trends. Desalination. https://doi.org/10.1016/j.desal.2021.115183
18. Williams J (2022) Desalination in the 21st century: a critical review of trends and debates. Water Altern 15(2):193–217
19. Ministry of Environment Water and Agriculture (2024) Innovation in the water sector in Saudi Arabia, technology adaption roadmap. Homepage, https://www.mewa.gov.sa/en/Ministry/Agencies/AgencyLivestock/Topics/PublishingImages/Pages/InnovationintheWatersector/EN-MEWA-Water_High-resolution%20for%20reading.pdf. Accessed 1 Aug 2024
20. Saudi water partnership company, company portfolio and profile. Homepage, https://www.swpc.sa/en/, Accessed 1 Aug 2024

Understanding the Leaching of Antimony from Commercially Bottled Water: Perspective on the Effect of Use Conditions and Associated Risks

Mahmoud Bataineh, Sohayla El Mokadem, Anika Fajutagana, and Ravi Rangarajan

Abstract Antimony (Sb), commonly used as a catalyst in the production of polyethylene terephthalate (PET) plastics, poses potential health risks when leached into drinking water. This study analyzed Sb concentrations in PET bottled water, focusing on samples stored from local and international brands for varying durations. Analysis results of the PET bottle materials showed Sb concentrations ranging from 106.4 to 248.8 mg/kg, with Sb levels in new water bottles ranging from 0.22 to 1.09 ppb, increasing to 0.26–1.77 ppb after eight months. International brands noted elevated concentrations due to structural requirements for shipping and environmental factors affecting leaching. Additionally, sunlight exposure further exacerbated Sb leaching, highlighting climate change's role in water quality degradation. These findings underscore the urgent need for enhanced monitoring of bottled water quality and implementing sustainable practices in PET production and distribution.

Keywords Antimony · Bottled water · PET · Health · Leaching

M. Bataineh (✉)
College of General Education, University of Doha for Science and Technology, Doha, Qatar
e-mail: mahmoud.bataineh@udst.edu.qa

S. El Mokadem
College of Engineering and Technology, University of Doha for Science and Technology, Doha, Qatar

A. Fajutagana
College of Health Sciences, University of Doha for Science and Technology, Doha, Qatar

R. Rangarajan
Department of Environmental Health and Safety, College of Health Sciences, University of Doha for Science and Technology, Doha, Qatar

S. Al-Naemi et al. (eds.), *Water and Food Security in the Face of Climate Change: Challenges and Opportunities for Resilience*, World Sustainability Series,
https://doi.org/10.1007/978-3-032-00098-9_59

Table 1 Bottled water samples with their representative characteristics collected in Qatar

Sample ID	Brand material	Bottle volume	Bottle color	Bottle pH	PET mass (g)	Production country
C1	PET	330 mL	Blue	7.1	14.02	Switzerland
C2	PET	500 mL	Transparent	7.8	12.15	Scotland
C3	Glass	375 mL	Transparent	6.4	Glass	Norway
C4	Glass	330 mL	Transparent	7.03	Glass	France
C5	PET	330 mL	Transparent	7.2	16.60	France
C6	PET	250 mL	Transparent	7.7	17.54	USA
C7	Glass	330 mL	Transparent	8	Glass	Italy
C8	PET	330 mL	Transparent	7.75	11.1	Qatar
C9	PET	330 mL	Transparent	8	19.6	Italy
C10	PET	500 mL	Transparent	7.5	13.67	Qatar
C11	PET	500 mL	Transparent	7.5	12.31	Qatar
C12	PET	1.5 L	Transparent	7.5	27.87	Qatar
C13	PET	200 mL	Transparent	7.7	8.15	Qatar
C14	PET	200 mL	Transparent	7.5	-	Qatar
C15	PET	200 mL	Transparent	6.5	9.44	Qatar
C16	PET	330 mL	Transparent	6.7	10.36	Qatar
A1	PET	330 mL	Light blue	7.0	10.52	Qatar
A2	PET	200 mL	Transparent	7.3	8.27	Qatar
A3	PET	330 mL	Light blue	7.4	10.47	Qatar
A4	PET	500 mL	Purple	7.06	12.45	Qatar
A5	PET	200 mL	Transparent	7.2	8.82	Qatar
M1	PET	250 mL	Transparent	7.02	9.68	Qatar
M2	PET	500 mL	Transparent	7.74	10.01	France
M3	Glass	250 mL	Transparent	6.8	Glass	Qatar
M4	PET	330 mL	Transparent	7.7	10.5	Qatar
M5	PET	250 mL	Transparent	7.5	9.59	Qatar
M6	PET	330 mL	Transparent	6.7	9.48	Qatar
M7	Glass	330 mL	Transparent	7.62	Glass	Scotland
M8	PET	500 mL	Transparent	7.38	14.12	Lebanon
M9	PET	330 mL	Green	5.0	22.18	Lebanon

There were discrepancies in the water pH on the bottle label and the measured pH, hence the measured values are only given in this table

2.2 Analytical Method Development and Sb Concentration Measurements

A sensitive, quantitative analytical method for Sb analysis to less than parts-per-billion (ppb; μg/L) level is needed to determine Sb leaching out of bottled water effectively. We used an analytical method based on APHA 3125B with the use of ICP-MS (Agilent 7800; Mass hunter 4.3) and ICP-OES (Thermo Scientific, 1CAP6300 Duo) at the Arab Center for Engineering Studies (ACES), Doha. The method achieved a detection limit of 0.12 μg/L, using calibration standards from 0 to 10 ppb, and quality control standards at 1 and 5 ppb.

2.3 Determination of Sb Concentration in the Bottling Material Used

PET bottling materials belonging to 24 different brands of water were selected and cut into small pieces, as shown in Fig. 1, to determine the concentration of Sb. The samples were digested using an Open Vessel Wet Digestion following a detailed digestion procedure by weighing 0.5 g of the sample and placing it into a beaker. Next, 10 mL of a 1:1 mixture of HNO_3 and water was added, and the beaker was covered with a watch glass before being heated to reflux for 10 min. Subsequently, after cooling, 10 mL of concentrated HNO_3 was added, and the mixture was refluxed for an additional 20 min. If brown fumes persisted, indicating incomplete oxidation, further HNO_3 was added until no more fumes were observed. Once cooled, 3 mL of water and 7 mL of H_2O_2 were introduced, and the mixture was warmed until bubbling ceased, reducing the volume to 5 mL. Finally, the solution was diluted to 50 mL with de-ionized water, filtered, and then analyzed using ICP-OES described in the previous section.

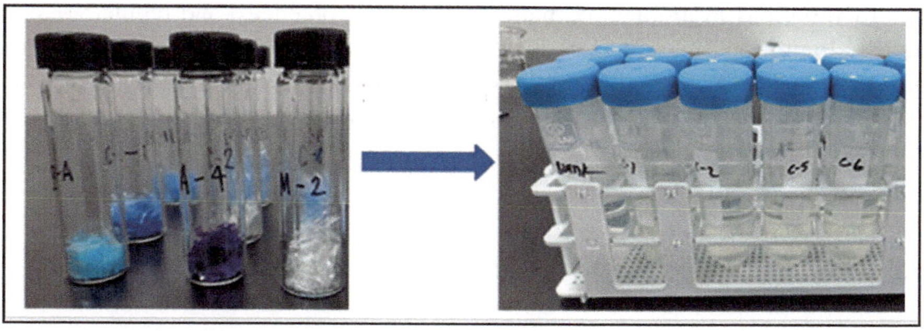

Fig. 1 PET bottling materials for 24 brands digested in Open Vessel Wet Digestion to analyze the concentration of Sb in the PET plastic material by ICP-OES

2.4 Understanding the Diffusion Rate of Sb Under Typical and Stressful Storage Conditions

The key conditions tested for Sb leaching from PET in the bottles were storage at room temperature, sunlight exposure, pH, and bottle characteristics (i.e., bottle mass, volume, the ratio of empty bottle volume to PET mass, and color). For sunlight irradiation, two bottle brands (C2 and C5) were tested by being placed outdoors in the ambient environment (Open Terrace, Doha) for 32 days between the period of August 22, 2023, and September 23, 2023. During this period, the average daytime temperature ranged from 35 to 46 °C, with humidity levels of ~ 60%. To assess the impact of bottle mass, various 330 mL water bottle brands were analyzed to determine Sb concentration. The tare mass of all empty 330 mL bottles was measured.

3 Results and Discussion

3.1 Sb Concentration in PET Bottles and Bottles of Water

Following the digestion process outlined in Display 1, the analysis of PET bottling materials was carried out. The obtained Sb concentration of 24 PET bottled water bottles ranged from 106.4 to 248.8 mg/kg (Fig. 2), considering the dilution factor. Notably, the measured Sb concentration falls within the industry-reported range of 80.6 to 352.7 mg Sb/kg PET [17], suggesting consistency in the manufacturing processes across different samples in different regions. It also indicates that in PET materials used for food and beverage packaging, stable use of Sb as a catalyst in its production exists.

The narrow-observed variation in Sb levels supports the idea that standardized manufacturing practices are being followed. However, future research is needed to understand better the regional differences in Sb concentration that can occur due to variations in production methods, recycling practices, and regulatory standards. Monitoring Sb levels in both newly produced and recycled PET bottles could also provide key insights into the lifecycle of Sb in these materials and its potential accumulation in recycled products, an area of sustainability that needs to be considered.

3.2 Storage Effect on Sb Concentration of Bottled Water

A comparison was made between new water samples and those stored for eight months at room temperature. The concentrations in the newly acquired water bottles ranged from 0.22 to 1.09 ppb (Fig. 3). In contrast, the results from the stored water bottles indicated varying concentration levels after an 8-month storage period at

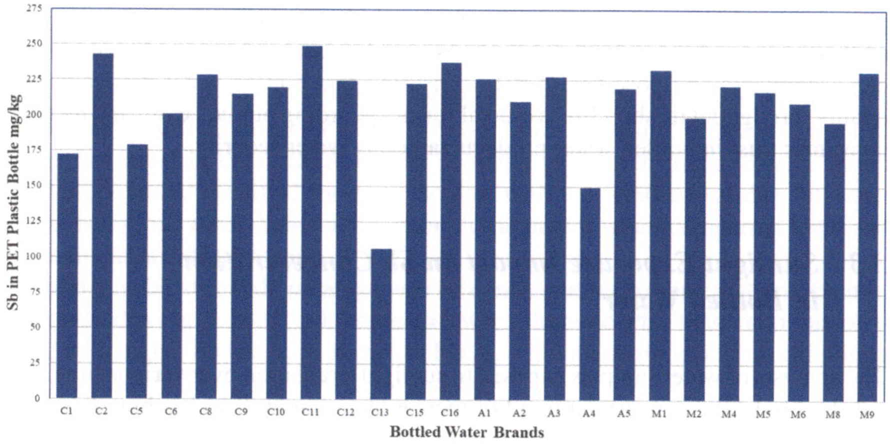

Fig. 2 Sb concentrations in PET bottling material of 24 brands analysed

room temperature. The obtained Sb values in the water ranged from 0.26 to 1.77 ppb (Fig. 3). Although the increments in Sb concentration remained within permissible and regulated limits, the stored water bottles exhibited elevated levels of Sb compared to the new samples.

This suggests that prolonged storage, even at room temperature, can increase Sb leaching from PET bottles, which may be influenced by factors such as material composition or storage conditions. The gradual rise in Sb concentration on storage suggests potential leaching from the packaging material over time. This

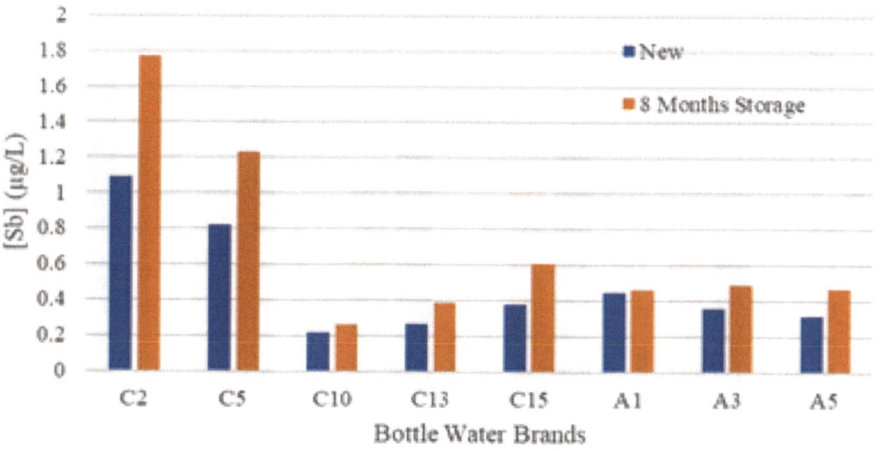

Fig. 3 Antimony concentration in new and 8-month-stored bottled water samples from supermarkets in Qatar, stored at room temperature

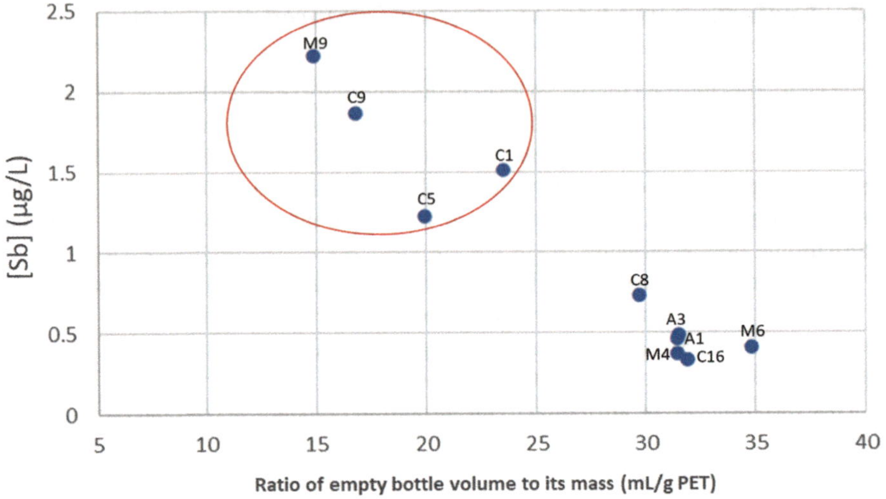

Fig. 6 Daily cloud percentage/fraction versus precipitation $\delta^{18}O$ in the collected samples from all the three sites

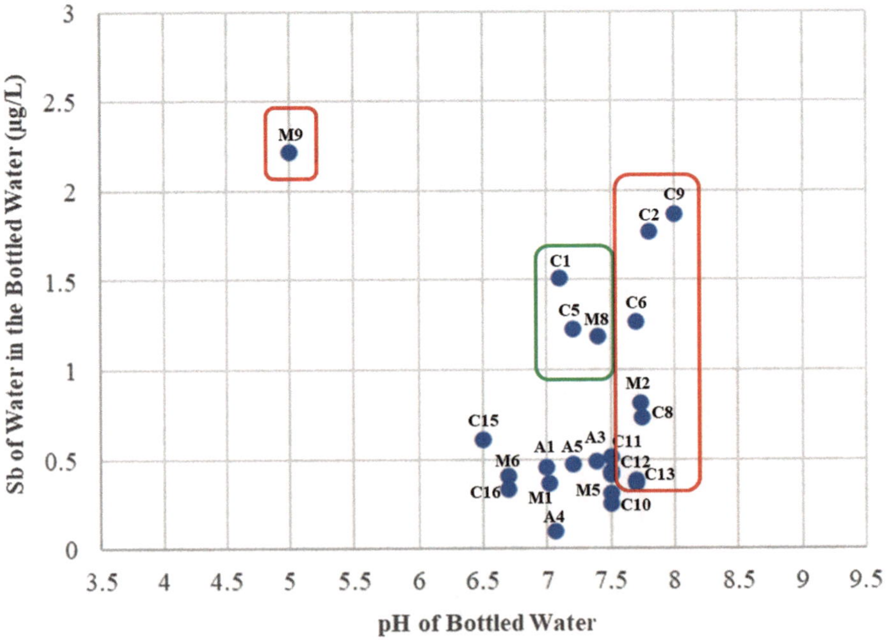

Fig. 7 Profile illustrating the correlation between antimony concentration in selected bottle brands and water pH values. The concentrations were measured in the 8-month stored bottled water samples

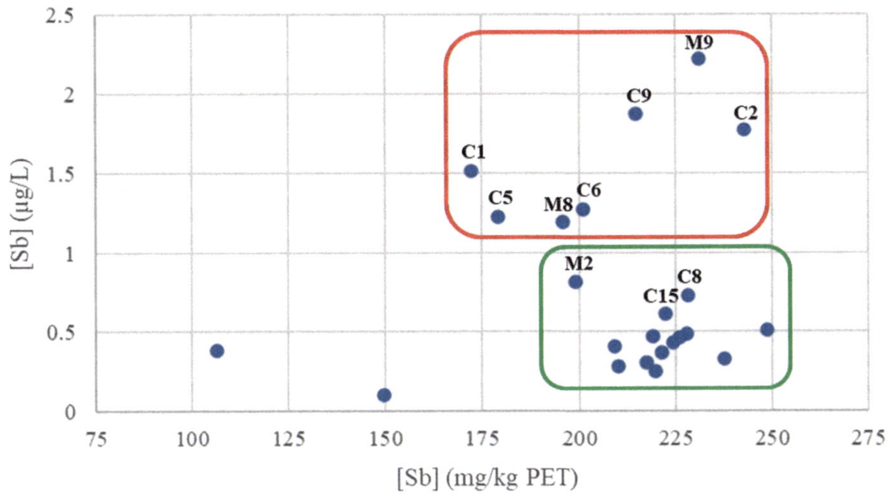

Fig. 8 Sb concentration profile in selected brands and its correlation with PET material levels from 8-month stored samples

during extended storage. From a sustainability perspective, the increased use of PET material for international shipments could be problematic, as it not only contributes to higher Sb leaching but also increases the environmental footprint of these bottles. The extended shipping durations and exposure to diverse environmental conditions further exacerbate this issue, as bottles experience a variety of temperatures and pH levels that may accelerate Sb migration into the water. This raises questions about the sustainability of long-distance bottled water transport and the safety implications of extended storage [9, 17].

To address whether the concentrations of Sb in PET materials impact Sb leaching, we compared Sb concentration profile in selected brands and its correlation with PET material levels from 8-month stored samples (Fig. 8). Brands highlighted within the green box exhibit slightly lower concentrations in bottled water, whereas brands highlighted within the red box show relatively high Sb concentrations after eight months of storage. All brands in the red box correspond to international brands with either low or high pH values. This observation can also be linked to the previously discussed (mL/PET mass) ratio.

4 Conclusion

The findings of this study reveal critical insights into the leaching behavior of antimony from PET bottled water, particularly under extended storage and environmental exposure. Although the observed Sb concentrations remained within permissible limits, the increases raise significant concerns regarding long-term water safety

Assessment of Satellite-Derived Bathymetry (SDB) Using Landsat-8 and Sentinel-2 Multispectral Images Along the North Coast of Cherchell, Algeria

Lamine Salim, Lamri Nacef, Merdjaoui Abdelkadir, and Bachari Nour El Islam

Abstract SDB is often employed in planning hydrographic surveys for marine regions that have not been surveyed or where bathymetric data is outdated. This method has gained significant popularity in recent years, relying on analytical modeling of light penetration through the water column in visible and infrared bands. In this study, the SDB method was applied using free-of-charge Landsat-8 and Sentinel-2 satellite images to obtain bathymetric data along the North Coast of Cherchell, Algeria. The objective of this research is to compare the accuracy of SDB derived from Landsat-8 and Sentinel-2 imagery after selecting and validating the best correlation model, with the implementation of the Stumpf Empirical Algorithm. The data analysis and SDB processing procedures are described in detail. Furthermore, the processed satellite data and the resulting bathymetric maps were compared through a comprehensive statistical analysis for SDB calibration and model validation. Echosounder measurements were collected from September 15–26, 2019, with a total of 1131 depth points recorded and geolocated. Two satellite images were utilized: a Landsat-8 image from September 24, 2019, and a Sentinel-2 image from February 24, 2020. Statistical indices such as R^2, RMSE, and MAE were calculated, with results indicating that Sentinel-2 images provided a better correlation between SDB-predicted values and in-situ depth data (Sentinel-2: $R^2 = 0.7892$; Landsat-8: $R^2 = 0.7233$, RMSE = 2.176, MAE = 4.735). Overall, the generated models demonstrate the potential to predict bathymetry on a large scale using only satellite images, highlighting the importance of spatial resolution for accurate SDB estimation.

L. Salim (✉) · L. Nacef · M. Abdelkadir · B. N. E. Islam
Laboratory of Biological Oceanography and the Marine Environment (LOBEM), University of Science and Technology of Houari Boumedien (USTHB), El Alia, Bab Ezzouar, Algiers, Algeria
e-mail: salim.lamine@gmail.com

L. Salim
Higher School of Saharan Agriculture Adrar, Adrar, Algeria

S. Al-Naemi et al. (eds.), *Water and Food Security in the Face of Climate Change: Challenges and Opportunities for Resilience*, World Sustainability Series,
https://doi.org/10.1007/978-3-032-00098-9_60

Keywords Remote sensing · Hydrography · Satellite derived bathymetry · Landsat-8 · Sentinel-2 · Stumpf algorithm

1 Introduction

Bathymetry information is one of the important parameters which plays a major role in planning near-shore structure activities such as engineering work, port management, pipeline laying, fishing, dredging operation. Seafloor bathymetry is essential for safety of navigation and for establishing the limits of the extended continental shelf (ECS) under the United Nations Convention on Law of the Sea (UNCLOS) [1]. This exemplifies that a detailed knowledge of a nation's coastal bathymetry is also vital for political and commercial purposes. Marine geohazards are not only of concern to coastal communities, but also to industries dealing with marine infrastructure. Geohazard assessments in the marine realm are mainly based on bathymetric data. Although only a snapshot in time, bathymetric data deepens the understanding of the seafloor fabric and helps to identify potential risks linked to hazardous processes, such as slope failures or turbidity currents, and with repeat surveys, can be used to monitor seafloor changes over time [2]. Generally, bathymetric data represent a fundamental dataset for addressing the growing challenges associated with climate change [3]. Bathymetric data are also fundamental to our ability to model ocean circulation, with the predicted location of key circulation features, such as the separation point of the Gulf Stream from the United States' coast, being critically dependent on accurate topography representation in the model in question [4]. Similarly, accurate ocean models can have a major impact on the ability of climate models to simulate global phenomena such as El Niño events [5]. Conventionally, bathymetric data is collected via various methods such as echosounding and Light Detection and Ranging (LIDAR) devices. For sounding based techniques, single beam echosounders (SBES) or multi-beam echosounders (MBES) may be alternatively used. As for the LIDAR, either airborne LIDAR bathymetry (ALB) or LIDAR installed on vessel devices are available tools [6]. Depth samples can be precisely generated in case of utilizing SBES with cost less than MBES. For both sounding types, the constraints are the cost, time and labor consuming and the sensitivity to the sea climate [7]. In addition, some difficulties in accessibility for shallow water, rocky seabed or coral reefs areas. ALB provides quick and accurate bathymetric information especially for quietly clear waters and in nearshore where sounding is least effective. In the present work, we will try to answer the following questions:

- What is the best multispectral optical sensor to retrieve the SDB?
- How the spatial resolution is affecting the accuracy of the bathymetric measures?

Our study aims to assess the potential of Landsat-8 and Sentinel-2 multispectral-satellite images in retrieving the SDB data on the North coast of Cherchell in Algeria.

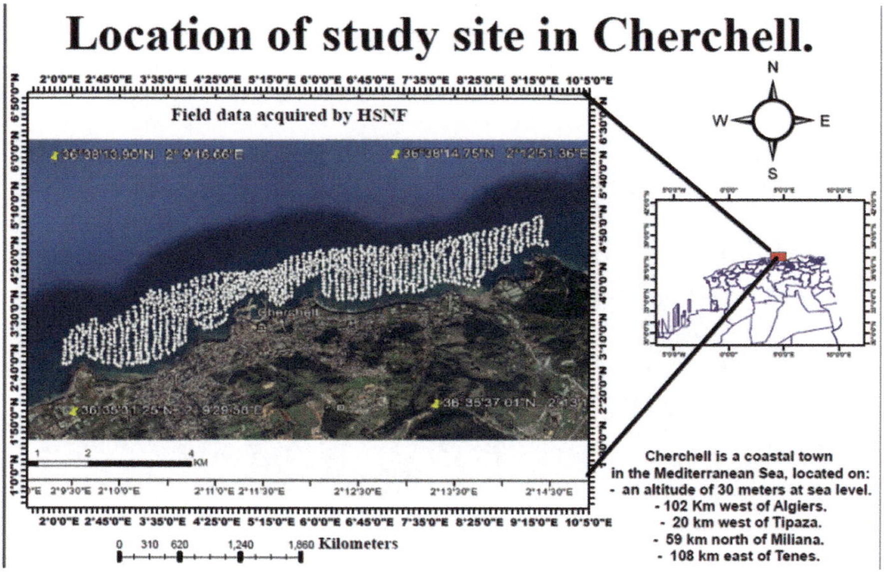

Fig. 1 Study area

2 Materials and Method

2.1 Study Area

The current study area is bounded by (36°38′19″, 36°35′15″) in northing and (2°09′03″, 2°14′46″) in easting and located at the North coast of Cherchell as presented in Fig. 1.

The geomorphology of this area is highly dynamic due to the pro-activity in sediment transport processes. A large amount of sediment is transported from west to the east. The area includes several features such as sand bars, shelf parts and ridges while the bottom is mostly composed of sand and rocks. The water body is almost clear at this region and no coral reefs formation exist; the tidal range is small: less than 0.5 m.

2.2 Single Beam Echosounder In-Situ Data

Bathymetric data were collected from the Northern coast of Algeria, in the Cherchell coastal region by Hydrographic Service of Naval Forces (HSNF, Ministry of National Defense). The echosounder measurements were collected during the period from 15-September-2019 to 26-September-2019, where a total of 1131 depth points were collected and geo-localized (Fig. 2).

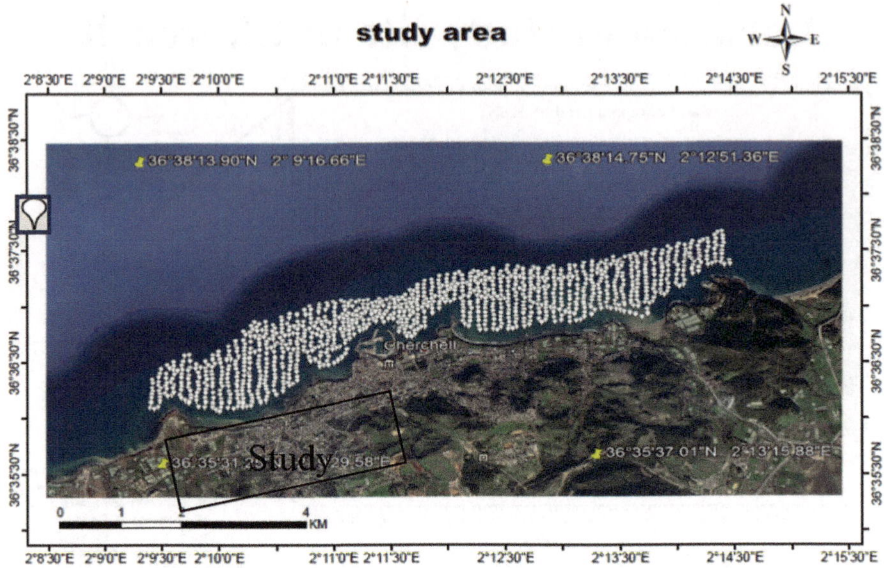

Fig. 2 Geographical distribution of the in-situ bathymetric data in the study area

The area covered by the bathymetric measures extends with about 8 km alongshore starting from Ras Rouge in the West to Ras Blanc in the East, and has an area of about 14 km^2. The water depths in the study region are varying from 1.23 to 19.95 m.

2.3 *Multispectral Satellite Images*

The satellite derived bathymetry (SDB) method was implemented in our study after the acquisition of bathymetric data by the non-survey vessels in the coastal area of Cherchell. Herein, we have used two multispectral satellite images taken inside the period of the echosounder field measurements (15 to 25/09/2019); Landsat 8 image of 24-Sep-2019 and Sentinel 2 image of 25-Sep-2019. The main advantage of using Landsat and sentinel 2 is the free access and the spatiotemporal resolution.

2.4 *Methods Used for Processing Satellites Images*

For this study Blue, Green and Infrared bands of Landsat 8 and Sentinel 2 satellite imagery are used. The reason for applying blue and green band in the coastal environment is that radiance in the blue band decreases more rapidly with depth

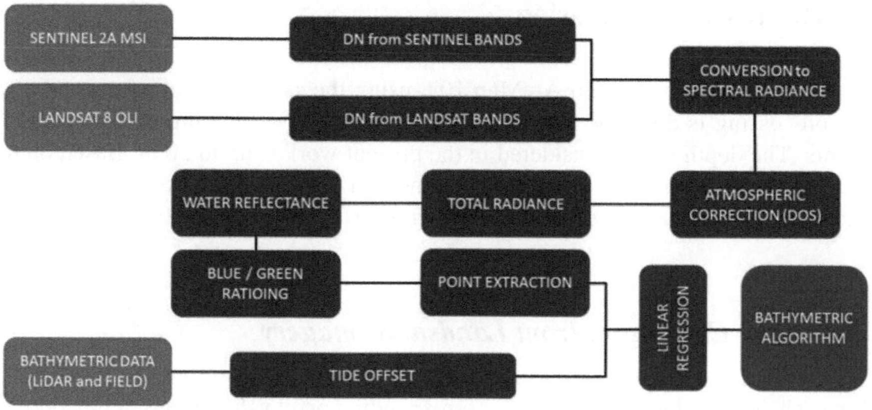

Fig. 3 Flowchart for deriving bathymetric maps from Landsat-8 images

than radiance in the green band. Generally, light at wavelengths above 700 nm has a very low transmittance in sea water. Therefore, water appears dark and the land appears bright [8]. For this reason, the infrared band (845–885 nm) is used for distinguishing water from land. For bathymetry values from our products, we must follow the flowchart of this study can be seen in (Fig. 3).

2.5 *Ratio Transform Algorithm*

Stumpf et al. developed a ratio transform algorithm to determine the bathymetry of the ocean [9, 10]. The ratio transform algorithm makes use of two bands in order to reduce the number of parameters to determine the bathymetry of the ocean. The algorithm is capable of retrieving depths greater than 25 m in clear water coastal environments and it is also can determine the depth of the turbid coastal environment efficiently. The following equation is used to determine the depth of the near shore coastal area:

$$z = m_1\left(\frac{\ln(L_{obs}(Band_i))}{\ln\left(L_{obs}\left(Band_j\right)\right)}\right) - m_0 \tag{1}$$

where, L_{obs} are observed radiance of bands, m_1 and m_0 are offset and gain which are determined empirically, and i refers to blue band and j refers to green band, z is depth in meter. The advantage of ratio transform algorithm is that it takes very few parameters to determine the depth and it is also easier for marine surveyors to put it into practice.

3 Results and Discussion

The processing was done in ArcMap 10.8 using the same Control points and after the processing is done the results are shown in Fig. 4—14 using all data of 1131 points. The depth range considered in the present work is up to 20 m. Based on the field data, the depth values have been obtained for depth ranging from 1 to 20 m, and each depth series has fractional values ranging from 2 to 4 up to 20 m.

3.1 SDB Estimation from Landsat 8 Imagery

For each depth value of in-situ data, corresponding pixel value from satellite derived bathymetry (SDB) image is obtained and averaged and the same process is continued for obtaining the remaining depth values. In order to validate the model, the statistical index (R^2) is computed between the algorithm derived value and the hydrographic chart sounding. Figure 4—6 shows R^2 of 0.723 between the satellite derived bathymetry value versus in-situ data value (Z).

In order to compare the estimated (derived) bathymetry with the in-situ data. Figure 4—6 shows a good correlation between SDB from Landsat 8 and in-situ data. The noise was observed between depths 0 to 5m and that might be due to the wave collapse on the shoreline and the break waters and rocks which are located close to the shoreline.

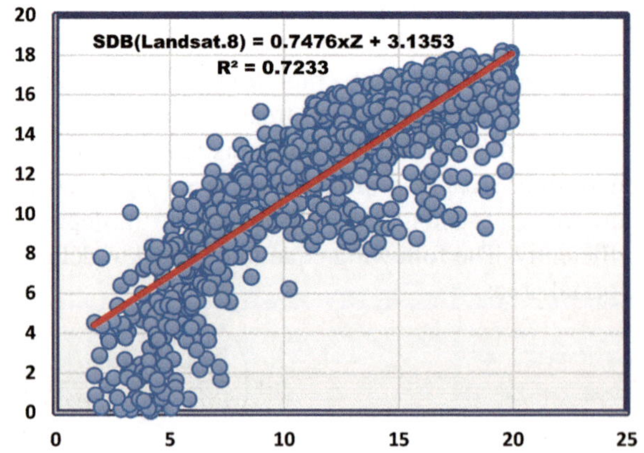

Fig. 4 Correlation between SDB from Landsat 8 and in-situ bathymetric data

3.2 *SDB Estimation from Sentinel 2 Satellite Imagery*

Several satellite images have been processed to select the best Sentinel-2 image that was captured in 24-02-2020 and used in the present study. The coordinates of each point have been loaded from in-situ depth to a text file and then imported in SNAP as pins, this procedure allowed us to extract the SDB values from Sentinel 2 corresponding to each of the 1131 points (Fig. 5). Later on, the points were plotted in Excel and the produced results are shown in Fig. 4—7.

The presented results registered from using Sentinel-2 data indicate strong correlation between SDB and the sounding data. Figure 4—14 shows $\mathbf{R}^2$ of 0.7892 between the satellite derived bathymetry value versus in-situ data value. The result shows a root mean square error (RMSE) of 1.982 and a mean absolute error (MAE) of 1.047 (see Table 1). After comparing the obtained coefficient of determination R^2 of Landsat-8 image with sentinel-2, we find a clear improvement in R^2 of Sentinel-2 image with 0.789 comparing to 0.723 for the Landsat-8 image. Therefore, Sentinel-2 provided the best correlation results (Table 1).

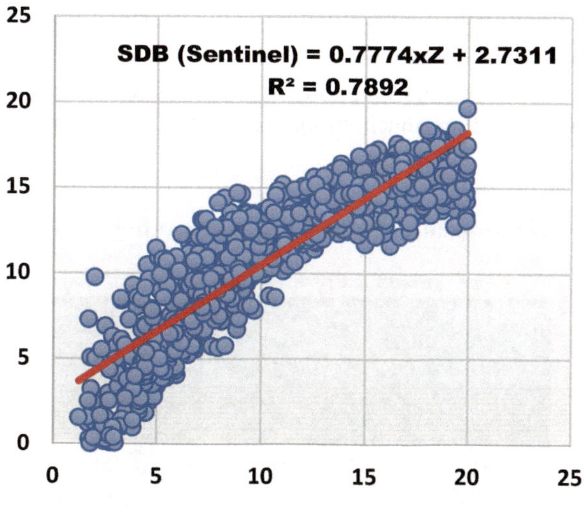

Fig. 5 Correlation between in-situ and SDB depths obtained from sentinel 2 image sensed of 24/02/2020

Table 1 Comparison the results obtained between Landsat 8 image and sentinel 2

Satellite	Coefficient of determination R^2	RMSE	MAE
Landsat 8	0.7233	2.176	4.735
Sentinel 2	0.7892	1.982	3.930

3.3 Mapping

The in-situ echosounder measurements have received an interpolation analysis using ArcMap 10.8 to generate a 2D interpolation map of the bathymetric data (Fig. 4—2).

From Fig. 6, the graduation in colors from green to blue represents the changes in bathymetric depths. What can be noticed at first glance is a rapid change in the depth of the sea, especially on the western side of the coast. The Fig. 4—2 clearly shows that big depths greater than 5 m are more dangerous in the western part of the coast. However, in the eastern part of the coast depths are relatively less risky, where the depth increases gradually to be approximately 20 m. The result of image processing of sentinel 2 was from 0.08 to 20 m. Then the value of the depth was grouped into 7 classes from 0.86 to 20 m to be used as depth map (Fig. 4—15). The result of image processing of sentinel 2 was from 0.08 to 20 m. Then the value of the depth was grouped into 7 classes from 0.86 to 20 m to be used as depth map (Fig. 7).

Figure 8 shows a more detailed view of the calculated depths, on Landsat 8 image (left) and on Sentinel 2 image (right). Figure 8 shows that the depths shown on the right-hand side image are much better because Sentinel 2 resolution is three times better than Landsat 8 resolution.

To make comparison easier we loaded the images next to each other. The resulted map is shown in Fig. 9.

From Fig. 9, Comparing to Landsat image, it can be clearly seen that results from Sentinel-2 are much similar to the bathymetric map of the echosounder data. The

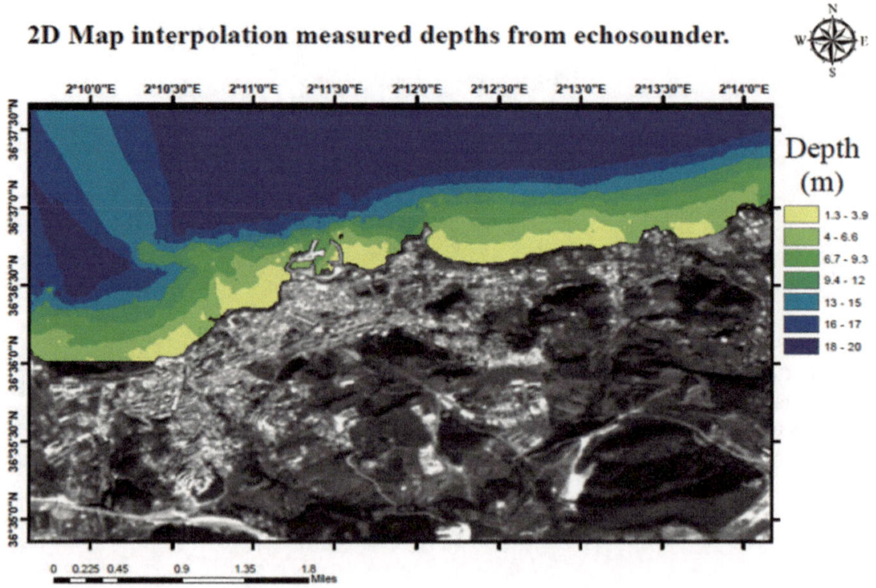

Fig. 6 2D interpolation map of the bathymetric echosounder in-situ data

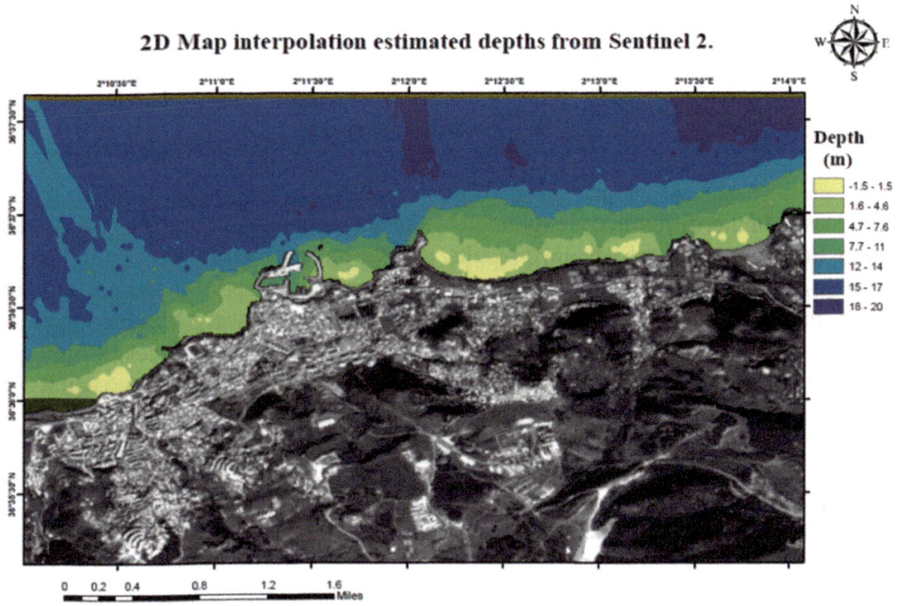

Fig. 7 2D Map interpolation estimated depths from Sentinel 2

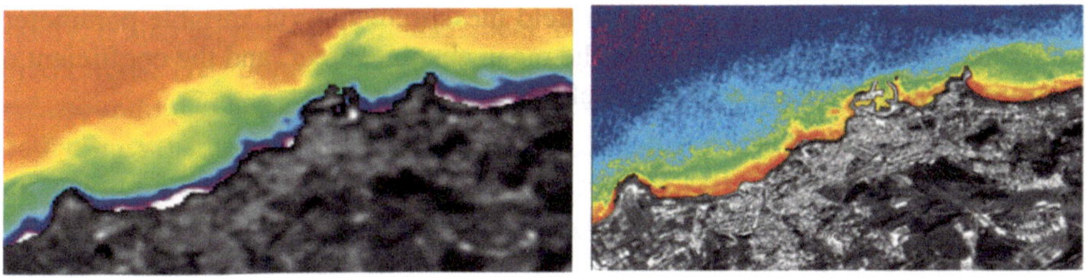

Fig. 8 More detailed view of satellite-derived bathymetry obtained from Landsat 8 (left) and from Sentinel 2 (right)

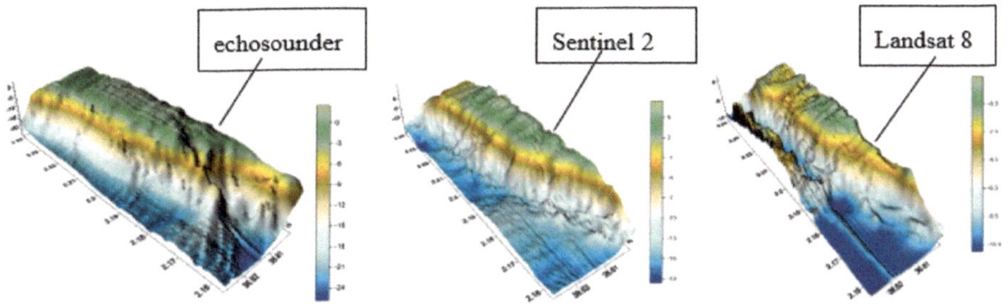

Fig. 9 3D view in Surfer echosounder in the (left) and satellite-derived bathymetry obtained from Sentinel 2 (center) and Landsat 8 (right)

1 Introduction

Achieving climate neutrality is the major goal of international organizations and governments around the world, this goal can only be achieved by establishing a circular economy based on minimizing the intensive exploitation of resources in various economic sectors [1]. In this context, of the sustainable economy, scientists have recently been focusing on the valorization of waste and natural resources in manufacturing and treatment processes.

As part of sustainability, water resources are crucial to development, human health and a basic principle for ensuring the vitality of all living beings. Hence, in response to the escalating issue of water contamination the development of biobased, low-cost and sustainable materials has emerged as a necessity to replace chemical products which are often detrimental and unsafe for human health, so the use of natural materials represents a promising solution for the remediation of emergent contaminants, including pharmaceuticals, dyes and heavy metals [2].

Adsorption remain the most studied processes for water remediation due to its numerous advantages such as inexpensive, simple to operate, prevents secondary contamination. However, industrial textile wastewater commonly contain a variety of dyes. the best approach for treating this type of effluent is to remove all dye molecules simultaneously by non-selective adsorption [3, 4].

Biochar is one of the most effective and promising materials which is characterized by its porous structure and derived from natural biomass. Activated carbon (AC) and biomass-based biochar are commonly used to decolorize wastewater. especially in adsorption process [5]. Biochar is a form of charcoal that is produced by pyrolysis of organic materials, such as crop residues, agricultural wastes and wood, in a oxygen limited atmosphere resulting in high specific surface area and low production costs. Moreover, biochar can easily be removed through simple coagulation sedimentation or filtration, which further illustrates biochar as a promising alternative and green technology for water purification [6].

Opuntia ficus-indica (OFI) is a plant native to Mexico that can grow in various soil types and withstand with very critical climatic conditions like drought, salinity and high temperatures. that is why it is widely cultivated in the Arab region. it has been used for various applications and in different industries like cosmetics, food, pharmaceuticals and others. it can also play an important role in restoring ecosystems and resilience against desertification [7]. The quantity of dry matter per hectare is estimated at 20 tons per year. OFI cladodes, in addition to its use in food and feed industry, it is also utilized as bio-coagulants, bio-flocculants and biosorbents in several research studies [8].

In the current study, the pyrolysis process was optimized for the production of highly effective adsorbent for both anionic MO and cationic MG dye. The biochar produced at optimal conditions were then used to investigate the impact of adsorption parameters such as adsorbent dose, solution pH and reaction duration in the adsorption efficiency of both MG and MO dyes. In addition, the adsorption process was studied using different kinetic and isotherm models. Therefore, the detailed analysis presented in this study provides a solid support for future research on biochar production and the use of *Opuntia ficus-indica* in water treatment processes.

2 Experimental Procedure

2.1 Chemical Reagents

All chemicals procured were used without modification and met standard purity specifications. Malachite green (MG) (CAS 2437-29-8), Methyl orange (MO) (CAS 547-58-0), Hydrochloric acid solution 37% (CAS 7647-01-0), Sodium hydroxide (CAS 1310-73-2).

2.2 Adsorbent Preparation

OFI cladodes were collected from Oued Al Athmania in the east of Algeria, the cladodes were thoroughly cleaned using tap water and then distilled water, cut to small 1 cm/1 cm pieces, dried at 80 °C during 24 h, crushed and sieved to a size of micrometers.

OFI powder was then placed in a limited oxygen atmosphere and pyrolyzed at temperatures ranges 300–700 °C during 1 h. in order to optimize the pyrolysis temperature.

After setting the optimum pyrolysis temperature, the duration was optimized by varying the carbonization time between 5 min and 2 h.

2.3 Biochar Characterization

Point of Zero Charge (pHpzc)

The point of zero charge (pHpzc) of OFI derived biochar was determined following a reported method (9). Briefly, 0.1 g of the adsorbent was added to 12 flasks containing 100 ml of 0.1 M NaCl solution with pH ranging from 2 to 12 which was adjusted with 0.1 M HCl and NaOH solutions. After shaking using magnetic stirrer during 48 h, the solution was filtered and the final pH were measured, and a plot of the initial pH against ΔpH (pH initial–pH final) was constructed to ascertain the point corresponding to the zero charge on the surface of the adsorbent.

adsorption efficiency of MG cationic dye is 30% higher than that of MO anionic dye which shows that OFI biochar can eliminate both dyes, but with a higher efficiency for the cationic dye.

Effect of pH

pH is considered as the most critical factor in adsorption process because the functional groups on the surface of the adsorbent and the degree of ionization can be influenced by solution pH [17]. The effect of pH variation was studied at constant initial concentration of 200 mg/L, constant time of 60 min, adsorbent dose of 1 g/L, variable pH [3, 5, 7, 9, 11].

The obtained results shown in Fig. 5b demonstrates a slight variation in the removal of MG cationic dye with a negligible decrease in removal efficiency in acidic solution pH, while the variation in pH for the anionic dye has a considerable effect on the removal efficiency and was promoted at acidic pH.

According to our findings, the adsorption in this case is a chemisorption-based process because the two dyes used in this study have shown the opposite effect with the same adsorbent and in the same conditions.

Effect of Adsorbent Dose

The effect of biochar dose was conducted using (0.25, 0.5, 1, 1.5, 2 g/L) at natural pH and initial dye concentration of 200 mg/L.

Our findings shown in Fig. 5c revealed that the adsorption capacity increased by increasing to biochar dose, this increase can be attributed to the a high number of surface area available for dye adsorption, the optimal biochar dose can be determined at 1 g/L because at this adsorbent amount the MG removal efficiency reached a constant level and a slight increase was noted in MO adsorption efficiency, this can be explained by reaching the point where the number of binding sites corresponds to the number of dye species [18].

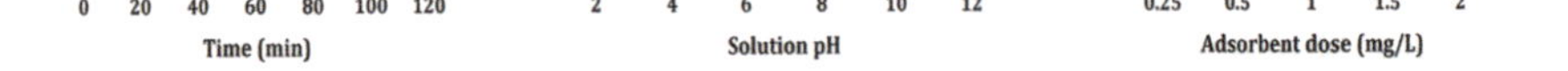

Fig. 5 **a** Contact time effect, **b** pH effect, **c** Adsorbent amount effect

3.3 Adsorption Isotherms

Langmuir, Freundlich and Temkin isotherm models provided by Eqs. (3), (4), (5) were employed to elucidate the relation between adsorption and dye concentration.

$$Q_e = \frac{Q_{\max} k_L C_e}{1 + k_L C_e} \tag{3}$$

$$Q_e = k_F C_e^n \tag{4}$$

$$Q_e = \frac{RT}{b} \ln k_T + \frac{RT}{b} \ln C_e \tag{5}$$

The isotherm model curves and fitting parameters of MG and MO on OFI biochar are shown in Fig. 6, (a) for MG and (b) for MO and Table 1.

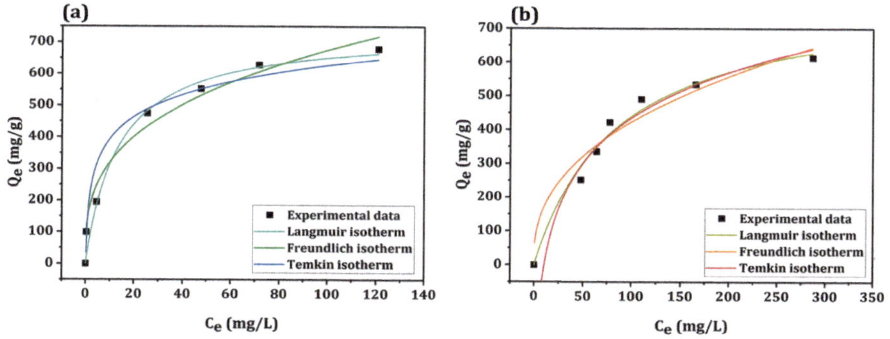

Fig. 6 Langmuir, Freundlich and Dubinin-Radushkevich isotherm models for **a** MG and **b** MO dye adsorption by OFI biochar

Table 1 Isotherm parameters for MG and MO adsorption

Isotherm model	Parameters	MG dye	MO dye
Langmuir	q_m (mg/g)	738.310	811.780
	K_L (L/mg)	0.074	0.012
	R_L	0.016–0.119	0.095–0.457
	R^2	0.982	0.984
Freundlich	K_f (L/mg)	152.630	68.089
	n	0.323	0.397
	R^2	0.976	0.890
Temkin	B_j	4.990	0.093
	K_T (L/mg)	100.934	194.917
	R^2	0.953	0.980

3. Sun Z, Zhao X, Chen Z, Jin Y, Ren F, Ren P (2024) Facile synthesis of carbon nitride-doped cellulose aerogels with high selective adsorption of cationic dyes from wastewater. Ind Crops Prod 221:119384
4. Nguyen VT, Ha LQ, Nguyen TDL, Ly PH, Nguyen DM, Hoang D (2022) Nanocellulose and graphene oxide aerogels for adsorption and removal methylene blue from an aqueous Environment. ACS Omega. 7(1):1003–1013
5. Hashemi E, Norouzi MM, Sadeghi-Kiakhani M (2024) Magnetic biochar as a revolutionizing approach for diverse dye pollutants elimination: a comprehensive review. Environ Res 261:119548
6. Tan Q, Chen Z, Shen J, Yan P, Kang J, Wang B et al (2025) Preparation of biomass-derived porous biochar for efficient removal of iodized X-ray contrast media in water: adsorption performance and underlying mechanism. Separation and Purification Technol 354:128751
7. Alzharani AA (2024) Activated Opuntia ficus-indica as an efficient adsorbent for iodine and heavy metals Cr(III), Cu(II) and Ni(II) ions uptake for wastewater treatment. JUmm Al-Qura Univ Appll Sci [Internet]. 9 févr 2024 [cité 3 sept 2024]; Disponible sur: https://doi.org/10.1007/s43994-023-00120-y
8. Barbera M, Indelicato S, Bongiorno D, Censi V, Saiano F, Piazzese D (2023) Untreated Opuntia ficus indica for the Efficient Adsorption of Ni(II), Pb(II), Cu(II) and Cd(II) Ions from Water. Molecules. 28(9):3953
9. Paluch D, Bazan-Wozniak A, Nosal-Wiercińska A, Pietrzak R (2024) Efficient dye removal by biocarbon obtained by chemical recycling of waste from the herbal industry. Ind Crops Prod 220:119254
10. Cui S, Ke Y, Fu Q, Hough R, Zhang Z, Shen Z et al (2024) Optimization preparation of biochar from garden waste and quantitative analysis for Cd2+ adsorption mechanism in aqueous solution. Biomass Conv Bioref 14(12):12761–12773
11. Veiga PADS, Schultz J, Matos TTDS, Fornari MR, Costa TG, Meurer L et al (2020) Production of high-performance biochar using a simple and low-cost method: Optimization of pyrolysis parameters and evaluation for water treatment. J Analytical Appl Pyrolysis 148:104823
12. Yuan X, Wang Q, Wang Z, Wu S, Zhai Y, Zhang H et al (2023) Optimization of mixed-based biochar preparation process and adsorption performance of lead and cadmium. Sustainability 15(15):11579
13. Doondani P, Gomase V, Saravanan D, Jugade RM (2022) Chitosan coated cotton-straw-biochar as an admirable adsorbent for reactive red dye. Results in Eng 15:100515
14. Jafari Zadegan MS, Moosaei R, Choopani L, Salehi MM, Maleki A, Zare EN (2024) Remediation of Safranin-O and acid fuchsin by using Ti_3C_2 MXene /rGo-Cu_2O nanocomposite: preparation, characterization, isotherm, kinetic and thermodynamic studies. Environ Res 258:119469
15. Sun X, Yin S, Zhao L, Yang W, You Y (2024) Adsorption properties of methylene blue and Cu(II) on magnetically oxidized tannic acid cross-linked carboxymethyl chitosan gels. Int J Biol Macromol 278:134709
16. Naribi Z, Nechchadi B, Mouratib R, Ghazzaf H, Salhi A, El Krati M et al (2024) Activated carbon from hydrothermal carbonization of ice cream factory wastewater with H_3PO_4 activation. Synthesis, characterisation and adsorption properties. Sustainable Chem Pharmacy 41:101697
17. Khandelwal R, Keelka S, Jain N, Jain P, Kumar Sharma M, Kaushik P (2024) Biosorption of arsenic (III) from aqueous solution using calcium alginate immobilized dead biomass of Acinetobacter sp. strain Sp2b. Sci Rep. 14(1):9972
18. Abd-Elhamid AI, Mostafa AG, Nayl AA, Akl MA (2024) Novel sulfonic groups grafted sugarcane bagasse biosorbent for efficient removal of cationic dyes from wastewater. Sci Rep 14(1):19129

The Water Future of the Kingdom of Saudi Arabia: Short- and Medium-Term Challenges

Christopher M. Fellows and Ali A. Al-Hamzah

Abstract Over the next few decades, over-use of groundwater is likely to place increasing pressure on the agricultural sector in the Kingdom of Saudi Arabia. This will necessitate increasing application of brackish water desalination, with associated challenges in maintaining water quality and in brine disposal. While continuing technical advances have secured low-cost water for domestic and industrial use, a different set of challenges are arising due to the shift from thermal desalination to reverse osmosis as the principal desalination method. The higher total dissolved solids content in water produced by reverse osmosis presents quality challenges related to corrosivity, disinfection by-products, and boron. These can all be addressed, but in many cases significant additional expenditure will be required. In this report we consider potential future water quality issues through the light of the results of an extensive water quality assessment carried out across the Kingdom of Saudi Arabia in 2020–2022.

Keywords Desalination · Groundwater · Boron · Bromate · Heavy metals · Trihalomethanes

1 Introduction

The long term prospective for water supply in Saudi Arabia is positive. We know from multiple paleoclimatic evidences that warmer global temperatures correspond to increased rainfall in the Arabian peninsula, especially in the south (Fig. 1) [1]. In the shorter term, however, Saudi Arabia is dependent on 'fossil' groundwater sources that are being consumed many times faster than they are being replenished. Most of this water is used for agriculture and in areas remote from any alternative

C. M. Fellows (✉) · A. A. Al-Hamzah
Water Technologies Innovation Institute and Research Advancement, Saudi Water Authority, Jubail, Kingdom of Saudi Arabia
e-mail: cfellows@une.edu.au

School of Science and Technology, The University of New England, Armidale, Australia

S. Al-Naemi et al. (eds.), *Water and Food Security in the Face of Climate Change: Challenges and Opportunities for Resilience*, World Sustainability Series,
https://doi.org/10.1007/978-3-032-00098-9_63

water supplies. The location, nature, and intensity of agricultural activities will all need to change for the agricultural sector to be sustainable. This process is already all well underway, with a significant drop in the irrigated land area over the past three decades and a shift from broad acre agriculture with inefficient centre-pivot irrigation to more intensive horticulture [2]. While groundwater also contributes to domestic and industrial consumption, this is being phased out, with the majority of water consumed in the urban centres of Saudi Arabia being obtained by desalination of seawater. Given the size of the nation and the distance of many populated areas from the ocean, it is not practicable to achieve 100% provision of domestic consumption by seawater desalination. Over the next few decades therefore we can expect to see a continuing decline in quantity and quality of groundwater resources, with an increasing requirement for brackish water desalination for both agricultural and domestic consumption. The disposal of brine generated by brackish water desalination will require careful management to prevent further degradation of groundwater resources.

Historically, the relatively concentrated seawaters of the Red Sea and Arabian Gulf (c. 45,000 mg/L cf. 35,000 mg/L for standard seawater) have been desalinated by thermal processes [3]. Technical advances in Reverse Osmosis (SWRO) desalination and rising energy costs have made SWRO less expensive than thermal desalination, and the Kingdom of Saudi Arabia is in the process of switching over from thermal desalination to SWRO. Approximately half of the Kingdom's desalinated water is now provided by each process. This shift has important implications for the maintenance of water quality in the Kingdom: current membrane-based desalination processes are more vulnerable to changes in water quality and to supply chain disruptions and provide water at a higher total dissolved solids content than thermal processes [4].

Finally, the Kingdom of Saudi Arabia has limited surface water resources, primarily in the southeast of the country, and water impounded on non-permanent watercourses is currently applied for both agriculture and potable water supply. As

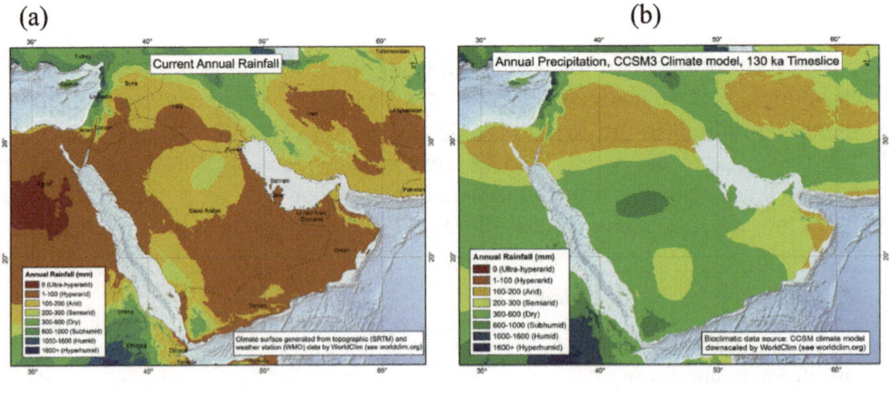

Fig. 1 Rainfall on the Arabian peninsula **a** currently; **b** at the time of the last interglacial climatic maximum [1]

the long-term transition to a wetter climate begins, more frequent and intense rain events will lead to a need for additional dams for flood control and improvement of the existing infrastructure, leading to a greater availability of relatively low-quality surface water [5].

In 2020–2022 we coordinated a major survey of water quality across the production and transmission system of the Saline Water Conversion Corporation (SWCC; now the Saudi Water Authority, SWA) [6]. The SWCC produced at the time approximately 6 million m^3 of desalinated water while transporting approximately 11 million m^3 through its network of pipelines (including surface water, ground water, and desalinated water from private producers). This paper will discuss the implications of the results of that survey in relation to the expected trends outlined above.

2 Methodology

2.1 Sampling

Water samples were collected from late 2019 through 2022 from 72 sites throughout the Kingdom of Saudi Arabia (Fig. 2). Full details of the sampling program can be found in our previous work [6].

For the majority of the samples, analysis of the parameters of interest was carried out within two weeks of collection in the laboratories of the Water Technologies Innovation Institute and Research Advancement (WTIIRA) of the SWA in Al Jubail, Kingdom of Saudi Arabia, using standard methods as follows.

Total dissolved solids (TDS): ASTM D 1193 (Gravimetry). For a number of samples collected in 2022 for which TDS measurements were not made, estimates of TDS are reported using cation concentrations determined by ICP-OES

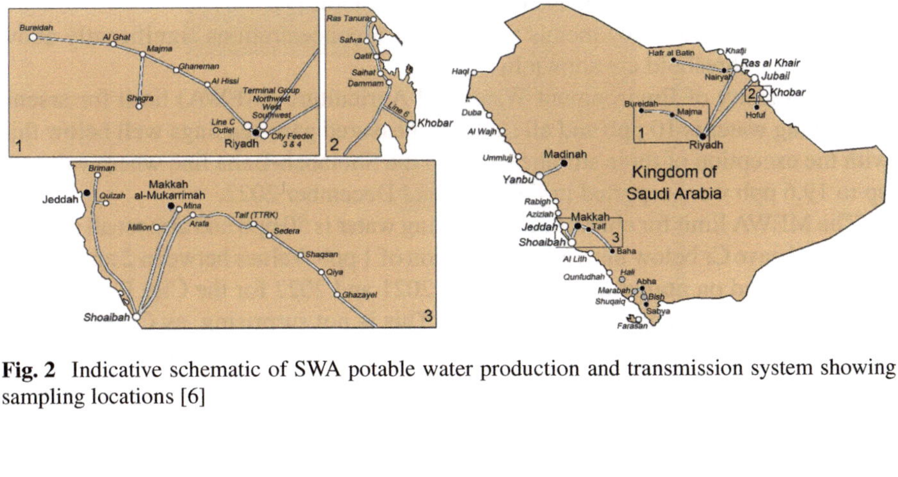

Fig. 2 Indicative schematic of SWA potable water production and transmission system showing sampling locations [6]

Copper values were also always well below the MEWA limit of 2 ppm, with a maximum value of 340 ppb reported anywhere in the system.

The MEWA limit for iron in drinking water is 300 ppb. Values above this were reported only at one occasion at the Qunfudah SWRO plant (550 ppb) and on a number of occasions for the Riyadh City Feeder 3&4, a ground water source, in June-November 2022.

The MEWA limit for lead in drinking water is 10 ppb. Only one sample measured gave a lead concentration above the detection limit of 1 ppb, and at 2 ppb this was far below the MEWA limit of 10 ppb for Pb.

The maximum manganese value (8 ppb), well below the MEWA limit of 400 ppb, was reported from a surface water (Marabah dam).

Insignificant levels of nickel were reported within the SWA system, well below the MEWA limit of 70 ppb, with a maximum value of 59 ppb reported on one occasion at Shoaibah Phase 2 with almost all values < 5 ppb.

The MEWA limit for selenium in drinking water is 10 ppb. All samples measured gave Se below the detection limits of 1 ppb with the exception of a cluster of results reported from the City Feeder 3 and 4 measuring points (groundwater), where 1.8–2.2 ppb was reported in January–February 2022, and a single measurement at Marabah (surface water) of 1.4 ppb.

All samples measured gave cadmium below the limit of detection of 1 ppb and mercury below the detection limits of 0.15 ppb (with AAS, used from January to June 2021) or 1 ppb (with ICP-OES, used from July 2021 onwards). It should be noted that the levels of heavy metals found in this study for the water supplied to Riyadh are consistently below the levels reported at households in Riyadh in previous studies [9, 10]. This suggests progressive improvement in water quality over time and/or a major contribution of corrosion within the municipal delivery system to the heavy metal load in water delivered to the consumer.

Arsenic, the only species found at any site at levels above the mandated standard, appears to be associated with surface water use. It may become a more significant issue if greater use is made of surface water resources and appropriate monitoring and mitigation procedures will need to be in place.

Elevated levels of chromium, iron, and selenium are primarily associated with groundwater usage. While major urban centres are transitioning away from groundwater, it is likely that more remote areas will remain dependent on this water source and with the degradation of groundwater resources, concentration of heavy metals in these waters is likely to increase. Risks for these communities in the short to medium term could therefore arise from delays in implementing brackish water desalination where groundwater quality is declining, and from failures in installed brackish water desalination systems. Corrosion of metallic components is also likely to be responsible for some instances of increased iron, copper, and nickel in produced water: trends in water corrosivity are discussed later on in this report.

Deterioration in groundwater quality also presents a potential hazard from radioisotopes, which were not considered in the 2019–2022 survey. While some surveys of groundwater quality across the Kingdom of Saudi Arabia have found no radioactive hazards of concern [11] others have found that some sites have groundwater with alpha and beta emission above international guidelines [12]. It is therefore necessary to introduce a monitoring program for radioisotopes in groundwater in identified susceptible areas.

3.2 Boron

An upper limit of 2.4 ppm boron is mandated for potable water in the Kingdom of Saudi Arabia to give a wide margin of safety to avoid potential adverse effects on human health [13]. Boron is present at a concentration of upwards of 5 ppm in Red Sea and Arabian Gulf waters and largely exists as the uncharged boric acid (H_3BO_3) species which is unaffected by RO membranes. Thus, if the pH of the permeate is not adjusted to near the upper range of the quality limit (~8.5) after the initial SWRO process and subjected to an additional brackish water reverse osmosis (BWRO) process, it is probably that boron levels above the limit will be delivered by a SWRO system.

Many crops show declines in yield and significant toxicity effects at boron concentrations well below 2.4 ppm and an optimal concentration of no more than 0.3 ppm is recommended for agricultural use [14]. This applies a significant constraint on the expansion of desalination for agricultural use.

Boron levels measured at representative locations in the SWA system are shown in Fig. 4. It can be seen that a number of large SWRO plants consistently deliver water with boron levels in excess of the 2.4 ppm limit, presumably due to avoidance of a pH-adjusted BWRO stage. Jeddah RO3 is the only currently operating plant producing water with an average level above the threshold value over the time period of the study (2.7 ppm). While the Ras al Khair results are well below 2.4 ppm, analysis was performed on combined water produced from approximately 30% SWRO and 70% MSF, suggesting that similarly high levels of boron are being passed by the Ras al Khair SWRO plant. Values reported above 1 ppm at the Yanbu Phase 2 plant suggest that the Yanbu RO plant is also not consistently producing water with acceptable boron levels.

Boron values in excess of 2.4 ppm are reported on a few occasions at the smaller SWRO plants at Farasan and Qunfudah. The majority of the smaller plants in Fig. 4b, c were built to the same plan and introduced at the same time to replace thermal (Multiple Effect Distillation, MED) plants on the Red Sea. It is most likely that the plants with low boron values are adjusting the pH in the second stage RO to ensure boron is present as dihydrogenborate ion, while those with high boron levels are not. The similar trend over the year for the plants with higher boron concentrations indicates the strong seasonal effect of temperature on boron passage [15]. RO membranes have lower selectivity at higher temperatures, which impacts performance of SWRO plants on the Red Sea and Arabian Gulf [16, 17]. Khafji, which is located in waters which have a very large change in temperature from winter to summer (~17–37 °C) has low boron levels in the cooler months, and the plants in the southern part of the

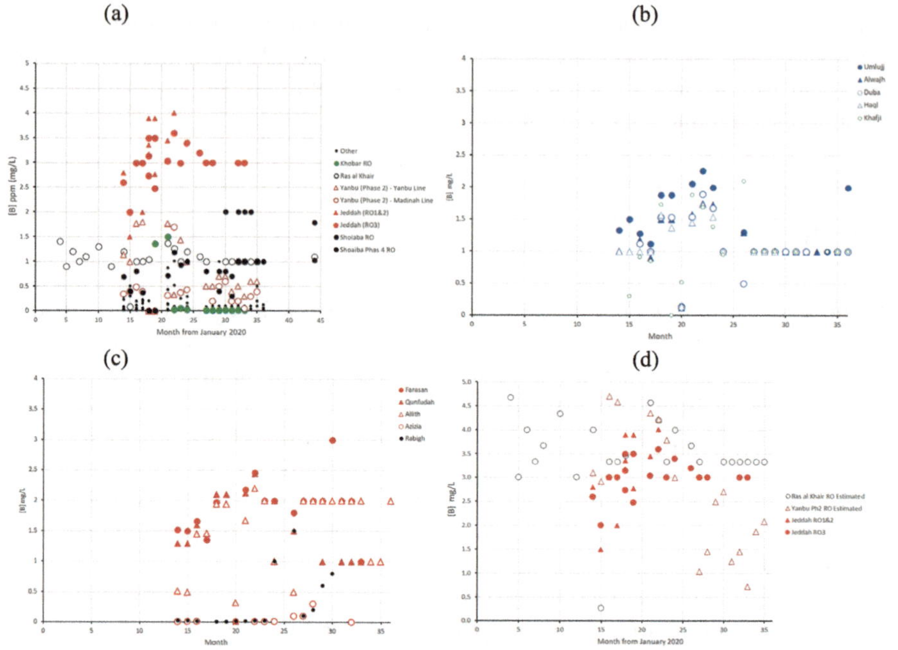

Fig. 4 **a** Boron Levels for large SWA desalination plants, 2020–2023; **b**, **c** Boron levels for small SWA desalination plants, 2020–2022. **d** Large RO plants giving elevated boron levels, 2020–2022 [6]

Red Sea (Farasan, Qunfudah, Allith) with maximum water temperatures of 33–34 °C have boron levels greater than plants in the northern part of the Red Sea (Haql, Duba, Alwajh) with maximum water temperatures of 28–30 °C [18]. It is thus evident that any increase in seawater temperatures will lead to an increase in boron concentrations in water delivered to consumers if no changes are made to the production system.

Within the distribution system, mixing of water from various sources ensures that the water delivered to the consumer is below the mandated limit in almost all cases. The elevated levels coming from Ras al Khair can be clearly traced through the levels in water delivered to Riyadh and Hafr al Batin; these can be used together with the Ras al Khair values and the relative capacity of the thermal and RO units to estimate the level of boron in the water produced by the Ras al Khair SWRO plant (Fig. 4d). From the relative capacity of the thermal and RO units at Yanbu phase 2 the boron production at that plant can also be estimated. It is clear that most major SWRO plants apparently produce water with levels of boron above the mandated limit at almost all times, and quality to the consumer is currently met only by mixing with water produced by thermal desalination. Note that the Jeddah RO1&2 plant was decommissioned in 2020 and no longer provides water to the SWA system.

It is apparent that RO plants across the SWA system are producing water with excessive boron levels, a trend that becomes more pronounced in warmer months.

Addressing these elevated boron levels by appropriate changes to operations should be a priority to ensure that the water delivered to consumers meets quality standards as a the SWA system shifts to a greater dependence on SWRO. Both the additional capital cost of a second-stage BWRO system to remove boron from desalinated water and the additional operating expenditure of chemicals for pH adjustment will impact the cost of water for agricultural use. As agricultural water use is extremely price sensitive, pursuing alternative technologies for producing low-boron water should be a high priority for the water and food security of Saudi Arabia. Restricting agriculture with desalinated water to crops that show a high boron tolerance is another mitigation strategy, though this is not optimal.

3.3 Bromate

Bromate is a toxic disinfection by-product which has been found to be carcinogenic in animal studies [19]. The bromate limit specified by MEWA regulations and SWA guidelines is 10 ppb. While bromate is traditionally associated with ozonation of surface waters and most research on its control is done within that context [20], it has been reported in levels of concern in some seawater desalination systems. For example, a recent survey of tap water quality in Kuwait which found bromate levels of between 10.7 and 50.3 ppb at all sites with an overall concentration of 19.6 ppb of bromate in drinking water [21]. Bromate production is encouraged by higher pH, higher T, the presence of catalytic amounts of transition metals such as copper, and higher concentrations of bromide and oxidising agent [22, 23]. n water produced in SWA desalination plants in 2020–2021, the target was exceeded several times in 2020, while only one instance of the target being exceeded was recorded in 2021 (Fig. 5).

The dependence of bromate levels due to changes in temperature is likely to be more prominent than apparent in Fig. 5, as the minimum bromate levels are dominated by points recorded as below the 2 ppb limit of detection. While bromate formation is expected to be greater at elevated pH, no clear correlation between plant produce water pH and bromate concentration was apparent in the data.

Excursions above 10 ppm bromate were reported at a number of locations across the SWA transmission system over the period of the study. Values in excess of the MEWA limit were reported in the pipeline from Riyadh to Buraidah (14 instances), from Yanbu to Madinah-al-Munawarah (6 instances) from Khobar to Ras Tanura (3 instances) and Shuqaiq to Abha (1 instance). Excessive bromate levels were also reported in a storage tank at Yanbu City for several months in early 2021, where the single highest value of 29.4 ppb was reported, which was associated with a period of abnormally high pH (8.5–8.8). Values above 10 ppb bromate were also reported from tanks in Yanbu and Medinah in September–October 2020, which was not a time associated with elevated pH. It seems likely that the elevated bromate levels in the Riyadh-Buraidah line can be attributed at least partly to the higher pH values in this line which in turn depends on the relatively high pH water produced at the Jubail

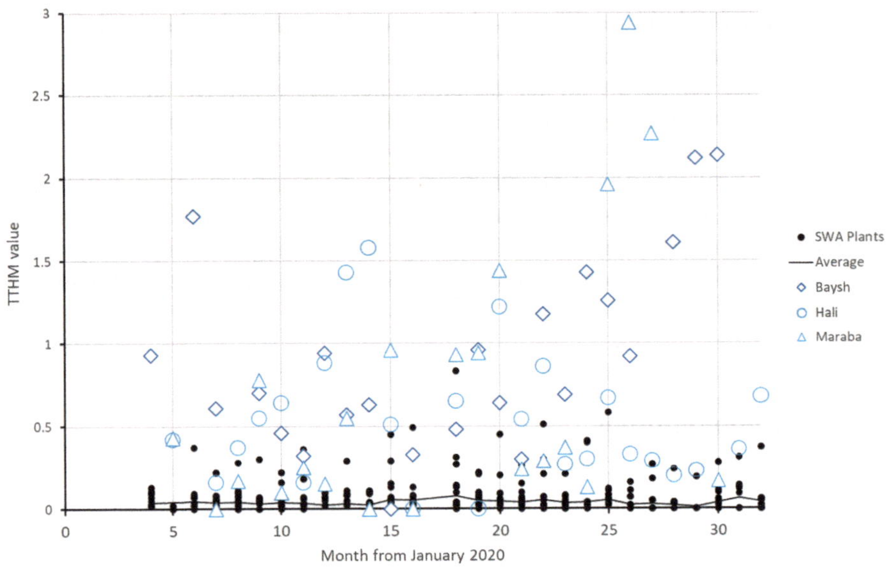

Fig. 7 TTHM values obtained from desalination plant and dam product waters, 2020–2022 [6]

4 Conclusion

In the short to medium term, Saudi Arabia faces challenges in maintaining water quality in the face of declining groundwater quality, increasing reliance on reverse osmosis rather than thermal desalination to supply domestic water, and a potential increased contribution of surface water to the national water budget. These challenges include potentially greater risks of heavy metal contamination, higher boron levels, increased corrosivity, and increased levels of halogenated organic by-products. These issues can all be addressed, but not without associated costs. It is necessary to expand research efforts now to determine the cost and timeframe of mitigation strategies. A cluster of emerging issues are related to the higher TDS obtained with SWRO compared to the historical thermal desalination processes. This higher TDS is not constant, but rises with temperature and with membrane lifetime. Rejection performance of current RO membranes is significantly worse at prevailing summer seawater temperatures of > 35 °C. Development of membranes that can operate more effectively under these conditions, already a priority for the industry, will become more urgent if these warm water temperatures persist for a greater fraction of the year. Development of membranes that are more robust to oxidative degradation, which will extend the period of time during which they give optimal rejection, is even more important for technical control of water quality challenges at the source. Other high priorities for ensuring water quality in the face of environmental challenges are extending the use of brackish water desalination in inland areas, (re)implementing pH adjustment and second stage RO for removal of boron, and pivoting away from chlorine-based disinfection to less problematic disinfectants such as chlorine dioxide and chloramine.

References

1. Jennings RP, Singarayer J, Stone EJ, Krebs-Kanzow U, Khon V, Nisancioglu KH, Pfeiffer M, Zhang X, Parker A, Parton A, Groucutt HS, White TS, Drake NA, Petraglia MD (2015) The greening of Arabia: Multiple opportunities for human occupation of the Arabian Peninsula during the Late Pleistocene inferred from an ensemble of climate model simulations'. Quatern Int 382:181–199
2. Ouda OKM (2014) Impacts of agricultural policy on irrigation water demand: a case study of Saudi Arabia. Int J Water Resour Dev 30(2):282–292
3. Fellows CM, Al-Hamzah A (2015) Thermal desalination: current challenges. In: Amjad Z, Demadis K (eds) Mineral scales and deposits: scientific and technological approaches. Elsevier, New York, pp 583–602
4. Fellows CM Polymer International, in press
5. Almazroui M (2020) Rainfall trends and extremes in Saudi Arabia in recent decades, Atmosphere 11(9):964
6. Al-Hamzah AA, Fellows CM (2024) Drinking water quality in the Kingdom of Saudi Arabia. Water 16(13):1810
7. Al-Hamzah AA, Aldowis GAH, Mahmoodur Rahman M, Fellows CM, Kurup PK, Alfefei JHM (2022) Monitoring of inorganic and organic pollutants in the desalinated water from thermal desalination plants, Desalination and Water Treatment 264:247–264
8. Alabdula'aly AI, Al Zarah AI, Khan MA (2011) Assessment of trace metals in groundwater sources used for drinking purposes in Riyadh Region. Int J Water Resources and Arid Environ 1(1):5–9
9. Al-Omran A, Al-Barakah F, Altuquq A, Aly A, Nadeem M (2015) Drinking water quality assessment and water quality index of Riyadh Saudi Arabia. Water Qual Res J 50(3):287–296
10. El-Rehaili A, Misbahuddin M (1995) Levels of trace metals in Riyadh drinking water at the consumer taps. J King Saud Univ 7(1):1–23
11. Alotaibi MF, Alharbi KN, Alosime EM, Alhawali LH, Albarqi MM, Alsulami RA (2024) Natural radioactivity in soil and water of Saudi Arabia: a mixed-studies review. J Radiation Res Appl Sci 17(2):100897
12. Almasoud FI, Ababneh ZQ, Alanazi YJ, Khandaker MU, Sayyed MI (2020) Assessment of radioactivity contents in bedrock groundwater samples from the northern region of Saudi Arabia. Chemosphere 242:125181
13. Nielsen FH (1997) Boron in human and animal nutrition. Plant Soil 193(1–2):199–208
14. Yermiyahu U, Tal A, Ben-Gal A, Bar-Tal A, Tarchitzky J, Lahav O (2007) Rethinking desalinated water quality and agriculture. Science 318(5852):920–921
15. Redondo J, Busch M, De Witte J-P (2003) Boron removal from seawater using FILMTEC™ high rejection SWRO membranes, Desalination 1256(1): 229–238
16. Jin X, Jawor A, Kim S, Hoek EMV (2009) Effects of feed water temperature on separation performance and organic fouling of brackish water RO membranes. Desalination 239(1):346–359
17. Kim J, Hong S (2023) Pilot study of emerging low-energy seawater reverse osmosis desalination technologies for high-salinity, high-temperature, and high-turbidity seawater. Desalination 565:116871
18. Fine M, Cinar M, Voolstra CR, Safa A, Rinkevich B, Laffoley D, Hilmi N, Allemand D (2019) Coral reefs of the Red Sea—Challenges and potential solutions. Reg Stud Marine Sci 25:100498
19. Kurokawa Y, Maekawa A, Takahashi M, Hayashi Y (1990) Toxicity and carcinogenicity of potassium bromate—a new renal carcinogen. Environ Health Perspect 87(7):309–335

20. Xie L, Shang C (2006) A review on bromate occurrence and removal strategies in water supply. Water Supply 6(6):131–136
21. Alomirah HF, Al-Zenki SF, Alaswad MC, Alruwaih NA, Wu Q, Kannan K (2020) Elevated concentrations of bromate in Drinking water and groundwater from Kuwait and associated exposure and health risks. Environ Res 181:108885
22. Alhamzah AA, Alofi AS, Abid AA, Fellows CM (2023) Control of bromate formation in desalinated seawater production and transmission with ammoniation. Water 15(21):3858
23. Liu C, von Gunten U, Croué J-P (2013) Chlorination of bromide-containing waters: enhanced bromate formation in the presence of synthetic metal oxides and deposits formed in drinking water distribution systems. Water Res 47(14):5307–5315
24. Al-Hamzah AA, Mahmoodur Rahman M, Kurup P, Barnawi A, Ghannam B, Musharraf I, Al Najjar F, Obeidallah A, Palmer N (2019) Use of chlorine dioxide as alternative to chlorination in reverse osmosis product water. Desalin Water Treat 163:57–66
25. Larson TE, Skold RV (1958) Laboratory studies relating mineral quality of water to corrosion of steel and cast iron. Corrosion 14:43–46
26. Shemer H, Hasson D, Semiat R, Priel M, Nadav N, Shulman A, Gelman E (2013) Remineralization of desalinated water by limestone dissolution with carbon dioxide. Desalination Water Treatment 51:877–881
27. Bajahlan AS, Wie J-M (2013) Post-treatment of desalinated water and water quality characteristics in Yanbu Industrial City'. Desalin Water Treat 51:1790–1803
28. Lahav O, Voutchkov N, Birnhack L (2012) Post-treatment of desalinated water. Desalination Publications. Hopkinton, Massachusetts, USA
29. Li X-F, Mitch WA (2018) Drinking water Disinfection Byproducts (DBPs) and human health effects: Multidisciplinary challenges and opportunities'. Environ Sci Technol 52(4):1681–1689
30. Sinha R, Gupta AK, Ghosal PS (2021) A review on trihalomethanes and haloacetic acids in drinking water: Global status, health impact, insights of control and removal technologies. J Environ Chem Eng 9(6):106511

From Desalination Waste to Wealth: Pioneering Brine Management for Sustainable Water Futures

Luma Abuorabi and Mohd Basheer

Abstract This research explores innovative brine management strategies for desalination plants, focusing on mineral recovery and renewable energy integration to enhance sustainability in water-scarce regions like Qatar. Mineral analysis of desalination brine revealed significant concentrations of magnesium (1290–1410 mg/L) and calcium (450–490 mg/L), demonstrating the feasibility of economically viable recovery. Using AI-powered optimization, the study improved mineral recovery efficiency by 10%, resulting in an estimated 5% reduction in energy consumption. Furthermore, integrating waste heat recovery for solar evaporation boosted system efficiency to 65–72%, depending on plant conditions, significantly lowering operational energy requirements. However, several challenges remain. Low lithium concentrations (0.30–0.40 mg/L) restrict the economic feasibility of lithium recovery, while seasonal variations in solar energy availability reduce energy efficiency during low-radiation periods. To overcome these limitations, future research should prioritize the development of advanced lithium extraction technologies, the exploration of hybrid renewable energy systems, and the expansion of AI models with larger datasets to improve adaptability and predictive accuracy. The findings have broad implications for global desalination practices, offering strategies to transform brine from a waste product into a valuable resource, reduce environmental impact, and enhance economic sustainability. This study provides actionable insights for policymakers and industry stakeholders, advocating for the adoption of eco-friendly brine management techniques to advance sustainable desalination.

Keywords Smart brine management · Mineral recovery · Renewable energy · AI optimization · Desalination sustainability

---

L. Abuorabi
Food and Agriculture Organization (FAO), Amman, Jordan
e-mail: luma.Abuorabi@fao.org

M. Basheer (✉)
University of Jordan, Amman, Jordan
e-mail: Mhm8220041@ju.edu.jo

S. Al-Naemi et al. (eds.), *Water and Food Security in the Face of Climate Change: Challenges and Opportunities for Resilience*, World Sustainability Series,
https://doi.org/10.1007/978-3-032-00098-9_64

1 Introduction

1.1 Overview of Current Research

In arid regions like Qatar, where freshwater is scarce, desalination has become an indispensable solution for ensuring water security. Among desalination technologies, reverse osmosis (RO) is widely used for producing potable water. However, desalination processes generate substantial amounts of brine, a byproduct with high salinity that poses significant environmental challenges, particularly for marine ecosystems [1]. Qatar's desalination plants produce over 200 million cubic meters of brine annually, and untreated brine discharge leads to elevated salinity levels, disrupting marine biodiversity and causing ecological imbalances [2].

Conventional brine disposal practices, which often involve discharging brine directly into the ocean, have raised environmental concerns, prompting the exploration of sustainable brine management solutions [3]. Researchers are increasingly focusing on innovative strategies like mineral recovery and Zero Liquid Discharge (ZLD) systems. These solutions aim to convert brine from a waste product into a source of economically valuable materials, such as magnesium, lithium, and calcium. However, these methods are constrained by high-energy demands and operational costs, which pose challenges, particularly in regions heavily reliant on desalination like Qatar. Given Qatar's near-total dependence on desalination for its freshwater needs, it serves as a pivotal case study for developing advanced, environmentally friendly brine management strategies. Success in implementing these approaches could provide a scalable model for sustainable brine management in arid regions globally, contributing to international efforts in climate adaptation and water security [4, 5].

1.2 Emerging Technologies

Advancements in brine management technologies have shifted the focus toward transforming brine into a valuable resource through mineral recovery, which offers both environmental and economic advantages. Materials such as lithium, a critical component in battery production, are particularly valuable, strengthening the economic feasibility of sustainable brine management solutions [6]. Additionally, waste heat recovery and renewable energy integration play vital roles in reducing the energy intensity of brine concentration processes. For example, solar thermal systems leverage Qatar's abundant sunlight, providing a sustainable energy source for desalination, while hybrid systems that combine solar and wind energy offer a reliable solution to seasonal variations in energy availability [7, 8].

Furthermore, the integration of real-time monitoring systems and AI-driven optimization enhances operational efficiency in desalination plants. Real-time data enables adaptive adjustments to brine management processes, while machine-learning algorithms predict mineral recovery rates and optimize resource utilization. These advancements align with the principles of a circular economy, where waste products are repurposed into valuable resources, reducing dependency on conventional disposal methods and supporting more sustainable desalination practices [9, 10].

1.3 Research Question and Motivation

This study addresses the critical research question: How can smart brine management strategies be implemented in Qatar to reduce environmental impacts and enhance desalination sustainability through mineral recovery and renewable energy integration? Qatar's heavy reliance on desalination, coupled with the significant environmental challenges posed by brine disposal, underscores the urgency of this research. By developing innovative and economically viable brine management methods, this study aligns with Qatar's National Vision 2030, contributing to long-term water security and environmental sustainability. Furthermore, these findings are relevant to other water-scarce regions worldwide, offering a scalable framework for sustainable desalination practices and advancing global efforts toward resource efficiency and climate resilience.

2 Methodology

2.1 Research Design and Approach

This research employs a comprehensive mixed-method approach, combining brine composition analysis, AI-driven optimization, and renewable energy integration to improve brine management in desalination facilities in Qatar. This approach allows for precise assessment of mineral recovery potential, operational efficiency, and sustainability improvements, aligning with Qatar's environmental goals.

2.2 Brine Composition Analysis

To evaluate the mineral recovery potential, brine samples were collected from Qatar's Ras Abu Fontas desalination plant. These samples were analyzed using Inductively Coupled Plasma Mass Spectrometry (ICP-MS), chosen for its high sensitivity and precision in detecting trace minerals. The analysis targeted key minerals of economic interest, including magnesium, lithium, and calcium, assessing their concentrations in the brine.

The Mineral Recovery Potential (MRP) was calculated to determine the economic feasibility of extraction using the following equation:

$$\text{MRP} = \frac{C_b * V_b}{C_s} \quad (1)$$

where:

C_b concentration of specific minerals in brine (mg/L),
V_b volume of brine produced (L),
C_s threshold concentration necessary for economically viable extraction [1, 2].

This calculation enabled a thorough evaluation of each mineral's recovery potential. While magnesium and calcium exhibited concentrations that supported economically viable recovery, lithium recovery presented unique challenges due to its lower concentrations and higher extraction costs. Given the high C_s values required for lithium profitability, further technological advancements are essential to enhance its economic feasibility.

2.3 Renewable Energy Integration

To reduce the energy demands of brine concentration, the study investigated the integration of renewable energy sources, focusing on waste heat recovery and solar energy. Waste heat generated during the desalination process was captured and repurposed for solar evaporation of brine, thereby enhancing energy efficiency. The system efficiency (η\etaη) was calculated using the equation:

$$\eta = \frac{\text{Quseful}}{\text{Qinput}} \quad (2)$$

where:

Q_{useful}: amount of heat recovered and effectively used for brine concentration (kJ)
Q_{input}: total heat available from the desalination process (kJ) [3]

In addition to waste heat, solar and wind energy were considered as complementary sources to reduce reliance on conventional power. Qatar's high solar irradiance provides an opportunity for solar thermal systems, which can deliver consistent energy during peak sun hours. However, seasonal variations in solar energy availability, such as reduced sunlight during winter months, impact energy input and system efficiency. To address these fluctuations, hybrid renewable systems that combine solar and wind energy are being explored. These systems offer a reliable energy supply throughout the year, mitigating the limitations of single-source renewable systems and ensuring uninterrupted brine management operations.

2.4 AI-Driven Optimization

An AI-driven optimization model was developed to enhance the efficiency of mineral recovery processes. The model utilized Random Forest algorithms, selected for their robustness in handling complex datasets and high predictive accuracy. Trained on over 1,000 historical data points from desalination plant operations, the model predicts brine composition and dynamically adjusts recovery processes in real time.

The accuracy of the model was evaluated using the error rate ER:

$$ER = \frac{1}{n}\sum_{i=1}^{n}(yi - y_i^{\wedge})^2 \quad (3)$$

where:

y_i: actual mineral concentration value in brine,

$y_i^{\wedge}$: predicted mineral concentration value by the model,

n: total number of data points [4].

By dynamically adjusting brine recovery parameters, the AI model achieved a 10% improvement in overall recovery efficiency. This optimization reduced operational variability and minimized resource wastage, showcasing the potential of AI in improving desalination processes. However, the model's performance was occasionally limited by the availability of diverse data under extreme operational conditions. Expanding the dataset to include a wider range of scenarios could further enhance the model's adaptability and predictive accuracy.

2.5 Terminology Definitions

Brine: A high-salinity byproduct produced during the desalination process, often viewed as waste but with potential for mineral recovery [5].

Mineral Recovery: The process of extracting valuable minerals, such as magnesium, lithium, and calcium, from desalination brine for economic benefit. Lithium extraction is particularly challenging due to its lower concentrations and higher costs [6].

Renewable Energy Integration: The use of renewable energy sources, such as solar, wind, and waste heat, to enhance brine concentration processes, improve operational efficiency, and reduce environmental impact [7].

Table 1 Mineral concentrations in brine samples

Plant	Magnesium (mg/L)	Lithium (mg/L)	Calcium (mg/L)
Ras Abu Fontas	1,320	0.35	450
Umm Al Houl	1,410	0.40	490
Dukhan	1,290	0.30	460

3 Results and Discussion

3.1 Mineral Recovery from Brine

The brine composition analysis (see Table 1) highlights significant potential for mineral recovery, particularly for magnesium and calcium, with concentrations ranging from 1290 to 1410 mg/L and 450 to 490 mg/L, respectively. While lithium levels were comparatively lower (0.30 to 0.40 mg/L), recovery remains feasible given the high volume of brine produced at Qatar's desalination plants. This underscores brine's potential as a valuable resource, shifting its perception from waste to an economic asset.

Umm Al Houl demonstrated the highest mineral recovery efficiency due to its optimized heat recovery infrastructure, which enhances brine concentration processes. The improved methodology used in this study reflects advancements in sustainable brine management and supports a circular economy model, where waste is transformed into valuable resources. This approach aligns with recent literature emphasizing resource recovery as a critical component of sustainable desalination practices [1, 2].

3.2 Renewable Energy Integration

The integration of renewable energy sources into desalination processes (see Fig. 1) improved system efficiency significantly, ranging from 65 to 72%, depending on plant conditions. This improvement was achieved through the effective use of waste heat for solar evaporation, demonstrating its potential to reduce both environmental impact and operational costs.

However, seasonal variations in solar energy, particularly during winter months, posed challenges to system consistency. Reduced sunlight availability during these periods decreased energy input and system efficiency. To mitigate these limitations, the study recommends hybrid renewable systems, combining solar and wind energy, to ensure consistent energy supply throughout the year. Such systems have been validated in recent studies, which highlight their potential to stabilize energy availability in desalination and brine management applications [3]. By diversifying energy

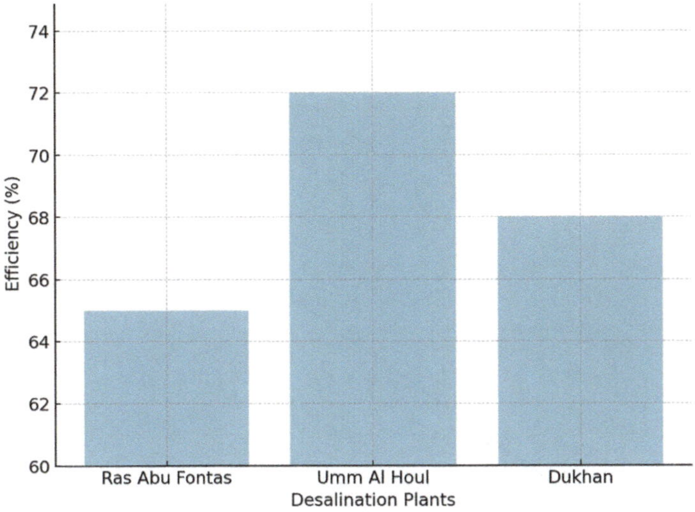

Fig. 1 Efficiency of renewable energy integration across desalination plants

inputs, hybrid systems could address seasonal fluctuations and improve overall sustainability.

3.3 AI-Driven Optimization Results

The AI-driven optimization model achieved an impressive 85% accuracy, with a 10% improvement in mineral recovery efficiency compared to conventional methods (see Fig. 2). This advancement reduced overall energy consumption by an estimated 5%, translating to substantial cost savings for desalination operations.

The AI model dynamically adjusted brine recovery parameters in real time, optimizing resource utilization and reducing waste. This approach exemplifies how machine learning can enhance complex operational processes, making them more efficient and sustainable. Recent literature supports the integration of AI in desalination, particularly for its ability to improve decision-making and predict outcomes with high accuracy [4].

3.4 Limitations of the Study

Several limitations were identified during this research:

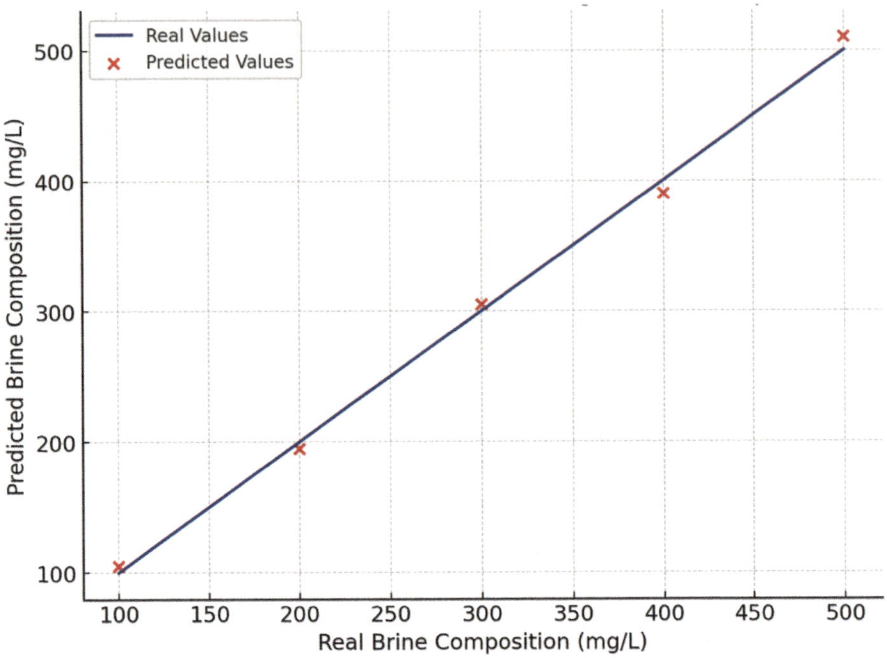

Fig. 2 AI model performance in predicting brine composition compared to real values

Lithium recovery remains economically challenging due to its low concentrations and high extraction costs. Further technological innovations are necessary to improve the efficiency and feasibility of lithium extraction.

Seasonal variations in solar energy impacted renewable energy integration, reducing system efficiency during periods of low sunlight. Hybrid renewable systems could address these fluctuations by providing a more reliable energy source.

The AI model's performance was occasionally constrained by the dataset, particularly during extreme operational conditions. Expanding the dataset to include more diverse scenarios would enhance the model's adaptability and robustness.

3.5 Interpretation of Findings

This research demonstrates that implementing smart brine management strategies can significantly improve the sustainability of desalination processes. The results emphasize the potential to transform brine from a waste product into a valuable resource, reducing its environmental impact while creating economic benefits through mineral recovery.

While the findings are promising, additional work is needed to address the challenges of lithium recovery, enhance the AI model's capabilities, and optimize renewable energy integration. These advancements would further align desalination practices with global sustainability goals and Qatar's National Vision 2030, providing a scalable framework for other water-scarce regions. This study offers valuable insights for policymakers and industry stakeholders aiming to improve desalination efficiency and sustainability worldwide.

4 Conclusion

4.1 Key Findings and Implications

This research highlights the transformative potential of smart brine management strategies in improving the sustainability of desalination processes. The findings demonstrate that the recovery of magnesium and calcium from brine is both feasible and economically viable, with concentrations that align with industry benchmarks. The integration of AI-based optimization enhanced mineral recovery efficiency by 10%, translating into measurable reductions in energy consumption and operational costs. These advancements underline the potential to reframe brine from a waste product into a valuable resource, supporting circular economy principles and reducing the environmental footprint of desalination [1].

This study's findings also carry broader implications for resource management and sustainability goals, particularly in water-scarce regions like Qatar. By incorporating renewable energy sources and advanced analytics, desalination can become a more environmentally friendly and economically sustainable solution to global water challenges [2].

4.2 Benefits and Limitations

The research underscores several key benefits, including the advantages of integrating state-of-the-art technologies for mineral recovery and renewable energy utilization. These technologies align with the transition toward a circular economy, where waste is minimized, and resources are maximized.

However, certain limitations were identified:

Lithium recovery remains challenging due to its lower concentrations in brine and higher extraction costs, which require innovative methods to improve economic viability.

Seasonal fluctuations in renewable energy availability, particularly for solar power, impact energy efficiency during low-radiation periods, such as winter months.

The AI optimization system, while effective, would benefit from an expanded dataset to increase its accuracy and adaptability under diverse operational scenarios, including extreme weather conditions [3].

These limitations highlight the need for continued research and technological development to fully realize the potential of sustainable brine management.

4.3 Suggestions for Future Research

Future research efforts should focus on:

1. Improving Lithium Extraction Methods:
   - Develop advanced methods or materials, such as selective adsorption techniques or membrane-based separation, to enhance the recovery of lithium from low-concentration brine.
2. Exploring Hybrid Renewable Energy Systems:
   - Investigate the integration of solar and wind energy systems to address seasonal variability and ensure consistent energy supply for brine management processes.
3. Expanding AI Models:
   - Incorporate larger and more diverse datasets to enhance the predictive capabilities of AI models, enabling them to perform reliably under a wide range of operational conditions, including extreme weather.

References

1. Safia A, Salima B, Ibrahim B (2024) Zero liquid discharge management: optimizing sustainable reverse osmosis desalination practices. Stud Eng Environ Sustain 4(2):99–110
2. Moharram NA, Konsowa AH, Shehata AI (2024) Advances in zero liquid discharge multigeneration plants: a novel approach for integrated power generation, supercritical desalination, and salt production. Energy Manage 15(6):133–147
3. Elazab SA, de Melo AB, Nechakh A (2024) Utilizing extremely high salinity waters for unconventional hydraulic fracturing in the UAE: a chemistry-oriented solution. In: Water Lifecycle management conference proceedings 2024. OnePetro, pp 43–52
4. Kumar S, Gupta R, Das S (2023) AI-based optimization of renewable energy integration in desalination plants. Energy AI 6:100115
5. Saleem F, Ahmed Z, Nazir M (2022) Hybrid renewable energy systems for desalination plants: a review of sustainability applications. J Clean Prod 367:132561
6. Mohsin F, Qadir M, Farooq U (2022) Environmental impact assessment of brine discharge from RO desalination plants: a meta-analysis. Sci Total Environ 825:153846
7. Alotaibi S, Alharbi Y (2023) Advances in solar thermal energy for sustainable brine management. Sol Energy Mater Sol Cells 249:112058

8. Nguyen TT, Le TH, Hoang T (2023) Predictive modeling for mineral recovery in desalination brine using machine learning. J Water Proc Eng 50:102237
9. Rahman F, Osman M, Sultan A (2023) Cost-benefit analysis of circular economy approaches in brine management. J Environ Manage 336:117497
10. Khalid S, Basu S, Ahmed A (2023) Hybrid wind-solar systems for brine management: a simulation-based study. Renew Energy 198:1207–1216
11. Lee CH, Yip NY (2022) Brine mining and zero-liquid discharge desalination: status and research needs. Desalination 520:115345
12. Ghosh P, Mitra D, Pal R (2023) Techno-economic feasibility of mineral recovery from brine using novel separation techniques. Chem Eng J Adv 15:100400

Innovative Nanomaterial Solutions for Shipboard Wastewater Management

Nourhan I. Ghoneim, Maathir Alhasani, Alzulfa Albulushi, Abdulaziz Al Shandoudi, Seyed Behbood Issa-Zadeh, Salim Alhashmi, Rashid Alsawai, Naser Alhakmani, Anteneh Mesfin Yeneneh, and Mohamed R. Shouman

Abstract Recently, using nanomaterials in treating marine wastewater has become a hot topic. In this paper, authors have investigated the practicality of nanoparticle-based technologies in wastewater treatment and examined the complicated systems at the connection points between nanoparticulate and wastewater. Authors have found that nanomaterial-based methods present some advantages, including improved removal efficiency of other contamination materials, lower treatment time, and fewer environmental concerns than traditional methods. Moreover, the nanomaterial's potential to specifically target toxins and assist in their removal is an up-and-coming method to mitigate the challenges of marine pollution. This study uses plasma from the arc discharge technique, which passes a high-voltage electric current on electrodes that treat wastewater, to treat it as part of modern science's effort to create treatment methods. This paper aims to improve recycled water quality and reduce marine pollution, thus achieving the global goal of providing access to clean water and safe sanitation for all, which is aligned with goals 6 and 14 of SDG.

Keywords Marine wastewater treatment · Marine pollution · Marine life

---

N.I. Ghoneim (✉) · M. Alhasani · A. Albulushi · A. Al Shandoudi · S. Alhashmi · R. Alsawai · N. Alhakmani · A.M. Yeneneh
International Maritime College Oman, National University of Science and Technology, Suhar, Oman
e-mail: norhan@imco.edu.om

S.B. Issa-Zadeh
School of Maritime Science, University of Gibraltar, Campus Europa Point, Gibralta GX11 1AA, UK

M.R. Shouman
Faculty of Maritime Transport and Technology, Arab Academy for Science, Technology and Maritime Transport, Alexandria, Egypt

S. Al-Naemi et al. (eds.), *Water and Food Security in the Face of Climate Change: Challenges and Opportunities for Resilience*, World Sustainability Series,
https://doi.org/10.1007/978-3-032-00098-9_65

1 Introduction

Urbanization, industrialization, and human activities have increased water consumption and led to contamination by heavy metals, fluoride, organic dyes, and biomolecules, posing a considerable environmental challenge [1]. Global water pollution caused by biological, inorganic, organic, and toxic metal contaminants obstructs access to clean drinking water, underscoring the urgent need for innovative solutions such as nanotechnology in integrated water management. Nanotechnology, which involves the manipulation of matter at the atomic and molecular scale, has significantly impacted disciplines such as physics, chemistry, and environmental science. Nanoparticles (1–100 nm) possess distinctive characteristics, including a high surface area-to-volume ratio and adjustable reactivity, making them exceptionally efficient for environmental applications, particularly in the treatment of water and wastewater [2]. Nanotechnology offers effective and sustainable solutions for water treatment utilizing nano-adsorbents, nano-membranes, and nano-photocatalysts. These advanced materials tackle water pollution, particularly in the marine sector, where traditional wastewater treatment techniques struggle to handle heavy metals, organic compounds, and pathogens adequately [1, 2]. Untreated wastewater worsens marine pollution from shipping, aquaculture, and tourism. Nanotechnology offers advanced solutions by utilizing nanoparticles to remove contaminants and reduce environmental impact. This research focuses on developing and optimizing nanoparticles for marine wastewater treatment, assessing their feasibility, and ensuring regulatory compliance. The study aims to advance sustainable marine engineering and wastewater management practices by leveraging an interdisciplinary approach.

2 Literature Review

Nanomaterials with dimensions less than 100 nm demonstrate significant efficacy in marine wastewater treatment, attributed to their elevated reactivity, adjustable surface chemistry, and extensive surface area. Different forms, such as nanoparticles, nanotubes, and nanofibers, effectively absorb, degrade, and stabilize pollutants, including organic molecules, heavy metals, and pathogens (Zhang et al. [3]). Graphene-based nanomaterials and metal oxide nanoparticles, such as TiO_2 and Fe_3O_4, exhibit high surface areas and photocatalytic properties, rendering them effective for degrading organic pollutants. Their small size facilitates compact and efficient wastewater treatment systems, suitable for space-constrained shipboard applications. Practical applications of graphene (Yang et al. [4]). Nanomaterial-based systems are advantageous for maritime applications owing to their compact dimensions, low energy requirements, minimal waste generation, cost-effectiveness, and decreased environmental impact (Yang et al. [4]). Nanomaterials provide efficient treatment for ship wastewater; nevertheless, their implementation encounters obstacles due to possible threats to marine ecosystems and human health (Gottschalk et al. [5]). Specific nanomaterials can adversely affect aquatic organisms and bioaccumulate throughout the food chain, prompting worries over their long-term environmental consequences (Gottschalk et al. [5]). Comprehensive risk evaluation is crucial for safely utilizing nanomaterials in maritime applications. Despite promising laboratory results, the scalability and dependability of nanomaterial-based wastewater treatment systems continue to pose substantial obstacles (Yang et al. [4]). Nanomaterial-based treatment systems have potential for marine wastewater treatment owing to their efficacy, compactness, and environmental sustainability. Nonetheless, obstacles, including pollution, maintenance, durability, environmental issues, scalability, and dependability, must be resolved to ensure effectiveness. Subsequent research should concentrate on surmounting these challenges and creating flexible marine ecosystem systems [6–9].

3 Methodology

Researching the application of nanoparticles for wastewater treatment aboard constitutes a compelling technological challenge. This research necessitates evaluating several factors, including the efficacy, efficiency, safety, and availability of nanomaterials, alongside the limitations presented by the maritime and vessel environment.

3.1 *The Steps of the Experiments*

- ***Initial needs assessment***: Before developing wastewater treatment technology, evaluate the vessel's needs and decide whether a temporary, minor, or permanent, efficient system for continuous use is necessary.
- ***Searching for suitable nanomaterials and electrodes***: Nanomaterials, such as mineral, organic, and composite types, are effective for water treatment due to their high surface area, reactivity, and adaptable properties. They benefit ship wastewater treatment, where space and efficiency are crucial. Different nanomaterials can be chosen to remove common wastewater contaminants.
  - *Graphene Oxide (GO)*: has a high adsorption capacity for water contaminants, including heavy metals and organic compounds, thanks to its two-dimensional structure and large surface area [[10]. Graphene oxide is also effective in removing heavy metals from ship ballast water.
  - *Titanium dioxide (TiO_2)*: Nanoparticles are known for their photocatalytic ability to degrade organic pollutants under UV light. When mounted on conveyors, they can be easily separated from water. A 2019 study by Johnson &

Brown showed that TiO_2-coated glass beads effectively removed organic pollutants from marine wastewater.
- *Carbon nanotubes (CNTs)*: Their high aspect ratio and large surface area are ideal for adsorption and filtration. Functionalized carbon nanotubes can selectively absorb specific pollutants. A recent study by Garcia et al. showed that they efficiently removed oil and fat from bilge water (2023).
- *Nanocomposite membranes*: Incorporating nanoparticles such as zeolite and metal oxides enhances the filtration performance of conventional membranes, effectively removing contaminants while maintaining high flow rates. (Chen and Wang 2020) A significant improvement in filtration efficiency for ship wastewater was achieved using nanocomposite membranes containing zeolite nanoparticles.
- *Silver nanoparticles (Ag NPs)*: have strong antimicrobial properties that help prevent bacterial growth in ship wastewater systems. They can be integrated into filters and coatings to reduce microbial contamination, as shown in a 2018 study by Robinson et al. However, challenges such as economic efficiency, environmental impact, and health risks need further investigation. A few nanomaterials have been commercialized, and future research should focus on improving affordability and assessing toxicity.

- ***Developing* suitable techniques:**

Using nanomaterials in ship wastewater treatment requires the integration of nanomaterials into an efficient and space-saving system.

- *Selection of nanomaterials*: Choice of nanomaterials depends on contaminant type, treatment objectives, and operational constraints. For example, graphene oxide (GO) provides high absorption capacity, and titanium dioxide (TiO_2) nanoparticles allow photodegradation under UV light [10, 11].
- *System design*: The system design and nanomaterials are integrated into the processing system, considering space constraints and operational requirements. Nanomaterials can be incorporated into membranes, filters, and coatings to increase treatment efficiency [12].
- *Incorporation of nanomaterials*: Nanomaterials can be immobilized on carriers or dispersed in solutions for effective processing. Functional carbon nanotubes have efficiently removed oils and fats from wastewater [13].
- *Technological challenges*: Key challenges for nanomaterial-based systems include scalability, stability, and cost-effectiveness. Reliable synthesis methods must be developed to enable large-scale use. Effective ship wastewater treatment using nanomaterials requires innovative technology and multidisciplinary collaboration.

  - Performance testing and optimization: The developed system should be experimentally tested using natural industrial wastewater to determine or assess its removal of pollutants. It can also be enhanced and adjusted by experimental testing of applicable nanomaterials.

– Safety and compliance with standards: Treating wastewater from ships using nanomaterials must safeguard the marine environment and public health while complying with regional and international safety, environmental standards, and maritime regulations. Adhering to the outlined steps will facilitate successful implementation.

3.2 The Design of the Project:

This paper involves conducting laboratory experiments using an arc discharge device. The discharge arc device uses the principle of producing an electric discharge inside a tank of water (see Fig. 1).

The plasma produced in water through this process has a wide range of uses aimed at increasing quality and efficiency, such as:

1. ***Plasma generation***: In an arc discharge device, high-voltage electric discharge ionizes water molecules, producing plasma of ions, reactive species, and photons. This plasma exhibits characteristic properties and interacts with surrounding environment.
2. ***Chemical reactions***: Arc discharge in water can generate reactive oxygen species like ozone, hydrogen peroxide, and hydroxyl radicals. These species oxidize and decompose organic pollutants and pathogens, aiding in water treatment.
3. ***Water treatment***: The plasma produced by the arc discharge device can deactivate or eliminate microorganisms such as bacteria, viruses, and algae while also breaking down organic pollutants in water, making it practical for general disinfection and water treatment [14].
4. ***Wastewater treatment***: Arc discharge provides an alternative to traditional wastewater treatment methods like UV radiation and chlorine, removing the need for external radiation or chemicals. However, challenges such as energy efficiency, electrode fouling, scalability, by-product generation, and environmental impact must be addressed. Careful design and control are essential to ensure sustainability and minimize adverse effects [15].

3.3 Experiment's Details

In this paper, an experiment for wastewater treatment using an arc discharge device was conducted as per the following details (Fig. 1):

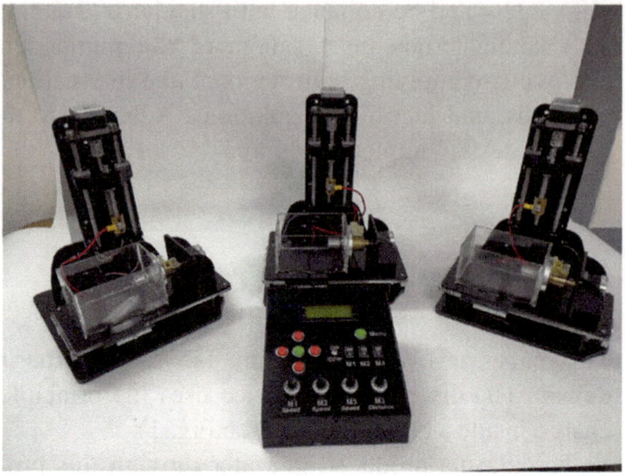

Fig. 1 A laboratory device specially designed for sample testing (arc discharge)

3.3.1 Instruments and Equipment

The following Instruments and equipment were used in performing the experiments:

- Arc discharge device (see Fig. 2.).
- Constant electrical current.
- Electrodes (anode and cathode).
- A wastewater sample.
- Hydrometer and Current measuring devices.
- Electrical wires and connectors.
- Water Quality Meter.
- Safety and security equipment like gloves, protective masks, safety shoes & a suit.

Fig. 2 Arc discharge device [16]

3.3.2 Steps of the Experiment

- Prepare a suitable wastewater sample containing all the typical specifications, organic substances, and microorganisms.
- Using electrodes such as graphite as the anode and tantalum as the cathode, installing them in parallel in the arc discharge device, immersing them with the sample of water to be treated, and ensuring the distance between the electrodes facilitates the arc discharge process.
- Provide a power source capable of providing a suitable voltage and current difference to start the arc discharge process, ensure its continuity and maintenance, and ensure electrical connections and proper polarity.
- Measurement of the initial parameters of the water sample to be treated and recorded, which includes the PH, conductivity, and any pollutants present and used as a reference at the end of the experiment.
- The arc discharge process is started by applying a voltage to the electrodes in the water sample. The potential difference gradually increased during the experiment while monitoring the discharge behavior.
- Set a certain period to ensure sufficient contact for discharge with the sample to be processed. Mixing and stirring the sample can facilitate the process.
- Measure the final parameters after processing, compare them with the initial parameters, and record them. In addition, the pH and conductivity of the final sample will be recorded to evaluate the effectiveness of the arc-emptying process.
- Analyzing the data to be obtained, evaluating the effectiveness of treatment and efficiency of process performance, improving water quality, and reducing pollutants and microorganisms, considering the energy consumption and the time required for optimal treatment.
- Follow all safety procedures and guidelines when using electrical equipment and sewage samples. Wear protective clothing, gloves, and goggles when entering the laboratory and throughout the device's use.

3.4 Data Collection

The experiment was carried out by fixing the voltage difference and the change of the current passing through the electrodes to observe the extent of the electric current's effect on the arc discharge and the changes that would occur on the sample used. The samples were obtained from the college laboratory. Two samples were used; one was prepared in the college laboratory, and the other was obtained from OQ, a factory wastewater company. This experiment measured the parameters of Connectivity (cond.) μs, total dissolved salts (TDS) ppm, and the hydrogen number (PH). There was a device (water quality meter) used to measure the parameters (see Fig. 3).

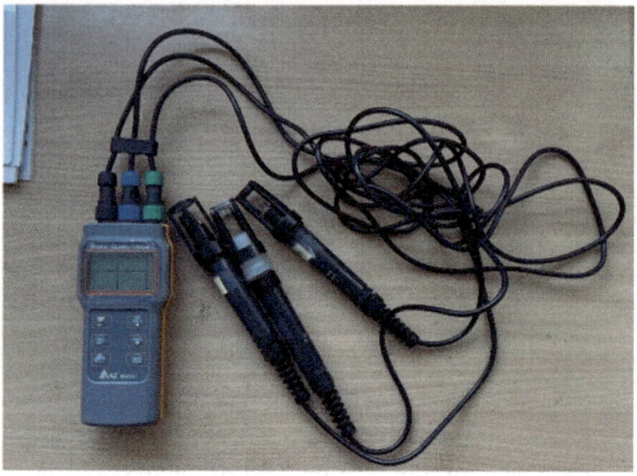

Fig. 3 Water quality meter [16]

Table 1 The initial parameters of the OQ wastewater sample

(TDS) ppm	(cond.) μs	(PH)
500	1006	7.3

3.4.1 OQ Wastewater Sample

The following figures and table show the initial parameters of the sample (see Table 1 and Fig. 4). Also, the following table shows the final parameters of the sample after performing the steps mentioned above (see Table 2 and Fig. 4):

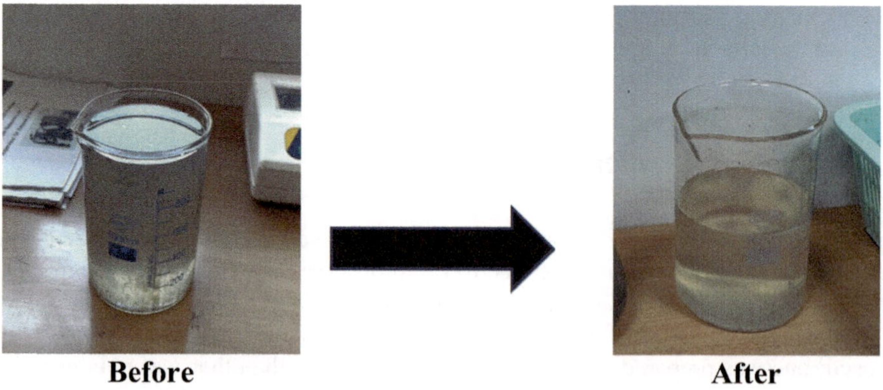

Fig. 4 OQ wastewater sample [16]

Table 2 The final parameters of the OQ wastewater sample

No	(V) v	(I) A	(d) mm	(t) min	(TDS) ppm	(cond.) µs	(PH)
1	100	15	2–3	15	500	1006	7.3
2	100	25	2–3	15	520	1060	7.6
3	100	35	2–3	15	470	917	7.5
4	100	45	2–3	15	354	720	8

3.4.2 International Maritime College Oman (IMCO) Wastewater Sample

The following figure and table show the initial parameters of the sample (Table 3):

The following table shows the final parameters of the sample after performing the above-mentioned steps (see Table 4 and Fig. 5):

Table 3 The initial parameters of the IMCO wastewater sample

(TDS) ppm	(cond.) µs	(PH)
394	760	7.3

Table 4 The final parameters of the IMCO wastewater sample

	(V) v	(I) A	(d) mm	(t) min	(TDS) ppm	(cond.) µs	(PH)
1	100	35	2–3	10	364	744	7.7
2	100	45	2–3	10	309	730	8.04

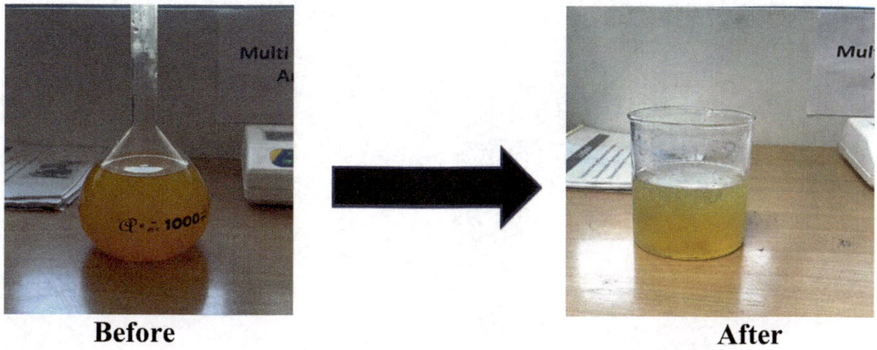

Fig. 5 IMCO wastewater sample [16]

Table 5 Freshwater (distilled water) parameters

(TDS) ppm	(cond.) μs	(PH)
33	66.5	7.6

3.4.3 Reference for Freshwater (Distilled Water) Parameters

The parameters of distilled water have been compared with a reference to verify the ideal values for freshwater production, and the results are shown in the following table (see Table 5):

4 Results and Discussion

4.1 Observations During the Experiment

When an electric current above 15 amperes is applied to the electrodes, arc discharge occurs, initially causing a rise in temperature without visible changes. As the temperature increases, white foam forms, and the sample changes to white (see Fig. 6). After about 10 min, the sample's color becomes more apparent, with white deposits forming beneath the pelvis.

Fig. 6 Foam formed with the color of the sample [16]

4.2 *Finding Results*

4.2.1 The Relationship Between Current and Arc Discharge

The experiment showed that increasing the electric current enhances the productivity of arc discharge between electrodes. At the lowest current of 15 amperes, arc discharge did not occur due to insufficient current to trigger ionization. The best results were observed at the highest current of 45 amperes.

4.2.2 The Relationship Between the Distance Between Electrodes and Arc Discharge

The smaller the distance between the electrodes, the greater the arc discharge efficiency, thereby improving the processing process. In the experiment, the distance was not fixed due to defects in the device, and the instability of the electrodes was suitable, as the distance was between 1–2 mm; it was observed that when approaching 1 mm, the appearance of the discharge arc was better.

4.2.3 The Relationship Between the Duration of the Experiment and the Efficiency of the Treatment Process

In the experiment, time played an essential role in improving the parameters of the treated water. The elapsed time of the treatment process increases the exposure of most of the water to be treated to arc discharge, better purification, and the appearance of better parameters.

4.2.4 The Difference Between the Initial and Final Parameters of the Experiment

1. *Total dissolved solids (TDS)*: At the beginning of the experiment, insoluble solids in the wastewater dissolve as the temperature rises, increasing the total dissolved solids (TDS) in the sample. These solids are removed during the treatment process, leading to a decrease in TDS by the end of the experiment. The results show that TDS remains high, and the value should be reduced to 33 ppm, which is the optimal level.
2. *Conductivity*: Initially, an increase in electric current raises the conductivity of the sample due to electrode ionization and the melting of conductive metals in water from high temperatures. After a while, the conductivity decreases as the treatment process progresses, aiming for optimal conductivity. The desired value for conductivity is 66.5 μs, which is considered the best for the process.

Fig. 7 Wastewater samples after 4 days [16]

3. *Power of hydrogen (PH)*: The power of hydrogen increased throughout the experiments. After several days of the experiment, deposits appeared in the treated sample, and a pronounced discoloration was noticed (see Fig. 7).

4.3 Problems that Appeared in the Device and Experiment

1. *Installation of electrodes*: The device has manufacturing defects. Since the electrodes are installed on two opposite sides only, not from four sides, they were not installed better, causing them to move circularly when the motor is running.
2. *The distance between the electrodes*: The failure to install the electrodes correctly resulted in the inability to stabilize the distance between them, which oscillated while producing the arc discharge throughout the experiment.
3. *Water quality measuring device*: The results from the device were not entirely accurate, so multiple devices were used to verify and confirm consistent findings.
4. *The current-conducting brush for electrodes*: The main issue encountered with the device was the brush burning when the electric current was increased. Upon investigation, it was found that the carbon material used in the brush could only tolerate 4–7 amperes. It was recommended to use high-current carbon graphite brushes, which can withstand higher currents (see Fig. 8).

4.4 Analysis

Water treatment is essential for providing safe, high-quality water and involves various methods, including physical, chemical, biological, and nanotechnological processes to remove contaminants. Experiments with arc discharge devices show

Fig. 8 Comparison of the brush before and after the experiments [16]

that increasing current enhances electrode ionization, improving the treatment process. Reducing the distance between electrodes boosts ionization and discharge efficiency, enhancing treatment quality. Total dissolved solids (TDS) changes reflect the treatment's impact on water quality. Electrodes like graphite, titanium, and tantalum are used in electrochemical processes, including nanomaterial synthesis and water treatment, to improve water quality through ionization and electroplating. The ionization of graphite or tantalum can be applied in the following cases:

1. *Production of nanomaterials.* Graphite and tantalum anodes are used in electrolysis to create nanostructured materials like carbon nanotubes and metal nanoparticles. By passing an electrical current through a chemical solution, water serves as the raw material, and process conditions are adjusted to promote the formation of these nanostructures. (Al-Halfi et al. 2014).
2. *Water Cleaning.* Graphite and tantalum electrodes are used in electrolysis to remove impurities from wastewater. When a constant current is passed through this solution, the impurity is hydrolyzed, and the water is soon purified and improved.

    - *The pH effect of using electrodes to produce nanomaterials in wastewater depends on the composition and the reaction process;* nanomaterials can vary depending on the targeted material, with different methods used based on the electrodes. The effect of pH can influence the process, and various factors must be considered during production to achieve the desired nanomaterial.

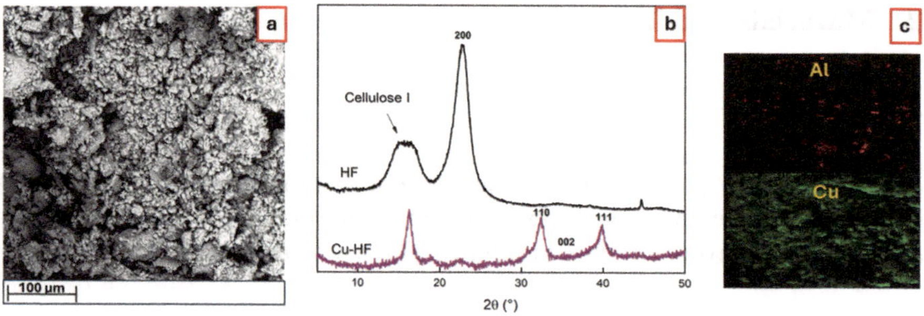

Fig. 1 **a** SEM image of Cu-HF; **b** XRD spectra of HF and Cu-HF; **c** EDX maps of Cu and Al on Cu-HF surface

3 Results and Discussion

3.1 Cu-HF Characterization

The produced adsorbent was characterized through SEM, XRD and EDX analyses (Fig. 1).

Figure 1a reports the typical morphology of Cu-HF. After the treatment and the functionalization, aggregates of Cu-based structures on the surface were evident. Figure 1b reports the XRD diffractograms of pristine and modified HF. The two peaks at $2\theta = 14°$ and $2\theta = 16°$ in the HF spectrum are typical of cellulose I [21]. For Cu-HF sample, diffraction peaks are present at $2\theta = 32.6$, 35.6, 39.4 which are related to the (110), (002), and (111) planes of CuO, respectively. Figure 1c shows that the distribution of Al and Cu on the surface of HF after the modification is quite homogeneous.

3.2 Adsorption Tests: Effect of Process Parameters

Figure 2 reports the experimental data obtained by carrying out different adsorption tests.

Figure 2a reports the impact of contact time on the adsorption of MO by the Cu-HF. Contact time optimization was achieved by varying contact times (1–96 h). The initial MO concentration, adsorbent dosage, temperature and pH were set at 50 ppm, 1 g L^{-1}, pH = 7 and T = 293 K, respectively. The presence of Cu-based particles contributes to greatly increasing the adsorption capacity changing from 3.8 to 48.4 mg/g. Initially, on the first stage, empty sites are accessible for adsorption. However, due to their saturation, as time proceeds, the unavailability of free sites and the repulsive interactions between the adsorbed MO and the MO in the bulk phase reduce further adsorption. The increase in adsorption capacity for the Cu modified

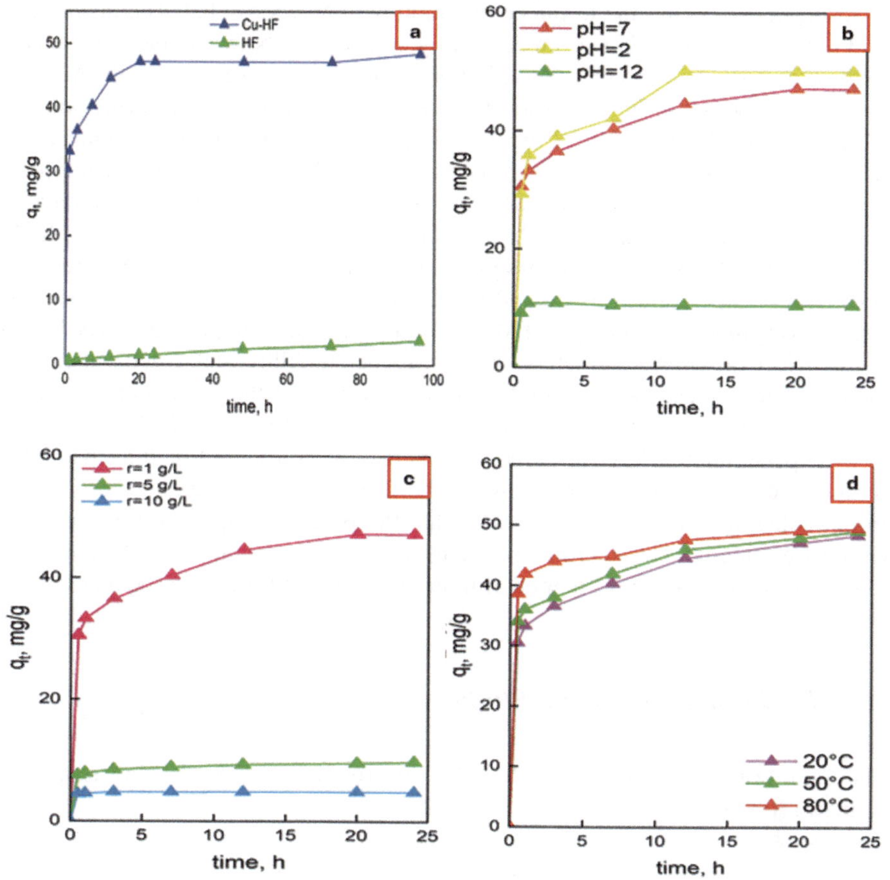

Fig. 2 **a** Adsorption capacity of HF and Cu-HF; **b** effect of pH; **c** effect of solid–liquid ratio; **d** effect of temperature

HF might be even ascribed to the improvement of specific surface area. A possible mechanism of adsorption could refer to the electrostatic interactions between the positive sites of the adsorbent and the negative groups of MO. Besides, even H bondings with N and S groups of MO could be responsible for the MO uptake. Figure 2b reports the effect of pH on the removal of MO ($c_0 = 50$ ppm, adsorbent dosage of 1 g L^{-1}, and T = 293 K). The maximum adsorption capacity changed from roughly 49 to 47 and 10 mg/g for pH = 2, 7 and 12, respectively. In order to understand the effect of pH on adsorption capacity, the point of zero charge was calculated as 7.26. As $pH < pH_{PZC}$, an excess of positive charge on the surface exists and the excess of H_3O^+ leads to a positively charged surface and as such, it can interact with the negative sulfonic acid group ($-SO_3^-$) of MO. In alkaline conditions, as $pH > pH_{PZC}$, an excess of negative charge on the adsorbent exists. MO assumes the anionic form and the presence of a negative charge and OH^- groups also led to a decrease in adsorption capacity due to the competition with MO ions [19]. The decrease in MO removal at high pHs is even caused by competition between the OH^- ions in solution and MO anions [11]. The effect of the adsorbent dosage (r) on the MO dye removal is reported in Fig. 2c. The MO removal increased as the adsorbent dosage increased from 1 g/L to 10 g/L while the q_t decreased. The improvement in MO removal percentage was ascribed to the high available sorption sites causing more MO to attach when the adsorbent amount increases [7]. Moreover, a low amount of adsorbent (5 g/L) is needed to fully decolorize a 50 ppm MO solution since the R, % is almost 100%. Figure 2d shows the adsorption profiles referred to MO removal at different temperatures (20, 50 and 80 °C). The q_t parameter rises as temperature rises proving that the adsorption of MO is favored at higher temperatures. The higher T can affect different parameters such as: higher mobility and kinetic energy of the adsorbate molecules, increased diffusion rate toward the adsorbent, decreased viscosity of the solution and internal swelling of the adsorbent which might favor the MO penetration. Moreover, other factors may be considered such as the higher solubility of dye at higher temperature. The thermodynamic parameters, obtained from isothermal studies, allow to state that the process is spontaneous (negative Gibb's free energy), endothermic (positive enthalpy) and characterized by enhanced randomness which suggests a good affinity of MO towards the adsorbent [13].

3.3 Isotherm and Kinetic Studies

The equilibrium adsorption data for MO adsorption were fitted using Langmuir isotherm (Eq. 3) [14]:

$$1/q_e = 1/q_m + 1/\left(b * q_e * c_e\right) \tag{3}$$

where C_e is the equilibrium dye concentration (mg/L), q_e is the equilibrium amount of dye adsorbed (mg/g), q_{max} is the maximum adsorption capacity (mg/g), b is the Langmuir constant (L/mg). Langmuir isotherm refers to the formation of a monomolecular layer without any interaction between the adsorbed molecules [14]. Langmuir's model parameters are reported in Table 1.

The Langmuir model fits the adsorption data with $R^2 \geq 0.98$. The q_{max} and k_L parameters are 338.98 mg/g and 0.31 L/mg, respectively. The feasibility of the adsorption process can be estimated by the calculation of the $R_L = 1/(1 + K_L{*}c_0)$. The calculated R_L values vary between 0.06–0.011, changing the c_0 from 50 to 300 ppm, proving the favourable uptake of MO. Ho's pseudo-second-order (Eq. 4) was applied to model the kinetic data [33]:

$$t/q_t = 1/\left(k_2 * q_e^2\right) + t/q_e \tag{4}$$

where q_e (mg g^{-1}) and q_t (mg g^{-1}) are the adsorption capacity at equilibrium and at time t, respectively, k_2 (g mg^{-1} min^{-1}) represents the pseudo second-order rate

Table 1 Parameters of Langmuir and pseudo second order models

Langmuir	
q_m (mg g^{-1})	338.98
b (L mg^{-1})	0.31
R^2	0.989
Pseudo second order	
q_e, (mg/g)	49.50
k_2, (g/mg*h)	0.024
R^2	0.999

constant. Finally, the adsorption data were described by the pseudo-second-order model ($R^2 > 0.99$). The calculated adsorption capacity (q_e) was very close to the experimental value (Table 1).

3.4 *Effect of Salinity and Regeneration Studies*

Figure 3a reports the effects of different salt solutions on the adsorption of MO to simulate the real wastewater scenario.

Figure 3a shows that the R% is affected by the salt presence. It can be hypothesized that the decrease in R,% could be ascribed to the competition effect between the MO ions and salt ones to occupy the adsorbent surface sites. For example, chloride ions could access the adsorbent surface more easily than MO, even due to their low sterical hindrance, obstructing the adsorption of the dye and decreasing the decolorization efficiency [12]. Besides, other factors such as hydrated radii and repulsive interactions

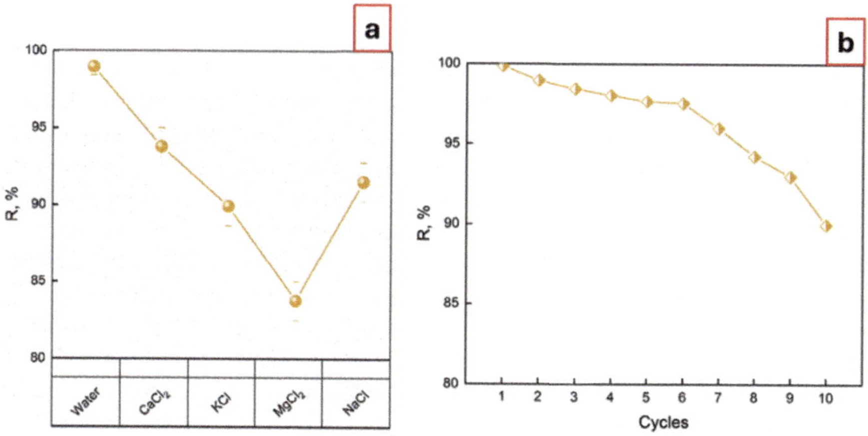

Fig. 3 **a** Effect of salinity on MO removal efficiency; **b** Regeneration studies

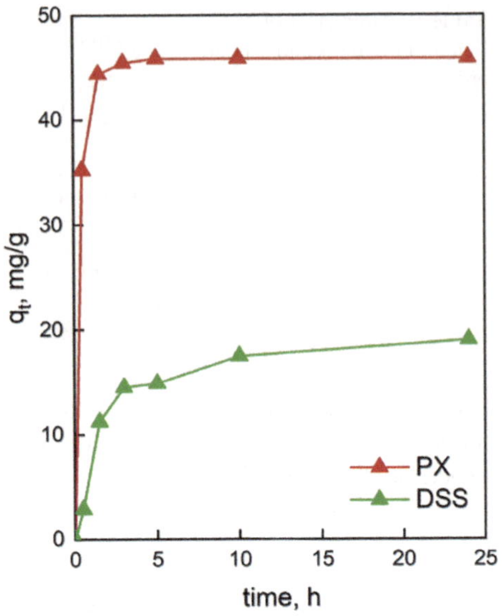

Fig. 4 q_t versus t for PX and DSS

should be considered. Generally speaking, the negative effect of electrolytes on decolorization advises the occurring of an ion exchange mechanism [18]. Figure 3b shows the regeneration studies of MO after 10 adsorption/desorption cycles. The results show that a drop of 9% occurs after 10 cycles of utilization demonstrating that the Cu-HF adsorbent can be used repeatedly without great loss in its performance.

3.5 Adsorption Tests of Two Pharmaceuticals

Figure 4 reports the q_t versus t profiles for PX and DSS removal.

It can be observed that the equilibrium capacities are 49 and 19 mg/g for PX and DSS which correspond to an R,% of 91.5% for PX and 34.7% for DSS corroborating the great performance of this novel adsorbent to remove different kinds of pollutants. The adsorption performances might be associated with the accessibility of free adsorption sites which are occupied in a few hours while the equilibrium regime is achieved in a few hours. H bonding interactions between the adsorbent surface and the PX are assumed to occur. The coordination between Cu and the nitrogen donor units in PX can then contribute to the removal. It follows that electrostatic interactions could occur between the two drugs and the positive sites of the adsorbent. Moreover, the metal cation (Cu)–π interactions and the H-bonding between chloride and the H atoms of the OH groups may also contribute to the binding of the drugs on the surface of Cu–HF.

4 Advantages, Limits and Industrial Applicability

Recently, there has been a notable increase in the application of natural fiber as a primary resource, driven by the growing concern for environmental sustainability, for producing novel materials for water remediation applications. Besides, the application of the adsorption technique allowed for benefiting of cost effectiveness adaptability, simplicity, profitability and ease of operation [24] which make the adsorption technique through bio-based adsorbent a more promising and effective way to remove pollutant from wastewaters than other various methodologies such as reverse osmosis, ion exchange, membrane separation [20], adsorption on inorganic materials [26] and chemical precipitation [10]. The efficiency of adsorption is strictly related to the selection of a good adsorbent with specific properties. For this reason, the proposed functionalization based on metal oxide modification of hemp fibers aims for improving the adsorption features of agrowaste resources with low adsorption capacity in order to confer the appropriate functional groups to remove different classes of pollutants. Besides, compared to other powdered materials such as synthetic resin, Mxene, zeolite, activated carbon, biochar, metal organic framework, which showed detrimental effects on the environment, lack of bioactivity, biodegradability, biocompatibility, cost prohibitive production and regeneration, the use of natural fibers-based materials can overcome these drawbacks due to their cost effectiveness, adaptability for cleanup, high availability and ease of recovery. The great performances of the adsorbents produced are highlighted in Table 2 which reports a comparison with other adsorbents to remove the target pollutants.

By analyzing the data reported in the table, the adsorption performances of the Cu-HF are higher than those of the reported systems, proving that the designed adsorbent has a great capability for the adsorption of different pollutants from water effluents. Economic analysis is essential for measuring the feasibility of application of an adsorbent at the industrial scale. In order to give an idea of scalability and industrial

Table 2 Comparison of adsorption capacity of Cu−HF with other adsorbents

Adsorbent	q_m, mg/g	References
Cu@HF	338	This work
Chitosan based polymer membrane Modified with CuO/Graphene Oxide NPs	96	[2]
Iron-copper metal–organic framework	294.43	[36]
Copper-benzimidazole coordination polymer	15.01	[4]
Copper/zinc oxides NPs	68.199	[30]
Orange and lemon peels activated carbons	38	[22]
Chitosan/montmorillonite composite	154.4	[3]
Activated carbon/polyaniline	192.52	[6]
Anchote peel	103.03	[16]
L-arginine treated populous charcoal powder	90.44	[29]

application of the designed adsorbent, the design of a single stage batch adsorbent was carried out by applying the Vadivelan and Kumar method [31] and Sen's method [27]. Starting from the equilibrium mass balance for the MO dye in the single-stage operation and describing the isothermal data through the Langmuir isotherm model, Eq. 5 was obtained which allows for evaluating the mass of adsorbent (m) required to obtain a specific removal efficiency (R) of a set volume of dye solution (V):

$$\frac{m}{V} = \frac{\left(\frac{1}{1-R/100} - 1\right) * \left(1 + b * C_{eq}\right)}{Q_m * b} \tag{5}$$

For instance, the amount of Cu-HF needed to remove 1000L of 50 ppm MO solution was 12.7, 102.7, 125.4, 156.1 and 218.4 g for a R, % of 50%, 60%, 70%, 80% and 90%, respectively. It follows that Cu-modified hemp fibers could be applied as a novel, effective and versatile adsorbent. A low amount of adsorbent could totally decolourize an aqueous solution of MO and remove a great amount of drugs after 2–3 h. Finally, it should be claimed that the performances of natural fibers-based adsorbents have been investigated under batch conditions with small volumes of wastewaters. The possibility of using natural fibers as practical biosorbents implementable in greener technologies for the removal, recycling, and reuse of pollutants should consider extensive research focusing on a complete clarification and quantification of the relationships between chemical composition and properties, evaluation of competitive biosorption, analysis of a continuous adsorption systems, lifecycle and long-term performance, disposal of exhausted biosorbents, scalability and economic analysis in real world scenario.

5 Conclusions

This study concerns the fabrication of a biomass-derived novel adsorbent modified with CuO to be applied for removing the azo dye MO and two pharmaceuticals, diclofenac sodium salt and piroxicam. The removed dye amount increases as T and solid/liquid ratio increases while decreasing in the basic medium compared to acidic one. The maximum adsorption capacity evaluated through the Langmuir model is about 339 mg/g while the kinetic data were modeled through the pseudo second order model. After 10 cycles of re-utilization, the drop in efficiency was only 9%. Besides, the produced adsorbent was even tested by adsorbing piroxicam and diclofenac with $q_{max} = 49$ and 19 mg/g for PX and DSS, respectively. It follows that Cu-modified hemp fibers could be applied as a novel, effective and versatile adsorbent.

References

1. Abbaz A, Arris S, Viscusi G, Ayat A, Aissaoui H, Boumezough Y (2023) Adsorption of safranin o dye by alginate/pomegranate peels beads: kinetic, isotherm and thermodynamic studies. Gels 9(11):916. https://doi.org/10.3390/GELS9110916
2. Afzal S, Hassan S, Imran Z, Aminullah S (2024) Chitosan based polymer membrane modified with CuO/graphene oxide nanoparticles: novel synthesis, characterization and enhanced methyl orange removal. J Inorganic Organomet Polymers Mater 1–18. https://doi.org/10.1007/S10904-024-03008-4/TABLES/6
3. Bahrudin NN, Nawi MA, Jawad AH, Sabar S (2020) Adsorption characteristics and mechanistic study of immobilized chitosan-montmorillonite composite for methyl orange removal. J Polym Environ 28(7):1901–1913. https://doi.org/10.1007/S10924-020-01734-7/FIGURES/12
4. Bamgbose JT, Elaigwu SE, Adimula VO, Okeowo HO, Olayemi VT, Ameen OA, Oyediran AO, Odunola OA, Adeyeni EG, Adeyemi AD, Tella AC (2023) Green route synthesis and adsorption studies of copper-benzimidazole coordination polymer for removal of methyl orange from water. Chemistry Africa 6(5):2581–2592. https://doi.org/10.1007/S42250-023-00675-7/TABLES/4
5. Baroncini EA, Kumar Yadav S, Palmese GR, Stanzione JF (2016) Recent advances in bio-based epoxy resins and bio-based epoxy curing agents. J Appl Polymer Sci 133(45. https://doi.org/10.1002/app.44103
6. Bekhoukh A, Moulefera I, Zeggai FZ, Benyoucef A, Bachari K (2022) Anionic methyl orange removal from aqueous solutions by activated carbon reinforced conducting polyaniline as adsorbent: synthesis, characterization, adsorption behavior, regeneration and kinetics study. J Polym Environ 30(3):886–895. https://doi.org/10.1007/S10924-021-02248-6/FIGURES/5
7. Chen S, Zhang J, Zhang C, Yue Q, Li Y, Li C (2010) Equilibrium and kinetic studies of methyl orange and methyl violet adsorption on activated carbon derived from Phragmites australis. Desalination 252(1–3):149–156. https://doi.org/10.1016/J.DESAL.2009.10.010
8. Cheng X, Jiang D, Chen H, Barati B, Yuan C, Li H, Wang S (2023) Multi-stage adsorption of methyl orange on the nitrogen-rich biomass-derived carbon adsorbent: DFT and MD evaluation. Chemosphere 338:139218. https://doi.org/10.1016/J.CHEMOSPHERE.2023.139218
9. Collado N, Rodriguez-Mozaz S, Gros M, Rubirola A, Barceló D, Comas J, Rodriguez-Roda I, Buttiglieri G (2014) Pharmaceuticals occurrence in a WWTP with significant industrial contribution and its input into the river system. Environ Pollut 185:202–212. https://doi.org/10.1016/J.ENVPOL.2013.10.040
10. Djedidi Z, Bouda M, Souissi MA, Cheikh RB, Mercier G, Tyagi RD, Blais JF (2009) Metals removal from soil, fly ash and sewage sludge leachates by precipitation and dewatering properties of the generated sludge. J Hazard Mater 172(2–3):1372–1382. https://doi.org/10.1016/J.JHAZMAT.2009.07.144
11. Dutta SK, Amin MK, Ahmed J, Elias M, Mahiuddin M (2022) Removal of toxic methyl orange by a cost-free and eco-friendly adsorbent: mechanism, phytotoxicity, thermodynamics, and kinetics. S Afr J Chem Eng 40:195–208. https://doi.org/10.1016/J.SAJCE.2022.03.006
12. Fan J, Guo Y, Wang J, Fan M (2009) Rapid decolorization of azo dye methyl orange in aqueous solution by nanoscale zerovalent iron particles. J Hazard Mater 166(2–3):904–910. https://doi.org/10.1016/J.JHAZMAT.2008.11.091
13. Fan S, Wang Y, Wang Z, Tang J, Tang J, Li X (2017) Removal of methylene blue from aqueous solution by sewage sludge-derived biochar: adsorption kinetics, equilibrium, thermodynamics and mechanism. J Environ Chem Eng 5(1):601–611. https://doi.org/10.1016/j.jece.2016.12.019
14. Farghali AA, Bahgat M, El Rouby WMA, Khedr MH (2013) Preparation, decoration and characterization of graphene sheets for methyl green adsorption. J Alloy Compd 555:193–200. https://doi.org/10.1016/J.JALLCOM.2012.11.190
15. Forgacs E, Cserháti T, Oros G (2004) Removal of synthetic dyes from wastewaters: a review. Environ Int 30(7):953–971. https://doi.org/10.1016/j.envint.2004.02.001

16. Hambisa AA, Regasa MB, Ejigu HG, Senbeto CB (2023) Adsorption studies of methyl orange dye removal from aqueous solution using Anchote peel-based agricultural waste adsorbent. Appl Water Sci 13(1):1–11. https://doi.org/10.1007/S13201-022-01832-Y/TABLES/4
17. Majadleh M, Shahwan T, Ahmed RB, Anjass M (2022) Application of Fe and Cu nanoparticles for methyl orange removal from water and water-ethanol mixtures under various experimental conditions. Water Res Ind 28:100189. https://doi.org/10.1016/J.WRI.2022.100189
18. Maurya NS, Mittal AK, Cornel P, Rother E (2006) Biosorption of dyes using dead macro fungi: effect of dye structure, ionic strength and pH. Biores Technol 97(3):512–521. https://doi.org/10.1016/J.BIORTECH.2005.02.045
19. Munagapati VS, Kim DS (2016) Adsorption of anionic azo dye Congo Red from aqueous solution by cationic modified orange peel powder. J Mol Liq 220:540–548. https://doi.org/10.1016/J.MOLLIQ.2016.04.119
20. O'Connell DW, Birkinshaw C, O'Dwyer TF (2008) Heavy metal adsorbents prepared from the modification of cellulose: a review. Biores Technol 99(15):6709–6724. https://doi.org/10.1016/J.BIORTECH.2008.01.036
21. Perel J (1990) An X-ray study of regain-dependent deformations in cotton crystallites. J Text Inst 81(3):241–244. https://doi.org/10.1080/00405009008658707
22. Ramutshatsha-Makhwedzha D, Mavhungu A, Moropeng ML, Mbaya R (2022) Activated carbon derived from waste orange and lemon peels for the adsorption of methyl orange and methylene blue dyes from wastewater. Heliyon 8(8). https://doi.org/10.1016/J.HELIYON.2022.E09930/ASSET/215F869D-D759-41A1-A742-6AF5755B2DD5/MAIN.ASSETS/GR9.JPG
23. Rastogi A, Tiwari MK, Ghangrekar MM (2021) A review on environmental occurrence, toxicity and microbial degradation of non-steroidal anti-inflammatory drugs (NSAIDs). J Environ Manage 300:113694. https://doi.org/10.1016/J.JENVMAN.2021.113694
24. Sadanand Giri Y, Subash A, Kandasubramanian B (2024) Green threads of progress: natural fibers reshaping wastewater cleanup strategies, a review. Hybrid Adv 6:100237. https://doi.org/10.1016/J.HYBADV.2024.100237
25. Sarioglu M, Atay ÜA, Sarioglu M, Atay UA (2006) Removal of methylene blue by using biosolid. Global NEST J 8(2). https://www.researchgate.net/publication/228482263
26. Sen Gupta S, Bhattacharyya KG (2011) Kinetics of adsorption of metal ions on inorganic materials: a review. Adv Coll Interface Sci 162(1–2):39–58. https://doi.org/10.1016/J.CIS.2010.12.004
27. Sen TK, Afroze S, Ang HM (2011) Equilibrium, kinetics and mechanism of removal of methylene blue from aqueous solution by adsorption onto pine cone biomass of Pinus radiata. Water Air Soil Pollut 218(1–4):499–515. https://doi.org/10.1007/s11270-010-0663-y
28. Shafqat SS, Zafar B, Masood SA, Shafqat SR, Khan HU, Syed A, Bahkali AH, Mutahir S, Khan MA, Xu G, Zafar MN (2024) Nitrophenylfurfural grafted amino functionalized silica nanoparticles for adsorptive removal of tartrazine dye from water. Mater Adv 5(18):7296–7311. https://doi.org/10.1039/D4MA00618F
29. Shah SS, Sharma T, Dar BA, Bamezai RK (2021) Adsorptive removal of methyl orange dye from aqueous solution using populous leaves: Insights from kinetics, thermodynamics and computational studies. Environ Chem Ecotoxicol 3:172–181. https://doi.org/10.1016/J.ENCECO.2021.05.002
30. Taha M, Elsayed A, Abbas M, Fakhry H, Ali EM (2024) Bio-sorption of Methyl blue from Synthetic wastewater onto copperzinc oxides bimetallic nanoparticles synthesized by fusarium oxysporum: equilibrium isotherms, kinetic models, process optimization, and antibacterial activity. Egypt J Chem 67(2):513–526. https://doi.org/10.21608/EJCHEM.2023.213597.8029
31. Vadivelan V, Vasanth Kumar K (2005) Equilibrium, kinetics, mechanism, and process design for the sorption of methylene blue onto rice husk. J Colloid Interface Sci 286(1):90–100. https://doi.org/10.1016/j.jcis.2005.01.007
32. Viscusi G, Lamberti E, Galluzzi A, Polichetti M, Gorrasi G (2022) Fabrication of novel hybrid materials based on iron-aluminum modified hemp fibers: Comparison between two proposed methodologies. Colloids Surf, A 642:128683. https://doi.org/10.1016/J.COLSURFA.2022.128683

33. Viscusi G, Lamberti E, Gorrasi G (2021) Hemp fibers modified with graphite oxide as green and efficient solution for water remediation: application to methylene blue. Chemosphere 132614. https://doi.org/10.1016/J.CHEMOSPHERE.2021.132614
34. Weber WJ Jr, Morris JC (1963) Kinetics of adsorption on carbon from solution. J Sanit Eng Div 89(2):31–59
35. Yang C, Xu W, Nan Y, Wang Y, Chen X (2020) Novel negatively charged nanofiltration membrane based on 4,4′-diaminodiphenylmethane for dye removal. Sep Purif Technol 248:117089. https://doi.org/10.1016/j.seppur.2020.117089
36. Yang X, Wang C, Zhou B, Cheng S (2023) Characterization of an iron-copper bimetallic metal-organic framework for adsorption of methyl orange in aqueous solution. J Anal Methods Chem 2023(1):9985984. https://doi.org/10.1155/2023/9985984
37. Yu C, He J, Tan M, Hou Y, Zeng H, Liu C, Meng H, Su Y, Qiao L, Lookman T, Bai Y (2022) Selective enhancement of photo-piezocatalytic performance in $BaTiO_3$ via heterovalent ion doping. Adv Func Mater 32(52):2209365. https://doi.org/10.1002/ADFM.202209365
38. Zhao B, Zhang X, Dou C, Han R (2015) Adsorption property of methyl orange by chitosan coated on quartz sand in batch mode. Desalin Water Treat 55(6):1598–1608. https://doi.org/10.1080/19443994.2014.925834
39. Zhou Y, Lu J, Zhou Y, Liu Y (2019) Recent advances for dyes removal using novel adsorbents: a review. Environ Pollut 252:352–365. https://doi.org/10.1016/J.ENVPOL.2019.05.072

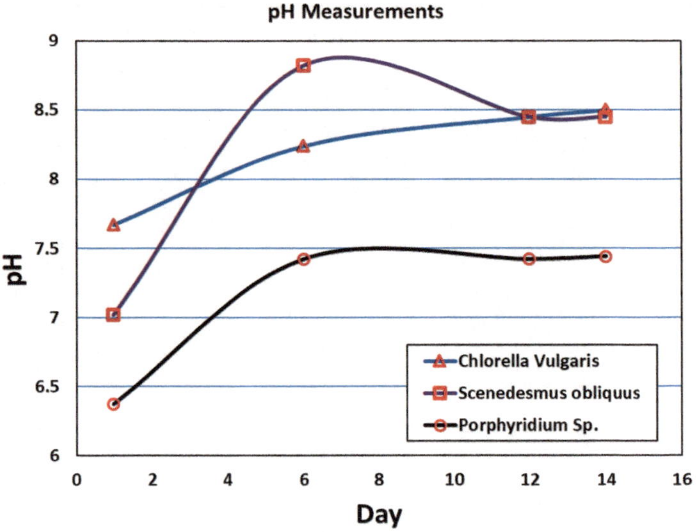

Fig. 2 The pH of growing media for three selected species

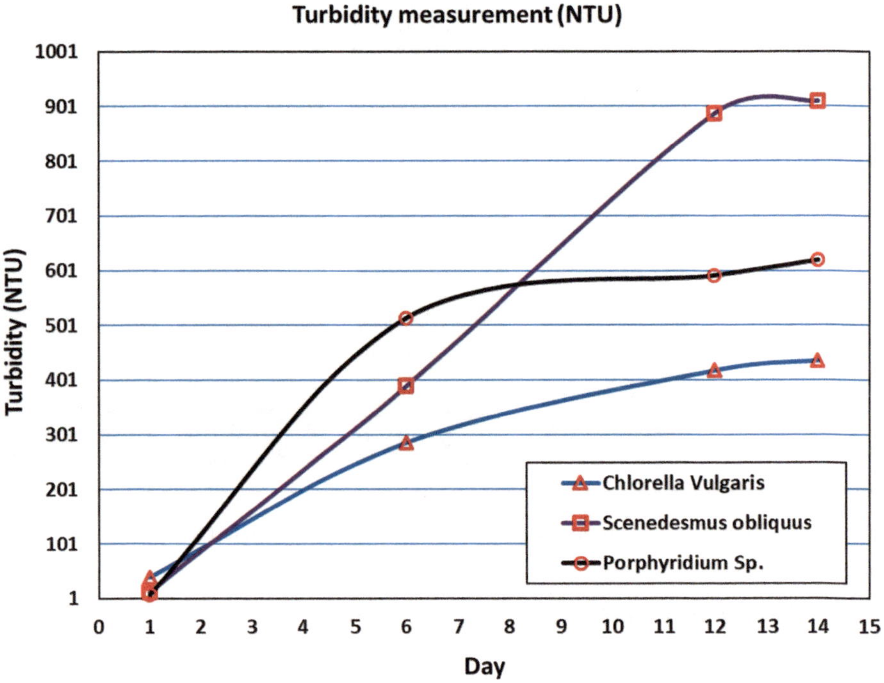

Fig. 3 The turbidity of growing media for three selected species

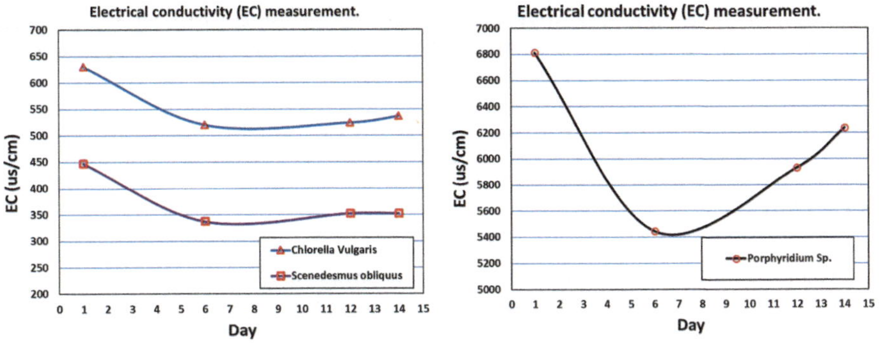

Fig. 4 The electrical conductivity of growing media for three species

EC decrease percentage of about 9% and 21%, respectively. This decrease might be attributed to the fact that over time, the ions are consumed by microalgae as nutrition for their growth. Hence, this leads to a trend depicted in Fig. 4. Moreover, the trend for Porphyridium Sp. is a bit different both in initial and final conductivity values, which implies its totally different behaviour with the total EC decrease percentage of about 4.8%. This might be a good issue for further investigations.

Moreover, the EC data can be used to calculate the Total dissolved solids (TDS) is an indicator of the amount of solids that are dissolved in the liquid, including soluble hydrogen carbonate ions, chloride salts, sulphates, calcium, magnesium, sodium, potassium, volatile solids, and non-volatile solids. The EC and TDS have a linear relation, and the trend for TDS will be the same as that of EC.

5 Conclusion

In this work, we discussed how to grow microalgae for desalination and supply fresh water in more eco-friendly methods; the tools needed to conduct the experiment using various types of microalgae were discussed. Three microalgae species were selected to run the experiments, including Chlorella Vulgaris, Scenedesmus obliquus, and Porphyridium Sp. in which Chlorella Vulgaris and Scenedesmus obliquus showed a good performance in reducing the salinity of the growing media with the total EC decrease percentage of about 9% and 21%, respectively. We emphasize that this research aimed to examine the use of microalgae and CO_2 for saline water desalination as a sustainable method with higher energy efficiency to desalinate brackish and seawater sources for agricultural and municipal water use, boosting the commercial microalgae sector by improving the production process.

Acknowledgements "This work was made possible by a UREP award [UREP29-255-1-056] from the Qatar National Research Fund (a member of the Qatar Foundation) and QRDI. The statements made herein are solely the responsibility of the authors." We also acknowledge Ms. Sara Soussy

(Lab Manager - Analytical Chemistry) at the Central Analysis Lab at UDST for supporting the research team during the experimental work and for conducting UV-Vis spectroscopy and FTIR analyses.

References

1. "World Population Clock: 8 Billion People (LIVE, 2023) - Worldometer." Accessed: Jun. 26, 2023. [Online]. Available: https://www.worldometers.info/world-population/#:~:text=World Population Clock%3A 8 Billion
2. "Water Facts - Worldwide Water Supply | ARWEC| CCAO | Area Offices | California-Great Basin | Bureau of Reclamation." Accessed: Jun. 26, 2023. [Online]. Available:https://www.usbr.gov/mp/arwec/water-facts-ww-water-sup.html#:~:text=97%25 of the earth
3. Clarke SF, Nawaz W, Skelhorn C, Amato A (2017) Towards a more sustainable waste management in Qatar: Retrofitting mindsets and changing behaviours. QScience Connect, vol. 2017, no. 1, Special Issue on Shaping Qatar's Sustainable Built Environment-Part I, p 4, May 2017. https://doi.org/10.5339/CONNECT.2017.QGBC.4
4. "Towards sustainable desalination." Accessed: Jun. 26, 2023. [Online]. Available: https://www.unep.org/news-and-stories/story/towards-sustainable-desalination
5. Esmaeili A, Moghadam HA, Golzary A (2023) Application of marine microalgae in biodesalination and CO_2 biofixation: a review. Desalination 567:116958. https://doi.org/10.1016/j.desal.2023.116958
6. El Sergany FAG, El Hosseiny OM, El Nadi MH (2019) The optimum algae dose in water desalination by algae ponds. Int Res J Adv Eng Sci 4(2):152–154
7. Zhu C et al (2018) Seawater desalination concentrate for cultivation of Dunaliella salina with floating photobioreactor to produce β-carotene. Algal Res 35:319–324. https://doi.org/10.1016/J.ALGAL.2018.08.035
8. Moayedi A, Yargholi B, Pazira E, Babazadeh H (2019) Investigated of desalination of saline waters by using Dunaliella Salina algae and its effect on water ions. Civ Eng J 5(11):2450–2460. https://doi.org/10.28991/CEJ-2019-03091423

Hybrid Deep Learning Model for Rainfall Prediction

Monther Tarawneh, Faisal Alzyoud, Khalaf Khatatneh, and Omar Tarawneh

Abstract Accurate rainfall forecasting addresses many issues, including energy production, agriculture, and most importantly disaster planning and water resource management. Many regions, including Jordan, suffer from water scarcity, which is a major concern for governments. Relying on traditional solutions is no longer possible with the ongoing climate change and weather fluctuations, so it has become necessary to rely on data analysis using artificial intelligence to increase the accuracy of rainfall forecasting. The variability in data from one place to another and over time periods leads to inconsistencies in the accuracy of rainfall prediction in models. This research presents a deep learning hybrid model for rainfall prediction. The model combines convolutional neural networks (CNN) and long short-term memory (LSTM) networks. It utilizes CNN's capabilities in analysing short-term spatial patterns and the ability to model long-term temporal dependencies in LSTM. The proposed model outperforms traditional machine learning and other deep learning approaches. The performance was examined on a dataset from the Ministry of Water and Irrigation, which includes daily rainfall amounts and some climate variables for the period 1987–1918. The study compared the performance of CNN-LSTM with other machine learning and deep learning models It obtained good values in the common key measures such as 9.98 Root Mean Squared Error, 6.8 Mean Absolute Error, and 6.24% Mean Absolute Percentage Error. Despite the complexity

M. Tarawneh (✉)
Department of IT, Faculty of IT and Communications, Tafila Technical University, Tafila, Jordan
e-mail: mtarawneh@ttu.edu.jo

F. Alzyoud
Department of Computer Science, Faculty of IT, Isra University, Amman, Jordan
e-mail: faisal.alzyoud@iu.edu.jo

K. Khatatneh
Prince Abdullah bin Ghazi for Communication and IT College, AlBalqa Applied University, Amman, Jordan
e-mail: Dr.khalf@bau.edu.jo

O. Tarawneh
Software Engineering Department, Amman Arab University, Amman, Jordan

S. Al-Naemi et al. (eds.), *Water and Food Security in the Face of Climate Change: Challenges and Opportunities for Resilience*, World Sustainability Series,
https://doi.org/10.1007/978-3-032-00098-9_68

and long-term rainfall [24]. K-Nearest Neighbour (KNN) is used also for rainfall predictions [24–26]. It works well with small datasets, but computationally expensive for large datasets and sensitive to noise. Machine learning techniques have been used to predict rainfall in many countries and cities and have been successful in areas with highly variable rainfall patterns such as the tropics. However, they must consider the availability of data, quality, computational resources, and simplicity.

Recently, deep learning (DL) has been widely utilized in rainfall forecasting [27]. The most popular deep learning models are CNNs and LSTMs. CNNs are suitable for obtaining spatial features, while LSTMs are constructed to process sequential data, making them ideal for time series forecasts such as rainfall. A study compared the performance of DL models in predicting monthly rainfall [28]. The study concluded that the LSTM algorithm is the most suitable model for predicting monthly rainfall. This is due to its ability to manage sequential data and retain long-term dependencies, making it the most suitable model for predicting complex time series such as rainfall. This proves that LSTM can improve the accuracy over other methods and combining it with other models will enhance the prediction accuracy. Therefore, combine CNN and LSTMs have demonstrated remarkable effects in rainfall forecasting by capturing spatial and sequential patterns in the data [29].

There is hybrid models integrate machine learning techniques [2, 30–32] or combine machine learning with physics-based models [33] to explain all possible patterns in the data. The main goal of hybrid models is to improve accuracy and reliability by combining the strengths of different methods and overcoming the limitations of individual methods. Also, the ability of DL makes AI techniques more effective in weather and rainfall forecasting than or complements traditional numerical weather prediction methods. A hybrid neural network approach combining wavelet transform and neural networks [34]. It shows effective result for short-term rainfall forecasting [35]. Genetic algorithms are combined with different machine learning algorithms to enhance accuracy such as SVMs [36] or ANNs [37], CNNs [38] and others to improve the model. Try all possible combinations. Ensemble method is also used to increase the accuracy of forecasts [39, 40]. Multiple models are employed to obtain a more robust final forecast. Random forests and gradient boosting machines (GBM) are well-known methods that have been used in rainfall forecasting [41]. The advantage of this method is to reduce the forecast gap and improve the accuracy of rainfall forecasts.

3 Methodology

This study aims to develop a hybrid model that combines CNN and LSTM networks to predict rainfall in Jordan based on rainfall records between 1987 and 2018. The proposed methodology consists of several stages starting with data processing, then training the hybrid model consisting of several levels to obtain the results and evaluate the model performance as shown in Fig. 1.

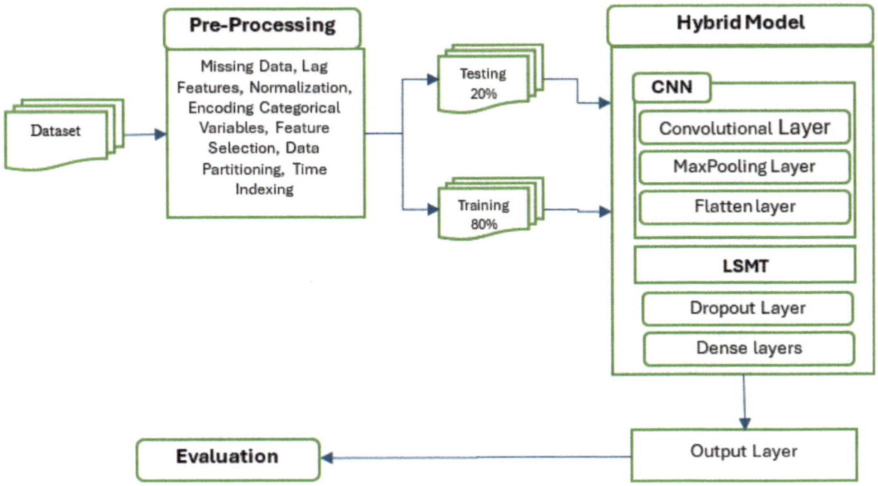

Fig. 1 Architecture of the hybrid CNN-LSTM model for rainfall prediction

3.1 Data Set

The dataset was obtained from the Jordanian Ministry of Water and Irrigation (Jordan Valley Authority). The dataset contains daily rainfall and weather patterns over three decades between 1987 and 2018. The dataset includes a range of features such as rainfall, water levels in dams, year, day, month, and climate variables such as temperature, wet-bulb and dry-bulb temperatures, and vapor pressure. The dataset contains more than 35,000 records. Some additional lagged variables were added to capture the temporal dependencies of daily measurements. The presence of day, day, and year features helps in conducting temporal analysis across different seasons and years. The comprehensiveness of the dataset also helps in studying the relationships between many variables to aid in water resource management and climate analysis. Table 1 describes the key features in the dataset. Figure 2 summarizes the average seasonal rainfall at different stations for the period 2015–2020. The data is classified by the four seasons. This gives us an idea of the seasonal rainfall patterns at each station.

3.2 Preprocessing

To ensure the dataset was properly prepared for analysis and machine learning, the following preprocessing steps were conducted:

- Handling Missing Data: Important rows with unnecessary missing data were eliminated, while other missing values were filled using forward and backward filling

- CNN Layers: In CNN model, the 1D Convolutional Layer filters the entered data over time to identify short-term patterns in the series, such as variations in rainfall or temperature. By examining the data, these layers enable the model to discover important developments and dependencies that might be invisible. The kernel size determines how many steps the filter covers at a time, allowing the model to capture local patterns over a given period. The ReLU function is used after the convolution to insert a bit of nonlinearity to the model, helping it to discover more complex patterns. To reduce noise and further increase the efficiency of the model, the Max Pooling layer focuses on the small parts of the input and only keeps the most important information (mostly the highest value).
- Flatten Layer: To ensure that data is passed to the LSTM layer that receives serial inputs in a specific format, the output is normalized to a one-dimensional vector.
- LSTM Layers: LSTMs are used for forecasting because they can record and study past patterns over a period and can be used to predict water levels or rainfall. Each LSTM cell has an internal memory that is updated from the input and the state of the previous cell, allowing the model to track information over time. There are instances where the LSTM replays the entire sequence, and some instances where a final output is required. In LSTM cells, tanh or sigmoid activations are utilized.
- Dropout Layer: After the LSTM layer, to avoid overfitting, add a dropout layer. The advantage of this is that the dropout layer generalizes the model better by randomly "dropping" a portion of the units (neurons) during the training phase, relying heavily on any one neuron. Thus, there is no dependence on any one neuron. Common dropout rates range from 0.2 to 0.5.
- Fully Connected (Dense) Layer: A fully connected dense layer is used to connect all the high-level patterns learned by the model and map them to the final output. In this layer, each input is connected to an output, allowing the model to understand the complex relationships between unique features. In the middle layers, a ReLU function is applied to help the model learn more effectively. For the final output layer, especially in regression tasks, there is usually no activation function, or a simple linear activation is used to provide continuous predictions.
- Output Layer: The result is a single value representing the amount of rainfall in millimetres.

4 Model Performance

To measure the performance of the proposed hybrid model combining CNN and LSTM, machine learning and deep learning algorithms were implemented on the data as shown in Table 2. The result shows that the traditional machine learning models (decision tree and random forest) performed poorly. The values of both MAPE and R2 proves that these models have difficulty overseeing complex data. Therefore, they should not be used for rainfall forecasting. DL models have an advantage over conventional machine learning models. The performance of DNN, RNN and LSTM was superior. DNN reduced the prediction errors as per the RMSE value (9.94).

Table 2 Comparative results between machine learning and deep learning models

Model	RMSE	MAE	MAPE	R^2
SVM	10.681789	6.245914	328.769417	– 0.12
DT	13.203792	9.360255	805.654174	– 0.68
RF	11.475742	8.231040	788.362856	– 0.30
RNN	11.887698	6.929927	175.929254	0.04
LSTM	12.408994	7.508246	122.048584	0.04
DNN	9.935443	6.795800	677.661413	0.03
CNN_LSTM	9.9799	6.7880	6.240	0.0259

Although the MAPE value is high and indicates that it has difficulty in handling scaling errors in some instances.

The hybrid model CNN_LSTM has obtained a low MAPE, which means that it can manage percentage-based prediction errors more effectively, especially in rainfall data, which can be highly variable. Combining convolutional layers that capture spatial dependencies with deep learning models that capture temporal dependencies combines the two to provide better predictive power for time series data such as rainfall. This shows how powerful deep learning models are in obtaining highly accurate rainfall predictions, especially when combined with a hybrid model like CNN_LSTM.

More optimization of the hybrid model may improve the ability to predict accurate rainfall. Some improvements can be applied, such as using bidirectional LSTM layers so that the model learns dependencies in both forward and backward directions and captures detailed temporal information. The model's ability to analyse complex time sequences can be improved by using stacked LSTM layers. Also, to avoid overfitting by randomly deactivating units during training, discrete regularization can be used. Increasing the number of convolutional filters in the CNN layer gives the ability to capture more spatial features, and by tuning the Adam optimizer, learning can be accelerated, and convergence can be improved.

5 Conclusion

Water is the backbone of all aspects of life on Earth, we use it for drinking, agriculture, industry, energy production, etc. Water scarcity has become a major problem in most parts of the world, and Jordan is one of the most concerned countries with providing water. Therefore, appropriate solutions must be used to predict rainfall amounts and manage water resources in a balanced manner.

Traditional models have become ineffective in this regard, so technology, especially artificial intelligence, has been employed. The study proposed a hybrid model for predicting rainfall amounts in Jordan that combines CNN and LSTM networks. The dataset from the Jordanian Ministry of Water and Irrigation was used, which includes rainfall records from 1987–1918. The model is remarkably effective in capturing both spatial and temporal dependencies in the data, and the ability to predict rainfall with high accuracy. The results showed that the proposed model outperformed machine learning models in predicting rainfall amounts. It was found that machine learning models have difficulty managing the complexity of the dataset. The output of this study opens the door to using innovative methods to improve the CNN-LSTM model. Bidirectional and stacked LSTM layers can be combined, hyper-parameters can be tuned, and additional climate features can be introduced. Applying the model to other regions or introducing climate variables increases the power and accuracy of the application and the use of its results in water resources management, disaster management, and agricultural decision-making in Jordan and similar arid regions. Incorporating additional weather variables improves the model's ability to predict weather in real time. Also, including advanced techniques, such as bidirectional LSTM layers or attention mechanisms, increases the model's ability to capture complex dependencies in the data.

References

1. Trinh TA (2018) The impact of climate change on agriculture: findings from households in Vietnam. Environ Resource Econ 71(4):897–921
2. Dotse S-Q et al (2024) A review of the application of hybrid machine learning models to improve rainfall prediction. Model Earth Syst Environ 10(1):19–44
3. Sethi SS, Sharma P (2023) New developments in the implementation of IoT in agriculture. SN Comput Sci 4(5):503
4. Drogkoula M, Kokkinos K, Samaras N (2023) A comprehensive survey of machine learning methodologies with emphasis in water resources management. Appl Sci 13(22):12147
5. Rajabi N, Rajabi K, Rajabi F (2023) Forecasting and management of disasters triggered by climate change. Visualization techniques for climate change with machine learning and artificial intelligence. Elsevier, pp 181–207
6. Luo Y et al (2023) Technology roadmap for flexible sensors. ACS Nano 17(6):5211–5295
7. Bauer P, Thorpe A, Brunet G (2015) The quiet revolution of numerical weather prediction. Nature 525(7567):47–55
8. Li W et al (2024) Deepphysinet: bridging deep learning and atmospheric physics for accurate and continuous weather modeling. ArXiv preprint arXiv:2401.04125
9. Nelson BK (1998) Time series analysis using autoregressive integrated moving average (ARIMA) models. Acad Emerg Med 5(7):739–744
10. Rezaiy R, Shabri A (2024) Enhancing drought prediction precision with EEMD-ARIMA modeling based on standardized precipitation index. Water Sci Technol 89(3):745–770
11. Saleh MA, Rasel H, Ray B (2024) A comprehensive review towards resilient rainfall forecasting models using artificial intelligence techniques. Green Technol Sustain, 100104
12. Baig F et al (2024) How accurate are the machine learning models in improving monthly rainfall prediction in hyper arid environment? J Hydrol 633:131040
13. Pham BT et al (2020) Development of advanced artificial intelligence models for daily rainfall prediction. Atmos Res 237:104845
14. Luk KC, Ball JE, Sharma A (2001) An application of artificial neural networks for rainfall forecasting. Math Comput Model 33(6–7):683–693
15. Amr H et al (2011) Artificial neural network technique for rainfall forecasting applied to Alexandria Egypt. Int J Phys Sci 6(6):1306–1316

16. Hung NQ et al (2009) An artificial neural network model for rainfall forecasting in Bangkok Thailand. Hydrol Earth Syst Sci 13(8):1413–1425
17. Lee J et al (2018) Application of artificial neural networks to rainfall forecasting in the Geum River Basin, Korea. Water 10(10):1448
18. Lin GF et al (2009) Effective forecasting of hourly typhoon rainfall using support vector machines. Water Res Res 45(8)
19. Yu P-S et al (2017) Comparison of random forests and support vector machine for real-time radar-derived rainfall forecasting. J Hydrol 552:92–104
20. Wang W-C et al (2013) Improved annual rainfall-runoff forecasting using PSO–SVM model based on EEMD. J Hydroinf 15(4):1377–1390
21. Sivapragasam C, Liong S-Y, Pasha M (2001) Rainfall and runoff forecasting with SSA–SVM approach. J Hydroinf 3(3):141–152
22. Samantaray S et al (2020) Rainfall forecasting through ANN and SVM in Bolangir Watershed, India. In: Smart intelligent computing and applications: proceedings of the third international conference on smart computing and informatics, vol 1. Springer (2020)
23. Hill AJ, Schumacher RS (2021) Forecasting excessive rainfall with random forests and a deterministic convection-allowing model. Weather Forecast 36(5):1693–1711
24. Diez-Sierra J, del Jesus M (2019) Subdaily rainfall estimation through daily rainfall downscaling using random forests in Spain. Water 11(1):125
25. Cristian M (2018) Average monthly rainfall forecast in Romania by using k-nearest neighbors regression. Annals of'Constantin Brancusi'University of Targu-Jiu. Economy Series/Analele Universității'Constantin Brâncuşi'din Târgu-Jiu Seria Economic, 2018(4)
26. Biruntha S et al (2022) Rainfall prediction using kNN and decision tree. In 2022: International conference on electronics and renewable systems (ICEARS). IEEE
27. Hernández E et al (2016) Rainfall prediction: a deep learning approach. In: Hybrid artificial intelligent systems: 11th international conference, HAIS 2016, Seville, Spain, April 18–20, 2016, Proceedings 11. 2016. Springer
28. Kanchan P, Kumar Bakkappa Shardoor N (2021) Rainfall analysis and forecasting using deep learning technique. J Inf Elect Electron Eng (JIEEE), 2(2):1–11
29. Sato R, Fujimoto Y (2024) Rainfall forecasting with LSTM by combining cloud image feature extraction with CNN and weather information. IEEJ J Ind Appl 13(1):24–33
30. Basha CZ et al (2020) Rainfall prediction using machine learning & deep learning techniques. In: 2020 international conference on electronics and sustainable communication systems (ICESC). IEEE
31. Singh G, Kumar D (2019) Hybrid prediction models for rainfall forecasting. In: 2019 9th International conference on cloud computing, data science & engineering (Confluence). IEEE
32. Di Nunno F et al (2022) Precipitation forecasting in Northern Bangladesh using a hybrid machine learning model. Sustainability 14(5):2663
33. Anuradha T, Formal PAS, RamaDevi J (2024) Hybrid model for rainfall prediction with statistical and technical indicator feature set. Expert Syst Appl 249:123260
34. Partal T, Cigizoglu HK (2009) Prediction of daily precipitation using wavelet—neural networks. Hydrol Sci J 54(2):234–246
35. Taie Semiromi M, Koch M (2024) Statistical downscaling of precipitation in northwestern Iran using a hybrid model of discrete wavelet transform, artificial neural networks, and quantile mapping. Theoret Appl Climatol, pp 1–31
36. Yang TC et al (2018) Predictor selection method for the construction of support vector machine (SVM)-based typhoon rainfall forecasting models using a non-dominated sorting genetic algorithm. Meteorol Appl 25(4):510–522
37. Nasseri M, Asghari K, Abedini MJ (2008) Optimized scenario for rainfall forecasting using genetic algorithm coupled with artificial neural network. Expert Syst Appl 35(3):1415–1421
38. Ngan TT et al (2023) A hybrid of RainNet and genetic algorithm in nowcasting prediction. Earth Sci Inf 16(4):3885–3894
39. Sani NS et al (2020) Ensemble learning for rainfall prediction. Int J Adv Comput Sci Applications 11(11)

Fig. 2 Tarragona roman aqueduct. *Source* Worldstrides

In the twentieth century, the development of satellite technology revolutionized elevation measurement. Satellites equipped with altimeters can measure the distance between the satellite and the Earth's surface. By combining these measurements with precise satellite positioning data, scientists can create detailed digital elevation models (DEMs) that provide highly accurate elevation information for the entire planet.

Modern elevation data is essential for various applications, including cartography, urban planning, environmental studies, and natural disaster management. By understanding the Earth's topography, scientists can better understand and predict geological processes, climate patterns, and the distribution of natural resources. Figure 3 illustrates the elevation map of Syria and Lebanon.

3 The Proposed Algorithm

By utilizing precise elevation maps to identify the optimal route for aggregation tunnels, we can eliminate the need for energy-intensive pumping systems to elevate water levels. This not only reduces electricity consumption but also enhances water security in remote areas. Additionally, the increased efficiency of water delivery can significantly promote irrigated agriculture across vast tracts of land. The fundamental principle guiding this approach is to select the path that minimizes the elevation drop between the water source, such as a spring, reservoir, or dam, and the intended destination.

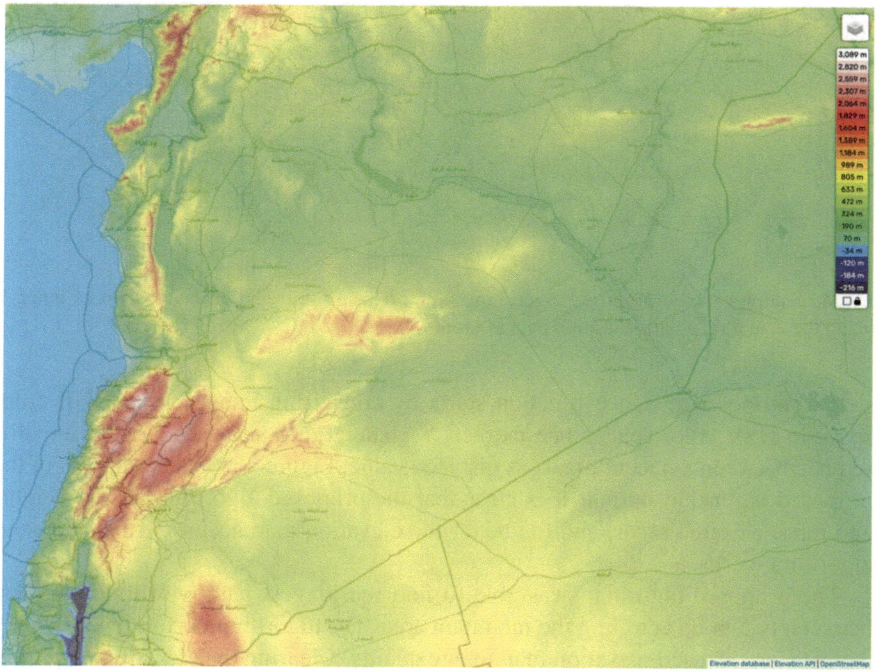

Fig. 3 The elevation map of Syria and Lebanon. *Source* The world topographic map

To achieve our task, we used Dijkstra's algorithm with the elevation maps. Dijkstra's algorithm is a classic graph search algorithm used to find the shortest path between nodes in a graph. It works by iteratively exploring nodes closest to the starting point, ensuring that the path to each visited node is optimal [3, 4]. We used Dijkstra's algorithm with elevation map images as follows.

1. Each pixel in the elevation map image was treated as a node in Dijkstra's algorithm.
2. Assign a distance of 0 for the starting pixel, and infinity for all other pixels (since they are not visited yet)
3. Calculate the distance between the start pixel and neighbor pixels, the used distance function reflects the color difference which aligns with the area elevation difference.
4. Move to the pixel with the least distance from the starting pixel.
5. We repeat steps 3 and 4 with the unvisited pixels until we reach our target pixel.

Dijkstra's algorithm has a time complexity of $O((|V| + |E|) \log |V|)$, where $|V|$ is the number of pixels (nodes) and $|E|$ is the number of edges in the graph (each pixel is connected with all adjacent pixels). This efficiency is achieved by using a priority queue to efficiently select the node with the smallest distance.

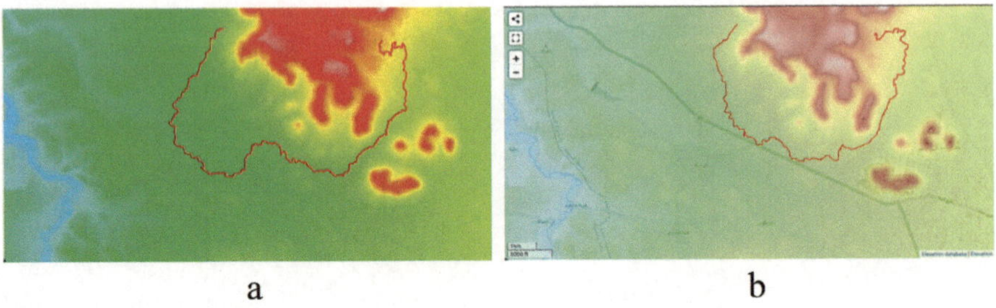

Fig. 4 Samples of the proposed algorithm output, **a** using only the elevation map as input, **b** using the actual map with elevation data as input

The proposed distance function starts by converting images from RGB color space to HSV color space, then extracts V channel info and uses the absolute difference between adjacent pixels as distance value. Figure 4 presents samples of the proposed algorithm output. It is clear that the proposed algorithm avoids heights and more elevated areas, and the better the elevation map accuracy, the better our algorithm works.

The proposed algorithm was able to find the longest path with the least drop between two points circling the mountain area. The algorithm performance heavily depends on the used elevation map image quality. Clearing the elevation map from metadata (area labels and roads) provides a better and more accurate path as illustrated in Fig. 4a.

4 Case Study: Lover's Tunnel

To illustrate our theoretical framework, we will examine the 'Lover's Roman Tunnel' as a case study. This historical tunnel, located in Syria, connects the Ayn az Zarqa spring in Salamiyah-Hama to the city of Afamia. Spanning approximately 80 km between the two locations, the tunnel's estimated length is around 150 km.

The Lover's Roman Tunnel not only served the city of Afamia but also provided a vital water source for numerous villages and towns along its route. Constructed approximately 2000 years ago, around the year 25. The channel was buried between one and three meters underground and was built using a combination of bricks and local stones. To navigate rocky terrain, the engineers skillfully curved the tunnel and incorporated twelve bridges. The tunnel's remarkable durability is evident, as it remained in operation until an earthquake struck Afamia in 1157, causing significant damage, Table 1 lists the towns where ruins of the tunnel were found. [5–7]

Towns' names align with Google Maps names for reference. We weren't able to find an accurate map of the tunnel path online, Fig. 5b draws a line between

Table 1 The known path of Lover's Tunnel [6]

Order	Area name	Elevation (m)
1	Salamiyah	475
2	Taldara	385
3	Jebrin	365
4	Around Zayn al-Abidin mountain	365
5	Maardes	360
6	Suran	335
7	Ltamenah	295
8	Morek	385
9	Lahaya	339
10	Latmeen	333
11	Kafr Zita	300

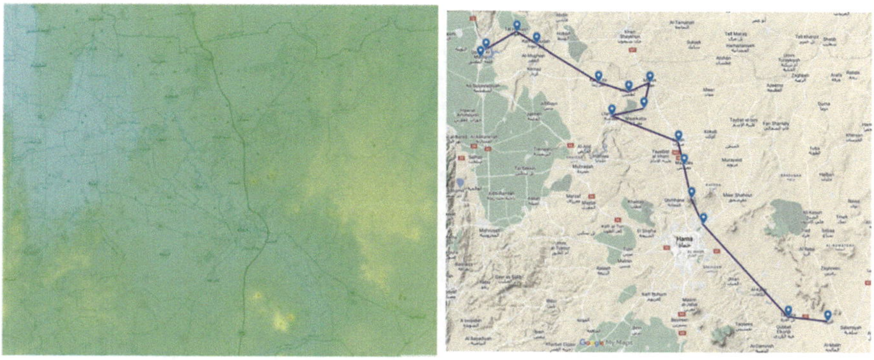

a b

Fig. 5 The known path of Lover's Tunnel, **a** The elevation map of the area, **b** the map with labels

the towns where ruins of the tunnel were found, while Fig. 5a shows the elevation map of the same area.

To apply our algorithm to the mentioned case, we start by getting the elevation map for the area (Fig. 6a), we used "The World Topographic Map" as the elevation map source for this paper.

After selecting the tunnel area in the map, we extract the accurate elevation map (without labels or road, Fig. 6b) for better accuracy. Then we extracted the V channel info for the least drop calculation (Fig. 6c), the suggested path of the algorithm is illustrated in Fig. 6c.

Figure 7 compares the proposed algorithm output with the actual tunnel path, the resemblance is clear, yet the proposed algorithm suggested lengthening the tunnel between Suran and Kafr Zita due to the low elevation difference for these areas.

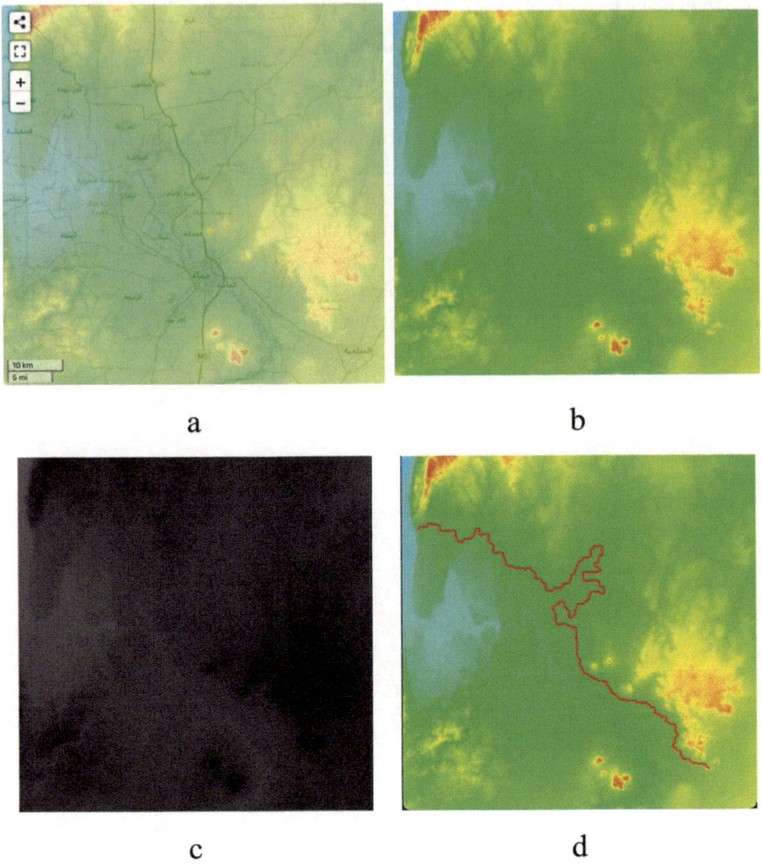

Fig. 6 Applying the proposed algorithm to the Lover's Tunnel case study, **a** The elevation map with labels, **b** the cleared elevation map, **c** The information from the V channel of the elevation map, **d** The suggested tunnel path

5 Conclusion

This research introduces a novel algorithm designed to optimize the path selection for water tunnels. By utilizing satellite-derived elevation data in conjunction with Dijkstra's algorithm, the proposed method effectively identifies the route between two points that minimizes the total elevation change. To evaluate the effectiveness of the proposed algorithm, we applied it to elevation data sourced from The World Topographic Map. The results obtained from this testing were highly promising. Figure 8 shows a few samples of the proposed algorithm outputs.

To assess the performance of the proposed algorithm, we conducted a case study using the Lover's Roman Tunnel. By applying our algorithm to the same geographic area and comparing the resulting optimal path to the historical route of

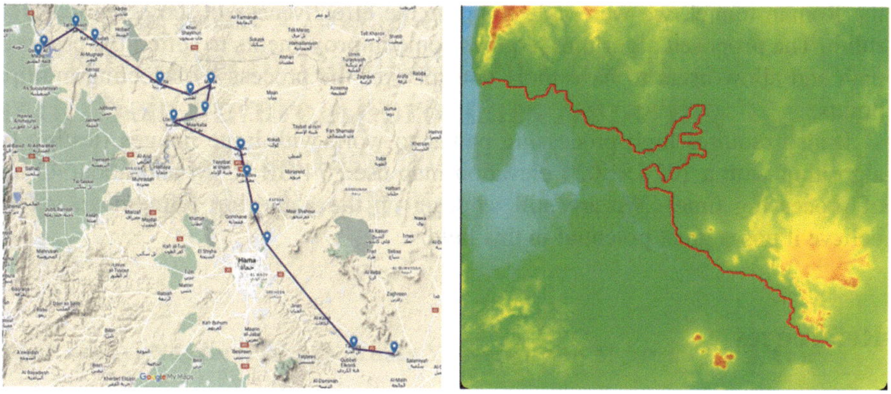

Fig. 7 Comparing the proposed algorithm output with the actual path of Lover's Tunnel

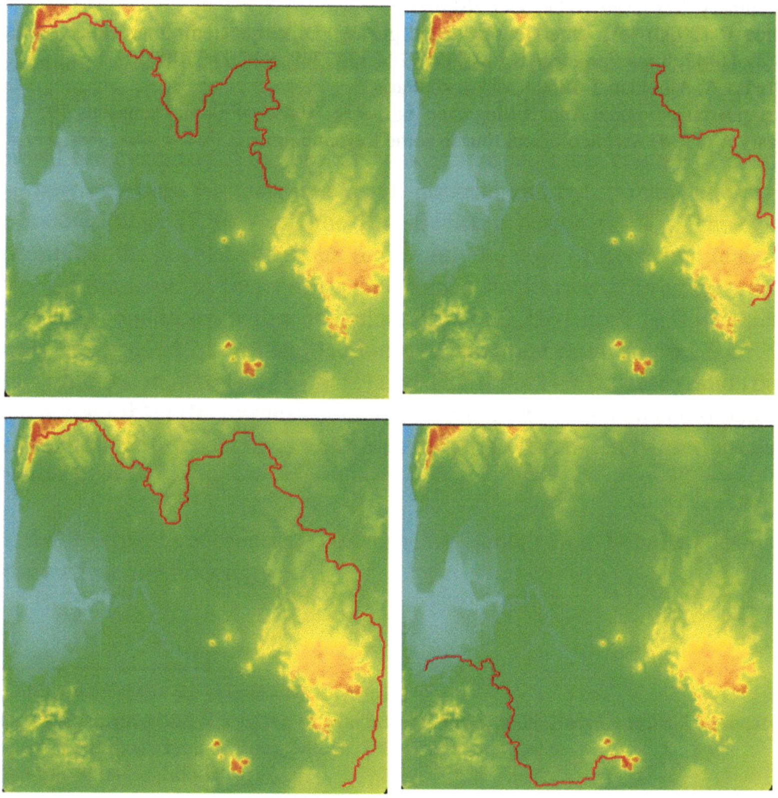

Fig. 8 Samples of the proposed algorithm outputs for different sources and destinations

1 Introduction

Water resources are a real wealth, both precious, fragile and limited, which must be the subject of coherent and rational management. Essential to life and any economic activity, water is used for domestic, industrial or agricultural purposes which all require an abundance of quality water. In fact, the most industrial operations consume a lot of water and release significant quantities of wastewater into the environment [1–3].

Water is said to be polluted when its balance is permanently modified by the contribution of very large quantities of more or less toxic substances of natural origin or from human activities. One of the main pollutants is dyes, which can be hazardous to human health and an annoyance to the environment once they dissolve in water. This is because they are synthetic and have a complex molecular structure that makes them more stable and recalcitrant to the conventional process traitment [4].

In principle, these waters are collected to be treated, in order to discharge them into waterways or into the environment does not degrade these natural environments, and does not cause serious damage (partial or total destruction of fauna and flora) [5]. Most of the conventional processes require the introduction of chemicals or additives (acids, bases, chlorines, etc.), the addition of these substances is quite complex and requires extensive dosage tests because it also leads to the generation of a large volume of hazardous and toxic sludge requiring additional treatment. Other processes generate by-products that are harmful to humans, animals and aquatic life ("DBPs" disinfection by-products; case of chlorination). In this work, it was focused on the novel process of biocoagulation/bioflocculation, which uses several coagulation-flocculation agents of biomaterials, in order to protect the environment and human health from all harmful chemicals such as $FeCl_3$, $Al_2(SO_4)_3$ which represent a source of diseases such as Alzheimer's, generate toxic and non-biodegradable sludge, expensive, require additional treatment, their delicate implementation sensitive to the physicochemical parameter (pH, alkalinity). The substitution of conventional coagulants by biomaterials which are effective and represent an actual plan for sustainable development and respect the principles of green chemistry, less expensive and produce less voluminous and less harmful residual sludge. This innovative process is therefore a promising alternative for the treatment of wastewater containing dyes [6–8].

The objective of this paper is to treat a synthetic reject assimilated to a real reject wastewater containing Congo Red as a principal pollutant, applying the response surface methodology (RSM) based on central composite faced design (CCFD) to optimize the bioprocess's major operational parameters and implementing the novel bioprocess, which is the biocoagulation/bioflocculation process. This study looked into two independent factors that could reduce the effectiveness of supernatant dye and COD removal: pH and carob powder dosage. Enhancing process optimization and determining the relative importance of multiple influencing elements, especially in the face of intricate interactions, is another goal. The diagnostic checking tests

Fig. 1 The preparation of carob powder (biocoagulant)

offered by the analysis of variance (ANOVA) are then used to assess the suitability of the given model [9].

2 Material and Methods

2.1 Biomaterial Preparation

In this study, it was used carob powder as a biocoagulant/bioflocculant. This is an abundant fruit in Algeria and it does not have a particular use in the national economy. Carob was selected on the market at a price of 800 DA/kg. The preparation steps were as follows:

Cleaning the pods, drying in the open air. Then, cutting the pods and removing the grains, grinding the carob pods with a domestic grinder.

Finally, sieving the powder through a 63 μm diameter and putting the powder (Car) in a bottle for use as showed in Fig. 1.

2.2 Preparation of Synthetic Release Assimilated to a Real Release

Congo Red is a synthetic dye ($C_{32}H_{22}N_6Na_2O_6S_2$), obtained by the action of Diaz benzidine on a naphthalene sulfonic acid, which turns blue in acid solution. It is a dye that is part of the polyazo category because it has two chromophores (region of the molecule that is mainly responsible for the color) of the azo type, that is to say, each formed of two doubly bonded nitrogen atoms, and variously substituted. Physicochemical characteristics of Congo red (CR) were presented in the Table 1.

Fig. 3 SEM images of MOF-303 (3000 and 10,000 × magnifications)

after the process was recorded. Finally, calculating the difference in weight of the sample before and after being placed in the vacuum oven will indicate the amount of water absorption.

3 Preliminary Results and Discussion of Feasibility Studies

3.1 Surface Morphology and FTIR Analysis of MOF-303

Electron microscopy images have been prepared to investigate the physical changes in the surface of wool fibers to compare the morphological structure of the samples. Microscopic images and FTIR spectra of MOF-303 are presented in Figs. 3 and 4.

Strong peaks in the frequency position of 1394 and 1606 and weak peaks in the frequency position of 1529 show the -COO-Al band and amide II (N–H bending C–N stretching), respectively, that form the backbone of the MOF-303.

3.2 Surface Morphology Analysis MOF-PAN Fiber Composite

Electron microscopy images have been prepared to investigate the physical changes in the surface of samples to compare their morphological structure. Microscopic images of control PAN12% and MOF-PAN fiber composites are presented in Fig. 5.

As seen in the microscopic images, nanofibers web or mat from a 12% PAN solution by the SBS method show a very smooth and uniform surface. Also, the microscopic images taken from the MOF-PAN fiber composite samples exhibit uniform dispersion of the MOF particles and the production of smooth and uniform nanofibers. It can also be seen that increasing the MOF concentration in MOF50%-PAN fiber

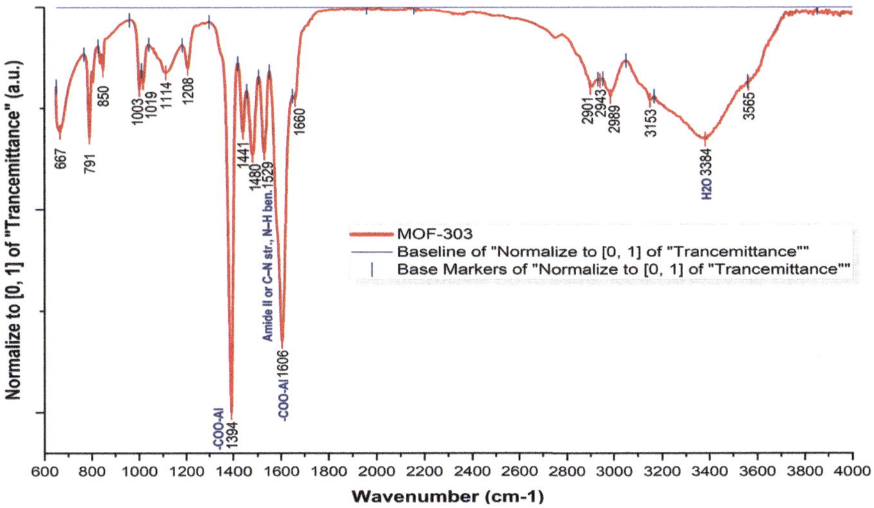

Fig. 4 Fourier transform infrared spectroscopy (FTIR) of MOF-303

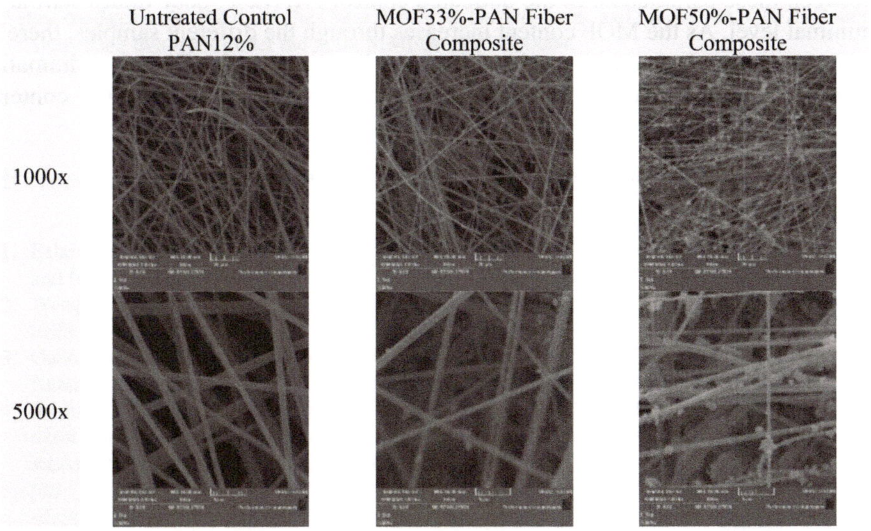

Fig. 5 SEM images of control PAN12% and MOF303-PAN fiber composite

maintains uniform distribution. The nanofiber diameter shows a normal distribution with a mean of 746.53 nm and a standard deviation of 279.42 nm, indicating a narrow and well-controlled spinning process.

1 Introduction

1.1 Background

Qatar relies heavily on desalinated water from the Arabian Gulf as its main source of fresh water [1]. The country has limited and overexploited groundwater resources, which are mainly used for irrigation purposes [2]. Qatar's high-water consumption per capita, rapid urbanization, and increased farming activities have put tremendous pressure on groundwater resources [3]. To address the need for alternative water sources, researchers have investigated the potential of using industrial low-grade waste heat to produce freshwater using membrane distillation [4].

In the arid expanses of Qatar, freshwater is a scarce and precious resource. The nation's reliance on desalination attests to the scarcity, with a significant portion of its freshwater needs met through technologically intensive processes. Official statistics reflect this deficiency; natural freshwater resources are limited to scarce seasonal rainfalls and confined aquifer systems with very low replenishment rates. In response to these constraints, agricultural entrepreneurs have sought autonomy over water supply through the excavation of artesian wells. These wells, often unlicensed, are proliferating across the countryside, especially around the capital city of Doha. They provide a crucial lifeline for irrigation and livestock, sustaining farms outside the reach of urban infrastructure (Fig. 1).

Yet, this approach comes with an unintended consequence: the potential for contamination. The lack of regulation and oversight that accompanies unlicensed well digging leaves these water sources vulnerable to bacterial contamination, among other pathogens. The ramifications of using such contaminated water are severe, extending beyond agricultural productivity to encompass both human and animal health. Evidence suggests that the absence of routine quality checks can lead to the proliferation of pathogenic bacteria, which, if used for irrigation or as drinking water, can introduce a host of waterborne diseases into the food supply and broader ecosystem.

Recognizing the gravity of this issue, our research targets the development and implementation of accessible biotechnological testing methods. These methods are designed to identify bacterial contaminants in well water, providing farm owners with essential tools to monitor and ensure the safety of their independent water supplies. Through this initiative, we aim to bridge the gap between the need for resource autonomy and the imperative of maintaining public and environmental health.

1.2 Objectives

- Assess the level of bacterial contamination in well water from unlicensed artesian wells.

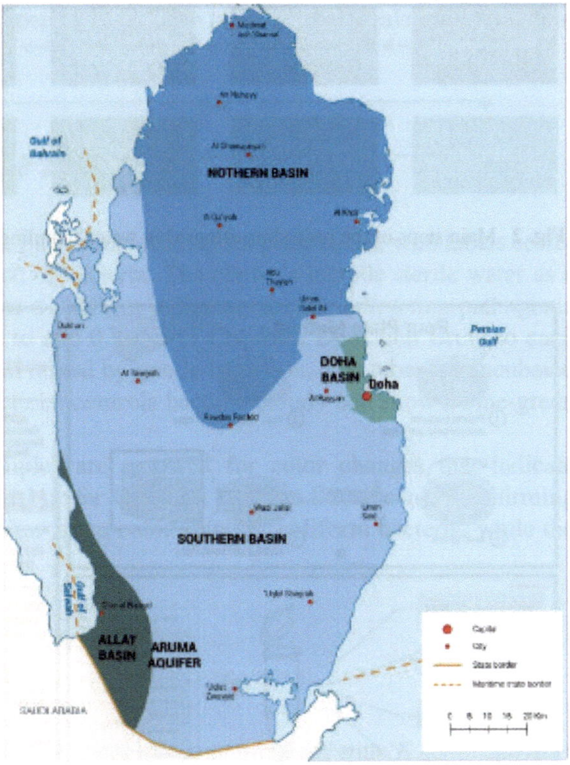

Fig. 1 Groundwater basins in Qatar

- Develop and implement biotechnological testing methods, including PCR, pour plating, and chromogenic assays, to detect and confirm bacterial contaminants.
- Provide farm owners with a set of reliable tests for routine monitoring of biological contaminants in well water.

2 Methods

2.1 Collecting Water Samples

See Fig. 2.

Figure 3 shows the main steps of the procedure adopted in water sampling for biological, physical, and chemical investigation. Below is the different testing techniques used.

Table 1 (continued)

Identifier	Geographic coordinates	Date	Time	Environmental conditions
SWW2	25.161680, 51.595524	22/1/2024	11:30 AM	Al-Wakrah
SWW3	25.161680, 51.595525	22/1/2024	12:00 PM	Al-Wakrah
ALKHOR C 1	25.686849, 51.509458	25/1/2024	4:30 PM	Al Khor
ALKHOR C 2	25.686849, 51.509459	23/1/2024	3:15 PM	Al Khor
ALKHOR C 3	25.686849, 51.509460	23/1/2024	3:45 PM	Al Khor
BFW 1	25.178206, 51.591447	4/2/2024	10:30 AM	Near WTP (Al-Wakrah)
BFW 2	25.178206, 51.591448	4/2/2024	10:40 AM	Near WTP (Al-Wakrah)
BFW 3	25.178206, 51.591449	4/2/2024	11:00 AM	Near WTP (Al-Wakrah)
SALWA BEACH 1	25.686849, 51.509458	12/1/2024	2:30 PM	Salwa Beach
SALWA BEACH 2	25.686849, 51.509459	12/1/2024	2:50 PM	Salwa Beach
SALWA BEACH 3	25.686849, 51.509460	12/1/2024	3:15 PM	Salwa Beach
DELMN BIOREACTOR 1	25.161680, 51.595520	14/1/2024	9:20 AM	Near WTP (Al-Wakrah)
DELMN BIOREACTOR 2	25.161680, 51.595521	14/1/2024	9:45 AM	Near WTP (Al-Wakrah)
DELMN BIOREACTOR 3	25.161680, 51.595522	14/1/2024	10:30 AM	Near WTP (Al-Wakrah)
Corniche Doha	25.298157, 51.520084	10/1/2024	3:30 PM	Doha City
Corniche Doha	25.298157, 51.520085	10/2/2024	3:40 PM	Doha City
Corniche Doha	25.298157, 51.520086	10/3/2024	4:00 PM	Doha City

3.4 Multiplex PCR Test Results

Samples that were tested positive for the presence of coliform bacteria were further tested using multiplex PCR to identify the type of bacteria (Fig. 7). The results are summarized below.

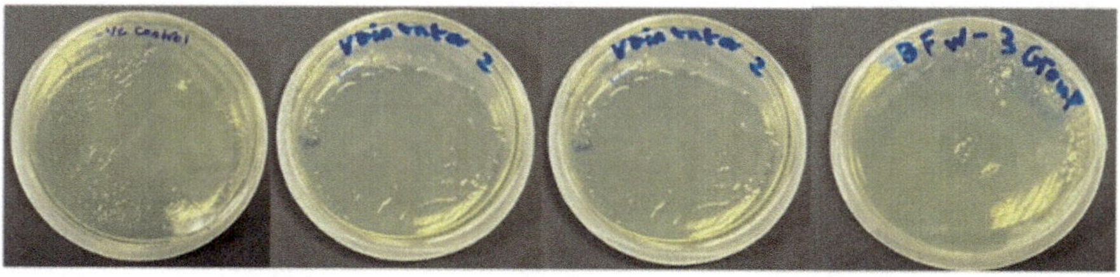

Fig. 4 Samples of negative results

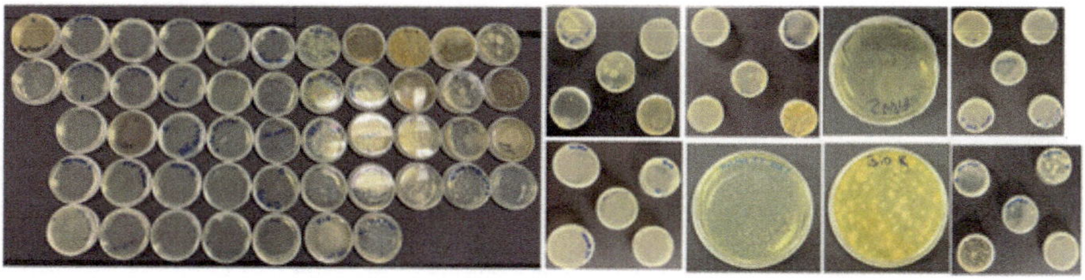

Fig. 5 Samples of the positive results

3.5 Comparing Contamination Levels Across Different Locations

The previous figures indicate that groundwater near animal farms is likely to be exposed to fecal contamination due to runoff from the farm. Also, the marine environments near the coast are influenced by human activities, waste runoff, and environmental factors such as salinity. In addition to that, the recreational pools contamination can still occur in poorly maintained pools or through human activity. Wastewater from industrial sources contains a diverse microbial community influenced by both human and industrial waste (Fig. 8).

4 Discussion

The exploration into the bacterial contamination of unlicensed artesian wells in Qatar reveals a substantial presence of coliform bacteria, including *E. coli*, which poses health risks for agricultural use and human consumption. The study's biotechnological testing methods, including PCR, pour plating, and chromogenic assays, have proven effective in detecting and quantifying these bacterial contaminants. The presence of *E. coli*, indicated by chromogenic assays, was confirmed through the indole test, showcasing the efficiency of the methods used.

The research methodology adopted in the study on water contamination involved several well-defined steps:

Table 2 Samples of the pour plate test results

Identifier	Result	Identifier	Result
Negative control -1	No growth	DEOILER OUTLET 2	No growth
Negative control -2	No growth	DEOILER OUTLET 3	No growth
Negative control -3	No growth	SW1	Growth
Positive control (*E. coli*)	Growth	SW2	Growth
Positive control (*S. merscenes*)	Growth	SW3	Growth
Positive control (*B. subtilis*)	Growth	SWW1	Growth
DNF 1	No growth	SWW2	Growth
DNF 2	No growth	SWW3	Growth
DNF 3	No growth	ALKHOR C 1	Growth
T-23251 1	No growth	ALKHOR C 2	Growth
T-23251 2	Growth	ALKHOR C 3	Growth
T-23251 3	No growth	BFW 1	Growth
OUTFALL 1	No growth	BFW 2	Growth
OUTFALL 2	No growth	BFW 3	Growth
OUTFALL 3	Growth	SALWA BEACH 1	No growth
RAIN WATER 1	No growth	SALWA BEACH 2	No growth
RAIN WATER 2	Growth	SALWA BEACH 3	No growth
RAIN WATER 3	Growth	DELMN BIOREACTOR 1	Growth
FARM 1	Growth	DELMN BIOREACTOR 2	Growth
FARM 2	Growth	DELMN BIOREACTOR 3	Growth
FARM 3	Growth	Corniche—Doha	Growth
DEOILER OUTLET 1	No growth	Corniche—Doha	Growth
		Corniche—Doha	No growth

Table 3 Predicted DNA bands sizes based on the DNA ladder

	Sample name	Molecular weight (bp)
	EdvoQuick^{TM} DNA ladder	2640, 1400, 1100, 700, 600, 400, 200
	Extracted DNA—*Serratia marcescens*	629
	Extracted DNA—*E. coli*	479
	Extracted DNA—*Bacillus subtilis*	893

Sample Collection: Water samples were successfully gathered from various locations, including unlicensed artesian wells and effluents from different sites around Doha, Qatar. This ensured a broad and representative sample pool, critical for the reliability of the study.

Initial Screening with Pour Plating Method: Following collection, an initial screening for bacterial contamination was conducted using the pour plating method.

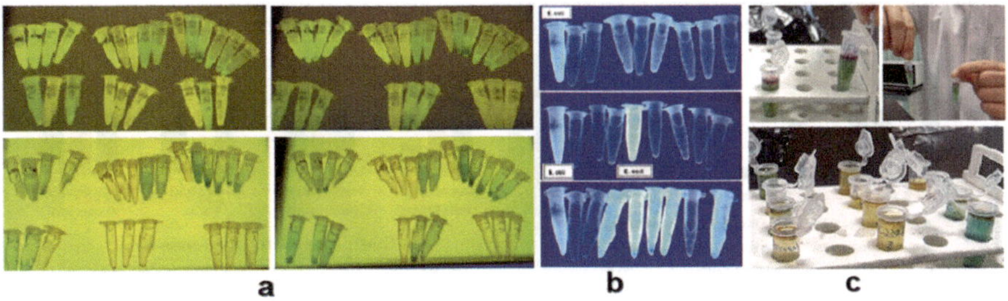

Fig. 6 **a** First chromogenic test results. **b** Second chromogenic test results. **c** Third chromogenic test results

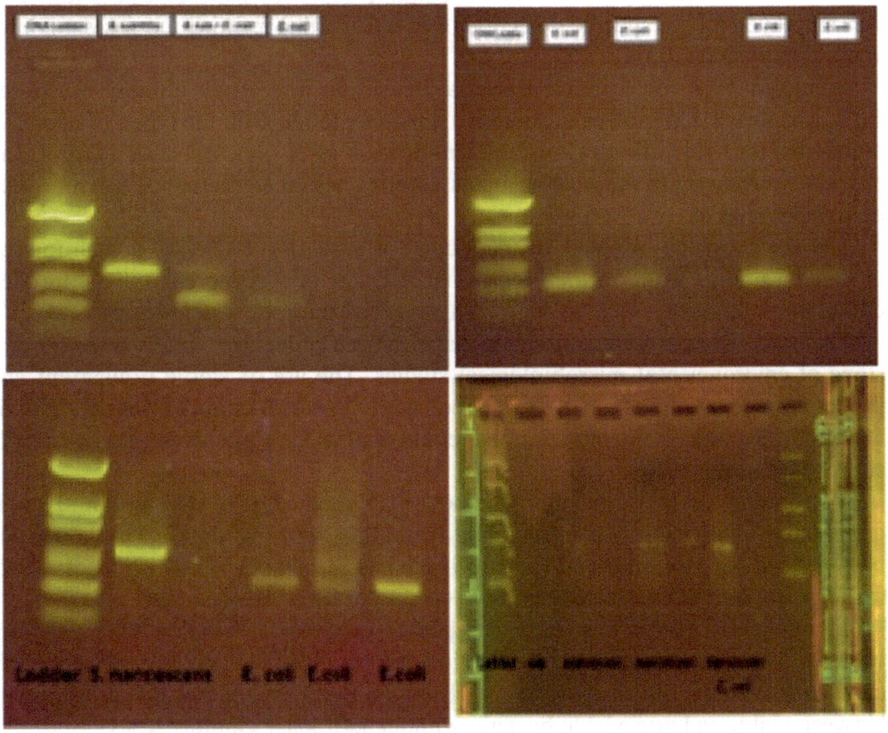

Fig. 7 The electrophoresis profile of PCR analysis of waterborne bacteria

This traditional microbiological approach allowed for the detection and enumeration of bacteria present in the water samples.

Detection of Coliform Bacteria with Chromogenic Assay: Chromogenic assays were utilized to specifically detect coliform bacteria, which are key indicators of fecal contamination and potential waterborne pathogens.

1 Introduction

Groundwater is a vital freshwater resource, especially in arid regions where available surface water is limited. It plays a key role in agriculture, serves as a strategic reserve, and supplements water needs in cases where desalination of seawater is not feasible. Ensuring its sustainable management is essential for maintaining environmental balance and supporting economic stability, while enabling water security [1]. However, excessive extraction, declining recharge rates, and increasing salinity threaten long-term groundwater sustainability [2]. Qatar, where groundwater is heavily used for agriculture and industrial purposes, faces severe depletion of this resource due to unregulated abstraction, increasing seawater intrusion, and lack of recharge [3]. The geology of Qatar is complex comprising karst limestone aquifers, characterized by sinkholes and underground channels, further complicating groundwater modeling and management [4]. Hence, assessing groundwater vulnerability in these aquifers is essential for sustainable water resource planning [5].

Numerical groundwater models such as MODFLOW are widely used for simulating aquifer behavior, but their accuracy relies on well-calibrated parameters such as hydraulic conductivity, recharge rates, and transmissivity [6, 7]. Traditional calibration methods, which manually tune parameters or apply computational optimization, struggle to capture the spatial variability of karst systems and are highly resource-intensive [8]. Studies have attempted pilot-point calibration to refine groundwater parameters, but they often fail to account for the extreme heterogeneity observed in Qatar, where hydraulic conductivity can range from less than 1 m/day in compact limestone to over 200 m/day in fractured formations [6]. Another major challenge is the scarcity of high-quality hydrogeological datasets. Historical hydrogeological maps, a primary data source, often contain overlaid text, legends, and inconsistent colors, making geospatial digitization difficult [9]. Traditional interpolation techniques, such as inverse distance weighting and kriging, struggle to fully capture the spatial complexity of aquifers [10]. Groundwater geochemistry studies further show that both natural and human activities influence groundwater chemistry, adding another layer of complexity to numerical modeling [5].

Machine learning (ML) has shown promise in hydrogeological data analysis and groundwater model calibration. ML algorithms such as Support Vector Regression (SVR), Random Forest (RF), and K-Nearest Neighbors (KNN) have been successfully applied to interpolate groundwater quality parameters, estimate recharge rates, and predict water table fluctuations [11]. Recent work has demonstrated how ML techniques can improve hydrogeological parameter extraction from historical hydrology maps, significantly enhancing numerical groundwater modeling [12]. However, most ML applications in hydrogeology focus on groundwater level prediction rather than direct calibration of hydraulic conductivity and recharge rates [13]. Studies have also highlighted the effectiveness of ML-based groundwater monitoring, including cost-effective irrigation models and aquifer recharge estimation, suggesting broader ML applications in hydrogeology [14, 15].

This study introduces a hybrid iterative calibration approach that integrates ML and zonation-based optimization to improve MODFLOW model accuracy for Qatar's aquifers. First, an ML-based preprocessing workflow extracts and refines hydraulic parameters from historical hydrogeological maps using RF regression to address color distortions and missing data. A zonation-based calibration method is then applied, where k-means clustering groups spatially homogenous regions, and gradients are computed for a single representative point per zone to balance computational efficiency and accuracy. The iterative calibration process updates hydraulic conductivity and recharge parameters based on gradient-based learning and modified bisection, reducing the mean squared error (MSE) between observed and simulated groundwater levels [13]. By integrating ML with zonation-based calibration, this study provides a scalable, computationally efficient method for groundwater modeling in data-scarce regions, advancing sustainable groundwater management strategies.

2 Hydrogeology of the Study Region

The MODFLOW-6 model used in this study simulates the 1958 steady-state groundwater conditions in Qatar when extraction rates were minimal [6], allowing for an accurate representation of natural recharge processes. Natural groundwater recharge in 1958 was primarily occurring through localized rainfall infiltration in depressions and fractures. The model grid consists of 208 columns and 420 rows, with a uniform cell size of 500×500 m. It incorporates three primary geological layers that define the hydrogeological framework of the region [6, 16]. The uppermost unit, the Dam and Dammam Formation, serves as the primary groundwater storage zone, consisting of limestone and dolomite with thicknesses reaching up to 50 m. Below this, the Russ Formation functions as a semi-confining layer, composed of chalky limestone interbedded with gypsum and shale, restricting vertical groundwater movement. The deepest unit, the Umm Al-Radhuma Formation, is a thick dolomitic limestone layer extending beyond 300 m, acting as a regional aquifer with limited recharge but contributing to lateral groundwater inflows from Saudi Arabia [16].

The karstic nature of the system results in highly variable hydraulic conductivity, ranging from less than 1 m per day in dense limestone to over 200 m per day in fractured zones. The piezometric map indicates that groundwater head across the entire aquifer varies between 0 m above sea level (a.s.l) near the coastal areas of the Qatar Peninsula and up to 12 m a.s.l in inland regions such as Al Jumaliyah and Ashahaniyah. Groundwater heads of approximately 8 m are also observed further south, near the Kar'anah region. These inland areas serve as the primary recharge zones of the aquifer, with the main groundwater flow direction moving from these higher-head regions toward the coastal areas. Given the complexity of Qatar's aquifer system, numerical modeling requires an approach that accurately captures spatial heterogeneity while maintaining computational efficiency. The selected grid resolution balances numerical stability with spatial accuracy, ensuring proper representation of hydraulic conductivity variations and recharge patterns.

3 Data Extraction and Preprocessing

Groundwater modeling is often limited by data scarcity, where high-resolution hydrogeological datasets are unavailable or fragmented. In particular, in Qatar, available groundwater records include scanned hydrogeological maps [6, 16], some of which contain overlaid text, legends, or color distortions such as faded contour lines, inconsistent shading, or blending of features, making it difficult to extract numerical data directly. Traditional digitization approaches, which rely on threshold-based segmentation and color conversion to numerical values might be ineffective in such scenarios. In particular, such techniques might struggle to differentiate hydrogeological features from non-relevant map elements and fail to recover missing or unclear data. To address these challenges, this study proposes ML based approach for hydrogeological data preprocessing, addressing data clarity, consistency, and geospatial accuracy. This methodology is tested in Qatar's aquifer system, by focusing on processing hydrogeological data collected from [6, 16] including calibrated hydraulic conductivities and recharge levels in addition to elevation of layers. These datasets were selected as they capture key stages in Qatar's groundwater history.

3.1 Hydrological and Geological Dataset

The initial phase in developing an accurate MODFLOW model, referred to as "Phase 1: Data Preparation and Processing," focuses on transforming scanned maps into structured raster datasets. As illustrated in Fig. 1, the preprocessing workflow in this phase is divided into two key steps. The first step involves data preparation, processing, georeferencing and digitization. The second step focuses on model training and prediction, where ML techniques are applied to extract and refine hydrogeological parameters for integration into the numerical model.

3.2 Data Preparation

The first step in processing these images is Data Preparation (P1.1 in Fig. 1), which involves extracting historical hydrogeological maps from scanned records to ensure their suitability for numerical modeling. These maps are then subjected to georeferencing, aligning them with modern coordinate systems to maintain spatial accuracy and consistency across datasets. Following georeferencing, the maps undergo automatic detection of overlaid text, legends, or extraneous features, which may obscure

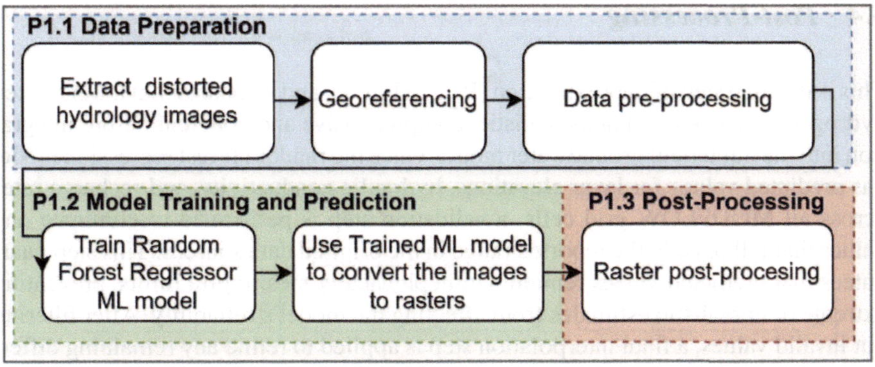

Fig. 1 ML-based data preprocessing workflow

critical hydrogeological information. These elements are identified and marked as regions with missing data, ensuring that they do not interfere with subsequent processing. To further refine the dataset, a Gaussian filter is applied to smooth the images by reducing noise and minimizing the influence of any undetected artifacts. This filtering process enhances the visibility of hydrogeological parameters, boundaries, and spatial variations, allowing for more accurate extraction of groundwater parameters.

3.3 Model Training and Prediction

The next step trains a random forest regression model to extract hydrogeological parameters from processed maps (P1.2 in Fig. 1). Using color pixels from map legends as reference points, the model learns to predict layer elevations, hydraulic conductivity, and recharge values with improved accuracy. Training on legend colors ensures robustness against image distortions, allowing the model to infer hydrogeological attributes even in faded or inconsistent regions. By leveraging a broad dataset of color-to-value mappings, the model generalizes across different map sections while maintaining consistency with the original data. Once trained, the model estimates hydrogeological parameters for each MODFLOW cell and evaluates the most probable values for layer elevation, hydraulic conductivity, and recharge based on valid input colors, ensuring spatial coherence and seamless integration into the groundwater model.

5 Results and Discussion

The calibration of the groundwater model was conducted to refine the hydraulic parameters and improve the accuracy of simulated water levels by minimizing the variation between observed and computed groundwater heads. The initial dataset extracted in phase 1 provided a preliminary understanding of groundwater distribution; however, due to the limitations associated with data Digitization and interpolation, significant deviations were observed. By implementing an iterative hybrid calibration approach, adjustments were made to the hydraulic conductivity and recharge parameters, leading to a substantial reduction in model error and a more realistic representation of groundwater flow dynamics. The following sections discuss the observed groundwater head distribution, the impact of calibration on reducing the Mean Square Eror (MSE), and the final spatial distribution of the calibrated hydraulic parameters.

The maps in Fig. 3 illustrate key hydrogeological parameters, including elevation, hydraulic conductivity, and recharge distribution across Qatar. The first row shows topography and formation elevations, while the second and third rows highlight hydraulic conductivity, active MODFLOW cells, and recharge zones, providing insights into groundwater flow dynamics.

5.1 Data Extraction—Phase 1

The extracted data after phase 1, represented in Fig. 3, shows the observed head distribution derived from digitized and interpolated historical groundwater maps. The observed groundwater levels exhibit significant spatial variability, with high water head values concentrated in the northern and central regions of Qatar, while lower values are observed in the southern and coastal areas. These variations reflect natural groundwater flow patterns influenced by karst features and localized recharge zones. However, due to data preprocessing limitations, initial dataset variations required further calibration, leading to an uncalibrated MSE of approximately 14 when compared to available field measurements. This difference highlighted the need for iterative calibration to refine the model parameters and improve the accuracy of the simulated groundwater levels.

5.2 Mean Squared Error Evolution

The calibration process, as detailed in Fig. 4, significantly reduced the mismatch between the observed and simulated groundwater heads. Through iterative adjustments of hydraulic conductivity and recharge values, the calibrated MSE was reduced to from 14 to 0.6, demonstrating a substantial improvement in model accuracy of

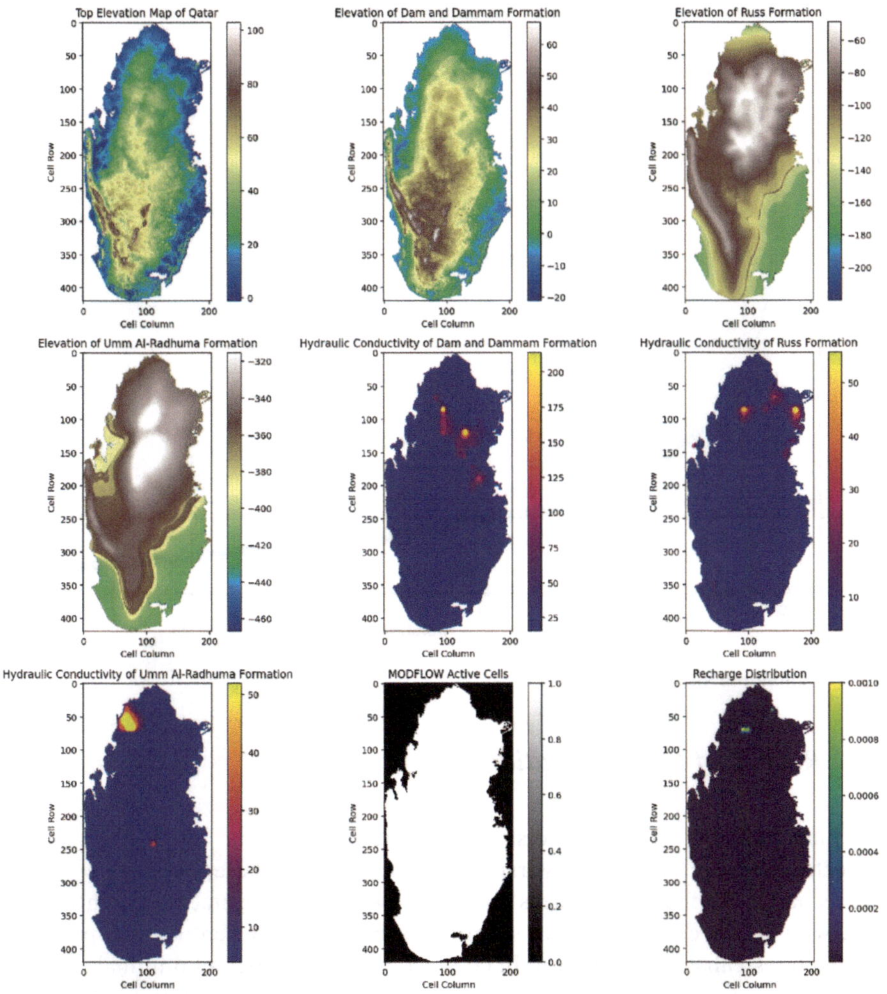

Fig. 3 Extracted hydrogeological parameters, including topography, formation elevations, and hydraulic conductivity

around 96% compared to the initially extracted images. The calibrated head distribution closely aligns with the observed values, indicating that the model effectively captures regional groundwater dynamics. Notably, the difference in southern Qatar is minimized, suggesting that the refined hydraulic parameters better account for the lower permeability formations in that region. The calibration also improved the representation of the hydraulic gradient, ensuring a more realistic simulation of groundwater flow.

Polices, Legislations and Society

an individual level since they are part of a complex system that interacts in a multitude of ways. Qatar has the opportunity to take a lead role in the field of sustainable development, by aggressively implementing a comprehensive plan for the development of the country. This plan would include strictly managed natural areas, innovative industrial development and urban design, promoting research and product design while giving Qatar a market advantage as a "first mover".

Survey on environmental perceptions and attitudes in the State of Qatar has been carried out and found that an overwhelming majority 80% know something about climate change but only 34% said they know something about sustainable development [13]. A majority of the public (92%) backed the statement that the quality of life depends on the quality of the environment.

Qatar is making significant efforts to achieve sustainable development and address SDGs. Knowledge of sustainable development and raising awareness play a crucial role in achieving sustainable development and addressing the SDGs. Therefore, this study aims at assessing the level of awareness, knowledge of, and attitudes towards sustainable development and SDGs amongst a sample of university students in the State of Qatar.

2 Methodology

The investigation was conducted using a random sample of 40 university students (20 males and 20 females). The investigation focused on the following:

(a) Sustainable Development Concepts: Use questionnaires with both open-ended and closed-ended questions to assess awareness and understanding among different demographics.
(b) SDGs Awareness: Utilize surveys' questionnaire to measure awareness of SDGs by students.
(c) Social Development Pillars: Develop survey questionnaires to understand public awareness and the impact on health, and education.
(d) Environmental Development Pillars: Gather data through survey questions on water restrictions, energy transition, and technology transfer.
(e) Green Economy: Collect data through survey questions to measure indicators of green growth.

The data analysis of the various themes of the survey was conducted as follows:

(a) Understanding the Concept of Sustainable Development
    - Analyze survey responses to identify common themes and levels of understanding.
    - Use descriptive statistics to quantify awareness levels.
    - Apply content analysis for open-ended responses to understand qualitative aspects of knowledge.

(b) Awareness of Sustainable Development Goals (SDGs)

- Perform frequency analysis to measure awareness levels of SDGs.
- Analyze the extent and nature of international collaborations and partnerships with the private sector.

(c) Social Development as a Pillar of Sustainable Development

- Use statistical software to perform descriptive analysis on health and education data.
- Perform thematic analysis for qualitative data from surveys.

(d) Environmental Development as a Pillar of Sustainable Development

- Use descriptive and inferential statistics to analyze data on water restriction, energy transition, and technology transfer.

(e) Green Economy

- Use descriptive and inferential statistics to analyze data on green economy indicators.
- Perform time series analysis to evaluate trends in green economic activities.
- Apply economic analysis tools such as input–output analysis to study the green economy.

3 Results and Discussions

3.1 Sustainable Development Concepts

Knowledge of Sustainable Development

In terms of the knowledge of the concept of sustainable development, a total of 21 students (11 males and 10 females) answered correctly in terms of balancing economic growth with social and environmental development (Fig. 1). This represents 50% of the sample of students. There are very few studies that investigated the knowledge of sustainable development concepts, but Sayeed in 2017 carried out an investigation of a considerable number of respondents who were above 17 years of age, and it appeared that only 34% stated that they know something about sustainable development [13]. Since our respondents were university students, the assumption is that they are more educated than the general population.

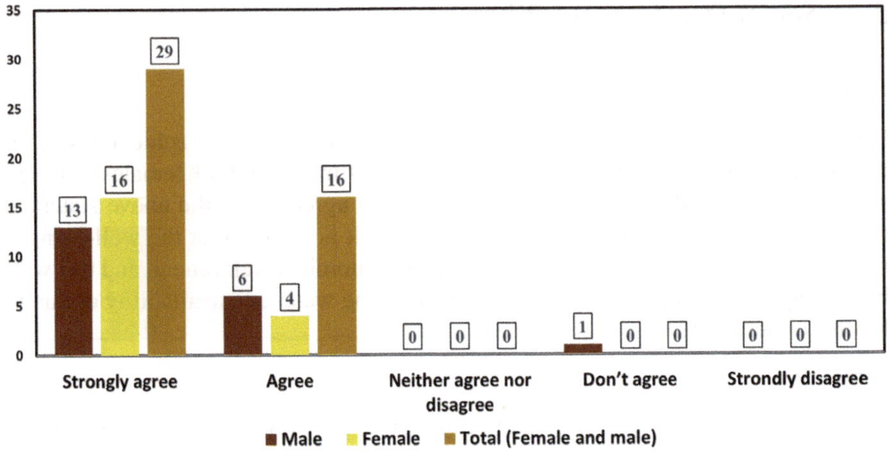

Fig. 4 Knowledge of social pillars of sustainable development: education

3.4 Environmental Development Pillars

Water Restrictions

Twenty-two students (9 males and 13 females) strongly agree while 13 students (7 males and 6 females) agree (Fig. 5) on having water restriction for water use in agriculture. This represents 87% of the sample. This result is logical since students are familiar with the initiative of the local water/electricity National company in urging consumers to reduce the per capita consumption of electricity and water. In addition, Kahramaa has set up the Awareness Park which encourages practices linked to sustainable development [17].

Energy Transition

Six students (2 males and 6 females) strongly agree while 8 students (4 males and 4 females) agree on the need to switch to electric cars (Fig. 6). This represents 35% of the sample. In the country, the concept of an electric car is not widespread, and the abundance of natural gas makes it difficult to adopt the concept of an electric car for the time being. However, it must be stated that the major dealers of international corporations have introduced a limited number of hybrid cars. This may be destined to change in the future as the survey conducted by Sayeed in 2014 found that traffic congestion and air pollution are the two most significant environmental concerns of Qataris [13].

Technology Transfer

Twenty-one students (11 males and 10 females) strongly agree while 14 students (5 males and 9 females) agree on the statement that developed countries should transfer technology to underdeveloped nations to protect the environment and achieve

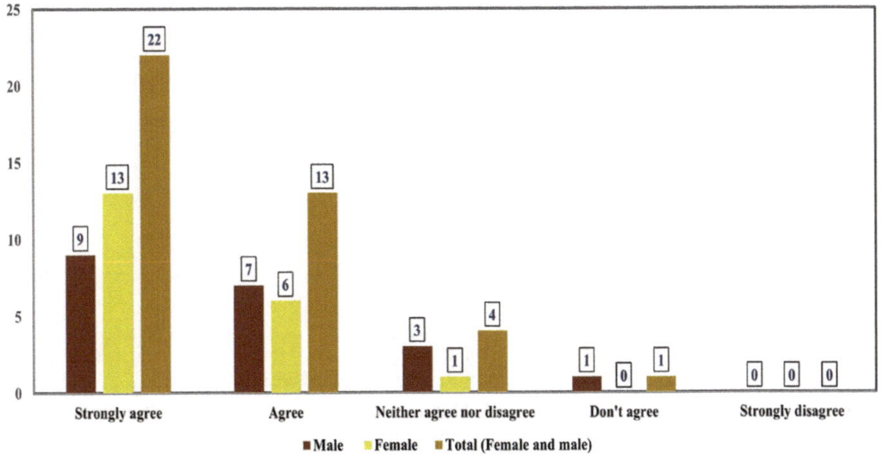

Fig. 5 Assessment of pillars of sustainable development: water restrictions

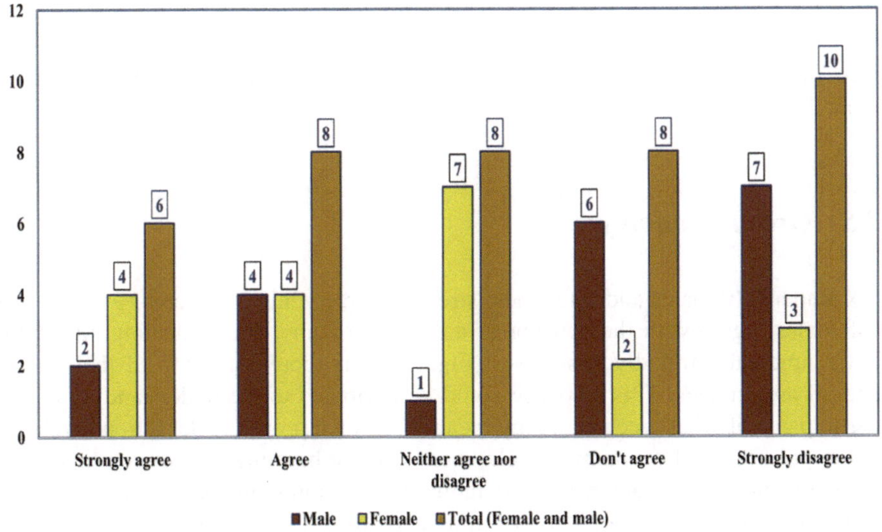

Fig. 6 Awareness of energy transition and the use of electric cars: is it necessary to switch to electric cars?

sustainable development (Fig. 7). This represents 85% of the sample, which is a clear majority. It has been reported that the transfer of clean technologies to developing countries is crucial to support the overall goals of reducing greenhouse gases but there are four issues that must be taken into account and these are the type of technology needs and its harmony with other development goals, the appropriate technology to

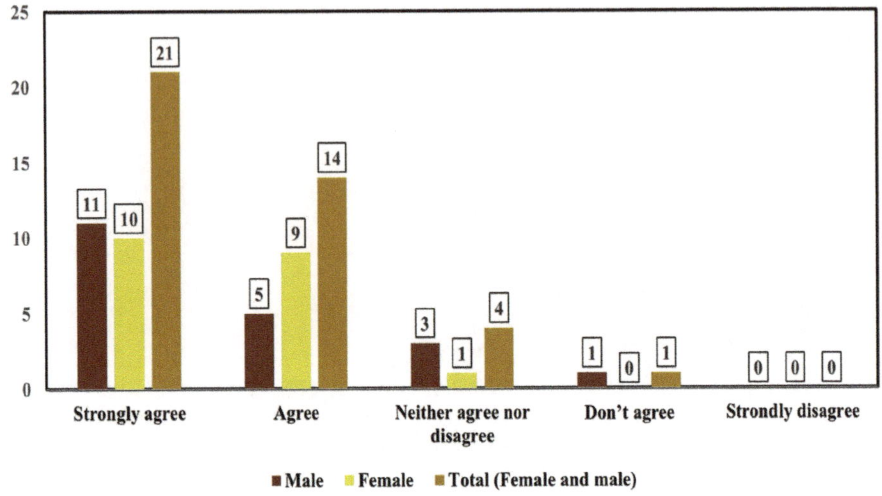

Fig. 7 Technology transfer: do you think that developed countries should transfer technology to developing countries to achieve sustainable development?

meet those needs, availability of expertise to master the technology, and other factors related to assimilation of the technology [18].

3.5 *Green Economy*

Six students (0 males and 6 females) strongly agree while 6 students (0 males and 6 females) agree with the concept of a green economy, that is dealing only with environmentally friendly businesses (Fig. 8). This represents 15% of the sample and all are females. This response could be attributed to the understanding of the participants of what green economy means. Our own proposed definition is related to the promotion of economic development without harming the environment. It is likely that males are more involved in industrial operations and businesses compared to females. Therefore, they may see it as unrealistic to promote a green economy when there is still enormous reliance on non-renewable resources in a country with huge reserves of fossil fuels.

4 Conclusions

The results of this study have shown that that 50% of the sample of university students in the State of Qatar have knowledge of the concept of sustainable development while none of them was able to identify the correct number of SDGs. Other results showed

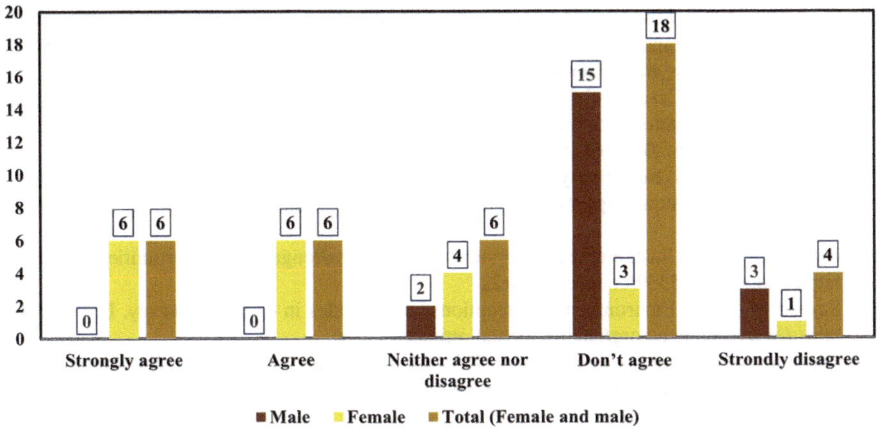

Fig. 8 Awareness of the concept of green economy

that more than 80% of the students are familiar with the social development pillars, water restriction and technology transfer. However, most students are unfamiliar with the concepts of energy transition and green economy. It is recommended that universities and schools across the country can allocate some units/courses in their programs to empower learners through climate change and sustainable development education.

Acknowledgements **Disclosure of Interests** The authors have no competing interests to declare that are relevant to the content of this article.

References

1. Brundtland GH (1987) Our common future: report of the world commission on environment and development. https://1987: Brundtland Report (admin.ch)
2. United Nations: Sustainable Development Goals (2016). https://www.un.org/en/academic-impact/page/sustainable-development-goals
3. Rosen MA (2019) Advances in sustainable development research. Eur J Sustain Dev Res. https://www.researchgate.net/publication/331540647_Advances_in_Sustainable_Development_Research
4. OECD and the Sustainable Development Goals: Delivering universal goals and targets. https://www.oecd.org/dac/sustainable-development-goals.htm
5. World Economic Forum. https://www.weforum.org/press/2019/09/global-survey-shows-74-are-aware-of-the-sustainable-development-goals/
6. KPMG. https://kpmg.com/xx/en/home/insights/2022/09/survey-of-sustainability-reporting-2022/sdg.html
7. IPSOS. https://www.ipsos.com/en-dk/global-opinion-about-un-sustainable-development-goals-survey-priorities-and-stakeholders-commitment

includes the Gulf Cooperation Council (GCC) countries, averaged 110 kg per capita annually, making it the highest level of household FW in the world [67]. China topped total household waste with 91.65 million metric tons each year, reporting a FW rate of 64 kg per person, while Qatar produced 95 kg of FW which was modestly reduced to 93kg per capita by 2023 [69]. Nair [43] highlighted that around thirty percent of food is wasted in Qatar, emphasizing that this issue poses a national security concern as the country has begun investing in local agriculture in recent years. An example of national efforts is the 2022 study conducted in Doha which assessed household FW during both Ramadan and non-Ramadan periods, following the UNEP FW Index methodology to monitor progress on SDG 12.3 [67]. The study found that 25% of food prepared in Qatar is wasted during Ramadan [67]. Its findings were included in the FW Index Report of 2024 and were classified as having high confidence for tracking purposes [69]. Moreover, the research highlights the importance of focused programs, like the "Sustainable Ramadan" initiative, aimed at fostering behavioral changes within households and the hospitality sector [69].

Qatar introduced its Vision 2030 in 2008, focusing on achieving balanced development incorporating environmental factors of air, land, water, and biodiversity (Planning and Statistics Authority State of Qatar (PSA) [47]). Qatar has unveiled its third National Development Strategy 2024–2030 outlining the four pillars of its national vision 2030: human development, social development, economic development and environmental development [47]. The fourth pillar of environmental development addresses the key issues of controlling national consumption, expanding recycling initiatives and bolstering biodiversity [47]. Its Strategic National Outcome six of Environmental Sustainability asserts that Qatar aims to reduce household consumption by 33% and water consumption by 40% per crop produced thereby promoting sustainable water conservation behavior in society [47].

This research aims to identify effective practices and key performance indicators (KPIs) suggested by policymakers and implemented by countries that have successfully reduced food waste. The findings will identify effective approaches to improve the handling of food waste and decrease its volume. These strategies are expected to be in harmony with Qatar's national goals and policies on FS. The outcomes will contribute to the government's broader objectives in this area, ensuring that waste reduction efforts align with the country's overarching food management policies. Upon outlining the main objective of this study, the subsequent Research Questions (RQ) are presented below.

RQ1: What best practices and KPIs have been adopted by countries and policy makers that effectively reduce food waste?

RQ2: How can public awareness and education impact food waste reduction in Qatar?

RQ3: What recommendations and KPIs can improve Qatar's food waste management keeping its food security goals?

2 Theoretical Framework

To accomplish the objectives of identifying best practices, key performance indicators and addressing the research questions, this study adopts secondary data collection on best practices that have effectively reduced FW, identifies initiatives to increase public awareness and education while maximizing the benefits food waste reduction can bring to the environment and the society, and addresses the present situation of Qatar with its food and water security concerns. Figure 1 presents the theoretical framework supporting this study.

The development of the framework brings forth best practices governments and stakeholders can enforce and implement to support the effective functioning of food systems to strengthen FS and reduce FW. The 2024 Science-Policy Brief for the Multistakeholder Forum on Science, Technology and Innovation (MFSTI) for the SDGs stresses the crucial need in aligning trade policies with sustainability goals to bolster FS while reducing adverse effects on ecological systems and communities to create a balanced relationship economic interests, environmental preservation, and social well-being [42]. Tomicic [64] emphasizes the essential adoption of innovative and impactful strategies to improve traditional educational approaches related to advanced food security and sustainability and public awareness. Moreover, encouraging farming partnerships in sharing resources and activities of production, marketing, and management decisions can enhance stronger community relationships, and improve food security [2, 64]. Educational programs and awareness campaigns establish alliances between governments, non-profit organizations, and local communities encouraging open dialogue for discussion, information exchange and joint problem-solving [64].

The research comprises three parts. The first part—the literature review—focuses on FW and benefits in reducing it and strengthening FS to help combat Climate Change. To achieve this an effective strategy will be designed to involve private and

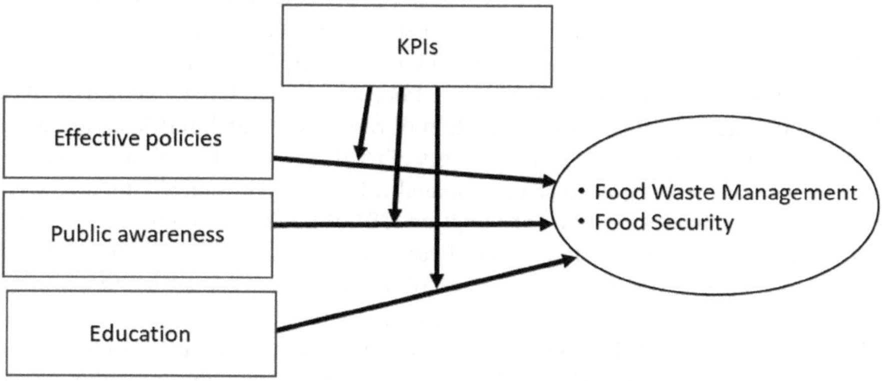

Fig. 1 Theoretical framework on best practices for managing FW and FS

public entities, as well as communities in fostering awareness and cultivating responsible practices. The second part of the research involves the analysis of secondary data, while the last part addresses recommendations, discussions, and future work.

3 Literature Review

3.1 Food Waste (FW)

There are different approaches to classifying waste in general. Well-known classifications include organic and inorganic solid waste (left-over food, cans, bottles, metal pieces, plastics, glass, paper, wrappers, ashes, etc.) [14, 35], gaseous, toxic, or non-toxic [14] and domestic and industrial waste [59]. The United Nations Food and Agriculture Organization (UN FAO) defines food waste as the loss of food quantity and food quality resulting from the irresponsible behavior practiced by retailers, food service providers and consumers [70].

Wasted food significantly exacerbates climate change, accounting for an estimated share of global greenhouse gas emissions (GHGEs) of 8–10% [51]. Additionally, FW production represents 30% of the world's agricultural land [19]. Sobaih [57, 58], Baig et al. [8], Mir et al. [40] posits that a substantial share of FW results from shifting consumer tendencies and lifestyle choices. Van Geffen et al. [73] mention behaviors and habits such as over-buying, inadequate food storage, and poor food planning. In Qatar, high household FW per capita reflects the challenges of balancing economic prosperity with sustainable consumption practices. Leal Filho et al. [38] identified four primary factors contributing to FW, namely insufficient planning, over-preparation of meals, challenges in preserving food for future use, and neglect of food safety regulations.

The Food Supply Chain (FSC) as shown in Fig. 2, comprises a chain of players containing the first three stages of food loss—agriculture, transportation and industrial food processing— as well as the two stages of food waste—retail, consumption and disposal. The phenomenon around food loss and FW represents significant sources of damage to the environment [25], Food and Agriculture Organization of the United Nations [70]. Besides the depletion of resources—land, fertilizer and water, FW also impacts the quantity and the quality of food [39]. A 2019 UN study predicted that 14% of all food produced at an estimated value of 400 billion US dollars, would be lost between harvesting and consumption [70]. While inefficiency, lack of infrastructure, equipment and training contribute to food loss, FW is mostly caused by lack of awareness, poor planning and lack of training, classified as intentional waste [67]. According to Goodwin [25], at the agricultural stage, it is estimated annual global food loss of 1.2 Bn tonnes, while in the household stage, FW reaches 2.5 Bn tonnes. In food processing, the third stage of the supply chain, the amount of perishable food loss varies by category -30% for cereals,40–50% for fruit and vegetables and 35% for fish (Global Food Waste Report [79]). Todd and Faour-Klingbeil [63]

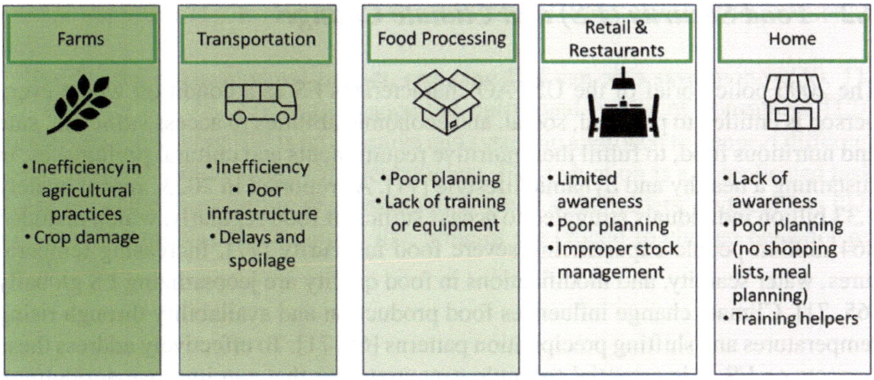

Fig. 2 Food loss and waste. Adapted from [25], World Resources Institute (WRI)

refer to the significant problem of FW and food surplus in retail and foodservice sectors, emphasizing the resulting economic and social burdens for both businesses and consumers.

The increasing importance of FW has led to initiatives to measure it in different parts of the world and the development of policies to manage it. The 2021 United Nations Environment Programme (UNEP) Food Waste Index (FWI) report [67] was the first attempt to collect and analyze FW data by country to track nations' efforts in fighting FW [67]. The data collected concentrated on the household, food service and retail sectors, which contributed to 61%, 26% and 13% of FW, respectively [67]. The methodology used, adopted the World Bank country income taxonomy classifyinng countries in low, lower middle, upper middle, and high-income groups. Results revealed that FW grows more rapidly amongst low-and-middle-income countries (LMICs) [9, 49, 53]. Within the European Union (EU), approximately 70% of FW occurs at household, food service, wholesale, and retail levels, of which wholesale and retail accounts for 5%, while the remaining FW is created during the production and processing stages [16].

FW is particularly harmful in developed countries, where it exceeds six hundred million tons annually [32]. In contrast, a significant share of the global population struggle with food scarcity and the inaccessibility of food to meet their basic needs. Similarly, World Health Organization (WHO) statistics confirmed that 2021 would see 29.3% of the global population being moderately or severely affected by food insecurity [76], and having limited means to acquire adequate, safe and nutritious food sources (Office of Disease Prevention and Health Promotion) [45]. Contrary to SDG 2-Zero hunger, the world has experienced an increase in hunger, food insecurity and malnutrition. According to the latest report by the UN State of Food Security and Nutrition in the World (SOFI), hunger plagued 733 million people worldwide in 2023 [77], which further indicates that hunger and food insecurities will still be ongoing by the year 2030 [71] if the necessary action are not taken.

results indicate low concern with FW, which points to the need of increasing public awareness and education campaigns on the matter. Aschemann-Witzel et al. [6] assessed consumers' motivations in minimizing waste, and the trade-offs consumers experience when deciding on food provisioning.

Apart from FW reduction, Prescott et al. [50] indicate that educational interventions targeting adolescents, can improve dietary behaviors in schools. Rondoni and Grasso [54] highlight the relevance of carbon footprint labeling on food products, to enhance consumers' awareness of the impact their choices can have on the environment and climate change. However, this labeling must be supported by a campaign to educate consumers on how to verify this information [54]. Furthermore, education plays a crucial role in behavioral change. According to Kushner and Nunes [36], increasing public awareness and education can lead to changes in individual behaviors by adopting habits such as responsible purchasing, food planning, mindful consumption, better food storage, and efficient food preparation and waste management. Besides, the authors mention the importance of engaging various stakeholders—governments, manufacturers, retailers, and educational institutions to create a unified approach.

New technologies can also help achieve improvements in FS. Onyeaka et al. [46] identified several technological applications to promote a circular economy. Machine Learning (ML) and AI-Driven Smart Agriculture can be adopted to predict food purchase and optimize food production, Blockchain with AI combined and Reinforcement Learning for improving supply chain efficiency, AI Image Recognition to assist in identifying only high-quality products for packaging and sale, and Natural Language Processing (NLP) for analyzing customer feedback and identifying areas for improvement to reduce FW. UNEP's report on green and digital technologies adoption to reduce FW proposes several technologies to prevent FW, like thermal preservation, solar-powered cold storage, biological and bio-chemical preservation, smart packing and labelling and monitoring of food quality [68].

3.5 Qatar Food and Water Security

Qatar is among the top four countries in income per capita in the world and is the richest country in the GCC, with a GDP per capita of 87,480 US dollars in 2022 [61]. According to The World Bank [62], within the period 2000–2023, Qatar's population grew from 645,937 thousand to 2,716,391 million, an increase of over 300%. This rapid growth brought challenges to a country with limited water resources [62]. Qatar is highly dependent on groundwater extraction and seawater desalination to meet water requirements, which includes agriculture, horticulture, and recreational activities [28].

Historically, Qatar relied significantly on imported products primarily from Saudi Arabia, to supply its residents. In 2008, the country launched the Qatar National Food Security Program (QNFSP) aiming to reduce its dependency on imports, by boosting self-sufficiency in agricultural and food production [56]. By 2010, 90% of Qatar's food was imported [7]. During the years of the embargo (blockade) from 2017 to 2021, Qatar faced an unprecedented diplomatic crisis when its neighboring GCC countries of Saudi Arabia, Bahrain, Egypt, and the UAE terminated all relations, despite its reliance on their food produce. As a response, Qatar developed efforts to improve its FS, pushing domestic production and enhancing its self-sufficiency, particularly in the agriculture sector.

The Ministry of Municipality and Environment (MME) [60] in Qatar released an updated Qatar National Food Security Strategy Plan (QNFSP), valid for 2018–2023, based on four pillars: (1) International Trade and Logistic, whose goal was to diversify the trade routes to reduce the risks associated with limited supply; (2) Domestic Market, adopting regulations to foster competition and encourage food safety; (3)Reserves Management that included measures to prevent water, seeds, fertilizer and food shortage; and (4) Domestic Self-Sufficiency, focusing on reaching cost-competitive crops, meat and fish [60]. According to Monroe [41] this was a turning point to make food an important geopolitical axis, as Qatar re-organized trade networks and promoted agricultural production to reduce food insecurity. Moreover, consumer culture changed, prioritizing a "buy local" movement, which forged nationalist sentiments among residents [41]. Agriculture played a significant role in the political landscape by boosting the FS agenda, and Qatari local product consumption became a way to express nationalist attitudes among the Qatar population [4]. However, the increased agricultural activity necessary for enhancing food production, puts considerable strain on Qatar's already limited water resources, potentially exacerbating water scarcity issues in the long term. [30].

According to the 2021 Water Statistics report from the National Planning Council, Qatar (NPC) [44], groundwater extraction remained constant between 2016 and 2021, despite a 5% population growth. The report also revealed that agriculture consumed 92% of the water supply during this period [44]. Water production resulted from water desalination and the re-use of Treated Sewage Effluent (TSE) with a substantial increase from 914 million m3 to 1,104 million m3 (almost 21%), while water loss from the desalinated water process increased from 4 to 6% [44]. The importance of reducing water consumption is highlighted in Qatar's most recent Food Security plan, part of the Third National Development Strategy for 2024–2030 [47]. According to this plan, Qatar strives to accomplish an average annual economic growth of 4% by adopting sustainable agriculture technologies, diversifying trade, expanding storage reserves, and reducing water consumption, while enhancing FS [47].

4 Methodology

The data collected was analyzed through an inductive approach, building concepts and theories from the data [55] by consolidating, ranking, and establishing relationships among the material collected [15]. Themes pertinent to the study's primary focus areas were categorized, enabling the recognition of the best practices and key performance indicators (KPIs), which are outlined in the theoretical framework and consider Qatar's present circumstances regarding food and water security.

5 Recommendations and Key Performance Indicators (KPIs)

The analysis of the literature review on best practices and the current situation of Qatar resulted in the following recommendations:

1. Improving Supply Chain Efficiency [5, 27, 33, 39, 50].
2. Public Awareness and Education campaigns to improve consumer engagement [1, 3, 6, 31, 36, 40, 48, 50, 54, 58].
3. FW, food processing and food transport mitigating measures [37].
4. FW Composting and Recycling practices [1, 35].
5. Legislation and Policy Interventions [9, 52, 75].
6. Green and Digital Technologies such as thermal preservation, solar-powered cold storage, biological and bio-chemical preservation, smart packing and labelling to monitor food quality to prevent FW [68].

Based on a detailed evaluation of FW Management and FS practices in Qatar and Qatar National Vision 2030, Table 1 presents additional recommendations to decrease FW, strengthen FS and minimize its impact on water usage.

The authors recommend that future research expand the investigation of opportunities to reduce FW, and enhance FS amid climate change in Qatar. The first recommendation suggests educational interventions aimed at reducing FW, intended for different age groups and socio-economic segments to influence consumer behavior. The second recommendation is to evaluate the potential of AI-driven solutions in Qatar's food supply chain to optimize waste reduction and improve FS. Additionally, interventions addressing FW behavior during Ramadan and other significant events should be developed. The last recommendation would be to include key stakeholders such as the Ministry of Municipality and Environment (MME); Ministry of Commerce and Industry (MOCI); Ministry of Public Health (MOPH); Ministry of Education and Higher Education, as well as local agricultural producers, food manufacturers, suppliers, retailers, hotels and restaurants in devising policies and fostering international collaborations, to achieve sustainable FS and waste reduction goals.

6 Conclusions and Future Research

The pressure imposed by the blockade on Qatar's economy increased the relevance of FS for the country. This research assessed the complex and significant relationship between FS and FW. To address this challange, Qatar should pursue two contradictory objectives. Firstly, increasing local food supply while implementing

Table 1 KPIs to reduce FW and enhance the Qatar Food Security Strategy 2024–2030

Recommendations	KPI	Description
1. Adoption of advanced agricultural technologies	Total food supply met by domestic production %	Employ cutting-edge techniques such as hydroponics, vertical farming, and precision agriculture
2. Investment in agricultural research and development	Research Projects Funded (number)	From an educational perspective, collaborate with universities and research bodies to create sustainable and suitable farming practices
3. Develop efficient supply chain management systems	Delivery time from production to consumers (days)	Use technology to improve logistics and ensure timely delivery
4. Support local farmers through marketing initiatives	Local product sales growth (%)	Effectively promote local products, including farmer's markets and online platforms for direct sales to consumers
5. Implement eco-conscious farming techniques	Sustainable farms %	Effectively promote organic farming, crop rotation, and pest control to enhance soil health and reduce environmental impact
6. Develop food processing infrastructure	Processing capacity increase %	Invest in infrastructure that improves food preservation, prolonging shelf life and minimizing waste
7. Utilize climate-resilient crop varieties	Climate-Resilient crops %	Research and develop crop varieties that can withstand extreme weather conditions and water scarcity
8. Launch public awareness campaigns on sustainable practices	Number of awareness campaigns	Educate consumers in supporting local agriculture and reducing FW through workshops and outreach programs

(continued)

Table 1 (continued)

Recommendations	KPI	Description
9. Establish a regional FS network	Annual joint GCC projetcs (number)	Collaborate with neighboring GCC countries to share resources, knowledge, and strategies for enhancing FS
10. Conduct regular agricultural censuses and surveys	Data-Informed policies %	Implement ongoing data collection to monitor agricultural production trends, consumer preferences, and market dynamics for informed policymaking

Source Authors' own work

water conservation and secondly, to mitigate the risks associated with limited water resources by reducing FW and promoting sustainable consumption among its residents. Addressing these objectives requires an integrated approach involving key stakeholders across the food supply chain, since different actors have different impacts on the UNSDGs in contributing to climate change [37] which includes technological innovations, increased efforts to educate the population and increase public awareness, to promote changes to consumer behavior.

Efforts to achieve environmental sustainability must not only come from governments, but also from companies who want to foster positive environmental change. It is noteworthy that as of 2023, among the 230 global members of the World Business Council for Sustainable Development (WBCSD), an organization focused on promoting a sustainable world, there is no Qatari company [74]. Qatar is also absent from other global initiatives involving companies that want to contribute to an environmentally sustainable cause, such as the Ellen MacArthur Foundation, whose goal is to promote a circular economy and work with business, academia, policymakers, and institutions to contribute to a global impact [17].

References

1. Abdelaal AH, McKay G, Mackey HR (2019) Food waste from a university campus in the Middle East: drivers, composition, and resource recovery potential. Waste Manage 98:14–20. https://doi.org/10.1016/j.wasman.2019.08.007
2. Agarwal B, Dorin B (2019) Group farming in France: why do some regions have more cooperative ventures than others? Environ Plan A: Econ Space 51(3):781–804
3. Aktas E, Sahin H, Topaloglu Z, Oledinma A, Huda AKS, Irani Z, Sharif AM, van't Wout T, Kamrava M (2018) A consumer behavioural approach to food waste. J Enterp Inf Manag 31(5):658–673. https://doi.org/10.1108/jeim-03-2018-0051
4. Al-Mal H, Saad AF (2021) Did the Qatar blockade work? Evidence from trade and consumer welfare three years after the blockade. Econ Peace Secur J 16(1). https://doi.org/10.15355/epsj.16.1.66

5. Al-Rumaihi A, McKay G, Mackey HR, Al-Ansari T (2020) Environmental impact assessment of food waste management using two composting techniques. Sustainability 12(4):1595. https://doi.org/10.3390/su12041595
6. Aschemann-Witzel J, de Hooge I, Amani P, Bech-Larsen T, Oostindjer M (2015) Consumer-related food waste: causes and potential for action. Sustainability 7(6):6457–6477. https://doi.org/10.3390/su7066457
7. Babar Z, Kamrava M (2014) Food security and food sovereignty in the Middle East. Oxford University Press EBooks, pp 1–18. https://doi.org/10.1093/acprof:oso/9780199361786.003.0001
8. Baig MB, Alotaibi BA, Alzahrani K, Pearson D, Alshammari GM, Shah AA (2022) Food waste in Saudi Arabia: causes, consequences, and combating measures. Sustainability 14(16):10362. https://doi.org/10.3390/su141610362
9. Barrera EL, Hertel T (2020) Global food waste across the income spectrum: implications for food prices, production and resource use. Food Policy 101874. https://doi.org/10.1016/j.foodpol.2020.101874
10. Bilal H, Govindan R, Al-Ansari T (2021) Investigation of groundwater depletion in the state of qatar and its implication to energy water and food nexus. Water 13(18):2464. https://doi.org/10.3390/w13182464
11. Buzby JC, Hyman J (2012) Total and per capita value of food loss in the United States. Food Policy 37(5):561–570. https://doi.org/10.1016/j.foodpol.2012.06.002
12. Carroll N, Sadowski A, Parizeau K, von Massow M, Wallace A, Jewell K, Ma DWL, Buchholz AC, Duncan AM, Chan B, Haines J (2021) Food skills: associations with diet quality and food waste among Canadian parents. J Nutr Educ Behav 53(5):371–379. https://doi.org/10.1016/j.jneb.2020.10.018
13. Crenna E, Sinkko T, Sala S (2019) Biodiversity impacts due to food consumption in Europe. J Clean Prod 227:378–391. https://doi.org/10.1016/j.jclepro.2019.04.054
14. Dias JM, Alvim-Ferraz MCM, Almeida MF, Rivera-Utrilla J, Sánchez-Polo M (2007) Waste materials for activated carbon preparation and its use in aqueous-phase treatment: a review. J Environ Manage 85(4):833–846. https://doi.org/10.1016/j.jenvman.2007.07.031
15. Dufour IF, Richard MC (2019) Theorizing from secondary qualitative data: a comparison of two data analysis methods. Cogent Educ 6(1). Tandfonline. https://doi.org/10.1080/2331186x.2019.1690265
16. Eičaitė O, Baležentis T, Ribašauskienė E, Morkūnas M, Melnikienė R, Štreimikienė D (2022) Food waste in the retail sector: a survey-based evidence from Central and Eastern Europe. J Retail Consum Serv 69:103116. https://doi.org/10.1016/j.jretconser.2022.103116
17. Ellen MacArthur Foundation (2018) Circular Economy—UK, USA, Europe, Asia and South America—The Ellen MacArthur Foundation. Ellenmacarthurfoundation.org. https://www.ellenmacarthurfoundation.org/
18. Eriksson M, Malefors C, Callewaert P, Hartikainen H, Pietiläinen O, Strid I (2019) What gets measured gets managed—or does it? Connection between food waste quantification and food waste reduction in the hospitality sector. Resour Conserv Recycl X 4:100021. https://doi.org/10.1016/j.rcrx.2019.100021
19. Fleck A (2024) Infographic: the enormous scale of global food waste. Statista Infographics. https://www.statista.com/chart/24350/total-annual-household-waste-produced-in-selected-countries/
20. Food and Agriculture Organization of the United Nations (2006) Food security policy brief. www.fao.org. https://www.fao.org/fileadmin/templates/faoitaly/documents/pdf/pdf_Food_Security_Cocept_Note.pdf
21. Food and Agriculture Organization of the United Nations (2022) Food loss and food waste|Policy support and governance gateway|Food and agriculture organization of the United Nations|Policy support and governance food and agriculture organization of the United Nations. www.fao.org. https://www.fao.org/policy-support/policy-themes/food-loss-food-waste/en/
22. Food and Agriculture Organization of the United Nations (2024) The State of food security and nutrition in the world 2024|Financing to end hunger, food insecurity and malnutrition in all its forms. https://openknowledge.fao.org/handle/20.500.14283/cd1254en

23. FAO, IFAD, UNICEF, WFP and WHO (2024) The state of food security and nutrition in the world 2024—financing to end hunger, food insecurity and malnutrition in all its forms. Rome. https://doi.org/10.4060/cd1254en
24. Gjerris M, Gaiani S (2013) Household food waste in Nordic countries: estimations and ethical implications. Etikk I Praksis - Nordic J Appl Ethics 7(1):6–23. https://doi.org/10.5324/eip.v7i1.1786
25. Goodwin L (2023) The global benefits of reducing food loss and waste, and how to do it. World Resources Institute. https://www.wri.org/insights/reducing-food-loss-and-food-waste
26. Gustavsson J, Cederberg C, Sonesson U, van Otterdijk R (2011) Global food losses and food waste. In: Meybeck A (ed) Food and agriculture organization of the United Nations. https://www.fao.org/3/mb060e/mb060e00.pdf
27. Hales J, Kemper J, White SK, Veer E (2024) Reflections on food policy in the context of healthy and sustainable diets. Food Policy 128:102708–102708. https://doi.org/10.1016/j.foodpol.2024.102708
28. Hansen L, Ribaudo M (2008) Economic measures of soil conservation benefits: regional values for policy assessment. In: SSNR. USDA Technical Bulletin No. 1922. https://papers.ssrn.com/sol3/papers.cfm?abstract_id=2014840
29. Hebrok M, Boks C (2017) Household food waste: drivers and potential intervention points for design—an extensive review. J Clean Prod 151:380–392. https://doi.org/10.1016/j.jclepro.2017.03.069
30. Hussein H, Lambert LA (2020) A rentier state under blockade: Qatar's water-energy-food predicament from energy abundance and food insecurity to a silent water crisis. Water 12(4):1051. https://doi.org/10.3390/w12041051
31. Jagau HL, Vyrastekova J (2017) Behavioral approach to food waste: an experiment. British Food J 119(4):882–894. https://doi.org/10.1108/bfj-05-2016-0213
32. Jeon D, Chung K, Shin J, Min Park C, Gu Shin S, Mo Kim Y (2020) Reducing food waste in residential complexes using a pilot-scale on-site system. Biores Technol 311:123497. https://doi.org/10.1016/j.biortech.2020.123497
33. Kasavan S, Ali NIBM, Ali SSBS, Masarudin NAB, Yusoff SB (2021) Quantification of food waste in school canteens: a mass flow analysis. Resour Conserv Recycl 164:105176. https://doi.org/10.1016/j.resconrec.2020.105176
34. Keating BA, Herrero M, Carberry PS, Gardner J, Cole MB (2014) Food wedges: framing the global food demand and supply challenge towards 2050. Glob Food Sec 3(3–4):125–132
35. Khan AH, López-Maldonado EA, Khan NA, Villarreal-Gómez LJ, Munshi FM, Alsabhan AH, Perveen K (2021) Current solid waste management strategies and energy recovery in developing countries—state of art review. Chemosphere 133088. https://doi.org/10.1016/j.chemosphere.2021.133088
36. Kushnir I, Nunes A (2022) Education and the UN development goals projects (MDGs and SDGs): definitions, links, operationalisations. J Res Int Educ 21(1):3–21. https://doi.org/10.1177/14752409221088942
37. Leal Filho W, Freitas Setti AF, Azeiteiro UM, Lokupitiya E, Donkor FK, NsikakAbasi Etim N, Matandirotya N, Olooto FM, Sharif A, Nagy GJ, Djekic I (2022) An overview of the interactions between food production and climate change. Sci Total Environ 838:156438. https://doi.org/10.1016/j.scitotenv.2022.156438
38. Leal Filho W, Ribeiro P, Setti A, Azam F, Abubakar I, Castillo-Apraiz J, Tamayo U, Ozuvar P, Frizzo K, Borsari B (2024) Toward food waste reduction at universities. Environ Dev Sustain 26:16585–16606. https://doi.org/10.1007/s10668-023-03300-2
39. Lourenco CE, Porpino G, Araujo CML, Vieira LM, Matzembacher DE (2022) We need to talk about infrequent high volume household food waste: a theory of planned behaviour perspective. Sustain Prod Consum 33:38–48. https://doi.org/10.1016/j.spc.2022.06.014
40. Mir MA, Chang SK, Hefni D (2024) A comprehensive review on challenges and choices of food waste in Saudi Arabia: exploring environmental and economic impacts. Environ Syst Res 13(40). https://doi.org/10.1186/s40068-024-00364-5

41. Monroe KV (2020) Geopolitics, food security, and imaginings of the state in Qatar's desert landscape. Cult Agric Food Environ 42(1):25–35. https://doi.org/10.1111/cuag.12243
42. Multistakeholder Forum on Science, Technology and Innovation for SDGs (STI Forum) (2024) https://sdgs.un.org/tfm/STIForum2024
43. Nair G (2021) The food we waste: antecedents of food wastage management behaviour. Int J Soc Econ, ahead-of-print(ahead-of-print). https://doi.org/10.1108/ijse-01-2020-0029
44. National Planning Council (2022) Water Statistics in the state of Qatar 2021. In National Planning Council. https://www.psa.gov.qa/en/statistics/Statistical%20Releases/Environmental/Water/2021/Water_Statistics_2021_EN.pdf
45. Office of Disease Prevention and Health Promotion (2024) Healthy People 2030. https://Odphp.health.gov/Healthypeople/Priority-Areas/Social-Determinants-Health/Literature-Summaries/Food-Insecurity
46. Onyeaka H, Tamasiga P, Nwauzoma UM, Miri T, Juliet UC, Nwaiwu O, Akinsemolu AA (2023) Using artificial intelligence to tackle food waste and enhance the circular economy: maximising resource efficiency and minimising environmental impact: a review. Sustainability 15(13):10482
47. Planning and Statistic Authority State of Qatar (2024) Third Qatar National Development Strategy 2024–2030. https://www.psa.gov.qa/en/nds1/nds3/Documents/QNDS3_EN.pdf
48. Ponis ST, Papanikolaou PA, Katimertzoglou P, Ntalla AC, Xenos, Konstantinos I (2017) Household food waste in Greece: a questionnaire survey. J Clean Prod 149:1268–1277. https://doi.org/10.1016/j.jclepro.2017.02.165
49. Porter SD, Reay DS, Higgins P, Bomberg E (2016) A half-century of production-phase greenhouse gas emissions from food loss and waste in the global food supply chain. Sci Total Environ 571:721–729. https://doi.org/10.1016/j.scitotenv.2016.07.041
50. Prescott MP, Burg X, Metcalfe JJ, Lipka AE, Herritt C, Cunningham-Sabo L (2019) Healthy planet, healthy youth: a food systems education and promotion intervention to improve adolescent diet quality and reduce food waste. Nutrients 11(8):1869. https://doi.org/10.3390/nu11081869
51. Ramísio PJ, Pinto LMC, Gouveia N, Costa H, Arezes D (2019) Sustainability strategy in higher education institutions: lessons learned from a nine-year case study. J Clean Prod 222:300–309. https://doi.org/10.1016/j.jclepro.2019.02.257
52. Reynolds C, Goucher L, Quested T, Bromley S, Gillick S, Wells VK, Evans D, Koh L, Carlsson Kanyama A, Katzeff C, Svenfelt Å, Jackson P (2019) Review: consumption-stage food waste reduction interventions—what works and how to design better interventions. Food Policy 83:7–27. https://doi.org/10.1016/j.foodpol.2019.01.009
53. Rolker H, Eisler M, Cardenas L, Deeney M, Takahashi T (2022) Food waste interventions in low-and-middle-income countries: a systematic literature review. Resour Conserv Recycl 186:106534. https://doi.org/10.1016/j.resconrec.2022.106534
54. Rondoni A, Grasso S (2021) Consumers behaviour towards carbon footprint labels on food: a review of the literature and discussion of industry implications. J Clean Prod 301:127031
55. Rosairo HSR (2023) Thematic analysis in qualitative research. J Agric Sci (JASS) 18(3):10526. https://doi.org/10.4038/jas.v18i3.10526
56. Semsia AAM (2017) Growing food pyramids in the sand: how sustainable are Qatar's self-sufficiency and foreign agro-investment policies? J Agric Environ Int Dev JAEID 2017(2):409–424. https://doi.org/10.12895/jaeid.20172.699
57. Sobaih AEE (2022) Excessive food buying in Saudi Arabia amid COVID-19: examining the effects of perceived severity, religiosity, consumption culture and attitude toward behavior. Int J Environ Res Public Health 20(4):3126. https://doi.org/10.3390/ijerph20043126
58. Sobaih AEE (2023) Saudi zero food waste certification: a novel approach for food waste management in Saudi Arabia. Agronomy 13(6):1654. https://doi.org/10.3390/agronomy13061654
59. Tezkan B, Hördt A, Gobashy M (2000) Two-dimensional radiomagnetotelluric investigation of industrial and domestic waste sites in Germany. J Appl Geophys 44(2–3):237–256. https://doi.org/10.1016/s0926-9851(99)00014-2

60. The Ministry of Municipality and Environment, Qatar (2020) Qatar National Food Security Strategy 2018–2023. (2020, January). https://www.mme.gov.qa/pdocs
61. The World Bank Group (2024) World Development Indicators|DataBank. Databank.worldbank.org. https://databank.worldbank.org/reports.aspx?source=2&series=NY.GDP.PCAP.CD&country=QAT
62. The World Bank Group (2024b) Population, total—Qatar|Data. Data.worldbank.org. https://data.worldbank.org/indicator/SP.POP.TOTL?locations=QA
63. Todd ECD, Faour-Klingbeil D (2024) Impact of food waste on society, specifically at retail and foodservice levels in developed and developing countries. Foods 13(13):2098. https://doi.org/10.3390/foods13132098
64. Tomicic A (2023) from awareness to action: empowering communities for sustainable food systems. https://feast2030.eu/from-awareness-to-action-empowering-communities-for-sustainable-food-systems
65. United Nations Food and Agriculture Organization (FAO) (2016) The State of Food and Agriculture 2016: climate change, agriculture and food security. https://reliefweb.int/report/world/state-food-and-agriculture-2016-climate-change-agriculture-and-food-security
66. United Nations (2022) The sustainable development agenda. United Nations Sustainable Development; United Nations. https://www.un.org/sustainabledevelopment/development-agenda/
67. United Nations Environmental Programme (2021) UNEP Food Waste Index Report 2021. UNEP—UN Environment Programme. https://www.unep.org/resources/report/unep-food-waste-index-report-2021
68. United Nations Environment Programme (2022) Reducing consumer food waste using green and digital technologies. In https://www.unep.org/resources/publication/reducing-consumer-food-waste-using-green-and-digital-technologies
69. United Nations Environmental Programme (2024) Food waste index report 2024. Think eat save: tracking progress to halve global food waste. In UN Environmental Program. https://wedocs.unep.org/handle/20.500.11822/45230
70. United Nations Food and Agriculture Organization (FAO) (2022) Tackling food loss and waste: a triple win opportunity. https://www.fao.org/newsroom/detail/FAO-UNEP-agriculture-environment-food-loss-waste-day-2022/en
71. United Nations Food and Agriculture Organization (FAO) (2024) The State of Food Security and Nutrition in the World 2024—financing to end hunger, food insecurity and malnutrition in all its forms. https://openknowledge.fao.org/items/09ed8fec-480e-4432-832c-5b56c672ed92
72. United Nations Environmental Programme (2024) UNEP Food Waste Index report 2024 features Qatar's national study. https://www.greenpolicyplatform.org/news/unep-food-waste-index-report-2024-features-qatars-national-study
73. Van Geffen L, van Herpen E, van Trijp H (2019) household food waste—how to avoid it? An integrative review. Food Waste Manag 27–55. https://doi.org/10.1007/978-3-030-20561-4_2
74. World Business Council for Sustainable Development (WBCSD). (2023, August 4). Who we are. www.wbcsd.org. https://www.wbcsd.org/who-we-are/
75. World Economic Forum (2023) WEF Report: 7 ways to bolster food security in the face of climate change. World Economic Forum. https://www.weforum.org/stories/2023/09/7-ways-to-boost-food-security-in-the-face-of-climate-change/
76. World Health Organization (2022) UN Report: global hunger numbers rose to as many as 828 million in 2021. World Health Organization. https://www.who.int/news/item/06-07-2022-un-report--global-hunger-numbers-rose-to-as-many-as-828-million-in-2021
77. World Health Organization (2024) UNICEF, 2024: hunger numbers stubbornly high for three consecutive years as global crises deepen: UN report. https://www.who.int/news/item/24-07-2024-hunger-numbers-stubbornly-high-for-three-consecutive-years-as-global-crises-deepen--un-report
78. World Resources Institute (WRI) (2023) The Global benefits of reducing food loss and waste, and how to do it. https://www.wri.org/insights/reducing-food-loss-and-food-waste

79. WWF-UK (2021) Driven to waste: the global impact of food loss and waste on farms. https://wwf.panda.org/discover/our_focus/food_practice/food_loss_and_waste/driven_to_waste_global_food_loss_on_farms/

Development of a Code of Practice for Offshore Aquaculture Vessels: Enhancing Safety and Efficiency in Seafood Production

Saeed Mohajernasab, Nagi Abdussamie, Andrew Harris, Nick Tighe, Chris Shearer, and Chris Chin

Abstract The Code of Practice (CoP) for Aquaculture Vessels provides guidance on the planning, construction, surveying, and operation of aquaculture vessels in diverse maritime environments, including Exclusive Economic Zones (EEZs). It was developed through: (i) Regulatory Review: Analysis of national and international standards, classification society rules, and aquaculture research from leading nations like Norway. (ii) Stakeholder Engagement: Input from workshops and interviews with regulators, designers, operators, and researchers, focusing on safety, vessel qualifications, and sustainable practices. (iii) Incident Analysis: Examination of fishing and aquaculture vessel incidents to identify key risks and accidents. The CoP addresses challenges in design, regulations, safety, food management, and fish welfare, including vessel stability, regulatory gaps, and humane live fish transport. Solutions include multi-purpose vessel designs, risk-based frameworks, advanced training, and stricter welfare protocols. Organized into five themes—vessel design and operation, regulatory compliance, human safety, training, and live species welfare—the CoP offers a global framework to improve safety, efficiency, and sustainability in offshore aquaculture operations.

S. Mohajernasab
Delmar Systems, Perth, WA, Australia

N. Abdussamie (✉)
College of Engineering and Technology, University of Doha for Science and Technology (UDST), Doha, Qatar
e-mail: nagi.abdussamie@udst.edu.qa

A. Harris
BMT Australia, Adeliade, SA, Australia

N. Tighe · C. Shearer
BMT Australia, Melbourne, VIC, Australia

C. Chin
Australian Maritime College, Launceston, TAS, Australia

S. Al-Naemi et al. (eds.), *Water and Food Security in the Face of Climate Change: Challenges and Opportunities for Resilience*, World Sustainability Series,
https://doi.org/10.1007/978-3-032-00098-9_79

Keywords Seafood · Offshore aquaculture · Support vessels · Risk and safety · Best practices

1 Introduction

Aquaculture is shifting from traditional shorelines to challenging offshore locations, offering both opportunities and complexities. The move to larger spaces with pristine waters requires robust structures to handle rough seas. Offshore farming reduces sea lice issues and minimizes ecosystem impacts in case of fish escapes, demonstrating the industry's adaptability and commitment to sustainable practices.

Designing aquaculture vessels is not a one-size-fits-all task. The American Bureau of Shipping (ABS) recently introduced guidelines for self-propelled aquaculture service vessels, covering structures, stability, fire safety, and machinery [1]. In Norway, regulations vary by vessel size and tonnage, with larger vessels following more comprehensive standards [2–4].

Aquaculture is among the most hazardous industries globally, with harsh conditions and high vessel motion posing significant risks. Small vessels, common in aquaculture, have the highest fatality rates, emphasizing the need for safety measures. A dedicated Code of Practice (CoP) for aquaculture vessels is vital to address these challenges, supporting safety, environmental compliance, and effective operations for stakeholders across the finfish and shellfish industries.

2 Aquaculture Vessels

Aquaculture vessels are broadly categorized into two types: specialized and multi-purpose. Specialized vessels are optimized for specific tasks, delivering high efficiency but limited versatility. In contrast, multi-purpose vessels can perform various operations, offering flexibility but typically being larger, less efficient, and less mission specific. This trend is not unique to aquaculture; naval ships also favor multi-role designs to minimize fleet size despite reduced single-task efficiency [5].

Three main vessel classes serve aquaculture sites: wellboats, feed carriers, and service boats. Wellboats transport and care for marine species, featuring live storage wells with water recirculation systems. These large, complex vessels perform tasks like transporting, treating, and resupplying species but can stress moorings due to their size [6]. Feed vessels transport feedstock to farms, often using cranes or hoses to transfer feed to barges. Modern designs incorporate dynamic positioning and wave-resistant hulls for stability in rough seas [7].

Service boats handle daily site operations and support larger vessels. Typically, under 15 m long, they feature high-capacity cranes, shallow drafts, and reduced freeboards for easy site access. Catamaran hulls dominate for their stability, maneuverability, and ample deck space [8].

Fig. 1 Different types of aquaculture service vessels

Additional vessel types, like flat-deck designs, serve short-term tasks such as mooring, net changing, and stock grading. Examples include Galician timber vessels for shellfish and flat-bottomed steel crafts in Scotland and Norway, which deliver substantial feed volumes to local aquaculture sites. Figure 1 presents examples of different types of aquaculture vessels.

3 Development Phases

The development of the code of practice unfolded across five planned phases. These phases served as the framework for exploration of the aquaculture vessels.

3.1 Phase 1: Literature Review

The first phase of the study focused on a literature review to understand the current landscape of aquaculture vessels and their operations. This desktop-based investigation examined national and international regulations, research studies, and industry guidelines:

- Regulations: Reviewed frameworks like the NSCV, marine orders, AMSA exemptions, DNV rules, ABS guidelines, and new IMO resolutions for insights into design and safety standards.
- Research: Included studies from Norway's SINTEF and NTNU, which explored exposed aquaculture operations, along with related topics like alternative fuels and crew transfer vessels.
- Global Standards: Highlighted varying regulations, such as Norway's rules for vessels under 24 m or 500 GT, and ABS guidelines addressing structural and operational elements for aquaculture vessels.

Despite ABS introducing classification standards, there remains a lack of industry-specific guidelines, leading to reliance on existing frameworks. Norway's 2015 revisions now require all vessels over 8 m to meet safety standards, reflecting growing safety concerns. Internationally, new IMO regulations under SOLAS Chapter XV and the IP Code establish minimum safety standards for industrial activities, including aquaculture. In Australia, the Navigation Act 2012 and Marine Safety National Law govern vessel compliance. AMSA oversees enforcement, with unique or large vessels requiring adherence to classification society standards and Recognized Organization guidelines. The outcomes of the literature review were presented at the 42nd International Conference on Ocean, Offshore and Arctic Engineering (OMAE) [9].

An analysis of over 160 vessel incidents, identifying key risks and accidents relevant to aquaculture vessels, ranging from minor operational issues to severe accidents involving loss of life. Table 1 summarizes the number of incidents that occurred on board fishing vessels, based on monthly reports from the AMSA website [10]. These incidents, covering the period from 2018 to the end of 2022, total 167 cases. The data includes a wide range of fishing vessel mishaps, from minor operational issues to more severe accidents involving property damage, personal injuries, or even fatalities. Analyzing these incidents offers valuable insights into the hazards and risks associated with fishing vessel operations, providing a deeper understanding of the safety challenges within the maritime industry. Figure 2 presents a detailed breakdown of the factors contributing to these incidents, highlighting the key variables involved. This analysis helps to identify the critical challenges faced by the fishing industry, pointing to areas that require immediate attention to improve safety standards.

The most common incidents reported were stability issues (49 cases) and slips, trips, and falls (52 cases). Stability problems directly threaten vessel safety, increasing the risk of capsizing and loss of life or property. Addressing these issues requires thorough stability assessments, adherence to guidelines, and design improvements. Slips, trips, and falls, which disrupt operations and pose injury risks, call for proactive

Table 1 Number of incidents by cause

#	Cause/Situation	Number of incidents
1	Adequate equipment	1
2	Compatibility of interaction	7
3	Considering shipping routes	6
4	Crew transfer between vessel and cage	2
5	Crossover of operations	8
6	Endurance	3
7	Maneuverability	2
8	Mooring	7
9	Sea state	10
10	Stability	49
11	Sustainability (Environmental)	6
12	Vessel's size	1
13	Industrial personnel	10
14	Lack of underwater regulation	6
15	Debris	2
16	Emergency services	7
17	Marking cages (Poor visibility)	4
18	Other users	6
19	Rope handling	5
20	Assets contact	22
21	Slips trips and fall	52

measures like non-slip surfaces, proper footwear, safety protocols, and crew training. By prioritizing these areas, stakeholders can significantly improve vessel safety.

Collaborative efforts are essential to analyze root causes of stability issues and implement targeted safety interventions. Similarly, understanding contributing factors for slips, trips, and falls can guide preventive measures, enhancing onboard safety.

Though these challenges primarily affect fishing vessels, their relevance extends to the aquaculture sector. Addressing them can improve safety practices across the industry, fostering safer and more sustainable operations through shared knowledge and best practices.

3.2 Phase 2: Industry Engagement

The second phase of our study marked a critical juncture where we actively engaged with key stakeholders in the aquaculture industry. This phase comprised two integral

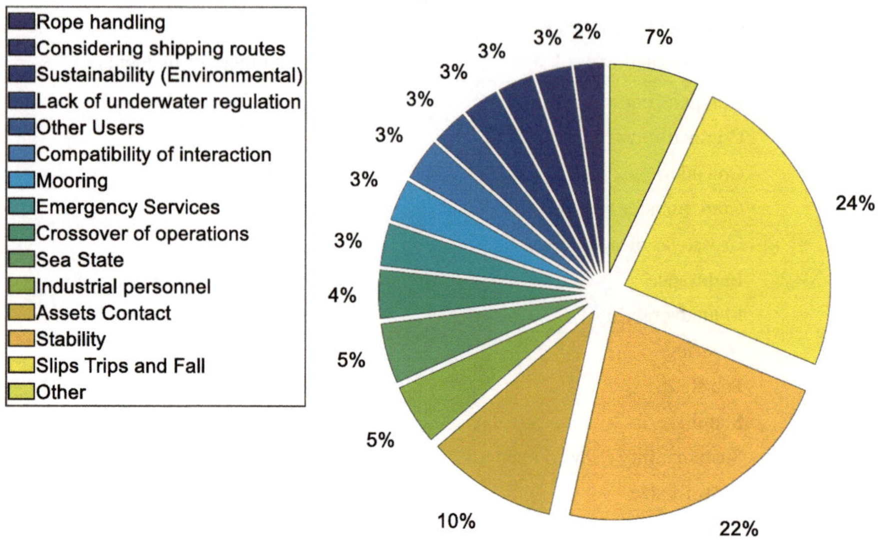

Fig. 2 Breakdown of contributing variables of incidents

parts, each contributing to our understanding of the challenges and dynamics within the sector.

Part One: Stakeholder Interviews

In the initial part of this phase, the focus was on engaging with stakeholders across four distinct aquaculture sectors, namely authorities, designers, operators, and researchers. Through an interview process, we conversed with 11 individuals representing a spectrum of expertise from eight different organisations. The list of organisations that participated in the interviews includes the Australian Maritime Safety Authority, AMC Search, Tasmanian Oyster Research Council, Green Aqua, AEXGroup, DNV Norway, BMT, and Tassal Group. This study was approved by the University of Tasmania Human Research Ethics Committee (H002739).

Part Two: Incident Review and Risk Assessment

The second part reviewed past incidents related to aquaculture alongside a risk assessment, analyzing 167 fishing vessel incidents reported on the AMSA website between 2018 and 2022. Incidents were classified under fishing vessels, as aquaculture vessels are currently grouped within this broader category due to the overlap between the two industries. However, distinguishing aquaculture-specific events from general fishing incidents was not always straightforward, as some cases were clearly linked to aquaculture, while others were less definitive. Given this interconnection, all incidents were included in the risk assessment, covering a range of severities—from minor operational issues to serious accidents involving property damage, injuries, or loss of life. This holistic approach acknowledges the shared risks between aquaculture and fishing sectors.

Table 2 provides an overview of the importance of each variable based on their mention during interviews and their corresponding score (from 1 to 4). The variables are related to different aspects of fishing vessel incidents, and their significance is assessed by different authorities, designers, operators, and researchers. The importance degree assigned to each cause/situation indicates the level of risk associated with the activity. It highlights variables that are commonly recognized as significant across multiple roles (e.g., situation number 10 and 21 with a degree of 4 by designers and operators). Additionally, it identifies variables that may not have been given much attention by certain stakeholders but are still regarded as important by others (e.g., situation number 9 recognized by researchers).

Table 2 The degree of importance assigned by experts for each cause/situation associated with incidents onboard fishing and aquaculture vessels

#	Cause/Situation	Authority	Design	Operator	Research
1	Adequate equipment	1	4	4	1
2	Compatibility of interaction	1	1	1	4
3	Considering shipping routes	1	1	1	4
4	Crew transfer between vessel and cage	1	1	1	4
5	Crossover of operations	4	1	1	1
6	Endurance	1	3	1	1
7	Maneuverability	1	3	3	1
8	Mooring	1	3	1	3
9	Sea state	1	1	1	2
10	Stability	1	4	4	1
11	Sustainability (Environmental)	2	2	1	2
12	Vessel's size	1	2	2	1
13	Industrial personnel	4	1	1	1
14	Lack of underwater regulation	1	4	1	1
15	Debris	1	1	4	1
16	Emergency services	1	1	4	1
17	Marking cages (Poor visibility)	1	1	1	4
18	Other users	1	1	3	1
19	Rope handling	1	1	1	3
20	Assets contact	1	1	2	1
21	Slips trips and fall	1	4	4	1

3.3 Phase 3: Gap Analysis Outcomes

This phase covered the outcomes of interviews conducted with aquaculture stakeholders from four different sectors, namely research, authority, design, and operator. A total of 11 interviews were conducted, with 2 from the research sector, 4 from the authority sector, 4 from the design sector, and 3 from the operator sector. The interview data was analyzed using the NVIVO software with the University of Tasmania license.

3.4 Phase 4: Guidance Drafting

As outlined in Table 3, the CoP guidance consists of eight chapters. In the first chapter, the guide lays the foundation by defining its scope and applications. Chapter 2 delves into the structural aspects of vessels and their stability. This section provides guidelines for designing and maintaining the structural integrity of aquaculture service vessels. Focusing on the machinery and technical aspects of vessels, Chapter 3 explores the various systems and components that are integral to the functioning of aquaculture service vessels. This includes propulsion systems, and other machinery critical for the vessel's operation. Chapter 4 concentrates on the specific operations related to aquaculture. Drawing from guidelines applicable to tugs, tows, oil and gas, and offshore wind operations, this section addresses crane operation, lifting, hoisting, towing, and work health and safety considerations. Chapter 5 is dedicated to the electrical control and monitoring systems onboard aquaculture service vessels. It provides guidance on the design, installation, and maintenance of electrical systems to ensure reliability and safety. Chapter 6 outlines the procedures for conducting surveys and obtaining certificates for aquaculture service vessels. It details the inspection processes and compliance requirements based on established codes and regulations, including NSCV, USL, AMSA, and COLREGs. Chapter 7 shifts the focus to food safety management, addressing the unique considerations for aquaculture operations. The final chapter centres on fish welfare, acknowledging the ethical and humane treatment of aquatic species in aquaculture operations.

4 Key Findings and Discussion

The responses have been categorized into five themes, namely Design and Operation, Regulations, Human Safety, Training and Qualifications, Food Safety Management Systems and Live Fish Health and Welfare During Transport.

Table 3 Draft guidance overall map

Code of practice	Content			References
Chapter 1: General	Scope	Definitions	Application	NSCV, USL codes, AMSA exemptions, Marine orders IMO resolutions ABS, Guide for building and classing aquaculture service vessels DNV, Rules for classification: high speed and light craft MLC 2006
Chapter 2: Structures and stability	Hull	Supporting structures	Stability	
Chapter 3: Systems and components	Operation Equipment	Machinery	Fire Safety	
Chapter 4: Aquaculture operations	Towing	Lifting	WHS	The ship owners' club, tugs and tows—a practical safety and operations guide G+/DROPS: reliable securing booklet for offshore wind IMCA guidelines for lifting operations IOGP, RP 376: lifting and hoisting safety recommended practice

(continued)

Table 3 (continued)

Code of practice	Content			References
Chapter 5: Electrical, control and monitoring systems	Navigation	Electrical	Control—monitoring	NSCV, USL codes, AMSA exemptions, Marine orders IMO resolutions COLREGs
Chapter 6: Surveys and certificates	Pre-construction	Post-construction	Crew Competency	
Chapter 7: Food safety management	Process	Policies	Managing	Food standards Australia New Zealand, Safe food Australia: a guide to the food safety standards
Chapter 8: Live fish welfare	Factors	Strategies	Certifications	Best Aquaculture Practice (BAP)

4.1 *Design and Operation*

- Stability is a key concern for designers and operators, with catamarans being preferred due to their stability in sea motions.
- One issue in aquaculture vessel design and operation is material handling, especially lifting heavy weights for mooring.
- Mooring challenges are not limited to cages only but also for mooring well boats to feed barges, causing potential damage to vessels.
- Aquaculture vessels are designed to be small and versatile to reduce operational costs, but this can lead to stability limitations.
- Various challenges were identified across different sectors of the industry, including the need for adequate equipment, compatibility of interactions, shipping routes, crew transfer, crossover of operations, deck space, design purpose, dry docking, endurance, environmental sustainability, and maneuverability.
- The use of multi-purpose vessels and dynamic positioning, as well as considering vessels' size, were suggested as potential solutions to some challenges.

4.2 *Regulations*

- The main challenge with regulations for aquaculture vessels is service notation, which results in these vessels being classified as fishing vessels, leading to issues with compliance and safety.

- The lack of specific regulations for aquaculture vessels has led to the need for vessels to be developed with specific purposes, such as catamaran hull forms, to accommodate the unique requirements of aquaculture operations.
- Operators demand a risk-based approach for smaller vessels, separate from larger ships, to ensure optimal design and avoid blanket regulations that may not be suitable for all types of operations.
- Challenges and recommendations include the development of hybrid fishing-aquaculture vessels that can accommodate both types of operations, the separation of fishing and aquaculture operations to enhance safety and efficiency, and the need for regulations related to underwater operations.
- Novel vessels present a challenge as they may not fit neatly into existing regulatory frameworks designed for traditional vessel types.
- Separating industrial personnel on vessels and addressing issues related to safety, housing, and transportation are important topics addressed by the International Maritime Organization (IMO).
- The importance of service notation for fishing and aquaculture vessels is recognized by both the authority and designer sectors for ensuring compliance with safety and environmental regulations.

4.3 Human Safety, Training and Qualifications

- Transitioning to offshore locations poses significant safety challenges for aquaculture vessels, requiring a complete rethink of operational practices and safety protocols.
- Rope handling is identified as the most hazardous offshore operation, and participants suggest improving training and working practices to minimize risks.
- Additional optional training courses are recommended to familiarize aquaculture workers with offshore operations and safety procedures.
- Stability and seakeeping issues are highlighted as important factors affecting human safety during offshore operations.
- Challenges and recommendations in human safety, training, and qualifications domain include aquaculture management courses, managing assets contact, debris management, licensing of deckhands, proximity to emergency services, creating an association for aquaculture, and proper marking of cages and equipment.
- Other concerns involve the number of crew members required for operations, managing interactions with other waterway users, risk management procedures, specific certificates for vessel inductions, equipment training, and surveying for accurate charts and navigational aids.
- Slips, trips, and falls are identified as major safety concerns that need attention during aquaculture operations.

4.4 Food Safety Management Systems

- Maintaining the cool chain for aquaculture products is crucial to ensure their quality and safety, as they are considered high-risk products.
- Vessels need specific features to prevent contamination, such as self-draining decks, proper fuel storage, and stringent hygiene practices.
- Food safety management challenges include the use of blockchain technology for traceability, maintaining the cool chain with temperature controls and proper packaging, and adhering to food safety standards and certifications.
- The authority sector is responsible for developing regulations related to blockchain technology and food safety, while the operator sector focuses on day-to-day operations for maintaining the cool chain.
- The designer sector's role involves designing facilities with food safety in mind and implementing systems for tracking and verifying food product safety.
- Food safety standards and certifications like Friends of the Sea, Natural Organics, Best Aquaculture Practice (BAP), and Aquaculture Stewardship Council (ASC) are used to promote safe and high-quality operations in the aquaculture industry.

4.5 Live Fish Health and Welfare During Transport

- The welfare of live fish during transport is crucial and requires gentle handling to minimize stress.
- Monitoring water quality parameters such as oxygen, temperature, ammonia, carbon dioxide, salinity, and ozone are essential during the transport process to ensure fish health and welfare.
- Challenges related to fish health and welfare fall under the responsibility of the designer and operator sectors.
- The designer sector is responsible for designing and managing fish farms, including addressing issues related to fish density and water quality standards.
- The operator sector manages day-to-day fish farm operations, including proper handling and feeding of fish and maintaining water quality.
- The designer sector is also responsible for developing systems for fish transportation, considering transport speed and suitable conditions to minimize stress and maintain good health.
- The authority sector plays a role in developing and enforcing regulations related to the use of ozone in aquaculture facilities to prevent the growth of pathogenic microorganisms in water treatment.

5 Concluding Remarks

Addressing challenges in aquaculture vessel operations requires a comprehensive approach, considering design, regulations, human safety, food safety, and fish welfare. Stakeholder collaboration and proactive measures are essential to navigate these complexities and ensure the sustainability and safety of aquaculture operations. The industry's commitment to continuous improvement and shared best practices will contribute to creating a safer and more sustainable future for aquaculture vessels.

Recognizing this, the objective of creating a Code of Practice (CoP) is underway. This code seeks to enhance ship design and operations by incorporating insights from crew members, addressing issues from design to crew competency. Given the unique challenges faced by aquaculture vessels, especially small-scale ones, a dedicated Code of Practice is crucial. This code would serve as a comprehensive guide for operators, vessel owners, regulators, researchers, and industry associations. The aim is to establish best practices for offshore aquaculture vessels, providing a user-friendly approach and addressing various areas for improvement, including Design, Audit, Equipment, Machinery, Stability, Shape, Size, Operation, Maintenance, Safety, and Crew Competency.

Acknowledgements The authors acknowledge the financial support of the Blue Economy Cooperative Research Centre (CRC), established and supported under the Australian Govern-ment's CRC Program, grant number CRC-20180101. The CRC Program supports industry led collaborations between industry, researchers and the community. Our gratitude extends to all project contributors (in alphabetic order): AEXGroup, AMC Search, Australian Maritime Safety Authority, BMT Pty Ltd, DNV, Green Aq-ua, Tasmanian Oyster Research Council Limited, Tassal Group Limited, The Unive sity of Queensland, and Universidad Austral de Chile. This study was approved by the University of Tasmania Human Research Ethics Committee (H002739).

References

1. ABS (2020) Guide for building and classing aquaculture service vessels. American Bureau of Shipping
2. Standards Norway (2016) Regulations on construction and supervision of smaller cargo ships
3. Standards Norway (2016) Regulations on shipbuilding
4. Standards Norway (2016) Ship Safety Act (Ship Safety Act)
5. Andrews D (2019) The multi role combatant-jack of all trades, master of none? In: Warship 2019 conference. RINA
6. Nekstad OJ (2017) Modularization of aquaculture service vessels—an approach for the implementation of operational flexibility. NTNU
7. Hornsletten H (2017) Optimization model aimed for the aquaculture industry for fleet composition and routing of Wellboats. In: Department of Marine Technology. NTNU
8. Paleo JB, Muir J, Turner R (2000) Offshore mariculture: workboats. Options Méditerranéennes: Série B. Etudes et Recherches (30):179–190
9. Mohajernasab S, Abdussamie N, Harris A, Shearer C, Tighe N, Bhaskar P, Chin C, Islam R, Binns J, Enshaei H (2023) Towards developing a code of practice for offshore aquaculture vessels. In: ASME 2023 42nd international conference on ocean, offshore and arctic engineering. American Society of Mechanical Engineers Digital Collection

10. AMSA. Monthly domestic incident reports. Available from: https://www.amsa.gov.au/marine-incident-reporting/monthly-domestic-incident-reports

Enhancing Awareness and Education on Food Security for Youths in Qatar: A Comparative Study

Hasna Ashraf, Abitha John, Amin Esmaeili, and Rachid Benlamri

Abstract Food security remains one of the most pressing global challenges, with Qatar making significant strides in realizing its National Food Security Strategy to achieve sustainability and resilience. However, despite the nation's strong commitment, higher education institutions in Qatar have yet to comprehensively integrate food security education into their academic curricula. This paper examines the gap in both the quantitative and qualitative aspects of food security education within Qatar's universities and compares existing educational frameworks in countries with well-established food security curricula. Through this comparative analysis, the study highlights the absence of structured programs, courses, and research opportunities related to food security within Qatar's higher education sector. Furthermore, the paper evaluates the current state of food security education in the Gulf Cooperation Council (GCC) countries, identifying various levels of implementation and challenges in fostering food security awareness. A key concern raised is the low participation of youth in food security matters, despite the growing need for their engagement in developing sustainable solutions. The study explores the age distribution and interest levels of university students in food security education, emphasizing the need to cultivate a new generation of experts equipped with knowledge and skills to contribute to Qatar's food security goals. Based on the findings, this research proposes strategic recommendations to strengthen food security education in Qatar. These include the introduction of well-structured, relevant academic programs tailored to the country's specific needs, the active involvement of key stakeholders—including government entities, educational institutions, and private sector players—and the promotion of youth engagement in food security initiatives. By fostering a deeper understanding of Integrated Sustainable Food Systems (ISSF) and related food security challenges, this study seeks to advance national efforts toward achieving long-term food sustainability. Strengthening food security education within Qatar's higher education system will not only support the country's strategic vision but also empower future generations to address critical food security challenges on both national and global scales.

H. Ashraf (✉) · A. John · A. Esmaeili · R. Benlamri
University of Doha for Science and Technology, Doha, Qatar
e-mail: 60100778@udst.edu.qa

S. Al-Naemi et al. (eds.), *Water and Food Security in the Face of Climate Change: Challenges and Opportunities for Resilience*, World Sustainability Series,
https://doi.org/10.1007/978-3-032-00098-9_80

Keywords Food security · Student awareness · Sustainability education · Youth engagement · Educational framework · Sustainable practices · Curriculum development

1 Introduction

Availability, accessibility, and appropriate use of food to meet the needs of a certain population have become more important today considering the global environmental crises, economic depression, and some political upheavals in the world [1]. While these factors have unfolded and escalated global food systems' challenges, the need to address food insecurity has emerged more forcefully, especially among countries that depend on imported foods. Qatar is a good example in this aspect since like other countries it does appreciate the importance of food security going by its National Food Security Strategy. This strategic framework seeks to build up the nation's food security; however, the teaching of food security in the higher education system is oddly missing [2] (Fig. 1).

This paper will argue that while food security has been accorded high priority in Qatar as a strategic resource for development, higher education institutions have not emphasized food security as a major feature in the development of academic curricula especially for the youth. This focused lack of educational attention constitutes a major social and political void for raising appreciation and activity among the younger generation, which will lead to future causes of food security and sustainability.

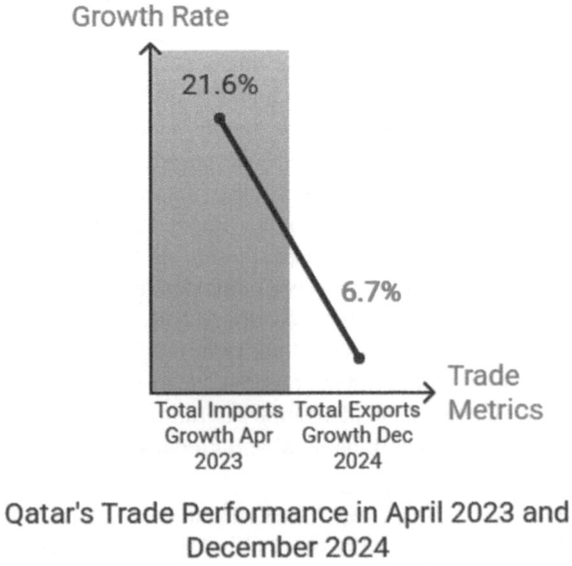

Fig. 1 Qatar's total imports increased by **5.1% year-on-year** in April 2023, reflecting a consistent growth trend since 1999, with an average growth rate of **7.0%** [3]

This paper is meant to establish why food security should be taught and why Qatar should take examples from other countries that include food security as a subject of study in their university curriculums. Identifying the levels of engagement and implementation of food security education developed in this study shows the differences in awareness among the GCC countries. Bilali and Hassen [4], suggest a clear systemic approach to how and where Qatar should introduce food security courses into the education system [4]. It also highlights the gap that currently exists in this aspect. Studies indicate that community-university partnerships enhance the effectiveness of educational programs on complex issues like sustainability and food security. Collaborative initiatives, such as those explored at universities focusing on sustainability education, show how community-based learning environments provide practical experiences that are crucial for grasping the multifaceted nature of food security [5]. In conclusion, this paper aims to stress the need for teaching, as well as ensuring, youth's food security knowledge as the essential component of youth's preparation for the appropriate engagement and coping with food sustainability issues in the twenty-first century.

2 Literature Review

A review of the literature on food security education has drawn interest from scholars in agriculture, nutrition, public policy, and the environment [6]. The literature review in this area shows an emerging appreciation of the importance of education in increasing food security knowledge and effective actions among university students. Several types of research have touched on the links between education/food security/sustainability, signaling the need to include topics related to food security in the curriculum since learners will need to competently deal with current issues.

Despite the availability of numerous studies about food security in the context of Qatar, there is a lack of empirical studies more specifically on the university education to food security. Although the Qatari government has declared their National Food Security Strategy that intends to raise the degree of food self-sufficiency and buffer stock, inadequate stimuli to promote food security competencies among the youthful population is the lack of a distinct scholarly line of study or study programs/courses on food security in Qatar [7]. This gap is even more shocking because youths will be the primary force to drive food security into the future in the country.

In other countries similarly, several universities have mainstreamed food security education in their higher learning institutions, with courses meant to address issues of food production, distribution, policies, and the like. Such educational endeavors have been found to enhance the student's knowledge about food systems plus the attitude and practice of food security and sustainability. For example, strategies such as using model-based activities, combined subjects, and external practices show a positive impact on developing students' reasoning and problem-solving competencies about the food security issue [8]. For example, Qatar's youth-focused food security programs could benefit from a model that combines academic learning with community-based experiences, like successful sustainability programs in other regions. This approach promotes both knowledge acquisition and a stronger sense of responsibility toward local food systems.

In addition, an analysis of the status of food security education in various GCC countries shows that there exist varying extents of commitment to implementing the food security themes into university curricula. Some countries have started implementing policies about food insecurity hoping to solve this issue through education, however, their education systems record no sign of the importance of this subject in their systems [4]. Inequity is why Qatar must look at neighboring countries' models and translate these paradigms to the specific Qatari setting. Engaging Qatari youth in local and regional initiatives focused on food security could benefit from encouraging an activist mindset. This approach, inspired by international sustainability and food security forums, can help students feel empowered to address food security both at the community level and within broader regional policies.

2.1 Global Perspectives on Food Security Education

In areas where people still struggle to gain access to adequate food, education, and training in this area also became one of the most successful interventions to promote awareness among the young generation of people in the field of food security, as well as to train a new generation of professionals in the problem field. Several countries present successful cases in this area which shows how important it is to involve issues related to food security in higher education.

United States: They argue that these approaches make sense in distinctions such as between Cornell University and the University of California where specialized programs in food security and sustainable agriculture have been set [9]. These programs integrate theories and cases and allow learners to analyze both national and international food security issues. Students acquire practical skills through practical assignments and ICT-oriented educational processes that help them to elaborate effective solutions to counteract the problem of food insecurity.

United Kingdom: In the UK alone, universities such as the University of London and Warwick University have incorporated Food policy, food safety, and food security courses in their agricultural sciences and sustainability track [10]. These courses are designed to prepare students for the vast knowledge about the food systems and the challenges involved, the legislations and policies in place, and the impact on human health. By highlighting these connections between food security, policy, and sustainability, these programs equip graduates for future discourse and practice about food.

India: In India at the federal level government has engaged universities to incorporate food security subjects within agriculture and Environmental Science faculties [11]. These initiatives underline commitments to sustainable rural direction, agriculture practices, and the global value chain. Through promoting a culture of learning that deems it essential to educate society on the lack of food security these partnerships seek to prepare students to produce sustainable food systems most especially in rural regions where food insecurity is prevalent.

These experiences can be used to demonstrate the need to incorporate food security education into the university curricula of these countries. In doing so, educational institutions can foster qualified human capital sufficient to innovate sustainable food systems. Not only does this approach directly deal with the short-term consequences of food insecurity but it also at the same time incorporates some of the longer-term solutions for food security contingency [6]. Since countries are progressively failing to adequately address issues such as population, and economic volatility, the endeavor of installing food security knowledge as a course option in higher learning institutions becomes more important to help foster leadership and capacity to effectively address these emergent questions.

The literature shows that critical to the success of addressing these issues that are intricately linked with food insecurity and climatic change is youth participation. According to the FAO, informed youths can encourage sustainable activities as well as champion change about food security and sovereignty both in their communities and at the policy level [6]. It is effective to tap young people's energy and creative thinking and to enable them to be policymakers and promote practices for improving food system resilience.

Many initiated organizations among them are the Global Youth Action Network a collective of young people from the developed world fighting food insecurity through project implementation in their communities [12]. Moreover, such collaborations as universities, NGOs, and governmental authorities can enhance the effectiveness of these types of activities by providing materials, tutoring, or having platforms that give opportunities to young people to share their opinions and ideas on the problems. It remains apparent that these global efforts to involve youths in delivering sustainable future assets to the world contribute to a broader understanding of food insecurity hence a generation armed to deal with these problems.

2.2 *Regional Context*

The Qatar National Food Security Strategy appreciates the imperative of filling this need through specialty-focused information sharing [13]. In the case of food security, GCC countries face similar challenges because of their climate, reliance on imports, and youth bulge. However, awareness of food security differs across the region depending on education levels. Table 1 provides an overview of the number of universities offering food security courses in GCC countries:

From Table 1 it can be deduced that Qatar is the only member of the GCC where no university currently offers courses in food security. On the other hand, food security education has made considerable progress in Saudi Arabia and UAE higher learning institutions.

Table 1 Food security in GCC countries

GCC country	Number of major universities	Universities with food security courses	Percentage (%)
Saudi Arabia	42	10	23.8
United Arab Emirates	30	8	26.7
Kuwait	10	2	20
Oman	22	5	22.7
Bahrain	12	3	25
Qatar	10	0	0

Qatar has laid a Food Security strategy spanning five years from 2018 to 2023 where food self-sufficiency via sustainable agriculture and innovation is seen as a major goal toward this end. However, to date, there has not been the development of specific academic courses or programs on food security within the higher education system that this strategic framework envisages, according to Abusin and Mandikiana [8].

There are many other areas of the future challenge which the students convey due to the lack of formal education about food security, and the lack of awareness of the shortcomings of the nation's food system and probable future careers in this exceedingly important field. Youth disengagement from issues of food security adds to the problem in this sense, seeing that most people active in food security-related activities in Qatar are over 45 years of age [14]. This demographic imbalance means that young students are either ignorant or do not care about the field, not least because they have never been introduced to it. In Qatar, fostering inclusive food security education programs that support both local and international students can lead to a more comprehensive and diverse understanding of global food security challenges. This inclusivity will enable international students to contribute their unique perspectives, enhancing the collective learning experience and potentially offering innovative solutions to local food security issues.

2.3 *Comparative Analysis: Countries with Education but no National Food Security Strategy*

Notably, some countries that do not have a coherent national food security policy have been able to design and implement educational programs geared toward food security. Finland and Sweden are good examples in this respect in that they included food security education in universities and proved that sound food security education can be attained with no link to formal national strategies [15].

Finland: People in Finland do not have extensive country-specific food security plans, but universities are important in mentoring future generations on food matters.

Our mappings show that all Finnish universities provide course offerings related to food systems, sustainability, and climate change. The programs offered by these universities provide the necessary knowledge to enable students to grasp and deal with food security issues. Due to the systematically rounded approach of the Finnish institutions reflecting considerations arising from the interdisciplinary triangle of agriculturally produced food, environmental protection, and population health, the issues underlying the notion of the food security strategy in question are regarded as comprehensive enough not to demand a formal strategy [16]. It positions graduates to support the global and local drive toward food security found in research, practice, and policymaking.

Sweden: Likewise, Sweden does not have an individual-identified national food security strategy. Thus, food security is included in the more general sustainability courses in some universities like the Swedish University of Agricultural Sciences. According to Gisslevik et al. [17], this approach is simply the keeper of food production and environmental and social sustainability where the food sustainability approach presents an integrated learning and understanding of how food systems function and perform within the climate change and resource opportunities [17]. Again, by training students to appreciate the different problems of food security, including the effect of farming on Biodiversity and social justice, the education institutions of Sweden are preparing leaders who will be able to look at issues of food security in a balanced way [18].

2.4 *Role of Education in Food Security*

The importance of education in improving food security is appreciated and more so in the public and higher education sectors. Public education brings awareness in the initial stages while a university trains and produces people with appropriate skills to address issues relating to food security. As institutions for learning and research, universities afford the students an inter-disciplinary view of food systems, and their relation to sustainability and climate change. This paper finds that by offering these topics in curricula, universities prepare students to enhance food security endeavors [19].

The role of universities as key players in the food security discourse is obvious based on the universities' function as enablers of research. By education in areas of sustainability, agriculture, and food systems, universities equip students with the development of high-utility approaches to current food security issues [20]. Also, there are good relations between higher education institutions and governmental and non-governmental organizations to research policies to enhance food security in the country [21].

The history and current state of food security in Saudi Arabia show that constructive collaboration between sustainability and food security should be achieved in the education system. The country's educational policies correlate with sustainable development goals (SDG 2) such as ending hunger and improving food production and land productivity [22]. Through such principles instilled in university programs, Saudi Arabia is seeking to solve food security issues. This approach is supported by the advent of the sectorial curricula in agriculture, water resources, and the production of food while still maintaining a sustainable environment.

However, the implementation of food security in education continues to pose great difficulty up to the present. However, most of the universities still have challenges in integrating education on food security into other related disciplines in the universities. Teaching methods that are commonly used, especially in agricultural sciences, are inclined toward passing more knowledge than skills; this hampers the further capability of students to solve food security issues adequately [23]. Moreover, students experience food insecurity inside and around the universities, which affects their learning outcomes and experiences; thus, there is a growing interest in the multifaceted approaches to educating students on food security.

Education, especially university education, a vital in the enhancement of food security. Universities are not only centers of awareness but also generators of new knowledge and ideas about food systems as students become productive participants in food security in their countries and globally [24]. Nonetheless, increasing focus on practical knowledge acquisition and issues of food insecurity while incorporated into learning institutions is vital to leverage these institutions [25].

3 Policy and Practice Recommendation

To address these two issues in the specific context of Qatar, it is vital to adopt a comprehensive educational process that integrates theoretical knowledge with practical experience. There is a great need for policymakers and educational institutions to join hands to improve the presence and understanding of these issues with the university youths. Several key recommendations can guide the development of more effective educational strategies:

3.1 Curriculum Integration

Among the areas that require urgent solutions, there is a requirement to include food security issues in the general education curricula for different disciplines. By including such topics in subjects like agriculture, environmental studies, economics, and social studies, the universities can enhance the standard education delivery systems. Such courses as 'Climate Change and Food System' or 'Sustainable Development and the Environment,' which are linked to both the Department of Food Science and the Department of Environment, would assist students in understanding the deeper significance of global environmental issues [26]. This would also encourage critical thinking and critical thinking skills so that the students would be in a better position to dissect various aspects of food security issues considering environmental and socio-economic factors in the country [27].

3.2 Community Engagement

In addition to formal academic engagement, there is an even greater potential in trying to establish community engagement projects in which students go out and participate in food security projects within their communities. Examples of such cooperation can involve local farms, community gardens, as well as NGOs when promoting principles of food security and Campus Food Pantry [28]. Through engaging in such practical-based projects, the students would not only be learning theory in food production and distribution but also attaining skills we hope they will require in future jobs. A few practical and engaging learning activities that would help students understand climate change's impact on food security and act as the basis for the latter community project could be involved in would be: Further, the proposed engagement can help develop the students' responsibility and empower them as such their actions result in enhancing local food security [29]. Drawing on community-university partnership models, this paper suggests that such collaborations be prioritized in Qatar's food security education initiatives. These partnerships can provide students with hands-on experience, grounding theoretical knowledge in practical applications that foster a sense of ownership and responsibility toward local food systems.

3.3 Research Opportunities

Promoting student research activities on food security and climate change issues is another major step towards high activity and engagement. Universities can offer the best chance for students to study important local problems, for instance, water shortage, soil erosion, or the susceptibility of the Qatari food chains. Such research projects might be included as part of a curriculum or could be taken as credit-recovery or special study programs [27]. By so doing, the students can have mastery of several areas, acquire research skills, and design practical solutions to reduce climate change impacts on food security. According to Abusin and Mandikiana [8], students undertaking research projects that include stakeholders from the government, or the private sector would contribute to addressing how locally generated knowledge can be applied [8].

Similarly, if the University were to provide grant or scholarship funding for student research, such issues of food security could be explored in greater depth. This, in turn, would enhance the buildup of locally produced scholarship to infuse the knowledge gaps in the state of Qatar policies and procedures.

3.4 Food and Agriculture Organization (FAO) Involvement

Food and Agriculture Organization (FAO) is a key factor in the efforts to fight food insecurity and is instrumental in the development of Qatar's approaches to addressing its food security issues, especially by involving youth [30]. The FAO has played a continually active role in Qatar's National Food Security plan, which is in line with FAO's global mandate to achieve food security for all Thus, the inclusion of Youth in the Agrifood initiatives cannot be overemphasized. This integration not only prepares a future human Force which can competently work and solve food Security issues but also build the research and a sustainable development agenda.

The most important effect of such cooperation with the FAO is the invention of educational activities and programs that may support the Qatar National Food Security Strategy. Many of these endeavors are aimed towards the engagement of youth in agriculture and food systems so that young people are well prepared to deal with challenges that surround the food chain. Concerning research, innovation, and sustainability, FAO's support will help Qatar to counteract the challenges of climate change and all the other adverse outside factors. To repeat, this approach provides solutions not only for near-term food security issues but at the same time it builds the capacity for sustainability in the agrifood sector eventually [25].

In addition, partnership with the Ministry of Municipality and other stakeholders enables to creation of frameworks of good practice for application in Qatar. This type of integrated work ensures that best practices and lessons learned from global and regional experiences are effectively conveyed to the local context. According to Stevens et al. (2019), through its large connections and experience in youth engagement mechanisms such as workshops, research and developments, and training activities [31]. Such platforms are crucial for building the right human capital to drive sustainable agrifood systems in Qatar. Encourage universities to formalize partnerships with local food producers, NGOs, and government agencies, creating structured pathways for students to engage directly with food security challenges.

4 Conclusion

Qatar has in the recent past tried to address food security issues through the formulation of the National Food Security Strategy. However, unlike other nations of the region, it has not set the machinery of formal education that directly faces the problem of food insecurity at a university level. These disparities imply that there is potential for the country to strengthen its interventions by integrating a food security curriculum into schools. Studying the educational activities of the neighboring countries of the GCC, including Saudi Arabia and the UAE, Qatar should consider a more all-encompassing way of inscribing the principles of food security into the university education system. For instance, Saudi Arabia has launched programs in agriculture, water, and sustainable development that help shed light on the existing food security questions. Likewise, the UAE has stuck to modern technology and local farming and Universities to pursue a specialized course in agriculture.

Food security education in the higher education system in Qatar will ensure that the nation gets a workforce that can support the vision of long-term food security. Such initiatives can range from cross-disciplinary programs that combine environmental science agriculture economics and technology which challenges the students to produce policies and solutions to food insecurity in the context of the Qatari environmental and economic policies. Further, research opportunities and cooperation with industrial companies and international organizations would enhance Qatar's preparedness for future food security threats.

To achieve this set goal, strategies like public education campaigns, and engagement of the youthful populace through youth-based organizations are equally significant at the initial stages of the students. Communications aimed at spreading information on the importance of food security among high school and university students help build a strong education and input new generations of professionals to lead and develop this important sector.

References

1. Karunasagar I, Karunasagar I (2016) Challenges of food security—need for interdisciplinary collaboration. Procedia Food Sci 6:31–33. https://doi.org/10.1016/j.profoo.2016.02.005
2. Ben Hassen T, El Bilali H (2022) Impacts of the Russia-Ukraine war on global food security towards more sustainable and resilient food systems foods, 11, Article 2301. References—Scientific Research Publishing (n.d.). https://www.scirp.org/reference/referencespapers?refereenceid=3554489
3. Qatar total imports growth 1999–2024|Monthly|%|CEIC Data (n.d.) CEIC data. Retrieved from https://www.ceicdata.com/en/indicator/qatar/total-imports-growth
4. Bilali HE, Hassen TB (2020) Food waste in the countries of the Gulf Cooperation Council: a systematic review. Foods 9(4):463. https://doi.org/10.3390/foods9040463
5. Johnson WC (2009) Preparing to appear: a case study of student activism (Doctoral dissertation, The University of British Columbia
6. Huambachano M, Arulingam I, Bowness E, Korzenszky A, Mungai C, Termine P, Wittman H (2022) Knowledge networks to support youth engagement in sustainable food systems. Front Sustain Food Syst 6. https://doi.org/10.3389/fsufs.2022.867344
7. Food Security Department (2020) Qatar National Food Security Strategy. In the Food Security Department
8. Abusin SA, Mandikiana BW (2020) Towards sustainable food production systems in Qatar: assessment of the viability of aquaponics. Glob Food Sec 25:100349. https://doi.org/10.1016/j.gfs.2020.100349
9. Degrees and programs (n.d.) CALS. https://cals.cornell.edu/food-science/degrees-programs
10. Food Research Institutes in the UK—Food Research Collaboration (2022). Food Research Collaboration. https://foodresearch.org.uk/food-research-institutes-in-the-uk/
11. Post A (2022) India to include agriculture curriculum in school education. Agriculture Post. https://agriculturepost.com/agri-research/india-to-include-agriculture-curriculum-in-school-education/
12. Youthinarn—Yarn (n.d.). https://youthinarn.com/
13. Ajibade I, Egge M, Pallathadka A (2020) Climate change and the sustainable development goal on food security: barriers and opportunities for progressive realization in Qatar and Nigeria. J Sustain Develop Law Policy (The) 10(2):158. https://doi.org/10.4314/jsdlp.v10i2.2

14. Sujarwo S, Prasetyaningrum DI, Fajar Y, Aini EK, Aprilia A, Setyowati PB, Laili F (2021) Agricultural education: investing basic agri-food education and agripreneurship knowledge to early age students. Agric Soc Econ J 21(1):33–40. https://doi.org/10.21776/ub.agrise.2021.021.1.5
15. Raheem D (2018) Food and nutrition security as a measure of resilience in the Barents region. Urban Sci 2(3):72. https://doi.org/10.3390/urbansci2030072
16. Rikkonen P, Rimhanen K, Aro K, Aakkula J (2023) The determinants of a resilient food system for Finland in the 2020s—three opinion polls for improvements based on a Delphi study among food system experts. Eur J Futures Res 11(1). https://doi.org/10.1186/s40309-023-00215-z
17. Gisslevik E, Wernersson I, Åberg H, Larsson C (2016) Food about sustainable development expressed in Swedish syllabuses of home and consumer studies: at present and past. J Educ Sustain Dev 10(1):68–87. https://doi.org/10.1177/0973408215627402
18. Tälle M, Wiréhn L, Ellström D, Hjerpe M, Huge-Brodin M, Jensen P, Lindström T, Neset T, Wennergren U, Metson G (2019) Synergies and trade-offs for sustainable food production in sweden: an integrated approach. Sustainability 11(3):601. https://doi.org/10.3390/su11030601
19. Kőmíves PM, Pilishegyi P, Novák N, Nagy AS, Körösparti P (2019) The role of higher education in the development of agriculture. Int J Inf Educ Technol 9(9):607–612. https://doi.org/10.18178/ijiet.2019.9.9.1275
20. Kawabata M, Berardo A, Mattei P, De Pee S (2020) Food security and nutrition challenges in Tajikistan: opportunities for a systems approach. Food Policy 96:101872. https://doi.org/10.1016/j.foodpol.2020.101872
21. Ashraf Z, Hussain M, Majeed I, Afzal M, Parveen K, Gilani S (2019) Effectiveness of health education on knowledge and practice regarding the importance of well-balanced nutrition among school students. J Health Med Nursing. https://doi.org/10.7176/jhmn/69-10
22. Unesco (2017) Education and the other SDGs—Unesco. https://gem-report-2017.unesco.org/en/chapter/education-and-the-other-sdgs/
23. Modebelu M, Nwakpadolu G (2013) Effective teaching and learning of agricultural science for food security and national sustainability. J Educ Soc Res. https://doi.org/10.5901/jesr.2013.v3n4p161
24. Google Scholar (n.d.) https://scholar.google.com/scholar_lookup?title=The+reality+of+educational+policies+related+to+gifted+education+programs+in+the+Kingdom+of+Saudi+Arabia&author=Abunasser,+F.&author=Al-Jughyman,+A.&publication_year=2012&journal=Jordanian+J.+Educ.+Sci.&volume=8&pages=193%E2%80%93214
25. Abunasser F, AlAli R, Al-Qahtani M (2022) KSA Universities' role in promoting the sustainability of food security: faculty opinions. Sustainability 14(16):10257. https://doi.org/10.3390/su141610257
26. Kumar A, Kumar VA (2014) Pedagogy in higher education of agriculture. Procedia Soc Behav Sci 152:89–93. https://doi.org/10.1016/j.sbspro.2014.09.160
27. Karanisa T, Ouammi A, Hamdi H, Saadaoui I, Fetais N, Sayadi S (2021) A design-led FEWW Nexus approach for Qatar University. Build Resil Univ Role Innov Entrep 11. https://doi.org/10.29117/quarfe.2021.0011
28. IJHAF|Influence of Agricultural Education Programme in Revamping Young Farmers Club for Food Security in Rivers State (n.d.) https://aipublications.com/ijhaf/detail/influence-of-agricultural-education-programme-in-revamping-young-farmers-club-for-food-security-in-rivers-state/
29. Nikolaus CJ, An R, Ellison B, Nickols-Richardson SM (2019) Food insecurity among college students in the United States: a scoping review. Adv Nutr 11(2):327–348. https://doi.org/10.1093/advances/nmz111
30. Mekouar MA (2019) 15. Food and agriculture organization of the United Nations (FAO). Yearb Int Environ Law 30(1):544–562. https://doi.org/10.1093/yiel/yvaa059
31. Stevens CJ, Whittle R, Davies WJ, Taylor JE (2019) Raising awareness about food security using a massive open online course. Plants People Planet 2(2):140–143. https://doi.org/10.1002/ppp3.10069

Farmer Training and Micro-credentialing for Food Security Upskilling: A Research Review

Abitha John, Amin Esmaeili, Rachid Benlamri, and Hasna Ashraf

Abstract This paper aims to present a review of the current literature on farmer training programs and interdisciplinary triage for food security workforce capacity building with emphasis on Qatar and comparable environments. These regions experience coma agricultural operations because of scarce water, unfavorable conditions for farming, and poor production as most of them depend on imported foods, hence realizing the importance of food security in their development plans. Indeed, in Qatar's National Vision 2030 and the Third National Development Strategy (NDS3), improved local agricultural productivity and sustainability are emphasized. Self-development training is important in the realization of these goals, especially in ensuring that the human capital in agricultural practices is professionally empowered with new knowledge and self-skill incorporation in modern practices in farming. Previous literature reviews show the need to use language in training given the complexity of Qatar's agricultural labor force which is comprised of people from countries like Nepal, India, and Bangladesh. Training with material in multiple languages enables cross-barriers and enhances understanding of new cultivation technologies. Apart from language-related interventions, the current major development has been interdisciplinary micro-credentialing as a way of building workforce resilience. From the information acquired from disciplines like environmental science, economics, and technology, the micro-credentials provide a coherent model of acquiring specific competencies that might be required to solve dynamic agricultural problems. Studies for other desert areas like Jordan and Egypt reveal that subjects like interdisciplinary training resulted in a remarkable increase in both output and water utilization. These programs are essential for enhancing crop yields and making agri-food systems more climate resilient—a critical need given the climate change risks confronting food systems across the region.

Keywords Training · Micro-credentials · Resilience · Education · Sustainability

A. John · A. Esmaeili · R. Benlamri · H. Ashraf (✉)
University of Doha for Science and Technology, Doha, Qatar
e-mail: 60100778@udst.edu.qa

S. Al-Naemi et al. (eds.), *Water and Food Security in the Face of Climate Change: Challenges and Opportunities for Resilience*, World Sustainability Series,
https://doi.org/10.1007/978-3-032-00098-9_81

1 Introduction

Food security is a major concern in Qatar and the GCC region due to dependency on imported foodstuffs and 'food insecurity' is caused by unsuitable climate for agriculture in the region [1]. Such factors such as high temperatures, low rainfall, and water stress work against the region's ability to produce food locally. Accordingly, efforts in the attainment of food security have been directed towards the improvement of production capacity and implementation of appropriate working agriculture. The importance of building resilient food systems has further gained importance as global value chains contracted and food security threats rose from climate change [2].

One of the significant improvement strategies in Qatar's food security system is the human capital development of agriculture workers consisting of employees from the NRIs including Nepalese, Indians, Bangladesh, and Egyptians. Just as most of the workers are no farming professionals with limited formal education in agriculture, most of them face language difficulties that restrict them from incorporating the new generation of farming practices [3]. Hence, there is a need to design and implement center-specific farmer training programs that incorporate multilingual and culturally appropriate education to increase the workforce's ability to apply state-of-the-art agricultural technologies and research findings.

Recent studies also confirm the significance of an interdisciplinary approach to micro-credentialing—a form of learning that is particularly suited for staff development doesn't bind workers into rigid programs but rather offers bite-sized, focused modules that enable them to obtain the skills necessary to meet the emerging demands of the agricultural sector [4]. From environmental science, economics, technology, and sustainable agriculture those micro-credentials equip the workers with a management strategy that can address most of the complex issues associated with farming in arid regions. This approach produces a flexible and well-informed human resource asset that can help realize Qatar's food security vision in the Qatar National Vision 2030 and the Third National Development Strategy.

This review synthesizes a critical analysis of the literature on the impact of Farmer Learning and current interdisciplinary micro-credentialing systems on farmer training programs about food security. It also discusses how these programs fit into other approaches toward sustainability, resilience, and the future vision of agriculture in Qatar. Drawing on the literature on case studies in Qatar and similar arid territories, this paper offers clues on the possibilities of training and workforce development in supporting a sustainable and improved Qatari food security system.

1.1 Food Security Challenges in Qatar

Food security remains a major concern in Qatar as in other countries within the Gulf Cooperation Council (GCC) because of its import dependency on food [5].

This makes the supply prone to risks such as the current global supply chain disruptions, price changes, and geopolitical. Qatar is endowed with several geographical constraints that predetermine its unfavorable conditions for agriculture, lack of water resources, and extreme climate [6]: the cultivated area constitutes 0.2% of the total land area of the country, and available water per capita is around 60 ^{m3} per year. The water stress is less than 1700 m^3 per year.

These are some challenges that the Qatari government has come to appreciate and include food security as one of the pillars within its Qatar National Vision 2030 [5]. This strategic initiative seeks to ensure the development and attainment of a convenient Food security system through participation in research on agriculture, promotion of technology, and training of human resources in the food sector. The National Development Strategy (NDS) 2018–2022 also defines the following major goals: increase food resilience and improve domestic agricultural capacities [7].

1.2 Interdisciplinary Approaches to Agricultural Training

Multidisciplinary perspectives in agricultural education include knowledge drawn from the following disciplines agricultural science, Food Systems economics, environmental science, and technology. This way educational programs can provide such a synthesis of different disciplines to farmers, which will prepare them for the challenges of modern farming. For example, knowledge about economic conditions affecting farming can be used in the improvement of production for agricultural businessmen.

The studies prove that interdisciplinary training is most suitable for closing the skill and knowledge deficits that prevail among Qatar's agriculture employees. According to Ridha et al. [8], some farm workers have diverse education levels and cultural endowments and are socially deprived of knowledge concerning sophisticated technological techniques in farming activities [8]. An interdisciplinary training approach enables the formulation of curricula that address these needs and whereby farmers acquire the appropriate skills for effective agricultural practices.

For instance, when learning about efficient water use, or the use of climate-smart agriculture in training, it is easier for the farmers to learn in the case of farming in arid regions. Precision practices of farming like using appropriate technology in irrigation and better crop tackling technology help to enhance yields without wastage of resources [9]. According to FAO (2018), smart agricultural technologies can help raise farm productivity to as much as 20% in areas prone to drought [10]. In addition, engagement between universities and farmers can enhance relationships; thereby knowledge transfer and innovative methods to enhance agricultural activities [11].

The convergence of these disciplines also creates opportunities for innovativeness in practices of raising crop and animal production. For instance, connecting IoT devices to AI farming practices revolves around the use of resources in production keyway of enhancing the capacity for those in the agricultural workforce. Through these programs, the farm workers are not only able to get the skills that would enable them to increase the productivity of their farms but also get to empower the nation to achieve its long-term vision of food security.

2.2 *Comparative Analysis: Farmer Training Programs in Other Arid Regions*

Qatar is not the only nation from the group of arid countries that suffers from the severe problems of agriculture. Literature reviews of comparative studies with other arid areas, for example, Jordan, Egypt, and Saudi Arabia give us indications as to the success of targeted and local farmer training initiatives in similar climatic conditions. For example, Egypt has paid a lot of attention to education for improvement in water use efficiency by smallholder farmers in the country, and this has gone a long way in helping them use water on crops effectively, and increase Crop Water Productivity, CWP [5]. Such programs highlight the principles of sustainable water use that may be a key aspect unlike in areas where water is not scarce.

The training programs conducted in Saudi Arabia focused on extension messages on the practical use of up-to-date and efficient farming practices such as the use of greenhouses and hydroponics in the area due to high temperature and poor soil fertility. All these causes have ensured that Saudi farmers boost their production capability for their farming produce, and at the same time, they reduce their negative impact on the environment. Likewise, Jordan has been strengthening Sustainable Agricultural practices through seminars/training such as on soil health and crop cropping patterns increasing the ability of farmers towards environmental shock.

The following case studies will demonstrate how similar training environments could bring similar improvements to Qatar. To improve its population's agricultural output, Qatar can make improvements to its training programs in areas relevant to local farming and farmers specifically through increased interdisciplinarity. Furthermore, the effectiveness of such programs in other similar environments is used to compare the current training activity in Qatar and outline the course of action to increase the size of such programs given local food security problems.

Therefore, it may be concluded that the comparative analysis proves the necessity of developing context-based training initiatives in arid surroundings. Apart from enhancing the links between research and practice, these programs also advance the dynamics of sustainable and resilient agricultural systems.

3 Interdisciplinary Micro-credentialing and Its Impact on Food Security

Micro-credentialing outside of the traditional discipline-specific modes of learning is gradually being embraced as an important strategy in the improvement of agricultural output and vulnerability. This transmitted and unique skill-based structural approach enhances specialized competencies among Qatar's agricultural workforce in a relatively shorter period and does so by combining various fields including environmental sciences, economics, and technology. These fields when integrated make micro-credentialing facilitate farm workers on how to respond to these changes that confront the agriculture industry, hence enhancing food production and utilization of the available resources [28].

The EU's 'Agri-Resilience' project, for instance, indicates that the blends of interdisciplinary information into training programs build up climate resilience [29]. Experience in Qatar shows that initial pilot projects, collaborative with LCA and TEA, can indicate acceptable results in raising consciousness of sustainable farming practices [24].

However, challenges exist as to how interprofessional content is delivered and how it is made more accessible to a workforce comprised of bi- and multilingual employees in Qatar. In addition, more synergy between universities and colleges, governmental and non-governmental organizations, as well as other corporations is required to enhance the relevance and extendibility of such programs [27].

4 Gender Diversity in the Workforce and Inclusive Training Programs

Attracting both genders to agricultural work creates essential conditions for innovative approaches as well as high productivity rates and sustainable agricultural development. Existing obstacles prevent complete female sector involvement because women encounter restricted access to training programs together with essential resources and leadership advancement possibilities [30]. Women face additional barriers regarding their agricultural development participation in Qatar and comparable regions because of cultural norms together with existing structural inequalities. Programs that train new workers should be improved to support diverse learning preferences and then eliminate all obstacles preventing employees from joining the workforce.

Training methods should have gender-responsive programs for women that acknowledge their dual work-family pressures. Women can access training programs more easily when learning options include flexible delivery methods such as part-time education alongside online resources. Training programs designed to be gender-conscious by showing female agricultural role models and handling gender-specific issues in educational materials will attract more women to farming careers

[31]. Industry leaders and experienced professionals connect women to mentorship programs that empower them through access to these critical relationships. Training programs that promote inclusive learning and equal opportunities enable agriculture to gain more women employees which creates stronger food security and economic strength.

5 Integration of AI and Technology in Agricultural Training Programs

The implementation of artificial intelligence along with digital tools in agricultural training programs improves productivity together with learning achievement and the flexibility of employees. Electronic learning platforms supported by artificial intelligence together with virtual simulation features and smart farming technology deliver personalized and interactive training experiences to workers who need specialized competence indicators while maintaining flexibility in their education. The austere agricultural conditions of Qatar with limited water availability and harsh climate make it necessary for AI-driven precision farming tools to teach proper resource-management methods to trainees. Machine learning systems monitor soil conditions by evaluating current data about moisture content and nutrient concentrations which helps farmers implement reasoned actions according to analysis results [32]. Machine learning algorithms can project crop disease risks along with providing suitable planting methods for incorporation in educational training materials to boost practical agricultural competence. The technological advancements enable workers to get ongoing learning experiences because they receive immediate decision assistance through real-time guidance which enhances agricultural productivity while promoting sustainability [33].

Various AI-based agricultural training programs have started to develop in Qatar which seeks to update the agricultural sector. The Qatar Computing Research Institute (QCRI) underlines its work in precision agriculture through research that focuses on developing automated irrigation systems for enhanced water management [34]. The educational institutions in Qatar created digital learning environments that use AI-based tracking systems for student progression as well as AI-driven adaptive content delivery. AI-powered virtual reality simulation systems should be developed to let agricultural workers train their techniques safely before implementing them at the actual farming sites. AI chatbots based on mobile technology enable instant responses to farmers when they seek information about best farming practices pest management and climate adaptation methods. Digital and artificial intelligence technologies allow Qatar to train an able agricultural workforce that resolves food security threats while maintaining environmental sustainability efforts in the sector.

6 Conclusion

This review highlights the significance of farmer training programs and interdisciplinary triage in strengthening food security workforce capacity, particularly in Qatar and comparable regions facing harsh agricultural conditions. The emphasis on self-development training, language accessibility, and interdisciplinary micro-credentialing demonstrates a structured approach to addressing the challenges of local agricultural productivity and sustainability. Studies indicate that providing training in multiple languages, given Qatar's diverse agricultural labor force, enhances knowledge transfer and adoption of modern farming practices [2]. Additionally, the integration of interdisciplinary micro-credentials, which draw from environmental science, economics, and technology, is a key strategy in equipping the workforce with relevant skills to address dynamic agricultural challenges [35]. Research on similar desert environments, such as Jordan and Egypt, suggests that structured interdisciplinary training significantly improves agricultural output and water efficiency, thereby enhancing climate resilience [36].

To operationalize these findings, a structured work plan should be implemented. The first step involves conducting a comprehensive needs assessment to identify key competency gaps within Qatar's agricultural workforce. Following this, multilingual training programs should be designed to accommodate the linguistic diversity of laborers. The third step focuses on integrating interdisciplinary micro-credentialing to provide specialized training in sustainable agriculture and resource-efficient practices.

A pilot program should then be launched to assess the effectiveness of these training interventions. Based on pilot outcomes, successful training methodologies should be scaled up to benefit a larger agricultural workforce. Additionally, institutional and policy support from government agencies, research institutions, and industry stakeholders is crucial in aligning training programs with national food security strategies. Finally, continuous monitoring and evaluation mechanisms should be established to track progress, refine training curricula, and incorporate emerging agricultural innovations [24]. By implementing this structured approach, Qatar can enhance its agricultural workforce, strengthen food security, and build resilience against climate change.

References

1. Al-Kandari D, Al-Abdeen J, Sidhu J (2019) Food safety knowledge, attitudes and practices of food handlers in restaurants in Kuwait. Food Control 103:103–110. https://doi.org/10.1016/j.foodcont.2019.03.040
2. Karanisa T, Amato A, Richer R, Majid SA, Skelhorn C, Sayadi S (2021) Agricultural production in Qatar's hot arid climate. Sustainability 13(7):4059. https://doi.org/10.3390/su13074059
3. Namany S, Govindan R, Di Martino M, Pistikopoulos EN, Linke P, Avraamidou S, Al-Ansari T (2022) Developing intelligence in food security: an agent-based modelling approach of Qatar's

food system interactions under socio-economic and environmental considerations. Sustain Prod Consum 32:669–689. https://doi.org/10.1016/j.spc.2022.05.017
4. Al-Kuwari MM, Du X, Koç M (2021) Performance assessment in education for sustainable development: a case study of the Qatar education system. Prospects 52(3–4):513–527. https://doi.org/10.1007/s11125-021-09570-w
5. Food Security Department (2020) Qatar National Food Security Strategy. In: Food Security Department
6. Abusin SA, Mandikiana BW (2020) Towards sustainable food production systems in Qatar: assessment of the viability of aquaponics. Glob Food Sec 25:100349. https://doi.org/10.1016/j.gfs.2020.100349
7. https://www.mme.gov.qa/pdocs/cview?siteID=2&docID=19772&year=2020
8. Ridha B. (2020) Agricultural extension system industrial managers in the light of their training needs to face job stress. ResearchGate. https://www.researchgate.net/publication/345920136_Agricultural_extension_system_industrial_managers_in_the_light_of_their_training_needs_to_face_job_stress_-_A_field_study
9. Adesipo A, Fadeyi O, Kuca K, Krejcar O, Maresova P, Selamat A, Adenola M (2020) Smart and climate-smart agricultural trends as core aspects of smart village functions. Sensors 20(21):5977. https://doi.org/10.3390/s20215977
10. https://openknowledge.fao.org/server/api/core/bitstreams/f5019ab4-0f6a-47e8-85b9-15473c012d6a/content
11. Aluko YA, Okuwa OB (2018) Innovative solutions and women empowerment: implications for sustainable development goals in Nigeria. https://econpapers.repec.org/article/tafrajsxx/v_3a10_3ay_3a2018_3ai_3a4_3ap_3a441-449.htm
12. Al Shammari F, Al-Kuwari M, Hassan S (2020) Food security and agricultural resilience in the GCC: challenges and opportunities. Arab J Agric Stud 8(3):112–130
13. Al-Otaibi H (2021) Sustainable agriculture in arid regions: exploring hydroponics and vertical farming in the GCC. J Desert Agric 15(2):78–94
14. Othman M, Azzam M, Al-Busaidi F (2019) Innovations in sustainable farming practices: solutions for food security in the GCC countries. J Agric Innov 22(1):54–69
15. Al-Mazrouei F, Al-Kaabi M (2022) The role of water-efficient farming in GCC food security: a strategic overview. Water Resour Manage 37(5):1453–1465
16. Faris H, Al-Habsi S (2021) Training the agricultural workforce in the GCC: addressing challenges and enhancing skills. Int J Agric Educ Dev 13(4):188–202
17. World Bank (2022) Agricultural Land (% of Total Land Area). Retrieved from https://data.worldbank.org/indicator/AG.LND.AGRI.ZS
18. United Nations Environment Programme (UNEP) (2020) Water scarcity and availability in the GCC. Retrieved from https://www.unenvironment.org/resources/report/water-scarcity-arab-region
19. Food and Agriculture Organization (FAO) (2021) Food security and food import dependency in the gulf countries. Retrieved from http://www.fao.org/3/y5686e/y5686e01.htm
20. International Labour Organization (ILO) (2021) Labour statistics by country (Agriculture Sector). Retrieved from https://www.ilo.org/ilostat
21. Food and Agriculture Organization (FAO) (2021) Agricultural production and trade by country. Retrieved from http://www.fao.org/faostat
22. GCC Agricultural Development Reports (2022) Sustainable agricultural practices in GCC countries. Retrieved from https://www.gcc-sustainability.org
23. Davey A (2022) Qatar turns to smart irrigation technology amid climate threat. H_2O Global News. https://h2oglobalnews.com/qatar-turns-to-smart-irrigation-technology-amid-climate-threat/
24. Tadesse KT (2024) The role of post-harvest management in ensuring food security in a changing world: review article. J Clin Res Case Stud 1–14. https://doi.org/10.61440/jcrcs.2024.v2.41
25. New farming method develops sustainable crop system suitable for Qatari climate|QAFCO|QATAR FERTILISER COMPANY (n.d.) https://www.qafco.qa/content/new-farming-method-develops-sustainable-crop-system-suitable-qatari-climate

26. B A, Rohilla S, Mishra A (2024) Communicating agricultural science: bridging the gap and fostering sustainability. ResearchGate. https://www.researchgate.net/publication/378964286_Communicating_Agricultural_Science_Bridging_the_Gap_and_Fostering_Sustainability
27. Okoth JR, Nalyongo W, Alexis Bonte (2010) Facilitators' Guide for running a farmer field school: an adaptation for a post emergency recovery programme. In: FAO Uganda. https://www.fao.org/fileadmin/user_upload/fsn/docs/Farmer_Field_School.pdf
28. Middle East Council on Global Affairs (2024) Food insecurity in the Middle East and North Africa—Middle East Council on Global Affairs. https://mecouncil.org/publication/food-insecurity-in-the-middle-east-and-north-africa/
29. Clayton R, Amaral A, Shafique A, (2021) Developing local learning ecosystems in Qatar to advance equity, inclusion and social cohesion. In: WISE. https://www.wise-qatar.org/app/uploads/2021/12/unnamed-file.pdf
30. FAO (2021) The role of women in agriculture: closing the gender gap for development. Food and Agriculture Organization
31. UN Women (2020) Gender-responsive agricultural training: a framework for inclusion and empowerment. United Nations Women
32. Al-Kuwari H, Al-Sulaiti F, Ahmed S (2022) AI-driven precision farming for arid regions: a case study from Qatar. Int J Smart Agric 10(1):45–60
33. Yadav AK, Reddy CA, Oraon S, Bharti SD (2024) Advancing disease management in agriculture: a review of plant pathology techniques. ResearchGate. https://doi.org/10.5147/PSA.2024.9.1.16
34. Qatar Computing Research Institute (QCRI) (2021) Artificial intelligence applications in agriculture: enhancing sustainability in Qatar. QCRI Research Reports
35. Smith J, Ahmed R (2020) Interdisciplinary education and workforce resilience in sustainable agriculture. Int J Agric Sustain 18(4):112–130
36. Hassan A, El-Sayed H, Othman M (2019) Climate resilience and interdisciplinary training in desert agriculture: case studies from Jordan and Egypt. J Environ Sci Agric 9(2):67–83

GPSR Compliance

The European Union's (EU) General Product Safety Regulation (GPSR) is a set of rules that requires consumer products to be safe and our obligations to ensure this.

If you have any concerns about our products, you can contact us on ProductSafety@springernature.com

In case Publisher is established outside the EU, the EU authorized representative is:

Springer Nature Customer Service Center GmbH
Europaplatz 3
69115 Heidelberg, Germany

Batch number: 10755329

Printed by Printforce, the Netherlands